Developments in Mathematics

Volume 57

More information about this series at http://www.springer.com/series/5834

Anthony B. Evans

Orthogonal Latin Squares Based on Groups

Springer

Anthony B. Evans
Mathematics and Statistics
Wright State University
Dayton, OH, USA

ISSN 1389-2177 ISSN 2197-795X (electronic)
Developments in Mathematics
ISBN 978-3-030-06850-9 ISBN 978-3-319-94430-2 (eBook)
https://doi.org/10.1007/978-3-319-94430-2

Mathematics Subject Classification: 05-02, 05B15, 05B25, 05E18, 12E20, 20F99, 20N05, 11T06, 11T22

Printed on acid-free paper

This Springer imprint is published by the registered company Springer Nature Switzerland AG
The registered company address is: Gewerbestrasse 11, 6330 Cham, Switzerland

To my wife Alma and to Leyla, Umair,
Jasmina, and Ariya

Preface

Latin squares and mutually orthogonal sets of Latin squares (MOLS) have an old history predating Euler's work in the late 1700s. With the emergence of the abstract concept of a group in the 1800s, Cayley observed that the multiplication/addition tables (Cayley tables) of groups are Latin squares. Latin squares and MOLS are used in several constructions of designs, notably nets, transversal designs, and affine and projective planes. In many constructions of designs from Latin squares and MOLS based on groups, the design is characterized by the action of the group on the design.

The focus of this book is on orthomorphisms and complete mappings of finite groups. Orthomorphisms are permutations of the columns of the Cayley table of a group that yield Latin squares that are orthogonal to the Cayley table of the group, and complete mappings of a group correspond to transversals of the Cayley table of the group. Pairwise orthogonal sets of orthomorphisms of a group correspond to MOLS based on the group.

I am writing this book to encourage others to work in this area. There are a number of problems in this area that remain unsolved. This book should serve as a reference for anyone wishing to tackle any of these problems, as well as for anyone who is curious about this subject, or who might wish to apply results from this area in their own disciplines. As the inspiration needed to tackle problems in this book may well come from the techniques used to obtain results, rather than from the results themselves, whenever two or more different proofs are available for the same result I will tend to give multiple proofs instead of selecting a proof.

In this book, I am expanding and updating my 1992 monograph *Orthomorphism graphs of groups* that was published in Springer's Lecture Notes in Mathematics series. This work also expands on material that can be found in other books, such as Dénes and Keedwell's books on Latin squares and Beth, Jungnickel, and Lenz's books on design theory. These books should be referred to for more general information on these topics.

This book is organized into four parts. Part I, the introduction, consists of two chapters. Chapter 1 introduces the concepts, orthomorphisms and complete mappings, and fundamental results that will be used in most of the book. Chapter 2, which is independent of the rest of the book, covers tests that have been used to determine whether a Latin square is based on a group or not. Part II consists of five chapters describing the work done on the existence problem for complete mappings of groups, culminating in a proof of the Hall-Paige conjecture. Part III, on orthomorphism graphs of groups, consists of six chapters and is the heart of the book. Chapter 8 introduces the topic. The other chapters in this part can, for the most part, be read independently of each other, except that Chapter 9 should be read before Chapter 10. Part IV consists of three chapters, covering projective planes constructed from complete sets of orthomorphisms, concepts closely related to orthomorphisms and complete mappings, and a list of problems. The last of these three chapters updates and expands on the list of problems in my 1992 monograph. Readers interested only in orthomorphism graphs can safely skip Part II.

For the most part, this book should be accessible to anyone with a basic understanding of finite group theory, finite field theory, linear algebra, and elementary number theory. In Part II, on the existence problem for complete mappings, I will make use of the classification of finite simple groups. However, I do not expect the reader to be familiar with the theory of these groups: relevant material will be introduced in the text as needed. Proofs involving simple groups of Lie type will usually be either omitted or outlined. While some applications to design theory will be included, prior knowledge of design theory will not be assumed.

I would like to thank Donald Keedwell and Rebecca Stones who read and commented on portions of this work and Ian Wanless who read this manuscript and suggested changes and problems for inclusion. I would also like to thank the reviewers of this manuscript for their helpful comments. These comments and suggestions have helped to improve this work. I would like to thank Wright State University for its support: portions of this book were written while on professional development leave from Wright State University. I would especially like to thank Michael Aschbacher for allowing me to include material from lectures he gave at NSA in the summer of 1990. Lastly, I would like to thank my wife, Alma, whose encouragement helped me to finish this work.

Dayton, OH, USA Anthony B. Evans

Contents

Part I
Introduction

Chapter 1
Latin Squares Based on Groups

Latin squares and orthogonal Latin squares have been used in the construction of many classes of designs: nets, affine planes, projective planes, and transversal designs, in particular.

When orthogonal sets of Latin squares are obtained from the multiplication table of a finite group by permuting columns, each square is determined by its first row which is a permutation of the elements of the group. This enables us to describe and study orthogonality from a purely algebraic point of view, using difference matrices, complete mappings, and orthomorphisms. The nets, affine planes, projective planes, and transversal designs constructed in this way are characterized by the action of the group on these designs.

1.1 Latin Squares

A *Latin square of order n* is an $n \times n$ matrix with entries chosen from an n-set of *symbols*, such that each symbol appears exactly once in each row and exactly once in each column. While the rows and columns of a Latin square of order n are usually indexed with the integers $1, \ldots, n$, other sets of order n can be used. We will usually use as indexing sets the integers $1, \ldots, n$, or $0, \ldots, n-1$, or the elements of a group of order n, and our most commonly used symbol sets for Latin squares of order n will be the element sets of groups of order n. When it becomes necessary to display the row and column indices of a latin square, we may *border* the latin square by writing the column indices above the columns and the row indices to the left of the rows. Figure 1.1 gives three examples of Latin squares of order 5.

In 1854 Cayley ([67] and [68]) pointed out that the multiplication/addition table of a group is a Latin square, and so we refer to such squares as Cayley tables. More formally, if $G = \{g_1, \ldots, g_n\}$ is a finite group, then the *Cayley table* of G is the $n \times n$ matrix with ijth entry equal to $g_i g_j$. It follows from the cancellation properties of

A. B. Evans, *Orthogonal Latin Squares Based on Groups*, Developments in Mathematics 57, https://doi.org/10.1007/978-3-319-94430-2_1

groups that this is a Latin square. With the natural border, the square, L_1, can be viewed as the Cayley table of the group $\mathbb{Z}_5$, as in Figure 1.2.

$$L_1 = \begin{bmatrix} 0\,1\,2\,3\,4 \\ 1\,2\,3\,4\,0 \\ 2\,3\,4\,0\,1 \\ 3\,4\,0\,1\,2 \\ 4\,0\,1\,2\,3 \end{bmatrix}, \quad L_2 = \begin{bmatrix} 0\,1\,2\,3\,4 \\ 1\,0\,3\,4\,2 \\ 2\,4\,0\,1\,3 \\ 3\,2\,4\,0\,1 \\ 4\,3\,1\,2\,0 \end{bmatrix}, \quad L_3 = \begin{bmatrix} 0\,1\,2\,3\,4 \\ 1\,4\,3\,0\,2 \\ 2\,3\,1\,4\,0 \\ 3\,0\,4\,2\,1 \\ 4\,2\,0\,1\,3 \end{bmatrix}$$

Fig. 1.1 Three Latin squares of order 5

Fig. 1.2 L_1 as a Cayley table for $\mathbb{Z}_5$

	0	1	2	3	4
0	0	1	2	3	4
1	1	2	3	4	0
2	2	3	4	0	1
3	3	4	0	1	2
4	4	0	1	2	3

With the natural border, the square, L_3, is another Cayley table for the group $\mathbb{Z}_5$, though this may not be immediately apparent. It is clear that the square, L_2, with the natural border cannot be the Cayley table of a group, as no group of odd order can have a nonidentity element of order 2. Is there any way to border L_2 so that it too can be viewed as a Cayley table for $\mathbb{Z}_5$? That is, can we find two orderings, $\{a_1, \dots, a_5\}$ and $\{b_1, \dots, b_5\}$ of $S = \{0, \dots, 4\}$, so that, by defining $a_i \cdot b_j$ to be the ijth entry of L_2, the set S becomes a group? We will return to this question in Chapter 2.

1.1.1 *Latin Squares and Quasigroups*

A *quasigroup* $(Q, \cdot)$ is a set Q with a binary operation $\cdot$ such that for each $b, c \in Q$, there exists a unique $x \in Q$ such that $x \cdot b = c$, and for each $a, c \in Q$, there exists a unique $y \in Q$ such that $a \cdot y = c$. It follows that a quasigroup satisfies both left and right cancellation laws. A quasigroup Q is called a *loop* if it contains an identity element e satisfying $a \cdot e = e \cdot a = a$ for all $a \in Q$. A *group* can be defined as a loop in which the associative law holds. As is the usual practice, we will write Q instead of $(Q, \cdot)$ and ab instead of $a \cdot b$, except when this might lead to ambiguity.

No matter how any Latin square in Figure 1.1 is bordered, it can be viewed as the multiplication table of a quasigroup, and with the natural bordering, it can be viewed as the multiplication table of a loop. There is, in fact, a natural correspondence between Latin squares of order n and quasigroups of order n.

Theorem 1.1. *There is a one-to-one correspondence between quasigroups with element set $Q = \{q_1, \ldots, q_n\}$ and Latin squares with symbol set Q.*

Proof. Let $\mathcal{Q}$ be the set of quasigroups with element set Q, and let $\mathcal{L}$ be the set of Latin squares with symbol set Q. Define f: $\mathcal{Q} \to \mathcal{L}$ by

$$\mathrm{f}(Q, \cdot) = L = \{l_{ij}\} = \{q_i \cdot q_j\}.$$

This mapping is well-defined as the cancellation properties of quasigroups ensure that the image of a quasigroup is a Latin square.

The mapping f is one to one as $\mathrm{f}(Q, \cdot) = \mathrm{f}(Q, \star)$ if and only if $q_i \cdot q_j = q_i \star q_j$ for all $q_i, q_j \in Q$, if and only if $(Q, \cdot) = (Q, \star)$.

To show that f is onto, let $L = \{l_{ij}\} \in \mathcal{L}$ and define a binary operation $\cdot$ on Q by $q_i \cdot q_j = l_{ij}$. If $x \cdot q_j = q_k$, then, as there exists a unique i for which $l_{ij} = q_k$, $x = q_i$ is uniquely determined; and if $q_i \cdot y = q_k$, then, as there exists a unique j for which $l_{ij} = q_k$, $y = q_j$ is uniquely determined. Hence, $(Q, \cdot)$ is a quasigroup, $\mathrm{f}(Q, \cdot) = L$, and f is onto. It follows that f is a bijection, proving the result. □

We should note that, in Theorem 1.1, to establish the one-to-one correspondence between quasigroups with element set Q and Latin squares with symbol set Q, we needed to have an order imposed on the elements of Q, and we needed the rows and columns of the Latin squares to be indexed by the integers $1, 2, \ldots, n$.

It is natural to extend the definition of a Cayley table to include quasigroups. If $Q = \{q_1, \ldots, q_n\}$ is a quasigroup then the *Cayley table* of Q is the $n \times n$ matrix with ijth entry equal to $q_i q_j$. By Theorem 1.1, the Cayley table of a quasigroup is a Latin square. We will sometimes find it more convenient to state and prove our results in a quasigroup setting rather than a Latin square setting, as Theorem 1.1 makes it clear that any result about quasigroups can be translated into a result about Latin squares.

1.1.2 Isotopism

There are several ways to obtain new Latin squares from old Latin squares. If we permute the rows of a Latin square, then the result is a Latin square; if we permute the columns of a Latin square, then the result is a Latin square; and if we permute the symbols of a Latin square, or replace the symbol set by another symbol set of the same cardinality, then the result is again a Latin square. Any Latin square obtained from a Latin square, L, by any combination of these operations is said to be isotopic to L. Formally, if $L = \{l_{ij}\}$ is a Latin square of order n with symbol set S and $L' = \{l'_{ij}\}$ is a Latin square of order n on a symbol set S', then L' is *isotopic* to L if there exist bijections $\alpha, \beta\colon \{1, \ldots, n\} \to \{1, \ldots, n\}$ and $\gamma\colon S \to S'$ such that $\gamma(l_{ij}) = l'_{\alpha(i),\beta(j)}$ for all i, j: the ordered triple of mappings (α, β, γ) is called an *isotopism* from L to L' and an *autotopism* if $L' = L$. It is routine to show that isotopy is an equivalence relation on the set of Latin squares. As Latin squares can be thought of as multiplication tables for quasigroups, we can translate the concept of isotopy for Latin squares into the concept of isotopy for quasigroups. The quasigroup $(Q', \star)$ is said to be *isotopic* to the quasigroup $(Q, \cdot)$ if there exist bijections $\alpha, \beta, \gamma\colon Q \to Q'$ such that $\gamma(a \cdot b) = \alpha(a) \star \beta(b)$ for all $a, b \in Q$. An *autotopism* of the quasigroup $(Q, \cdot)$ is an isotopism from $(Q, \cdot)$ to itself. In the special case $\alpha = \beta = \gamma$, this equation becomes $\alpha(a \cdot b) = \alpha(a) \star \alpha(b)$. A bijection $\alpha\colon Q \to Q'$ is an *isomorphism* from $(Q, \cdot)$ to $(Q', \star)$ if (α, α, α) is an isotopism from $(Q, \cdot)$ to $(Q', \star)$: it is an *automorphism* if $(Q, \cdot) = (Q', \star)$. As with Latin squares, isotopism is an equivalence relation on the set of quasigroups. Every isotopism class of quasigroups contains loops.

Theorem 1.2. *Any quasigroup is isotopic to a loop.*

Proof. Let $(Q, \cdot)$ be a quasigroup and pick $e \in Q$. Define $\alpha\colon Q \to Q$ by $\alpha(q) \cdot e = q$. Next define $\beta\colon Q \to Q$ by $\alpha(e) \cdot \beta(q) = q$. Note that $\beta(e) = e$ as

$$\alpha(e) \cdot \beta(e) = e = \alpha(e) \cdot e;$$

α is injective as if $\alpha(q_1) = h = \alpha(q_2)$, then

$$q_1 = \alpha(q_1) \cdot e = h \cdot e = \alpha(q_2) \cdot e = q_2;$$

and β is injective as if $\beta(q_1) = h = \beta(q_2)$, then

$$q_1 = \alpha(e) \cdot \beta(q_1) = \alpha(e) \cdot h = \alpha(e) \cdot \beta(q_2) = q_2.$$

α is surjective as if $q \in Q$, then $q \cdot e = q'$ for some $q' \in Q$, and, hence, $\alpha(q') = q$; and β is surjective as if $q \in Q$, then $\alpha(e) \cdot q = q'$ for some $q' \in Q$, and, hence, $\beta(q') = q$. Let γ be the identity mapping, and define a binary operation $\star$ on Q by $q_1 \star q_2 = \alpha(q_1) \cdot \beta(q_2)$. If $x \star q_j = q_k$, then $\alpha(x) \cdot \beta(q_j) = q_k$, which, as $(Q, \cdot)$ is a quasigroup, uniquely determines $\alpha(x)$, which, as α is a bijection, uniquely determines x. Similarly, if $q_i \star y = q_k$, then $\alpha(q_i) \cdot \beta(y) = q_k$, which, as $(Q, \cdot)$ is a quasigroup, uniquely determines $\beta(y)$, which, as β is a bijection, uniquely determines y. Thus $(Q, \star)$ is a quasigroup and is, in fact, a loop, as, for all $q \in Q$,

$$q \star e = \alpha(q) \cdot \beta(e) = \alpha(q) \cdot e = q, \quad \text{and} \quad e \star q = \alpha(e) \cdot \beta(q) = q.$$

To complete the proof, we need to show that (α, β, γ) is an isotopy from $(Q, \star)$ to $(Q, \cdot)$: this is true as

$$\gamma(a \star b) = a \star b = \alpha(a) \cdot \beta(b).$$ □

For Latin squares this tells us that any Latin square is isotopic to a Latin square with the entries of the first row, as well as the first column, in some standard order.

The proof of Theorem 1.2 describes algebraically a simple algorithm for finding such a standard square: first permute the rows so that the first column is in standard order, and then permute the columns so that the first row is in standard order.

We will say that a Latin square is *based on the group* G if it is isotopic to the Cayley table of G. While the isotopism class of a quasigroup may contain nonisomorphic loops, we will see in Chapter 2 that, if a quasigroup is based on a group G, then every loop in its isotopism class will be a group isomorphic to G.

1.2 Orthogonality

A pair of Latin squares, $L = \{l_{ij}\}$ and $L' = \{l'_{ij}\}$, of the same order is said to be *orthogonal* if for each a in the symbol set of L and each b in the symbol set of L', there exists a unique pair i and j for which $l_{ij} = a$ and $l'_{ij} = b$. We also say that L' is an *orthogonal mate* of L. There is evidence that orthogonal Latin squares were known in the *thirteen*th century. Information on the early history of Latin squares and orthogonal Latin squares can be found in two articles by Ullrich [313] in 1999 and [314] in 2002, as well as in a 2007 article on the history of design theory by Anderson, Colbourn, Dinitz, and Griggs [20]. The modern study of orthogonality can be traced back to a 1779 paper by Euler [110].

1.2.1 MOLS

An example of an orthogonal pair of Latin squares of order 4 is given in Figure 1.3.

Fig. 1.3 An orthogonal pair of Latin squares of order 4

$$L_1 = \begin{bmatrix} 0\,1\,2\,3 \\ 1\,0\,3\,2 \\ 2\,3\,0\,1 \\ 3\,2\,1\,0 \end{bmatrix}, \quad L_2 = \begin{bmatrix} 0\,2\,3\,1 \\ 1\,3\,2\,0 \\ 2\,0\,1\,3 \\ 3\,1\,0\,2 \end{bmatrix}$$

The square, L_1, with the natural border, is the Cayley table of $\mathbb{Z}_2 \times \mathbb{Z}_2$, and L_2 is obtained from L_1 by permuting columns. Thus each of L_1, L_2 is based on the group $\mathbb{Z}_2 \times \mathbb{Z}_2$. As a further example, consider the four Latin squares of order 5 in Figure 1.4. Each pair of Latin squares in this set can be seen to be orthogonal: we call a set of Latin squares, each pair of which is orthogonal, a *mutually orthogonal set of Latin squares* (MOLS).

Fig. 1.4 A set of four MOLS of order 5

$$M_1 = \begin{bmatrix} 0&1&2&3&4 \\ 1&2&3&4&0 \\ 2&3&4&0&1 \\ 3&4&0&1&2 \\ 4&0&1&2&3 \end{bmatrix}, \quad M_2 = \begin{bmatrix} 0&2&4&1&3 \\ 1&3&0&2&4 \\ 2&4&1&3&0 \\ 3&0&2&4&1 \\ 4&1&3&0&2 \end{bmatrix}$$

$$M_3 = \begin{bmatrix} 0&3&1&4&2 \\ 1&4&2&0&3 \\ 2&0&3&1&4 \\ 3&1&4&2&0 \\ 4&2&0&3&1 \end{bmatrix}, \quad M_4 = \begin{bmatrix} 0&4&3&2&1 \\ 1&0&4&3&2 \\ 2&1&0&4&3 \\ 3&2&1&0&4 \\ 4&3&2&1&0 \end{bmatrix}$$

We may informally define orthogonality by saying that two Latin squares of the same order are orthogonal if when they are superimposed each ordered pair of symbols appears exactly once. Figure 1.5 shows that the two squares of order 4 in Figure 1.3 are orthogonal. We can similarly show that each pair of Latin squares in Figure 1.4 is orthogonal. Each square in this set is easily seen to be based on the group $\mathbb{Z}_5$. We will say that a set of MOLS is *based on* the group G if each square in the set is based on the group G. A somewhat different definition is given in [91], where an orthogonal pair of Latin squares is said to be based on the group G if one of the squares is the Cayley table of G. Thus the set of 2 MOLS in Figure 1.3 is based on the group $\mathbb{Z}_2 \times \mathbb{Z}_2$ and the set of 4 MOLS in Figure 1.4 is based on the group $\mathbb{Z}_5$. A number of constructions of MOLS in the literature are based on groups: many of these are described in Chapter 13.

Fig. 1.5 L_1 and L_2 superimposed

$$\begin{bmatrix} 00&12&23&31 \\ 11&03&32&20 \\ 22&30&01&13 \\ 33&21&10&02 \end{bmatrix}$$

A set of MOLS is *maximal* if it cannot be extended to a larger set of MOLS. An important problem is that of determining pairs k, n, for which there exists a maximal set of k MOLS of order n. The set of 2 MOLS of order 4 in Figure 1.3 is not maximal as the square

$$\begin{bmatrix} 0&3&1&2 \\ 1&2&0&3 \\ 2&1&3&0 \\ 3&0&2&1 \end{bmatrix}$$

is orthogonal to each of these two squares. However, the set of 4 MOLS of order 5 in Figure 1.4 is maximal. Note that each M_i has the same 0th column. Suppose that L is a Latin square on the same symbol set and with the same 0th column as each

M_i. Then the 0,1th entry of L cannot be 0, as 0 can only occur once in any row, and cannot be i, for $i = 1, \ldots, 4$, as i already occurs opposite i in the 0th column of M_i. Thus L cannot be orthogonal to each of the M_i. For this to be a proof of the maximality of $M_1, \ldots, M_4$, we need to show that the assumptions that we made about the structure of L can be made without loss of generality. This is the case as isotopisms preserve orthogonality as long as we use the same permutation for the rows as we use for the columns.

Theorem 1.3. *Let L_1 and L_2 be two Latin squares of the same order, and let $(\alpha, \beta, \gamma_1)$ be an isotopism from L_1 to L_1' and $(\alpha, \beta, \gamma_2)$ an isotopism from L_2 to L_2'. Then L_1 is orthogonal to L_2 if and only if L_1' is orthogonal to L_2'.*

Proof. Let $L_k = \{l_{k,ij}\}$, where $k = 1, 2$, and $L_k' = \{l'_{k,ij}\}$, where $k = 1, 2$. It is routine to prove that, for a in the symbol set of L_1 and b in the symbol set of L_2, there exists a unique pair i, j for which $l_{1,ij} = a$ and $l_{2,ij} = b$ if and only if, for $a' = \gamma(a)$ in the symbol set of L_1' and $b' = \gamma(b)$ in the symbol set of L_2', there exists a unique pair $i' = \alpha(i)$, $j' = \beta(j)$ for which $l'_{1,i'j'} = a'$ and $l'_{2,i'j'} = b'$, hence the result. □

The following is an easy corollary:

Corollary 1.4. *Let $L_1, \ldots, L_r$ be a set of MOLS, and let $(\alpha, \beta, \gamma_i)$ be an isotopism from L_i to L_i' for $i = 1, \ldots, r$. Then $L_1, \ldots, L_r$ is a set of MOLS if and only if $L_1', \ldots, L_r'$ is a set of MOLS.*

Thus we may standardize a set of MOLS. We often assume, without loss of generality, that for a set of MOLS, the symbol set is the same for each square, and each square in the set has the same first row, or each square in the set has the same first column. This justifies our restrictions on the choice of Latin square L, assumed to be orthogonal to each of $M_1, \ldots, M_4$, completing the proof that the set of MOLS in Figure 1.4 is maximal.

1.2.2 The Number of Squares in a Set of MOLS

A natural question to ask is, how many Latin squares can we have in a set of MOLS of order n? We use $N(n)$ to denote the maximum number of squares possible in a set of MOLS of order n. We set $N(1) = \infty$ as any Latin square of order 1 must be orthogonal to itself and to any other Latin square of order 1. Determining $N(n)$ is a central problem in the study of Latin squares.

Theorem 1.5. *If $n \geq 2$, then $N(n) \leq n - 1$.*

Proof. Let $L_1, L_2, \ldots, L_r$ be a set of MOLS of order n. By Theorem 1.3 we may, without loss of generality, assume that the entries of each square are $\{1, 2, \ldots, n\}$ and that the i,1th entry of L_j is i for $i = 1, 2, \ldots, n$. Now the 1,2th entries of L_1,

$L_2,\ldots,L_r$ form a set of r distinct elements of $\{1, 2, \ldots, n\}$, none of which can equal 1. Hence, $r \leq n-1$ as claimed. □

For $n \geq 2$, a set of $n-1$ MOLS of order n is said to be a *complete set* of MOLS of order n. Such sets are known to exist when n is a prime power.

Theorem 1.6. *If q is a prime power, then there exists a complete set of MOLS of order q.*

Proof. Let $\mathrm{GF}(q) = \{x_1, x_2, \ldots, x_q\}$ be the field of order q, and define a $q \times q$ matrix L_b to have ijth entry $x_i + bx_j$. Then $\{L_b \mid b \in \mathrm{GF}(q) \setminus \{0\}\}$ is a complete set of MOLS of order q. □

It should be noted that the construction given in Theorem 1.6 is of complete sets of MOLS based on additive groups of finite fields. No example of a complete set of MOLS of non-prime power order is known, and it is a long-standing conjecture that no such examples exist.

Conjecture 1.7 (The Prime Power Conjecture). For $n \geq 2$ a complete set of MOLS of order n exists if and only if n is a prime power.

For no non-prime power value of n, other than 1 and 6, is the value of $N(n)$ known. We know that $1 \leq N(n) \leq n-1$ for $n \geq 2$. Much work has gone into improving the lower bound. See the 2007 article by Abel, Colbourn, and Dinitz [6] for a list of lower bounds for $N(n)$.

One recursive construction of MOLS is the Kronecker product construction. For this construction we will use ordered m-tuples to index rows and columns and as symbol sets. If $L_k = \{l_{ij}^k\}$ is a Latin square of order n_k for $k = 1, \ldots, m$, then the *Kronecker product* of $L_1, \ldots, L_m$ is the square $L_1 \times \cdots \times L_m$, of order $n_1 n_2 \cdots n_m$, with $(i_1, \cdots, i_m)(j_1, \cdots, j_m)$th entry equal to $(l_{i_1 j_1}^1, \ldots, l_{i_m j_m}^m)$.

Lemma 1.8 (The Kronecker Product Construction). *If, for $k = 1, \ldots, m$, the squares $L_1^k, \ldots, L_r^k$ form a set of r MOLS of order n_k, then $L_1^1 \times \cdots \times L_1^m, \ldots, L_r^1 \times \cdots \times L_r^m$ is a set of r MOLS of order $n_1 n_2 \cdots n_m$.*

Proof. The $(i_1, \cdots, i_m)$th row of $L_t^1 \times \cdots \times L_t^m$ contains the symbol $(a_1, \ldots, a_m)$ exactly once as, for $k = 1, \ldots, m$, there is a unique j_k for which $a_k = l_{i_k j_k}^t$. Thus each symbol appears exactly once in each row of $L_t^1 \times \cdots \times L_t^m$, for $t = 1, \ldots, r$. Similarly $(a_1, \ldots, a_m)$ appears exactly once in each column of $L_t^1 \times \cdots \times L_t^m$, for $t = 1, \ldots, r$. Hence, $L_t^1 \times \cdots \times L_t^m$ is a Latin square, for $t = 1, \ldots, r$.

If $1 \leq s < t \leq r$ and we are given symbols $(a_1, \ldots, a_m)$ and $(b_1, \ldots, b_m)$, then there is a unique row $\underline{i} = (i_1, \cdots, i_m)$ and column $\underline{j} = (j_1, \cdots, j_m)$ for which the $\underline{i}\,\underline{j}$th entry of $L_s^1 \times \cdots \times L_s^m = (a_1, \ldots, a_m)$ and the $\underline{i}\,\underline{j}$th entry of $L_t^1 \times \cdots \times L_t^m = (b_1, \ldots, b_m)$, as, for $k = 1, \ldots, m$, the equalities $a_k = l_{i_k j_k}^s$ and $b_k = l_{i_k j_k}^t$ uniquely determine i_k and j_k. Thus $L_s^1 \times \cdots \times L_s^m$ is orthogonal to $L_t^1 \times \cdots \times L_t^m$. The result follows. □

There is a close relationship between Kronecker products and direct products of groups.

Corollary 1.9. *If $L_1^k, \ldots, L_r^k$ is a set of r MOLS based on the group G_k for $k = 1, \ldots, m$, then $L_1^1 \times \cdots \times L_1^m, \ldots, L_r^1 \times \cdots \times L_r^m$ is a set of r MOLS based on the group $G_1 \times \cdots \times G_m$.*

Proof. It suffices to prove that, if $L_k = \{l_{ij}^k\}$ is a Latin square based on the group G_k for $k = 1, \ldots, m$, then $L_1 \times \cdots \times L_m$ is based on the group $G_1 \times \cdots \times G_m$. Let $n_k = |G_k|$. As L_k is based on G_k, for $k = 1, \ldots, m$, there exist orderings $G_k = \{(g_k)_1, \ldots, (g_k)_{n_k}\} = \{(g_k')_1, \ldots, (g_k')_{n_k}\}$ of the elements of G_k for which the ijth entry of L_k is $(g_k)_i(g_k')_j$.

Then the $(i_1, \cdots, i_m)(j_1, \cdots, j_m)$th entry of $L_1 \times \cdots \times L_m$ is equal to

$$\begin{aligned}(l_{i_1 j_1}^1, \ldots, l_{i_m j_m}^m) =&((g_1)_{i_1}(g_1')_{j_1}, \ldots, (g_m)_{i_m}(g_m')_{j_m}) \\ =&((g_1)_{i_1}, \ldots, (g_m)_{i_m})((g_1')_{j_1}, \ldots, (g_m')_{j_m}),\end{aligned}$$

and, hence, $L_1 \times \cdots \times L_m$ is based on the group $G_1 \times \cdots \times G_m$. □

From Lemma 1.8 we derive a lower bound for $N(n)$, discovered by MacNeish [235] in 1922.

Theorem 1.10 (The MacNeish Bound). *If $n = q_1 q_2 \cdots q_m$, where $q_1 < q_2 < \cdots < q_m$ are powers of distinct primes, then $N(n) \geq q_1 - 1$.*

Proof. By Theorem 1.6 there exist a set of $q_1 - 1$ MOLS of order q_i for $i = 1, \ldots, m$, and so, by Lemma 1.8, there exists a set of $q_1 - 1$ MOLS of order $q_1 q_2 \cdots q_m$. □

Based on this result, and an erroneous proof of Euler's conjecture, MacNeish conjectured the following.

Conjecture 1.11 (The MacNeish Conjecture). If $n = q_1 q_2 \cdots q_m$, where $q_1 < q_2 < \cdots < q_m$ are powers of distinct primes, then $N(n) = q_1 - 1$.

This conjecture generalizes Euler's conjecture [110], one of the oldest conjectures on orthogonal Latin squares. Euler conjectured in 1779 that if $n \equiv 2 \pmod 4$, then no Latin square of order n can have an orthogonal mate: MacNeish's conjecture reduces to Euler's conjecture when $q_1 = 2$. Euler's conjecture is true for the smallest cases, $n = 2$ and $n = 6$. Euler was unable to solve the case $n = 6$, which he posed as the "36 officers problem". This case was finally settled in 1900 by Tarry [307]. Many "proofs" of Euler's conjecture have been given, all of which are incorrect as Euler's conjecture is, in fact, false. The first hint that Euler's conjecture might actually be false came with a construction in 1959 of a counterexample to MacNeish's conjecture by Parker [264]. Parker constructed a set of 4 MOLS of order 21: as $21 = 3 \times 7$, MacNeish's conjecture states that $N(21) = 2$! Interestingly, Parker's set of 4 MOLS of order 21 is based on the group $\mathbb{Z}_{21}$. Subsequently, also in 1959, Bose and Shrikhande [45] constructed a pair of orthogonal Latin squares of order 22, the first counterexample to Euler's conjecture. This was followed in 1960 by the celebrated result of Bose, Shrikhande, and Parker [46] in which orthogonal pairs of Latin squares of order n for each $n \equiv 2 \pmod 4$, $n > 6$, were constructed. A further counterexample to MacNeish's conjecture was found independently by Bose, Chakravarti, and Knuth [43] in 1960 and by Johnson, Dulmage, and Mendelsohn [196] in 1961: they constructed 5 MOLS of order 12. MacNeish's conjecture states that $N(12) = 2$! The values of $n \leq 200$ for which MacNeish's conjecture has not been settled are 63, 72, 77, 88, 99, 104, 117, 119, 136, 143, 152, 153, 171, 175, 184, and 187: see the 2007 table of values of $N(n)$ by Abel, Colbourn, and Dinitz [6]. Many histories of Euler's conjecture and its disproof can be found in the literature: see the 1981 paper by Frisinger [148] or the 2006 paper by Klyve and Stemkosi [222], for instance. In none of the constructions of counterexamples to Euler's conjecture were any of the Latin squares based on groups. In fact, we shall see in Chapter 3 that Latin squares based on groups satisfy Euler's conjecture.

1.2.3 Transversals

Let us revisit Figure 1.3 to get some clues as to when a Latin square might have an orthogonal mate. For the cells of L_2 that have entry 0, the corresponding entries in L_1 are shown in Figure 1.6.

Fig. 1.6 Cells of L_1 for which the corresponding entries in L_2 are 0

$$\begin{bmatrix} 0 & \cdot & \cdot & \cdot \\ \cdot & \cdot & \cdot & 2 \\ \cdot & 3 & \cdot & \cdot \\ \cdot & \cdot & 1 & \cdot \end{bmatrix}$$

We note that as the 0 entries of L_2 appear once in each row and once in each column, so do the corresponding cells of L_1. Further the entries of these cells in L_1 must be distinct as when L_1 and L_2 are superimposed no ordered pair of symbols can occur more than once. A set of cells in a Latin square, exactly one in each row and exactly one in each column, whose entries are distinct, is called a *transversal* of the Latin square. The cells in L_1 whose corresponding entries in L_2 are 0 form a transversal of L_1. Similarly for the cells in L_1 whose corresponding entries in L_2 are 1, 2, or 3. These four transversals partition the cells of L_1. It is the existence of transversals that partition the cells of a Latin square that determines the existence of an orthogonal mate.

Theorem 1.12. *A Latin square possesses an orthogonal mate if and only if its cells can be partitioned by a set of transversals.*

Proof. Let $L = \{l_{ij}\}$ be a Latin square of order n. If $L' = \{l'_{ij}\}$ is a Latin square orthogonal to L and a is a symbol of L', then set $T_a = \{(i, j) | l'_{ij} = a\}$. The set $\{T_a | a \text{ a symbol of } L'\}$ forms a set of transversals of L that partitions the cells of L.

If $\{T_1, \ldots, T_n\}$ is a set of transversals of L that partitions the cells of L, then define $L' = \{l'_{ij}\}$ by $l'_{ij} = k$ if $(i, j) \in T_k$. Then L' is well-defined and is a Latin square on the symbol set $\{1, \ldots, n\}$ that is an orthogonal mate to L. □

For Latin squares based on groups, a single transversal determines the existence of an orthogonal mate.

Theorem 1.13. *A Latin square based on a group possesses an orthogonal mate if and only if it possesses a transversal.*

Proof. If a Latin square based on a group possesses an orthogonal mate, then it must possess a transversal by Theorem 1.12. Conversely suppose that $G = \{g_1, \ldots, g_n\}$ is a group, and let L be a Latin square based on G. By Theorem 1.3, L has an orthogonal mate if and only if the Cayley table of G has an orthogonal mate. Thus, by Theorem 1.12, the cells of L can be partitioned by a set of transversals if and only if the cells of the Cayley table of G can be partitioned by a set of transversals. Hence, we may assume without loss of generality that L is the Cayley table of G. If $T = \{(1, j_1), \ldots, (n, j_n)\}$ is a transversal of L, then set $T_k = \{(1, j_{k,1}), \ldots, (n, j_{k,n})\}$, where $j_{k,i}$ is determined by the equation $g_i g_{j_{k,i}} = g_i g_{j_i} g_k$. Then the set $\{T_k | k = 1, \ldots, n\}$ is a set of transversals of L that partitions the cells of L, and thus L possesses an orthogonal mate by Theorem 1.12. □

1.2.4 Complete Mappings

Let us suppose that $L = \{l_{ij}\}$ is the Cayley table of a quasigroup $Q = \{q_1, \ldots, q_n\}$ and that $T = \{(1, j_1), \ldots, (n, j_n)\}$ is a transversal of L. Then the mapping $\theta\colon Q \to Q$ defined by $\theta(q_i) = q_{j_i}$ is a bijection. Further, as the entries of T are distinct, the mapping $q_i \mapsto q_i\theta(q_i) = q_i q_{j_i}$ must also be a bijection. If Q is a quasigroup, then we say that a bijection $\theta\colon Q \to Q$ is a *complete mapping* of Q if the mapping $q \mapsto q\theta(q)$ is also a bijection. There is a natural correspondence between complete mappings of a quasigroup and transversals of its Cayley table.

Theorem 1.14. *Let $Q = \{q_1, \ldots, q_n\}$ be a quasigroup and let L be its Cayley table. For $T = \{(1, j_1), \ldots, (n, j_n)\}$, a transversal of L, define θ_T by $\theta_T(q_i) = q_{j_i}$. Then the mapping $T \mapsto \theta_T$ defines a one-to-one correspondence between the transversals of L and the complete mappings of Q. Further, if S and T are two transversals of L, then S and T are disjoint if and only if $\theta_S(q) \neq \theta_T(q)$ for all $q \in Q$.*

Proof. Define $\phi\colon Q \to \{1, \ldots, n\}$ by $\phi(q_i) = i$. If T is a transversal of L, then $T = \{(i, \phi(\theta_T(q_i))) \mid i = 1, \ldots, n\}$. It follows that if S and T are transversals of L, then $\theta_S = \theta_T$ if and only if $S = T$. Thus the mapping $T \mapsto \theta_T$ is one to one. If

θ is a complete mapping of Q, then the set $T = \{(i, \phi(\theta(q_i))) \mid i = 1, \ldots, n\}$ is a transversal of L, and so the mapping $T \mapsto \theta_T$ is onto and hence a bijection.

If S and T are transversals of L, then S and T are disjoint if and only if $(i, \phi(\theta_S(q_i))) \neq (i, \phi(\theta_T(q_i)))$ for all $i = 1, \ldots, n$, if and only if $\theta_S(q_i) \neq \theta_T(q_i)$ for all $i = 1, \ldots, n$. □

There is a simple relationship between the existence problem for orthogonal mates of Latin squares based on groups and the existence problem for complete mappings of finite groups.

Theorem 1.15. *If L is a Latin square based on a group G, then L possesses an orthogonal mate if and only if G admits a complete mapping.*

Proof. By Theorem 1.3, L possesses an orthogonal mate if and only if the Cayley table of G possesses an orthogonal mate. So, without loss of generality, we will assume that L is the Cayley table of G. By Theorem 1.13, L possesses an orthogonal mate if and only if L possesses a transversal. By Theorem 1.14, this happens if and only if G admits a complete mapping. □

The existence problem for complete mappings will be studied in Chapters 3 through 7.

A related problem is the problem of determining all n for which there exists a Latin square of order n that has no orthogonal mate. We call such squares *bachelor squares*, a term coined by van Rees [315] in 1990. In 1779 Euler [110] proved that cyclic Latin squares of even order have no orthogonal mates. Euler's cyclic Latin squares are really Cayley tables of cyclic groups. Thus there exist bachelor squares of order n for all even n. The case $n \equiv 1 \pmod 4$ was dealt with by Mann [238] in 1944, and the problem was completely solved in 2006 by Wanless and Webb [326] and independently by Evans [126]. If $n \neq 1, 3$, then there exists a bachelor square of order n. The bachelor squares constructed by Wanless and Webb have an additional property. There is at least one cell in each of their squares through which no transversal passes: they called such squares *confirmed bachelor squares*. We shall see in Chapter 3 that no bachelor square of odd order can be based on a group as the Cayley table of a group of odd order always has an orthogonal mate.

Readers interested in more information on Latin squares may consult the 1991 and 2015 books by Dénes and Keedwell [93, 95], the 1998 book by Laywine and Mullen [223], or the 2007 article by Abel, Colbourn, and Dinitz [6].

1.3 Difference Matrices and Orthomorphisms

Let us look at the known lower bounds on $N(n)$ for small values of n. We know that $N(n) = n - 1$ if n is a prime power and that in this case a complete set of MOLS, of order n, based on the additive group of a field of order n exists. We will also see in Chapter 3 that if $n \equiv 2 \pmod 4$, then we cannot even construct an orthogonal pair of Latin squares of order n based on a group. The values of $N(n)$

for the remaining values of $n \leq 48$ are given in Table 1.1: this data is from a 2007 article by Abel, Colbourn, and Dinitz [6]. We see from this table that all but one of these constructions is a construction of MOLS based on groups.

In each of the constructions of MOLS based on groups in Table 1.1, the set of MOLS can be obtained from the Cayley table of the group by either permuting rows or permuting columns. If a set of MOLS is constructed from the Cayley table of a group by permuting columns, then each square in the set is determined by its first row: treating these first rows as the rows of a matrix yields a difference matrix. We shall see that, from a set of MOLS constructed from the Cayley table of a group by permuting rows, we can obtain a set of MOLS constructed from the Cayley table of a group by permuting columns. Each of the constructions of MOLS based on groups in Table 1.1 is obtained from a difference matrix.

Table 1.1 Some lower bounds on $N(n)$

n	$N(n) \geq$	Construction
12	5	Based on $GF(3)^+ \times GF(4)^+$
15	4	Based on $GF(3)^+ \times GF(5)^+$
20	4	Quasi-difference matrix
21	5	Based on $GF(3)^+ \times GF(7)^+$
24	7	Based on $GF(3)^+ \times GF(8)^+$
28	5	Based on $GF(4)^+ \times GF(7)^+$
33	5	Based on $GF(3)^+ \times GF(11)^+$
35	5	Based on $GF(5)^+ \times GF(7)^+$
36	8	Based on $GF(4)^+ \times GF(9)^+$
39	5	Based on $GF(3)^+ \times GF(13)^+$
40	7	Based on $GF(5)^+ \times GF(8)^+$
44	5	Based on $GF(4)^+ \times GF(11)^+$
45	6	Based on $GF(5)^+ \times GF(9)^+$
48	8	Based on $GF(3)^+ \times GF(16)^+$

1.3.1 Difference Matrices

If a set of MOLS is based on a group G, then, as permuting the symbol set of any given square does not affect orthogonality, we may assume without loss of generality that each square is obtained from the Cayley table of G by permuting rows and/or columns. Most of the constructions of MOLS based on groups in this book will be obtained by permuting the columns of Cayley tables of groups. Let us look more closely at this option. The set of 4 MOLS of order 5 in Figure 1.4 is based on the group $\mathbb{Z}_5$. The square, M_1, is the Cayley table of $\mathbb{Z}_5$, and each of M_2, M_3, and M_4 is obtained from M_1 by permuting columns. Thus each square is uniquely determined by its first row. The matrix F in Figure 1.7 consists of the first rows of $M_1, \ldots, M_4$. Observe that the difference of any two distinct rows of F contains each element of $\mathbb{Z}_5$ exactly once. For G a group of order n an $(n, r; \lambda)$-*difference matrix* over G is an $r \times \lambda n$ matrix $D = (d_{ij})$ with entries from G such that for any $i, k \in \{1, \ldots, r\}$, $i \neq k$, each element of G appears λ times in the form $d_{ij}^{-1} d_{kj}$. We call λ the *index* of D. The matrix F in Figure 1.7 is a $(5, 4; 1)$-difference matrix over $\mathbb{Z}_5$. By adding a row of zeroes, we can turn this into a $(5, 5; 1)$-difference matrix over $\mathbb{Z}_5$. A difference matrix is *maximal* if no row can be added to create a larger difference matrix. Clearly, the difference matrix F in Figure 1.7 is not maximal.

Fig. 1.7 A $(5, 4; 1)$-difference matrix over $\mathbb{Z}_5$

$$F = \begin{bmatrix} 0 & 1 & 2 & 3 & 4 \\ 0 & 2 & 4 & 1 & 3 \\ 0 & 3 & 1 & 4 & 2 \\ 0 & 4 & 3 & 2 & 1 \end{bmatrix}$$

There are a number of operations that we can perform on a difference matrix that yield other difference matrices.

Theorem 1.16. *Let D be an $(n, r; \lambda)$-difference matrix over G. If D' is obtained from D by any combination of permuting rows, permuting columns, multiplying all the elements of a row on the right by an element of G, and multiplying all the elements of a column on the left by an element of G, then D' is an $(n, r; \lambda)$-difference matrix over G.*

Proof. It is clear that if D' is obtained from D by permuting rows and/or columns, then D' is a difference matrix.

If D' is obtained from D by multiplying all the entries in row i on the right by a, then

$$d'^{-1}_{kj} d'_{lj} = \begin{cases} d_{kj}^{-1} d_{lj} & \text{if } k, l \neq i, \\ a^{-1} d_{ij}^{-1} d_{lj} & \text{if } k = i, \\ d_{kj}^{-1} d_{ij} a & \text{if } l = i. \end{cases}$$

In each case, as j runs through $1, \ldots, \lambda n$, the quotients $d'^{-1}_{kj} d'_{lj}$ run through the element set of G exactly λ times.

If D' is obtained from D by multiplying all the entries in column j on the left by a, then $d'^{-1}_{ij} d'_{kj} = d_{ij}^{-1} a^{-1} a d_{kj} = d_{ij}^{-1} d_{kj}$ and the row quotients are unchanged. □

By applying the transformations in Theorem 1.16, we may transform any difference matrix into a *normalized* difference matrix, that is, one in which every entry in the first row and first column is the identity.

From difference matrices we can construct MOLS based on groups, and from MOLS based on groups, we can construct difference matrices.

Theorem 1.17. *There exists an $(n, r; 1)$-difference matrix over a group G if and only if there exists a set of $r - 1$ MOLS based on G, obtained by permuting the columns of the Cayley table of G.*

Proof. Let $G = \{g_1, \ldots, g_n\}$ be a group, $g_1 = 1$, and let L be the Cayley table of G.

If $D = (d_{ij})$ is an $(n, r; 1)$-difference matrix over G, then, by Theorem 1.16, we may assume without loss of generality that D is normalized and that $d_{2i} = g_i$ for all i. For $k = 1, \ldots, r-1$, let L_k be the $n \times n$ matrix with ijth entry equal to $g_i d_{k+1,j}$. Clearly L_1 is the Cayley table, L, of G and, hence, is a Latin square. As $d_{k+1,1}, \ldots, d_{k+1,n}$ is a permutation of the elements of G, each L_k is a Latin square obtained from L_1 by permuting columns. Let $k, t \in \{1, \ldots, r-1\}$, $t \neq k$, and let $a = g_i d_{k+1,j}$ and $b = g_i d_{t+1,j}$. Then $a^{-1}b = d_{k+1,j}^{-1} d_{t+1,j}$, which, by the defining property of difference matrices, uniquely determines j, and so i is uniquely determined by $g_i = a d_{k+1,j}^{-1}$. Thus $L_1, \ldots, L_{r-1}$ are pairwise orthogonal.

Let $L_1, \ldots, L_{r-1}$ be a set of MOLS obtained from L by permuting columns, and let the $1j$th entry of L_k be denoted l_j^k: the ijth entry of L_k is $g_i l_j^k$. Define an $r \times n$ matrix $D = (d_{ij})$ by

$$d_{ij} = \begin{cases} 1 & \text{if } i = 1, \\ l_j^{i-1} & \text{if } i \neq 1. \end{cases}$$

We claim that D is an $(n, r; 1)$-difference matrix over G. As the L_ks are Latin squares, $\{d_{sj}^{-1} d_{tj} \mid j = 1, \ldots, n\} = G$ if either of s or t is 1, $s \neq t$. If neither s nor t is 1, and $s \neq t$, then

$$S = \{d_{sj}^{-1} d_{tj} \mid j = 1, \ldots, n\} = \{(l_j^{s-1})^{-1} l_j^{t-1} \mid j = 1, \ldots, n\},$$

a set we claim equals G. If $g \in G$, then, as L_s is orthogonal to L_t, for any $a \in G$, there exists an i and j for which $a = g_i l_j^{s-1}$ and $ag = g_i l_j^{t-1}$, from which it follows that $g = (l_j^{s-1})^{-1} l_j^{t-1} \in S$. Thus $S = G$ and D is an $(n, r; 1)$-difference matrix over G. □

In particular, if G is nontrivial, as the existence of an $(n, r; 1)$-difference matrix over G implies the existence of a set of $r - 1$ MOLS of order n, then $r - 1 \leq n - 1$, that is, $r \leq n$. From a similar argument, using transversal designs, a bound can be obtained for difference matrices of any index.

Theorem 1.18. *If G is a group of order $n > 1$ and there exists an $(n, r; \lambda)$-difference matrix over G, then $r \leq \lambda n$.*

Proof. See Corollary **VIII**.3.7 in [41], for instance. □

Note that, if G is trivial, then the bound of Theorem 1.18 does not apply as any matrix with entries from G is then a difference matrix over G.

Difference matrices for which the bound in Theorem 1.18 is attained are of special significance. If G is of order $n > 1$, then an $(n, \lambda n; \lambda)$-difference matrix over G is called a *generalized Hadamard matrix* and denoted $\mathrm{GH}(n, \lambda)$. The existence and nonexistence of generalized Hadamard matrices will prove important in Chapter 14. While we are interested in difference matrices of index 1, results for difference matrices of higher index are still useful to us as they may be homomorphic images of difference matrices of lower index.

Theorem 1.19. *Let* $\mathrm{f}\colon G \to H$ *be an epimorphism,* $|G| = n$ *and* $|H| = m$*. If there exists an* $(n, r; \lambda)$*-difference matrix over* G*, then there exists an* $(m, r; (n/m)\lambda)$*-difference matrix over* H*.*

Proof. Let $D = (d_{ij})$ be an $(n, r; \lambda)$-difference matrix over G and let $\mathrm{f}(D) = (\mathrm{f}(d_{ij}))$. For any $i, k \in \{1, \ldots, r\}$, $i \neq k$, any element of G appears λ times in the form $d_{ij}^{-1}d_{kj}$. Thus, if $h \in H$ and $i \neq k$, then

$$|\{j \mid \mathrm{f}(d_{ij}^{-1}d_{kj}) = h\}| = |\{g \in G \mid \mathrm{f}(g) = h\}||\{j \mid d_{ij}^{-1}d_{kj} = g\}| = \left(\frac{n}{m}\right)\lambda.$$

The result follows. □

1.3.2 Orthomorphisms

There is another way to view the construction of MOLS, based on a group G, obtained by permuting the columns of the Cayley table of G: we have seen how the first rows of these squares form rows of a difference matrix. These first rows can also be thought of as permutations of the element set of the group: MOLS based on groups, obtained by permuting columns, can be studied through these permutations and relationships between them. Let $G = \{g_1, \ldots, g_n\}$ be a group of order n and let L be the Cayley table of G: the ijth entry of L is $g_i g_j$. For a mapping $\theta\colon G \to G$, define L_θ to be the square with ijth entry $g_i\theta(g_j)$. The relationship between properties of the mapping $\theta\colon G \to G$ and properties of the corresponding matrix L_θ is described in the next result.

Theorem 1.20. *Let* G *be a finite group, and let* $\theta, \phi\colon G \to G$ *be mappings. Then the following hold.*

1. L_θ *is a Latin square if and only if* θ *is a bijection.*
2. *If* θ *is a bijection, then* L_θ *is orthogonal to* L *if and only if the mapping* $x \mapsto x^{-1}\theta(x)$ *is a bijection.*
3. *If* θ *and* ϕ *are bijections, then* L_θ *is orthogonal to* L_ϕ *if and only if the mapping* $x \mapsto \phi(x)^{-1}\theta(x)$ *is a bijection.*

Proof.

1. By the cancellation property of groups, every element of G appears exactly once in each column of L_θ. Further, every element of G appears exactly once in the ith row of L_θ if and only if the mapping $g_j \mapsto g_i\theta(g_j)$ is a bijection, if and only if θ is a bijection. Hence L_θ is a latin square if and only if θ is a bijection.

2. If the mapping $x \mapsto x^{-1}\theta(x)$ is a bijection, $a = g_i g_j$, and $b = g_i\theta(g_j)$, then $a^{-1}b = g_j^{-1}\theta(g_j)$ which uniquely determines g_j. Then $g_i = ag_j^{-1}$ is uniquely determined, and so L and L_θ are orthogonal.

   If the mapping $x \mapsto x^{-1}\theta(x)$ is not a bijection, then there exists $g_k, g_{k'} \in G$, $g_k \neq g_{k'}$, for which $g_k^{-1}\theta(g_k) = g_{k'}^{-1}\theta(g_{k'})$. Let $g_t, g_{t'} \in G$ satisfy $g_t g_k = g_{t'} g_{k'}$, and set $a = g_t g_k$ and $b = g_t\theta(g_k)$. Then the system of equations $a = g_i g_j$ and $b = g_i\theta(g_j)$ has distinct solutions corresponding to $i = t$ and $j = k$, and to $i = t'$ and $j = k'$, and so L and L_θ are not orthogonal.
3. Similar to the proof of (2). □

If G is a group and $\theta\colon G \to G$ is a bijection, then, by Theorem 1.20, L_θ is a Latin square based on G. Further L_θ is obtained from L, the Cayley table of G, by permuting columns, and any Latin square, obtained from L by permuting columns, is equal to L_θ for a unique bijection θ. For a group G, we will say that a bijection $\theta\colon G \to G$ is an *orthomorphism* of G if the mapping $x \mapsto x^{-1}\theta(x)$ is a bijection and that two orthomorphisms, θ, ϕ of G, are *orthogonal*, written $\theta \perp \phi$, if the mapping $x \mapsto \phi(x)^{-1}\theta(x)$ is a bijection. Note that $\theta \perp \phi$ if and only if $\phi \perp \theta$, as the mapping $x \mapsto \phi(x)^{-1}\theta(x)$ is a bijection if and only if the mapping $x \mapsto (\phi(x)^{-1}\theta(x))^{-1} = \theta(x)^{-1}\phi(x)$ is a bijection.

If θ is an orthomorphism of G and we define the mapping $\mathrm{r}_a\colon G \to G$ by $\mathrm{r}_a(x) = xa$, then the mapping $\mathrm{r}_a\theta$ is also an orthomorphism of G as the mappings $x \mapsto \mathrm{r}_a\theta(x) = \theta(x)a$ and $x \mapsto x^{-1}\mathrm{r}_a\theta(x) = x^{-1}\theta(x)a$ are clearly both bijections. Further, if $a = \theta(1)^{-1}$, then $\mathrm{r}_a\theta(1) = 1$. We will say that an orthomorphism of a group is *normalized* if it fixes the identity. Normalization does not affect orthogonality as if θ and ϕ are orthogonal orthomorphisms of a group G, then $\mathrm{r}_a\theta \perp \mathrm{r}_b\phi$ as the mapping $x \mapsto \mathrm{r}_a\theta(x)^{-1}\mathrm{r}_b\phi(x) = a^{-1}\theta(x)^{-1}\phi(x)b$ is clearly a bijection.

From Theorem 1.20, we can derive a simple relationship between orthogonal sets of orthomorphisms, difference matrices, and MOLS obtained from Cayley tables of groups by permuting columns.

Theorem 1.21. *Let G be a group of order n. The following are equivalent.*

1. *There exists a set of r pairwise orthogonal orthomorphisms of G.*
2. *There exists a set of $r+1$ MOLS, based on G, obtained by permuting the columns of the Cayley table of G.*
3. *There exists an $(n, r+2; 1)$-difference matrix over G.*

Proof. By Theorem 1.20, the existence of r pairwise orthogonal orthomorphisms of G implies the existence of a set of $r+1$ MOLS, based on G, obtained by permuting the columns of the Cayley table of G.

By Theorem 1.17, the existence of a set of $r+1$ MOLS, based on G, obtained by permuting the columns of the Cayley table of G implies the existence of an $(n, r+2; 1)$-difference matrix over G.

If $D = (d_{ij})$ is an $(n, r+2; 1)$-difference matrix over G, then, by Theorem 1.16, we may assume without loss of generality that D is normalized and, hence, that

$d_{1j} = 1$ for all j. Then $\{d_{2j} \mid j = 1, \ldots, n\} = G$. For $i = 1, \ldots, r$, define $\theta_i : G \to G$ by $\theta_i(d_{2j}) = d_{i+2,j}$. As each of rows $3, \ldots, r+2$ of D is a permutation of the elements of G, each of $\theta_1, \ldots, \theta_r$ is a bijection. Further, for $i = 1, \ldots, r$,

$$\{x^{-1}\theta_i(x) \mid x \in G\} = \{d_{2j}^{-1} d_{i+2,j} \mid j = 1, \ldots, n\} = G,$$

and hence each of $\theta_1, \ldots, \theta_r$ is an orthomorphism. Similarly, if $i, k \in \{1, \ldots, r\}$, $i \neq k$, then

$$\{\theta_i(x)^{-1}\theta_k(x) \mid x \in G\} = \{d_{i+2,j}^{-1} d_{k+2,j} \mid j = 1, \ldots, n\} = G,$$

and so, $\theta_1, \ldots, \theta_r$ are pairwise orthogonal. □

The following is an immediate corollary:

Corollary 1.22. *Let G be a group of order $n > 1$. If there exists a set of r pairwise orthogonal orthomorphisms of G, then $r + 1 \leq N(n)$.*

In a 1995 survey of orthomorphisms, Bedford [35, 36] gave an alternative, though equivalent, construction of mutually orthogonal Latin squares from pairwise orthogonal sets of orthomorphisms of finite groups.

Corollary 1.22 yields a bound on the number of orthomorphisms possible in a set of pairwise orthogonal orthomorphisms.

Corollary 1.23. *Let G be a group of order $n > 1$. If there exists a set of r pairwise orthogonal orthomorphisms of G, then $r \leq n - 2$.*

Proof. By Theorem 1.5, if $n > 1$, then $N(n) \leq n - 1$. The result then follows from Corollary 1.22. □

If $|G| = n > 1$, then we will call a set of $(n - 2)$ pairwise orthogonal orthomorphisms of G a *complete set* of orthomorphisms of G. Theorem 1.21 yields a simple relationship between complete sets of orthomorphisms, generalized Hadamard matrices of index one, and complete sets of MOLS obtained from the Cayley tables of groups by permuting columns.

Corollary 1.24. *Let G be a group of order $n > 1$. The following are equivalent.*

1. *There exists a complete set of orthomorphisms of G.*
2. *There exists a* $\mathrm{GH}(n, 1)$ *over G.*
3. *There exists a complete set of MOLS obtained from the Cayley table of G by permuting columns.*

Here are some examples of pairwise orthogonal sets of orthomorphisms.

Example 1.25. Let $G = \mathrm{GF}(q)^+$, the additive group of the field of order q, and let $a \in \mathrm{GF}(q)$. Define $\theta\colon G \to G$ by $\theta(x) = ax + b$. It is easily seen that θ is a bijection if and only if $a \neq 0$, and the mapping $x \mapsto \theta(x) - x = (a - 1)x + b$ is a bijection if and only if $a \neq 1$. Thus θ is an orthomorphism if and only if $a \neq 0, 1$.

If $\theta' \colon G \to G$ is defined by $\theta'(x) = a'x + b'$, then $\theta \perp \theta'$ if and only if $a \neq a'$ as $\theta'(x) - \theta(x) = (a' - a)x + (b' - b)$. Thus the set $\{x \mapsto ax \mid a \neq 0, 1\}$ is a complete set of orthomorphisms of $\mathrm{GF}(q)^+$. Pairwise orthogonal sets of orthomorphisms of $\mathrm{GF}(q)^+$, q a prime power, will be studied in Chapters 9 and 10.

Example 1.26. Let $G = \mathrm{GF}(11)^+ \cong \mathbb{Z}_{11}$, the additive group of the field of order 11. Let $[a, b]$ denote the mapping $\mathrm{GF}(11)^+ \to \mathrm{GF}(11)^+$ defined by

$$[a,b](x) = \begin{cases} 0 & \text{if } x = 0, \\ ax & \text{if } x \in \{1,3,4,5,9\}, \\ bx & \text{if } x \in \{2,6,7,8,10\}. \end{cases}$$

It is easy to check that $[2, 6]$, $[4, 5]$, and $[9, 3]$ form a pairwise orthogonal set of orthomorphisms of $\mathrm{GF}(11)^+$. Note that the set $\{1, 3, 4, 5, 9\}$ is the set of nonzero squares of $\mathrm{GF}(11)$, while $\{2, 6, 7, 8, 10\}$ is the set of nonsquares of $\mathrm{GF}(11)$. Orthomorphisms of $\mathrm{GF}(11)^+$ of the form $[a, b]$ are quadratic orthomorphisms of $\mathrm{GF}(11)^+$. Quadratic orthomorphisms and their generalizations, cyclotomic orthomorphisms, will be studied in Chapter 10.

Example 1.27. Let G be a group of odd order n, and define $\theta \colon G \to G$ by $\theta(x) = x^2$. Clearly the mapping $x \mapsto x^{-1}\theta(x) = x$ is a bijection. Further, the mapping $x \mapsto \theta(x) = x^2$ is onto, and hence one to one, as $\theta(x^{(n+1)/2}) = (x^{(n+1)/2})^2 = x^{n+1} = x$. Thus θ is an orthomorphism of G. If n is also not divisible by 3, then the mapping $x \mapsto x^3$ is also an orthomorphism of G and is orthogonal to the mapping $x \mapsto x^2$. Pairwise orthogonal sets of orthomorphisms of the form $x \mapsto x^k$ will be studied in Section 8.2.

Example 1.28. Let $G = \mathbb{Z}_{21}$ and $D = \{d_1, \ldots, d_5\} = \{3, 6, 7, 12, 14\}$. As each nonzero element of G is uniquely expressible in the form $d_i - d_j$, D is a difference set in G. Let $M_1 = \left(m^1_{ij}\right)$, $M_2 = \left(m^2_{ij}\right)$, and $M_3 = \left(m^3_{ij}\right)$, $i, j = 1, \ldots, 5$, be defined by

$$M_1 = \begin{pmatrix} 1&5&4&3&2 \\ 3&2&1&5&4 \\ 5&4&3&2&1 \\ 2&1&5&4&3 \\ 4&3&2&1&5 \end{pmatrix}, \quad M_2 = \begin{pmatrix} 1&3&5&2&4 \\ 5&2&4&1&3 \\ 4&1&3&5&2 \\ 3&5&2&4&1 \\ 2&4&1&3&5 \end{pmatrix}, \quad \text{and} \quad M_3 = \begin{pmatrix} 1&4&2&5&3 \\ 4&2&5&3&1 \\ 2&5&3&1&4 \\ 5&3&1&4&2 \\ 3&1&4&2&5 \end{pmatrix}.$$

Note that M_1, M_2, and M_3 form a mutually orthogonal set of Latin squares in which $m^k_{ii} = i$ for $k = 1, 2, 3$ and $i = 1, \ldots, 5$. Let mappings $\theta_1, \theta_2, \theta_3 \colon G \to G$ be defined by $\theta_k(0) = 0$ and $\theta_k(d_i - d_j) = d_{m^k_{ij}} - d_j$ for $i \neq j$. As D is a difference set, θ_1, θ_2, and θ_3 are well-defined. These three mappings are shown in Table 1.2. It is routine to check that θ_1, θ_2, and θ_3 are pairwise orthogonal orthomorphisms of $\mathbb{Z}_{21}$. Pairwise orthogonal orthomorphisms constructed from difference sets, difference families, relative difference sets, and relative difference families will be studied in Section 8.3.

Table 1.2 Pairwise orthogonal orthomorphisms of $\mathbb{Z}_{21}$

x	0	1	2	3	4	5	6	7	8	9	10	11	12	13	14	15	16	17	18	19	20
$\theta_1(x)$	0	6	12	4	11	7	18	20	1	3	13	9	16	19	10	2	15	5	8	14	17
$\theta_2(x)$	0	18	16	11	9	20	8	17	6	4	19	3	15	14	13	12	2	7	1	10	5
$\theta_3(x)$	0	8	15	9	3	17	1	5	18	11	14	4	2	10	19	16	12	20	6	13	7

Example 1.29. Let α be a fixed-point-free automorphism of a finite group G, i.e., an automorphism of G that fixes only the identity element of G. We will see in Section 8.4 that any fixed-point-free automorphism of G is an orthomorphism of G and that two fixed-point-free automorphisms α and β of G are orthogonal orthomorphisms of G if and only if $\alpha\beta^{-1}$ is a fixed-point-free automorphism of G.

Let us extend the definition of orthogonality to cover mappings in general. For any two mappings $\theta, \phi \colon G \to G$, we say that θ is *orthogonal* to ϕ, written $\theta \perp \phi$, if the mapping $x \mapsto \theta^{-1}(x)\phi(x)$ is a bijection. As with orthomorphisms $\theta \perp \phi$ if and only if $\phi \perp \theta$. Now $\theta \colon G \to G$ is a bijection if orthogonal to the mapping $x \mapsto 1$, an orthomorphism if orthogonal to the mappings $x \mapsto 1$ and $x \mapsto x$, and a complete mapping if orthogonal to the mappings $x \mapsto 1$ and $x \mapsto x^{-1}$. A simple characterization of orthogonality of bijections follows:

Theorem 1.30. *Let $\theta, \phi \colon G \to G$ be bijections. Then $\theta \perp \phi$ if and only if the mapping $x \mapsto x^{-1}\theta\phi^{-1}(x)$ is a bijection $G \to G$.*

Proof. The mapping $x \mapsto x^{-1}\theta\phi^{-1}(x)$ is a bijection $G \to G$ if and only if the mapping

$$y = \phi^{-1}(x) \mapsto x = \phi(y) \mapsto \phi(y)^{-1}\theta(y)$$

is a bijection $G \to G$ if and only if $\theta \perp \phi$. □

As an easy corollary, we obtain a characterization of orthogonality of orthomorphisms.

Corollary 1.31. *Let θ and ϕ be orthomorphisms of G. Then $\theta \perp \phi$ if and only if $\theta\phi^{-1}$ is an orthomorphism of G.*

Let us consider Latin squares obtained by permuting the rows of the Cayley table of a group G rather than the columns. If we define L_θ to be the square with ijth entry $\theta(g_i)g_j$, then analogously to Theorem 1.20, we find that L_θ is a Latin square if and only if θ is a bijection; if θ is a bijection, then L_θ is orthogonal to L if and only if the mapping $x \mapsto \theta(x)x^{-1}$ is a bijection; and if θ and ϕ are bijections, then L_θ is orthogonal to L_ϕ if and only if the mapping $x \mapsto \theta(x)\phi(x)^{-1}$ is a bijection. This yields an alternative definition of orthomorphisms and orthogonality. To distinguish between the two types of orthomorphisms, let us say that a bijection $\theta \colon G \to G$ is a *right orthomorphism* if the mapping $x \mapsto \theta(x)x^{-1}$ is a bijection and a *left orthomorphism* if the mapping $x \mapsto x^{-1}\theta(x)$ is a bijection; and let us say that two left orthomorphisms θ, ϕ are *left orthogonal* if the mapping $x \mapsto \phi(x)^{-1}\theta(x)$ is a bijection, and two right orthomorphisms θ, ϕ are *right orthogonal* if the mapping

$x \mapsto \theta(x)\phi(x)^{-1}$ is a bijection. The "right handed" and "left handed" theories are equivalent as, if to each right orthomorphism θ, we associate the mapping θ' defined by $\theta'(x) = \theta(x^{-1})^{-1}$, then the mapping $\theta \mapsto \theta'$ establishes a one-to-one correspondence between the set of right orthomorphisms of G and the set of left orthomorphisms of G. Further θ is right orthogonal to ϕ if and only if θ' is left orthogonal to ϕ'. From now on we will use the term orthomorphism to refer to left orthomorphisms only and the term orthogonal to refer to left orthogonality only.

In Section 1.2, we defined a complete mapping of G to be a bijection $\theta\colon G \to G$ for which the mapping $x \mapsto x\theta(x)$ is a bijection. Orthomorphisms and complete mappings are closely related as a mapping $\theta\colon G \to G$ is an orthomorphism of G if and only if the mapping $x \mapsto x^{-1}\theta(x)$ is a complete mapping of G and a complete mapping of G if and only if the mapping $x \mapsto x\theta(x)$ is an orthomorphism of G.

1.3.3 Maximal Sets of MOLS

We will say that a set of pairwise orthogonal orthomorphisms of G is *maximal* if it cannot be extended to a larger set of pairwise orthogonal orthomorphisms of G. We know that from a set of r pairwise orthogonal orthomorphisms of G, we can construct a set of $r + 1$ MOLS based on G. The question is, if the set of orthomorphisms is maximal, is the set of MOLS also maximal? This question was answered in the affirmative, in the language of nets, by Ostrom [258] in 1966. The proof that we give here is a Latin square proof.

Theorem 1.32 (Ostrom, 1966). *Let G be a finite group and let L be its Cayley table. If $\theta_1, \ldots, \theta_r$ is a maximal set of pairwise orthogonal orthomorphisms of G, then $L, L_{\theta_1}, \ldots, L_{\theta_r}$ is a maximal set of MOLS.*

Proof. Let $\theta_1, \ldots, \theta_r$ be a maximal set of pairwise orthogonal orthomorphisms of G. Let the rows and columns of $L, L_{\theta_1}, \ldots, L_{\theta_r}$ be indexed by the elements of G. Suppose that M is orthogonal to each of $L, L_{\theta_1}, \ldots, L_{\theta_r}$, and, for s a symbol of M, let $T = \{(g_1, h_1), \ldots, (g_n, h_n)\}$ be the cells of M containing s. Then, T is a common transversal of $L, L_{\theta_1}, \ldots, L_{\theta_r}$. There exists a bijection $\phi\colon G \to G$ for which $g_k = \phi(h_k)^{-1}$. Then, the entries of T in L are $\{\phi(h_k)^{-1}h_k \mid k = 1, \ldots, n\}$. Thus, the mapping $x \mapsto x^{-1}\phi(x) = (\phi(x)^{-1}x)^{-1}$ is a bijection, implying that ϕ is an orthomorphism of G. The entries of T in L_{θ_i} are $\{\phi(h_k)^{-1}\theta_k(h_k) \mid k = 1, \ldots, n\}$. Thus, the mapping $x \mapsto \phi(x)^{-1}\theta_k(x)$ is a bijection, implying that $\phi \perp \theta_k$. Hence, $\{\theta_1, \ldots, \theta_r, \phi\}$ is a pairwise orthogonal set of orthomorphisms of G, contradicting the maximality of $\{\theta_1, \ldots, \theta_r\}$. Thus $L, L_{\theta_1}, \ldots, L_{\theta_r}$ is a maximal set of MOLS. □

Theorem 1.32 might tempt one to think that if there can exist at most k pairwise orthogonal orthomorphisms of a group G, then any set of $k+1$ MOLS, each isotopic to the Cayley table of G, would be maximal. This was shown to be false in 2016 by Egan and Wanless [107]. In the process of enumerating MOLS of small order, they created an unpublished catalogue of such MOLS. The following two examples are taken from that catalogue.

Example 1.33. Figure 1.8 shows three MOLS of order eight. In Figure 1.8, L_1 is the Cayley table of D_8, and L_2 is isotopic to the Cayley table of D_8. We will see in Section 13.2 that no pair of orthogonal orthomorphisms of D_8 exists. Hence, by Theorem 1.32, if L_2 were obtained from L_1 by permuting columns (or by permuting rows), then L_1 and L_2 would form a maximal set of two MOLS of order eight. However, L_3 is orthogonal to both L_1 and L_2. Note that L_3 in Figure 1.8 is isotopic to the Cayley table of $\mathbb{Z}_2 \times \mathbb{Z}_4$.

$$L_1 = \begin{pmatrix} a & b & c & d & e & f & g & h \\ b & c & d & a & f & g & h & e \\ c & d & a & b & g & h & e & f \\ d & a & b & c & h & e & f & g \\ e & h & g & f & a & d & c & b \\ f & e & h & g & b & a & d & c \\ g & f & e & h & c & b & a & d \\ h & g & f & e & d & c & b & a \end{pmatrix}, \quad L_2 = \begin{pmatrix} a & b & c & d & e & f & g & h \\ f & e & h & g & d & c & b & a \\ h & g & f & e & b & a & d & c \\ c & d & a & b & g & h & e & f \\ g & f & e & h & c & b & a & d \\ b & c & d & a & h & e & f & g \\ d & a & b & c & f & g & h & e \\ e & h & g & f & a & d & c & b \end{pmatrix}, \quad L_3 = \begin{pmatrix} a & b & c & d & e & f & g & h \\ h & g & f & e & b & a & d & c \\ e & h & g & f & c & b & a & d \\ b & c & d & a & f & g & h & e \\ d & a & b & c & h & e & f & g \\ g & f & e & h & a & d & c & b \\ f & e & h & g & d & c & b & a \\ c & d & a & b & g & h & e & f \end{pmatrix}.$$

Fig. 1.8 Three MOLS of order eight

Example 1.34. Figure 1.9 shows four MOLS of order nine. In Figure 1.9, M_1 is the Cayley table of $\mathbb{Z}_9$, and M_2 is isotopic to the Cayley table of $\mathbb{Z}_9$. We will see in Section 13.3 that no pair of orthogonal orthomorphisms of $\mathbb{Z}_9$ exists. Hence, by Theorem 1.32, if M_2 were obtained from M_1 by permuting columns (or by permuting rows), then M_1 and M_2 would form a maximal set of two MOLS of order nine. However, M_3 and M_4 extend M_1 and M_2 to a set of four MOLS of order nine. Note that neither M_3 nor M_4 in Figure 1.9 is isotopic to the Cayley table of a group.

$$M_1 = \begin{pmatrix} a & b & c & d & e & f & g & h & i \\ b & c & d & e & f & g & h & i & a \\ c & d & e & f & g & h & i & a & b \\ d & e & f & g & h & i & a & b & c \\ e & f & g & h & i & a & b & c & d \\ f & g & h & i & a & b & c & d & e \\ g & h & i & a & b & c & d & e & f \\ h & i & a & b & c & d & e & f & g \\ i & a & b & c & d & e & f & g & h \end{pmatrix}, \quad M_2 = \begin{pmatrix} a & b & c & d & e & f & g & h & i \\ f & a & b & i & d & e & c & g & h \\ e & f & a & h & i & d & b & c & g \\ g & h & i & a & b & c & d & e & f \\ c & g & h & f & a & b & i & d & e \\ b & c & g & e & f & a & h & i & d \\ d & e & f & g & h & i & a & b & c \\ i & d & e & c & g & h & f & a & b \\ h & i & d & b & c & g & e & f & a \end{pmatrix},$$

$$M_3 = \begin{pmatrix} a & b & c & d & e & f & g & h & i \\ d & f & e & g & i & h & a & c & b \\ i & a & h & c & d & b & f & g & e \\ h & i & a & b & c & d & e & f & g \\ b & d & f & e & g & i & h & a & c \\ g & e & i & a & h & c & d & b & f \\ c & g & b & f & a & e & i & d & h \\ f & h & d & i & b & g & c & e & a \\ e & c & g & h & f & a & b & i & d \end{pmatrix}, \quad M_4 = \begin{pmatrix} a & b & c & d & e & f & g & h & i \\ e & g & f & h & a & i & b & d & c \\ h & i & d & b & c & g & e & f & a \\ b & f & g & e & i & a & h & c & d \\ i & h & a & c & b & d & f & e & g \\ c & d & e & f & g & h & i & a & b \\ f & a & h & i & d & b & c & g & e \\ d & c & b & g & f & e & a & i & h \\ g & e & i & a & h & c & d & b & f \end{pmatrix}.$$

Fig. 1.9 Four MOLS of order nine

There is also a natural correspondence between maximal set of orthomorphisms of a group G and maximal difference matrices of index 1 over G.

Theorem 1.35. *The existence of a maximal set of r orthomorphisms of G is equivalent to the existence of a maximal $(|G|, r+2; 1)$-difference matrix over G.*

Proof. Let $G = \{g_1, \ldots, g_n\}$ be a group of order n, and let $\theta_1, \ldots, \theta_r$ be a maximal set of orthomorphisms of G. Form a matrix D whose ijth entry is

$$d_{ij} = \begin{cases} 1 & \text{if } i = 1, \\ g_j & \text{if } i = 2, \\ \theta_{i-2}(g_j) & \text{if } 2 < i \leq r+2. \end{cases}$$

The matrix D is an $(n, r+2; 1)$-difference matrix over G, and if we can add a row $d_1, \ldots, d_n$ to D, then the mapping $g_i \mapsto d_i$ is an orthomorphism of G orthogonal to each of $\theta_1, \ldots, \theta_r$, contradicting the maximality of $\theta_1, \ldots, \theta_r$. Hence, D is a maximal $(n, r+2; 1)$-difference matrix over G.

Let $D = \{d_{ij}\}$ be a maximal $(n, r+2; 1)$-difference matrix over G. Without loss of generality, we may assume that D is normalized, $d_{1j} = 1$ and $d_{2j} = g_j$ for $j = 1' \ldots, n$. For $i = 1, \ldots, r$, define $\theta_i : G \to G$ by $\theta_i(g_j) = d_{i+2,j}$. Then, $\theta_1, \ldots, \theta_r$ is a pairwise orthogonal set of orthomorphisms of G. If ϕ is an orthomorphism of G orthogonal to each of $\theta_1, \ldots, \theta_r$, then adding a row $\phi(g_1), \ldots, \phi(g_n)$ extends D to an $(n, r+3; 1)$-difference matrix over G, contradicting the maximality of D. Hence, $\theta_1, \ldots, \theta_r$ is a maximal set of orthomorphisms of G. □

1.3.4 Quotient Group Constructions

In 1999, for H a normal subgroup of G, Quinn [273] gave a construction of pairwise orthogonal orthomorphisms of G from pairwise orthogonal orthomorphisms of H and pairwise orthogonal orthomorphisms of G/H: this is a special case of a theorem on constructions of difference matrices. The proof we will give is Quinn's proof for the special case in which each difference matrix is of index one.

Theorem 1.36 (Quinn, 1999). *If H is a normal subgroup of G and there exists a $(|H|, r; 1)$-difference matrix over H and a $(|G/H|, r; 1)$-difference matrix over G/H, then there exists a $(|G|, r; 1)$-difference matrix over G.*

Proof. Let $|H| = m$ and $|G/H| = n$. Let $A = (a_{ij})$ be an $(m, r; 1)$-difference matrix over H and $C = (c_{ij}H)$ be an $(n, r; 1)$-difference matrix over G/H. Let T be a system of distinct coset representatives for H in G, and let t_{ij} denote the unique element of T for which $t_{ij}H = c_{ij}H$. Let $B_k = (a_{ij}t_{ik})$ be an $r \times m$ matrix, for $k = 1, \ldots, n$, and set $E = (B_1 \cdots B_n)$. We will call B_k the kth block of E.

We will show that E is an $(mn, r; 1)$-difference matrix over G. To do this we will need to show for $i_1 \neq i_2$ that the quotients $(a_{i_1 j}t_{i_1 k})^{-1}(a_{i_2 j}t_{i_2 k})$, $j = 1, \ldots, m$, $k = 1, \ldots, n$, are distinct. We first note that quotients from the same block must be distinct as

$$(a_{i_1 j_1} t_{i_1 k})^{-1}(a_{i_2 j_1} t_{i_2 k}) = (a_{i_1 j_2} t_{i_1 k})^{-1}(a_{i_2 j_2} t_{i_2 k})$$

if and only if

$$a_{i_1 j_1}^{-1} a_{i_2 j_1} = a_{i_1 j_2}^{-1} a_{i_2 j_2},$$

which, as A is a difference matrix of index one, occurs if and only if $j_1 = j_2$.

It remains to show that quotients from different blocks are distinct. This follows as, if

$$(a_{i_1 j_1} t_{i_1 k_1})^{-1}(a_{i_2 j_1} t_{i_2 k_1}) = (a_{i_1 j_2} t_{i_1 k_2})^{-1}(a_{i_2 j_2} t_{i_2 k_2}),$$

then

$$t_{i_1 k_1}^{-1} t_{i_2 k_1} (t_{i_2 k_1}^{-1} (a_{i_1 j_1}^{-1} a_{i_2 j_1}) t_{i_2 k_1}) H = t_{i_1 k_2}^{-1} t_{i_2 k_2} (t_{i_2 k_2}^{-1} (a_{i_1 j_2}^{-1} a_{i_2 j_2}) t_{i_2 k_2}) H,$$

and so

$$t_{i_1 k_1}^{-1} t_{i_2 k_1} H = t_{i_1 k_2}^{-1} t_{i_2 k_2} H,$$

which implies that

$$(t_{i_1 k_1} H)^{-1}(t_{i_2 k_1} H) = (t_{i_1 k_2} H)^{-1}(t_{i_2 k_2} H),$$

from which, as D is a difference matrix of index one, it follows that $k_1 = k_2$. □

Quinn's result was stronger and is given as the following corollary:

Corollary 1.37. *If H is a normal subgroup of G and there exists a $(|H|, r; \lambda)$-difference matrix over H and a $(|G/H|, r; 1)$-difference matrix over G/H, then there exists a $(|G|, r; \lambda)$-difference matrix over G.*

Proof. This is similar to the proof of Theorem 1.36 except that the blocks B_k are $r \times m\lambda$ matrices, and if g appears as a quotient in a block, then it appears λ times as a quotient in the same block. □

The following are immediate corollaries of Theorem 1.36: the first of these was proved by Hall and Paige [171] in 1955.

Corollary 1.38 (Hall and Paige, 1955). *Let H be a normal subgroup of G. If H and G/H admit orthomorphisms (resp. complete mappings), then G admits orthomorphisms (resp. complete mappings).*

Hall and Paige's proof of Corollary 1.38 will be given in Section 4.1.

Corollary 1.39. *Let H be a normal subgroup of G. If H and G/H admit a set of r pairwise orthogonal orthomorphisms, then G admits a set of r pairwise orthogonal orthomorphisms.*

1.4 Incidence Structures Related to MOLS

An *incidence structure* $(\mathbb{P}, \mathbb{L}, \mathbb{I})$ consists of a set $\mathbb{P}$ of *points*, a set $\mathbb{L}$ of *lines*, and a subset $\mathbb{I}$ of $\mathbb{P} \times \mathbb{L}$. If $(P, l) \in \mathbb{I}$, then we say that the point P is *incident* with the line l and the line l is *incident* with the point P. We will often make use of more geometric language to describe incidence such as, the point P is on the line l, or the line l passes through the point P, etc. Readers interested in the many classes of incidence structures and what is known about them can consult the 1999 book Design Theory by Beth, Jungnickel, and Lenz [41] or the 2007 CRC Handbook of Combinatorial Designs [79]. Four classes of incidence structures are very closely related to MOLS: nets, transversal designs, affine planes, and projective planes.

1.4.1 Nets and Transversal Designs

A *net* is an incidence structure, in which the lines are partitioned into equivalence classes called *parallel classes* and which satisfies the following:

1. Two distinct points are incident with at most one line.
2. Each point is incident with exactly one line of each parallel class.
3. Lines in different parallel classes have exactly one point in common.
   We also impose two nondegeneracy conditions.
4. There are at least three parallel classes.
5. Some line is incident with more than one point.

We leave it to the reader to determine all degenerate nets, i.e., nets in which at least one of the last two conditions does not hold. We say that two lines in the same parallel class are *parallel* and so any line is parallel to itself. As each point is incident with exactly one line of each parallel class, distinct lines from the same parallel class do not intersect. It follows that the lines of a parallel class induce a partition of the point set. For a finite net, a net with finitely many points and lines, the *degree* is the number of parallel classes. A net of degree k is also called a k-*net*.

Theorem 1.40. *Let $\mathcal{N}$ be a finite k-net. If some line is incident with n points, then the following hold:*

1. *Each line is incident with n points.*
2. *Each point is incident with k lines.*
3. *Each parallel class contains n lines.*
4. *$\mathcal{N}$ has n^2 points.*
5. *$\mathcal{N}$ has nk lines.*

Proof. Let l be a line of $\mathcal{N}$ incident with n points and let l' be a line distinct from l. As $k \geq 3$, there exists a parallel class π not containing either l or l'. We define a mapping f from the points incident with l to the points incident with l' as follows. If P is a point incident with l and P' is a point incident with l', then $\mathrm{f}(P) = P'$ if there is a line of π incident with both P and P'. Now f is well-defined as if P is a point incident with l, then P is incident with exactly one line, m, of π, and m meets l' in exactly one point, P'. Similarly we can define a mapping g from the points incident with l' to the points incident with l using the same parallel class π. It is clear that g is the inverse of f. Hence, f is a bijection from which it follows that every line of $\mathcal{N}$ is incident with n points.

Let π be a parallel class of $\mathcal{N}$ and let $m \notin \pi$ be a line of $\mathcal{N}$. As each point incident with m is incident with exactly one line of π and each line of π meets m in exactly one point, there are as many lines in π as points incident with m. Hence each parallel class of $\mathcal{N}$ contains n lines. As there are k parallel classes, $\mathcal{N}$ has nk lines and, as each point is incident with exactly one line of each parallel class, each point is incident with exactly k lines.

Let π and π' be distinct parallel classes of $\mathcal{N}$. Number the lines of π and the lines of π' from 1 to n, and assign the ordered pair (i, j) to a point of $\mathcal{N}$ if the point is incident with the ith line of π and the jth line of π'. As each point of $\mathcal{N}$ is incident with exactly one line of π and exactly one line of π', each point is represented by an ordered pair and, as each line of π intersects each line of π' in exactly one point, each ordered pair represents exactly one point of $\mathcal{N}$. Hence, as there are n^2 ordered pairs, $\mathcal{N}$ has n^2 points. □

If any, and hence every, line of a net is incident with n points, then we say that the net has *order* n. Note that, as a line of a net is uniquely determined by the points that it is incident with, we will frequently describe lines as sets of points: a parallel class then partitions the point set. Here are three examples of nets of order 4, a 3-net and two 4-nets.

Example 1.41. Consider the Latin square

$$L = \begin{pmatrix} 0\,1\,2\,3 \\ 1\,2\,3\,0 \\ 2\,3\,0\,1 \\ 3\,0\,1\,2 \end{pmatrix}$$

of order 4. We can construct an incidence structure from L by taking the points to be the cells of L and the lines to be the rows, columns, and sets of cells with the same entries. Thus, labeling the rows and columns 0, 1, 2, 3, and writing ij instead of (i, j), the lines are

$$\begin{array}{lll} r_0 = \{00, 01, 02, 03\}, & c_0 = \{00, 10, 20, 30\}, & s_0 = \{00, 13, 22, 31\}, \\ r_1 = \{10, 11, 12, 13\}, & c_1 = \{01, 11, 21, 31\}, & s_1 = \{01, 10, 23, 32\}, \\ r_2 = \{20, 21, 22, 23\}, & c_2 = \{02, 12, 22, 23\}, & s_2 = \{02, 11, 20, 33\}, \\ r_3 = \{30, 31, 32, 33\}, & c_3 = \{03, 13, 23, 33\}, & s_3 = \{03, 12, 21, 30\}. \end{array}$$

The incidence structure that we have constructed satisfies the properties of a 3-net of order 4. The parallel classes of this net are $\{r_0, r_1, r_2, r_3\}$, $\{c_0, c_1, c_2, c_3\}$, and $\{s_0, s_1, s_2, s_3\}$.

Example 1.42. Consider the orthogonal pair of Latin squares

$$L_1 = \begin{pmatrix} 0\,1\,2\,3 \\ 1\,0\,3\,2 \\ 2\,3\,0\,1 \\ 3\,2\,1\,0 \end{pmatrix} \quad \text{and} \quad L_2 = \begin{pmatrix} 0\,2\,3\,1 \\ 1\,3\,2\,0 \\ 2\,0\,1\,3 \\ 3\,1\,0\,2 \end{pmatrix}$$

of order 4. We can construct an incidence structure from L_1 and L_2 by taking the points to be the cells of these squares. Labeling the rows and columns 0, 1, 2, 3, and writing ij instead of (i, j), the cells are ij, $i, j = 0, 1, 2, 3$. The lines are the rows, the columns, sets of cells in L_1 with the same entries, and sets of cells in L_2 with the same entries. Thus the lines are

$$\begin{aligned} r_0 &= \{00, 01, 02, 03\}, & c_0 &= \{00, 10, 20, 30\}, \\ r_1 &= \{10, 11, 12, 13\}, & c_1 &= \{01, 11, 21, 31\}, \\ r_2 &= \{20, 21, 22, 23\}, & c_2 &= \{02, 12, 22, 22\}, \\ r_3 &= \{30, 31, 32, 33\}, & c_3 &= \{03, 13, 23, 33\}, \end{aligned}$$

$$\begin{aligned} s_0 &= \{00, 11, 22, 33\}, & t_0 &= \{00, 13, 21, 32\}, \\ s_1 &= \{01, 10, 23, 32\}, & t_1 &= \{03, 10, 22, 31\}, \\ s_2 &= \{02, 13, 20, 31\}, & t_2 &= \{01, 12, 20, 33\}, \\ s_3 &= \{03, 12, 21, 30\}, & t_3 &= \{02, 11, 23, 30\}. \end{aligned}$$

The incidence structure that we have constructed satisfies the properties of a 4-net of order 4. The parallel classes of this net are $\{r_0, r_1, r_2, r_3\}$, $\{c_0, c_1, c_2, c_3\}$, $\{s_0, s_1, s_2, s_3\}$, and $\{t_0, t_1, t_2, t_3\}$.

Example 1.43. Let $\mathrm{GF}(4) = \{a + b\alpha \mid a, b \in \mathrm{GF}(2), \alpha^2 = \alpha + 1\}$, and let D be the incidence structure whose points are the elements of $\mathrm{GF}(4) \times \mathrm{GF}(4)$ and whose lines are

$$r_a = \{(a, y) \mid y \in \mathrm{GF}(4)\}, a \in \mathrm{GF}(4);$$

$$c_b = \{(x, b) \mid x \in \mathrm{GF}(4)\}, b \in \mathrm{GF}(4);$$

$$s_b = \{(x, x + b) \mid x \in \mathrm{GF}(4)\}, b \in \mathrm{GF}(4);$$

and

$$t_b = \{(x, \alpha x + b) \mid x \in \mathrm{GF}(4)\}, b \in GF(4).$$

Thus the lines are represented by the equations $x = a$, $y = b$, $y = x + b$, and $y = \alpha x + b$.

The incidence structure D is a 4-net of order 4, the four parallel classes being lines represented by equations of the form $x = a$, lines represented by equations of the form $y = b$, lines represented by equations of the form $y = x + b$, and lines represented by equations of the form $y = \alpha x + b$.

The constructions of Examples 1.41 and 1.42 illustrate how we can construct nets from Latin squares and MOLS. In fact any k-net of order n can be constructed from a set of $k - 2$ MOLS of order n, and we can also construct $k - 2$ MOLS of order n from k-nets of order n.

Theorem 1.44. *A k-net of order n exists if and only if a set of $k - 2$ MOLS of order n exists.*

Proof. Let $L_1, \ldots, L_{k-2}$ be a set of $k - 2$ MOLS of order $n \geq 2$ on the symbol set S, and let the rows and columns of these squares be indexed $1, \ldots, n$. Form an incidence structure $\mathscr{N}$ whose points are the ordered pairs (i, j), $i, j = 1, \ldots, n$ and whose lines are

$$r_a = \{(a, y) \mid y = 1, \ldots, n\}, a = 1, \ldots, n;$$

$$c_b = \{(x, b) \mid x = 1, \ldots, n\}, b = 1, \ldots, n;$$

and

$$s_{ib} = \{(x, y) \mid \text{ the entry in the } xy\text{th cell of } L_i \text{ is } b\}, i = 1, \ldots, k-2 \quad \text{and} \quad b \in S.$$

Each line of $\mathscr{N}$ contains $n \geq 2$ points, and the $k \geq 3$ parallel classes of $\mathscr{N}$ are $R = \{r_1, \ldots, r_n\}$, $C = \{c_1, \ldots, c_n\}$, and $S_i = \{s_{ib} \mid b \in S\}$ for $i = 1, \ldots, k - 2$. Clearly $\mathscr{N}$ satisfies the nondegeneracy properties of a net, and distinct lines from the same parallel class have no points in common. The point (i, j) is on only one line of the parallel class R, r_i; only one line of the parallel class C, c_j; and, for $i = 1, \ldots, k - 2$, only one line in the parallel class S_k, s_{kb}, where b is the ijth entry of L_k. Any line from R intersects any line from C in exactly one point as any row intersects any column in one cell. As each L_k is a Latin square, s_{kb} contains one entry in each row and each column and hence meets each line of R and each line of C in exactly one point. If $k \neq k'$, as L_k and $L_{k'}$ are orthogonal, for each $a, b \in S$, there is exactly one choice for i and j for which the ijth entry of L_k is a and the ijth entry of $L_{k'}$ is b. Thus each line s_{ka} meets each line $s_{k'b}$ in exactly one point. Hence lines in different parallel classes have exactly one point in common. As distinct lines from the same parallel class have no points in common, two distinct points can be contained in at most one line from a given parallel class, and, as lines from different parallel classes have exactly one point in common, two distinct points can be contained in at most one line. Thus $\mathscr{N}$ is a k-net of order n.

Let $\mathcal{N}$ be a k-net of order n, let the parallel classes of $\mathcal{N}$ be $\pi_1, \dots, \pi_k$, and suppose that the lines of each parallel class are numbered $1, \dots, n$. Label a point P with the ordered pair (i, j) if P is on the ith line of π_1 and the jth line of π_2. We next form n by n arrays $L_1, \dots, L_{k-2}$ by setting the ijth entry of L_m equal to t if the point (i, j) is on the tth line of π_{m+2}. For each $i = 1 \dots, n$, $m = 1, \dots, k-2$, and $t \in S$, symbol t cannot appear twice in the ith row of L_m as the ith line of π_1 would then meet the tth line of π_{m+2} in at least two points. Similarly, for each $j = 1 \dots, n$, $m = 1, \dots, k-2$, and $t \in S$, the symbol t cannot appear twice in the jth column of L_m. Thus each of $L_1, \dots, L_{k-2}$ is a Latin square. If $a, b \in S$ and $m, m' \in \{1, \dots, k-2\}$, $m \neq m'$, then the ath line of π_m meets the bth line of $\pi_{m'}$ in exactly one point, (i, j). It follows that L_m and $L_{m'}$ are orthogonal. Hence $L_1, \dots, L_{k-2}$ is a set of $k-2$ MOLS of order n. □

As a corollary we obtain a bound on the degree of a finite net.

Corollary 1.45. *If there exists a k-net of order $n \geq 2$, then $k \leq n+1$.*

Proof. By Theorem 1.44, if there exists a k-net of order $n \geq 2$, then there exists a set of $k-2$ MOLS of order n. Hence, by Theorem 1.5, $k-2 \leq n-1$ and so $k \leq n+1$. □

The construction of nets from MOLS, given in the proof of Theorem 1.44, was illustrated in Examples 1.41 and 1.42. Let us use Example 1.43 to illustrate the construction of MOLS from nets given in the proof of Theorem 1.44. From the 4-net of order 4 in Example 1.43, we will construct a pair of orthogonal Latin squares of order 4. As the symbols used are the elements of $GF(4) = \{a\alpha + b \mid a, b \in GF(2)\}$, let the lines and columns of these squares be labelled in order $\{0, 1, \alpha, 1+\alpha\}$. A point P will be labelled (x, y) if it is on the lines $\{(x, y) \mid y \in GF(4)\}$ and $\{(x, y) \mid x \in GF(4)\}$. We can then construct squares L_1 and L_2 by setting the xyth entry of L_1 equal to w if $y = x + w$ and the xyth entry of L_2 equal to w if $y = \alpha x + w$. Thus the xyth entry of L_1 is $x + y$, while the xyth entry of L_2 is $y + \alpha x$. Hence

$$L_1 = \begin{pmatrix} 0 & 1 & \alpha & 1+\alpha \\ 1 & 0 & 1+\alpha & \alpha \\ \alpha & 1+\alpha & 0 & 1 \\ 1+\alpha & \alpha & 1 & 0 \end{pmatrix} \quad \text{and} \quad L_2 = \begin{pmatrix} 0 & 1 & \alpha & 1+\alpha \\ \alpha & 1+\alpha & 0 & 1 \\ 1+\alpha & \alpha & 1 & 0 \\ 1 & 0 & 1+\alpha & \alpha \end{pmatrix}.$$

It is easy to see that L_1 and L_2 are orthogonal Latin squares.

The incidence structures that are most often used to construct MOLS are actually the duals of nets. If $(\mathbb{P}, \mathbb{L}, \mathbb{I})$ is an incidence structure, then its *dual* is the incidence structure $(\mathbb{P}', \mathbb{L}', \mathbb{I}')$, where $\mathbb{P}' = \mathbb{L}$, $\mathbb{L}' = \mathbb{P}$, and $\mathbb{I}' = \{(l, P) \mid (P, l) \in \mathbb{I}\}$. The dual of a net of order n and degree k is a *transversal design* of *order* n and *linesize* k, denoted $TD(k, n)$. In the duals of properties of a net, parallel classes of lines become *point classes*, and the following hold:

1. Two distinct lines meet in at most one point.
2. Each line is incident with exactly one point of each point class.

3. Points in different point classes are incident with exactly one line.
   The nondegeneracy conditions become the following:
4. There are at least three point classes.
5. Some point is incident with more than one line.

As nets and transversal designs are duals of each other, constructions of MOLS from nets are equivalent to constructions of MOLS from transversal designs.

Corollary 1.46. *If $n \geq 2$ and $k \geq 3$, then a* TD(k, n) *exists if and only if a set of $k-2$ MOLS of order n exists.*

Transversal designs are used more than nets in the construction of MOLS because there are many constructions of transversal designs from other classes of designs. The extra step of dualizing is superfluous. For a survey of the constructions of MOLS using transversal designs as well as other classes of designs, see the 2007 paper of Colbourn and Dinitz [78].

A natural question to ask is, if a set of MOLS is based on a group, what does this tell us about the net constructed from the set of MOLS? A *collineation* of a net $\mathcal{N} = \{\mathbb{P}, \mathbb{L}, \mathbb{I}\}$ is a bijection $\phi\colon \mathbb{P} \to \mathbb{P}$ that maps lines of $\mathcal{N}$ to lines of $\mathcal{N}$ and hence induces a bijection $\mathbb{L} \mapsto \mathbb{L}$. Under composition the collineations of a net form a group, the *collineation group* of the net. Of particular interest to us are the *strict translations*, described by Ostrom [258] in 1966: these are collineations that fix no points and fix all lines of a given parallel class. Strict translations fix all parallel classes.

Corollary 1.47. *Any strict translation of a net fixes all parallel classes of a net.*

Proof. Let τ be a strict translation fixing all lines of a given parallel class π. Suppose l to be a line that is not parallel to $\tau(l)$. Let P be the unique point on both l and $\tau(l)$. Then $P = \tau(Q)$ for some point Q. Let m be the unique line of π through Q. As m is fixed, Q, P, and hence $\tau(P)$ must all be on m. As the line joining P and $\tau(P)$ is unique, $\tau(l) = m$ and so $l = \tau(l)$, a contradiction from which the result follows. □

The nets arising from orthogonal sets of orthomorphisms are precisely those with a group of collineations, consisting of the identity and strict translations fixing the lines of one parallel class that acts regularly on the points of a line in this class.

Theorem 1.48. *Let G be a group of order $n \geq 2$ and suppose that $k \geq 3$. There exists a k-net of order n that admits a collineation group, isomorphic to G, consisting of the identity and strict translations fixing the lines of one parallel class, that acts regularly on the points of a line in this class if and only if G admits a set of $k-3$ pairwise orthogonal orthomorphisms.*

Proof. Let us first assume that $k > 3$.

Let $\theta_1, \ldots, \theta_{k-3}$ be a set of pairwise orthogonal orthomorphisms of G. We will define an incidence structure $\mathcal{N}$ as follows. The point set of $\mathcal{N}$ is $G \times G$, and the lines are partitioned into k parallel classes, $\pi_1, \ldots, \pi_k$. For $g \in G$, the gth line of π_1 is $\{(g, y) \mid y \in G\}$, the gth line of π_2 is $\{(x, g) \mid x \in G\}$, the gth line of π_3 is $\{(x, xg) \mid x \in G\}$, and, for $m = 4, \ldots, k$, the gth line of π_m is

$\{(x, \theta_{m-3}(x)g) \mid x \in G\}$. Thus the lines are described by equations of the form $x = g$, $y = g$, $y = xg$, and, for $m = 4, \ldots, k$, $y = \theta_{m-3}(x)g$. $\mathcal{N}$ clearly satisfies the nondegeneracy conditions of a net: each line contains $n \geq 2$ points and there are $k \geq 3$ parallel classes. The point (x, y) is on the xth line only of π_1, the yth line only of π_2, the $x^{-1}y$th line only of π_3, and, for $m = 4, \ldots, k$, as θ_{m-3} is a bijection, the $\theta_{m-3}(x)^{-1}y$th line only of π_m. Thus any point is on exactly one line of each parallel class. The gth line of π_1 meets the hth line of π_2 in exactly one point, the point (g, h); the gth line of π_1 meets the hth line of π_3 in exactly one point, the point (g, gh); and, for $m = 4, \ldots, k$, the gth line of π_1 meets the hth line of π_m in exactly one point, the point $(g, \theta_{m-3}(g)h)$. The gth line of π_2 meets the hth line of π_3 in exactly one point, the point (gh^{-1}, g); and, for $m = 4, \ldots, k$, the gth line of π_2 meets the hth line of π_m in exactly one point, the point $(\theta_{m-3}^{-1}(gh^{-1}), g)$. For $m = 4, \ldots, k$, the gth line of π_3 meets the hth line of π_m in the point $(x, xg) = (x, \theta_{m-3}(x)h)$: this point is uniquely determined by the equation $x^{-1}\theta_{m-3}(x) = gh^{-1}$ and the fact that θ_{m-3} is an orthomorphism of G. For $m, m' = 4, \ldots, k$, $m \neq m'$, the gth line of π_m meets the hth line of $\pi_{m'}$ in the point $(x, \theta_{m-3}(x)g) = (x, \theta'_{m-3}(x)h)$: this point is uniquely determined by the equation $\theta_{m-3}(x)^{-1}\theta_{m'-3}(x) = gh^{-1}$ and the fact that $\theta_{m-3} \perp \theta_{m'-3}$. Thus lines from distinct parallel classes meet in exactly one point. As the lines of any parallel class partition the points, two distinct points are connected by at most one line. Hence, $\mathcal{N}$ is k-net of order n. The mapping $(x, y)^g = (x, yg)$ defines an action of G on the points of $\mathcal{N}$. If $g \neq 1$, then $(x, y)^g \neq (x, y)$, and hence, under this action, the nonidentity elements of G fix no points of $\mathcal{N}$. As this action fixes every line of π_1, the nonidentity elements of G are strict translations. Further G acts regularly on the points of any line of π_1.

Let $\mathcal{N}$ be a k-net of order n with parallel classes $\pi_1, \ldots, \pi_k$. Let G be a collineation group that acts regularly on the points of a line l of π_1 and whose nonidentity elements are strict translations that fix the lines of π_1. Let P be a point on l. First note that, as G acts regularly on the points of l, for each point Q on l, there exists a unique element $g \in G$ that moves P to Q. We will label the lines of each parallel class as follows. First assign to each point Q of l the unique element $g \in G$ that moves P to Q. As, for $i = 2, \ldots, k$, each line of π_i meets l in exactly one point Q, we will label this line with the unique element $g \in G$ assigned to Q. We next need to label the lines of π_1. A line l' of π_1 meets the line of π_3 labelled 1 in a unique point Q which must be on a unique line of π_2, labelled g say. We will label l' with g. Next label a point of $\mathcal{N}$ (x, y) if it is on the xth line of π_1 and the yth line of π_2. By Lemma 1.47, each parallel class is fixed by each element of G. For $g \in G$ the xth line of π_1 is fixed by g and the yth line of π_2 is moved to the ygth line of π_2 by g. Hence g moves the point (x, y) to the point (x, yg). The gth line of π_1 is $\{(x, g) \mid x \in G\}$, the gth line of π_2 is $\{(g, y) \mid y \in G\}$, and, as the action of G on $\mathcal{N}$ is given by $(x, y)^g = (x, yg)$, the gth line of π_3 is $\{(x, xg) \mid x \in G\}$. For $m = 4, \ldots, k$, let l' be the line of π_m through P and define a mapping $\theta_{m-3}\colon G \to G$ by $\theta_{m-3}(x) = y$ if (x, y) is the point of intersection of l' with the xth line of π_2: θ_{m-3} is well-defined as this intersection point is uniquely determined. θ_{m-3} is onto and hence, by finiteness one to one, as the intersection of l' with the yth line of π_1 is uniquely determined. For each $m = 4, \ldots, k$, the mapping $x \mapsto x^{-1}\theta_{m-3}(x)$ is a bijection as the intersection $(x, \theta_{m-3}(x)) = (x, xg)$ of l' with the gth line of π_3 is uniquely determined, and so x is uniquely determined by the equation $x^{-1}\theta_{m-3}(x) = g$. Hence, the mapping $x \mapsto x^{-1}\theta_{m-3}(x)$ is 1–1 and hence, by finiteness, is a bijection. It follows that $\theta_1, \ldots, \theta_{k-3}$ are orthomorphisms of G. For each $m, m' \in \{4, \ldots, k\}$, $m \neq m'$, the mapping $x \mapsto \theta_{m'-3}(x)^{-1}\theta_{m-3}(x)$ is a bijection as the intersection $(x, \theta_{m-3}(x)) = (x, \theta_{m'-3}(x)g)$ of the 1st line of π_m with the gth line of $\pi_{m'}$ is uniquely determined, and so x is uniquely determined by the equation $\theta_{m'-3}(x)^{-1}\theta_{m-3}(x) = g$. Hence the mapping $x \mapsto \theta_{m'-3}(x)^{-1}\theta_{m-3}(x)$ is 1–1 and hence, by finiteness, is a bijection. It follows that $\theta_1, \ldots, \theta_{k-3}$ are pairwise orthogonal orthomorphisms of G.

We leave it to the reader to verify that this result still holds when $k = 3$. □

1.4.2 Affine Planes

An *affine plane* is an incidence structure, in which the lines are partitioned into equivalence classes called *parallel classes*, which satisfies the following:

1. Two distinct points are incident with exactly one line.
2. Each point is incident with exactly one line of each parallel class.
3. Lines in different parallel classes have exactly one point in common.
   We also impose a nondegeneracy condition.
4. There exist three noncollinear points.

An affine plane is a net. From the single nondegeneracy condition for an affine plane, we can derive both nondegeneracy conditions for a net. As any two points are joined by a line, the existence of three noncollinear points implies the existence of a line incident with more than one point, and, as each pair of points in a set of three noncollinear points determines a unique line and the three lines so determined intersect each other, these lines must belong to three distinct parallel classes. The *order* of an affine plane is its order as a net, i.e., the number of points incident with a line. An affine plane is actually a net with the maximum number of parallel classes.

Theorem 1.49. *If $\mathscr{A}$ is a finite affine plane of order n, then the following hold.*

1. *Each line of $\mathscr{A}$ is incident with n points.*
2. *Each point of $\mathscr{A}$ is incident with $n + 1$ lines.*
3. *Each parallel class of $\mathscr{A}$ contains n lines.*
4. *$\mathscr{A}$ has n^2 points.*
5. *$\mathscr{A}$ has $n^2 + n$ lines.*

Proof. As $\mathscr{A}$ is a net of order n, by Theorem 1.40, each line of $\mathscr{A}$ contains n points, each parallel class of $\mathscr{A}$ contains n lines, and $\mathscr{A}$ has n^2 points.

As any pair of points is joined by a unique line, each line is incident with n points, and as $\mathscr{A}$ has n^2 points, the number of lines incident with any given point must be exactly $(n^2 - 1)/(n - 1) = n + 1$. Hence, as distinct lines from the same parallel class do not intersect, and each point is incident with exactly one line of each parallel class, there must be $n + 1$ parallel classes, and so, by Theorem 1.40, $\mathscr{A}$ has $n(n + 1) = n^2 + n$ lines. □

As a direct consequence of Theorem 1.44, we obtain a simple relationship between the existence of affine planes of order n and complete sets of mutually orthogonal Latin squares of order n.

Theorem 1.50. *An affine plane of order n exists if and only if a complete set of mutually orthogonal Latin squares of order n exists.*

The smallest affine plane, of order 2, is depicted in Figure 1.10. The points of the plane are the four dots and the lines are the six straight line segments.

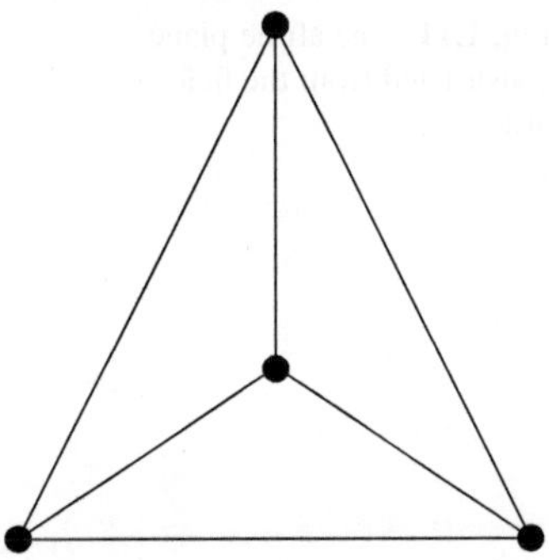

Fig. 1.10 The affine plane of order 2

From any field we can construct an affine plane. Let F be a field, and let $\mathscr{A}$ be the incidence structure whose points are the elements of $F \times F$ and whose lines are solutions to equations: $x = c$, $c \in F$; and $y = mx + b$, $m, b \in F$. The lines $x = c$, $c \in F$ form one parallel class, and, for each $m \in F$, the lines $y = mx + b$, $b \in F$ form a parallel class. Let (x_1, y_1) and (x_2, y_2) be two distinct points of $\mathscr{A}$. If $x_1 = x_2$, then the only line through both points is the line $x = x_1$. If $x_1 \neq x_2$, then the line $y = mx + b$ through both points is uniquely determined by solving the system of equations

$$y_1 = mx_1 + b$$

$$y_2 = mx_2 + b$$

for m and b. Hence two distinct points are joined by a unique line. A given point (x_1, y_1) is incident with exactly one line of the parallel class $\{x = c \mid c \in F\}$, the line $x = x_1$, and, for $m \in F$, incident with exactly one line of the parallel class $\{y = mx + b \mid b \in F\}$, the line $y = mx + (y_1 - mx_1)$. Hence, each point is incident with exactly one line of each parallel class. For $c, m, b \in F$, the line $x = c$ intersects the line $y = mx + b$ in exactly one point, the point $(c, mc + b)$, and, for $m_1, m_2, b_1, b_2 \in F$, $m_1 \neq m_2$, the point of intersection of the lines $y = m_1x + b_1$ and $y = m_2x + b_2$ is uniquely determined by solving the system of equations

$$y = m_1x + b_1$$
$$y = m_2x + b_2$$

for x and y. Hence lines in different parallel classes have exactly one point in common. The points $(0, 0)$, $(0, 1)$, and $(1, 0)$ are noncollinear, and so $\mathscr{A}$ satisfies the nondegeneracy condition of an affine plane. It follows that $\mathscr{A}$ is an affine plane. These affine planes are called *Desarguesian* affine planes. The affine plane of order 2 depicted in Figure 1.10 is the Desarguesian affine plane constructed using the field GF(2), as is demonstrated in Figure 1.11.

Fig. 1.11 The affine plane constructed from the field of order 2

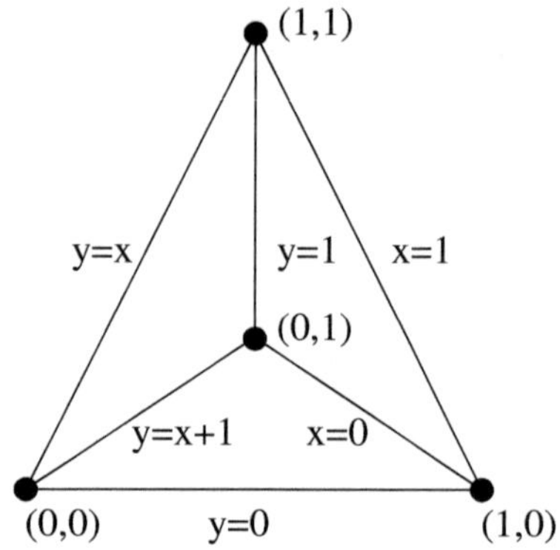

A *translation* of an affine plane is a collineation that fixes all parallel classes and fixes all the lines of a given parallel class.

Corollary 1.51. *Nonidentity translations of affine planes are strict translations.*

Proof. Let τ be a translation of an affine plane $\mathscr{A}$, let π be a parallel class all of whose lines are fixed by τ, let P be a fixed point of τ, and let l be the line of π through P. If Q is a point not incident with l, then the line m joining P and Q must be fixed by τ as $\tau(m)$ and m are parallel and incident with the same point P which is only possible if $\tau(m) = m$. As Q is the intersection point of the fixed line m with the fixed line in π through Q, Q must be fixed. Hence, each point of $\mathscr{A}$ not incident with l is fixed. Repeating this argument with some point not incident with l in place of P shows that all points of l are also fixed. Thus, τ is the identity. It follows that any nonidentity translation is a strict translation. □

From Lemma 1.51 and Theorem 1.48, we obtain a relationship between affine planes and orthogonal sets of orthomorphisms.

Theorem 1.52. *An affine plane admits a group G of translations that fixes all lines of a given parallel class and is transitive on the points of a line of this parallel class if and only if G admits a complete set of orthomorphisms.*

Proof. By Lemma 1.51, all nonidentity translations are strict translations, and so the result follows from Theorem 1.48. □

1.4.3 Projective Planes

A *projective plane* is an incidence structure, satisfying the following:

1. Two distinct points are incident with exactly one line.
2. Two distinct lines meet in exactly one point.
   We also impose a nondegeneracy condition.
3. There exist four points, no three of which are collinear.

Let P, Q, R, and S be four points in a projective plane, no three of which are collinear, let l be the unique line incident with P and Q, and let m be the unique line incident with R and S. The lines l and m are distinct and meet in a unique point. Hence any projective plane has a line that is incident with at least three points. In fact, we can say more.

Theorem 1.53. *If $\mathscr{P}$ is a finite projective plane, and some line of $\mathscr{P}$ is incident with $n+1$ points, $n \geq 2$, then the following hold:*

1. *Each line of $\mathscr{P}$ is incident with $n+1$ points.*
2. *Each point of $\mathscr{P}$ is incident with $n+1$ lines.*
3. *$\mathscr{P}$ has n^2+n+1 points.*
4. *$\mathscr{P}$ has n^2+n+1 lines.*

Proof. Let l be a line of $\mathscr{P}$ incident with $n+1$ points, $n \geq 2$. Let P be a point not incident with l. Each point incident with l is joined by a unique line to P and, as each of these lines intersect l in exactly one point, these $n+1$ lines are distinct. As any line through P must intersect l, there are exactly $n+1$ lines through P. Similarly, if m is a line not through P, then, as each of the $n+1$ lines through P intersect m in exactly one point and each point incident with l is joined by a unique line to P, m is incident with exactly $n+1$ points. Thus we have shown that every line not through P is incident with $n+1$ points, and every point not incident with l is incident with $n+1$ lines. Let P' be a point incident with l. Four points, no three of which are collinear, determine six lines, no three of which can meet in a common point. Hence there exist at least three lines that do not pass through P', and at least one of these, l', will not pass through P. Repeating the above arguments with l' instead of l and P' instead of P shows that P is incident with $n+1$ lines, and every line through P is incident with $n+1$ points. Hence every line is incident with $n+1$ points and every point is incident with $n+1$ lines.

As there are $n+1$ lines incident with P, each pair of lines intersecting in P only, each line incident with n points other than P, and each point, other than P, incident with exactly one of these lines, there are a total of $n(n+1)+1 = n^2+n+1$ points.

Similarly, as there are $n + 1$ points incident with l, each pair of points joined by l only, each point incident with n lines other than l, and each line, other than l, meeting exactly one of these points, there are a total of $n(n + 1) + 1 = n^2 + n + 1$ lines. □

A projective plane is of *order n* if each line is incident with $n + 1$ points. The smallest projective plane, of order 2, is depicted in Figure 1.12. The points of the plane are the seven dots and the lines are the six straight line segments and the circle. This plane is called the *Fano plane*.

Fig. 1.12 The Fano plane

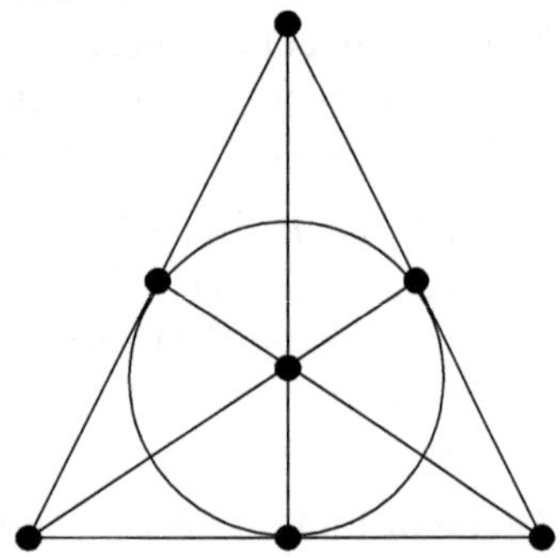

From any field we can construct a projective plane. Let F be a field, and let V be a *three*-dimensional vector space over F. We form an incidence structure $\mathscr{P}$ whose points are the *one*-dimensional subspaces of V, i.e., $\langle(x, y, z)\rangle$, x, y, and z not all zero, and whose lines are also the *one*-dimensional subspaces of V, i.e., $\langle(A, B, C)\rangle$, A, B, and C not all zero, a point $\langle(x, y, z)\rangle$ being incident with a line $\langle(A, B, C)\rangle$ if $Ax+By+Cz = 0$. If $\langle(x_1, y_1, z_1)\rangle$ and $\langle(x_2, y_2, z_2)\rangle$ are two distinct points of $\mathscr{P}$, then there is a unique line $\langle(A, B, C)\rangle$ incident with both points which is determined by solving

$$Ax_1 + By_1 + Cz_1 = 0$$
$$Ax_2 + By_2 + Cz_2 = 0$$

for A, B, and C. Hence two distinct points are joined by exactly one line. Similarly, if $\langle(A_1, B_1, C_1)\rangle$ and $\langle(A_2, B_2, C_2)\rangle$ are two distinct lines of $\mathscr{P}$, then there is a unique point of intersection $\langle(x, y, z)\rangle$ which is determined by solving

$$A_1x + B_1y + C_1z = 0$$
$$A_2x + B_2y + C_2z = 0$$

for x, y, and z. Hence two distinct lines meet in exactly one point. We leave it to the reader to verify that no three of the four points $\langle(0, 0, 1)\rangle$, $\langle(0, 1, 1)\rangle$, $\langle(1, 0, 0)\rangle$, and $\langle(1, 1, 0)\rangle$ are collinear. It follows that $\mathscr{P}$ is a projective plane. These projective planes are called *Desarguesian* projective planes. The projective plane of order 2 depicted in Figure 1.12 is the Desarguesian projective plane constructed using the field GF(2) as is demonstrated in Figure 1.13: in this figure we shorten $\langle(i, j, k)\rangle$ to ijk and use italics to signify the lines.

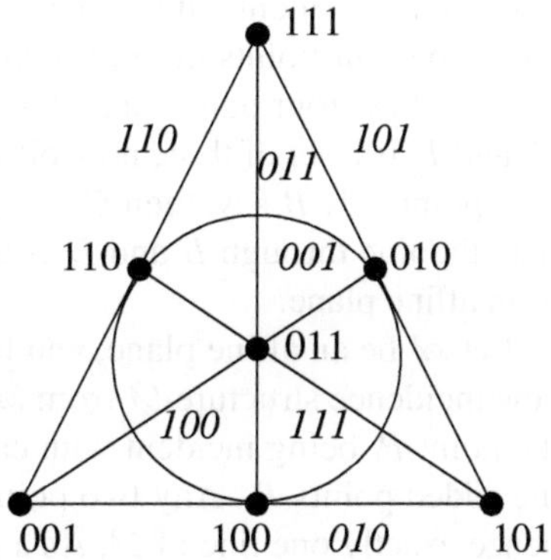

Fig. 1.13 The Fano plane constructed from the field of order 2

The dual of a projective plane is a projective plane. The axioms, "Two distinct points are incident with in exactly one line" and "Two distinct lines meet in exactly one point," are duals of each other, and it is easy to see that the axiom, "There exist four points, no three of which are collinear," and its dual, "There exist four lines, no three of which meet in a common point," are equivalent. As the description of the Desarguesian projective planes is symmetric with respect to points and lines, these planes are *self-dual*, i.e., isomorphic to their duals.

Comparing Figures 1.10 and 1.12, it is easy to see that when we remove one line from the Fano plane, the line represented by a circle, and all the points incident with it, we obtain the affine plane of order 2. Let us remove from the Desarguesian projective plane $\mathscr{P}$, constructed from the field F, the line $\langle(0, 0, 1)\rangle$ and all points incident with it, i.e., all points of the form $\langle(x, y, 0)\rangle$. In the new incidence structure, the points are $\langle(x, y, 1)\rangle$, $x, y \in F$, and the lines are $\langle(A, B, C)\rangle$, $A, B, C \in F$, A, and B not both zero. As the points incident with the line $\langle(A, B, C)\rangle$ are determined by the solutions to the equation $Ax + By + C = 0$, this incidence structure is the affine Desarguesian plane constructed from the field F. The relationship between affine Desarguesian planes and projective Desarguesian planes applies more generally.

Theorem 1.54. *If we remove any line from a projective plane along with all the points incident with the line, then the resulting incidence structure is an affine plane, and any affine plane can be obtained from a projective plane by removing a line and all points incident with it.*

Proof. Let $\mathscr{P}$ be a projective plane, let l be a line of $\mathscr{P}$, and let $\mathscr{D}$ be the incidence structure obtained from $\mathscr{P}$ by removing l and all points incident with l. Any two points of $\mathscr{D}$ are joined by a unique line in $\mathscr{P}$, and, as this line cannot be l, they are also joined by a unique line in $\mathscr{D}$. Let $\{P_i \mid i \in I\}$ be the points incident with l. Then the ith parallel class of $\mathscr{D}$, $i \in I$, consists of the lines of $\mathscr{D}$ incident with P_i in $\mathscr{P}$. Each point P of $\mathscr{D}$ is joined in $\mathscr{P}$ to any point P_i incident with l by a unique line, and hence, any point of $\mathscr{D}$ is incident with exactly one line of each parallel class. Two lines from distinct parallel classes intersect in a unique point of $\mathscr{P}$, and, as this point is not incident with l, they also intersect in a unique point of $\mathscr{D}$. Let A, B, C, and D be four points in $\mathscr{P}$, no three of which are noncollinear. If l passes through none of these four points or l passes through one of these four points, A say, then B, C, and D is a set of three noncollinear points in $\mathscr{D}$. If l passes through two of these four points, A, B say, then C, D, and the intersection of the line through A and C with the line through B and D is a set of three noncollinear points in $\mathscr{D}$. Hence, $\mathscr{D}$ is an affine plane.

Let $\mathscr{A}$ be an affine plane, and let π_i, $i \in I$, be the parallel classes of $\mathscr{A}$. Form a new incidence structure $\mathscr{D}$ from $\mathscr{A}$ by adding a point P_i for each parallel class π_i, the point P_i being incident with each line in π_i and a line l_∞ incident with each of the added points P_i. Any two points of $\mathscr{A}$ are joined by exactly one line of $\mathscr{A}$ and hence exactly one line of $\mathscr{D}$. As a point P of $\mathscr{A}$ is incident with exactly one line of each parallel class, for each $i \in I$, P and P_i are joined by exactly one line of $\mathscr{D}$. For $i, j \in I$, $i \neq j$, the points P_i and P_j are joined by exactly one line if $\mathscr{D}$, l_∞. Hence, any two points of $\mathscr{D}$ are joined by exactly one line of $\mathscr{D}$. Two lines of $\mathscr{A}$ from different parallel classes intersect in a point of $\mathscr{A}$ only and so exactly one such point. Two distinct parallel lines of $\mathscr{A}$ intersect in no points of $\mathscr{A}$ and exactly one point of l_∞ and so exactly one point. Any line of $\mathscr{A}$ belongs to a unique parallel class, π_i say, and so intersects l_∞ in exactly one point, P_i. Hence, any two lines of $\mathscr{D}$ intersect in exactly one point of $\mathscr{D}$. Let A, B, and C be three noncollinear points of $\mathscr{A}$, let P be the point, incident with l_∞, collinear with A and B, and let Q be the point, on l_∞, collinear with A and C. Then B, C, P, and Q is a set of four points in $\mathscr{D}$, no three of which are collinear. It follows that $\mathscr{D}$ is a projective plane and $\mathscr{A}$ can be obtained from a projective plane by removing one line and all points incident with it. □

When adding a line to an affine plane to obtain a projective plane, the added line is called the *line at infinity* and the added points are called *points at infinity*. Affine and projective planes related to complete sets of orthomorphisms will be studied in Chapter 14. Readers interested in more information on affine and projective planes may consult the 1968 book by Dembowski [88], the 1973 book by Hughes and Piper [190], or the 1982 book by Kallaher [210].

Chapter 2
When Is a Latin Square Based on a Group?

A question that presents itself to us is, given a Latin square, is it isotopic to the Cayley table of a group? That is, is it based on a group? The case in which the Latin square is the Cayley table of a loop might seem to be the easiest to tackle, but surprisingly it turns out that permuting the rows, columns, and symbols of the square does not cause any significant difficulties. Several approaches have been proposed for answering this question, dating back to the 1800s. Some of the answers that we will present will be based on configurations in the given square or given bordered square, the quadrangle criterion for Cayley tables of quasigroups in general and its variants for loops and the Thomsen condition characterizing Latin squares based on abelian groups. We will present the rectangle rule, a criterion for the normal multiplication table of a loop to be based on a group; and we will present approaches that are based on permuting rows and/or columns of the given square, or that treat rows and/or columns of the given square as permutations. We will describe Keedwell's tests that determine whether an element of a loop is in the left, right, or middle nucleus of the loop: these tests can be used to determine whether a loop is a group or not and requires less work to show that a loop is in fact a group. We will also present Keedwell's test to determine whether a loop is an abelian group or not.

Our emphasis will be on particular properties of Latin squares that characterize Latin squares based on groups. We will not emphasize algorithmic solutions, or algebraic properties of loops or quasigroups that characterize loops and quasigroups that are isotopic to groups: interested readers may consult the 1991 and 2015 books by Dénes and Keedwell [93, 95] on latin squares, or for more information on quasigroups and loops, the 1958 book by Bruck [56] on binary systems, the 1990 book by Pflugfelder [267] on quasigroups and loops, or the papers in [71], a 1990 volume on quasigroups and loops. For more information on identities that imply that a loop is a group, interested readers may consult the 2007 paper by Fiala [143] or the 2009 paper and its corrigendum by Keedwell [218, 219].

A. B. Evans, *Orthogonal Latin Squares Based on Groups*, Developments in Mathematics 57, https://doi.org/10.1007/978-3-319-94430-2_2

2.1 The Quadrangle Criterion

The oldest criterion for determining whether a Latin square is based on a group or not is called the *quadrangle criterion* in Dénes and Keedwell's 1991 and 2015 books on Latin squares [93, 95]. Frolov [149] is credited with discovering the quadrangle criterion in 1890 and proving that it is a necessary condition for a quasigroup to be isotopic to a group: this condition is also sufficient. The quadrangle criterion has been rediscovered at least three times. It was rediscovered by Reidemeister [277] in 1929, and is called the *Reidemeister condition* in Aczél's 1964, 1965, and 1969 papers [10–12]. It was also rediscovered by Watson [328] in 1967 and by Tecklenburg [308] in 1990.

In proving the quadrangle criterion, it is important to observe that any equation of the form $ab = cd$ in a quasigroup is preserved by isotopisms.

Lemma 2.1. *If* (α, β, γ) *is an isotopism from* $(Q, \cdot)$ *to* $(Q', \star)$ *and* $a \cdot b = c \cdot d$ *in* Q*, then* $\alpha(a) \star \beta(b) = \alpha(c) \star \beta(d)$ *in* Q'.

Proof.

$$\alpha(a) \star \beta(b) = \gamma(a \cdot b) = \gamma(c \cdot d) = \alpha(c) \star \beta(d). \quad \square$$

As the quadrangle criterion is described entirely in terms of equations of the form $ab = cd$, which by Lemma 2.1 are preserved by isotopisms, in proving the quadrangle criterion for an arbitrary quasigroup, we are free to work with any quasigroup isotopic to our given quasigroup.

Theorem 2.2 (The Quadrangle Criterion). *A quasigroup* Q *is isotopic to a group if and only if for all* $a_i, b_i, c_i, d_i \in Q$, $i = 1, 2$, *the equalities* $a_1c_1 = a_2c_2$, $a_1d_1 = a_2d_2$, *and* $b_1c_1 = b_2c_2$ *imply the equality* $b_1d_1 = b_2d_2$.

Proof. We will first show that the quadrangle criterion holds for groups and hence, by Lemma 2.1, for any quasigroup isotopic to a group. Let G be a group, and let $a_i, b_i, c_i, d_i \in G$, $i = 1, 2$, satisfy the equalities $a_1c_1 = a_2c_2$, $a_1d_1 = a_2d_2$, and $b_1c_1 = b_2c_2$. Then

$$b_1d_1 = (b_1c_1)(a_1c_1)^{-1}(a_1d_1) = (b_2c_2)(a_2c_2)^{-1}(a_2d_2) = b_2d_2.$$

To complete the proof, we will show that any quasigroup that satisfies the quadrangle criterion is isotopic to a group. If Q satisfies the quadrangle criterion, then, by Lemma 2.1, as any isotope of Q also satisfies the quadrangle criterion, we may assume without loss of generality, by Theorem 1.2, that Q is a loop with identity e. Let $a, b, c \in Q$ and set $a_1 = e$, $a_2 = b$, $b_1 = a$, $b_2 = ab$, $c_1 = b$, $c_2 = e$, $d_1 = bc$, and $d_2 = c$. Then

$$a_1c_1 = eb = be = a_2c_2,$$

$$a_1d_1 = e(bc) = bc = a_2d_2,$$

$$b_1c_1 = ab = (ab)e = b_2c_2,$$

and so by the quadrangle criterion

$$a(bc) = b_1 d_1 = b_2 d_2 = (ab)c.$$

Hence Q is an associative loop and, thus, a group. □

The quadrangle criterion is best understood pictorially, as in Figure 2.1. If the quadrangle criterion holds, then the "?" in Figure 2.1 stands for a d. Stripping away the borders, we obtain a Latin square depiction of the quadrangle criterion. As an example of its application, recall that we described a Latin square, L_2, in Figure 1.1: at that time, we did not know whether it was based on a group or not. In Figure 2.2, we present L_2 in three different ways: in two of these ways, some entries are replaced by a "·" to suppress unnecessary detail. It is now clear, by the quadrangle criterion, that L_2 cannot based on a group.

Fig. 2.1 Pictorial depiction of the quadrangle criterion

$$\begin{array}{c|ccccccc}
 & c_1 & \cdots & d_1 & c_2 & \cdots & d_2 \\ \hline
a_1 & a & \cdots & b & & & \\
\vdots & \vdots & \ddots & \vdots & & & \\
b_1 & c & \cdots & d & & & \\
 & & & & & & \\
a_2 & & & & a & \cdots & b \\
\vdots & & & & \vdots & \ddots & \vdots \\
b_2 & & & & c & \cdots & ?
\end{array}$$

Fig. 2.2 How L_2 fails the quadrangle criterion

$$L_2 = \begin{bmatrix} 0&1&2&3&4 \\ 1&0&3&4&2 \\ 2&4&0&1&3 \\ 3&2&4&0&1 \\ 4&3&1&2&0 \end{bmatrix}, \quad \begin{bmatrix} 0&1&\cdot&\cdot&\cdot \\ 1&0&\cdot&\cdot&\cdot \\ \cdot&\cdot&0&1&\cdot \\ \cdot&\cdot&\cdot&\cdot&\cdot \\ \cdot&\cdot&1&2&\cdot \end{bmatrix}, \quad \text{or} \quad \begin{bmatrix} 0&1&\cdot&\cdot&\cdot \\ 1&0&\cdot&\cdot&\cdot \\ \cdot&\cdot&\cdot&\cdot&\cdot \\ \cdot&\cdot&4&\cdot&1 \\ \cdot&\cdot&1&\cdot&0 \end{bmatrix}$$

In the course of proving Theorem 2.2, we showed that the quadrangle criterion for loops is equivalent to the associative law for loops.

Corollary 2.3. *A loop satisfies the quadrangle criterion if and only if it satisfies the associative law, in which case the loop is a group.*

The result of Corollary 2.3 is not true for quasigroups in general. A quasigroup may satisfy the quadrangle criterion, but not be associative. In fact we will see, in Theorem 2.14, that a quasigroup that satisfies the associative law must be a group. In Figure 2.3 we describe the smallest nonassociative quasigroup. In this quasigroup, which is isotopic to the group $\mathbb{Z}_3$ and so satisfies the quadrangle criterion,

$$(1 \cdot 2) \cdot 1 = 2 \cdot 1 = 1 \quad \text{but} \quad 1 \cdot (2 \cdot 1) = 1 \cdot 1 = 0,$$

and so the associative law does not hold.

Fig. 2.3 The smallest nonassociative quasigroup that satisfies the quadrangle criterion

	0	1	2
0	0	2	1
1	1	0	2
2	2	1	0

2.1.1 Loops Isotopic to Groups

Corollary 2.3 implies that for a loop to be isotopic to a group, it must be a group, but actually a stronger result is known and can be found in a 1943 paper by Albert [14].

Theorem 2.4 (Albert, 1943). *A loop L is isotopic to a group G if and only if L is a group isomorphic to G.*

Proof. Let (α, β, γ) be an isotopy from the group $(G, \cdot)$ to the loop $(L, \star)$, and let e be the identity for L. Let g, h be the unique elements of G that satisfy $\alpha(g) = \beta(h) = e$. Then, for all $x \in G$,

$$\gamma(x \cdot h) = \alpha(x) \star \beta(h) = \alpha(x) \star e = \alpha(x)$$

and

$$\gamma(g \cdot x) = \alpha(g) \star \beta(x) = e \star \beta(x) = \beta(x).$$

Let us define $\mathrm{f}\colon G \to L$ by $\mathrm{f}(x) = \gamma(g \cdot x \cdot h)$.

Now $\mathrm{f}(x) = \mathrm{f}(y)$ if and only if $\gamma(g \cdot x \cdot h) = \gamma(g \cdot y \cdot h)$, if and only if $g \cdot x \cdot h = g \cdot y \cdot h$, if and only if $x = y$. Thus f is injective. Further, f is surjective and hence bijective, as if $l \in L$, then there exists $x \in G$ for which $\gamma(x) = l$; but then,

$$\mathrm{f}(g^{-1} \cdot x \cdot h^{-1}) = \gamma(g \cdot g^{-1} \cdot x \cdot h^{-1} \cdot h) = \gamma(x) = l.$$

The bijection f is, in fact, an isomorphism from $(G, \cdot)$ to $(L, \star)$ as

$$\begin{aligned} \mathrm{f}(x \cdot y) &= \gamma(g \cdot (x \cdot y) \cdot h) \quad = \gamma((g \cdot x) \cdot (y \cdot h)) \\ &= \alpha(g \cdot x) \star \beta(y \cdot h) = \gamma(g \cdot x \cdot h) \star \gamma(g \cdot y \cdot h) \\ &= \mathrm{f}(x) \star \mathrm{f}(y). \end{aligned}$$

The result follows. □

The following are immediate corollaries:

Corollary 2.5. *If G_1 and G_2 are groups, then G_1 is isotopic to G_2 if and only if G_1 is isomorphic to G_2.*

Corollary 2.6. *If the quasigroup Q_1 is isotopic to the group G_1 and the quasigroup Q_2 is isotopic to the group G_2, then Q_1 is isotopic to Q_2 if and only if G_1 is isomorphic to G_2.*

As a consequence of Corollary 2.6, if a Latin square is isotopic to the Cayley table of a group, then that group is uniquely determined up to isomorphism.

2.1.2 The Quadrangle Criterion for Loops

The quadrangle criterion has a variant form when the quasigroup being tested is a loop. This version of the quadrangle criterion was observed in 1927 by Brandt [49] in the context of a class of groupoids, called Brandt groupoids in Clifford and Preston's 1961 book on semigroups [76]: see also [297]. It was called the rectangle rule by Hammel [173] in 1968, a name which we reserve for the test in Theorem 2.11.

Theorem 2.7 (Brandt, 1927). *A loop L with identity e is isotopic to a group if and only if for all $a_i, b_i, c_i, d_i \in L$, $i = 1, 2$, the equalities $a_1c_1 = e = a_2c_2$, $a_1d_1 = a_2d_2$, and $b_1c_1 = b_2c_2$ imply the equality $b_1d_1 = b_2d_2$.*

Proof. We need only prove that this test is equivalent to the quadrangle criterion.

If the quadrangle criterion holds, then this test will be satisfied as it is a special case of the quadrangle criterion.

Next, let us suppose that this test is satisfied and that $a_i, b_i, c_i, d_i \in L$, $i = 1, 2$ satisfy $a_1c_1 = a_2c_2$, $a_1d_1 = a_2d_2$ and $b_1c_1 = b_2c_2$. We will show that $b_1d_1 = b_2d_2$, i.e., the quadrangle criterion holds, by applying the test of this theorem twice.

Choose A_i to satisfy $A_ic_i = e$ for $i = 1, 2$, and choose D_2 to satisfy $A_2D_2 = A_1d_1$. Set $B_1 = a_1$, $B_2 = a_2$, $C_1 = c_1$, $C_2 = c_2$ and $D_1 = d_1$. Now

$$\begin{aligned} A_1C_1 = &\ A_1c_1 = e = A_2c_2 = A_2C_2, \\ A_1D_1 = &\ A_1d_1 = A_2D_2, \quad \text{and} \\ B_1C_1 = &\ a_1c_1 = a_2c_2 = B_2C_2. \end{aligned}$$

Thus, by the test, $B_1D_1 = B_2D_2$. But $B_1D_1 = a_1d_1 = a_2d_2$ and $B_2D_2 = a_2D_2$ and so $D_2 = d_2$.

Next set $A'_1 = A_1$, $A'_2 = A_2$, $B'_1 = b_1$, $B'_2 = b_2$, $C'_1 = C_1 = c_1$, $C'_2 = C_2 = c_2$, $D'_1 = D_1 = d_1$ and $D'_2 = D_2 = d_2$. Then

$$\begin{aligned} A'_1C'_1 = &\ A_1c_1 = e = A_2c_2 = A'_2C'_2, \\ A'_1D'_1 = &\ A_1D_1 = A_2D_2 = A'_2D'_2, \\ B'_1C'_1 = &\ b_1c_1 = b_2c_2 = B'_2C'_2. \end{aligned}$$

Thus, by the test,

$$b_1d_1 = B'_1D'_1 = B'_2D'_2 = b_2d_2,$$

and the quadrangle criterion is satisfied. □

Pictorially, this test is easier to apply than the quadrangle criterion when the quasigroup is a loop, as in Figure 2.1 we need only look at those "quadrangles" in which $a = e$. We can prove a similar restricted version of the quadrangle criterion for quasigroups by copying the proof of Theorem 2.7 with e replaced by an arbitrary element of the quasigroup.

Corollary 2.8. *Let Q be a quasigroup and let $a \in Q$. Then Q is isotopic to a group if and only if for all $a_i, b_i, c_i, d_i \in Q$, $i = 1, 2$, the equalities $a_1c_1 = a = a_2c_2$, $a_1d_1 = a_2d_2$, and $b_1c_1 = b_2c_2$ imply the equality $b_1d_1 = b_2d_2$.*

2.1.3 Aczél's Test

If we are given the multiplication table of a *groupoid*, a set with a binary operation, how can we determine if the groupoid is associative or not? In 1969 Aczél [12] described a test to answer this question. This test was also described by Ferrero and Ferrero Cotti [142] in 1975, where it was suggested that it could be taught in secondary schools. Applied to loops, Aczél's test is a variant of the quadrangle criterion.

Theorem 2.9 (Aczél's Test for Loops). *A loop L is a group if and only if for all $x, b_i, d_i \in L$, $i = 1, 2$, the equalities $xd_1 = d_2$ and $b_2x = b_1$ imply the equality $b_1d_1 = b_2d_2$.*

Proof. This test is really the special case of the quadrangle criterion in which $a_2 = c_1 = e$ and $a_1 = c_2 = x$, where e is the identity for L. Hence this test does give a necessary condition for L to be a group.

To show that this condition is also sufficient, assume that L passes this test, let $a, b, c \in L$, and set $x = b$, $b_1 = ab$, $b_2 = a$, $d_1 = c$, and $d_2 = bc$. Then $xd_1 = bc = d_2$ and $b_2x = ab = b_1$ and so, as the test holds,

$$a(bc) = b_2(xd_1) = b_2d_2 = b_1d_1 = (ab)c,$$

that is, the associative law holds, and hence L is a group. □

This test is depicted pictorially in Figure 2.4. If Aczél's test holds, then the "?" in Figure 2.4 stands for a y: note that the entries b_2 and d_1 can be suppressed.

As examples, consider the loops L and L', of order 9, described in Figure 2.5. Are either of these loops groups? Figure 2.6 shows three instances of L passing Aczél's test: to suppress unnecessary detail, most of the entries are either removed or replaced with a "·". This is, of course, inconclusive. However, the reader can verify that L is isomorphic to the elementary abelian group of order 9. Figure 2.7

Fig. 2.4 Aczél's test

	d_1	x	$\cdots$	d_2
x	$\cdots\ d_2$	$\vdots$		$\vdots$
$\vdots$	$\vdots$	$\vdots$		$\vdots$
b_1	$\cdots\ y$	$\vdots$		$\vdots$
		$\vdots$		$\vdots$
b_2		b_1	$\cdots$	?

shows three instances of L' failing Aczél's test: again to suppress unnecessary detail, most of the entries are either removed or replaced with a "·". It follows that L' is not a group.

Fig. 2.5 Aczél's test: two squares to be tested

$L =$

	0	1	2	3	4	5	6	7	8
0	0	1	2	3	4	5	6	7	8
1	1	5	3	7	8	0	4	2	6
2	2	3	6	4	1	7	0	8	5
3	3	7	4	8	5	2	1	6	0
4	4	8	1	5	7	6	3	0	2
5	5	0	7	2	6	1	8	3	4
6	6	4	0	1	3	8	2	5	7
7	7	2	8	6	0	3	5	4	1
8	8	6	5	0	2	4	7	1	3

, $L' =$

	0	1	2	3	4	5	6	7	8
0	0	1	2	3	4	5	6	7	8
1	1	2	3	0	5	6	7	8	4
2	2	3	0	1	6	7	8	4	5
3	3	0	1	2	7	8	4	5	6
4	4	5	6	7	8	0	1	2	3
5	5	6	7	8	1	4	0	3	2
6	6	7	8	4	2	3	5	0	1
7	7	8	4	5	3	1	2	6	0
8	8	4	5	6	0	2	3	1	7

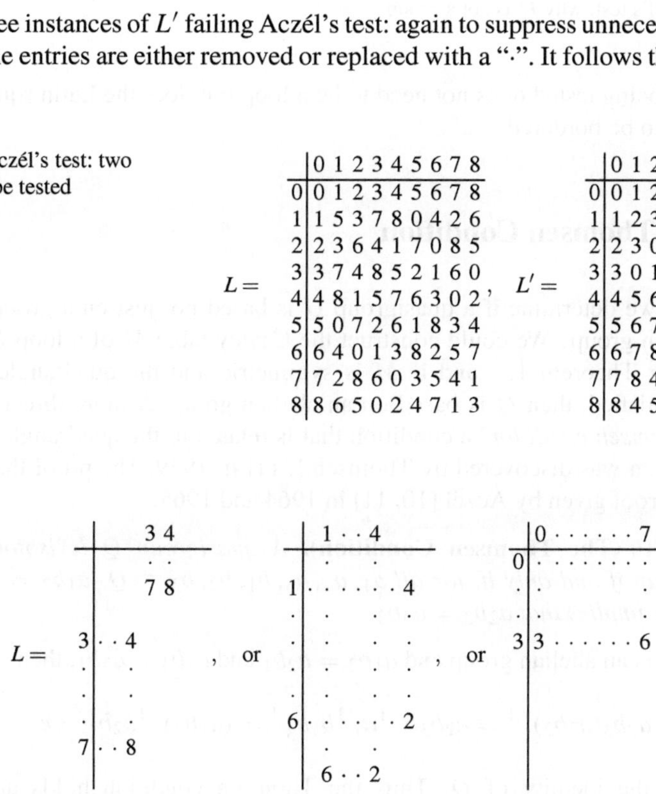

Fig. 2.6 Aczél's test: is L a group?

Jonathan Hall has pointed out that Aczél's test can be applied to quasigroups and unbordered Latin squares. In the Cayley table of a loop, x is the first entry of the row indexed x, and y is the first entry of the column indexed y. Hence the first row and column can be used instead of borders, and, when we make this change, the

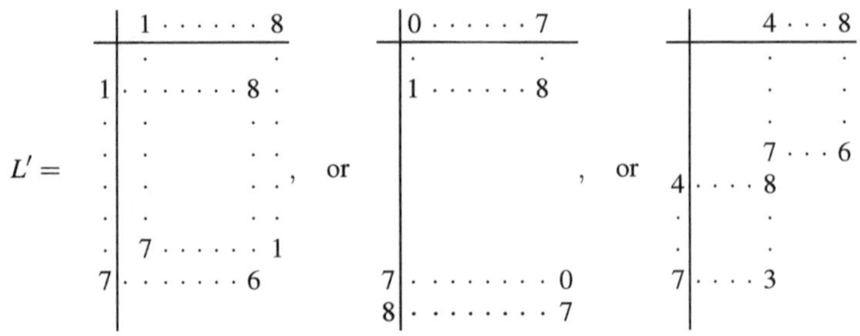

Fig. 2.7 Aczél's test: why L' is not a group

quasigroup being tested does not need to be a loop nor does the Latin square being tested need to be bordered.

2.2 The Thomsen Condition

How might we determine if a quasigroup Q is based not just on a group but also on an abelian group? We could construct the Cayley table M of a loop L isotopic to Q, as per Theorem 1.2, and if M is symmetric and the quadrangle criterion criteria is satisfied, then Q is based on an abelian group. A more direct approach uses the *Thomsen condition*, a condition that is related to the quadrangle criterion. This condition was discovered by Thomsen [311] in 1939. The proof that we will give is the proof given by Aczél [10, 11] in 1964 and 1965.

Theorem 2.10 (The Thomsen Condition). *A quasigroup Q is isotopic to an abelian group if and only if, for all $a_1, a_2, a_3, b_1, b_2, b_3 \in Q$, $a_1b_2 = a_2b_1$ and $a_1b_3 = a_3b_1$ implies that $a_2b_3 = a_3b_2$.*

Proof. If Q is an abelian group and $a_1b_2 = a_2b_1$ and $a_1b_3 = a_3b_1$, then

$$a_2b_3(a_3b_2)^{-1} = a_2b_3b_2^{-1}a_1^{-1}b_1b_3^{-1} = (a_1b_2)^{-1}a_2b_1 = e,$$

where e is the identity of Q. Thus the Thomsen condition holds and so, by Lemma 2.1, will hold in any quasigroup isotopic to an abelian group.

Next assume that the Thomsen condition holds for Q. By Lemma 2.1, this condition must hold for all isotopes of Q, and so, without loss of generality, by Theorem 1.2, we may assume Q to be a loop. We will first show that Q is a commutative loop. Let $a, b \in Q$ and set $a_1 = b_1 = e$, $a_2 = b_2 = a$ and $a_3 = b_3 = b$. Then

$$a_1b_2 = a = a_2b_1 \quad \text{and} \quad a_1b_3 = b = a_3b_1,$$

and so, by the Thomsen condition, $ab = a_2b_3 = a_3b_2 = ba$ implying that Q is commutative.

We will next show that the Thomsen condition implies the quadrangle criterion. We will do this by applying the Thomsen condition three times. Suppose that $a_i, b_i, c_i, d_i \in Q$ satisfy $a_1c_1 = a_2c_2$, $a_1d_1 = a_2d_2$ and $b_1c_1 = b_2c_2$.

Set $A_1 = a_2$, $A_2 = a_1$, $B_1 = d_1$, $B_2 = d_2$, $B_3 = c_1$, and choose A_3 to satisfy $A_1B_3 = A_3B_1$. Then

$$A_1B_2 = a_2d_2 = a_1d_1 = A_2B_1 \quad \text{and} \quad A_1B_3 = A_3B_1$$

and so, by the Thomsen condition, $A_2B_3 = A_3B_2$.

Next, set $A'_1 = b_2$, $A'_2 = b_1$, $A'_3 = A_1 = a_2$, $B'_1 = B_3 = c_1$, $B'_2 = c_2$, and choose B'_3 to satisfy $A'_1B'_3 = A'_3B'_1$. Then

$$A'_1B'_2 = b_2c_2 = b_1c_1 = A'_2B'_1 \quad \text{and} \quad A'_1B'_3 = A'_3B'_1$$

and so, by the Thomsen condition, $A'_2B'_3 = A'_3B'_2$.

For the final application of the Thomsen condition, set $A''_1 = A_3$, $A''_2 = A'_2 = b_1$, $A''_3 = A'_1 = b_2$, $B''_1 = B'_3$, $B''_2 = B_2 = d_2$, and $B''_3 = B_1 = d_1$. Then

$$A''_1B''_2 = A_3B_2 = A_2B_3 = a_1c_1 = a_2c_2 = A'_3B'_2 = A'_2B'_3 = A''_2B''_1$$

$$\text{and} \quad A''_1B''_3 = A_3B_1 = A_1B_3 = A'_3B'_1 = A'_1B'_3 = A''_3B''_1$$

and so, by the Thomsen condition, $A''_2B''_3 = A''_3B''_2$ or $b_1d_1 = b_2d_2$ and so the quadrangle criterion holds. □

The Thomsen condition is described pictorially in Figure 2.8. If the Thomsen condition holds, then the "?" in Figure 2.8 stands for a c. As with the quadrangle criterion, a Latin square depiction of the Thomsen condition can be obtained by stripping away the borders. The configuration in Figure 2.8, with the "?" replaced by $d \neq c$, was rediscovered in 2017 by Wanless and Webb [327] in their study of small partial Latin squares that cannot be embedded in the Cayley tables of groups. They proved that this partial Latin square could not be embedded in the Cayley table of an abelian group.

Fig. 2.8 Pictorial depiction of the Thomsen condition

	b_1	b_2	b_3
a_1		a ···	b
		⋮	⋮
a_2	a ···	···	c
	⋮	⋮	
a_3	b ···	?	

As examples, consider the two Latin squares, M and M', of order 9, described in Figure 2.9. Both of these Latin squares are based on groups. Figure 2.10 shows three instances of M satisfying the Thomsen condition: to suppress unnecessary detail, most of the entries are replaced with a "·" and some are replaced with a "—" or a "|". The reader can verify that M is based on the elementary abelian group of order 8, an abelian group. Figure 2.11 shows three instances of M' not satisfying the Thomsen condition: to suppress unnecessary detail, most of the entries are replaced with a "·" and some are replaced with a "—" or a "|". The reader can verify that M' is based on the dihedral group of order 8, a nonabelian group.

Fig. 2.9 The Thomsen condition: two squares to be tested

$$M = \begin{bmatrix} 0&4&5&2&6&1&7&3 \\ 2&1&3&0&7&4&6&5 \\ 1&2&6&4&5&0&3&7 \\ 4&0&7&1&3&2&5&6 \\ 3&6&2&5&4&7&1&0 \\ 5&7&0&3&1&6&4&2 \\ 7&5&4&6&2&3&0&1 \\ 6&3&1&7&0&5&2&4 \end{bmatrix}, \quad M' = \begin{bmatrix} 0&6&1&3&2&7&5&4 \\ 3&5&0&2&1&4&7&6 \\ 2&7&3&1&0&6&4&5 \\ 1&4&2&0&3&5&6&7 \\ 4&1&6&7&5&3&2&0 \\ 6&0&5&4&7&2&1&3 \\ 5&3&7&6&4&1&0&2 \\ 7&2&4&5&6&0&3&1 \end{bmatrix}$$

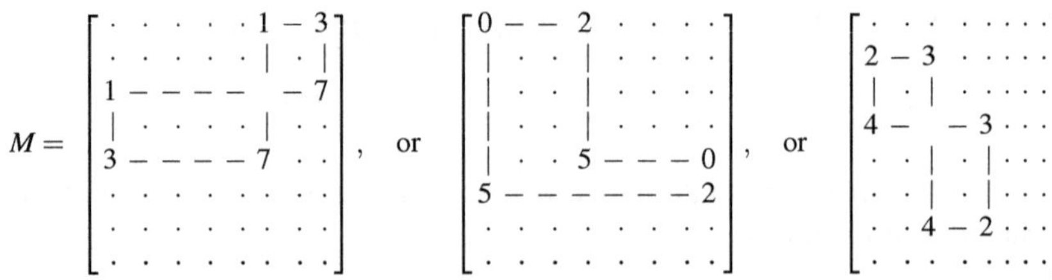

Fig. 2.10 The Thomsen condition: is M based on an abelian group?

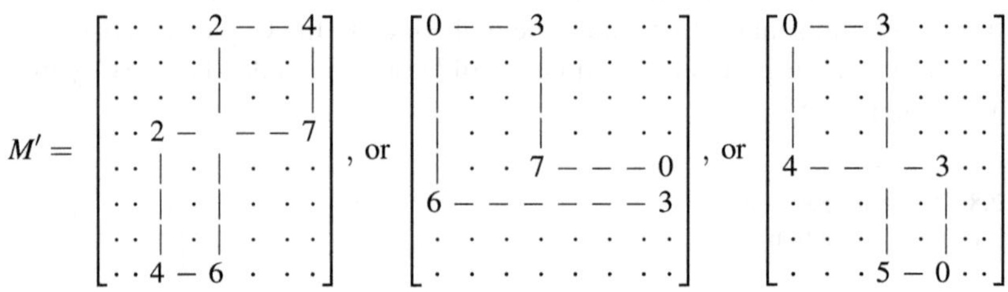

Fig. 2.11 The Thomsen condition: why M' is not based on an abelian group

2.3 The Rectangle Rule

The *rectangle rule* for loops uses the normal multiplication table of the loop instead of the usual multiplication table of the loop. In his 1958 book on group theory, Zassenhaus [338] credited Brandt [49] with discovering this rule in 1927. It was credited to Zassenhaus by Parker [265] in 1965 and was called the Zassenhaus rectangle rule by Watson [328] in 1967.

If $L = \{q_1, \ldots, q_n\}$ is a loop with identity e and, for each $i \in \{1, \ldots, n\}$, q_i^* is the unique element of L that satisfies $q_i q_i^* = e$, then the *normal multiplication table* of L is the $n \times n$ matrix with ijth entry $q_i q_j^*$. The normal multiplication table of a loop is an isotope of the multiplication table of the loop, obtained by permuting its columns so that each entry on the main diagonal is the identity element.

Theorem 2.11 (The Rectangle Rule). *Let $N = \{l_{ij}\}$ be the normal multiplication table of a loop L. Then L is a group if and only if $l_{ij}l_{jk} = l_{ik}$ for all i, j, k.*

Proof. Let $L = \{q_1, \ldots, q_n\}$ be a loop with identity e and normal multiplication table $N = \{l_{ij}\}$. If L is a group, then

$$l_{ij}l_{jk} = \left(q_i q_j^{-1}\right)\left(q_j q_k^{-1}\right) = q_i q_k^{-1} = l_{ik} \quad \text{for all } i, j, k.$$

If $l_{ij}l_{jk} = l_{ik}$ for all i, j, k then

$$(xy^*)(yz) = xz \quad \text{for all } x, y, z \in L. \tag{a}$$

Applying equation (a) with $x = z = e$ yields $y^*y = e = yy^*$ and so $(y^*)^* = y$ for all $y \in L$.

If we now set $x = a$, $y = b^*$ and $z = e$ in equation (a), we get

$$(ab)b^* = a \quad \text{for all } a, b \in L. \tag{b}$$

For all $a, b, c \in L$

$$\begin{aligned} a(bc) &= ((ab)b^*)(bc) && \text{by equation (b)} \\ &= (ab)c && \text{by equation (a).} \end{aligned}$$

Thus the associative law holds, and L is a group. □

The rectangle rule is described pictorially in Figure 2.12. Here we show the multiplication table of the loop L, and its normal multiplication table N with most entries suppressed: we need both the multiplication table and normal multiplication table of L to apply the rectangle rule. If the rectangle rule is satisfied, then the "?" in N stands for ab.

As an example, consider the loop L, of order 6, described in Figure 2.13. We show the multiplication table of L and its normal multiplication N: in the second depiction of N, we have suppressed unnecessary detail by replacing most of the entries with a "·" and some with a "–". If L had satisfied the rectangle rule, then the

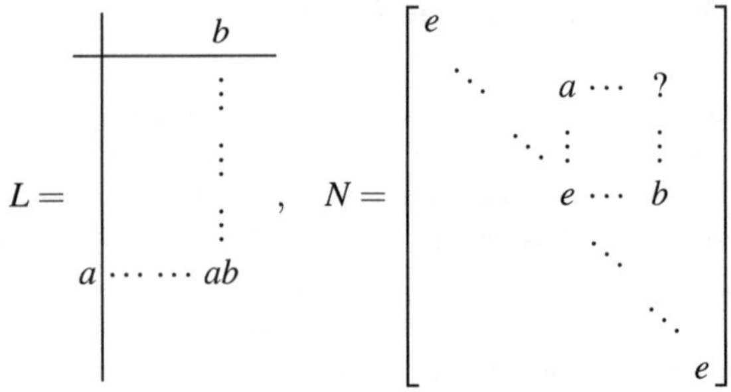

Fig. 2.12 Pictorial depiction of the rectangle rule

entry in parentheses in the second depiction of N would be 0 and not 4. Hence, L is not a group.

$$L = \begin{array}{c|cccccc} & 0 & 1 & 2 & 3 & 4 & 5 \\ \hline 0 & 0 & 1 & 2 & 3 & 4 & 5 \\ 1 & 1 & 3 & 5 & 2 & 0 & 4 \\ 2 & 2 & 4 & 1 & (0) & 5 & 3 \\ 3 & 3 & 5 & 0 & 4 & 1 & 2 \\ 4 & 4 & 0 & 3 & 5 & 2 & 1 \\ 5 & 5 & 2 & 4 & 1 & 3 & 0 \end{array}, \quad N = \begin{bmatrix} 0 & 4 & 3 & 2 & 1 & 5 \\ 1 & 0 & 2 & 5 & 3 & 4 \\ 2 & 5 & 0 & 1 & 4 & 3 \\ 3 & 1 & 4 & 0 & 5 & 2 \\ 4 & 2 & 5 & 3 & 0 & 1 \\ 5 & 3 & 1 & 4 & 2 & 0 \end{bmatrix} \quad \text{or} \quad \begin{bmatrix} 0 & \cdot & \cdot & \cdot & \cdot & \cdot \\ \cdot & 0 & 2 & - & - & (4) \\ \cdot & \cdot & 0 & - & - & 3 \\ \cdot & \cdot & \cdot & 0 & \cdot & \cdot \\ \cdot & \cdot & \cdot & \cdot & 0 & \cdot \\ \cdot & \cdot & \cdot & \cdot & \cdot & 0 \end{bmatrix}$$

Fig. 2.13 An example of the rectangle rule not holding

Based on the rectangle rule, in 1965 Parker [265] suggested a test for a loop to be isotopic to a group. He observed that if $N = \{l_{ij}\}$ is the normal multiplication table for a group G and if we proceed to multiply the second row on the left by $l_{1,2}$, then the second column on the right by $l_{2,1}$, then the third row on the left by $l_{1,3}$, then the third column on the right by $l_{3,1}$, etc., then all the entries of the final matrix will be the identity. He suggested that this property characterized loops isotopic to groups. While this property is indeed a necessary condition for a loop to be isotopic to a group, in 1968 Zassenhaus [339] showed it not to be sufficient by providing a class of counterexamples.

There is an alternative formulation of the rectangle rule. As in Theorem 2.11, let N be the normal multiplication table of a loop L. From N we form a Latin square X by replacing each entry g with x_g, where $\{x_g \mid g \in L\}$ are commuting indeterminates. In 1996 Ford and Johnson [145] observed that, when L is a group, X^2 is a Latin square that has the same pattern as N, i.e., there is a bijection from the entries of N to the entries of X^2 that induces an isotopy. By Theorem 2.11, this observation is actually a characterization of groups: L is a group if and only if X^2 is a Latin square that has the same pattern as N.

2.4 Tests Based on Row and Column Permutations

There are two tests that involve permuting the rows and/or columns of a Latin square or that regard the rows and/or columns of a Latin square as permutations. Light's test uses row and column permutations to test a groupoid for associativity and hence can be used to show that a loop is a group. In the row composition rule, the rows of a Latin square are treated as permutations, whose products tell us if the Latin square is the Cayley table of a group or not.

2.4.1 Light's Test

The first of these tests was suggested by F.W. Light in 1949 (see [76]). While *Light's test* was devised to test a groupoid for associativity, we will prove it only for quasigroups.

Theorem 2.12 (Light's Test). *Let Q be a quasigroup with Cayley table M, and, for $w \in Q$, let M_C^w be the matrix obtained from M by replacing the column headed y by the column headed wy for each $y \in Q$ and let M_R^w be the matrix obtained from M by replacing the row headed x by the row headed xw for each $x \in Q$. Q is associative if and only if $M_C^w = M_R^w$ for all $w \in S$, S a set of generators for Q.*

Proof. As the xyth entry of M_C^w is $x(wy)$ and the xyth entry of M_R^w is $(xw)y$, it follows that $M_C^w = M_R^w$ if and only if $(xw)y = x(wy)$ for all $x, y \in Q$. Thus Q is associative if and only if $M_C^w = M_R^w$ for all $w \in Q$.

If $M_C^s = M_R^s$ and $M_C^t = M_R^t$, then $(xs)y = x(sy)$ and $(xt)y = x(ty)$ for all $x, y \in Q$. But then

$$(x(st))y = ((xs)t)y = (xs)(ty) = x(s(ty)) = x((st)y),$$

and hence $M_C^{st} = M_R^{st}$. Thus, by induction, if $M_C^w = M_R^w$ for all $w \in S$, S a set of generators for Q, then $M_C^w = M_R^w$ for all $w \in Q$. The result follows. □

As groups are associative loops, the following is an immediate corollary.

Corollary 2.13. *If L is a loop, then L is a group if and only if L satisfies Light's test.*

In the test as originally suggested by Light, you had to check that $M_C^w = M_R^w$ for all $w \in Q$: it was pointed out by Clifford and Preston [76] in 1961 that we need only check this for all w in a set of generators of Q.

Figure 2.14 shows a loop L of order 9 and a quasigroup Q of order 8. The loop L is from Figure 2.5: in Figure 2.6 we showed three instances of its satisfying Aczél's test and left it to the reader to verify that this loop is isotopic to the elementary abelian group of order 9. We can use Light's test to verify that L is a group. L is generated by 1 and 2 as $0 = (1 \cdot 1) \cdot 1$, $1 = 1$, $2 = 2$, $3 = 1 \cdot 2$, $4 = (1 \cdot 2) \cdot 2$, $5 = 1 \cdot 1$, $6 = 2 \cdot 2$, $7 = 2 \cdot (1 \cdot 1)$, and $8 = (1 \cdot 2) \cdot (1 \cdot 2)$. A demonstration that the

Cayley tables L_R^1 and L_C^1 are equal is shown in Figure 2.15, and a demonstration that the Cayley tables L_R^2 and L_C^2 are equal is shown in Figure 2.16: we have left the borders in place as they show the exact permutations of rows and columns that we used. It follows that L is a group: that L is isomorphic to the elementary abelian group of order 9 follows from the relations $(1\cdot 1)\cdot 1 = 5\cdot 1 = 0$, $(2\cdot 2)\cdot 2 = 6\cdot 2 = 0$, and $1\cdot 2 = 3 = 2\cdot 1$.

$$L = \begin{array}{c|ccccccccc} & 0 & 1 & 2 & 3 & 4 & 5 & 6 & 7 & 8 \\ \hline 0 & 0 & 1 & 2 & 3 & 4 & 5 & 6 & 7 & 8 \\ 1 & 1 & 5 & 3 & 7 & 8 & 0 & 4 & 2 & 6 \\ 2 & 2 & 3 & 6 & 4 & 1 & 7 & 0 & 8 & 5 \\ 3 & 3 & 7 & 4 & 8 & 5 & 2 & 1 & 6 & 0 \\ 4 & 4 & 8 & 1 & 5 & 7 & 6 & 3 & 0 & 2 \\ 5 & 5 & 0 & 7 & 2 & 6 & 1 & 8 & 3 & 4 \\ 6 & 6 & 4 & 0 & 1 & 3 & 8 & 2 & 5 & 7 \\ 7 & 7 & 2 & 8 & 6 & 0 & 3 & 5 & 4 & 1 \\ 8 & 8 & 6 & 5 & 0 & 2 & 4 & 7 & 1 & 3 \end{array}, \quad Q = \begin{array}{c|cccccccc} & 0 & 1 & 2 & 3 & 4 & 5 & 6 & 7 \\ \hline 0 & 0 & 6 & 1 & 3 & 2 & 7 & 5 & 4 \\ 1 & 3 & 5 & 0 & 2 & 1 & 4 & 7 & 6 \\ 2 & 2 & 7 & 3 & 1 & 0 & 6 & 4 & 5 \\ 3 & 1 & 4 & 2 & 0 & 3 & 5 & 6 & 7 \\ 4 & 4 & 1 & 6 & 7 & 5 & 3 & 2 & 0 \\ 5 & 6 & 0 & 5 & 4 & 7 & 2 & 1 & 3 \\ 6 & 5 & 3 & 7 & 6 & 4 & 1 & 0 & 2 \\ 7 & 7 & 2 & 4 & 5 & 6 & 0 & 3 & 1 \end{array}$$

Fig. 2.14 A loop of order 9 and a quasigroup of order 8

$$L_R^1 = \begin{array}{c|ccccccccc} & 0 & 1 & 2 & 3 & 4 & 5 & 6 & 7 & 8 \\ \hline 1 & 1 & 5 & 3 & 7 & 8 & 0 & 4 & 2 & 6 \\ 5 & 5 & 0 & 7 & 2 & 6 & 1 & 8 & 3 & 4 \\ 3 & 3 & 7 & 4 & 8 & 5 & 2 & 1 & 6 & 0 \\ 7 & 7 & 2 & 8 & 6 & 0 & 3 & 5 & 4 & 1 \\ 8 & 8 & 6 & 5 & 0 & 2 & 4 & 7 & 1 & 3 \\ 0 & 0 & 1 & 2 & 3 & 4 & 5 & 6 & 7 & 8 \\ 4 & 4 & 8 & 1 & 5 & 7 & 6 & 3 & 0 & 2 \\ 2 & 2 & 3 & 6 & 4 & 1 & 7 & 0 & 8 & 5 \\ 6 & 6 & 4 & 0 & 1 & 3 & 8 & 2 & 5 & 7 \end{array}, \quad \text{and} \quad L_C^1 = \begin{array}{c|ccccccccc} & 1 & 5 & 3 & 7 & 8 & 0 & 4 & 2 & 6 \\ \hline 0 & 1 & 5 & 3 & 7 & 8 & 0 & 4 & 2 & 6 \\ 1 & 5 & 0 & 7 & 2 & 6 & 1 & 8 & 3 & 4 \\ 2 & 3 & 7 & 4 & 8 & 5 & 2 & 1 & 6 & 0 \\ 3 & 7 & 2 & 8 & 6 & 0 & 3 & 5 & 4 & 1 \\ 4 & 8 & 6 & 5 & 0 & 2 & 4 & 7 & 1 & 3 \\ 5 & 0 & 1 & 2 & 3 & 4 & 5 & 6 & 7 & 8 \\ 6 & 4 & 8 & 1 & 5 & 7 & 6 & 3 & 0 & 2 \\ 7 & 2 & 3 & 6 & 4 & 1 & 7 & 0 & 8 & 5 \\ 8 & 6 & 4 & 0 & 1 & 3 & 8 & 2 & 5 & 7 \end{array}.$$

Fig. 2.15 Verifying that $L_R^1 = L_C^1$

The quasigroup Q is obtained from the Latin square M' in Figure 2.9 by adding borders $0, 1, \ldots, 7$ in their natural order: we left it to the reader to verify that M' was isotopic to the Cayley table of a dihedral group of order 8. Q is generated by 1 as $5 = 1\cdot 1$, $4 = 1\cdot 5$, $3 = 4\cdot 5$, $2 = 1\cdot 3$, $6 = 4\cdot 2$, $7 = 5\cdot 4$, and $0 = 5\cdot 1$. Figure 2.17 shows that Q_R^1 and Q_C^1 are not equal, and thus Q, though isotopic to a group, is not associative. As Q is isotopic to a group, any loop isotopic to Q must be associative and so must satisfy Light's test. However, any quasigroup that is isotopic to Q, but which is not a loop, will not be associative. In fact, for a quasigroup to be associative, it must be a group.

Theorem 2.14. *An associative quasigroup is a group.*

$$L_R^2 = \begin{array}{c|ccccccccc} & 0&1&2&3&4&5&6&7&8 \\ \hline 2&2&3&6&4&1&7&0&8&5 \\ 3&3&7&4&8&5&2&1&6&0 \\ 6&6&4&0&1&3&8&2&5&7 \\ 4&4&8&1&5&7&6&3&0&2 \\ 1&1&5&3&7&8&0&4&2&6 \\ 7&7&2&8&6&0&3&5&4&1 \\ 0&0&1&2&3&4&5&6&7&8 \\ 8&8&6&5&0&2&4&7&1&3 \\ 5&5&0&7&2&6&1&8&3&4 \end{array}, \quad \text{and} \quad L_C^2 = \begin{array}{c|ccccccccc} & 2&3&6&4&1&7&0&8&5 \\ \hline 0&2&3&6&4&1&7&0&8&5 \\ 1&3&7&4&8&5&2&1&6&0 \\ 2&6&4&0&1&3&8&2&5&7 \\ 3&4&8&1&5&7&6&3&0&2 \\ 4&1&5&3&7&8&0&4&2&6 \\ 5&7&2&8&6&0&3&5&4&1 \\ 6&0&1&2&3&4&5&6&7&8 \\ 7&8&6&5&0&2&4&7&1&3 \\ 8&5&0&7&2&6&1&8&3&4 \end{array}.$$

Fig. 2.16 Verifying that $L_R^2 = L_C^2$

Proof. Let Q be an associative quasigroup and pick $a \in Q$. Let e be the unique element of Q satisfying $ae = a$. Then $ab = (ae)b = a(eb)$ for all $b \in Q$ and so $eb = b$ for all $b \in Q$. Further $(ce)b = c(eb) = cb$ for all $c \in Q$ and so $ce = c$ for all $c \in Q$. It follows that Q is a loop, with identity e, and hence a group. □

It is now clear why Q did not satisfy Light's test. Had we constructed instead a loop from M' in Figure 2.9 by bordering the rows with any column of M' and the columns with any row of M', then the constructed loop would satisfy Light's test. As an example, consider the loop L' in Figure 2.18: this is constructed from M' by using a column of M' to border the rows and a row of M' to border the columns. The identity of L' is 1, and L' is generated by 0 and 7: $3 = 0 \cdot 0$, $2 = 0 \cdot 3$, $1 = 0 \cdot 2$, $5 = 7 \cdot 0$, $6 = 5 \cdot 0$, and $4 = 6 \cdot 0$.

A demonstration that the Cayley tables $L_R'^0$ and $L_C'^0$ are equal is shown in Figure 2.19, and a demonstration that the Cayley tables $L_R'^7$ and $L_C'^7$ are equal is shown in Figure 2.20: we have left the borders in place as they show the exact permutations of rows and columns that we used. It follows that L' is a group: that L' is isomorphic to the dihedral group of order 8 follows from the relations $0^4 = 7^2 = 1$, and $7 \cdot 0 \cdot 7 = 0^3$.

$$Q_R^1 = \begin{array}{c|cccccccc} & 0&1&2&3&4&5&6&7 \\ \hline 6&5&3&7&6&4&1&0&2 \\ 5&6&0&5&4&7&2&1&3 \\ 7&7&2&4&5&6&0&3&1 \\ 4&4&1&6&7&5&3&2&0 \\ 1&3&5&0&2&1&4&7&6 \\ 0&0&6&1&3&2&7&5&4 \\ 3&1&4&2&0&3&5&6&7 \\ 2&2&7&3&1&0&6&4&5 \end{array}, \quad \text{and} \quad Q_C^1 = \begin{array}{c|cccccccc} & 3&5&0&2&1&4&7&6 \\ \hline 0&3&7&0&1&6&2&4&5 \\ 1&2&4&3&0&5&1&6&7 \\ 2&1&6&2&3&7&0&5&4 \\ 3&0&5&1&2&4&3&7&6 \\ 4&7&3&4&6&1&5&0&2 \\ 5&4&2&6&5&0&7&3&1 \\ 6&6&1&5&7&3&4&2&0 \\ 7&5&0&7&4&2&6&1&3 \end{array}.$$

Fig. 2.17 How Q fails Light's test

Fig. 2.18 A loop of order 8

$L' =$

	2	7	3	1	0	6	4	5
3	0	6	1	3	2	7	5	4
2	3	5	0	2	1	4	7	6
1	2	7	3	1	0	6	4	5
0	1	4	2	0	3	5	6	7
7	4	1	6	7	5	3	2	0
4	6	0	5	4	7	2	1	3
6	5	3	7	6	4	1	0	2
5	7	2	4	5	6	0	3	1

$L'^{0}_{R} =$

	2	7	3	1	0	6	4	5
2	3	5	0	2	1	4	7	6
1	2	7	3	1	0	6	4	5
0	1	4	2	0	3	5	6	7
3	0	6	1	3	2	7	5	4
5	7	2	4	5	6	0	3	1
7	4	1	6	7	5	3	2	0
4	6	0	5	4	7	2	1	3
6	5	3	7	6	4	1	0	2

, and $L'^{0}_{C} =$

	1	4	2	0	3	5	6	7
3	3	5	0	2	1	4	7	6
2	2	7	3	1	0	6	4	5
1	1	4	2	0	3	5	6	7
0	0	6	1	3	2	7	5	4
7	7	2	4	5	6	0	3	1
4	4	1	6	7	5	3	2	0
6	6	0	5	4	7	2	1	3
5	5	3	7	6	4	1	0	2

Fig. 2.19 Verifying that $L'^{0}_{R} = L'^{0}_{C}$

$L'^{7}_{R} =$

	2	7	3	1	0	6	4	5
6	5	3	7	6	4	1	0	2
5	7	2	4	5	6	0	3	1
7	4	1	6	7	5	3	2	0
4	6	0	5	4	7	2	1	3
1	2	7	3	1	0	6	4	5
0	1	4	2	0	3	5	6	7
3	0	6	1	3	2	7	5	4
2	3	5	0	2	1	4	7	6

, and $L'^{7}_{C} =$

	4	1	6	7	5	3	2	0
3	5	3	7	6	4	1	0	2
2	7	2	4	5	6	0	3	1
1	4	1	6	7	5	3	2	0
0	6	0	5	4	7	2	1	3
7	2	7	3	1	0	6	4	5
4	1	4	2	0	3	5	6	7
6	0	6	1	3	2	7	5	4
5	3	5	0	2	1	4	7	6

Fig. 2.20 Verifying that $L'^{7}_{R} = L'^{7}_{C}$

A variation on Light's test was proved by Bondesen [42] in 1969. It is an exercise to show that, if M is the Cayley table of a group, C a column of M, and the rows of M are permuted so as to put C in standard order, then the square obtained can also be obtained from M by permuting the columns of M. Bondesen proved the converse.

Theorem 2.15 (Bondesen, 1969). *Let M be a Latin square. For any column C of M, let M_C denote the square obtained from M by permuting the rows of M so as to put C in standard order. If, for any column C, M_C can be obtained from M by permuting the columns of M, then M is the Cayley table of a group.*

2.4.2 The Row Composition Rule

The next test that we will describe is the *row composition rule*. This is a test based on treating the rows of a Latin square as permutations: the Latin square is the Cayley table of a group if and only if this set of permutations is closed under composition. This was implicitly proved by Albert [14] in 1943.

Theorem 2.16 (The Row Composition Rule). *Let L be a loop and for each $a \in L$ define a permutation r_a by $\mathrm{r}_a\colon x \mapsto ax$, set $L_\mathrm{r} = \{\mathrm{r}_a \mid a \in L\}$, and let H be the permutation group generated by L_r. Then L is a group if and only if $H = L_\mathrm{r}$, in which case $L \cong H$.*

Proof. Define a mapping $\phi\colon L \mapsto L_\mathrm{r}$ by $\phi(a) = \mathrm{r}_a$. The mapping ϕ is one to one as if $\phi(a) = \phi(b)$, then $ax = bx$ for all $x \in L$, and so $a = b$.

If L is a group, then

$$\phi(ab)(x) = (ab)x = a(bx) = \phi(a)\phi(b)(x),$$

and so ϕ is an isomorphism, L_r is a group, $H = L_\mathrm{r}$, and $L \cong H$.

If $L_\mathrm{r} = H$ then for any $a, b \in L$, $\phi(a)\phi(b) = \phi(c)$ for some $c \in L$. Thus $a(bx) = cx$ for all $x \in L$, and setting $x = e$, the identity of e, yields $ab = c$, and so $\phi(a)\phi(b) = \phi(ab)$ and ϕ is an isomorphism, and L is a group isomorphic to H. □

To apply the row composition rule, treat the rows as permutations. L is then a group if and only if the set of row permutations is closed under composition. As an example, Figure 2.21 shows a loop of order 11. In Figure 2.22 we see that $\mathrm{r}_1 \circ \mathrm{r}_2$ is not in L_r, and so L_r is not closed under composition and is, therefore, not a group. It follows that L is not a group.

There are other proofs of the row composition rule in the literature. Hammel [173] gave a proof in 1968, and Siu [296] gave another proof in 1991. An application of the row composition rule was described by Burn [59] in 1978. In 1929, Suschkewitsch [305] implicitly proved that a quasigroup is a group if and only if both the row composition rule and the analogous column composition rule hold.

Theorem 2.17 (Suschkewitsch, 1929). *Let Q be a quasigroup and for each $a \in Q$ define permutations $\mathrm{r}_a\colon x \mapsto ax$ and $\mathrm{c}_a\colon x \mapsto xa$, set $Q_\mathrm{r} = \{\mathrm{r}_a \mid a \in Q\}$ and $Q_\mathrm{c} = \{\mathrm{c}_a \mid a \in Q\}$, and let H_r be the permutation group generated by Q_r and H_c the permutation group generated by Q_c. Then Q is a group if and only if $H_\mathrm{r} = Q_\mathrm{r}$ and $H_\mathrm{c} = Q_\mathrm{c}$, in which case $Q \cong H_\mathrm{r} \cong H_\mathrm{c}$.*

Fig. 2.21 A loop of order 11

$L =$

	0	1	2	3	4	5	6	7	8	9	10
0	0	1	2	3	4	5	6	7	8	9	10
1	1	0	3	2	5	4	7	6	9	10	8
2	2	3	4	5	6	7	8	9	10	0	1
3	3	2	5	4	7	6	9	10	1	8	0
4	4	5	6	7	8	9	10	0	2	1	3
5	5	4	7	6	9	10	1	8	0	3	2
6	6	7	8	9	10	0	2	1	3	4	5
7	7	6	9	10	1	8	0	3	5	2	4
8	8	9	10	0	2	1	3	4	6	5	7
9	9	10	1	8	0	3	5	2	4	7	6
10	10	8	0	1	3	2	4	5	7	6	9

Fig. 2.22 How L fails does not satisfy the row composition rule

x	0	1	2	3	4	5	6	7	8	9	10
$\mathrm{r}_1(x)$	1	0	3	2	5	4	7	6	9	10	8
$\mathrm{r}_2(x)$	2	3	4	5	6	7	8	9	10	0	1
$\mathrm{r}_1(\mathrm{r}_2(x))$	3	2	5	4	7	6	9	10	8	1	0

2.5 Keedwell's Criteria

In 2005 Keedwell [217] described tests to determine whether an element of a loop was in its left nucleus, its middle nucleus, or its right nucleus. As each nucleus is a group, these tests can be used to determine if a loop is a group or not. Keedwell introduced three properties that rows/columns of the Cayley table of a loop might have the Frolov property, the F^*-property, and the F_a^*-property. The Frolov property for rows or columns determines elements of the middle nucleus, the F^*-property for rows determines elements of the left nucleus, and the F^*-property for columns determines elements of the right nucleus. The F_a^*-property for columns is used to determine whether a Latin square is the Cayley table of an abelian group.

2.5.1 *The Frolov Property*

If M is the Cayley table of a loop L, we say that row i of M has the *Frolov property*, which we will abbreviate to the F-property; if when the columns of M are reordered to put row i in standard order, then the rows of the new square M^i are the same as the rows of M. Similarly column j of M has the F-property if when the rows of M are reordered to put column j in standard order, then the columns of the new square M' are the same as the columns of M. Bondesen's test (Theorem 2.15) can be expressed as "if every column of M has the F-property, then M is the Cayley table of a group."

For a loop L, $b \in L$ is in the *middle nucleus* of L if $a(bc) = (ab)c$ for all $a, c \in L$, Clearly L is a group if and only if its middle nucleus is L. It is an exercise to show that the middle nucleus is a group. In 2005 Keedwell [217] gave a test to determine whether an element of a loop is in the middle nucleus.

Theorem 2.18 (Keedwell, 2005). *If $L = \{a_1, \ldots, a_n\}$ is a loop with identity a_1 and $M = \{a_i a_j\}$ is the Cayley table of L, then a_i is in the middle nucleus of L if and only if the ith row of M has the F-property and a_j is in the middle nucleus of L if and only if the jth column of M has the F-property.*

Proof. Let us consider the ith row of M and permute the columns of M so that the new square M' has first row $\theta(a_1), \ldots, \theta(a_n)$ and ith row $a_1, \ldots, a_n$. Hence $a_i\theta(a_j) = a_j$.

If a_i is in the middle nucleus of L, which is a group, it has an inverse a_i^{-1}. Hence $a_i^{-1}(a_i\theta(a_j)) = a_i^{-1}a_j$ from which it follows, as a_i is in the middle nucleus, that $\theta(a_j) = (a_i^{-1}a_i)\theta(a_j) = a_i^{-1}a_j$. The kth row of M' is $a_k\theta(a_1), \ldots, a_k\theta(a_n)$ which equals $a_k(a_i^{-1}a_1), \ldots, a_k(a_i^{-1}a_n)$. As a_i^{-1} is in the nucleus of L, this row is $(a_k a_i^{-1})a_1, \ldots, (a_k a_i^{-1})a_n$ which equals $a_m a_1, \ldots, a_m a_n$ for some m. Hence the kth row of M' is the mth row of M. As k is arbitrary, the ith row of M has the F-property.

If the ith row of M has the F-property, then the kth row of M' is the mth row of M for some m and so $a_k\theta(a_j) = a_m a_j = a_m(a_i\theta(a_j))$ for all j. Now, as $\theta(a_j) = a_1$ for some j, $a_k = a_m a_i$. Thus $(a_m a_i)\theta(a_j) = a_m(a_i\theta(a_j))$ for all m, j. It follows that a_i is in the middle nucleus of L.

To prove that a_j is in the middle nucleus of L if and only if the jth column of M has the F-property, let L^T be the loop with Cayley table M^T, the transpose of M. The jth column of M has the F-property if and only if the jth row of M^T has the F-property, if and only a_j is in the middle nucleus of L^T, if and only a_j is in the middle nucleus of L. □

Note that, to show that M is the Cayley table of a group using Theorem 2.15, we need to test each column of M for the F-property. Theorem 2.18 allows us to test far fewer columns (rows) to determine that M is the Cayley table of a group: this is because the order of the middle nucleus of a loop divides the order of the loop. The proof that the order of the middle nucleus of a loop divides the order of the loop is similar to Lagrange's proof that the order of a subgroup divides the order of the group. If N is the middle nucleus of a loop L, then the left cosets $aN = \{ax \mid x \in L\}$ partition L and all have the same order: the proof is an exercise.

Corollary 2.19. *If L is a loop of order n, p the smallest prime divisor of n, M the Cayley table of L, and $(n/p) + 1$ columns (rows) of L have the F-property, then L is a group.*

Proof. If N is the middle nucleus of L, then $|N|$ divides n. By Theorem 2.18, $|N| \geq (n/p) + 1$. It follows that $|N| = n$ and hence that $L = N$, a group. □

Applying Corollary 2.19 when $n = p$ yields a shorter test.

Corollary 2.20. *If L is a loop of prime order p, M the Cayley table of L, and two columns (rows) of have the F-property, then L is isomorphic to the group $\mathbb{Z}_p$.*

In Figure 1.1 we gave three examples of Latin squares of order 5. The third of these, L_3, we claimed to be the Cayley table of $\mathbb{Z}_5$. This square is depicted in Figure 2.23. It is clear that the first column has the F-property. If we take the second column, highlighted in bold, and rearrange the rows of L_3 so that the second column is in standard order, then we obtain the square L_3' shown in Figure 2.23. As the columns of L_3 and L_3' are the same, though in a different order, both the first and second columns of L_3 have the F-property. From Corollary 2.20, it follows that L_3 is the Cayley table of $\mathbb{Z}_5$ as claimed.

Fig. 2.23 A Latin square that is the Cayley table of $\mathbb{Z}_5$

$$L_3 = \begin{bmatrix} 0\,\mathbf{1}\,2\,3\,4 \\ 1\,\mathbf{4}\,3\,0\,2 \\ 2\,\mathbf{3}\,1\,4\,0 \\ 3\,\mathbf{0}\,4\,2\,1 \\ 4\,\mathbf{2}\,0\,1\,3 \end{bmatrix}, \quad L_3' = \begin{bmatrix} 3\,0\,4\,2\,1 \\ 0\,1\,2\,3\,4 \\ 4\,2\,0\,1\,3 \\ 2\,3\,1\,4\,0 \\ 1\,4\,3\,0\,2 \end{bmatrix}$$

Figure 2.5 showed a loop of order 9. We showed three instances of its satisfying Aczél's test and later used Light's test to show that this loop was a group. Let us apply Theorem 2.18 to the Cayley table of this loop. In Figure 2.24 we show the Cayley table M of the loop and squares M_2, M_3, and M_4: square M_i is obtained from M by permuting the rows so as to put column i in standard order. In each case the columns of M_i are the columns of M. Hence columns 1, 2, 3, and 4 have the F-property. It follows from Corollary 2.19 that this loop is a group.

2.5.2 *The F*-Property*

If M is the Cayley table of a loop L, we say that row i of M has the F^**-property* if when the symbols of M are permuted to put row i in standard order, then the rows of the new square M' are the same as the rows of M. Similarly, we say that column j of M has the F^**-property* if when the symbols of M are permuted to put column j in standard order, then the columns of the new square M' are the same as the columns of M.

For a loop L, $a \in L$ is in the *left nucleus* of L if $a(bc) = (ab)c$ for all $b, c \in L$, and $c \in L$ is in the *right nucleus* of L if $a(bc) = (ab)c$ for all $a, b \in L$. Clearly L is a group if and only if its left nucleus is L, if and only if its middle nucleus is L, if and only if its right nucleus is L. It is an exercise to show that the left and right nuclei are groups. In 2005 Keedwell [217] gave tests to determine whether an element of a loop is in the right nucleus or the left nucleus.

Theorem 2.21 (Keedwell, 2005). *If $L = \{a_1, \ldots, a_n\}$ is a loop with identity a_1 and $M = \{a_i a_j\}$ is the Cayley table of L, then a_i is in the left nucleus of L if and* *only if the ith row of M has the F^*-property and a_j is in the right nucleus of L if and only if the jth column of M has the F^*-property.*

Fig. 2.24 A loop of order 9 that is a group

$$M = \begin{pmatrix} 0&1&2&3&4&5&6&7&8 \\ 1&5&3&7&8&0&4&2&6 \\ 2&3&6&4&1&7&0&8&5 \\ 3&7&4&8&5&2&1&6&0 \\ 4&8&1&5&7&6&3&0&2 \\ 5&0&7&2&6&1&8&3&4 \\ 6&4&0&1&3&8&2&5&7 \\ 7&2&8&6&0&3&5&4&1 \\ 8&6&5&0&2&4&7&1&3 \end{pmatrix}, \quad M_2 = \begin{pmatrix} 5&0&7&2&6&1&8&3&4 \\ 0&1&2&3&4&5&6&7&8 \\ 7&2&8&6&0&3&5&4&1 \\ 2&3&6&4&1&7&0&8&5 \\ 6&4&0&1&3&8&2&5&7 \\ 1&5&3&7&8&0&4&2&6 \\ 8&6&5&0&2&4&7&1&3 \\ 3&7&4&8&5&2&1&6&0 \\ 4&8&1&5&7&6&3&0&2 \end{pmatrix}.$$

$$M_3 = \begin{pmatrix} 6&4&0&1&3&8&2&5&7 \\ 4&8&1&5&7&6&3&0&2 \\ 0&1&2&3&4&5&6&7&8 \\ 1&5&3&7&8&0&4&2&6 \\ 3&7&4&8&5&2&1&6&0 \\ 8&6&5&0&2&4&7&1&3 \\ 2&3&6&4&1&7&0&8&5 \\ 5&0&7&2&6&1&8&3&4 \\ 7&2&8&6&0&3&5&4&1 \end{pmatrix}, \quad M_4 = \begin{pmatrix} 8&6&5&0&2&4&7&1&3 \\ 6&4&0&1&3&8&2&5&7 \\ 5&0&7&2&6&1&8&3&4 \\ 0&1&2&3&4&5&6&7&8 \\ 2&3&6&4&1&7&0&8&5 \\ 4&8&1&5&7&6&3&0&2 \\ 7&2&8&6&0&3&5&4&1 \\ 1&5&3&7&8&0&4&2&6 \\ 3&7&4&8&5&2&1&6&0 \end{pmatrix}.$$

Proof. Let us consider the ith row of M, let $\theta\colon L \to L$ be a bijection satisfying $\theta(a_i a_j) = a_j$, and let M' be the square obtained from M by replacing entry x with $\theta(x)$. The ith row of M' is $a_1, \ldots, a_n$ and the kth row of M' is $\theta(a_k a_1), \ldots, \theta(a_k a_n)$. The row of M with first entry $\theta(a_k a_1)$ is $\theta(a_k)a_1, \ldots, \theta(a_k)a_n$. Hence the ith row of M has the F^*-property if and only if $\theta(a_k a_j) = \theta(a_k)a_j$ for all j, k. Let a_i' be the unique element of L for which $a_i a_i' = a_1$. Then $\theta(a_1) = \theta(a_i a_i') = a_i'$.

If the ith row of M has the F^*-property, then $\theta(a_k a_j) = \theta(a_k)a_j$ for all j, k. Setting $k = 1$ yields $\theta(a_j) = a_i' a_j$ for all j and, hence, $a_i'(a_k a_j) = (a_i' a_k)a_j$ for all j, k. It follows that a_i' is in the left nucleus of L and, as the left nucleus of L is a group, a_i is in the left nucleus of L.

Suppose that a_i is in the left nucleus of L and, for each k, let a_k^* be the unique element of L for which $a_k = a_i a_k^*$. Then, for all j, k,

$$\theta(a_k a_j) = \theta((a_i a_k^*)a_j) = \theta(a_i(a_k^* a_j)) = a_k^* a_j = \theta(a_i a_k^*)a_j = \theta(a_k)a_j.$$

Hence the ith row of M has the F^*-property.

To prove that a_j is in the right nucleus of L if and only if the jth column of M has the F^*-property, let L^T be the loop with Cayley table M^T, the transpose of M. The jth column of M has the F^*-property if and only if the jth row of M^T has the F^*-property, if and only a_j is in the left nucleus of L^T, if and only if a_j is in the right nucleus of L. □

As the orders of the left and right nuclei divide the order of the loop, Corollaries 2.19 and 2.20 still hold with the "F-property" replaced by the "F^*-property." It should be noted that if M is the Cayley table of a group, then whether we permute the rows (or columns) or permute the symbols to put a given column (or row) into standard order, the resulting square will be the same.

2.5.3 The F_a^*-Property

If M is the Cayley table of a loop L, we say that column j of M has the *F_a^*-property* if when the symbols of M are permuted to put column j in standard order, then the columns of the new square M' are the same as the columns of M and the rows of M' are the same as the rows of M. Note that, if a column has the F_a^*-property, then it also has the F^*-property and hence, by Theorem 2.21, the F_a^*-property can be used to show that M is the Cayley table of a group. Keedwell [217] showed in 2005 that the F_a^*-property can be used to tell us whether M is the Cayley table of an abelian group.

Theorem 2.22 (Keedwell, 2005). *If $L = \{a_1, \ldots, a_n\}$ is a loop with identity a_1 and $M = \{a_i a_j\}$ is the Cayley table of L, then L is an abelian group if and only if every column of M has the F_a^*-property.*

Proof. If a column has the F_a^*-property, then it also has the F^*-property. By Theorem 2.21, L is a group if and only if each column of M has the F^*-property. Thus we may assume that L is a group and we need only show that L is abelian if and only if each column of M has the F_a^*-property.

The elements of the jth column of M are $a_1 a_j, \ldots, a_n a_j$. The permutation of the symbols that puts the jth column of M into standard order is $x \mapsto x a_j^{-1}$. Let M' be the square obtained from M by replacing each entry x by $x a_j^{-1}$. The ith row of M' is $a_i a_1 a_j^{-1}, \ldots, a_i a_n a_j^{-1}$. The row of M with the same first entry as M' is $a_i a_j^{-1} a_1, \ldots, a_i a_j^{-1} a_n$. It follows that the jth column of M has the F_a^*-property for all j if and only if $a_i a_k a_j^{-1} = a_i a_j^{-1} a_k$ for all i, j, k, if and only if $a_k a_j^{-1} = a_j^{-1} a_k$ for all j, k, if and only if L is an abelian group. □

Figure 2.25 depicts two Latin squares of order 6 with the second column in bold. For each square we permute the symbols to put the second column in standard order. The resultant squares are shown in Figure 2.26. The columns of L_1' are the same as those of L_1, and so the second column of L_1 has the F^*-property. However the rows of L_1' are not the same as those of L_1, and so L_1 cannot be the Cayley table of an abelian group. We leave it to the reader to show that L_1 is the Cayley table of the nonabelian group of order 6. The columns of L_2' are the same as those of L_2 and so the second column of L_2 has the F^*-property. Further, the rows of L_2' are the same as those of L_2, and so the second column of L_2 has the F_a^*-property. We leave it to the reader to show that every column of L_2 has the F_a^*-property and hence that L_2 is the Cayley table of the abelian group of order 6.

Fig. 2.25 Two Latin squares of order 6

$$L_1 = \begin{bmatrix} 0 & \mathbf{1} & 2 & 3 & 4 & 5 \\ 1 & \mathbf{2} & 0 & 4 & 5 & 3 \\ 2 & \mathbf{0} & 1 & 5 & 3 & 4 \\ 3 & \mathbf{5} & 4 & 0 & 2 & 1 \\ 4 & \mathbf{3} & 5 & 1 & 0 & 2 \\ 5 & \mathbf{4} & 3 & 2 & 1 & 0 \end{bmatrix}, \quad L_2 = \begin{bmatrix} 0 & \mathbf{1} & 2 & 3 & 4 & 5 \\ 1 & \mathbf{3} & 0 & 5 & 2 & 4 \\ 2 & \mathbf{0} & 4 & 1 & 5 & 3 \\ 3 & \mathbf{5} & 1 & 4 & 0 & 2 \\ 4 & \mathbf{2} & 5 & 0 & 3 & 1 \\ 5 & \mathbf{4} & 3 & 2 & 1 & 0 \end{bmatrix}$$

Fig. 2.26 The F_a^*-property for two Latin squares of order 6?

$$L_1' = \begin{bmatrix} 2 & 0 & 1 & 4 & 5 & 3 \\ 0 & 1 & 2 & 5 & 3 & 4 \\ 1 & 2 & 0 & 3 & 4 & 5 \\ 4 & 3 & 5 & 2 & 1 & 0 \\ 5 & 4 & 3 & 0 & 2 & 1 \\ 3 & 5 & 4 & 1 & 0 & 2 \end{bmatrix}, \quad L_2' = \begin{bmatrix} 2 & 0 & 4 & 1 & 5 & 3 \\ 0 & 1 & 2 & 3 & 4 & 5 \\ 4 & 2 & 5 & 0 & 3 & 1 \\ 1 & 3 & 0 & 5 & 2 & 4 \\ 5 & 4 & 3 & 2 & 1 & 0 \\ 3 & 5 & 1 & 4 & 0 & 2 \end{bmatrix}$$

Part II
Admissible Groups

Chapter 3
The Existence Problem for Complete Mappings: The Hall-Paige Conjecture

Given a group, does its Cayley table have an orthogonal mate? Equivalently, does a given group admit complete mappings? In this chapter we will introduce the existence problem for complete mappings of groups, a problem that we will solve for groups of odd order, countably infinite groups, and finite abelian groups.

For finite groups in general, we will present Hall and Paige's important result: if the Sylow 2-subgroup of a finite group is nontrivial and cyclic, then the group does not admit complete mappings. Hall and Paige conjectured the converse. Work on this conjecture will be described in Chapters 4 through 7. In Chapter 4 we will establish the admissibility of several classes of groups; in Chapter 5 we will specialize in work on the determination of the admissibility of the linear groups, $\mathrm{GL}(n, q)$, $\mathrm{SL}(n, q)$, $\mathrm{PGL}(n, q)$, and $\mathrm{PSL}(n, q)$; in Chapter 6 we will describe work on determining possible minimal counterexamples to the Hall-Paige conjecture; and a proof of this conjecture will be given in Chapter 7.

Much of the material in this and subsequent chapters can be skipped by readers who just want to see the proof of the Hall-Paige conjecture: the proof that the conjecture is true for solvable groups will be given in Sections 4.1 and 4.2, and the proof that the conjecture is true for nonsolvable groups will be given in Section 6.4 and Chapter 7.

3.1 Introduction

An old question is, when does a Latin square have an orthogonal mate? This problem dates back at least to Euler. Let us consider a class of examples that was studied by Euler [110] in 1779. A *cyclic* Latin square of order n is a Latin square of the form

A. B. Evans, *Orthogonal Latin Squares Based on Groups*, Developments in Mathematics 57, https://doi.org/10.1007/978-3-319-94430-2_3

$$\begin{pmatrix} a_0 & a_1 & \cdots & a_{n-1} \\ a_1 & a_2 & \cdots & a_0 \\ \vdots & \vdots & \ddots & \vdots \\ a_{n-1} & a_0 & \cdots & a_{n-2} \end{pmatrix}.$$

We may ask, for what values of n does this square have an orthogonal mate? As isotopisms preserve orthogonality, in answering this question we may, without loss of generality, assume that $a_i = i$ for each i and that the rows and columns are indexed by $0, 1, 2, \ldots, n-1$. This is clearly, then, the Cayley table of $\mathbb{Z}_n$. We will see that the cases n even and n odd have quite different answers. The following are the smallest cyclic Latin squares:

$$\begin{pmatrix} 0\,1 \\ 1\,0 \end{pmatrix}, \quad \begin{pmatrix} 0\,1\,2 \\ 1\,2\,0 \\ 2\,0\,1 \end{pmatrix}, \quad \begin{pmatrix} 0\,1\,2\,3 \\ 1\,2\,3\,0 \\ 2\,3\,0\,1 \\ 3\,0\,1\,2 \end{pmatrix}, \quad \begin{pmatrix} 0\,1\,2\,3\,4 \\ 1\,2\,3\,4\,0 \\ 2\,3\,4\,0\,1 \\ 3\,4\,0\,1\,2 \\ 4\,0\,1\,2\,3 \end{pmatrix}.$$

Let us first look at the squares of odd order. The entries on the main diagonal of the square of order 3 are, in order, 0, 2, and 1; and those on the main diagonal of the square of order 5 are, in order, 0, 2, 4, 1, and 3. Hence, in each of these cases, the main diagonal is a transversal, which, as these are the Cayley tables of $\mathbb{Z}_3$ and $\mathbb{Z}_5$, implies by Theorem 1.13 that each of these squares has an orthogonal mate. This applies more generally. If n is odd, then the main diagonal of a cyclic Latin square of order n is a transversal, which, as this is the Cayley table of $\mathbb{Z}_n$, implies by Theorem 1.13 that this square has an orthogonal mate. For the squares of even order, it is fairly easy to see that neither of them has a transversal. More generally, if n is even, then the cyclic Latin square of order n has no transversal, which implies by Theorem 1.12 that this square has no orthogonal mate. This is summarized in the following theorem, due to Euler [110].

Theorem 3.1 (Euler, 1779). *A cyclic Latin square has an orthogonal mate if and only if it is of odd order.*

Proof. Let L be a cyclic Latin square of order n. The ijth entry of L may be assumed to be $i + j \pmod n$. For L to have an orthogonal mate, by Theorem 1.12, L must have a transversal.

Let n be even, and let $(0, j_0)$, $(1, j_1)$, $\ldots$, $(n-1, j_{n-1})$ be a transversal of L. Then $0 + 1 + \cdots + (n-1) \equiv n/2 \pmod n$, and

$$j_0 + j_1 + \cdots + j_{n-1} \equiv 0 + 1 + \cdots + (n-1) \equiv n/2 \pmod n.$$

As the entries of these cells are distinct

$$(0 + j_0) + (1 + j_1) + \cdots + ((n-1) + j_{n-1}) \equiv 0 + 1 + \cdots + (n-1) \equiv n/2 \pmod n.$$

But adding in a different way, we obtain

$$\begin{aligned}&(0+j_0)+(1+j_1)+\cdots+((n-1)+j_{n-1})\\ \equiv&(0+1+\cdots+(n-1))+(j_0+j_1+\cdots+j_{n-1})\\ \equiv&n/2+n/2\equiv 0 \pmod{n},\end{aligned}$$

a contradiction. Hence, L cannot have a transversal and so cannot have an orthogonal mate.

If n is odd, then the entries on the main diagonal, $0, 2, \ldots, 2(n-1)$ modulo n, are all distinct as, if $2i \equiv 2j \pmod{n}$, then, 2 being relatively prime to n, $i = j$. Thus the main diagonal of L is a transversal, which, as this is the Cayley table of $\mathbb{Z}_n$, implies by Theorem 1.13 that L has an orthogonal mate. □

In proving the nonexistence of transversals of cyclic Latin squares of even order, we used the fact that $0+1+\cdots+(n-1) \equiv n/2 \pmod{n}$. But, if n is odd, then $0+1+\cdots+(n-1) \equiv 0 \pmod{n}$, and our nonexistence argument, used for n even, cannot be applied. We will see that the sum or product of all the elements of a group plays an important role in determining whether a Latin square based on the group has an orthogonal mate or not.

3.1.1 Cyclic Groups

By Theorem 1.15 a Latin square based on a group G has an orthogonal mate if and only if G admits complete mappings. We will call a group that admits complete mappings *admissible*, a term that was defined by Paige [263] in 1951, though he had previously used a different definition of admissibility in a 1949 paper [262]: there he defined a group to be admissible if it could be the multiplicative group of a neofield: neofields will be studied in Chapter 15. The problem of determining which finite groups have Cayley tables that have orthogonal mates is equivalent to the problem of determining which finite groups are admissible. Theorem 3.1 solves the admissibility problem for finite cyclic groups.

Corollary 3.2. *A finite cyclic group is admissible if and only if its order is odd.*

Rediscoveries of special cases of Theorem 3.1, equivalently Corollary 3.2, can be found in the literature. As an example, consider the following 1882 result of Hurwitz [193]: the proof that we give is from a 1948 paper of Vijayaraghavan and Chowla [318].

Theorem 3.3 (Hurwitz, 1882). *If p is an odd prime and $x_1, \ldots, x_p$ and $y_1, \ldots$, y_p are two complete sets of residues modulo p, then $x_1y_1, \ldots, x_py_p$ cannot be a complete set of residues modulo p.*

Proof. Assume to the contrary that $x_1y_1, \ldots, x_py_p$ is a complete set of residues modulo p. As $x_iy_i \equiv 0 \pmod p$ for exactly one i, we may assume without loss of generality that $x_1 \equiv y_1 \equiv 0 \pmod p$. Thus

$$\begin{aligned}
1 \times \cdots \times (p-1) &\equiv x_2 \cdots x_p \\
&\equiv y_2 \cdots y_p \\
&\equiv x_2y_2 \cdots x_py_p \\
&\equiv (1 \times \cdots \times (p-1))^2 \pmod p.
\end{aligned}$$

But $1 \times \cdots \times (p-1) \equiv -1 \pmod p$ by Wilson's theorem (see Theorem 5.4 in [60], for instance), and so $-1 \equiv (-1)^2 \equiv 1 \pmod p$, a contradiction. □

In Theorem 3.3 we proved that the mapping $x_i \mapsto y_i$, $i \neq 1$, can never be a complete mapping of the multiplicative group of a field of order p. As this is a cyclic group of order $p-1$, Theorem 3.3 is a special case of Corollary 3.2. In 1948 Vijayaraghavan and Chowla [318] extended Hurwitz's result to complete primitive residue sets modulo n. The sequence $x_1, \ldots, x_{\phi(n)}$, ϕ the Euler function, is a *complete primitive* residue set modulo n if each x_i is relatively prime to n and $x_i \not\equiv x_j \pmod n$ if $i \neq j$.

Theorem 3.4 (Vijayaraghavan and Chowla, 1948). *Let $n > 2$ have a primitive root, and let $x_1, \ldots, x_{\phi(n)}$ and $y_1, \ldots, y_{\phi(n)}$ be two complete primitive residue sets modulo n. Then $x_1y_1, \ldots, x_{\phi(n)}y_{\phi(n)}$ cannot be a complete primitive residue set modulo n.*

Proof. Assume to the contrary that $x_1y_1, \ldots, x_{\phi(n)}y_{\phi(n)}$ is a complete primitive residue set modulo n and that g is a primitive root of n. Then

$$\begin{aligned}
gg^2 \cdots g^{\phi(n)} &\equiv x_1 \cdots x_{\phi(n)} \\
&\equiv y_1 \cdots y_{\phi(n)} \\
&\equiv x_1y_1 \cdots x_{\phi(n)}y_{\phi(n)} \\
&\equiv (gg^2 \cdots g^{\phi(n)})^2 \pmod n.
\end{aligned}$$

But $gg^2 \cdots g^{\phi(n)} \equiv g^{\phi(n)(\phi(n)+1)/2} \pmod n$, and, as a simple argument shows that $\phi(n)$ is even, $g^{\phi(n)(\phi(n)+1)/2} \equiv g^{\phi(n)/2} \pmod n$. As g is a primitive root of n, $g^{\phi(n)/2} \equiv -1 \pmod n$. Thus $-1 \equiv (-1)^2 \equiv 1 \pmod n$, a contradiction. □

Theorem 3.4 applies when n has a primitive root, and it is known that n has a primitive root if and only if n is one of 2, 4, p^t or $2p^t$, where p is an odd prime and t is a positive integer: see Theorem 8.10 of [60] for a proof. The case $n = 2$ is the trivial exception to the theorem. In Theorem 3.4 the mapping $x_i \mapsto y_i$ is potentially a complete mapping of the group of units of the ring of integers modulo n. As, if n has a primitive root, this group is a cyclic group of even order, Theorem 3.4 is another special case of Corollary 3.2.

A more recent example is to be found in a 1960 result of Singer [295]. Singer proved, using essentially the same argument as in Theorem 3.1, that if G is a cyclic group of even order, then there does not exist a bijection $\theta\colon G \to G$ for which the mapping $x \mapsto \theta(x) - x$ is a bijection. Singer called such bijections 1-permutations. We recognize these as orthomorphisms, which exist if and only if complete mappings exist, and so Singer's result is a special case of Corollary 3.2. More recently still, in 1969 Hedayat and Federer [175] proved that a Latin square based on a cyclic permutation group of order n and degree n does not possess a transversal if n is even. Their proof is also essentially the same as in Theorem 3.1.

3.1.2 Groups of Odd Order

In proving Theorem 3.1, we showed that the main diagonal of any cyclic Latin square of odd order is a transversal. As this also holds true for Cayley tables of groups of odd order, we can conclude that Latin squares based on groups of odd order have orthogonal mates.

Theorem 3.5. *Any Latin square based on a group of odd order has an orthogonal mate.*

Proof. Let $G = \{g_1, \ldots, g_n\}$, n odd, and let M be a Latin square based on G. Without loss of generality, we may assume M to be the Cayley table of G and the ijth entry of M to be $g_i g_j$. The entries on the main diagonal of M, $g_1^2, \ldots, g_n^2$, are all distinct, as if $g_i^2 = g_j^2$, 2 being relatively prime to n, $g_i = g_j$. Thus the main diagonal of M is a transversal, which, as M is the Cayley table of a group, implies by Theorem 1.13 that M has an orthogonal mate. □

This solves the admissibility problem for groups of odd order.

Corollary 3.6. *Any group of odd order is admissible.*

If M is the Cayley table of a finite group G, then Theorem 1.14 explicitly establishes a one-to-one correspondence between the set of transversals of M and the set of complete mappings of G. In this correspondence the main diagonal corresponds to the identity mapping. Thus the proof of Theorem 3.5 implies that the identity mapping $g \mapsto g$ is a complete mapping of any group of odd order. This observation was made by Paige [261] in 1947. In 1963, Sade [287] also proved the mapping $x \mapsto x$ to be a complete mapping of any finite group of odd order, as well as any commutative quasigroup of odd order.

3.1.3 Infinite Groups

For infinite groups the admissibility problem was solved in 1950 by Bateman [29] using transfinite induction. In the following theorem, we will give his admissibility proof for countably infinite groups: see Bateman's paper for the general proof.

Theorem 3.7 (Bateman, 1950). *Any countably infinite group is admissible.*

Proof. Let $G = \{g_1, g_2, g_3, \ldots\}$ be a countably infinite group. Pick $a, b \in G$ and set $x_1 = a$, $y_1 = b$, and $z_1 = ab$. Set $X_1 = \{x_1\}$, $Y_1 = \{y_1\}$, and $Z_1 = \{z_1\}$. We will recursively define sets X_n, Y_n, and Z_n, $n = 1, 2, 3, \ldots$, as follows. Having defined X_n, Y_n, and Z_n, let u_n be the element of G of least index that is not an element of $X_n \cap Y_n \cap Z_n$. If $u_n \notin X_n$, then set $x_{n+1} = u_n$, and set y_{n+1} equal to the element of G of least index which is not in Y_n and for which $x_{n+1}y_{n+1} \notin Z_n$. Set $z_{n+1} = x_{n+1}y_{n+1}$, $X_{n+1} = X_n \cup \{x_{n+1}\}$, $Y_{n+1} = Y_n \cup \{y_{n+1}\}$, and $Z_{n+1} = Z_n \cup \{z_{n+1}\}$. If $u_n \in X_n$ but $u_n \notin Y_n$, then set $y_{n+1} = u_n$, and set x_{n+1} equal to the element of G of least index which is not in X_n and for which $x_{n+1}y_{n+1} \notin Z_n$. Set $z_{n+1} = x_{n+1}y_{n+1}$, $X_{n+1} = X_n \cup \{x_{n+1}\}$, $Y_{n+1} = Y_n \cup \{y_{n+1}\}$, and $Z_{n+1} = Z_n \cup \{z_{n+1}\}$. If $u_n \in X_n \cap Y_n$ but $u_n \notin Z_n$, then set $z_{n+1} = u_n$, and set x_{n+1} equal to the element of G of least index which is not in X_n and for which $x_{n+1}^{-1}z_{n+1} \notin Y_n$. Set $y_{n+1} = x_{n+1}^{-1}z_{n+1}$, $X_{n+1} = X_n \cup \{x_{n+1}\}$, $Y_{n+1} = Y_n \cup \{y_{n+1}\}$, and $Z_{n+1} = Z_n \cup \{z_{n+1}\}$.

If we set $X = \{x_1, x_2, \ldots\}$, $Y = \{y_1, y_2, \ldots\}$, and $Z = \{z_1, z_2, \ldots\}$, then each element of G appears exactly once in each of the sequences X, Y, and Z. Hence, the mapping $x_n \mapsto y_n$ is a complete mapping of G. □

A special case of Theorem 3.7 had already been proved by Paige [262] in 1949, who showed that finitely generated abelian groups, that do not possess a unique involution, are admissible.

3.2 The Admissibility of Finite Abelian Groups

Having determined the admissibility of cyclic groups, groups of odd order, and infinite groups, we turn our attention to the admissibility of finite abelian groups. Admissible finite abelian groups were completely characterized by Paige [261] in 1947.

3.2.1 Paige's Theorem

For a finite group G, and a bijection $\theta\colon G \to G$, let η_θ be defined by $\eta_\theta(g) = g\theta(g)$ and set $O(\eta_\theta) = |\{\eta_\theta(g) \mid g \in G\}|$. It is clear that if $O(\eta_\theta) = |G|$, then θ is a complete mapping of G. Starting with a bijection $\theta\colon G \to G$, G a finite abelian group, Paige gave an existence proof for a bijection $\theta'\colon G \to G$ for which $O(\eta_{\theta'}) = |G|$ if G does not have a unique involution, and $|G| - 1$ if G does have a unique involution. Essential to Paige's proof is the value of the product of all the elements of a finite abelian group.

Lemma 3.8. *Let G be a finite abelian group. If G has a unique involution u, then*

$$\prod_{g\in G} g = u, \quad \textit{otherwise} \quad \prod_{g\in G} g = 1.$$

Proof. In the product $\prod_{g\in G} g$, each element $g \in G$ that is not 1 or an involution cancels with g^{-1}. Hence $\prod_{g\in G} g = \prod_{h\in H} h$, where H is the subgroup of G consisting of the identity and the involutions of G. The result is then immediate if G has at most one involution.

If G has more than one involution, then H is an elementary abelian 2-group. Let $\{u_1, \ldots, u_m\}$ be a minimal generating set of H. We can use the isomorphism of H with the additive group of an m-dimensional vector space over $GF(2)$ to show that each element of H can be written uniquely in the form $u_1^{\varepsilon_1} \cdots u_m^{\varepsilon_m}$, where $\varepsilon_1, \ldots, \varepsilon_m \in \{0, 1\}$. Thus

$$\prod_{g\in G} g = \prod_{h\in H} h = \prod_{\varepsilon_1,\ldots,\varepsilon_m=0,1} u_1^{\varepsilon_1} \cdots u_m^{\varepsilon_m} = \prod_{i=1}^{m} u_i^{2^{m-1}} = 1,$$

hence the result. □

It is pointed out in [94] that Lemma 3.8 had earlier been proved by Miller [242] in 1903 and had subsequently been reproved by Paige [261] and Ramanathan [274], both in 1947 and by Carlitz [61] in 1953. The admissibility of a finite abelian group depends on whether the product of its elements is the identity or not.

Theorem 3.9 (Paige, 1947). *A finite abelian group is admissible if and only if it does not have a unique involution.*

Proof. We will first prove that there exists a bijection $\theta\colon G \to G$ for which $O(\eta_\theta) \geq |G| - 1$. Let us suppose to the contrary that for all bijections $\theta\colon G \to G$, $O(\eta_\theta) \leq |G| - 2$, and let $\theta\colon G \to G$ be a bijection for which $O(\eta_{u\theta}) = r$ is maximum. Let $G = \{g_1, g_2, \ldots, g_n\}$ be ordered so that $\eta_\theta(g_1), \ldots, \eta_\theta(g_r)$ are distinct. There cannot exist $s, t > r$ for which $g_s\theta(g_t) \neq \eta_\theta(g_i)$ for any i, as if we define $\theta'\colon G \to G$ by

$$\theta'(g_i) = \begin{cases} \theta(g_t), & \text{if } i = s; \\ \theta(g_s), & \text{if } i = t; \\ \theta(g_i), & \text{if } i \neq s, t, \end{cases}$$

then $O(\eta_{\theta'}) > r$.

As $\eta_\theta(g_{r+1}) = \eta_\theta(g_i)$ for some $i = 1, \ldots, r$, we may assume, by relabeling if need be, that $\eta_\theta(g_{r+1}) = \eta_\theta(g_1)$. Now $g_1\theta(g_{r+2}) \neq \eta_\theta(g_1)$ as then $\theta(g_{r+2}) = \theta(g_1)$. But $g_1\theta(g_{r+2}) = \eta_\theta(g_i)$, for some $i = 2, \ldots, r$, as otherwise, if we set

$$\theta'(g_i) = \begin{cases} \theta(g_{r+2}), & \text{if } i = 1; \\ \theta(g_1), & \text{if } i = r+2; \\ \theta(g_i), & \text{if } i \neq 1, r+2, \end{cases}$$

then $O(\eta_{\theta'}) > r$. We may assume, by relabeling if need be, that $g_1\theta(g_{r+2}) = \eta_\theta(g_2)$.

Now $g_2\theta(g_1) \neq \eta_\theta(g_1)$ as then $g_2 = g_1$, and $g_2\theta(g_1) \neq \eta_\theta(g_2)$ as then $\theta(g_2) = \theta(g_1)$. But $g_2\theta(g_1) = \eta_\theta(g_i)$, for some $i = 3, \ldots, r$, as otherwise if we set

$$\theta'(g_i) = \begin{cases} \theta(g_{r+2}), & \text{if } i = 1; \\ \theta(g_1), & \text{if } i = 2; \\ \theta(g_2), & \text{if } i = r+2 \\ \theta(g_i), & \text{if } i \neq 1, 2, r+2, \end{cases}$$

then $O(\eta_{\theta'}) > r$. We may assume, by relabeling if need be, that $g_2\theta(g_1) = \eta_\theta(g_3)$.

Let us suppose that we have proved that

$$g_{i+1}\theta(g_i) = \eta_\theta(g_{i+2}), \tag{3.1}$$

for $i = 1, \ldots, k$. From this and $g_1\theta(g_{r+2}) = \eta_\theta(g_2)$, it follows that $\eta_\theta(g_1)\theta(g_{r+2}) = \eta_\theta(g_{i+1})\theta(g_i)$ for $i = 1, \ldots, k+1$ as

$$\begin{aligned} \eta_\theta(g_1)\theta(g_{r+2}) &= g_1\theta(g_{r+2})\theta(g_1) \\ &= \eta_\theta(g_2)\theta(g_1) \\ &= g_2\theta(g_1)\theta(g_2) \\ &= \eta_\theta(g_3)\theta(g_2) = \cdots = \eta_\theta(g_{k+2})\theta(g_{k+1}). \end{aligned}$$

Now $g_{k+2}\theta(g_{k+1}) \neq \eta_\theta(g_i)$ for any $i \leq k+2$ as otherwise

$$\eta_\theta(g_i)\theta(g_{k+2}) = g_{k+2}\theta(g_{k+1})\theta(g_{k+2}) = \eta_\theta(g_{k+2})\theta(g_{k+1}) = \eta_\theta(g_i)\theta(g_{i-1}),$$

from which it follows that $\theta(g_{k+2}) = \theta(g_{i-1})$. But $g_{k+2}\theta(g_{k+1}) = \eta_\theta(g_i)$ for some $i \leq r$ as otherwise if we set

$$\theta'(g_j) = \begin{cases} \theta(g_{r+2}), & \text{if } j = 1; \\ \theta(g_i), & \text{if } j = i+1,\ i = 1, \ldots, k+1; \\ \theta(g_{k+2}), & \text{if } j = r+2 \\ \theta(g_j), & \text{otherwise,} \end{cases}$$

then $O(\eta_{\theta'}) > r$. We may assume, by relabeling if need be, that $g_{k+2}\theta(g_{k+1}) = \eta_\theta(g_{k+3})$, and so equation 3.1 also holds for $i = k+1$. But $O(\eta_\theta)$ is finite, and so we must arrive at a j for which $g_j\theta(g_{j-1}) \neq \eta_\theta(g_i)$ for any $i = 1, \ldots, r$, and so we obtain a contradiction.

Thus there exists a bijection $\theta\colon G \to G$ for which $O(\eta_\theta) \geq |G| - 1$.

To complete the proof, there are two cases to consider, depending on whether G has a unique involution or not. If G has a unique involution u and θ is a complete mapping of G. Then

$$u = \prod_{g\in G} g = \prod_{g\in G} g\theta(g) = \prod_{g\in G} g \prod_{g\in G} \theta(g) = u^2 = 1,$$

a contradiction, and so G is not admissible. If G does not have a unique involution, then we know by Lemma 3.8 that $\prod_{g\in G} g = 1$. We may assume that $\eta_\theta(g_1), \ldots, \eta_\theta(g_{|G|-1})$ are distinct. Let $y = \eta_\theta(g_{|G|})$ and let $z \in G$ satisfy $z \neq \eta_\theta(g_i)$ for $i = 1, \ldots, |G| - 1$. Then

$$1 = \prod_{i=1,\ldots,|G|} g_i \prod_{i=1,\ldots,|G|} \theta(g_i) = \prod_{i=1,\ldots,|G|} \eta_\theta(g_i) = yz^{-1}.$$

Thus $z = y$, $O(\eta_\theta) = |G|$, and θ is a complete mapping of G. □

The proof of Theorem 3.9 tells us not only when there exists a bijection $\theta\colon G \to G$ for which $O(\eta_\theta) = |G|$ but also when there exists a bijection $\theta\colon G \to G$ for which $O(\eta_\theta) = |G| - 1$.

Corollary 3.10. *Let G be a finite abelian group. There exists a bijection $\theta\colon G \to G$ for which $O(\eta_\theta) = |G| - 1$ if and only if G has a unique involution.*

3.2.2 *Hall's Theorem*

Using a different approach, Hall [167] generalized Paige's result in 1952. If $a_1, \ldots, a_n$ and $c_1, \ldots, c_n$ are two orderings of the elements of an additively written finite abelian group G and $b_i = c_i - a_i$ for $i = 1, \ldots, n$, then $\sum_{i=1}^n b_i = \sum_{i=1}^n c_i - \sum_{i=1}^n a_i = \sum_{g\in G} g - \sum_{g\in G} g = 0$. The question arises, if we have a sequence $b_1, \ldots, b_n$ of the elements of an abelian group G of order n, and $\sum_{i=1}^n b_i = 0$, then do there exist orderings $a_1, \ldots, a_n$ and $c_1, \ldots, c_n$ of the elements of G for which $c_i - a_i = b_i$ for $i = 1, \ldots, n$? Hall answered this question in the affirmative. The crux of his proof is the following lemma.

Lemma 3.11. *Let $y_1, \ldots, y_n$ and $y'_1, \ldots, y'_n$ be orderings of the elements of a finite abelian group G satisfying $y'_i = y_i$, for $i = 1, \ldots, n$, $i \neq s, t$, and $y'_s + y'_t = y_s + y_t$. If there exist orderings $x_1, \ldots, x_n$ and $z_1, \ldots, z_n$ of the elements of G satisfying $y_i = z_i - x_i$ for all i, then there exist orderings $x'_1, \ldots, x'_n$ and $z'_1, \ldots, z'_n$ of the elements of G satisfying $y'_i = z'_i - x'_i$ for all i.*

Proof. To simplify this proof we may assume without loss of generality that $s = n - 1$ and $t = n$. Thus $y'_i = y_i$ for $i = 1, \ldots, n - 2$, and $y'_{n-1} + y'_n = y_{n-1} + y_n$. Let $x_1, \ldots, x_n$ and $z_1, \ldots, z_n$ be orderings of the elements of G satisfying $y_i = z_i - x_i$ for all i. We will construct bijections $f'\colon \{x_1, \ldots, x_n\} \to \{y'_1, \ldots, y'_n\}$ and

$g' : \{x_1, \ldots, x_n\} \to \{z_1, \ldots, z_n\}$ for which $f'(x_i) = g'(x_i) - x_i$ for all i. The result will then follow by setting $x_i' = f'^{-1}(y_i')$ and $z_i' = g'(x_i)$ for all i.

Define bijections f and g by $f(x_i) = y_i'$ and $g(x_i) = z_i$ for all i. Thus $f(x_i) = g(x_i) - x_i$ for $i = 1, \ldots, n-2$ and $x_{n-1} + x_n + f(x_{n-1}) + f(x_n) = g(x_{n-1}) + g(x_n)$. If there exist $k, t \in \{n-1, n\}$ for which $x_k + f(x_{n-1}) = g(x_t)$, then we can define bijections f′ and g′ by

$$f'(x_i) = \begin{cases} f(x_i) & \text{if } i \leq n-2, \\ f(x_{n-1}) & \text{if } i = k, \\ f(x_n) & \text{if } i \in \{n-1, n\} \setminus k, \end{cases}$$

and

$$g'(x_i) = \begin{cases} g(x_i) & \text{if } i \leq n-2, \\ g(x_t) & \text{if } i = k, \\ g(x_{n-1}) & \text{if } i \neq k \text{ and } t = n, \\ g(x_n) & \text{if } i \neq k \text{ and } t = n-1. \end{cases}$$

Then f′ and g′ are the bijections we need to prove the result.

If there exist no $k, t \in \{n-1, n\}$ for which $x_k + f(x_{n-1}) = g(x_t)$, then there exists a unique $r_1 \leq n-2$ satisfy $x_{r_1} + f(x_{n-1}) = g(x_{n-1})$. We define bijections f_1 and g_1 by

$$f_1(x_i) = \begin{cases} f(x_i) & \text{if } i \neq r_1, n-1, \\ f(x_{n-1}) & \text{if } i = r_1, \\ f(x_{r_1}) & \text{if } i = n-1, \end{cases}$$

and

$$g_1(x_i) = \begin{cases} g(x_i) & \text{if } i \neq r_1, n-1, n, \\ g(x_{n-1}) & \text{if } i = r_1, \\ g(x_n) & \text{if } i = n-1, \\ g(x_{r_1}) & \text{if } i = n. \end{cases}$$

If there exist $k, t \in \{n-1, n\}$ for which $x_k + f_1(x_{n-1}) = g_1(x_t)$, then we can define bijections f′ and g′ by

$$f'(x_i) = \begin{cases} f_1(x_i) & \text{if } i \leq n-2, \\ f_1(x_{n-1}) & \text{if } i = k, \\ f_1(x_n) & \text{if } i \in \{n-1, n\} \setminus k, \end{cases}$$

and

$$g'(x_i) = \begin{cases} g_1(x_i) & \text{if } i \leq n-2, \\ g_1(x_t) & \text{if } i = k, \\ g_1(x_{n-1}) & \text{if } i \neq k \text{ and } t = n, \\ g_1(x_n) & \text{if } i \neq k \text{ and } t = n-1. \end{cases}$$

Then f′ and g′ are the bijections we need to prove the result.

If there exist no $k, t \in \{n-1, n\}$ for which $x_k + f_1(x_{n-1}) = g_1(x_t)$, then there exists a unique $r_2 \leq n-2$ satisfy $x_{r_2} + f_1(x_{n-1}) = g_1(x_{n-1})$. Now, $r_2 \neq r_1$ as $x_{r_2} = g_1(x_{n-1}) - f_1(x_{n-1}) = g(x_n) - f(x_{r_1}) = (g(x_n) - g(x_{r_1})) + x_{r_1}$ and $g(x_n) - g(x_{r_1}) \neq 0$ as g is a bijection. We define bijections f_2 and g_2 by

$$f_2(x_i) = \begin{cases} f_1(x_i) & \text{if } i \neq r_2, n-1, \\ f_1(x_{n-1}) & \text{if } i = r_2, \\ f_1(x_{r_2}) & \text{if } i = n-1, \end{cases}$$

and

$$g_2(x_i) = \begin{cases} g_1(x_i) & \text{if } i \neq r_2, n-1, n, \\ g_1(x_{n-1}) & \text{if } i = r_2, \\ g_1(x_n) & \text{if } i = n-1, \\ g_1(x_{r_2}) & \text{if } i = n. \end{cases}$$

At the jth step, $j \geq 2$, f_j and g_j are defined by

$$f_j(x_i) = \begin{cases} f_{j-1}(x_i) & \text{if } i \neq r_j, n-1, \\ f_{j-1}(x_{n-1}) & \text{if } i = r_j, \\ f_{j-1}(x_{r_j}) & \text{if } i = n-1, \end{cases}$$

and

$$g_j(x_i) = \begin{cases} g_{j-1}(x_i) & \text{if } i \neq r_j, n-1, n, \\ g_{j-1}(x_{n-1}) & \text{if } i = r_j, \\ g_{j-1}(x_n) & \text{if } i = n-1, \\ g_{j-1}(x_{r_j}) & \text{if } i = n, \end{cases}$$

where $r_0 = n$.

If there exist $k, t \in \{n-1, n\}$ for which $x_k + f_j(x_{n-1}) = g_1(x_t)$, then we can define bijections f′ and g′ by

$$f'(x_i) = \begin{cases} f_j(x_i) & \text{if } i \le n-2, \\ f_j(x_{n-1}) & \text{if } i = k, \\ f_j(x_n) & \text{if } i \in \{n-1, n\} \setminus k, \end{cases}$$

and

$$g'(x_i) = \begin{cases} g_j(x_i) & \text{if } i \le n-2, \\ g_j(x_t) & \text{if } i = k, \\ g_j(x_{n-1}) & \text{if } i \ne k \text{ and } t = n, \\ g_j(x_n) & \text{if } i \ne k \text{ and } t = n-1. \end{cases}$$

Then f' and g' are the bijections we need to prove the result.

If there exist no $k, t \in \{n-1, n\}$ for which $x_k + f_j(x_{n-1}) = g_1(x_t)$, then there exists a unique $r_{j+1} \le n-2$ satisfying $x_{r_{j+1}} + f_j(x_{n-1}) = g_j(x_{n-1})$. Now, $r_{j+1} \notin \{r_1, \ldots, r_j\}$ as $x_{r_{j+1}} = g_{j-k}(x_{r_{j-k+1}}) - f_{j-k}(x_{r_j}) = (g_{j-k}(x_{r_{j-k+1}}) - f_{j-k}(x_{r_{j-k+1}})) + (f_{j-k}(x_{r_{j-k+1}}) - f_{j-k}(x_{r_j})) = x_{r_{j-k+1}} + (f_{j-k}(x_{r_{j-k+1}}) - f_{j-k}(x_{r_j}))$ for $k = 1, \ldots, j$, where $f_0 = f$ and $g_0 = g$, and $(f_{j-k}(x_{r_{j-k+1}}) - f_{j-k}(x_{r_j})) \ne 0$ as f_{j-k} is a bijection. As $x_{r_1}, \ldots, x_{r_j} \le n-2$ are distinct for all j, there must exist $k, t \in \{n-1, n\}$ for which $x_k + f_j(x_{n-1}) = g_1(x_t)$ for some j. The result follows.

□

We are now in a position to prove Hall's theorem.

Theorem 3.12 (Hall, 1952). *Let G be an additive abelian group of order n, and let $b_1, \ldots, b_n \in G$ satisfy $\sum_{i=1}^n b_i = 0$. Then there exist orderings $a_1, \ldots, a_n$ and $c_1, \ldots, c_n$ of the elements of G for which $c_i - a_i = b_i$ for $i = 1, \ldots, n$.*

Proof. If $y_i = 0$ for all i, then clearly there exist orderings $x_1, \ldots, x_n$ and $z_1, \ldots, z_n$ of the elements of G for which $y_i = z_i - x_i$: Let $x_1, \ldots, x_n$ be any ordering of the elements of G and set $z_i = x_i$ for all i. Set $y'_1, \ldots, y'_n = b_1, -b_1, 0, \ldots, 0$. By Lemma 3.11 there exist orderings $x'_1, \ldots, x'_n$ and $z'_1, \ldots, z'_n$ of the elements of G for which $y'_i = z'_i - x'_i$ for $i = 1, \ldots, n$. Next set $y''_1, \ldots, y''_n = b_1, b_2, -(b_1 + b_2), 0, \ldots, 0$. By Lemma 3.11 there exist orderings $x''_1, \ldots, x''_n$ and $z''_1, \ldots, z''_n$ of the elements of G for which $y''_i = z''_i - x''_i$ for $i = 1, \ldots, n$. We repeat this procedure. At the ith step $y_1^{(i)}, \ldots, y_n^{(i)} = b_1, \ldots, b_i, -\sum_{j=1}^i b_j, 0, \ldots, 0$. By Lemma 3.11 there exist orderings $x_1^{(i)}, \ldots, x_n^{(i)}$ and $z_1^{(i)}, \ldots, z_n^{(i)}$ of the elements of G for which $y_k^{(i)} = z_k^{(i)} - x_k^{(i)}$ for $k = 1, \ldots, n$. At the $(n-1)$th step $y_1^{(n-1)}, \ldots, y_n^{(n-1)} = b_1, \ldots, b_{n-1}, -\sum_{j=1}^{n-1} b_j = b_1, \ldots, b_n$. Setting $a_i = x_i^{(n-1)}$ and $c_i = z_i^{(n-1)}$ completes the proof. □

In Theorem 3.12, if the sequence $b_1, \ldots, b_n$ is an ordering of the elements of G, then the mapping $a_i \mapsto b_i$ is a complete mapping of G, and we obtain the following immediate corollary.

Corollary 3.13. *Let G be an additive abelian group of order n. If $\sum_{g \in G} g = 0$, then G is admissible.*

By Lemma 3.8, $\sum_{g \in G} g = 0$ if and only if G does not have a unique involution. Thus Corollary 3.13 gives us another proof of the admissibility of finite abelian groups that do not have unique involutions.

3.2.3 Complete Primitive Residue Sets

In 1948 Vijayaraghavan and Chowla [318] independently proved a special case of Theorem 3.9. They answered the question, for which n does there exist $x'_1, \ldots, x'_{\phi(n)}$ and $y'_1, \ldots, y'_{\phi(n)}$ complete primitive residue sets modulo n for which $x'_1 y'_1, \ldots, x'_{\phi(n)} y'_{\phi(n)}$ is a complete primitive residue set modulo n? Their proof that, for $n > 2$, if there exists a primitive root, then the answer is no, was given in Theorem 3.4. They proved the converse to be true also: their proof relies on a special case of Corollary 1.38.

Lemma 3.14. *If m and n are relatively prime and there exist $x_1, \ldots, x_{\phi(m)}$ and $y_1, \ldots, y_{\phi(m)}$, complete primitive residue sets modulo m, for which $x_1 y_1, \ldots, x_{\phi(m)} y_{\phi(m)}$ is a complete primitive residue set modulo m, and there exist $x'_1, \ldots, x'_{\phi(n)}$ and $y'_1, \ldots, y'_{\phi(n)}$, complete primitive residue sets modulo n, for which $x'_1 y'_1, \ldots, x'_{\phi(n)} y'_{\phi(n)}$ is a complete primitive residue set modulo n, then there exist $x''_1, \ldots, x''_{\phi(mn)}$ and $y''_1, \ldots, y''_{\phi(mn)}$, complete primitive residue sets modulo mn, for which $x''_1 y''_1, \ldots, x''_{\phi(mn)} y''_{\phi(mn)}$ is a complete primitive residue set modulo mn.*

Proof. As m and n are relatively prime, we may assume, without loss of generality, that $x_i \equiv 1 \pmod{n}$ and $y_i \equiv 1 \pmod{n}$ for $i = 1, \ldots, \phi(m)$, and $x'_i \equiv 1 \pmod{m}$ and $y'_i \equiv 1 \pmod{m}$ for $i = 1, \ldots, \phi(n)$. Then $\{x_i x'_j \mid i = 1, \ldots, \phi(m), j = 1, \ldots, \phi(n)\}$ and $\{y_i y'_j \mid i = 1, \ldots, \phi(m), j = 1, \ldots, \phi(n)\}$ are complete primitive residue sets modulo mn, as is $\{(x_i y_i)(x'_j y'_j) \mid i = 1, \ldots, \phi(m), j = 1, \ldots, \phi(n)\}$. □

To prove the converse of Theorem 3.4, using Lemma 3.14, we employ factorizations of integers into products $m_1 \ldots m_r$, where $m_1 \ldots m_r$ pairwise relatively prime, $m_i > 2$ for all i, and m_i has no primitive root for all i.

Theorem 3.15 (Vijayaraghavan and Chowla, 1948). *If $n > 2$ does not have a primitive root, then there exist complete primitive residue sets $x_1, \ldots, x_{\phi(n)}$ and $y_1, \ldots, y_{\phi(n)}$ modulo n for which $x_1 y_1, \ldots, x_{\phi(n)} y_{\phi(n)}$ is a complete primitive residue set modulo n.*

Proof. Recall that if there is a primitive root modulo n, then n is of the form 2; 4; p^t, p an odd prime; or $2p^t$, p an odd prime. Thus, by Lemma 3.14, it is sufficient to give constructions for n of the form 2^m, $m \neq 2$; $4q^s$, q an odd prime; $2^m r^t$, $m \geq 3$ and r an odd prime; $p^m q^s$, p and q distinct odd primes; $4q^s r^t$, q and r

distinct odd primes; and $p^m q^s r^t$, p, q, and r distinct odd primes. We will give these constructions in terms of mappings $\theta\colon x_i \mapsto y_i$.

If $n = p^m q^s$, p and q distinct odd primes, then let $g \equiv 1 \pmod{q^s}$ be a primitive root modulo p^m and $h \equiv 1 \pmod{p^m}$ be a primitive root modulo q^s, and set $2M = \phi(p^m)$ and $2N = \phi(q^s)$. Then set

$$\theta(g^i h^j) = \begin{cases} g^i h^j, & \text{if } 1 \le i \le M,\ 1 \le j \le N; \\ g^i h^{j-1}, & \text{if } M < i \le 2M,\ 1 \le j \le N; \\ g^{i+1} h^{j-1}, & \text{if } M \le i < 2M,\ N < j \le 2N; \\ g^{i+1} h^j, & \text{if } 0 \le i < M,\ N < j \le 2N. \end{cases}$$

This construction can be adapted to the case $n = 4q^s$ by setting $p = 2$ and $m = 2$, in which case $g \equiv 3 \pmod 4$.

If $n = 2^m$, $m \neq 2$, then, as the case $m = 1$ is trivial, we shall assume that $m \ge 3$. In this case set $M = 2^{m-3}$. Then $\{5^i(-1)^j \mid i = 0, \ldots, 2M-1, j = 0, 1\}$ is a complete primitive residue set modulo n. Then set

$$\theta(5^i(-1)^j) = \begin{cases} -5^i, & \text{if } 1 \le i \le M,\ j = 1; \\ 5^i, & \text{if } M < i \le 2M,\ j = 1; \\ -5^{i+1}, & \text{if } M \le i < 2M,\ j = 0; \\ 5^{i+1}, & \text{if } 0 \le i < M,\ j = 0. \end{cases}$$

If $n = p^m q^s r^t$, p, q, and r distinct odd primes, then let $g \equiv 1 \pmod{q^s}$ and $1 \pmod{r^t}$ be a primitive root modulo p^m, $h \equiv 1 \pmod{p^m}$ and $1 \pmod{r^t}$ be a primitive root modulo q^s, and $k \equiv 1 \pmod{p^m}$ and $1 \pmod{q^s}$ be a primitive root modulo r^t, and set $2M = \phi(p^m)$, $2N = \phi(q^s)$, and $2L = \phi(r^t)$. Then set

$$\theta(g^i h^j k^l) = \begin{cases} g^i h^j k^l, & \text{if } 1 \le i \le M,\ 1 \le j \le N,\ 1 \le l \le L; \\ g^i h^{j-1} k^l, & \text{if } M \le i \le 2M,\ 1 \le y \le N,\ 1 \le l \le L; \\ g^i h^{j-1} k^{l-1}, & \text{if } M < i \le 2M,\ N < j \le 2N,\ L < l \le 2L; \\ g^i h^j k^{l-1}, & \text{if } 1 \le i \le M,\ N < j \le 2N,\ L < l \le 2L; \\ g^{i+1} h^{j-1} k^{l-1}, & \text{if } M \le i < 2M,\ N < j \le 2N,\ 1 \le l \le L; \\ g^{i+1} h^j k^{l-1}, & \text{if } 0 \le i < M,\ N < j \le 2N,\ 1 \le l \le L; \\ g^{i+1} h^j k - 1^l, & \text{if } M \le i < 2M,\ 1 \le j \le N,\ L < l \le 2L; \\ g^{i+1} h^j k^l, & \text{if } 0 \le i < M,\ 1 \le j \le N,\ L < l \le 2L. \end{cases}$$

This construction can be adapted to the case $n = 4q^s r^t$ by setting $p = 2$ and $m = 2$, in which case $g \equiv 3 \pmod 4$. It can also be adapted to the case $n = 2^u r^t$, $u \ge 3$ by setting $p = 2$, $m = u/2$, $q = 2$, $s = 1$ and choosing g to also be congruent to 5 modulo 2^u and h to be congruent to -1 modulo 2. □

The smallest examples of complete primitive residue sets $x'_1, \ldots, x'_{\phi(n)}$ and $y'_1, \ldots, y'_{\phi(n)}$ modulo n for which $x'_1y'_1, \ldots, x'_{\phi(n)}y'_{\phi(n)}$ is a complete primitive residue set modulo n are given below.

Example 3.16. If $n = 2$, then $\phi(n) = 1$,

$$\begin{aligned} x_1 &= 1, \\ y_1 &= 1, \text{ and} \\ x_1y_1 &= 1. \end{aligned}$$

Example 3.17. If $n = 8$, then $\phi(n) = 4$, setting

$$\begin{aligned} x_1, \ldots, x_4 &= 1, 3, 5, 7, \text{ and} \\ y_1, \ldots, y_4 &= 1, 5, 7, 3, \text{ we see that} \\ x_1y_1, \ldots, x_4y_4 &= 1, 7, 3, 5. \end{aligned}$$

Example 3.18. If $n = 12$, then $\phi(n) = 4$, setting

$$\begin{aligned} x_1, \ldots, x_4 &= 1, 5, 7, 11, \text{ and} \\ y_1, \ldots, y_4 &= 1, 7, 11, 5, \text{ we see that} \\ x_1y_1, \ldots, x_4y_4 &= 1, 11, 5, 7. \end{aligned}$$

Example 3.19. If $n = 15$, then $\phi(n) = 8$, setting

$$\begin{aligned} x_1, \ldots, x_8 &= 1, 2, 4, 7, 8, 11, 13, 14, \text{ and} \\ y_1, \ldots, y_8 &= 1, 4, 14, 2, 11, 7, 13, 8, \text{ we see that} \\ x_1y_1, \ldots, x_8y_8 &= 1, 8, 11, 14, 13, 2, 4, 7. \end{aligned}$$

Vijayaraghavan and Chowla's results, Theorem 3.4 and Theorem 3.15, establish that the group of units of the ring of integers modulo n, $n > 2$, is admissible if and only if it is not cyclic. This is a special case of Theorem 3.9. Note that in the case $n = 2$, this group is trivially admissible. Some constructions of complete primitive residue sets modulo powers of 2 were given by Bambah [26] in 1946. There seems to have been little or no further work done on complete primitive residue sets.

3.2.4 *Carlitz's Proof*

Inspired by Vijayaraghavan and Chowla's result, Carlitz [61] rediscovered Theorem 3.9 in 1953. The nonexistence part of his proof is essentially the same as in Theorem 3.9. His existence proof is constructive and borrows from Vijayaraghavan and Chowla's constructions. Any finite abelian group with a noncyclic Sylow 2-subgroup can be written as a direct product of groups of odd order, groups of the form $\mathbb{Z}_{2^m} \times \mathbb{Z}_{2^n}$, and groups of the form $\mathbb{Z}_{2^m} \times \mathbb{Z}_{2^n} \times \mathbb{Z}_{2^l}$. Any group of odd order is admissible by Corollary 3.6. For the group $\mathbb{Z}_{2^m} \times \mathbb{Z}_{2^n}$, Carlitz adapts the construction in Theorem 3.15 for $n = p^m q^s$, replacing $g^i h^j$ with (i, j), M with 2^{m-1}, and N with 2^{n-1}. For the group $\mathbb{Z}_{2^m} \times \mathbb{Z}_{2^n} \times \mathbb{Z}_{2^l}$, Carlitz adapts the construction in Theorem 3.15 for $n = p^m q^s r^t$, replacing $g^i h^j k^l$ with (i, j, l), M with 2^{m-1}, N with 2^{n-1}, and L with 2^{l-1}. The admissibility of direct products follows from Corollary 1.31, completing the admissibility proof. The following is a shorter version of Carlitz's proof.

Theorem 3.20. *A finite abelian group is admissible if its Sylow 2-subgroup is trivial or noncyclic.*

Proof. Let G be a finite abelian group with a trivial or noncyclic Sylow 2-subgroup. Then G admits a subnormal series of the form $\{1\} = G_1 \subset G_2 \subset \cdots \subset G_m = G$, where G_{i+1}/G_i is isomorphic to either an elementary abelian 2-group of order at least 4, a group of odd order, or a group of the form $\mathbb{Z}_2 \times \mathbb{Z}_m$, $m = 2^n$, $n > 1$. By Corollary 1.31, it suffices to prove admissibility of each factor. Odd-order groups are admissible by Corollary 3.6, and elementary abelian 2-groups of order at least 4 are admissible by Example 1.25. Thus we need only show that $\mathbb{Z}_2 \times \mathbb{Z}_m$, $m = 2^n$, $n > 1$, is admissible. Let $\mathbb{Z}_2 = \langle h \rangle$ and $\mathbb{Z}_m = \langle g \rangle$. Set $S = \{1, \cdots, 2^{n-1}\}$. Then

$$\theta(x) = \begin{cases} hg^i & \text{if } x = hg^i \text{ and } i \in S \\ g^{i-1} & \text{if } x = g^i \text{ and } i \in S \\ g^{i-1} & \text{if } x = hg^i \text{ and } i \notin S \\ hg^i & \text{if } x = g^i \text{ and } i \notin S \end{cases}$$

is a complete mapping of $\mathbb{Z}_2 \times \mathbb{Z}_m$. This completes the proof. □

3.3 Some Admissibility Criteria

Having settled the admissibility question for infinite groups and finite abelian groups, we now turn our attention to conditions that might be used to determine whether a given finite group is admissible or not.

3.3.1 Euler's Conjecture and Cayley Tables of Groups

Euler's conjecture that a Latin square of order $4n + 2$ can have no orthogonal mate was shown to be false for all $n \geq 2$ in 1960 by Bose, Shrikhande, and Parker [46]. The inspiration for this result was the construction in 1959 by

Parker [264] of 4 MOLS of order 21, the first counterexample to the MacNeish conjecture. As the MacNeish conjecture is a generalization of Euler's conjecture and Parker's counterexample was based on the group $\mathbb{Z}_{21}$, it is reasonable to look for counterexamples to Euler's conjecture that are based on groups. Surprisingly, no such counterexamples can exist: Latin squares based on groups satisfy Euler's conjecture. The earliest proof of this is in Fleisher's [144] 1934 PhD thesis. The first published proof was given in 1942 by Mann [237]. We give Mann's proof, which requires the following simple lemma.

Lemma 3.21. *A group of order* $4n+2$ *has a normal subgroup of index* 2.

Proof. Let G be a group of order $4n+2$. As G acts on itself by right multiplication, G can be viewed as a subgroup of the symmetric group of degree $4n+2$. Let H be the subgroup of G consisting of the even permutations. It is well-known that H is a normal subgroup of G and is either equal to G if G contains no odd permutations or is of index 2 if G contains odd permutations. As G is of even order, G contains an involution u, and u must be a product of $2n+1$ disjoint transpositions. Thus, $u \in G$ is an odd permutation and hence H is of index 2. □

The existence of an odd-order subgroup of index 2 leads to a simple nonadmissibility argument.

Theorem 3.22 (Fleisher, 1934). *A group of order* $4n+2$ *cannot be admissible.*

Proof. Assume that $|G| = 4n+2 = 2m$ and that G admits a complete mapping θ. By Lemma 3.21, G has a normal subgroup H of odd-order m. Let H and Hg be the cosets of H and set

$$a = |\{x \mid x \in H, \theta(x) \in H\}|,$$

$$b = |\{x \mid x \in H, \theta(x) \in Hg\}|,$$

and

$$c = |\{x \mid x \in Hg, \theta(x) \in H\}|.$$

As $a+b = m = a+c$, it follows that $b = c$. Now $x\theta(x) \in Hg$ if and only if $x \in H$ and $\theta(x) \in Hg$ or $x \in Hg$ and $\theta(x) \in H$. Hence $m = |\{x \mid x\theta(x) \in Hg\}| = b+c = 2b$. Thus m is even, a contradiction from which the result follows. □

Mann's proof was rediscovered by Jungnickel [201] in 1980. In 1944 Mann [238] studied conditions on a Latin square that would imply that the square did not have an orthogonal mate. We present a special case of his result below.

Theorem 3.23 (Mann, 1944). *If a Latin square L of order 2n, n odd, has a Latin subsquare of order n, then L has no orthogonal mate.*

Proof. Let L be a Latin square of order $2n$, n odd, that has a Latin subsquare of order n. By reordering rows and columns, if need be, we may assume that $L = \begin{pmatrix} A & B \\ C & D \end{pmatrix}$, where A is a Latin square of order n on the symbol set S, and thus, D must also be a latin square of order n on the symbol set S. Assume that $L' = \begin{pmatrix} A' & B' \\ C' & D' \end{pmatrix}$ is a Latin square orthogonal to L, where A' is $n \times n$, and let x be a symbol of L'. Let m be the number of occurrences of x in A'. Then x occurs $n - m$ times in B', and so x occurs m times in D'. That is, x occurs an even number of times opposite cells of A and D. But x must occur exactly once opposite each symbol of S, and so x must occur n times opposite cells of A and D. As n is odd, we have arrived at a contradiction from which the result follows. □

This leads to another proof of Theorem 3.22.

Corollary 3.24. *A group of order* $4n + 2$ *cannot be admissible.*

Proof. If G is a group of order $4n + 2$, then G has a normal subgroup H of order $2n + 1$ by Lemma 3.21. Thus the Cayley table M of G contains a Latin subsquare of order $2n + 1$. Thus, by Theorem 3.23, M, and hence any Latin square based on G does not have an orthogonal mate. Hence, G is not admissible. □

3.3.2 Products of All Group Elements

We have already seen, in the case of finite abelian groups, the importance of the product of all the elements of the group. By Lemma 3.8 and Theorem 3.9, a finite abelian group is admissible if and only if the product of all of its elements is the identity. In 1951 Paige [263] proved this to be a necessary condition for groups in general.

Theorem 3.25 (Paige, 1951). *If G is a finite admissible group, then there is an ordering $g_1, \ldots, g_n$ of the elements of G for which $g_1 \cdots g_n = 1$.*

Proof. Let θ be a complete mapping of G. Then $\alpha\colon x \mapsto \theta(x)^{-1}$, as a permutation of G, can be represented as a product of disjoint cycles

$$\alpha = (x_1 \ldots x_{i_1})(x_{i_1+1} \ldots x_{i_2}) \cdots (x_{i_t+1} \ldots x_n).$$

Set $g_i = x_i\theta(x_i)$. Then

$$\begin{aligned} g_1 g_2 \cdots g_n &= x_1\theta(x_1) \cdots x_n\theta(x_n) \\ &= (x_1\theta(x_1) \cdots x_{i_1}\theta(x_{i_1})) \cdots (x_{i_t+1}\theta(x_{i_t+1}) \cdots x_n\theta(x_n)). \end{aligned}$$

If $\alpha(x) = y$, then $y = \theta(x)^{-1}$ and $x\theta(x)y\theta(y) = x\theta(y)$. Thus

$$(x_{i_k+1}\theta(x_{i_k+1}) \cdots x_{i_{k+1}}\theta(x_{i_{k+1}})) = x_{i_k+1}\theta(x_{i_{k+1}}) = \theta(x_{i_{k+1}})^{-1}\theta(x_{i_{k+1}}) = 1,$$

and so $g_1 g_2 \cdots g_n = 1$. □

From this we obtain another necessary condition for a finite group to be admissible.

Corollary 3.26 (Paige, 1951). *If G is a finite admissible group, then the product of the elements of G in any order is an element of the commutator subgroup of G.*

Proof. Let G' be the commutator subgroup of G. As G is admissible, by Theorem 3.25, there exists an ordering $g_1, \ldots, g_n$ of the elements of G for which $g_1 \cdots g_n = 1$. Let $g_{\sigma(1)}, \ldots, g_{\sigma(n)}$ be another ordering of the elements of G. Then, as G/G' is abelian,

$$\begin{aligned}(g_{\sigma(1)} \cdots g_{\sigma(n)})G' &= (g_{\sigma(1)}G') \cdots (g_{\sigma(n)}G') = (g_1G') \cdots (g_nG') \\ &= (g_1 \cdots g_n)G' = G',\end{aligned}$$

hence the result. □

As an easy consequence of the proof of Theorem 3.25, we see that if G is admissible, then there exists a partition of the elements of G into subsets such that, for each subset S, $|S| = m$ say, there is an ordering of the elements $g_1, \ldots, g_m$ of S such that the products $g_1 \cdots g_i$ are distinct for $i = 2, \ldots, m$ and $g_1 \cdots g_m = 1$. In the special case $S = G$, Paige [263] proved that this condition implies that G is admissible.

Theorem 3.27 (Paige, 1951). *If there exists an ordering $g_1, \ldots, g_n$, $g_1 = 1$, of the elements of G, such that the products $g_1 \cdots g_i$ are distinct for $i = 2, \ldots, n$ and $g_1 \cdots g_n = 1$, then G is admissible.*

Proof. Let h be the unique element of G that is not equal to $g_1 \cdots g_i$ for any $i \in \{2, \ldots, n\}$ and define $\theta\colon G \to G$ by

$$\theta(x) = \begin{cases} 1 & \text{if } x = 1, \\ g_2 & \text{if } x = h^{-1}, \\ g_{i+1} & \text{if } x = h^{-1}g_2 \cdots g_i \text{ and } i \in \{2, \ldots, n-1\}. \end{cases}$$

Then θ is a bijection. Further θ is a complete mapping as the products $x\theta(x)$ are $1, h^{-1}g_2$, and $h^{-1}g_2 \cdots g_{i+1}$ for $i = 2, \ldots, n-1$, i.e., all the elements of G. □

3.3.3 Some Nonadmissibility Criteria

Under certain circumstances, the nonadmissibility of a quotient group can imply the nonadmissibility of the group. Paige [263] established such a result in 1951.

Theorem 3.28 (Paige, 1951). *Let H be an odd-order normal subgroup of a finite group G. If G/H is a nonadmissible abelian group, then G is not admissible.*

Proof. As G/H is a finite nonadmissible abelian group, by Theorem 3.9 there is a unique involution uH in G/H. Thus, the product of the elements of G, in any order, must be an element of uH. But the commutator subgroup of G is a subgroup of H, and so the product of the elements of G, in any order, cannot be an element of the commutator subgroup of G. Hence G cannot be admissible by Corollary 3.26. □

A similar result was discovered by Bruck [53], also in 1951. Let $\mathcal{N}$ be a finite net (see Section 1.4 for the definition and properties of a net), and let f be a function from the points of $\mathcal{N}$ to the integers. We say that m is *represented on $\mathcal{N}$ by* f if for each line l of $\mathcal{N}$, $\sum_{P\in l} \mathrm{f}(P) = m$,. If in addition f takes on only nonnegative values, then we say that m is *represented positively on $\mathcal{N}$ by* f. Bruck used $\phi(\mathcal{N})$ to denote the smallest positive integer representable on $\mathcal{N}$.

Lemma 3.29. *A line can be adjoined to a finite net $\mathcal{N}$ if and only if* 1 *can be represented positively on $\mathcal{N}$.*

Proof. Suppose that a line l can be adjoined to $\mathcal{N}$. Then define f by

$$\mathrm{f}(P) = \begin{cases} 1, & \text{if } P \text{ is on } l; \\ 0, & \text{if } P \text{ is not on } l. \end{cases}$$

This yields a positive representation of 1 on $\mathcal{N}$.

If 1 is represented positively on $\mathcal{N}$ by f, then $\mathrm{f}(P)$ is either 0 or 1. Set $l = \{P \mid \mathrm{f}(P) = 1\}$. Then l is a line that can be adjoined to $\mathcal{N}$. □

We really only need the following special case of Lemma 3.29.

Corollary 3.30. *If a line can be adjoined to a finite net $\mathcal{N}$, then $\phi(\mathcal{N}) = 1$.*

For $\mathcal{N}$ a net of order n and degree k, we assign coordinates as follows: pick a point C, number the parallel classes $1, \ldots, k$, and number the lines in each parallel class $1, \ldots, n$ so that the line numbered 1 in each parallel class passes through C. Next we define the ith *indicator function* I_i by $I_i(P) = j$ if the jth line of the ith parallel class passes through P. Bruck used this coordinatization to characterize $\phi(\mathcal{N})$.

Lemma 3.31. *If $\mathcal{N}$ is a net of order n and degree k, then $\phi(\mathcal{N})$ is the smallest positive integer s with the property that for all functions $\mathrm{f}_1, \ldots, \mathrm{f}_k \colon \{1, \ldots, n\} \to \{1, \ldots, n\}$, if $\mathrm{f}_i(1) \equiv 0 \pmod{n}$ for $i = 1, \ldots, k$, and $\sum_{i=1}^{k} \mathrm{f}_i(I_i(P)) \equiv 0 \pmod{n}$ for all points P, then $s \sum_{i=1}^{n} \mathrm{f}_1(i) \equiv 0 \pmod{n}$.*

Proof. See Theorem 3 in [53]. □

When the net is constructed from the Cayley table of a group, the elements of the group can naturally be used to number the lines of a parallel class, enabling us to use the structure of the group in attempts to evaluate $\phi(\mathcal{N})$.

Theorem 3.32 (Bruck, 1951). *If a finite group G contains a normal subgroup H of odd order, such that the quotient group G/H is a nontrivial cyclic group of even order, then G is not admissible.*

Proof. Let $|G| = n = mt$, t odd and m even; and suppose that $|H| = t$ and G/H is cyclic. We will first construct a net $\mathcal{N}$ of order n and degree 3 from G. The points of $\mathcal{N}$ are the ordered pairs $\{(x, y) \mid x, y \in G\}$. In indexing the lines in the ith parallel class, we will find it more convenient to use elements of the group G rather than integers. Thus the range of each indicator function will be G, and the domain of each function f_i in Lemma 3.31 will also be G. Further, as the computations in Lemma 3.31 are all performed modulo n, we may, without loss of generality, assume that the domain of each f_i is $\mathbb{Z}_n$. The numbering of the lines is as follows:

1. The ath line of parallel class 1 is $\{(a, y) \mid y \in G\}$;
2. the bth line of parallel class 2 is $\{(x, b) \mid y \in G\}$;
3. and the cth line of parallel class 3 is $\{(x, y) \mid x, y \in G, xy = c\}$.

It follows that if $P = (x, y)$, then $I_1(P) = x$, $I_2(P) = y$, and $I_3(P) = xy$. Let $\mathrm{g}\colon G \to \mathbb{Z}_n$ be a homomorphism with kernel H, and set $\mathrm{f}_1 = \mathrm{f}_2 = \mathrm{g}$ and $\mathrm{f}_3 = -\mathrm{g}$. We saw in Section 1.4 that if the Cayley table of G has an orthogonal mate, then we can adjoin, not just a line but a parallel class of lines to $\mathcal{N}$. Thus, if we assume that G is admissible, then the Cayley table of G has an orthogonal mate and hence $\phi(\mathcal{N}) = 1$ by Corollary 3.30. It follows that $\sum_{x\in G} \mathrm{g}(x) = 0$ by Lemma 3.31, contradicting Lemma 3.8. □

Bruck proved this result in the more general setting of loops: if a loop G contains a normal subloop H of odd order, such that the quotient loop G/H is a nontrivial cyclic group of even order, then $\phi(\mathcal{N}) = 2$, where $\mathcal{N}$ is the net described in Theorem 3.32. Otherwise $\phi(\mathcal{N}) = 1$, which, it should be noted, does not imply that a line can be adjoined to $\mathcal{N}$ as this guarantees only that 1 can be represented on $\mathcal{N}$, not that 1 can be represented positively on $\mathcal{N}$. Information on loops can be found in [56].

3.3.4 The Hall-Paige Theorem and Conjecture

We have seen several nonadmissibility results: cyclic groups of even order are not admissible; finite abelian groups with unique involutions are not admissible; a finite group G with an odd-order normal subgroup H is not admissible if G/H is a nontrivial cyclic group of even order; and, more generally, a finite group G with an odd-order normal subgroup H is not admissible if G/H is a nonadmissible abelian group. Each of these results is a special case of a nonadmissibility result that was proved by Hall and Paige [171] in 1955. Before we can prove this result, we need a lemma due to Burnside.

Lemma 3.33 (Burnside). *If the Sylow 2-subgroup of a finite group is nontrivial and cyclic, then the group admits a homomorphism onto its Sylow 2-subgroup.*

Proof. See Chapter X, Lemma 12.1 of [41]. □

Using Lemma 3.33 and a generalization of Euler's proof that a cyclic Latin square of even order has no orthogonal mate (see Theorem 3.1), Hall and Paige established the nonadmissibility of finite groups with nontrivial, cyclic Sylow 2-subgroups.

Theorem 3.34 (Hall and Paige, 1955). *A finite group is not admissible if its Sylow 2-subgroup is nontrivial and cyclic.*

Proof. Let G be a group with a nontrivial, cyclic Sylow 2-subgroup, and let $|G| = mn$, n odd, m a power of 2. Then by Lemma 3.33, there exists an epimorphism $\phi\colon G \to \mathbb{Z}_m$.

If G admits a complete mapping θ, then:

$$\sum_{g\in G}\phi(g) = \sum_{g\in G}\phi(g\theta(g)) = \sum_{g\in G}(\phi(\theta(g)) + \phi(g)) = 2\sum_{g\in G}\phi(g)$$

From this it follows that $\sum_{g\in G}\phi(g) = 0$. But direct computation yields

$$\sum_{g\in G}\phi(g) = n\sum_{i=0}^{m-1} i = nm(m-1)/2 = nm/2 \neq 0.$$

A contradiction from which the result follows. □

Note that either Theorem 3.28 or Theorem 3.32 combined with Lemma 3.33 would have yielded an earlier proof of Hall and Paige's theorem. All earlier nonadmissibility results are special cases of Theorem 3.34: these are incorporated in Corollary 3.2, Theorem 3.9, and Theorem 3.22. Hall and Paige conjectured that the converse of their theorem would be true.

Conjecture 3.35 (Hall and Paige, 1955). A finite group with a trivial or noncyclic Sylow 2-subgroup is admissible.

In Theorem 3.25, we gave Paige's proof that the product of the elements of a finite admissible group in some order is the identity, and from Theorem 3.9, we know that the converse of this theorem holds for abelian groups. In 1951 Paige [263] conjectured the converse of Theorem 3.25 to be true.

Conjecture 3.36 (Paige, 1951). If there exists an ordering $g_1, \ldots, g_n$ of the elements of a group G for which $g_1 \cdots g_n = 1$, then G is admissible.

We will show that the Hall-Paige conjecture and Paige's 1951 conjecture are equivalent. Let us first note that if the Sylow 2-subgroup of a finite group G is nontrivial and cyclic, then the product of the elements of G cannot be the identity no matter the order. In this case, by Lemma 3.33, there exists a normal odd-order subgroup H of G for which G/H is a cyclic 2-group. Thus G/H has a unique involution uH, and it is easy to show, using Lemma 3.8, that the product of the elements of G in any order is an element of uH. If $|G|$ is odd, then G is admissible by Corollary 3.2, and so the product of the elements of G in some order is the identity by Theorem 3.25. If the Sylow 2-subgroup of a finite group G is noncyclic and G is solvable, then we will see in Theorem 4.22 that G is admissible, and hence the product of the elements of G in some order is the identity by Theorem 3.25. This leaves the nonsolvable groups. In 1989 Dénes and Keedwell [92] proved that for any nonsolvable group G, the product of the elements of G in some order is the identity. Their proof relies on a 1982 result of Dénes and Hermann [90]. For G a finite group of order n, let $p(G)$ denote the set of elements of G that can be expressed as a product $g_1 g_2 \ldots g_n$, for some ordering $g_1, g_2, \ldots, g_n$ of the elements of G. Fuchs asked if $p(G)$ might be a coset of G'. Fuchs question was answered positively for solvable groups by Rhemtulla [279] in 1969 and for groups in general by Dénes and Hermann in 1982.

Lemma 3.37 (Dénes and Hermann, 1982). *If G' is the commutator subgroup of a finite group G, then $p(G) = G'g$, for some $g \in G$.*

Proof. See Dénes and Hermann [90]. □

In proving the next lemma, the crux of Dénes and Keedwell's result, we need to apply Lemma 3.37 to the case, when the product of all the elements of the group in some order is an element of the commutator subgroup.

Lemma 3.38. *If G' is the commutator subgroup of a finite group G and $|G| \equiv |G'| \pmod 2$, then the product of the elements of G in some order is the identity.*

Proof. Let $G = \{g_1, \ldots, g_n\}$ and let σ be any permutation of $\{1, \ldots, n\}$. Then $\prod_{i=1}^{n} g_{\sigma(i)} G' = \prod_{i=1}^{n} g_i G'$. Thus the product of the elements of G in any order will be an element of the same coset K of G'. We will prove that under the conditions of this lemma, this coset is G' itself.

Clearly if $G = G'$, then $K = G'$. If $G \neq G'$, then there are two cases to consider, G/G' admissible, and G/G' not admissible. Now G/G' is abelian, and a finite abelian group is admissible if and only if the product of all the elements is the identity.

Case 1. G/G' is admissible.

As the product of the cosets of G' is G', any product of all the elements of G is an element of G'.

Case 2. G/G' is not admissible.

There is a unique involution $C \in G/G'$, and if an involution $a \in G$ is not an element of G', then $aG' \neq G'$ and $(aG')^2 = G'$. Thus $aG' = C$ for all involutions $a \notin G'$. As $|G|$ and $|G'|$ are both even, the number of elements of G not in G' is also even. As each non-involution b of G not in G' is paired with its inverse b^{-1}, m, the number of involutions in G but not in G' is even. Thus $\prod_{g \in G} gG' = C^m = G'$, and the product of the elements of G in any order is an element of G'.

To complete the proof, as the product of the elements of G in some order is an element of G', it follows that $p(G) = G'$, and hence the product of the elements of G in some order is the identity. □

Next we show that the conditions of Lemma 3.38 hold for all nonsolvable groups, yielding a proof of Dénes and Keedwell's result.

Theorem 3.39 (Dénes and Keedwell, 1989). *If G is a finite nonsolvable group, then the product of the elements of G in some order is the identity.*

Proof. By the Feit-Thompson theorem [141], any nonsolvable group is of even order. If G is nonsolvable, then so is G'. Thus $|G| \equiv |G'| \equiv 0 \pmod 2$. The result then follows from Lemma 3.38. □

Theorem 3.39 led Dénes and Keedwell to conjecture the following:

Conjecture 3.40 (Dénes and Keedwell, 1989). All nonsolvable groups are admissible.

Dénes and Keedwell's conjecture is a special case of the Hall-Paige conjecture as it is easy to establish, from the Feit-Thompson theorem and Lemma 3.33, that any finite nonsolvable group has a noncyclic Sylow 2-subgroup. In fact, these two conjectures are equivalent as we will see in Chapter 4 that the Hall-Paige conjecture is true for solvable groups.

Since 1955, all work on admissibility has revolved around attempts to settle this conjecture and to determine new classes of admissible groups. In Chapters 4 through 6, we will give a number of admissibility results, and in Chapter 7, we will prove the conjecture true: this proof will depend on the classification of finite simple groups.

Chapter 4
Some Classes of Admissible Groups

In Chapter 3 we called a group admissible if it admitted complete mappings, and we proved that a finite group cannot be admissible if its Sylow 2-subgroup is nontrivial and cyclic. We proved the converse, known as the Hall-Paige conjecture, for abelian groups and groups of odd order. In this chapter we will prove the Hall-Paige conjecture true for more classes of groups. In Section 4.1 we will introduce HP-systems, an important tool in the construction of complete mappings, and use these to prove the Hall-Paige conjecture true for alternating and symmetric groups. In Section 4.2 we will prove the Hall-Paige conjecture true for solvable groups, in Section 4.3 for Mathieu groups and Suzuki groups, and in Section 4.4 for unitary groups and groups whose Sylow 2-subgroups intersect trivially. For the linear groups $GL(n, q)$, $SL(n, q)$, $PGL(n, q)$, and $PSL(n, q)$, we will see many constructions of complete mappings in Chapter 5.

4.1 A Construction of Hall and Paige

In Section 3.2 we proved the Hall-Paige conjecture true for abelian groups. For nonabelian groups one of the most fruitful approaches has been to construct complete mappings of a group from complete mappings of a subgroup. The only example of such a construction that we have seen so far is the quotient group construction described in Corollary 1.38. In 1955 Hall and Paige [171] gave a construction of complete mappings of a group from complete mappings of a, not necessarily normal, subgroup.

A. B. Evans, *Orthogonal Latin Squares Based on Groups*, Developments in Mathematics 57, https://doi.org/10.1007/978-3-319-94430-2_4

4.1.1 HP-Systems

Let H be a subgroup of a finite group G. A *dual system of coset representatives* for H in G is a system of left coset representatives for H in G that is also a system of right coset representatives for H in G. For D a dual system of coset representatives for H in G, a bijection $\theta\colon D \to D$ will be called a *complete mapping* of D if there exists a bijection $\eta\colon D \to D$ such that $x\theta(x)H = \eta(x)H$ for all $x \in D$. We will say that D is *admissible* if D admits complete mappings, and we will then call (H, D, θ, η) an *HP-system*, a term coined by Aschbacher [21] in 1990. The admissibility of D and H implies the admissibility of G.

Theorem 4.1 (Hall and Paige, 1955). *Let G be a finite group, H a subgroup of G, and D a dual system of coset representatives for H in G. If H and D are admissible, then so is G.*

Proof. Let θ be a complete mapping of D and ϕ a complete mapping of H, and let $\eta\colon D \to D$ be defined by $\eta(d)H = d\theta(d)H$. As every element in G can be uniquely expressed as an element of DH and also as an element of HD, for all $h \in H$ and all $d \in D$, there exist functions $g_h\colon D \to D$ and $f_d\colon H \to H$ such that $dh = f_d(h)g_h(d)$.

If $f_d(h) = f_d(h')$ for some $h = h' \in H$, then $hg_h(d)^{-1} = h'g_{h'}(d)^{-1}$ from which it follows that $h^{-1}h' = g_{h'}(d)^{-1}g_h(d)$. As $h^{-1}h' \in H$ and $g_{h'}(d), g_h(d) \in D$, a dual system of coset representatives for H in G, $g_{h'}(d) = g_h(d)$, and $h' = h$. It follows that, for all $d \in D$, f_d is one to one and so, by finiteness, is a bijection. Similarly, if $g_h(d) = g_h(d')$, for some $d = d' \in D$, then $f_d(h)^{-1}dh = f_{d'}(h)^{-1}d'h$ and so $f_d(h)f_{d'}(h)^{-1} = d'd^{-1}$. As $f_d(h)f_{d'}(h)^{-1} \in H$ and D is a dual system of coset representatives for H in G, $d = d'$. Hence g_h is one to one and so, by finiteness, is a bijection.

We now know that each element of G can be written uniquely as $df_{\theta(d)}(h)$. Define $\theta'\colon G \to G$ by $\theta'(df_{\theta(d)}(h)) = g_h(\theta(d))\phi(h)$. We claim that θ' is a complete mapping of G.

If $\theta'(d'f_{\theta(d')}(h')) = \theta'(df_{\theta(d)}(h))$, then $g_h(\theta(d))^{-1}g_{h'}(\theta(d')) = \phi(h)\phi(h')^{-1}$. As $\phi(h)\phi(h')^{-1} \in H$ and $g_h(\theta(d)), g_{h'}(\theta(d')) \in D$, a dual system of coset representatives for H in G, $g_h(\theta(d)) = g_{h'}(\theta(d'))$, and $\phi(h) = \phi(h')$. As ϕ is a bijection, $h = h'$, as g_h is a bijection $\theta(d) = \theta(d')$, and as θ is a bijection, $d = d'$. Hence θ' is one to one and so, by finiteness, is a bijection.

If $df_{\theta(d)}(h)\theta'(df_{\theta(d)}(h)) = d'f_{\theta(d')}(h')\theta'(d'f_{\theta(d')}(h'))$, then

$$(d\theta(d))(h\phi(h)) = (d'\theta(d'))(h'\phi(h')).$$

As $d\theta(d) \in \eta(d)H$ and $d'\theta(d') \in \eta(d')H$ it must be that $\eta(d) = \eta(d')$, and so $d = d'$ from which it follows that $h\phi(h) = h'\phi(h')$ and hence that $h = h'$. Hence θ' is a complete mapping of G. □

As an example illustrating Theorem 4.1, from a normal subgroup H of a group G and a complete mapping of G/H, we can construct an HP-system.

Example 4.2. Let H be a normal subgroup of a group G and suppose that G/H is admissible. Let D be a system of dual coset representatives for H in G, let θ be a complete mapping of G/H, and let $\eta: G/H \to G/H$ be the mapping uniquely determined by $(dH)\theta(dH) = \eta(dH)$. The complete mapping θ induces a mapping $\theta': D \to D$ uniquely determined by $\theta(dH) = \theta'(d)H$. This in turn uniquely determines a mapping $\eta': D \to D$ by $(dH)\theta(dH) = (dH)\theta'(d)H = \eta'(d)H$. As θ' and η' are bijections, (H, D, θ', η') is an HP-system for G.

Example 4.2 together with Theorem 4.1 yields another proof of Corollary 1.38: if H is a normal subgroup of G, and H and G/H are both admissible, then G is admissible. The subgroup need not be normal for us to be able to construct an HP-system: the following examples illustrate this:

Example 4.3. Let G be the group A_4, and let

$$H = \{(1), (1\ 2\ 3), (1\ 3\ 2)\} \quad \text{and} \quad K = \{(1), (1\ 2)(3\ 4), (1\ 3)(2\ 4), (1\ 4)(2\ 3)\}.$$

Both H and K are subgroups of G and $\{G : H\} = |K|$. As the intersection of H and K is trivial, for all $k_1, k_2 \in K$, $k_1 \neq k_2$, $k_1k_2^{-1} \notin H$ and $k_1^{-1}k_2 \notin H$, and so K is a dual system of coset representatives for H in G. Let θ be a complete mapping of K: K is admissible by Theorem 3.9. Define $\eta: K \to K$ by $\eta(k) = k\theta(k)$. As θ is a complete mapping of K, η is a bijection, and so (H, K, θ, η) is an HP-system. Note that H is not a normal subgroup of G. Note also that H is admissible by Corollary 3.2, and hence A_4 is admissible by Theorem 4.1.

Example 4.4. Let G be the group S_4, and let

$$H = \langle (1\ 3), (1\ 2\ 3\ 4) \rangle \quad \text{and} \quad K = \langle (1\ 2\ 3) \rangle.$$

H and K are both subgroups of G and $\{G : H\} = |K|$. As the intersection of H and K is trivial, for all $k_1, k_2 \in K$, $k_1 \neq k_2$, $k_1k_2^{-1} \notin H$ and $k_1^{-1}k_2 \notin H$, and so K is a dual system of coset representatives for H in G. Let $\theta: K \to K$ be identity mapping and $\eta: K \to K$ be the mapping $k \mapsto k^2$. Then (H, K, θ, η) is an HP-system. Note that H is not a normal subgroup of G.

If we set $\alpha = (1\ 2\ 3\ 4)$ and $\beta = (1\ 3)$, then

$$H = \langle \alpha, \beta \mid \alpha^4 = \beta^2 = 1, \beta\alpha = \alpha^{-1}\beta \rangle$$

is the dihedral group of order 8. A complete mapping of H is described below.

x	1	α	α^2	α^3	β	$\alpha\beta$	$\alpha^2\beta$	$\alpha^3\beta$
$\theta(x)$	1	α	$\alpha\beta$	$\alpha^2\beta$	$\alpha^3\beta$	α^3	α^2	β
$x\theta(x)$	1	α^2	$\alpha^3\beta$	$\alpha\beta$	α	$\alpha^2\beta$	β	α^3

It follows that S_4 is admissible by Theorem 4.1.

4.1.2 The Symmetric and Alternating Groups

One of the first applications of HP-systems was to proving the Hall-Paige conjecture true for the alternating and symmetric groups. This was accomplished by Hall and Paige [171] in 1955. Just a few years earlier, in 1951 Paige [263] had actually conjectured that all symmetric groups were inadmissible. In our presentation of their proofs, it should be noted that products of permutations will be computed from left to right. As an example $(1\ 2)(1\ 3) = (1\ 2\ 3)$ as opposed to $(1\ 3\ 2)$.

As the alternating groups are either trivial, of odd order, or have noncyclic Sylow 2-subgroups, the truth of the Hall-Paige conjecture would imply that they are all admissible.

Theorem 4.5 (Hall and Paige, 1955). *A_n is admissible for all n.*

Proof. The groups A_1 and A_2 are trivial and hence admissible; the group A_3 is of order 3 and so is admissible by Corollary 3.6; and A_4 was proved admissible in Example 4.3.

To establish the result for $n > 4$, we will proceed by induction. Suppose that $n \geq 4$ and that A_n is admissible. Using the natural action of A_{n+1} on $\{1, \ldots, n+1\}$, we will take H to be the stabilizer of $n+1$ in A_{n+1}. Hence $H \cong A_n$. Let

$$D = \langle (1\ n\ n+1) \rangle \cup \{(i\ n+1)(1\ n) \mid i = 2, \ldots, n-1\}.$$

For $x \in A_{n+1}$, $x \in H$ if and only if x fixes $n+1$. Now

$$\{A_{n+1} : H\} = |A_{n+1}|/|A_n| = n+1 = |D|.$$

Further D is closed under inverses, and so D is a dual system of coset representatives for H in A_{n+1} if and only if D is a system of right coset representatives for H in A_{n+1} if and only if D is a system of left coset representatives for H in A_{n+1}. The elements xy^{-1}, x and y distinct elements of D, are nonidentity elements of D, elements of the form $(i\ n+1)(1\ n)(j\ n+1)(1\ n) = (i\ j\ n+1)$, $(i\ n+1)(1\ n)(1\ 2\ 3) = (1\ n\ 2\ 3)(i\ n+1)$, $(i\ n+1)(1\ n)(1\ 3\ 2) = (1\ n\ 3\ 2)(i\ n+1)$, $(1\ 2\ 3)(i\ n+1)(1\ n) = (1\ 2\ 3\ n)(i\ n+1)$, and $(1\ 3\ 2)(i\ n+1)(1\ n) = (1\ 3\ 2\ n)(i\ n+1)$, none of which fix $n+1$, and, hence, for all $x, y \in D$, $x \neq y$, $x^{-1}y \notin H$ and $xy^{-1} \notin H$. It follows that D is a dual system of coset representatives for H in A_{n+1}.

Let us define mappings $\theta, \eta\colon D \to D$ by

$$\theta(x) = \begin{cases} x & \text{if } x = 1,\ (1\ n\ n+1),\ \text{or } (1\ n+1\ n), \\ (i+1\ n+1)(1\ n) & \text{if } x = (i\ n+1)(1\ n),\ \text{for } i = 2, \ldots, n-2, \\ (2\ n+1)(1\ n) & \text{if } x = (n-1\ n+1)(1\ n), \end{cases}$$

and

$$\eta(x) = \begin{cases} 1 & \text{if } x = 1, \\ (1\; n+1\; n) & \text{if } x = (1\; n\; n+1), \\ (1\; n\; n+1) & \text{if } x = (1\; n+1\; n), \\ (i+1\; n+1)(1\; n) & \text{if } x = (i\; n+1)(1\; n), \text{ for } i = 2, \dots, n-2, \\ (2\; n+1)(1\; n) & \text{if } x = (n-1\; n+1)(1\; n). \end{cases}$$

Clearly θ and η are both bijections. Further

$$(x\theta(x))^{-1}\eta(x) = \begin{cases} 1 & \text{if } x = 1, (1\; n\; n+1), \text{ or } (1\; n+1\; n), \\ (i\; i+1)(1\; n) & \text{if } x = (i\; n+1)(1\; n), \text{ for } i = 2, \dots, n-2, \\ (2\; n-1)(1\; n) & \text{if } x = (n-1\; n+1)(1\; n). \end{cases}$$

It follows that $(x\theta(x))^{-1}\eta(x)$ does not move $n+1$ for any $x \in D$. Thus $x\theta(x)H = \eta(x)H$, and so (H, D, θ, η) is an HP-system, and, as $H \cong A_n$, the result then follows inductively by Theorem 4.1. □

A similar proof to the proof of Theorem 4.5 can be used to prove the Hall-Paige conjecture true for the symmetric groups, S_n. Note that S_1 is trivial, and hence admissible; and S_2 and S_3 have nontrivial, cyclic Sylow 2-subgroups and hence are not admissible. This leaves S_n, $n \geq 4$, to deal with.

Theorem 4.6 (Hall and Paige, 1955). *S_n is admissible for all $n \geq 4$.*

Proof. The group S_4 was shown to be admissible in Example 4.4. To establish the result for $n > 4$, we will proceed by induction. Suppose that $n \geq 4$ and, using the natural action of S_{n+1} on $\{1, \dots, n+1\}$, that H is the stabilizer of $n+1$ in S_{n+1}. Let

$$D = \{1, (1\; n+1), (2\; n+1), \dots, (n\; n+1)\}.$$

If $x \in S_{n+1}$, then $x \in H$ if and only if x fixes $n+1$. Thus $H \cong S_n$. Now

$$\{S_{n+1} : H\} = |S_{n+1}|/|S_n| = n+1 = |D|.$$

Further D is closed under inverses, and so D is a dual system of coset representatives for H in S_{n+1} if and only if D is a system of right coset representatives for H in S_{n+1} if and only if D is a system of left coset representatives for H in S_{n+1}. The elements xy^{-1}, x and y distinct elements of D, are nonidentity elements of D of the form $(i\; n+1)(j\; n+1) = (i\; j\; n+1)$, which moves $n+1$, and hence, for all $x, y \in D$, $x \neq y$, $x^{-1}y \notin H$ and $xy^{-1} \notin H$. It follows that D is a dual system of coset representatives for H in S_{n+1}.

Let us define $\theta, \eta\colon D \to D$ by

$$\theta(x) = \begin{cases} 1 & \text{if } x = 1, \\ (i+1\; n+1) & \text{if } x = (i\; n+1),\ \text{for } i = 2, \ldots, n-1, \\ (2\; n+1) & \text{if } x = (n\; n+1), \end{cases}$$

and

$$\eta(x) = \begin{cases} 1 & \text{if } x = 1, \\ (i+1\; n+1) & \text{if } x = (i\; n+1),\ \text{for } i = 2, \ldots, n-1, \\ (2\; n+1) & \text{if } x = (n\; n+1). \end{cases}$$

Clearly θ and η are both bijections. Further

$$(x\theta(x))^{-1}\eta(x) = \begin{cases} 1 & \text{if } x = 1, \\ (i\; i+1) & \text{if } x = (i\; n+1),\ \text{for } i = 2, \ldots, n-1, \\ (2\; n) & \text{if } x = (n\; n+1). \end{cases}$$

It follows that $(x\theta(x))^{-1}\eta(x)$ does not move $n+1$ for any $x \in D$. Thus $x\theta(x)H = \eta(x)H$, and so (H, D, θ, η) is an HP-system, and, as $H \cong S_n$, the result then follows inductively by Theorem 4.1. □

4.1.3 Factorable Groups

We say that a group G is *factorable* if G has nontrivial subgroups A and B such that $G = AB$ and $A \cap B = 1$. We call A and B *factors* of G. Recognizing that a finite group is factorable is easy if the potential factors have relatively prime orders.

Lemma 4.7. *If A and B are subgroups of a finite group G, $\gcd(|A|, |B|) = 1$, and $|G| = |A||B|$, then $A \cap B = \{1\}$ and $G = AB$.*

Proof. As $|A \cap B|$ divides both $|A|$ and $|B|$ and $\gcd(|A|, |B|) = 1$, $|A \cap B| = 1$. Hence $A \cap B = \{1\}$.

Suppose that $a_1, a_2 \in A$, $b_1, b_2 \in B$, and $a_1b_1 = a_2b_2$. Then

$$a_2^{-1}a_1 = b_2b_1^{-1} \in A \cap B = \{1\}.$$

It follows that $a_1 = a_2$ and $b_1 = b_2$. Thus there are $|A||B| = |G|$ distinct elements of G of the form ab, $a \in A$, $b \in B$. The result follows. □

In Example 4.3 we described two subgroups of A_4, $H = \{(1), (1\ 2\ 3), (1\ 3\ 2)\}$ and $K = \{(1), (1\ 2)(3\ 4), (1\ 3)(2\ 4), (1\ 4)(2\ 3)\}$. As $|H| = 3$ and $|K| = 4$, $\gcd(|H|, |K|) = 1$ and $|A_4| = 12 = |H||K|$, it follows from Lemma 4.7 that A_4 is factorable with factors H and K. Similarly in Example 4.4, we described two subgroups of S_4, $H = \langle(1\ 3), (1\ 2\ 3\ 4)\rangle$ and $K = \langle(1\ 2\ 3)\rangle$. As $|H| = 8$ and $|K| = 3$, $\gcd(|H|, |K|) = 1$ and $|S_4| = 24 = |H||K|$, it follows from Lemma 4.7 that S_4 is factorable with factors H and K.

If G is factorable with factors A and B, then from the admissibility of A and B, we can deduce the admissibility of G.

Theorem 4.8. *Let G be a factorable group with factors A and B. If A and B are admissible, then so is G.*

Proof. Let θ be a complete mapping of A, and define $\eta\colon A \to A$ by $\eta(x) = x\theta(x)$. Then $\{B, A, \theta, \eta\}$ is an HP-system, and the result follows from the admissibility of B and Theorem 4.1. □

Theorem 4.8 will prove crucial to proving the Hall-Paige conjecture true for solvable groups and will be used in Section 5.1 to prove the admissibility of some special linear groups. A useful class of factorable groups are Frobenius groups. A nontrivial, proper subgroup H of G is a *Frobenius complement* if $H \cap g^{-1}Hg = \{1\}$ for all $g \in G \setminus H$. A group G is a *Frobenius group* if it contains a Frobenius complement.

Theorem 4.9. *Let G be a Frobenius group with Frobenius complement H, and let F be the set of elements of G that are not conjugate to any nonidentity element of H. Then F is a normal subgroup of G.*

Proof. See Chapter 6, Theorem 2.3 in [306], for example. □

The group F in Theorem 4.9 is called the *Frobenius kernel* of G. Many Frobenius groups are factorable.

Theorem 4.10. *If G is a finite Frobenius group with Frobenius complement H and Frobenius kernel F, and there are precisely $\{G : H\}$ conjugates of H in G, then $G = FH$.*

Proof. See Chapter 6, Theorem 2.3 in [306], for example. □

In 1955 Hall and Paige [171] used Theorem 4.10 to prove the admissibility of a class of Frobenius groups.

Corollary 4.11 (Hall and Paige, 1955). *Let G be a finite group with a noncyclic Sylow 2-subgroup S. If the number of Sylow 2-subgroups of G is $\{G : S\}$ and distinct Sylow 2-subgroups intersect trivially, then G is admissible.*

Proof. As $N_G(S) = S$, the Sylow 2-subgroup S is a Frobenius complement of G. Let F be the Frobenius kernel. By Theorem 4.10, $G = FS$ and F is a normal subgroup of odd order. We will show S to be admissible in Theorem 4.20. Further, F is admissible by Corollary 3.6. Thus, G is admissible by Theorem 4.8. □

4.2 Solvable Groups

We saw in Corollary 3.6 that all odd-order groups are admissible and in Theorem 3.9 that all abelian groups with noncyclic Sylow 2-subgroups are admissible. Thus we have seen that many solvable groups satisfy the Hall-Paige conjecture. In this section we will prove the Hall-Paige conjecture true for all solvable groups. We will do this by first proving all noncyclic 2-groups admissible. For a solvable group with a noncyclic Sylow 2-subgroup, we will then use Theorem 4.8 to extend complete mappings of its Sylow 2-subgroup to complete mappings of the whole group.

4.2.1 Some Properties of 2-Groups

To prove the admissibility of noncyclic 2-groups, we will need some facts about 2-groups and, in particular, facts about maximal subgroups of 2-groups.

Lemma 4.12. *Let G be a group of order 2^n. If H is a subgroup of G of order 2^m, $1 \leq m < n$, then there exists a subgroup K of G of order 2^{m+1} that contains H.*

Proof. By a corollary to Theorem 1.9 in [306], there is a composition series of G through H, and, by $(1.9)'$ in [306], all factors in any composition series for G are isomorphic to $\mathbb{Z}_2$, hence the result. □

By Lemma 4.12, all maximal subgroups of 2-groups are of index 2.

Corollary 4.13. *If $|G| = 2^n$ and M is a maximal subgroup of G, then $|M| = 2^{n-1}$ and M is normal in G.*

Proof. By Lemma 4.12, any subgroup of G of order less than 2^{n-1} is not maximal. Hence any maximal subgroup must be of index 2. It is an exercise in many group theory texts to show that any subgroup of index 2 is normal. □

A cyclic group of order 2^n has a unique subgroup of order 2^m for each m, $1 \leq m \leq n$. While the converse is true, a stronger result is known.

Lemma 4.14. *Let G be a group of order 2^n. If, for some m, $1 \leq m < n$, there is a unique subgroup of order 2^m, then G is cyclic.*

Proof. See Theorem 16 in [338], for instance. □

An immediate corollary:

Corollary 4.15. *If G is noncyclic of order 2^n, then G contains more than one maximal subgroup.*

The intersections of distinct maximal subgroups of 2-groups are limited.

Lemma 4.16. *Let G be a group of order 2^n. If M_1 and M_2 are distinct maximal subgroups of G, then $|M_1 \cap M_2| = 2^{n-2}$ and $G/(M_1 \cap M_2) \cong \mathbb{Z}_2 \times \mathbb{Z}_2$.*

Proof. Set $A = M_1 \cap M_2$, $B = M_1 \setminus A$, $C = M_2 \setminus A$, and $D = G \setminus (M_1 \cup M_2)$. Now

$$|G| = 2^n = |A| + |B| + |C| + |D| = 2(2^{n-1} - |A|) + |A| + |D|,$$

from which it follows that $|D| = |A|$. Now $|A| \leq |B|$ and if $x \in M_2 \setminus M_1$, then $xA \subseteq C$ and $xB \subseteq D$. Hence $|B| \leq |D| = |A|$. It follows that $|B| = |A|$. Similarly $|C| = |A|$ and so $4|A| = 2^n$ and $|M_1 \cap M_2| = |A| = 2^{n-2}$.

As M_1 and M_2 are normal in G, $A = M_1 \cap M_2$ is normal in G, and G/A is isomorphic to either $\mathbb{Z}_4$ or $\mathbb{Z}_2 \times \mathbb{Z}_2$. The elements of G/A are the cosets A, B, C, and D. Clearly $A^2 = A$. As $B \subseteq M_1$, $B^2 \in M_1$ and, as $B \not\subseteq M_2$ and $G/M_2 \cong \mathbb{Z}_2$, $B^2 \in M_2$. Hence $B^2 = A$. Similarly $C^2 = A$ and $D^2 = A$ from which it follows that $G/(M_1 \cap M_2) \cong \mathbb{Z}_2 \times \mathbb{Z}_2$. □

We are particularly interested in those noncyclic 2-groups whose maximal subgroups are cyclic. A classification of these groups can be found in [23] and in [338].

Theorem 4.17. *If G is a nonabelian 2-group of order $4m$ that possesses a cyclic subgroup of order $2m$, then G is isomorphic to one of the following:*

1. $Q_{4m} = \langle a, b \mid a^{2m} = 1, b^2 = a^m, bab^{-1} = a^{-1}\rangle$, $m \geq 2$,
2. $D_{4m} = \langle a, b \mid a^{2m} = b^2 = 1, bab = a^{-1}\rangle$, $m \geq 2$,
3. $\mathrm{Mod}_{4m} = \langle a, b \mid a^{2m} = b^2 = 1, bab = a^{m+1}\rangle$, $m \geq 2$, *or*
4. $\mathrm{SD}_{4m} = \langle a, b \mid a^{2m} = b^2 = 1, bab = a^{m-1}\rangle$, $m \geq 4$.

Proof. See p. 150 in [338] or result (23.4) in [23]. □

The group Q_{4m} is called the *quaternion group* of order $4m$, D_{4m} is called the *dihedral group* of order $4m$, Mod_{4m} is called the *modular group* of order $4m$, and SD_{4m} is called the *semidihedral group* of order $4m$. Observe that Mod_8 and D_8 are the same group. The quaternion group is also called the generalized quaternion group in the literature, and the modular group is also called the quasi-dihedral group. The quaternion groups have a simple characterization which will prove useful: this characterization can be found in [338].

Theorem 4.18. *If G is a noncyclic 2-group with only one involution, then G is a quaternion group.*

Proof. See Theorem 15 in [338]. □

4.2.2 Admissible 2-Groups

We saw in Theorem 3.9 that noncyclic abelian 2-groups are admissible. To prove the admissibility of nonabelian 2-groups, we begin by proving the admissibility of 2-groups whose maximal subgroups are cyclic: these groups were classified in Theorem 4.17.

Lemma 4.19 (Hall and Paige, 1955). *If G is a nonabelian group of order 2^n that possesses a cyclic subgroup of order 2^{n-1}, then G is admissible.*

Proof. It must be that $n \geq 3$ as groups of order 2 and 4 are abelian. By Theorem 4.17, G must be one of Q_{4m}, D_{4m}, Mod_{4m}, or SD_{4m}, where $m = 2^{n-2}$.

Letting a and b be the generators of these groups, as in Theorem 4.17, in each of these groups, any element can be uniquely written in the form $a^i b^j$, $i = 0, \ldots, 2m-1$, $j = 0, 1$.

If G is Q_{4m}, D_{4m}, or SD_{4m}, then define $\theta\colon G \to G$ by

$$\theta(x) = \begin{cases} a^i & \text{if } x = a^i \text{ and } i = 0, \ldots, m-1, \\ a^{i-m}b & \text{if } x = a^i \text{ and } i = m, \ldots, 2m-1, \\ a^{-(i+1)} & \text{if } x = a^i b \text{ and } i = 0, \ldots, m-1, \\ a^{m-(i+1)}b & \text{if } x = a^i b \text{ and } i = m, \ldots, 2m-1. \end{cases}$$

As

$$\{a^{-(i+1)} \mid i = 0, \ldots, m-1\} = \{a^i \mid i = m, \ldots, 2m-1\},$$

$$\{a^{i-m}b \mid i = m, \ldots, 2m-1\} = \{a^i b \mid i = 0, \ldots, m-1\},$$

and

$$\{a^{m-(i+1)}b \mid i = m, \ldots, 2m-1\} = \{a^i b \mid i = m, \ldots, 2m-1\},$$

$$\{\theta(x) \mid x \in G\} = \{a^i \mid i = 0, \ldots, 2m-1\} \cup \{a^i b \mid i = 0, \ldots, 2m-1\} = G.$$

Hence θ is one to one and so, by finiteness, is a bijection.

If G is one of Q_{4m}, D_{4m}, or SD_{4m}, let the mapping $\eta : G \to G$ be defined by $\eta(x) = x\theta(x)$. There are three cases to consider.

Case 1: $G = Q_{4m}$.

If $i \in \{0, \ldots, 2m-1\}$ is even, then $a^i = \eta(a^{i/2})$. If $i \in \{0, \ldots, 2m-1\}$ is odd, then $a^i = \eta(a^{(i+m-1)/2}b)$. If $i \in \{0, \ldots, 2m-1\}$ is even, then $a^i b = \eta(a^{(i+m)/2})$. If $i \in \{0, \ldots, 2m-1\}$ is odd, then $a^i b = \eta(a^{(i-1)/2})$. Thus η is onto and, hence, by finiteness, is a bijection. It follows that θ is a complete mapping of G, and so G is admissible.

Case 2: $G = D_{4m}$.

If $i \in \{0, \ldots, 2m-1\}$ is even, then $a^i = \eta(a^{i/2})$. If $i \in \{0, \ldots, 2m-1\}$ is odd, then $a^i = \eta(a^{(i+m-1)/2}b)$. If $i \in \{0, \ldots, 2m-1\}$ is even, then $a^i b = \eta(a^{(i+m)/2}b)$. If $i \in \{0, \ldots, 2m-1\}$ is odd, then $a^i b = \eta(a^{(i-1)/2})$. Thus η is onto and, hence, by finiteness, is a bijection. It follows that θ is a complete mapping of G, and so G is admissible.

Case 3: $G = \mathrm{SD}_{4m}$.

If $i \in \{0, \ldots, 2m-1\}$ is even, then $a^i = \eta(a^{i/2})$. If $i \in \{0, \ldots, 2m-1\}$ is congruent to 1 modulo 4, then $a^i = \eta(a^{(i-1)/2}b)$. If $i \in \{0, \ldots, 2m-1\}$ is congruent to 3 modulo 4, then $a^i = \eta(a^{(i-m-1)/2}b)$. If $i \in \{0, \ldots, 2m-1\}$ is even, then $a^i b = \eta(a^{(i+m)/2})$. If $i \in \{0, \ldots, 2m-1\}$ is congruent to 1 modulo 4, then $a^i b = \eta(a^{(i+m-1)/2}b)$. If $i \in \{0, \ldots, 2m-1\}$ is congruent to 3 modulo 4, then $a^i b = \eta(a^{(i-1)/2}b)$. Thus η is onto and, hence, by finiteness, is a bijection. It follows that θ is a complete mapping of G and so G is admissible.

If G is Mod_{4m}, then define $\theta\colon G \to G$ by

$$\theta(x) = \begin{cases} a^{i-1} & \text{if } x = a^i \text{ and } i = 1, \ldots, m, \\ a^{i-1+m}b & \text{if } x = a^i \text{ and } i = m+1, \ldots, 2m, \\ a^{i+m} & \text{if } x = a^i b \text{ and } i = 0, \ldots, m-1, \\ a^i b & \text{if } x = a^i b \text{ and } i = m, \ldots, 2m-1. \end{cases}$$

$$\begin{aligned} \{\theta(x) \mid x \in G\} =& \{a^{i-1} \mid i = 1, \ldots, m\} \cup \{a^{i+m} \mid i = 0, \ldots, m-1\} \\ & \cup \{a^{i-1+m}b \mid i = m+1, \ldots, 2m\} \cup \{a^i b \mid i = m, \ldots, 2m-1\} \\ =& \{a^i \mid i = 0, \ldots, 2m-1\} \cup \{a^i b \mid i = 0, \ldots, 2m-1\} = G. \end{aligned}$$

Hence θ is one to one and so, by finiteness, is a bijection.

Let the mapping $\eta\colon G \to G$, be defined by $\eta(x) = x\theta(x)$. If $i \in \{0, \ldots, 2m-1\}$ is odd, then $a^i = \eta(a^{(i+1)/2})$. If $i \in \{0, \ldots, 2m-1\}$ is congruent to 0 modulo 4, then $a^i = \eta(a^{i/2}b)$. If $i \in \{0, \ldots, 2m-1\}$ is congruent to 2 modulo 4, then $a^i = \eta(a^{(i-m)/2}b)$. If $i \in \{0, \ldots, 2m-1\}$ is odd, then $a^i b = \eta(a^{(i-m+1)/2})$. If $i \in \{0, \ldots, 2m-1\}$ is congruent to 0 modulo 4, then $a^i b = \eta(a^{(i-m)/2}b)$. If $i \in \{0, \ldots, 2m-1\}$ is congruent to 2 modulo 4, then $a^i b = \eta(a^{i/2}b)$. Thus η is onto and, hence, by finiteness, is a bijection. It follows that θ is a complete mapping of G, and so G is admissible. □

We are now in a position to prove the admissibility of noncyclic 2-groups.

Theorem 4.20 (Hall and Paige, 1955). *Any noncyclic 2-group is admissible.*

Proof. Let us assume the result to be false and that G is a minimal counterexample. If G is abelian, then G is admissible by Theorem 3.9. Thus we will assume that G is nonabelian and that $|G| = 2^n$. If $n \leq 3$, then $n = 3$, and G contains a cyclic subgroup of order 4 and so is admissible by Lemma 4.19. If G contains only one involution, then it is a quaternion group by Theorem 4.18 and so is admissible by Lemma 4.19. It follows that $n \geq 4$.

Thus G must contain distinct involutions as otherwise G would be a quaternion group by Theorem 4.18 and hence admissible by Lemma 4.19. Further, as $Z(G)$ is nontrivial it contains an involution. Let v_1 and v_2 be distinct involutions in G, $v_1 \in Z(G)$, and let $V = \langle c, d\rangle \cong \mathbb{Z}_2 \times \mathbb{Z}_2$.

If V is contained in two distinct maximal subgroups M_1 and M_2, then $M_1 \cap M_2$, as it contains V, is noncyclic, as is $G/(M_1 \cap M_2)$ by Lemma 4.16. By the minimality of G, both $M_1 \cap M_2$ and $G/(M_1 \cap M_2)$ are admissible. It follows from Corollary 1.38 that G must be admissible, a contradiction.

Thus we may suppose that V is contained in a unique maximal subgroup M. Any maximal subgroup of G, in particular M, must be noncyclic, as, otherwise, G is admissible by Lemma 4.19. By Lemma 4.16 there must exist another maximal subgroup M' of G. If $M \cap M'$ is noncyclic, then, as $G/(M \cap M')$ is also noncyclic by Lemma 4.16, G is admissible as, by the minimality of G, both $M_1 \cap M_2$ and $G/(M_1 \cap M_2)$ are admissible. It follows from Corollary 1.38 that G must be admissible, a contradiction.

Thus we may assume that $K = M \cap M'$ is cyclic. Now $|M| = 2^{n-1}$ and $|K| = 2^{n-2}$ by Lemma 4.16. Thus, as M is noncyclic, either M is an abelian group isomorphic to $\mathbb{Z}_2 \times \mathbb{Z}_{2^{n-2}}$ or, by Theorem 4.17, a modular group, a dihedral group, or a semidihedral group: M cannot be a quaternion group by Theorem 4.18. Let us set $2^{n-3} = m$. If M is abelian, then we may assume that $M = \langle a, b \mid a^{2m} = b^2 = 1, bab = a\rangle$. If M is one of D_{4m}, Mod_{4m}, or SD_{4m}, then $M = \langle a, b\rangle$, where a and b are as in Theorem 4.17. In each case we are free to assume that $M \cap M' = K = \langle a\rangle$. Let c be an element of M' that is not in M. Now K is a normal subgroup of M and M', and M' is noncyclic. As K is a normal subgroup of M' of index 2, $c^2 = a^t$ for some integer t, and t cannot be odd as then $M' = \langle c\rangle$, contradicting M' noncyclic. Thus t is even. By the normalcy of K in M', $c^{-1}ac = a^s$ for some integer s, and s must be odd as a^s is a generator of K.

Set $H = \langle a^2, b\rangle$. Now $n \geq 4$. It is an exercise to show that, if M is abelian or Mod_{4m}, then $H \cong \mathbb{Z}_2 \times \mathbb{Z}_m$; if $n = 4$ and $M = D_{4m}$, then $H \cong \mathbb{Z}_2 \times \mathbb{Z}_2$; and if $n > 4$ and M is D_{4m} or SD_{4m}, then $M \cong D_{2m}$. In each case H is admissible by either Theorem 3.9 or Lemma 4.19. Set $D = \{1, a, c, ac\}$. As $|D| = \{G : H\}$ and $a, c, ac \notin H$, to show that D is a dual system of coset representatives for H in G, we need only show that neither xy^{-1} nor $x^{-1}y$ is an element of H for any distinct $x, y \in D$, $x, y \neq 1$. It is an exercise to verify that $a^{-1}c \in acH$, $a^{-1}(ac) \in cH$, $c^{-1}a \in acH$, $c^{-1}(ac) \in aH$, $(ac)^{-1}a \in cH$, and $(ac)^{-1}c \in aH$ and to verify that $ac^{-1} \in acH$, $a(ac)^{-1} \in cH$, $ca^{-1} \in acH$, $c(ac)^{-1} \in aH$, $(ac)a^{-1} \in cH$, and $(ac)c^{-1} \in aH$, thus establishing that D is a dual system of coset representatives for H in G as claimed.

Define mappings $\theta, \eta \colon D \to D$ by

$$\theta(x) = \begin{cases} 1 & \text{if } x = 1, \\ c & \text{if } x = a, \\ ac & \text{if } x = c, \\ a & \text{if } x = ac, \end{cases}$$

and

$$\eta(x) = \begin{cases} 1 & \text{if } x = 1, \\ ac & \text{if } x = a, \\ a & \text{if } x = c, \\ c & \text{if } x = ac. \end{cases}$$

Now $1\theta(1) = 1 = \eta(1), a\theta(a) = ac = \eta(a), c\theta(c) = cac = c^2a^s = a^{s+t} \in aH = \eta(c)H$, and $(ac)\theta(ac) = aca = ca^{s+1} \in cH = \eta(ac)H$. Hence D is admissible, and thus G is admissible by Theorem 4.1, a contradiction from which the result follows. □

Hall and Paige indicated that Theorem 4.20 was known. In 1951 Paige [263] credited Bruck with having proved this and claimed that arguing modulo the commutator subgroup enables one to establish this result for those 2-groups with noncyclic commutator subgroups and that Bruck had dealt with the remaining cases. This appears not to have been published.

4.2.3 Admissible Solvable Groups

In 1937 Philip Hall [172] proved finite solvable groups to be factorable. In fact, he showed that, if G is a finite solvable group, p is any prime divisor of $|G|$, and S is a Sylow p-subgroup of G, then G has a factorization in which one of the factors is S: this is Theorem 4.5 in [157]. We need only the special case in which $p = 2$.

Lemma 4.21. *If G is finite and solvable, then G is factorable with factors A and B, $|A|$ odd, and B a Sylow 2-subgroup of G.*

Combined with Theorem 4.8, Lemma 4.21 yields a proof that finite solvable groups satisfy the Hall-Paige conjecture.

Theorem 4.22 (Hall and Paige, 1955). *Any finite solvable group whose Sylow 2-subgroups are noncyclic is admissible.*

Proof. Let G be a finite solvable group with noncyclic Sylow 2-subgroups. By Lemma 4.21, G is factorable with factors A and B , $|A|$ odd, and B a Sylow 2-subgroup of G. A is admissible by Corollary 3.6, and B is admissible by Theorem 4.20. The result then follows from Theorem 4.8. □

4.3 The Admissibility of the Mathieu Groups and Suzuki Groups

In this section we will prove two classes of simple groups to be admissible, the Mathieu groups and the Suzuki groups.

4.3.1 Mathieu Groups

For a long time, the five Mathieu groups M_{11}, M_{12}, M_{22}, M_{23}, and M_{24}, discovered in the 1800s, were the only known sporadic simple groups. In 1990 Aschbacher [21] proved that these groups were not minimal counterexamples to the Hall-Paige conjecture, and in 1993 Dalla Volta and Gavioli [83] proved them to be admissible. Each of the Mathieu groups has a group action that is at least 3-transitive and Dalla Volta and Gavioli used the multiple transitivity of these groups to prove their admissibility.

Lemma 4.23. *Let G be a transitive permutation group acting on $X = \{1, \dots, n\}$, and suppose that a set $D = \{g_1, \dots, g_n\}$, satisfying $i^{g_i} = n$, is a dual system of coset representatives for G_n, the stabilizer of n in G. If $\pi\colon X \to X$ is a bijection satisfying $\pi(i)^{g_i} = \pi(i)$, then the mapping $g_i \mapsto g_{\pi(i)}$ is a complete mapping of D.*

Proof. As D is a dual system of coset representatives for G_n in G, there exists a unique mapping $\sigma\colon X \to X$ determined by the equation $g_i g_{\pi(i)} G_n = g_{\sigma(i)} G_n$. Now

$$\pi(i)^{g_i g_{\pi(i)}} = \pi(i)^{g_{\pi(i)}} = n$$

from which it follows that $g_i g_{\pi(i)} G_n = g_{\pi(i)} G_n$ and, thus, that $\sigma = \pi$, a bijection. Hence π is a complete mapping of D as claimed. □

Using Lemma 4.23, Dalla Volta and Gavioli [83] proved that a 3-transitive group with an admissible point stabilizer is admissible.

Lemma 4.24 (Dalla Volta and Gavioli, 1993). *Let G be a group that acts 3-transitively on $X = \{1, \dots, n\}$, $n \geq 3$, and let G_n be the subgroup of G fixing n. If G_n is admissible, then G is admissible.*

Proof. Let G_i denote the subgroup of G fixing i, and, for $i \neq j$, let G_{ij} denote the subgroup of G fixing both i and j. As G acts 3-transitively on X, G_i acts transitively on $X \setminus \{i\}$ and G_{ij} acts transitively on $X \setminus \{i, j\}$. It follows that $n - 2$ divides $|G_{ij}|$ and $n - 1$ divides $|G_i|$ and so $(n - 2)(n - 1)$ divides $|G_i|$. Hence $|G_i|$ is even for all i and so each G_i contains an involution. By renumbering the elements of X if needed, we may assume, without loss of generality, that G_2 contains an involution τ that swaps 1 and n. Let π be a permutation of X that fixes n and maps 1 to 2. As G is 3-transitive, for each $i = 1, \dots, n - 1$, there exists $\sigma_i \in G_n$ that maps 1 to i

and 2 to $\pi(i)$. Set $\tau_i = \sigma_i\tau\sigma_i^{-1}$ and $D = \{1, \tau_1, \ldots, \tau_{n-1}\}$. As $|D| = \{G : G_n\}$ and D consists of the identity and involutions only, D is a dual system of coset representatives for G_n in G if and only if D is a system of left coset representatives for G_n in G if and only if D is a system of right coset representatives for G_n in G. Now no nonidentity element of D fixes n and if $i, j \in \{1, \ldots, n-1\}$, $i \neq j$, then $\tau_i\tau_j^{-1}(n) = \sigma_i\tau\sigma_i^{-1}\sigma_j\tau\sigma_j^{-1}(n) = n$ if and only if $\sigma_i\tau(n) = \sigma_j\tau(n)$. Now $\sigma_i\tau(n) = \sigma_i(1) = i$. Similarly $\sigma_j\tau(n) = j$. Hence $\tau_i\tau_j^{-1}(n) = n$ if and only if $i = j$. It follows that D is a dual system of coset representatives for G_n in G.

As $\tau_i(\pi(i)) = \sigma_i\tau\sigma_i^{-1}(\pi(i)) = \sigma_i\tau(2) = \sigma_i(2) = \pi(i)$, the mapping $\theta\colon D \to D$, defined by

$$\theta(x) = \begin{cases} 1 & \text{if } x = 1, \\ \tau_{\pi(i)} & \text{for } i = 1, \ldots, n-1, \end{cases}$$

is a complete mapping of D by Lemma 4.23. Hence, by Theorem 4.1, if G_n is admissible, then G is admissible. □

In Chapter 7 we will strengthen Lemma 4.24 by showing that 2-transitivity is sufficient for finite nonabelian simple groups: see Corollary 7.30. To use Lemma 4.24 to prove the admissibility of the Mathieu groups requires properties of the Mathieu groups as permutation groups. The needed properties of the Mathieu groups are well-known and can be found in many books such as [266] or [284].

Theorem 4.25 (Dalla Volta and Gavioli, 1993). *The Mathieu groups are admissible.*

Proof. M_{11} acts 4-transitively on a set $X = \{1, \ldots, 10\}$, the subgroup H fixing 1 and 2 acts 3-transitively on $\{3, \ldots, 10\}$ and is a solvable group of order 72 with a noncyclic Sylow 2-subgroup. By Theorem 4.22 and by Lemma 4.24, the subgroup K of M_{11} fixing 1 is admissible. The admissibility of M_{11} follows from Lemma 4.24. M_{12} acts 5-transitively on a set of order 12, the point stabilizer being isomorphic to M_{11}. As M_{11} is admissible, M_{12} is admissible by Lemma 4.24. M_{22} acts 3-transitively on a set of order 22, the point stabilizer being isomorphic to PSL(3, 4). The admissibility of $PSL(3, 4)$ will be proved in Corollary 5.28. The admissibility of M_{22} follows from Lemma 4.24. M_{23} acts 4-transitively on a set of order 23, the point stabilizer being isomorphic to M_{22}. As M_{22} is admissible, M_{23} is admissible by Lemma 4.24. M_{24} acts 5-transitively on a set of order 24, the point stabilizer being isomorphic to M_{23}. As M_{23} is admissible, M_{24} is admissible by Lemma 4.24. □

Note that the group K in the proof of Theorem 4.25 is often denoted M_{10} and is not solvable, which is why we needed to prove its admissibility using Lemma 4.24, before we could apply Lemma 4.24 to prove the admissibility of M_{11}.

4.3.2 Suzuki Groups

The Suzuki groups $\mathrm{Sz}(2^{2n+1})$, $n \geq 1$, form an infinite class of finite simple groups. These were discovered by Suzuki in 1960 and were recognized to be groups of Lie type by Ree. These groups are also called Suzuki-Ree groups and are also denoted ${}^2B_2(2^{2n+1})$. In 1989 Di Vincenzo [96] proved these groups to be admissible. Descriptions of these groups can be found in [192].

Lemma 4.26. *For $G = \mathrm{Sz}(2^{2n+1})$, $n \geq 1$, the following hold:*

1. *The Sylow 2-subgroups of G are noncyclic.*
2. *Distinct Sylow 2-subgroups of G intersect trivially.*
3. *Any element of G of even order is a 2-element.*

Proof. If we set $q = 2^{2n+1}$, then the order of G is $(q^2 + 1)q^2(q - 1)$. The Sylow 2-subgroups of G are described in Lemma 3.1, Chapter XI of [192], where they are shown to be nonabelian.

Let F be a Sylow 2-subgroup of G. By Lemma 3.1, Chapter XI of [192], G has a subgroup H of order $q - 1$, and, by Theorem 3.10, Chapter XI of [192], G has a subgroup K of order $q + 2\sqrt{(q/2)} + 1$ and a subgroup L of order $q - 2\sqrt{(q/2)} + 1$. Further, by Theorem 3.10, Chapter XI of [192], the conjugates of $F \setminus \{1\}$, $H \setminus \{1\}$, $K \setminus \{1\}$, and $L \setminus \{1\}$ partition the nonidentity elements of G. It follows that distinct Sylow 2-subgroups of G intersect trivially. Further, as elements of H, K, or L are of odd order, the only elements of even order are 2-elements. □

It follows from Lemma 4.26 that the Suzuki groups are Frobenius groups, their Sylow 2-subgroups being Frobenius subgroups. From the partitions described in Lemma 4.26, we can construct complete mappings of the Suzuki groups.

Theorem 4.27 (Di Vincenzo, 1989). *The Suzuki groups, $\mathrm{Sz}(2^{2n+1})$, $n \geq 1$, are admissible.*

Proof. Let $G = \mathrm{Sz}(2^{2n+1})$, $n \geq 1$, let O be the set of odd-order elements of G, and let $H_1, \ldots, H_m$ be the Sylow 2-subgroups of G. By Lemma 4.26, the Sylow 2-subgroups of G are noncyclic, and so, by Theorem 4.20, each of $H_1, \ldots, H_m$ is admissible. For each $i = 1, \ldots, m$, let θ_i be a normalized complete mapping for H_i, and define $\theta\colon G \to G$ by

$$\theta(x) = \begin{cases} \theta_i(x) & \text{if } x \in H_i \setminus \{1\}, \\ x & \text{if } x \in O. \end{cases}$$

Let $x \in G$. By Lemma 4.26, if $x \notin O$, that is, x is of even order, then x must be in at least one of $H_i \setminus \{1\}$ and, as $H_1, \ldots, H_m$ intersect trivially, in exactly one of $H_i \setminus \{1\}$. It follows that θ is well-defined. Restricted to O, θ is clearly a bijection $O \to O$, and, as each θ_i induces a bijection $H_i \setminus \{1\} \to H_i \setminus \{1\}$, θ restricted to $H_i \setminus \{1\}$ is a bijection $H_i \setminus \{1\} \to H_i \setminus \{1\}$. It follows that θ is a bijection $G \to G$. Similarly, as $x \mapsto x\theta_i(x)$ is a bijection $H_i \setminus \{1\} \to H_i \setminus \{1\}$ for each $i = 1, \ldots, m$

and, by the argument used in the proof of Theorem 3.5, the mapping $x \mapsto x^2$ is a bijection $O \to O$, θ is a complete mapping of G. □

The use of partitions of a group to construct complete mappings will be revisited in Chapter 5.

4.4 Unitary Groups and Groups with Trivially Intersecting Sylow 2-Subgroups

In this section we describe the 1989 work of Di Vincenzo [96] on the admissibility of certain unitary groups. We will also describe his work on the admissibility of groups in which distinct Sylow 2-subgroups intersect trivially: this will improve on Corollary 4.11.

4.4.1 Unitary Groups

In 1989 Di Vincenzo [96] proved the unitary group, $SU(3, q)$, and the projective special unitary group, $PSU(3, q)$, to be admissible whenever q is even. These groups are described by their action on a vector space.

Let V be a vector space, consisting of row vectors, of dimension 3 over $GF(q^2)$, q even. We define a mapping ϕ: $V \times V \to GF(q^2)$ by

$$\phi((x_1, y_1, z_1), (x_2, y_2, z_2)) = x_1 z_2^q + y_1 y_2^q + z_1 x_2^q.$$

The mapping ϕ is called a *nondegenerate hermitian (or unitary) form*. The *special unitary group*, $SU(3, q)$, is the subgroup of $SL(3, q^2)$ consisting of those matrices A satisfying $\phi(v_1 A, v_2 A) = \phi(v_1, v_2)$. If aI is a scalar matrix in $SU(3, q)$, then $a^3 = 1$, and, as $\phi(v_1(aI), v_2(aI)) = a^{q+1}\phi(v_1, v_2)$, $a^{q+1} = 1$. Hence, the normal subgroup Z, of $SU(3, q)$, consisting of scalar matrices, has order $\gcd(3, q + 1)$. The *projective special unitary group*, $PSU(3, q)$, is the quotient group $SU(3, q)/Z$. There are alternative notations for these groups, using q^2 in place of q. Thus, the groups that we are denoting $SU(3, q)$ and $PSU(3, q)$ are sometimes denoted by $SU(3, q^2)$ and $PSU(3, q^2)$: see [96], for example. The group that we are denoting $PSU(3, q)$ is also denoted $U_3(q)$.

Let us define U to be the set of vectors $v \in V$ satisfying $\phi(v, v) = 0$. U is called a *unital*. As the mapping $v \mapsto v(aI)$ maps elements of U to elements of U for all $a \in GF(q^2)$, U is a union of one-dimensional subspaces of V. Recall, in Section 1.4, our description of the Desarguesian projective plane, π, of order q^2. The points of π are the one-dimensional subspaces of V, and the lines of π are the two-dimensional subspaces of V. In particular U can be regarded as a set of points of π, namely, the points $\langle(1, x, y)\rangle$, $x^{q+1} + y + y^q = 0$, and $\langle(0, 0, 1)\rangle$. It is clear that

$\mathrm{SU}(3, q)$ acts on U as well as the points and lines of π and, as, for all $a \in \mathrm{GF}(q^2)$, the mapping $\mathrm{v} \mapsto \mathrm{v}(aI)$ fixes all points and lines of π, the action of $\mathrm{SU}(3, q)$ on U induces a natural action of $\mathrm{PSU}(3, q)$ on U. Here are some properties of $\mathrm{SU}(3, q)$ and $\mathrm{PSU}(3, q)$.

Lemma 4.28.

1. $|\mathrm{SU}(3, q)| = q^3(q^3 + 1)(q^2 - 1)$.
2. $|\mathrm{PSU}(3, q)| = q^3(q^3 + 1)(q^2 - 1)/\gcd(q + 1, 3)$.
3. *If* $q > 2$, *then* $\mathrm{PSU}(3, q)$ *is a simple group.*
4. $\mathrm{SU}(3, q)$ *and* $\mathrm{PSU}(3, q)$ *act doubly transitively on the points of* U.

Proof. See [191], for instance. □

If l is a line of π, then l is a *tangent line* of U if it meets U in exactly one point of π and a *secant line* of U if it meets U in exactly $q + 1$ points of π. Here are some basic properties of U.

Lemma 4.29.

1. U *contains* $q^3 + 1$ *points of* π.
2. *If* l *is a line of* π, *then* l *is either a tangent line of* U *or a secant line of* U.
3. *There is exactly one tangent line of* U *through each point of* U.

Proof. The mapping $y \mapsto y^q + y$ is a homomorphism from $\mathrm{GF}(q^2)^+$ onto $\mathrm{GF}(q)^+$, and the mapping $x \mapsto x^{q+1}$ is a homomorphism from $\mathrm{GF}(q^2)^*$, the multiplicative group of $\mathrm{GF}(q^2)$, onto $\mathrm{GF}(q)^*$, the multiplicative group of $\mathrm{GF}(q)$. Hence, for each $c \in \mathrm{GF}(q)^*$, there are $q(q+1)$ pairs (x, y) for which $x^{q+1} = y^q + y = c$. There are also q pairs $(0, y)$ for which $y^q + y = 0$. Hence, there are $(q-1)q(q+1)+q = q^3$ pairs $(x.y)$ for which $x^{q+1} + y + y^q = 0$. It follows that U contains $q^3 + 1$ points of π.

Let l be a line of π. If l contains points $\langle(1, x, y)\rangle$ and $\langle(0, 0, 1)\rangle$ of U, then any point of l, other than $\langle(0, 0, 1)\rangle$, is of the form $\langle(1, x, y + w)\rangle$. Now $x^{q+1} + (y + w) + (y + w)^q = 0$ if and only if $w + w^q = 0$, if and only if $w \in GF(q)$. Hence l contains precisely $q + 1$ points of U. As, by Lemma 4.28, $\mathrm{SU}(3, q)$ acts doubly transitively on the points of U, any line that passes through at least two points of U must pass through exactly $q + 1$ points of U.

Let $P \in U$. As only one line can pass through two distinct points, exactly q^2 lines through P pass through another point of U. As there are $q^2 + 1$ lines through P, exactly one line must pass through P and be tangent to U.

The number of lines passing through points of U, secant and tangent lines, is

$$q^3 + 1 + \frac{(q^3 + 1)q^2}{q + 1} = q^4 + q^2 + 1,$$

which is the number of lines in π. Hence, every line of π is either a secant line or a tangent line. □

Set

$$A_{a,b} = \begin{pmatrix} 1 & a & b \\ 0 & 1 & a^q \\ 0 & 0 & 1 \end{pmatrix} \quad \text{and} \quad B_c = \begin{pmatrix} c^{-q} & 0 & 0 \\ 0 & c^{q-1} & 0 \\ 0 & 0 & c \end{pmatrix}.$$

It is easily verified that, if $a^{q+1} + b^q + b = 0$, then $A_{a,b} \in \mathrm{SU}(3, q)$ and, if $c \in \mathrm{GF}(q^2)$, $c \neq 0$, then $B_c \in \mathrm{SU}(3, q)$. Set

$$S = \{A_{a,b} \mid a, b \in GF(q^2), a^{q+1} + b^q + b = 0\},$$

and

$$T = \{B_c \mid c \in GF(q^2), c \neq 0\}.$$

S is a Sylow 2-subgroup of $\mathrm{SU}(3, q^2)$.

Lemma 4.30. *For q even, the following hold:*

1. *If $A_{a,b}, A_{c,d} \in S$, then $A_{a,b}A_{c,d} = A_{a+c,b+d+ac^q}$.*
2. *If $A_{a,b} \in S$, then $A^4_{a,b} = I$.*
3. *If $A_{a,b} \in S$, then $A_{a,b}$ is an involution if and only if $a = 0$ and $b \neq 0$ and of order 4 if and only if $a \neq 0$.*
4. *S is a Sylow 2-subgroup of $\mathrm{SU}(3, q)$.*
5. *$S^* = \{AZ \mid A \in S\}$ is a Sylow 2-subgroup of $\mathrm{PSU}(3, q)$.*

Proof. Simple computation shows that $A_{a,b}A_{c,d} = A_{a+c,b+d+ac^q}$. It follows that $A^2_{a,b} = A_{0,a^{q+1}}$ and $A^4_{a,b} = A_{0,0}$, and so $A_{a,b}$ is either the identity if $a = b = 0$, an involution if $a = 0$ and $b \neq 0$, or of order 4 if $a, b \neq 0$. It also follows, as S is closed under multiplication, that S is a 2-subgroup of $\mathrm{SU}(3, q)$.

Now the mapping $b \mapsto b^q + b$ is a homomorphism from $\mathrm{GF}(q^2)^+$ onto $\mathrm{GF}(q)^+$, and the mapping $a \mapsto a^{q+1}$ is a homomorphism from $\mathrm{GF}(q^2)^*$, the multiplicative group of $\mathrm{GF}(q^2)$, onto $\mathrm{GF}(q)^*$, the multiplicative group of $\mathrm{GF}(q)$. Hence, for each $c \in \mathrm{GF}(q)^*$, there are $q(q+1)$ pairs (a, b) for which $a^{q+1} = b^q + b = c$. There are also q pairs $(0, b)$ for which $b^q + b = 0$. Hence, $|S| = (q - 1)q(q + 1) + q = q^3$. As the Sylow 2-subgroup is of order q^3, it follows that S is a Sylow 2-subgroup of $\mathrm{SU}(3, q)$.

As $|Z|$ is odd, the mapping $A \mapsto AZ$ is a homomorphism from $\mathrm{SU}(3, q)$ onto $\mathrm{SU}(3, q)/Z = \mathrm{PSU}(3, q)$, which induces a bijection $S \to S^*$. Hence S^* is a Sylow 2-subgroup of $\mathrm{PSU}(3, q)$. □

Some information about the actions of bijections from the point set of π to the point set of π will be needed to establish the admissibility of $\mathrm{PSU}(3, q)$ and $\mathrm{SU}(3, q)$, q even.

Lemma 4.31. *Let σ be a bijection from the point set of π to the point set of π that maps lines into lines. If σ fixes all lines through a point P, then σ fixes all points on some line l of π. Further, if σ is not the identity, then σ fixes no other points.*

Proof. This is a special case of a result for collineations of projective planes: see [88, 190], or [210], for instance. □

The actions of even-order elements on points and lines of π are described in the following lemma:

Lemma 4.32. *Let $q > 4$ be even and let σ be a nonidentity element of* $\mathrm{PSU}(3, q)$.

1. *σ fixes exactly one point, P, of U.*
2. *If σ is an involution, then σ also fixes each line of π through P.*
3. *If σ is of order 4, then σ fixes exactly one line of π through P, the unique tangent line of U through P.*
4. *If σ is not a 2-element, then σ fixes exactly two lines of π through P, one of which is the unique tangent line of U through P.*

Proof. If σ is a 2-element, then, without loss of generality, by Lemma 4.30, we may assume that $\sigma = A_{a,b}Z$. It is easy to see that the only point in U fixed by σ is $\langle(0, 0, 1)\rangle$. If σ is an involution, then, by Lemma 4.30, $a = 0$, $b \neq 0$, and

$$(1, x, y)\sigma = (1, x, b + y) = (1, x, y) + b(0, 0, 1).$$

It follows that σ fixes all secant lines through $\langle(0, 0, 1)\rangle$, and, as σ is nonsingular, σ also fixes the other line through $\langle(0, 0, 1)\rangle$, the tangent line. If σ is of order 4, then, by Lemma 4.30, $a \neq 0$ and

$$(1, x, y)\sigma = (1, a + x, b + a^q x + y),$$

which cannot be expressed as a linear combination of $(1, x, y)$ and $(0, 0, 1)$. It follows that σ fixes no secant line through $\langle(0, 0, 1)\rangle$ but does permute the secant lines through $\langle(0, 0, 1)\rangle$, and so the remaining line through $\langle(0, 0, 1)\rangle$, the tangent line, must also be fixed by σ.

If σ is of even order but is not a 2-element, then the order of σ must be $2^n m$ for some $n \geq 1$ and some odd $m > 1$. Set $\alpha = \sigma^{2^{n-1}m}$. As α is an involution, α fixes exactly one point P of U and $(P\sigma)\alpha = (P\alpha)\sigma = P\sigma$. Hence, α fixes $P\sigma$ from which it follows that $P\sigma = P$. If $Q \in U$ is fixed by σ, then, as α is a power of σ, Q is fixed by α and so $Q = P$. Thus, σ fixes exactly one point of U.

Let p be a prime divisor of m and set $\beta = \sigma^{2^n m/p}$. As β acts on the q^3 points of $U \setminus \{P\}$ and is of prime order p, which is relatively prime to q^3, and each orbit is of length p or 1, β fixes a point Q of $U \setminus \{P\}$. As β permutes the secant lines through P and the secant lines through Q, β also fixes the point R that is the intersection of the tangent lines through P and through Q. As $\beta(\sigma(Q)) = \sigma(\beta(Q) = \sigma(Q)$, β fixes $\sigma(Q)$. Thus β fixes the four distinct points P, Q, R, and $\sigma(Q)$. As β is not the identity mapping, it is an exercise to show that three of these points must be on the same line (Recall that β is a linear transformation on the 3-dimensional vector space V, points are 1-dimensional subspaces of V, and lines are 2-dimensional subspaces of V.): the only possibility for these three points is P, Q, and $\sigma(Q)$. It follows that σ fixes the secant line through P and Q. As σ also fixes the tangent line through P, it remains to show that σ fixes no other line through P. Assume to the contrary that σ fixes a third line through P. Then, it is an exercise to show that σ fixes all lines through P, and so, by Lemma 4.31, σ must fix all points on some line through P. This line must be the tangent line through P. As β is a power of σ, β also fixes all lines through P and all points on the tangent line through P. Hence, as β is not the identity mapping, by Lemma 4.31, β only fixes points on the tangent line through P, a contradiction from which the result follows. □

An important fact is that the Sylow 2-subgroups of $\mathrm{PSU}(3, q)$ intersect trivially.

Lemma 4.33. *If q is even, then $\mathrm{PSU}(3, q)$ has $q^3 + 1$ Sylow 2-subgroups and these subgroups intersect trivially.*

Proof. By Lemma 4.30, $S^* = \{A_{a,b}Z \mid a, b \in \mathrm{GF}(q^2), a^{q+1} + b^q + b = 0\}$ is a Sylow 2-subgroup of $\mathrm{PSU}(3, q)$. It is also easy to show that S^* fixes the point $\langle(0, 0, 1)\rangle$ of U. If g maps the point $\langle(0, 0, 1)\rangle$ of U to the point Q of U, then $Q^{-1}S^*Q$ is a Sylow 2 subgroup of $\mathrm{PSU}(3, q)$ that fixes Q. Hence $\mathrm{PSU}(3, q)$ has at least $q^3 + 1$ Sylow 2-subgroups. Now

$$(ST)^* = \{A_{a,b}B_cZ \mid a, b \in \mathrm{GF}(q^2), a^{q+1} + b^q + b = 0, c \in \mathrm{GF}(q^2), c \neq 0\}$$

is a subgroup of $\mathrm{PSU}(3, q)$ of index $q^3 + 1$. As S^* is a normal subgroup of $(ST)^*$, $\mathrm{PSU}(3, q)$ has at most $q^3 + 1$ Sylow 2-subgroups and so has exactly $q^3 + 1$ Sylow 2-subgroups. By Lemma 4.32, no nonidentity two-element can fix more than one point of U, and so these Sylow 2-subgroups intersect trivially. □

One class of subgroups will play an important role in our proof of the admissibility of $\mathrm{PSU}(3, q)$.

Lemma 4.34. *Let $q > 4$ be even and let $d = \gcd(q + 1, 3)$. There exist a set*

$$\mathcal{G} = \{G_{ij} \mid i = 1, \ldots, q^3 + 1, j = 1, \ldots, q^2\}$$

of subgroups of $\mathrm{PSU}(3, q)$ *satisfying the following:*

1. *$G_{ij} \cong \mathrm{GF}(q)^+ \times \mathbb{Z}_{(q+1)/d}$, for $i = 1, \ldots, q^3 + 1$, $j = 1, \ldots, q^2$.*
2. *If $C \in \mathrm{PSU}(3, q)$ is of even order that is not a 2-element, then $C \in G_{ij}$ for a unique G_{ij}.*

Proof. The subgroup $(ST)^*$ of $\mathrm{PSU}(3, q)$ is the stabilizer of point $\langle(0, 0, 1)\rangle$ of U and the subgroup of $(ST)^*$, which also fixes the line through $\langle(0, 0, 1)\rangle$ and $\langle(1, 0, 0)\rangle$, is $K = \{A_{a,b}B_cZ \in (ST)^* \mid a = 0\}$. Now $K \cong \mathrm{GF}(q)^+ \times \mathbb{Z}_{(q+1)/d}$. Let $\{P_1, \ldots, P_{q^3+1}\}$ be the points of U, and, for $i = 1, \ldots, q^3 + 1$, let $\{l_{i,j} \mid j = 1, \ldots, q^2\}$ be the secant lines through P_i. For $i = 1, \ldots, q^3 + 1$ and $j = 1, \ldots, q^2$, let $Q_{i,j}$ be a point on $l_{i,j}$ distinct from P_i. By Lemma 4.28, for $i = 1, \ldots, q^3 + 1$ and $j = 1, \ldots, q^2$, there exists $\alpha_{i,j} \in \mathrm{PSU}(3, q)$ that maps $\langle(0, 0, 1)\rangle$ to P_i and $\langle(1, 0, 0)\rangle$ to $Q_{i,j}$. Set $G_{i,j} = \alpha_{i,j}^{-1}K\alpha_{i,j}$. Then $G_{i,j}$ fixes P_i and $l_{i,j}$. By Lemma 4.32, any even-order element g of $\mathrm{PSU}(3, q)$ that is not a two-element fixes precisely one point P_i of U and precisely one secant line $l_{i,j}$ of π. Hence $g \in G_{i,j}$ for precisely one choice of i and j. □

We now have all the ingredients we need to prove the admissibility of PSU(3, q).

Theorem 4.35 (Di Vincenzo, 1989). *If q is even, then* PSU(3, q) *is admissible.*

Proof. For the smallest case, $q = 2$, it is known that PSU(3, 4) is solvable with its Sylow 2-subgroup isomorphic to the quaternion group of order 8: see [81], for example. It follows from Theorem 4.22 that PSU(3, 4) is admissible.

Let us assume that $q > 2$. Set $\mathscr{E} = \{A \in \mathrm{PSU}(3, q) \mid |A| \text{ even, not a power of } 2\}$, $\mathscr{T} = \{A \in \mathrm{PSU}(3, q) \mid |A| \text{even, a power of } 2\}$, and $\mathscr{O} = \{A \in \mathrm{PSU}(3, q) \mid |A| \text{ odd}\}$. Clearly $\mathscr{E}$, $\mathscr{T}$, and $\mathscr{O}$ partition the element set of PSU(3, q).

Let $\{S_1, \ldots, S_{q^3+1}\}$ be the Sylow 2-subgroups of PSU(3, q). By Lemma 4.33, $S_i \cap S_j = \{1\}$ if $i \neq j$ and so $\mathscr{T}$ is partitioned by $\{S_i \setminus \{1\} \mid i = 1, \ldots, t\}$. By Lemma 4.30, each S_i is noncyclic. It follows from Theorem 4.20 that each S_i admits a normalized complete mapping θ_i. By Lemma 4.34, the sets $\{G_{ij} \cap \mathscr{E} \mid i = 1, \ldots, q^3 + 1, j = 1, \ldots, q^2\}$ partitions $\mathscr{E}$.

By Lemma 4.34, each $G_{ij} \cong \mathrm{GF}(q)^+ \times \mathbb{Z}_{(q+1)/d}$, where $d = \gcd(q + 1, 3)$, and hence

$$G_{ij} \cap \mathscr{E} \cong \{(a, b) \mid a \in \mathrm{GF}(q), b \in \mathbb{Z}_{(q+1)/d}, a, b \neq 0\}.$$

Note that $\mathrm{GF}(q)^+$ admits a normalized complete mapping α by Lemma 4.30, and $\mathbb{Z}_{(q+1)/d}$ admits a normalized complete mapping β by Corollary 3.6, and, hence, $\alpha \times \beta\colon (a, b) \mapsto (\alpha(a), \beta(b))$ is a normalized complete mapping of $\mathrm{GF}(q)^+ \times \mathbb{Z}_{(q+1)/d}$. It follows that G_{ij} admits a normalized complete mapping ϕ_{ij} for $i = 1, \ldots, q^3+1$, $j = 1, \ldots, q^2$. Let $\iota\colon \mathscr{O} \to \mathscr{O}$ be the identity mapping.

Define $\theta'\colon \mathrm{PSU}(3, q) \to \mathrm{PSU}(3, q)$ by

$$\theta'(g) = \begin{cases} \theta_i(g) & \text{if } g \in S_i,\ i = 1, \ldots, q^3 + 1, \\ \phi_{ij}(g) & \text{if } g \in G_{ij},\ i = 1, \ldots, q^3 + 1,\ j = 1, \ldots, q^2, \\ \iota(g) & \text{if } g \in \mathscr{O}. \end{cases}$$

It is routine to show that θ' is a complete mapping of PSU(3, q). □

The admissibility of SU(3, q) for q even is an easy corollary of Theorem 4.35.

Corollary 4.36. *If q is even, then* SU(3, q) *is admissible.*

Proof. The group SU(3, q) has a normal subgroup H of odd order, $\gcd(3, q + 1)$, for which $\mathrm{SU}(3, q)/H \cong \mathrm{PSU}(3, q)$. As H is admissible by Corollary 3.6 and PSU(3, q) is admissible by Theorem 4.35, SU(3, q) is admissible by Corollary 1.38. □

4.4.2 Groups with Trivially Intersecting Sylow 2-Subgroups

In 1955 Hall and Paige [171] proved that, if the Sylow 2-subgroup S of a finite group G is noncyclic and G has $\{G : S\}$ distinct Sylow 2-subgroups, which intersect pairwise trivially, then G is admissible (see Corollary 4.11). In 1989, in the course of removing the condition on the number of distinct Sylow 2-subgroups, Di Vincenzo [96] proved Theorem 4.27 on the admissibility of Suzuki groups and Theorem 4.35 on the admissibility of the projective special unitary group $\mathrm{PSU}(3, q)$, q even. In this section we will use these results to extend Corollary 4.11 to cover any group whose Sylow 2-subgroups are noncyclic and intersect trivially.

Theorem 4.37 (Hering, 1972). *Let G be a finite group and let H be a subgroup of G satisfying the following:*

i) $|H|$ *is even,*
ii) $N_G(H) \neq G$, *and*
iii) $N_G(H) \cap H^g = \{1\}$ *for all* $g \in G \setminus N_G(H)$.

If K is the smallest normal subgroup of G that contains H, then one of the following holds:

1. $K = H \cdot O(K)$ *and H is a Frobenius complement.*
2. $K \equiv \mathrm{SL}(2, q)$, *$q$ even,* $q \geq 4$.
3. $K \equiv \mathrm{Sz}(q)$, $q = 2^{2n+1}$, $n \geq 1$.
4. $K \equiv \mathrm{SU}(3, q)$, *$q$ even.*
5. $K \equiv \mathrm{PSU}(3, q)$, *$q$ even.*

Proof. See Theorem 1 in [177]. □

With Hering's classification, Di Vincenzo was able to improve on Corollary 4.11 by removing the condition on the number of Sylow 2-subgroups.

Theorem 4.38 (Di Vincenzo, 1989). *Let G be a finite group with noncyclic Sylow 2-subgroups. If distinct Sylow 2-subgroups intersect trivially, then G is admissible.*

Proof. Let S be a Sylow 2-subgroup of G. If S is normal in G, then S is admissible by Theorem 4.20, and G/S, being of odd order, is admissible by Corollary 3.6. It follows that G is admissible by Corollary 1.38.

Next, assume that S is not normal in G, and let H be a normal subgroup of G that contains S but contains no proper subgroup that is both normal in G and contains S. If $g \in G \setminus N_G(H)$, then $S \neq S^g$, and, if $s \in N_G(S) \cap S^g$, then, as $s \in S^g$, s must be a 2-element, and, as S is a normal subgroup of odd index in $N_G(S)$, $s \in S$, from which it follows that $s = 1$. Thus, by Theorem 4.37 there are four cases to consider.

1. $H \equiv KS$, K an odd-order subgroup of G. By Theorem 4.20, S is admissible, and by Corollary 3.6, K is admissible. Hence, by Corollary 1.38, H is admissible.
2. $H \equiv \mathrm{SL}(2, q)$, q even, $q \neq 2$. The admissibility of H will be proved in Theorem 5.5.

3. $H \equiv \mathrm{Sz}(q), q = 2^{2n+1}, n \geq 1$. H is admissible by Theorem 4.27.
4. $H \equiv \mathrm{PSU}(3, q), q$ even. H is admissible by Theorem 4.35.

In each case H is admissible, and, as G/H is of odd order, G/H is admissible by Corollary 3.6. It follows, from Corollary 1.38, that G is admissible. □

Chapter 5
The Groups $\mathrm{GL}(n, q)$, $\mathrm{SL}(n, q)$, $\mathrm{PGL}(n, q)$, and $\mathrm{PSL}(n, q)$

In this chapter we gather together admissibility proofs for the general linear group $\mathrm{GL}(n, q)$, the special linear group $\mathrm{SL}(n, q)$, the projective general linear group $\mathrm{PGL}(n, q)$, and the projective special linear group $\mathrm{PSL}(n, q)$.

Recall that a group is admissible if it admits complete mappings. In this chapter we will show that the groups $\mathrm{GL}(n, q)$, $\mathrm{SL}(n, q)$, $\mathrm{PGL}(n, q)$, and $\mathrm{PSL}(n, q)$ are admissible when q is even, except when n and q are both 2. In this exceptional case, each of these groups is isomorphic to the nonabelian group of order 6, a group with a nontrivial, cyclic Sylow 2-subgroup, which therefore is not admissible by Theorem 3.34. In this chapter we will also prove the admissibility of $\mathrm{GL}(n, q)$, $\mathrm{SL}(2, q)$, $\mathrm{PSL}(2, q)$, and $\mathrm{PGL}(n, q)$ when q is odd.

The group $\mathrm{SL}(2, q)$, $q \neq 2$, will be shown to be admissible in Section 5.1. For q even, complete mappings of $\mathrm{SL}(2, q)$ will be constructed using partitions of the element set of the group and, for q odd, using partitions of a dual system of coset representatives.

In Section 5.2 we will describe constructions of complete mappings of $\mathrm{PSL}(2, q)$ when q is odd; in Section 5.3 constructions of complete mappings of $\mathrm{SL}(n, q)$, $\mathrm{GL}(n, q)$, $\mathrm{PGL}(n, q)$, and $\mathrm{PSL}(n, q)$ when q is even; and in Section 5.4 constructions of complete mappings of $\mathrm{GL}(n, q)$ and $\mathrm{PGL}(n, q)$ when q is odd. These constructions yield proofs of special cases of the Hall-Paige conjecture that a finite group with a trivial or noncyclic Sylow 2-subgroup is admissible.

5.1 Complete Mappings of $\mathrm{SL}(2, q)$

A proof of the admissibility of $\mathrm{SL}(2, q)$, when q is even but not 2, was given by Saeli [288] in 1989. He obtained his proof by partitioning the element set of the group and constructing complete mappings for each set in the partition. In the same paper, he proved the admissibility of $\mathrm{SL}(2, q)$ when $q = 5$, 7, and

A. B. Evans, *Orthogonal Latin Squares Based on Groups*, Developments in Mathematics 57, https://doi.org/10.1007/978-3-319-94430-2_5

11 using factorizations of each of these groups. The proof of the admissibility of SL$(2, q)$ for odd q was given by Evans [123, 125] in two papers, the case $q \equiv 1 \pmod 4$ in 2001 and the case $q \equiv 3 \pmod 4$ in 2005. In each of these proofs, an admissible subgroup was constructed, a dual system of coset representatives was partitioned, and a complete mapping was constructed for each set in this partition. Thus partitions of the element set of a group and partitions of dual systems of coset representatives play an important role in admissibility proofs for SL$(2, q)$.

5.1.1 Partitions and Dual Systems of Coset Representatives

In constructing complete mappings of SL$(2, q)$, we will need to use partitions both of the element set of a group and of dual systems of coset representatives. Complete mappings of a group or dual systems of coset representatives can then be stitched together from complete mappings of subsets of a group or subsets of a dual system of coset representatives. To do this we first need to extend our definition of complete mappings of groups to complete mappings of subsets of groups and our definition of complete mappings of dual systems of coset representatives to complete mappings of subsets of dual systems of coset representatives.

If X is a subset of the element set of a group and $\theta\colon X \to X$ is a bijection, then θ is a *complete mapping* of X if the mapping $x \mapsto x\theta(x)$ is also a bijection $X \to X$. If $X_1, \ldots, X_r$ partition the element set of a group G, then complete mappings of G can be constructed from complete mappings of the X_is.

Lemma 5.1. *Let $X_1, \ldots, X_r$ partition the element set of a group G. If each X_i admits complete mappings, then G is admissible.*

Proof. For $i = 1, \ldots, r$, let θ_i be a complete mapping of X_i. Define $\theta\colon G \to G$ by $\theta(x) = \theta_i(x)$ if $x \in X_i$. It is routine to show that θ is a complete mapping of G. □

We will use Lemma 5.1 to construct complete mappings of SL$(2, q)$ when q is even.

Let H be a subgroup of a finite group G, and let D be a dual system of coset representatives for H in G. Recall from Section 4.1 that a bijection $\theta\colon D \to D$ is called a *complete mapping* of D if there exists a bijection $\eta\colon D \to D$ such that $x\theta(x)H = \eta(x)H$ for all $x \in D$, and we called D *admissible* if D admitted complete mappings. By Theorem 4.1, if H is admissible and D is admissible, then G is admissible.

A set X of elements of G will be called a *partial system of left coset representatives* for H in G if, for all $x, y \in X$, $xH = yH$ implies $x = y$. A partial system of left coset representatives, X, for H in G, is a system of left coset representatives for H in G if and only if $G = XH$. Partial systems of right coset representatives and partial dual systems of coset representatives are defined similarly. For X a partial system of left coset representatives for H in G, a bijection $\theta\colon X \to X$ will be called a *complete mapping* of X if, for each $x \in X$, there is exactly one element $\eta(x) \in X$ such that $x\theta(x)H = \eta(x)H$ and the mapping $\eta\colon X \to X$ is a bijection. Complete mappings of partial systems of right coset representatives and partial dual systems of coset representatives are defined similarly. As in Lemma 5.1, if $X_1, \ldots, X_r$ partitions D, a dual system of coset representatives for H in G, then complete mappings of D can be constructed from complete mappings of the X_is.

Lemma 5.2. *Let H be a subgroup of a finite group G, and let D be a dual system of coset representatives for H in G. If $X_1, \ldots, X_r$ partitions D, and each X_i admits complete mappings, then D is admissible.*

Proof. Similar to the proof of Lemma 5.1. □

We will use Lemma 5.2 to construct complete mappings of SL(2, q) and GL(n, q) when q is odd.

5.1.2 SL(2, q), q Even

The smallest of these groups, SL(2, 2), is isomorphic to S_3 and, hence, does not admit complete mappings. However, if $q > 2$ is even, then SL(2, q) admits complete mappings. This was proved by Saeli [288] in 1989. Saeli's proof is based on the distribution of the 2-elements in SL(2, q). To present Saeli's proof, we need to first establish some properties of the Sylow 2-subgroups of SL(2, q). The subgroups of PSL(2, q) are described in [191]. Note that PSL(2, q) $\cong$ SL(2, q) when q is even.

Lemma 5.3. *If q is even, then the following hold:*

1. *Each Sylow* 2*-subgroup of* SL(2, q) *is isomorphic to* GF$(q)^+$.
2. *There are $q + 1$ Sylow* 2*-subgroups of* SL(2, q).
3. *Distinct Sylow* 2*-subgroups of* SL(2, q) *intersect trivially.*
4. *If m is the order of a cyclic subgroup of* SL(2, q)*, then either $m = 2$, or m divides $q^2 - 1$.*

Proof. See Satz 8.2 and Satz 8.5 in [191]. □

As an immediate corollary, we see that any element of even order in SL(2, q), q even, is contained in a Sylow 2-subgroup of SL(2, q).

Corollary 5.4. *If q is even, then any element of* SL(2, q) *that is not a* 2*-element is of odd order.*

Proof. The order of a nonidentity element is either 2, making it a 2-element, or a divisor of $q^2 - 1$, which is odd. □

We are now in a position to prove the existence of complete mappings of SL(2, q), q even, $q > 2$, by partitioning the element set of SL(2, q).

Theorem 5.5 (Saeli, 1989). *If q is even, $q \neq 2$, then* SL(2, q) *is admissible.*

Proof. Let $G = \mathrm{SL}(2, q)$, q even, $q > 2$. By Lemma 5.3(2), G has $q + 1$ distinct Sylow 2-subgroups, $S_1, \ldots, S_{q+1}$. For $i = 1, \ldots, q + 1$, set $H_i = S_i \setminus \{I\}$. By Lemma 5.3(3), $H_i \cap H_j = \emptyset$ if $i \neq j$. Set $X = \{A \mid A \in G, A \neq I, A$ of odd order$\}$. By Lemma 5.3(4), any nonidentity element of G, that is not in H_i, for some i, is an element of X. Hence the sets $H_1, \ldots, H_{q+1}$, X, and $\{I\}$ partition the element set of G. We will show that each set in this partition admits complete mappings.

The mapping $I \mapsto I$ is a complete mapping of $\{I\}$. For $q > 2$, $\mathrm{GF}(q)^+$ admits normalized complete mappings, for example, $x \mapsto ax$ is a normalized complete mapping of $\mathrm{GF}(q)^+$ if $a \neq 0, 1$. By Lemma 5.3(1), each of $S_1, \ldots, S_{q+1}$ is isomorphic to $\mathrm{GF}(q)^+$, and so each of $S_1, \ldots, S_{q+1}$ admits normalized complete mappings. As these complete mappings fix I, each of $H_1, \ldots, H_{q+1}$ admits complete mappings.

For X, the identity mapping is a complete mapping. The proof of this is very similar to the proof of Theorem 3.5. Clearly $\iota\colon X \to X$, defined by $\iota(x) = x$, is a bijection. To see that $x \mapsto x\iota(x)$ is a bijection, let $x, y \in X$. If $x\iota(x) = y\iota(y)$, then $x^2 = y^2$. Let m be the lowest common multiple of the orders of x and y. Then m is odd and

$$x = (x^2)^{(m+1)/2} = (y^2)^{(m+1)/2} = y.$$

Thus each set in the partition of the element set of G admits complete mappings. The result then follows from Lemma 5.1. □

5.1.3 SL(2, q), $q = 5$, 7, *and* 11

In 1989, Saeli [288] proved that SL(2, 5), SL(2, 7), and SL(2, 11) are admissible. Saeli's proof is based on Theorem 4.8. Recall that a group G is factorable if G has nontrivial subgroups A and B, such that $G = AB$ and $A \cap B = \{1\}$. By Theorem 4.8, if A and B are admissible, then G is admissible. Lemma 4.7 gave us one way to determine the factorization of a group: given two subgroups A and B of a finite group G, if $|G| = |A||B|$ and $\gcd(|A|, |B|) = 1$, then $A \cap B = \{1\}$ and $G = AB$. Using Lemma 4.7 and some knowledge of the subgroups of SL(2, 5), SL(2, 7), and SL(2, 11), we can prove that SL(2, 5), SL(2, 7), and SL(2, 11) are admissible. The subgroups of SL(2, q) were first determined by Dickson [98] in 1901: they are also described in [306] . The facts about subgroups of SL(2, q) that we will use can also be verified using magma [47] or GAP [153].

Lemma 5.6. SL(2, 3) *and* GL(2, 3) *are admissible.*

Proof. SL(2, 3) is solvable and, as the matrices

$$\begin{pmatrix} 0 & 1 \\ 2 & 0 \end{pmatrix} \text{ and } \begin{pmatrix} 1 & 1 \\ 1 & 2 \end{pmatrix}$$

generate a subgroup of SL(2, 3) that is isomorphic to the quaternion group of order 8, its Sylow 2-subgroup, which is of order 8, is noncyclic. Hence, by Theorem 4.22, SL(2, 3) is admissible. As GL(2, 3) is also solvable and its Sylow 2-subgroup is also noncyclic, containing the Sylow 2-subgroup of SL(2, 3), GL(2, 3) is admissible by Theorem 4.22. □

Theorem 5.7 (Saeli, 1989). *The groups* $SL(2, 5)$, $SL(2, 7)$, *and* $SL(2, 11)$ *are admissible.*

Proof. If p is a prime, then

$$H = \left\{ \begin{pmatrix} 1 & x \\ 0 & 1 \end{pmatrix} \mid x \in GF(p) \right\}$$

is a cyclic subgroup of SL(2, p) of order p.

Now SL(2, 5) contains a subgroup $A \cong SL(2, 3)$ and, as noted, a subgroup $B \cong \mathbb{Z}_5$. As $|A| = 24$, $|B| = 5$, $\gcd(24, 5) = 1$, and $|SL(2, 5)| = 120$, by Lemma 4.7, $SL(2, 5) = AB$, and $A \cap B = \{1\}$. Now $\mathbb{Z}_5$ is admissible by Corollary 3.2, and SL(2, 3) is admissible by Lemma 5.6, and so SL(2, 5) is admissible by Theorem 4.8. The proofs for SL(2, 7) and SL(2, 11) are similar. SL(2, 7) can be written in the form $SL(2, 7) = AB$, where A and B are subgroups of SL(2, 7), $A \cap B = \{1\}$, $A \cong GL(2, 3)$, and $B \cong \mathbb{Z}_7$, and SL(2, 11) can be written in the form $SL(2, 11) = AB$, where A and B are subgroups of SL(2, 11), $A \cap B = \{1\}$, $A \cong SL(2, 5)$, and $B \cong \mathbb{Z}_{11}$. □

5.1.4 SL(2, q), $q \equiv 1 \pmod 4$

For q odd, constructions of complete mappings for SL(2, q) were given by Evans [123, 125]. The case $q \equiv 1 \pmod 4$ was settled in 2001 and the case $q \equiv 3 \pmod 4$ in 2005. In each of these cases, a subgroup H of SL(2, q), that admits complete mappings, was determined, a dual system of coset representatives, D, for H in SL(2, q) was constructed, and, piece by piece, a complete mapping of D was constructed. The fact that SL(2, q) admits complete mappings then followed from Lemma 5.2.

We will first deal with the case $q \equiv 1 \pmod 4$. Let $q \equiv 1 \pmod 4$ and let $G = SL(2, q)$. For $a, b, x, y \in GF(q)$, $a, b \neq 0$, let

$$A_a = \begin{pmatrix} a & 0 \\ 0 & 1/a \end{pmatrix}, \quad B_b = \begin{pmatrix} 0 & b \\ -1/b & 0 \end{pmatrix}, \quad \text{and} \quad D_{x,y} = \begin{pmatrix} 1 & x \\ y & 1+xy \end{pmatrix}.$$

The set $H = \{A_a \mid a \neq 0\} \cup \{B_b \mid b \neq 0\}$ is a subgroup of G that is admissible.

Lemma 5.8. *H is a subgroup of G, and H is admissible.*

Proof. We first note that $A_a A_b = A_{ab}$, $A_a B_b = B_{ab}$, $B_a A_b = B_{a/b}$, and $B_a B_b = A_{-a/b}$. It follows that H is closed under multiplication and hence is a subgroup of G.

As $|H| = 2(q-1)$, and $\{A_a \mid a \neq 0\}$ is a cyclic normal subgroup of H of order $q-1$, H is solvable. Further $\langle A_{-1}, B_{-1}\rangle \cong \mathbb{Z}_2 \times \mathbb{Z}_2$ is a subgroup of H. This implies that the Sylow 2-subgroup of H is noncyclic, and so H is admissible by Theorem 4.22. □

We now turn to the construction of a dual system of coset representatives for H in G. The mapping $x \mapsto -(1+x)$ is an involutory permutation of GF(q) that fixes only $(q-1)/2$. Let $P \subseteq \mathrm{GF}(q)$, $-1, (q-1)/2 \notin P$, have the property that each orbit of $x \mapsto -(1+x)$, of length 2, intersects P in exactly one element. We will assume that $0 \in P$ and, hence, $-1 \notin P$ as $\{0, -1\}$ is an orbit of $x \mapsto -(1+x)$. Similarly, the mapping $x \mapsto -x$ is an involutory permutation of GF(q) that fixes only 0. Let $T \subseteq GF(q)$, $0 \notin T$, have the property that each orbit of $x \mapsto -x$, of length 2, intersects T in exactly one element. As an example, if q is prime, then the orbits of $x \mapsto -(1+x)$ are

$$\{\{0, q-1\}, \{1, q-2\}, \ldots, \{(q-3)/2, (q+1)/2\}, \{(q-1)/2\}\}$$

and we could then choose P to be $\{0, 1, \ldots, (q-3)/2\}$. Similarly the orbits of $x \mapsto -x$ are

$$\{\{0\}, \{1, q-1\}, \ldots, \{(q-1)/2, (q+1)/2\}\}$$

and we could choose T to be $\{1, 2, \ldots, (q-1)/2\}$. In what follows we will only need to specify our choice of P for small values of q, and we will never need to specify our choice of T.

We define the set

$$D = \{D_{x,y} \mid xy \in P\} \cup \{D_{x,y} \mid xy = (q-1)/2, x \in T)\}.$$

The set D is a dual system of coset representatives for H in G.

Lemma 5.9. *D is a dual system of coset representatives for H in G.*

Proof. If $D_{x,y}, D_{w,z} \in D$ satisfy $D_{x,y}D_{w,z}^{-1} \in H$, then $D_{x,y} = A_a D_{w,z}$ for some $a \neq 0$, or $D_{x,y} = B_b D_{w,z}$ for some $b \neq 0$. If $D_{x,y} = A_a D_{w,z}$ for some $a \neq 0$, then

$$\begin{pmatrix} 1 & x \\ y & 1+xy \end{pmatrix} = \begin{pmatrix} a & 0 \\ 0 & 1/a \end{pmatrix} \begin{pmatrix} 1 & w \\ z & 1+wz \end{pmatrix} = \begin{pmatrix} a & aw \\ z/a & (1+wz)/a \end{pmatrix},$$

from which it follows that $a = 1$, $w = x$, and $z = y$. If $D_{x,y} = B_b D_{w,z}$ for some $b \neq 0$, then

$$\begin{pmatrix} 1 & x \\ y & 1+xy \end{pmatrix} = \begin{pmatrix} 0 & b \\ -1/b & 0 \end{pmatrix} \begin{pmatrix} 1 & w \\ z & 1+wz \end{pmatrix} = \begin{pmatrix} bz & b(1+wz) \\ -1/b & -w/b \end{pmatrix},$$

from which it follows that $z = 1/b = -y$, and $wz = -(1+xy)$. If $xy \in P$, then $wz \notin P$ and $wz \neq (q-1)/2$, from which it follows that $D_{w,z} \notin D$, a contradiction. If $xy = (q-1)/2$ and $x \in T$, then $wz = (q-1)/2$ and $w = (q-1)/(-2y) = -x \notin T$, from which it follows that $D_{w,z} \notin D$, a contradiction. Hence $D_{x,y}D_{w,z}^{-1} \in H$ if and only if $x = w$ and $y = z$ and D is a partial system of right coset representatives for H in G.

If $D_{x,y}, D_{w,z} \in D$ satisfy $D_{x,y}^{-1}D_{w,z} \in H$, then $D_{x,y}A_a = D_{w,z}$ for some $a \neq 0$, or $D_{x,y}B_b = D_{w,z}$ for some $b \neq 0$. If $D_{x,y}A_a = D_{w,z}$ for some $a \neq 0$, then

$$\begin{pmatrix} 1 & w \\ z & 1+wz \end{pmatrix} = \begin{pmatrix} 1 & x \\ y & 1+xy \end{pmatrix} \begin{pmatrix} a & 0 \\ 0 & 1/a \end{pmatrix} = \begin{pmatrix} a & x/a \\ ay & (1+xy)/a \end{pmatrix},$$

from which it follows that $a = 1$, $x = w$, and $y = z$. If $D_{x,y}B_b = D_{w,z}$ for some $b \neq 0$, then

$$\begin{pmatrix} 1 & w \\ z & 1+wz \end{pmatrix} = \begin{pmatrix} 1 & x \\ y & 1+xy \end{pmatrix} \begin{pmatrix} 0 & b \\ -1/b & 0 \end{pmatrix} = \begin{pmatrix} -x/b & b \\ -(1+xy)/b & yb \end{pmatrix},$$

from which it follows that $w = b = -x$, and $wz = -(1+xy)$. If $xy \in P$, then $wz \notin P$ and $wz \neq (q-1)/2$, from which it follows that $D_{w,z} \notin D$, a contradiction. If $xy = (q-1)/2$ and $x \in T$, then $wz = (q-1)/2$ and $w = -x \notin T$, from which it follows that $D_{w,z} \notin D$, a contradiction. Hence $D_{x,y}^{-1}D_{w,z} \in H$ if and only if $x = w$ and $y = z$ and D is a partial system of left coset representatives for H in G, and, as D is also a partial system of right coset representatives for H in G, D is a partial dual system of coset representatives for H in G.

As D is a partial dual system of coset representatives for H in G, D is a dual system of coset representatives for H in G if and only if $|D| = |G : H|$. Now $|G| = (q^2-1)(q^2-q)/(q-1)$ and $|H| = 2(q-1)$. Hence $|G : H| = q(q+1)/2$. Now D can be partitioned into three subsets $D_0 = \{D_{x,y} \mid xy = 0\}$, $D_T = \{D_{x,y} \mid xy = (q-1)/2, x \in T\}$, and $D_P = \{D_{x,y} \mid xy \in P \setminus \{0\}\}$. There are $2q-1$ pairs $(x, y) \in \mathrm{GF}(q) \times \mathrm{GF}(q)$ that satisfy $xy = 0$, and so $|D_0| = 2q-1$. As $|P \setminus \{0\}| = (q-3)/2$, for each of the $q-1$ choices of $x \in \mathrm{GF}(q) \setminus \{0\}$ there exist precisely $(q-3)/2$ choices of $y \in \mathrm{GF}(q)$ for which $xy \in P \setminus \{0\}$. Hence $|D_P| = (q-1)(q-3)/2$. For each $x \in T$, there is precisely one $y \in \mathrm{GF}(q)$ for which $xy = (q-1)/2$. Hence, as $|T| = (q-1)/2$, $|D_T| = (q-1)/2$. It follows that $|D| = (2q-1) + (q-1)(q-3)/2 + (q-1)/2 = q(q+1)/2$. As $|D| = |G : H|$ the result follows. □

In Lemma 5.9 we partitioned D into three subsets, $D_0 = \{D_{x,y} \mid xy = 0\}$, $D_T = \{D_{x,y} \mid xy = -1/2, x \in T\}$, and $D_P = \{D_{x,y} \mid xy \in P \setminus \{0\}\}$. If we can show that each of D_0, D_P, and D_T admits complete mappings, then we will have shown, by Lemma 5.2, that D is admissible. We will first construct complete mappings of D_0 and D_T.

Lemma 5.10. *D_0 admits complete mappings and D_T admits complete mappings.*

Proof. Let $\theta_0 \colon D_0 \to D_0$ be the identity mapping, and let $\eta_0 \colon D_0 \to D_0$ be defined by $\eta_0(D_{x,y}) = D_{2x,2y}$. Clearly θ_0 is a bijection. As $D_{x,y} = \eta_0(D_{x/2,y/2})$ and $xy = 0$ if and only if $(x/2)(y/2) = 0$, the mapping η_0 is onto and hence is a bijection. Now, if $xy = 0$, then

$$D_{x,y}\theta_0(D_{x,y}) = \begin{pmatrix} 1 & x \\ y & 1 \end{pmatrix}^2 = \begin{pmatrix} 1 & 2x \\ 2y & 1 \end{pmatrix} = \eta_0(D_{x,y}).$$

Hence θ_0 is a complete mapping of D_0.

Let $i^2 = -1$ and define $\theta_T, \eta_T \colon D_T \to D_T$ by

$$\theta_T\left(D_{x,-1/(2x)}\right) = \begin{cases} D_{xi/2,-1/(xi)} & \text{if } xi/2 \in T \\ D_{-xi/2,1/(xi)} & \text{if } xi/2 \notin T, \end{cases}$$

and

$$\eta_T\left(D_{x,-1/(2x)}\right) = \begin{cases} D_{xi,-1/(2xi)} & \text{if } xi \in T \\ D_{-xi,1/(2xi)} & \text{if } xi \notin T. \end{cases}$$

As the mapping

$$x \mapsto \begin{cases} xi/2 & \text{if } xi/2 \in T \\ -xi/2 & \text{if } xi/2 \notin T \end{cases}$$

is a bijection $T \to T$, the mapping θ_T is a bijection $D_T \to D_T$, and as the mapping

$$x \mapsto \begin{cases} xi & \text{if } xi \in T \\ -xi & \text{if } xi \notin T \end{cases}$$

is a bijection $T \to T$, the mapping $\eta_T \colon D_T \to D_T$ is also a bijection $D_T \to D_T$.

For $x \in T$

$$D_{x,-1/(2x)}\theta_T\left(D_{x,-1/(2x)}\right) = \begin{cases} D_{xi,-1/(2xi)}A_{1+i} = D_{-xi,1/(2xi)}B_{x(1+i)/2} & \text{if } xi/2 \in T \\ D_{-xi,1/(2xi)}A_{1-i} = D_{xi,-1/(2xi)}B_{x(1-i)/2} & \text{if } xi/2 \notin T. \end{cases}$$

It follows that $D_{x,-1/(2x)}\theta_T\left(D_{x,-1/(2x)}\right)H = \eta_T\left(D_{x,-1/(2x)}\right)H$, and so θ_T is a complete mapping of D_T. □

To complete the proof that D is admissible, we need to construct complete mappings of D_P. To achieve this end, we define two symmetric polynomials over $GF(q)$.

- $\mathrm{p}_1(a,b,c) = 1+2(a+b+c)+2(ab+ac+bc)+(a^2+b^2+c^2)+4abc$, and
- $\mathrm{p}_2(a,b,c) = (a^2+b^2+c^2)-2(ab+ac+bc)-4abc$.

These polynomials play a crucial role in determining the existence of complete mappings of D_P.

Lemma 5.11. *If there exist bijections* $\theta, \eta\colon P\setminus\{0\} \to P\setminus\{0\}$ *such that, for each* $x \in P\setminus\{0\}$, *at least one of* $\mathrm{p}_1(x,\theta(x),\eta(x))$ *or* $\mathrm{p}_2(x,\theta(x),\eta(x))$ *is a square in* GF(q), *then* D_P *admits complete mappings.*

Proof. Let $\theta, \eta\colon P\setminus\{0\} \to P\setminus\{0\}$ be bijections such that, for each $x \in P\setminus\{0\}$, at least one of $\mathrm{p}_1(x,\theta(x),\eta(x))$ or $\mathrm{p}_2(x,\theta(x),\eta(x))$ is a square in GF(q).

For $a \in P\setminus\{0\}$, set $D_a = \{D_{x,y} \mid xy = a\}$. Then the sets D_a, $a \in P\setminus\{0\}$, partition D_P. We will use this partition to construct complete mappings for D_P. For $a \in P\setminus\{0\}$, define polynomials $\mathrm{f}_{a,1}$ and $\mathrm{f}_{a,2}$ by

$$\mathrm{f}_{a,1}(x) = \eta(a)(1+\eta(a))x^2 - (1+a+\theta(a)+\eta(a)+2a\eta(a))x + a(1+a)$$

and

$$\mathrm{f}_{a,2}(x) = \eta(a)(1+\eta(a))x^2 - (a-\theta(a)+\eta(a)+2a\eta(a))x + a(1+a).$$

Now $\mathrm{p}_1(a,\theta(a),\eta(a))$ is the discriminant for $\mathrm{f}_{a,1}$, and $\mathrm{p}_2(a,\theta(a),\eta(a))$ is the discriminant for $\mathrm{f}_{a,2}$. Hence, as at least one of $\mathrm{p}_1(a,\theta(a),\eta(a))$, $\mathrm{p}_2(a,\theta(a),\eta(a))$ is a square, at least one of $\mathrm{f}_{1,a}$, $\mathrm{f}_{2,a}$ has a root in GF(q). Let α be a root of $\mathrm{f}_{1,a}$ if $\mathrm{p}_1(a,\theta(a),\eta(a))$ is a square or a root of $\mathrm{f}_{2,a}$ otherwise: thus $\alpha \in$ GF(q). As $a \in P\setminus\{0\}$ and $-1 \notin P\setminus\{0\}$, $\mathrm{f}_{a,1}(0) = \mathrm{f}_{a,2}(0) = a(1+a) \neq 0$, and so $\alpha \neq 0$. If α is a root of $\mathrm{f}_{1,a}$, then $\eta(a)-\theta(a)-(1+a)/\alpha \neq 0$, as if $\eta(a)-\theta(a)-(1+a)/\alpha = 0$ then, -1 not being in $P\setminus\{0\}$, $(1+a)/\alpha \neq 0$, which implies that $\eta(a)-\theta(a) \neq 0$, $\alpha = (1+a)/(\eta(a)-\theta(a))$, and

$$\mathrm{f}_{a,1}(\alpha) = \frac{\theta(a)(1+\theta(a))(1+a)^2)}{(\eta(a)-\theta(a))^2)} \neq 0:$$

a contradiction. Similarly, if α is a root of $\mathrm{f}_{2,a}$, then $-1-\eta(a)-\theta(a)+(1+a)/\alpha \neq 0$, as if $-1-\eta(a)-\theta(a)+(1+a)/\alpha = 0$, then, -1 not being in $P\setminus\{0\}$, $(1+a)/\alpha \neq 0$, which implies that $-1-\eta(a)-\theta(a) \neq 0$, $\alpha = (1+a)/(1+\eta(a)+\theta(a))$, and

$$\mathrm{f}_{a,2}(\alpha) = \frac{\theta(a)(1+\theta(a))(1+a)^2}{(1+\eta(a)+\theta(a))^2} \neq 0:$$

a contradiction.

For $a \in P \setminus \{0\}$, we define two mappings $\theta_a : D_a \to D_{\theta(a)}$ and $\eta_a : D_a \to D_{\eta(a)}$ by

$$\theta_a\left(D_{x,a/x}\right) = \begin{cases} D_{(\eta(a)-\theta(a)-(1+a)\alpha)x,\frac{\theta(a)}{((\eta(a)-\theta(a)-(1+a)/\alpha)x)}} & \text{if } \mathrm{f}_{1,a}(\alpha)=0, \\ D_{(-1-\eta(a)-\theta(a)+(1+a)/\alpha)x,\frac{\theta(a)}{((-1-\eta(a)-\theta(a)+(1+a)/\alpha)x)}} & \text{otherwise,} \end{cases}$$

and

$$\eta_a\left(D_{x,a/x}\right) = D_{x/\alpha,\alpha\eta(a)/x}.$$

We next define $\theta_P, \eta_P : D_P \to D_P$ by

$$\theta_P\left(D_{x,a/x}\right) = \theta_a\left(D_{x,a/x}\right) \quad \text{and} \quad \eta_P\left(D_{x,a/x}\right) = \eta_a\left(D_{x,a/x}\right) \quad \text{for } a \in P \setminus \{0\}.$$

As $\eta(a) - \theta(a) - (1+a)/\alpha \neq 0$ if $\mathrm{f}_{1,a}(\alpha) = 0$ and $-1 - \eta(a) - \theta(a) + (1+a)/\alpha \neq 0$ if $\mathrm{f}_{2,a}(\alpha) = 0$, θ_a is a bijection for all $a \in P \setminus \{0\}$, and further, as $\theta : P \setminus \{0\} \to P \setminus \{0\}$ is a bijection, θ_P is a bijection. Also, as $\alpha \neq 0$, η_a is a bijection, and further, as $\eta : P \setminus \{0\} \to P \setminus \{0\}$ is a bijection, η_P is a bijection.

If $\mathrm{f}_{1,a}(\alpha) = 0$, then

$$D_{x,a/x}\theta_P\left(D_{x,a/x}\right) = \eta_P\left(D_{x,a/x}\right) B_{(\eta(a)+1-((1+a)/\alpha))x}.$$

To see that this equation is well-defined, we need to show that $\eta(a)+1-(1+a)/\alpha \neq 0$. This is the case as, if $\eta(a)+1-(1+a)/\alpha = 0$, then $\alpha = (1+a)/(1+\eta(a))$ and $\mathrm{f}_{1,a}(\alpha) = -(1+a)(1+\theta(a))/(1+\eta(a)) \neq 0$, a contradiction.

If $\mathrm{f}_{1,a}(\alpha) \neq 0$, then $\mathrm{f}_{2,a}(\alpha) = 0$ and

$$D_{x,a/x}\theta_P\left(D_{x,a/x}\right) = \eta_P\left(D_{x,a/x}\right) A_{(-1-\eta(a)+(1+a)/\alpha)/(-1-\eta(a)-\theta(a)+(1+a)/\alpha)}.$$

To see that this equation is well-defined, as we have already shown that $-1-\eta(a)-\theta(a)+(1+a)/\alpha \neq 0$, we need to show that $-1-\eta(a)+(1+a)/\alpha \neq 0$. This is the case as, if $-1-\eta(a)+(1+a)/\alpha = 0$, then $\alpha = (1+a)/(1+\eta(a))$ and $\mathrm{f}_{2,a}(\alpha) = (1+a)\theta(a)/(1+\eta(a)) \neq 0$, a contradiction.

We have thus shown that $D_{x,a/x}\theta_P(D_{x,a/x})H = \eta_P(D_{x,a/x})H$ and hence that θ_P is a complete mapping of D_P. □

As an application of Lemma 5.11, we will show that SL(2, q) is admissible for $q = 5, 9, 13, 17, 25$, or 29: these are the smallest values of $q \equiv 1 \pmod 4$. The smallest case, $q = 5$, was established in Theorem 5.7.

Lemma 5.12. SL(2, q) *is admissible if* $q = 5, 9, 13, 17, 25,$ *or* 29.

Proof. For $q = 5, 9, 13, 17, 25$, and 29, we can choose P and define mappings θ_P and η_P, satisfying the conditions of Lemma 5.11.

For the smallest case, $q = 5$, choose $P \setminus \{0\} = \{1\}$ and set $\theta_P(1) = \eta_P(1) = 1$. Then $\mathrm{p}_1(1, 1, 1) = 0$ a square.

In the remaining cases, we will define θ and η using tables. The last two rows of the tables, labeled p_1 and p_2, show the values of $\mathrm{p}_1(a, \theta(a), \eta(a))$ and $\mathrm{p}_2(a, \theta(a), \eta(a))$ for $a \in P \setminus \{0\}$, the squares being shown in brackets.

For $q = 9$, setting $\mathrm{GF}(9) = \{x+yi \mid i^2+1 = 0, x, y \in GF(3)\}$, choose $P\setminus\{0\} = \{i, 2i, 1+i\}$. The squares are 0, 1, 2, i, and $2i$. The mappings $\theta, \eta\colon P\setminus\{0\} \to P\setminus\{0\}$ are defined in the following table:

a	i	$2i$	$1+i$
$\theta(a)$	i	$2i$	$1+i$
$\eta(a)$	$1+i$	$2i$	i
p_1	$[2i]$	$1+i$	$[1]$
p_2	$2+2i$	$[2i]$	$2+2i$

For $q = 13$, choose $P \setminus \{0\} = \{1, 2, 3, 4, 5\}$. The squares are 0, 1, 3, 4, 9, 10, and 12. The mappings $\theta, \eta\colon P \setminus \{0\} \to P \setminus \{0\}$ are defined in the following table:

a	1	2	3	4	5
$\theta(a)$	1	2	3	4	5
$\eta(a)$	2	1	3	4	5
p_1	7	[0]	[0]	[9]	2
p_2	[1]	[3]	8	8	[10]

For $q = 17$, choose $P \setminus \{0\} = \{1, \ldots, 7\}$. The squares are 0, 1, 2, 4, 8, 9, 13, 15, and 16. The mappings $\theta, \eta\colon P \setminus \{0\} \to P \setminus \{0\}$ are defined in the following table:

a	1	2	3	4	5	6	7
$\theta(a)$	1	2	3	4	5	6	7
$\eta(a)$	7	2	3	4	5	6	1
p_1	[9]	[1]	[4]	[0]	[8]	[1]	10
p_2	10	11	[1]	[2]	3	14	[15]

For $q = 25$, setting $GF(25) = \{x + y\alpha \mid \alpha^2 - 2 = 0, x, y \in \mathrm{GF}(5)\}$, choose $P \setminus \{0\} = \{1, \alpha, 2\alpha, 3\alpha, 4\alpha, 1 + \alpha, 1 + 2\alpha, 1 + 3\alpha, 1 + 4\alpha, 2 + \alpha, 2 + 2\alpha\}$. The squares are 0, 1, 2, 3, 4, $1 + \alpha$, $1 + 4\alpha$, $2 + 2\alpha$, $2 + 3\alpha$, $3 + 2\alpha$, $3 + 3\alpha$, $4 + \alpha$, and $4 + 4\alpha$. The mappings $\theta, \eta\colon P \setminus \{0\} \to P \setminus \{0\}$ are defined in the following table:

a	1	α	2α	3α	4α	$1+\alpha$	$1+2\alpha$	$1+3\alpha$	$1+4\alpha$	$2+\alpha$	$2+2\alpha$
$\theta(a)$	1	α	2α	3α	4α	$1+\alpha$	$1+2\alpha$	$1+3\alpha$	$1+4\alpha$	$2+\alpha$	$2+2\alpha$
$\eta(a)$	1	α	2α	$1+\alpha$	4α	$2+2\alpha$	$1+2\alpha$	$1+3\alpha$	3α	$2+\alpha$	$1+4\alpha$
p_1	$[0]$	$[4+4\alpha]$	$3+\alpha$	$[4]$	$[4+\alpha]$	$[3]$	$3+\alpha$	$3+4\alpha$	$[3+2\alpha]$	$[2+3\alpha]$	$[3]$
p_2	$[3]$	$4+2\alpha$	$[1+\alpha]$	$[2+3\alpha]$	$4+3\alpha$	$[2+2\alpha]$	$[3]$	$[3]$	2α	$[2+2\alpha]$	$2+4\alpha$

For $q = 29$, choose $P \setminus \{0\} = \{1, \ldots, 13\}$. The squares are 0, 1, 4, 5, 6, 7, 9, 13, 16, 20, 22, 23. 24, 25, and 28. The mappings $\theta, \eta\colon P \setminus \{0\} \to P \setminus \{0\}$ are defined in the following table:

a	1	2	3	4	5	6	7	8	9	10	11	12	13
$\theta(a)$	1	2	3	4	5	6	7	8	9	10	11	12	13
$\eta(a)$	1	2	3	9	5	6	7	8	10	4	11	12	13
p_1	[20]	[23]	[5]	[1]	2	[7]	[0]	[5]	21	21	[13]	[16]	[6]
p_2	[22]	14	10	[28]	[5]	14	18	[22]	[9]	[25]	26	[22]	14

We have thus shown that for $q = 5, 9, 13, 17, 25$, and 29, the conditions of Lemma 5.11 are satisfied. Hence, D_P admits complete mappings. By Lemma 5.10, D_T and D_0 also admit complete mappings, and, hence, by Lemma 5.2, D is admissible. As H is admissible by Lemma 5.8, by Theorem 4.1, $\mathrm{SL}(2, q)$ is admissible for $q = 5, 9, 13, 17, 25$, and 29. □

We will use C_0 to denote the set of nonzero squares of $\mathrm{GF}(q)$, C_1 to denote the set of nonsquares of $\mathrm{GF}(q)$, and $\mathrm{cycl}_2(i, j)$ to denote the number of solutions to $x_i+1 = x_j$, $x_i \in C_i$, $x_j \in C_j$. The numbers $\mathrm{cycl}_2(i, j)$ are the *cyclotomic numbers* of *index* 2: cyclotomic numbers will be discussed in Section 10.1. Cyclotomic numbers of index 2 have been determined.

Lemma 5.13. *If $q \equiv 3 \pmod 4$, then*

$$\mathrm{cycl}_2(0, 1) = \frac{q+1}{4}, \quad \textit{and} \quad \mathrm{cycl}_2(0, 0) = \mathrm{cycl}_2(1, 0) = \mathrm{cycl}_2(1, 1) = \frac{q-3}{4}.$$

If $q \equiv 1 \pmod 4$, then

$$\mathrm{cycl}_2(0, 0) = \frac{q-5}{4}, \quad \textit{and} \quad \mathrm{cycl}_2(0, 1) = \mathrm{cycl}_2(1, 0) = \mathrm{cycl}_2(1, 1) = \frac{q-1}{4}.$$

Proof. This will be proved later: see Theorem 10.5. □

In Lemma 5.11 we saw the importance of triples $(a, \theta(a), \eta(a))$, for which at least one of $p_1((a, \theta(a), \eta(a))$ or $p_2(a, \theta(a), \eta(a))$ is a square, in constructing complete mappings of D_P. We will call a triple (a, b, c) of elements of $GF(q)$ an *adoptable triple* if at least one of $p_1(a, b, c)$ or $p_2(a, b, c)$ is a square in $GF(q)$. Many of the triples $(a, \theta(a), \eta(a))$ used in Lemma 5.12 were of the form (a, a, a): there is a simple characterization of such adoptable triples.

Lemma 5.14. *If $a \in P \setminus \{0\}$, then (a, a, a) is an adoptable triple if and only if at least one of $4a + 1$, $4a + 3$ is a square. Further*

$$|\{a \in P \setminus \{0\} \mid (a, a, a) \text{ an adoptable triple}\}| \geq \frac{3q - 11}{8}.$$

Proof. As $p_1(a, a, a) = (a+1)^2(4a+1)$, $p_2(a, a, a) = -a^2(4a+3)$, and $a+1 \neq 0$, $p_1(a, a, a)$ is a square if and only if $4a + 1$ is a square and $p_2(a, a, a)$ is a square if and only if $4a + 3$ is a square.

Set

$$A = \{a \in \mathrm{GF}(q) \mid \text{ at least one of } 4a + 1, 4a + 3 \text{ square}\},$$

and

$$A_I = \{a \in \mathrm{GF}(q) \mid 4a + i \text{ a square for each } i \in I\}.$$

Including 0 there are $(q + 1)/2$ squares in GF(q), and so $|A_1| = |A_3| = (q + 1)/2$.

As a consequence of Lemma 5.13, if 2 is a square, then

$$\begin{aligned}
|A_{1,3}| &= |\{a \in \mathrm{GF}(q) \mid 4a + 1, 4a + 3 \text{ squares}\}| \\
&= |\{x \in \mathrm{GF}(q) \mid x, x + 2 \text{ squares}\}| \\
&= |\{x \in \mathrm{GF}(q) \mid x, x + 1 \text{ squares}\}| \\
&= |\{x \mid x, x + 1 \in C_0\} \cup \{0, -1\}| \\
&= \mathrm{cycl}_2(0, 0) + 2 \\
&= (q + 3)/4,
\end{aligned}$$

and, if 2 is a nonsquare, then

$$\begin{aligned}
|A_{1,3}| &= |\{a \in \mathrm{GF}(q) \mid 4a + 1, 4a + 3 \text{ squares}\}| \\
&= |\{x \in \mathrm{GF}(q) \mid x, x + 2 \text{ squares}\}| \\
&= |\{x \in \mathrm{GF}(q) \mid x, x + 1 \text{ nonsquares}\}| \\
&= |\{x \mid x, x + 1 \in C_1\} \\
&= \mathrm{cycl}_2(1, 1) \\
&= (q - 1)/4.
\end{aligned}$$

Thus,

$$|A| = |A_1| + |A_3| - |A_1 \cap A_3| = |A_1| + |A_3| - |A_{1,3}|$$
$$= \begin{cases} (3q+1)/4 & \text{if 2 is a square,} \\ (3q+5)/4 & \text{if 2 is a nonsquare.} \end{cases}$$

If $a \in A$, then either $4a+1$ or $4a+3$ is a square. If $4a+1$ is a square, then $4(-(1+a))+3 = -(4a+1)$is also a square. If $4a+3$ is a square, then $4(-(1+a))+1 = -(4a+3)$ is also a square. In either case if $a \in A$, then $-(1+a) \in A$ also, and so A is a union of orbits of $x \mapsto -(1+x)$. Thus, as $(q-1)/2 \in A$,

$$|A \cap P| = \begin{cases} (3q-3)/8 & \text{if 2 is a square,} \\ (3q+1)/8 & \text{if 2 is a nonsquare.} \end{cases}$$

Further $0 \in A \cap P$, and so

$$|A \cap (P \setminus \{0\})| = \begin{cases} (3q-11)/8 & \text{if 2 is a square,} \\ (3q-7)/8 & \text{if 2 is a nonsquare,} \end{cases}$$

and hence $|A \cap (P \setminus \{0\})| \geq (3q-11)/8$. □

We do not have a more general characterization of those triples (a, b, c) that are adoptable triples. We can, however, find useful bounds for the number of ways to extend an ordered pair (a, b) to an adoptable triple (a, b, c).

Lemma 5.15. *If $a, b \in P \setminus \{0\}$, then*

$$|\{c \in P \setminus \{0\} \mid (a, b, c) \text{ an adoptable triple}\}| \geq \frac{q-7}{4}.$$

Proof. Let us set $\alpha = 4ab + 2a + 2b$ and $\beta = (a-b)^2$. Then

$$\mathrm{p}_1(a, b, c) = c^2 + (\alpha+2)c + (\alpha+\beta+1)$$

and

$$\mathrm{p}_2(a, b, c) = c^2 - \alpha c + \beta.$$

Thus, for $r \in \mathrm{GF}(q)$, the equation $\mathrm{p}_1(a, b, c) = r$ is a quadratic equation in c which has a solution in GF(q) if and only if $(\alpha+2)^2 - 4(\alpha+\beta+1) + 4r = \alpha^2 - 4\beta + 4r$ is a square. Similarly $\mathrm{p}_2(a, b, c) = r$ is a quadratic equation in c which has a solution in GF(q) if and only if $\alpha^2 - 4\beta + 4r$ is a square.

Hence,

$$m = |\{c \in \mathrm{GF}(q) \mid \mathrm{p}_1(a,b,c) \text{ a square}\}|$$
$$= |\{c \in \mathrm{GF}(q) \mid \mathrm{p}_2(a,b,c) \text{ a square}\}|.$$

We claim that $m \geq (q-1)/2$. If $\alpha^2 - 4\beta = 0$, then $\alpha^2 - 4\beta + 4r$ is a square for each nonzero square r, and so $m \geq (q-1)/2$. If $\alpha^2 - 4\beta$ is a nonzero square and r is a square, then $\alpha^2 - 4\beta + 4r$ is a square if $r = 0$ or $-(\alpha^2 - 4\beta)/4$, or r and $((\alpha^2 - 4\beta)/4r) + 1$ are nonzero squares. In the latter case, there are $\mathrm{cycl}_2(0,0)$ values of r for which r and $((\alpha^2 - 4\beta)/4r) + 1$ are nonzero squares, and, for each of these values of r, there are two solutions for c to $\mathrm{p}_1(a,b,c) = r$. Hence, in this case, by Lemma 5.13, $m = 2\mathrm{cycl}_2(0,0) + 2 = (q+1)/2 \geq (q-1)/2$. If $\alpha^2 - 4\beta$ is a nonsquare, and r and $\alpha^2 - 4\beta + 4r$ are squares, then r and $((\alpha^2 - 4\beta)/4r) + 1$ are nonzero squares: this is true for $\mathrm{cycl}_2(0,1)$ values of r, and for each such r, there are two solutions for c to the equation $\mathrm{p}_1(a,b,c) = r$. Hence, in this case, by Lemma 5.13, $m = 2\mathrm{cycl}_2(0,1) = (q-1)/2$. Hence $m \geq (q-1)/2$ as claimed.

Next we set

$$K = \{c \in \mathrm{GF}(q) \mid \mathrm{p}_1(a,b,c) \text{ and/or } \mathrm{p}_2(a,b,c) \text{ a square}\}.$$

As $\mathrm{p}_1(a,b,-(1+c)) = \mathrm{p}_2(a,b,c)$ and $\mathrm{p}_2(a,b,-(1+c)) = \mathrm{p}_1(a,b,c)$, K is a union of orbits of $x \mapsto -(1+x)$, and, as 0, -1, and $(q-1)/2$ may be elements of K,

$$|(P \setminus \{0\}) \cap K| \geq (|K| - 3)/2.$$

Hence, as $|K| \geq m$, $|\{c \in P \setminus \{0\} \mid (a,b,c) \text{ an adoptable triple}\}| \geq (m-3)/2 \geq (q-7)/4$. □

We can now prove the admissibility of SL(2, q) for $q \equiv 1 \pmod 4$.

Theorem 5.16 (Evans, 2001). *SL(2, q) is admissible for all $q \equiv 1 \pmod 4$.*

Proof. We proved the admissibility of SL(2, q) for $q = 5, 9, 13, 17, 25$, and 29 in Lemma 5.12. Thus we may assume that $q \geq 37$.

For $q \geq 37$ we will construct mappings $\theta_P, \eta_P \colon P \setminus \{0\} \to P \setminus \{0\}$ that satisfy the conditions of Lemma 5.11. Set

$$M = \{a \in P \setminus \{0\} \mid (a,a,a) \text{ not an adoptable triple}\}.$$

By Lemma 5.14, $|(P \setminus \{0\}) \setminus M| \geq (3q-11)/8$, and as $|P \setminus \{0\}| = (q-3)/2$, it follows that

$$|M| \leq \frac{q-3}{2} - \frac{3q-11}{8} = \frac{q-1}{8}.$$

Set

$$N = \begin{cases} M & \text{if } |M| \text{ is even,} \\ M \cup \{d\} & \text{if } |M| \text{ is odd,} \end{cases}$$

where $d \in (P \setminus \{0\}) \setminus M$. Thus $|N|$ is even and $|N| \leq (q+7)/8$. As $|N|$ is even, N can be partitioned into 2-element subsets $\{a_1, b_1\}, \ldots, \{a_{|N|/2}, b_{|N|/2}\}$.

We claim that there exist distinct elements $c_1, \ldots, c_{|N|/2}$ of $(P\setminus\{0\})\setminus N$ for which each (a_i, b_i, c_i) is an adoptable triple, $i = 1, \ldots, |N|/2$. Assume this to be false, and let r be the largest integer for which there exist distinct elements $c_1, \ldots, c_r$ of $(P \setminus \{0\}) \setminus N$ satisfying (a_i, b_i, c_i) is an adoptable triple, $i = 1, \ldots, r$. Now, as $q \geq 37$,

$$|N \cup \{c_1, \ldots, c_r\}| \leq \frac{3|N|}{2} - 1 \leq \frac{3q+5}{16} < \frac{q-7}{4},$$

and so by Lemma 5.15, there exists $c_{r+1} \in (P \setminus \{0\}) \setminus N$, distinct from $c_1, \ldots, c_r$, such that $(a_{r+1}, b_{r+1}, c_{r+1})$ is an adoptable triple: a contradiction which proves the claim.

Define $\theta_P, \eta_P \colon P \setminus \{0\} \to P \setminus \{0\}$ by

$$\theta_P(a) = \begin{cases} b_i & \text{if } a = a_i,\ i = 1, \ldots, |N|/2, \\ c_i & \text{if } a = b_i,\ i = 1, \ldots, |N|/2, \\ a_i & \text{if } a = c_i,\ i = 1, \ldots, |N|/2, \\ a & \text{otherwise,} \end{cases}$$

and

$$\eta_P(a) = \begin{cases} c_i & \text{if } a = a_i,\ i = 1, \ldots, |N|/2, \\ a_i & \text{if } a = b_i,\ i = 1, \ldots, |N|/2, \\ b_i & \text{if } a = c_i,\ i = 1, \ldots, |N|/2, \\ a & \text{otherwise.} \end{cases}$$

As θ_P and η_P satisfy the conditions of Lemma 5.11, D_P admits complete mappings. Now by Lemma 5.10, D_0 and D_T admit complete mappings and, hence, D is admissible by Lemma 5.2. As H is admissible by Lemma 5.8, $\mathrm{SL}(2, q)$ is admissible for $q \geq 37$, $q \equiv 1 \pmod 4$, by Theorem 4.1. □

5.1.5 SL(2, q), $q \equiv 3 \pmod 4$

We now turn to the case $q \equiv 3 \pmod 4$. Let $q \equiv 3 \pmod 4$ and let $G = \mathrm{SL}(2, q)$. As with the case $q \equiv 1 \pmod 4$, we will construct a subgroup H of G that is admissible and a dual system of coset representatives of H in G that is admissible. The admissibility of G will then follow from Theorem 4.1. For $a, b, x, y \in \mathrm{GF}(q)$, $a, b \neq 0$, let

$$A_{a,b} = \begin{pmatrix} a & b \\ -b & a \end{pmatrix}, \ B_{a,b} = \begin{pmatrix} a & b \\ b & -a \end{pmatrix}, \text{ and } D_{x,y} = \begin{pmatrix} 0 & x \\ \frac{-1}{x} & y \end{pmatrix}.$$

Set

$$H = \{A_{a,b} \mid a^2 + b^2 = 1\} \cup \{B_{a,b} \mid a^2 + b^2 = -1\},$$

and

$$D = \{D_{x,y} \mid x \in C_0\},$$

where C_0 is the set of nonzero squares. If $a^2 + b^2 = 1$, then $\det(A_{a,b}) = 1$, and if $a^2 + b^2 = -1$, then $\det(B_{a,b}) = 1$, and so each element of H is an element of G. In fact the set H is a subgroup of G and is also admissible.

Lemma 5.17. *H is an admissible subgroup of G.*

Proof. Let $A_{a,b}, A_{c,d}, B_{a,b}, B_{c,d} \in H$. The following are the multiplication rules:

- $A_{a,b}A_{c,d} = A_{ac-bd,ad+bc}$, and, as $a^2 + b^2 = c^2 + d^2 = 1$, $(ac - bd)^2 + (ad + bc)^2 = (a^2 + b^2)(c^2 + d^2) = 1$, and so $A_{a,b}A_{c,d} \in H$.
- $A_{a,b}B_{c,d} = B_{ac+bd,ad-bc}$, and, as $a^2 + b^2 = 1$ and $c^2 + d^2 = -1$, $(ac + bd)^2 + (ad - bc)^2 = (a^2 + b^2)(c^2 + d^2) = -1$, and so $A_{a,b}B_{c,d} \in H$.
- $B_{a,b}A_{c,d} = B_{ac-bd,ad+bc}$, and, as $a^2 + b^2 = -1$ and $c^2 + d^2 = -1$, $(ac - bd)^2 + (ad + bc)^2 = (a^2 + b^2)(c^2 + d^2) = -1$, and so $B_{a,b}A_{c,d} \in H$.
- $B_{a,b}B_{c,d} = A_{ac+bd,ad-bc}$, and, as $a^2 + b^2 = c^2 + d^2 = -1$, $(ac + bd)^2 + (ad - bc)^2 = (a^2 + b^2)(c^2 + d^2) = 1$, and so $B_{a,b}B_{c,d} \in H$.

It follows that H is closed under multiplication and so is a subgroup of G. Set $A = \{A_{a,b} \mid a^2 + b^2 = 1\}$ and $B = \{B_{a,b} \mid a^2 + b^2 = -1\}$. The multiplication rules show that A is an abelian subgroup of H.

Now $|A|$ is equal to the number of solutions to $a^2 + b^2 = 1$. There are four solutions with one of a, b equal to zero, namely, $(a, b) = (0, \pm 1)$ or $(\pm 1, 0)$. If $a, b \neq 0$, then $a^2 = 1 + (-b^2)$, an equation which has $\mathrm{cycl}_2(1, 0)$ solutions for (a^2, b^2). Hence, by Lemma 5.13,

$$|A| = 4 + 4\mathrm{cycl}_2(1, 0) = 4 + 4\left(\frac{q-3}{4}\right) = q + 1.$$

Similarly $|B|$ is equal to the number of solutions to $a^2 + b^2 = -1$. As there are no solutions with one of a, b equal to zero, $1 + a^2 = (-b^2)$, an equation which has $\mathrm{cycl}_2(0, 1)$ solutions for (a^2, b^2). Hence, by Lemma 5.13,

$$|B| = 4\mathrm{cycl}_2(0, 1) = 4\left(\frac{q+1}{4}\right) = q + 1.$$

As $A \cap B = \emptyset$, $|H| = 2(q+1)$. Hence H has a normal abelian subgroup A of index 2, and so H is solvable. Further, H has a noncyclic Sylow 2-subgroup as, if we let $c, d \in GF(q)$ satisfy $c^2 + d^2 = -1$, then $\langle A_{0,1}, B_{c,d}\rangle$ is a 2-subgroup of H that is isomorphic to Q_8, the quaternion group of order 8. Hence H is admissible. □

Lemma 5.18. *D is a dual system of coset representatives for H in G.*

Proof. As $|G| = q(q^2 - 1)$ and $H = 2(q + 1)$, any dual system of coset representatives for H in G must be of order $|G|/|H| = q(q-1)/2$. As $D_{x,y} \in D$ if $x \in C_0$, there are $(q-1)/2$ choices for x and q choices for y. Hence $|D| = q(q-1)/2$, the right number for D to be a dual system of coset representatives for H in G.

Let $D_{x,y}, D_{w,z} \in D$ and suppose that

$$D_{x,y}D_{w,z}^{-1} = \begin{pmatrix} x/w & 0 \\ (y/w) - (z/x) & w/x \end{pmatrix} \in H.$$

If $D_{x,y}D_{w,z}^{-1} = A_{a,b}$, then $a = x/w$, $b = 0$, and $y/w = z/x$. As $a^2 = 1$, $x^2 = w^2$. As $x, w \in C_0$ and $-1 \notin C_0$, $w = x$, and so $z = y$. If $D_{x,y}D_{w,z}^{-1} = B_{a,b}$, then $a = x/w$, $b = 0$, and $y/w = z/x$. As $a^2 = -1$, $x^2 = -w^2$, an impossibility. It follows that $D_{x,y}D_{w,z}^{-1} \in H$ if and only if $x = w$ and $y = z$, and so D is a system of right coset representatives for H in G.

Next let

$$D_{x,y}^{-1}D_{w,z} = \begin{pmatrix} x/w & wy - xz \\ 0 & w/x \end{pmatrix} \in H.$$

If $D_{x,y}^{-1}D_{w,z} = A_{a,b}$, then $a = x/w$, $b = 0$, and $yw = zx$. As $a^2 = 1$, $x^2 = w^2$. As $x, w \in C_0$ and $-1 \notin C_0$, $w = x$, and so $z = y$. If $D_{x,y}^{-1}D_{w,z} = B_{a,b}$, then $a = x/w$, $b = 0$, and $yw = zx$. As $a^2 = -1$, $x^2 = -w^2$, an impossibility. It follows that $D_{x,y}^{-1}D_{w,z} \in H$ if and only if $x = w$ and $y = z$, and so D is a system of left coset representatives for H in G. Thus, D is both a system of left coset representatives for H in G, as well as a system of right coset representatives for H in G, and hence is a dual system of coset representatives for H in G, as claimed. □

We can now prove $\mathrm{SL}(2, q)$ to be admissible for $q \equiv 3 \pmod 4$.

Theorem 5.19 (Evans, 2005). $\mathrm{SL}(2, q)$ *is admissible for all* $q \equiv 3 \pmod 4$.

Proof. For a square x, we will use $\sqrt{x}$ to denote the unique square satisfying $(\sqrt{x})^2 = x$. Let us define $\theta, \eta\colon D \to D$ by

$$\theta(D_{x,y}) = \begin{cases} D_{x^2,0} & \text{if } y = 0,\\ D_{\sqrt{x},-y/(x\sqrt{x})} & \text{if } y \neq 0 \text{ and } x^2+y^2 \in C_0,\\ D_{\sqrt{x},y/(x\sqrt{x})} & \text{if } y \neq 0 \text{ and } x^2+y^2 \in C_1,\end{cases}$$

and

$$\eta(D_{x,y}) = \begin{cases} D_{\frac{1}{x},0} & \text{if } y = 0,\\ D_{\sqrt{(x^2+y^2)/x},(xy+yx^2+y^3)/(x\sqrt{x}\sqrt{x^2+y^2})} & \text{if } y \neq 0 \text{ and } x^2+y^2 \in C_0,\\ D_{\sqrt{-(x^2+y^2)/x},(xy-yx^2-y^3)/(x\sqrt{x}\sqrt{-(x^2+y^2)})} & \text{if } y \neq 0 \text{ and } x^2+y^2 \in C_1.\end{cases}$$

If $D_{w,z} \in D$ and we set

$$x = \begin{cases} \sqrt{w} & \text{if } z = 0,\\ w^2 & \text{if } z \neq 0,\end{cases}$$

and

$$y = \begin{cases} 0 & \text{if } z = 0,\\ -w^3 z & \text{if } z \neq 0 \text{ and } 1 + w^2z^2 \in C_0,\\ w^3 z & \text{if } z \neq 0 \text{ and } 1 + w^2z^2 \in C_1,\end{cases}$$

then $\theta(D_{x,y}) = D_{w,z}$, and so θ is onto and, hence, by finiteness, is a bijection.

If $D_{w,z} \in D$ and we set

$$x = \begin{cases} 1/w & \text{if } z = 0,\\ (1+w^2)^2w^2/\left(w^2z^2+(1+w^2)^2\right) & \text{if } z \neq 0 \text{ and } (1+w^2)^2 + w^2z^2 \in C_0,\\ -(1+w^2)^2w^2/\left(w^2z^2+(1+w^2)^2\right) & \text{if } z \neq 0 \text{ and } (1+w^2)^2 + w^2z^2 \in C_1,\end{cases}$$

and

$$y = \begin{cases} 0 & \text{if } z = 0,\\ (1+w^2)w^3z/\left(w^2z^2+(1+w^2)^2\right) & \text{if } z \neq 0 \text{ and } (1+w^2)^2 + w^2z^2 \in C_0,\\ -(1+w^2)w^3z/\left(w^2z^2+(1+w^2)^2\right) & \text{if } z \neq 0 \text{ and } (1+w^2)^2 + w^2z^2 \in C_1,\end{cases}$$

then $\eta(D_{x,y}) = D_{w,z}$ and so η is onto and, hence, by finiteness, is a bijection.

To see that $D_{x,y}\theta(D_{x,y}) \in \eta(D_{x,y})H$ for all $D_{x,y} \in D$, by routine computation, we find that

$$D_{x,y}\theta(D_{x,y}) = \begin{cases} \eta(D_{x,y})A_{0,1} & \text{if } y = 0, \\ \eta(D_{x,y})A_{\frac{-y}{\left(\sqrt{x^2+y^2}\right)}, \frac{x}{\left(\sqrt{x^2+y^2}\right)}} & \text{if } y \neq 0 \text{ and } x^2 + y^2 \in C_0, \\ \eta(D_{x,y})B_{\frac{-y}{\left(\sqrt{-(x^2+y^2)}\right)}, \frac{-x}{\left(\sqrt{-(x^2+y^2)}\right)}} & \text{if } y \neq 0 \text{ and } x^2 + y^2 \in C_1. \end{cases}$$

Hence $D_{x,y}\theta(D_{x,y}) = \eta(D_{x,y})H$ for each $D_{x,y} \in D$, and θ is a complete mapping of D. As H is admissible by Lemma 5.17, the admissibility of G follows from Theorem 4.1. □

5.2 The Group PSL(2, q), q Odd

The proof that PSL(2, q) is admissible when q is odd was given by Dalla Volta and Gavioli [83, 84] in 1993 and 1997; in 1993 they dealt with the case $q \equiv 3 \pmod 4$; and in 1997 with the case $q \equiv 1 \pmod 4$.

Let us begin with the case $q \equiv 3 \pmod 4$. We will regard SL(2, q) as the set of 2×2 matrices over GF(q) of determinant one, and we will let

$$P = \left\{ \begin{pmatrix} 1 & 0 \\ 0 & 1 \end{pmatrix}, \begin{pmatrix} -1 & 0 \\ 0 & -1 \end{pmatrix} \right\}.$$

Now P is a normal subgroup of SL(2, q) and PSL(2, q) $\cong$ SL(2, q)/P. We will write elements of PSL(2, q) in the form AP, $A \in$ SL(2, q). Let V denote the two-dimensional vector space of column vectors over GF(q). Then the mapping $\mathbf{v} \mapsto \mathbf{v}^A = A\mathbf{v}$ describes an action of SL(2, q) on V. It is easily seen that SL(2, q) permutes the one-dimensional subspaces of V, and, as each element of P fixes each one-dimensional subspace of V, this describes a natural action of PSL(2, q) on the one-dimensional subspaces of V. This action will prove to be crucial in establishing the admissibility of PSL(2, q). The action of PSL(2, q) is known to be doubly-transitive: it is an easy exercise to prove this from the fact that GL(2, q) is transitive on the set of ordered bases. In fact PSL(2, q) contains a subgroup that acts regularly on the set of one-dimensional subspaces of V: the action is transitive and only the identity fixes any one-dimensional subspace.

Lemma 5.20. PSL(2, q) *contains a dihedral subgroup of order* $q + 1$ *that acts regularly on the one-dimensional subspaces of* V.

Proof. See Satz 11.8.4 in [191], for example. □

We can now prove the admissibility of PSL$(2, q)$ when $q \equiv 3 \pmod 4$ by showing this group to be factorable with admissible factors.

Theorem 5.21 (Dalla Volta and Gavioli, 1993). *If* $q \equiv 3 \pmod 4$, *then* PSL$(2, q)$ *is admissible.*

Proof. Let $G = \mathrm{PSL}(2, q)$, $q \equiv 3 \pmod 4$. Let $\langle \mathbf{v} \rangle$ be a one-dimensional subspace of V, and let H be the stabilizer in G of $\langle \mathbf{v} \rangle$. Now $|H| = (q-1)q/2$, and, as $(q-1)q/2$ is odd, H is admissible by Corollary 3.6. By Lemma 5.20, G contains a dihedral subgroup K, of order $q+1$, that acts regularly on the one-dimensional subspaces of V. As K is a dihedral group of doubly even order, K is admissible by Theorem 4.22. As $\gcd(H, K) = 1$ and

$$|H||K| = \left(\frac{(q-1)q}{2}\right)(q+1) = |G|,$$

$G = HK$ by Lemma 4.7. As H and K are both admissible, G is admissible by Theorem 4.8. □

For the case $q \equiv 1 \pmod 4$, to prove that PSL$(2, q)$ is admissible, we will partition the elements of the group using stabilizers of subsets of the set of one-dimensional subspaces of V. To do this we need some known facts about these stabilizers.

Lemma 5.22. *Let* $G = \mathrm{PSL}(2, q)$, *let* p *be the characteristic of* $GF(q)$, *and let* $\langle \mathbf{u} \rangle$, $\langle \mathbf{v} \rangle$, $\langle \mathbf{w} \rangle$, *and* $\langle \mathbf{z} \rangle$ *be one-dimensional subspaces of* V.

1. *If* $\langle \mathbf{u} \rangle$, $\langle \mathbf{v} \rangle$, *and* $\langle \mathbf{w} \rangle$ *are distinct, then* $G_{\{\langle \mathbf{u} \rangle, \langle \mathbf{v} \rangle, \langle \mathbf{w} \rangle\}} = 1$.
2. *If* $\langle \mathbf{u} \rangle \neq \langle \mathbf{v} \rangle$, *then* $G_{\{\langle \mathbf{u} \rangle, \langle \mathbf{v} \rangle\}}$ *is a cyclic group of order* $(q-1)/2$, *and the normalizer of* $G_{\{\langle \mathbf{u} \rangle, \langle \mathbf{v} \rangle\}}$ *is a dihedral group of order* $q-1$, *each element of which either fixes both* $\langle \mathbf{u} \rangle$ *and* $\langle \mathbf{v} \rangle$ *or interchanges* $\langle \mathbf{u} \rangle$ *and* $\langle \mathbf{v} \rangle$.
3. *If* $\langle \mathbf{u} \rangle \neq \langle \mathbf{v} \rangle$ *and* $\langle \mathbf{w} \rangle \neq \langle \mathbf{z} \rangle$, *then* $G_{\{\langle \mathbf{u} \rangle, \langle \mathbf{v} \rangle\}} = G_{\{\langle \mathbf{w} \rangle, \langle \mathbf{z} \rangle\}}$ *if and only if* $\{\langle \mathbf{u} \rangle, \langle \mathbf{v} \rangle\} = \{\langle \mathbf{w} \rangle, \langle \mathbf{z} \rangle\}$.
4. $G_{\langle \mathbf{u} \rangle}$ *is a Sylow* p*-subgroup of* G.
5. *If* $\langle \mathbf{u} \rangle \neq \langle \mathbf{v} \rangle$, *then* $G_{\langle \mathbf{u} \rangle} \cap G_{\langle \mathbf{u} \rangle} = 1$.

Proof. See Section 11.8 in [191]. □

We need information about those elements of PSL$(2, q)$ that stabilize no one-dimensional subspaces of V.

Lemma 5.23. PSL$(2, q)$ *contains a cyclic subgroup* S *of order* $(q+1)/2$ *which has the following properties:*

1. *Any nonidentity element of* S *stabilizes no one-dimensional subspaces of* V.
2. *Distinct conjugates of* S *intersect trivially.*
3. *Any element of* PSL$(2, q)$ *that stabilizes no one-dimensional subspaces of* V *is contained in some conjugate of* S.

Proof. See Satz 8.4 and Satz 8.5 in [191]. □

Theorem 5.24 (Dalla Volta and Gavioli, 1997). *If* $q \equiv 1 \pmod 4$, *then* $\mathrm{PSL}(2, q)$ *is admissible.*

Proof. As $\mathrm{PSL}(2, 5) \cong A_5$ and $\mathrm{PSL}(2, 9) \cong A_6$ (see [81]), both are admissible by Theorem 4.5, and a complete mapping of PSL(2, 13) was found using Gap (see [84]). Thus we may assume that $q > 13$.

By Lemmas 5.22 and 5.23, the element set of $G = \mathrm{PSL}(2, q)$ is partitioned by $\{I\}$; $G_{\langle\mathbf{v}\rangle} \setminus \{I\}$, $\mathbf{v} \neq \mathbf{0}$; the conjugates of $S \setminus \{I\}$; and $H = \cup\{G_{\{\langle\mathbf{u}\rangle,\langle\mathbf{v}\rangle\}} \setminus \{I\} \mid \mathbf{u} \neq \mathbf{v}, \mathbf{u}, \mathbf{v} \neq 0\}$.

The identity mapping is a complete mapping of $\{I\}$. As the group $G_{\langle\mathbf{v}\rangle}$, $\mathbf{v} \neq \mathbf{0}$, is of odd order, it admits normalized complete mappings, and hence $G_{\langle\mathbf{v}\rangle} \setminus \{I\}$ admits complete mappings. Similarly, as S, and thus any conjugate of S, is of odd order, any conjugate of $S \setminus \{I\}$ admits complete mappings. It remains to show that H admits complete mappings.

It should be noted that if $\mathbf{u} \neq \mathbf{v}$ and $\mathbf{u}, \mathbf{v} \neq 0$, then $G_{\{\langle\mathbf{u}\rangle,\langle\mathbf{v}\rangle\}}$ does not admit complete mappings as $G_{\{\langle\mathbf{u}\rangle,\langle\mathbf{v}\rangle\}}$ is a cyclic group of even order, and hence we will need to find a more useful partition of H. From the partition of G, it is clear that any even-order element of G is contained in H. Now, by lemma 5.22, if $\mathbf{u} \neq \mathbf{v}$ and $\mathbf{u}, \mathbf{v} \neq 0$, then the normalizer of $G_{\{\langle\mathbf{u}\rangle,\langle\mathbf{v}\rangle\}}$, which we will denote $N_{\{\langle\mathbf{u}\rangle,\langle\mathbf{v}\rangle\}}$, is a dihedral group of order $2|G_{\{\langle\mathbf{u}\rangle,\langle\mathbf{v}\rangle\}}|$. Thus every element of $N_{\{\langle\mathbf{u}\rangle,\langle\mathbf{v}\rangle\}} \setminus G_{\{\langle\mathbf{u}\rangle,\langle\mathbf{v}\rangle\}}$ is of even order and, hence, is contained in H. It follows that $H = \cup\{N_{\{\langle\mathbf{u}\rangle,\langle\mathbf{v}\rangle\}} \setminus \{I\} \mid \mathbf{u} \neq \mathbf{v}, \mathbf{u}, \mathbf{v} \neq 0\}$. As each $N_{\{\langle\mathbf{u}\rangle,\langle\mathbf{v}\rangle\}}$ is a dihedral group of doubly even order, each admits normalized complete mappings: however, two such groups need not intersect trivially. Now if $\mathbf{u} \neq \mathbf{v}$ and $\mathbf{u}, \mathbf{v} \neq 0$, then $G_{\{\langle\mathbf{u}\rangle,\langle\mathbf{v}\rangle\}}$ is a cyclic group of even order. Hence $G_{\{\langle\mathbf{u}\rangle,\langle\mathbf{v}\rangle\}}$ contains a unique involution σ: we will set $G^{\dagger}_{\{\langle\mathbf{u}\rangle,\langle\mathbf{v}\rangle\}} = G_{\{\langle\mathbf{u}\rangle,\langle\mathbf{v}\rangle\}} \setminus \{I, \sigma\}$. Let $\langle\mathbf{u}\rangle$ be a one-dimensional subspace of V, and set

$$K = \cup\{N_{\{\langle\mathbf{u}\rangle,\langle\mathbf{v}\rangle\}} \setminus \{I\} \mid \langle\mathbf{v}\rangle \neq \langle\mathbf{u}\rangle, \langle\mathbf{0}\rangle\} \cup \{G^{\dagger}_{\{\langle\mathbf{z}\rangle,\langle\mathbf{w}\rangle\}} \mid \langle\mathbf{u}\rangle, \langle\mathbf{z}\rangle, \langle\mathbf{w}\rangle, \langle\mathbf{0}\rangle \text{ distinct}\}.$$

We claim that $K = H$ and further that the sets $N_{\{\langle\mathbf{u}\rangle,\langle\mathbf{v}\rangle\}} \setminus \{I\}$, $\langle\mathbf{v}\rangle \neq \langle\mathbf{u}\rangle, \langle\mathbf{0}\rangle$, and $G^{\dagger}_{\{\langle\mathbf{z}\rangle,\langle\mathbf{w}\rangle\}}$, $\langle\mathbf{u}\rangle, \langle\mathbf{z}\rangle, \langle\mathbf{w}\rangle, \langle\mathbf{0}\rangle$ distinct, partition H.

Clearly $K \subseteq H$, and if $\tau \in H \setminus K$, then τ is the unique involution of some $G_{\{\langle\mathbf{z}\rangle,\langle\mathbf{w}\rangle\}}$, $\langle\mathbf{z}\rangle$ and $\langle\mathbf{w}\rangle$ distinct one-dimensional subspaces of V that are distinct from $\langle\mathbf{u}\rangle$. But then, as any element of H fixes exactly two one-dimensional subspaces of V, τ must swap $\langle\mathbf{u}\rangle$ with some $\langle\mathbf{v}\rangle$, from which it follows that $\tau \in N_{\{\langle\mathbf{u}\rangle,\langle\mathbf{v}\rangle\}} \setminus \{I\} \subseteq K$: a contradiction. Hence $K = H$ as claimed.

Suppose that $\tau \in N_{\{\langle\mathbf{u}\rangle,\langle\mathbf{v}\rangle\}} \cap N_{\{\langle\mathbf{u}\rangle,\langle\mathbf{w}\rangle\}}$ for one-dimensional subspaces, $\langle\mathbf{v}\rangle$, $\langle\mathbf{w}\rangle$, of V. Then τ must stabilize $\langle\mathbf{u}\rangle$, $\langle\mathbf{v}\rangle$, and $\langle\mathbf{w}\rangle$ and hence either $\tau = I$ or $\langle\mathbf{v}\rangle = \langle\mathbf{w}\rangle$. It follows that $N_{\{\langle\mathbf{u}\rangle,\langle\mathbf{v}\rangle\}} \setminus \{I\} \cap N_{\{\langle\mathbf{u}\rangle,\langle\mathbf{w}\rangle\}} \setminus \{I\} = \emptyset$ if $\langle\mathbf{u}\rangle \neq \langle\mathbf{w}\rangle$. By Lemma 5.22, $G^{\dagger}_{\{\langle\mathbf{z}\rangle,\langle\mathbf{w}\rangle\}} \cap G^{\dagger}_{\{\langle\mathbf{x}\rangle,\langle\mathbf{y}\rangle\}} = \emptyset$ if $\{\langle\mathbf{z}\rangle, \langle\mathbf{w}\rangle\} \neq \{\langle\mathbf{x}\rangle, \langle\mathbf{y}\rangle\}$.

Suppose that $\tau \in N_{\{\langle\mathbf{u}\rangle,\langle\mathbf{v}\rangle\}} \cap G^{\dagger}_{\{\langle\mathbf{z}\rangle,\langle\mathbf{w}\rangle\}}$, $\langle\mathbf{z}\rangle, \langle\mathbf{w}\rangle \neq \langle\mathbf{u}\rangle$. Then $\tau \neq I$ and so τ cannot stabilize $\langle\mathbf{u}\rangle$, $\langle\mathbf{v}\rangle$, $\langle\mathbf{z}\rangle$, and $\langle\mathbf{w}\rangle$. Thus τ stabilizes $\langle\mathbf{z}\rangle$ and $\langle\mathbf{w}\rangle$ and swaps $\langle\mathbf{u}\rangle$ and $\langle\mathbf{v}\rangle$. It follows that τ^2 stabilizes $\langle\mathbf{u}\rangle$, $\langle\mathbf{v}\rangle$, $\langle\mathbf{z}\rangle$, and $\langle\mathbf{w}\rangle$ and so τ is an involution.

But $G^{\dagger}_{\{\langle \mathbf{z}\rangle,\langle \mathbf{w}\rangle\}}$ contains no involutions: a contradiction. Hence the sets $N_{\{\langle \mathbf{u}\rangle,\langle \mathbf{v}\rangle\}}\setminus\{I\}$, $\langle \mathbf{v}\rangle \neq \langle \mathbf{u}\rangle$, $\langle \mathbf{0}\rangle$, and $G^{\dagger}_{\{\langle \mathbf{z}\rangle,\langle \mathbf{w}\rangle\}}$, $\langle \mathbf{u}\rangle$, $\langle \mathbf{z}\rangle$, $\langle \mathbf{w}\rangle$, $\langle \mathbf{0}\rangle$ distinct, partition H as claimed.

Now, for each $\langle \mathbf{v}\rangle$ distinct from $\langle \mathbf{u}\rangle$, $N_{\{\langle \mathbf{u}\rangle,\langle \mathbf{v}\rangle\}}$ is a doubly even dihedral group which admits normalized complete mappings, and hence $N_{\{\langle \mathbf{u}\rangle,\langle \mathbf{v}\rangle\}} \setminus \{I\}$ admits complete mappings. For each pair of distinct one-dimensional subspaces, $\langle \mathbf{z}\rangle$, $\langle \mathbf{w}\rangle$, of V, $G_{\{\langle \mathbf{z}\rangle,\langle \mathbf{w}\rangle\}}$, being a cyclic group of even order, contains a unique involution σ and admits a bijection $\theta\colon G_{\{\langle \mathbf{z}\rangle,\langle \mathbf{w}\rangle\}} \setminus \{\sigma\} \to G_{\{\langle \mathbf{z}\rangle,\langle \mathbf{w}\rangle\}} \setminus \{I\}$ for which the mapping η defined by $\eta(x) = x\theta(x)$ is also a bijection $G_{\{\langle \mathbf{z}\rangle,\langle \mathbf{w}\rangle\}} \setminus \{\sigma\} \to G_{\{\langle \mathbf{z}\rangle,\langle \mathbf{w}\rangle\}} \setminus \{I\}$. Clearly θ restricted to $G^{\dagger}_{\{\langle \mathbf{z}\rangle,\langle \mathbf{w}\rangle\}}$ is a complete mapping of $G^{\dagger}_{\{\langle \mathbf{z}\rangle,\langle \mathbf{w}\rangle\}}$. Hence H admits complete mappings and the result then follows from Lemma 5.1. □

5.3 The Groups SL(n, q), GL(n, q), PSL(n, q), and PGL(n, q), q Even

In 1993 Dalla Volta and Gavioli [83] proved that SL(n, q) is admissible if q is even and n and q are not both 2. This was proved in two parts. For $q > 2$ the proof was by induction on n using the 1989 proof by Saeli [288] that SL($2, q$) is admissible when q is even, $q \neq 2$ (see Theorem 5.5), as the basis step. Their proof that SL($n, 2$) is admissible for $n > 2$ is similar, using the admissibility of SL(3, 2) as the basis step. The admissibility of GL(n, q) for q even, n and q not both 2, can then be deduced. Their proofs rely on a particular type of complete mapping of a dual system of coset representatives described in Lemma 4.23: given a transitive permutation group G acting on $X = \{1, \dots, n\}$ and a dual system of coset representatives for G_n, the stabilizer of n, in G, $D = \{g_1, \dots, g_n\}$, satisfying $i^{g_i} = n$. If $\pi : X \to X$ is a bijection satisfying $\pi(i)^{g_i} = \pi(i)$, then the mapping $g_i \mapsto g_{\pi(i)}$ is a complete mapping of D.

Let V_n denote the n-dimensional vector space of column vectors over GF(q). Then $\mathbf{v}^A = A\mathbf{v}$ defines an action of SL(n, q) on V_n, which induces a natural action of SL(n, q) on the set of one-dimensional subspaces of V_n. Involutions permuting the one-dimensional subspaces of a two-dimensional subspace of V_n will play a crucial role in establishing the admissibility of SL(n, q), q even, $n \geq 2$, n and q not both 2.

Lemma 5.25. *If U is a two-dimensional subspace of V_n and $\langle \mathbf{u}\rangle$, $\langle \mathbf{v}\rangle$, and $\langle \mathbf{w}\rangle$ are distinct one-dimensional subspaces of U, then there exists an involution in* SL(n, q) *that fixes $\langle \mathbf{u}\rangle$ and swaps $\langle \mathbf{v}\rangle$ and $\langle \mathbf{w}\rangle$.*

Proof. By an appropriate change of basis, we may assume that $\langle \mathbf{v}\rangle = \langle (1, 0, \dots, 0)^T\rangle$ and $\langle \mathbf{w}\rangle = \langle (0, 1, 0, \dots, 0)^T\rangle$. Then $\langle \mathbf{u}\rangle = \langle (a, 1, 0, \dots, 0)^T\rangle$ for some $a \neq 0$ and the action of

$$\begin{pmatrix} 0 & a & 0 & \cdots & 0 \\ a^{-1} & 0 & 0 & \cdots & 0 \\ 0 & 0 & 1 & \cdots & 0 \\ \vdots & \vdots & & \ddots & \vdots \\ 0 & 0 & \cdots & \cdots & 1 \end{pmatrix}$$

swaps $\langle \mathbf{v} \rangle$ and $\langle \mathbf{w} \rangle$ and fixes $\langle \mathbf{u} \rangle$. □

We can now prove SL(n, q) is admissible for q even, $n \geq 2$, n and q not both 2.

Theorem 5.26 (Dalla Volta and Gavioli, 1993). *If q is even, $n \geq 2$, n and q not both* 2, *then* SL(n, q) *is admissible.*

Proof. By Theorem 5.5, SL$(2, q)$ is admissible if q is even, $q \neq 2$. Now SL$(3, 2)$ is isomorphic to PSL$(2, 7)$: see [81]. Hence, as by Theorem 5.21, PSL$(2, 7)$ is admissible; SL$(3, 2)$ is admissible. We will prove that, for q even, if SL$(n-1, q)$, $n > 2$, is admissible, then so is SL(n, q). The result will then follow by induction.

Let us assume that q is even, that $n > 2$, and that SL$(n-1, q)$ is admissible. Let $\mathbf{v} = (1, 0, \ldots, 0)^T$ and set

$$H = \{\text{diag}(a, a^{-1}, 1, \ldots, 1) \mid a \in \text{GF}(q), a \neq 0\},$$

$$K = \left\{ \begin{pmatrix} 1 & v \\ O^T & I_{n-1} \end{pmatrix} \mid v \in V_{n-1} \right\}, \text{ and}$$

$$L = \left\{ \begin{pmatrix} 1 & O \\ O^T & A \end{pmatrix} \mid A \in SL(n-1, q) \right\},$$

where O^T is the zero vector in V_{n-1}. Now, each of H, K, and L is a subgroup of $G_{\langle \mathbf{v} \rangle}$. H is isomorphic to the multiplicative group of GF(q), which is of odd order, and thus is admissible by Corollary 3.6; K is isomorphic to the additive group of V_{n-1}, an elementary abelian group of order $q^{n-1} > 2$, which is admissible by Example 1.25; and L is isomorphic to SL$(n-1, q)$, which we have assumed to be admissible. $L \cap K = 1$ and L normalizes K. Hence $M = LK$ is a subgroup of $G_{\langle \mathbf{v} \rangle}$ of order $q^{n-1}|\text{SL}(n-1, q)|$. As L and K are admissible, M is admissible by Theorem 4.8. Further, as H and M are subgroups of $G_{\langle \mathbf{v} \rangle}$, $H \cap M = 1$, and

$$|M||H| = q^{n-1}(q-1)|\text{SL}(n-1, q)| = |G_{\langle \mathbf{v} \rangle}|,$$

$G_{\langle \mathbf{v} \rangle} = MH$. As M and H are admissible, it follows from Theorem 4.8 that $G_{\langle \mathbf{v} \rangle}$ is admissible.

Let $U_1, \ldots, U_m$, $m = (q^{n-1} - 1)/(q - 1)$ be the two-dimensional subspaces of V_n containing $\mathbf{v}$. For each i, $i = 1, \ldots, m$, let π_i be a permutation of the one-dimensional subspaces of U_i that fixes $\langle \mathbf{v} \rangle$ only: it is an exercise to show such permutations exist using the involution constructed in Lemma 5.25.

Define π, a permutation of the one-dimensional subspaces of V_n, by $\pi(\langle \mathbf{u} \rangle) = \pi_i(\langle \mathbf{u} \rangle)$ if $\langle \mathbf{u} \rangle \in U_i$. Then π is well-defined, and by Lemma 5.25, there exists, for each $\langle \mathbf{u} \rangle \neq \langle \mathbf{v} \rangle$, an involution $\sigma_{\langle \mathbf{u} \rangle} \in \mathrm{SL}(2, q)$ that swaps $\langle \mathbf{u} \rangle$ and $\langle \mathbf{v} \rangle$ and fixes $\pi(\langle \mathbf{u} \rangle)$. Set $\sigma_{\langle \mathbf{v} \rangle} = I$, and set $D = \{\sigma_{\langle \mathbf{u} \rangle} \mid \langle \mathbf{u} \rangle$ as a one-dimensional subspace of $V_n\}$. As $|D|$ is equal to the number of one-dimensional subspaces of V_n, D has the right number of elements to be a dual system of coset representatives for $G_{\langle x \rangle}$ in G. Now, as $\sigma_{\langle \mathbf{u} \rangle}^{-1} \sigma_{\langle \mathbf{w} \rangle} = \sigma_{\langle \mathbf{u} \rangle} \sigma_{\langle \mathbf{w} \rangle} = \sigma_{\langle \mathbf{u} \rangle} \sigma_{\langle \mathbf{w} \rangle}^{-1}$, D is a system of left coset representatives for $G_{\langle x \rangle}$ in G if and only if D is a system of right coset representatives for $G_{\langle x \rangle}$ in G if and only if D is a dual system of coset representatives for $G_{\langle x \rangle}$ in G if and only if $\sigma_{\langle \mathbf{u} \rangle} \sigma_{\langle \mathbf{w} \rangle} \notin G_{\langle \mathbf{v} \rangle}$ if $\sigma_{\langle \mathbf{u} \rangle} \neq \sigma_{\langle \mathbf{w} \rangle}$. Now $\langle \mathbf{v} \rangle^{\sigma_{\langle \mathbf{u} \rangle} \sigma_{\langle \mathbf{w} \rangle}} = \langle \mathbf{u} \rangle^{\sigma_{\langle \mathbf{w} \rangle}} = \langle \mathbf{v} \rangle$ if and only if $\langle \mathbf{u} \rangle = \langle \mathbf{w} \rangle$. Hence D is a dual system of coset representatives for $G_{\langle x \rangle}$ in G. Now $\langle \mathbf{u} \rangle^{\sigma_{\langle \mathbf{u} \rangle}} = \langle \mathbf{v} \rangle$ for each one-dimensional subspace $\langle \mathbf{u} \rangle$ of V_n, and so D satisfies the conditions of Lemma 4.23. By construction $\pi(\langle \mathbf{u} \rangle)^{\sigma_{\langle \mathbf{u} \rangle}} = \langle \mathbf{u} \rangle$ and so, by Lemma 4.23, D is admissible. Hence, by Theorem 4.1, G is admissible. The result then follows by induction. □

We obtain the admissibility of GL(n, q), $n \geq 2$ and q even, n and q not both 2, as a corollary.

Corollary 5.27 (Dalla Volta and Gavioli, 1993). GL(n, q) *is admissible for* $n \geq 2$ *and* q *even,* n *and* q *not both* 2.

Proof. If $n \geq 2$ and q are even, and n and q not both 2, then SL(n, q) is admissible by Theorem 5.26. Further GL(n, q)/SL(n, q) is isomorphic to the multiplicative group of GF(q), a group of odd order, and hence is admissible by Corollary 3.6. The admissibility of GL(n, q) then follows from Corollary 1.38. □

The proof that PSL(n, q) is admissible for $n \geq 2$ and q even, n and q not both 2, is similar to the proof that SL(n, q) is admissible for $n \geq 2$ and q even, n and q not both 2, except that we will not use induction but will instead rely on the already determined admissibility of SL(n, q).

Theorem 5.28 (Dalla Volta and Gavioli, 1993). PSL(n, q) *is admissible for* $n \geq 2$ *and* q *even,* n *and* q *not both* 2.

Proof. If q is even, then PSL(2, q) $\cong$ SL(2, q), and, hence, by Theorem 5.5, PSL(2, q) is admissible for q even, $q \neq 2$. Thus we will assume that $n \geq 3$. Let

$$Z = Z(\mathrm{SL}(n, q)) = \{aI \mid a^n = 1\},$$

and let ϕ: SL(n, q) $\to$ PSL(n, q) be the canonical homomorphism mapping A to AZ. We will let $G = \mathrm{SL}(n, q)$ and $G^* = \mathrm{PSL}(n, q)$.

Our proof of the admissibility will use elements of the proof of Theorem 5.26: $\mathbf{v}$, K, L, and M will be as in the proof of Theorem 5.26; but H will be redefined as

$$H = \{\mathrm{diag}(a, a^{1-n}, a, \ldots, a) \mid a \in \mathrm{GF}(q), a \neq 0\}.$$

Now $M \cap Z = 1$, and, hence, $\phi(M) \cong M$ and so is admissible whenever SL$(n-1, q)$ is admissible. As H is isomorphic to the multiplicative group of GF(q), which is of odd order, $\phi(H)$ is also of odd order and thus is admissible by Corollary 3.6. Further, as $\phi(H)$ and $\phi(M)$ are subgroups of $G^*_{\langle \mathbf{v} \rangle}$, $\phi(H) \cap \phi(M) = 1$, and

$$|\phi(M)||\phi(H)| = \frac{q^{n-1}(q-1)|SL(n-1, q)|}{|Z|} = |G^*_{\langle \mathbf{v} \rangle}|,$$

$G^*_{\langle \mathbf{v} \rangle} = \phi(M)\phi(H)$. As $\phi(M)$ and $\phi(H)$ are admissible, it follows from Theorem 4.8 that $G^*_{\langle \mathbf{v} \rangle}$ is admissible.

To complete the proof, let D be as in Theorem 5.26 and set $D^* = \{\tau Z \mid \tau \in D\}$. The proof that D^* is admissible is identical to the proof that D is admissible. The result then follows from Theorem 4.1. □

We obtain the admissibility of PGL(n, q), $n \geq 2$ and q even, n and q not both 2, as a corollary.

Corollary 5.29 (Dalla Volta and Gavioli, 1993). PGL(n, q) *is admissible for* $n \geq 2$ *and* q *even,* n *and* q *not both* 2.

Proof. Similar to the proof of Corollary 5.27. □

5.4 The Groups GL(n, q) and PGL(n, q), q Odd

In 1997, Dalla Volta and Gavioli [84] proved GL(n, q) and PGL(n, q) to be admissible when q is odd. The principal tool in establishing the admissibility of these groups is the 3-transitivity of GL$(2, q)$ and PGL$(2, q)$. Let V_n denote the n-dimensional vector space of column vectors over $GF(q)$. The mapping $x \mapsto Ax$ describes an action of GL(n, q) on V_n. In this action GL(n, q) permutes the one-dimensional subspaces of V_n. It is well-known that the action of GL$(2, q)$ on the one-dimensional subspaces of V_2 is 3-transitive: to prove this consider the action of GL$(2, q)$ on the set of ordered bases. From this we can deduce that, if U is a 2-dimensional subspace of V_n, then the stabilizer of U in GL(n, q) acts 3-transitively on the one-dimensional subspaces of U. Now $Z = \{aI \mid a \in \text{GF}(q), a \neq 0\}$ is a normal subgroup of GL(n, q) and PGL$(n, q) \cong$ GL$(n, q)/Z$. As each element of Z fixes each one-dimensional subspace of V_n, the action of GL(n, q) induces a natural action of PGL(n, q) on the one-dimensional subspaces of V_n. If $n = 2$, then this action is 3-transitive. Further, if U is a 2-dimensional subspace of V_n, then the stabilizer of U in PGL(n, q) acts 3-transitively on the one-dimensional subspaces of U.

For $q \equiv 3 \pmod 4$ the admissibility of PGL$(2, q)$ was proved by showing PGL$(2, q)$ to be factorable with admissible factors.

Theorem 5.30 (Dalla Volta and Gavioli, 1997). *If* $q \equiv 3 \pmod 4$, *then* $\mathrm{PGL}(2,q)$ *is admissible.*

Proof. For $a, b \in \mathrm{GF}(q)$, let

$$A_{a,b} = \begin{pmatrix} a & b \\ -b & a \end{pmatrix}, \; B_{a,b} = \begin{pmatrix} 1 & a \\ 0 & b \end{pmatrix}, \; C_a = \begin{pmatrix} a & 0 \\ 0 & a \end{pmatrix}, \; D = \begin{pmatrix} -1 & 0 \\ 0 & 1 \end{pmatrix},$$

and, for α a given nonsquare in $\mathrm{GF}(q)$, let

$$E = \begin{pmatrix} 0 & 1 \\ \alpha & 0 \end{pmatrix}.$$

Simple computation yields

$$A_{a,b}A_{c,d} = A_{ac-bd,ad+bc}, \; B_{a,b}B_{c,d} = B_{c+ad,bd},$$

$$C_aC_b = C_{ab}, \; DA_{a,b} = A_{a,-b}D, \text{ and } B_{0,-1}E = -EB_{0,-1}.$$

Let $Z = \{C_a \mid a \in \mathrm{GF}(q), a \neq 0\}$, the center of $\mathrm{GL}(2,q)$, and let $\phi\colon \mathrm{GL}(2,q) \to \mathrm{GL}(2,q)/Z \cong \mathrm{PGL}(2,q)$ be the canonical homomorphism. As $|\mathrm{GL}(2,q)| = (q^2-1)(q^2-q)$ and $|Z| = q-1$, it follows that $|\mathrm{PGL}(2,q)| = q(q^2-1)$.

We will show that $\mathrm{PGL}(2,q)$ is a factorable group with admissible factors.

Let $H = \{A_{a,b} \mid a, b \in \mathrm{GF}(q), a^2+b^2 \neq 0\}$. If $a^2+b^2 \neq 0$ and $c^2+d^2 \neq 0$, then $(ac-bd)^2+(ad+bc)^2 = (a^2+b^2)(c^2+d^2) \neq 0$. It follows that H is an abelian subgroup of $\mathrm{GL}(2,q)$. Let $K = \{B_{a,b} \mid a, b \in GF(q), b \text{ be a nonzero square}\}$, and $L = \langle D\rangle$. The sets K, L, K, and HD are also subgroups of $\mathrm{GL}(2,q)$. Now $\phi(K)$ is a subgroup of $\mathrm{PGL}(2,q)$ of odd order $q(q-1)/2$ and, hence, is admissible by Corollary 3.6, and the group $\phi(HL)$ is a group of order $2(q+1)$. As H is abelian and is a normal subgroup of HL of index 2, HL and, hence, $\phi(HL)$ is solvable. As DZ and $A_{0,1}Z$ are distinct involutions in $\phi(HL)$, the Sylow 2-subgroup of $\phi(HL)$ is noncyclic. It follows that $\phi(HL)$ is admissible by Theorem 4.22. As $|\phi(K)||\phi(HL)| = q(q^2-1) = |\mathrm{PGL}(2,q)|$ and $\gcd(|\phi(K)|, |\phi(HL)|) = 1$, by Lemma 4.7, $\mathrm{PGL}(2,q)$ is factorable with factors $\phi(K)$ and $\phi(HL)$. As both factors are admissible, $\mathrm{PGL}(2,q)$ is admissible by Theorem 4.8. □

In the proof of Theorem 5.30, Dalla Volta and Gavioli's construction of the group K differs from the construction that we have given here. We can extend the action of $\mathrm{PGL}(2,q)$ on π_1, the set of one-dimensional subspaces of V_2. Let $\psi\colon \mathrm{PGL}(2,q) \to \mathrm{PGL}(2,q)/\mathrm{PSL}(2,q) \cong \mathbb{Z}_2$ be the canonical homomorphism. Set $E = \pi_1 \times \mathbb{Z}_2$, and define an action of $\mathrm{PGL}(2,q)$ on E by $(p,e)^{AZ} = (Ap, e + \psi(AZ))$. If $p = \langle (0,1)^T\rangle$, then the stabilizer of $(p,0)$ is a subgroup of $\phi(N)$ of index 2, where $N = \{C_aB_{c,d} \mid a, b, c \in \mathrm{GF}(q), a, c \neq 0\}$: the subgroup $\phi(K)$, that we constructed, is a subgroup of $\phi(N)$ of index 2.

In 1997, Dalla Volta and Gavioli [84] proved $\mathrm{PGL}(2, q)$ to be admissible when $q \equiv 1 \pmod 4$ by partitioning the element set of the group. To prove this result, we need to define special near orthomorphisms. If G is a cyclic group with a unique involution g, then $G^{\#} = G \setminus \{1, g\}$, and a *special near orthomorphism* of G is a bijection $\theta\colon G^{\#} \to G^{\#}$ for which the mapping $\eta\colon x \mapsto x^{-1}\theta(x)$ is also a bijection $G^{\#} \to G^{\#}$. The mapping η is a complete mapping of $G^{\#}$. In 1996 Keedwell and Wang [220] proved that every cyclic group of even order admits a special near orthomorphism.

Lemma 5.31. *$\mathbb{Z}_n$, n even, admits a special near orthomorphism.*

Proof. See [220]. □

Note that Lemma 5.31 shows that, if n is even, then $\mathbb{Z}_n^{\#}$ admits complete mappings.

Theorem 5.32 (Dalla Volta and Gavioli, 1997). *If $q \equiv 1 \pmod 4$, then $\mathrm{PGL}(2, q)$ is admissible.*

Proof. Let the matrices $A_{a,b}$, $B_{a,b}$, C_a, D, and E be defined as in Theorem 5.30, and let Z be as in theorem 5.30.

As $\mathrm{PGL}(2, 5) \cong S_5$, $\mathrm{PGL}(2, 5)$ is admissible by Theorem 4.6. Thus we may assume that $q \geq 9$.

The group $\mathrm{GL}(2, q)$ acts on V, a vector space of dimension 2 over $\mathrm{GF}(q)$, by multiplication on the right. Under this action, the group Z fixes every one-dimensional subspace of V, and, hence, this action induces an action of $\mathrm{PGL}(2, q)$ on π_1, the set of one-dimensional subspaces of V. Now $\mathrm{PGL}(2, q)$ acts sharply 3-transitively of the elements of π_1. The stabilizer of $\langle(0, 1)\rangle$ is $\{B_{a,b} \mid a, b \in \mathrm{GF}(q), b \neq 0\}Z$, which contains the subgroup PZ, where $P = \{B_{a,1} \mid a \in \mathrm{GF}(q)\}$; and the point-stabilizer of $\{\langle(0, 1)\rangle, \langle(1, 0)\rangle\}$ is MZ where $M = \{B_{0,a} \mid a \in \mathrm{GF}(q), a \neq 0\}$. Clearly $\phi(P) \cong \mathrm{GF}(q)^+$ and $\phi(M) \cong \mathbb{Z}_{q-1}$. Let $F = \{aE + bI \mid a, b \in \mathrm{GF}(q), (a, b) \neq (0, 0)\}$, a subgroup of $\mathrm{GL}(2, q)$. As the set $\{aE + bI \mid a, b \in \mathrm{GF}(q)\}$ is a field of order q^2, $F \cong \mathbb{Z}_{q^2-1}$ and, as $Z \subseteq F$, $\phi(F) \cong \mathbb{Z}_{q+1}$.

There are precisely $q + 1$ distinct conjugates of $\phi(P)$ in $\mathrm{PGL}(2, q)$, and these intersect pairwise trivially: $\mathrm{PGL}(2, q)$ acts transitively on the elements of π_1, and the only eigenvalue of any nontrivial element of P is 1 with corresponding eigenspace $\langle(0, 1)\rangle$. Let $P_1, \dots, P_{q+1}$ denote the conjugates of $\phi(P) \setminus \{\phi(I)\}$ in $\mathrm{PGL}(2, q)$. There are precisely $q(q + 1)/2$ distinct conjugates of $\phi(M)$ in $\mathrm{PGL}(2, q)$, and these intersect pairwise trivially: $\mathrm{PGL}(2, q)$ acts sharply triply transitively on the elements of π_1, and the intersection of two distinct conjugates of $\phi(M)$ fixes at least three elements of π_1. Let $M_1, \dots, M_{q(q+1)/2}$ denote the conjugates of $\phi(M) \setminus \{\phi(I)\}$ in $\mathrm{PGL}(2, q)$. The normalizer of $\phi(F)$ in $\mathrm{PGL}(2, q)$ is the group $\langle F, \{B_{0,-1}\}\rangle$ of order $2(q + 1)$. Hence, there are $|\mathrm{PGL}(2, q)|/(2(q + 1))$ distinct conjugates of $\phi(F)$ in $\mathrm{PGL}(2, q)$. We leave it as an exercise to show that distinct conjugates of $\phi(F)$ intersect trivially. Let $F_1, \dots, F_{q(q-1)/2}$ be the conjugates of $\phi(F) \setminus \{\phi(I)\}$ in $\mathrm{PGL}(2, q)$. We will show that the sets $\{\phi(I)\}$, $P_1, \dots, P_{q+1}$, $\bigcup_{i=1}^{q(q+1)/2} M_i$, $\bigcup_{i=1}^{q(q-1)/2} F_i$ partition $\mathrm{PGL}(2, q)$.

As $\gcd(|\phi(P)|, |\phi(M)|) = 1$ and $\gcd(|\phi(P)|, |\phi(F)|) = 1$, $P_i \cap M_j = \emptyset$ and $P_i \cap F_j = \emptyset$ for all i and j. Further $M_i \cap F_j = \emptyset$ for all i and j: consider eigenvalues. Simple computation shows that

$$\begin{aligned}|\{\phi(I)\}| &+ \textstyle\sum_{i=1}^{q+1} |P_i| + \sum_{i=1}^{q(q+1)/2} |M_i| + \sum_{i=1}^{q(q-1)/2} |F_i| \\ &= 1 + (q+1)(q-1) + \tfrac{q(q+1)}{2}(q-2) + \tfrac{q(q-1)}{2}q \\ &= q(q^2-1) = |\mathrm{PGL}(2,q)|.\end{aligned}$$

Hence, the sets $\{\phi(I)\}$, $P_1, \ldots, P_{q+1}$, $\bigcup_{i=1}^{q(q+1)/2} M_i$, $\bigcup_{i=1}^{q(q-1)/2} F_i$ partition PGL$(2,q)$.

We next need to show that each of the sets $\{\phi(I)\}$, $P_1, \ldots, P_{q+1}$, $\bigcup_{i=1}^{q(q+1)/2} M_i$, $\bigcup_{i=1}^{q(q-1)/2} F_i$ is admissible. The set $\{\phi(I)\}$ is admissible: the mapping $\phi(I)\colon \phi(I) \to \phi(I)$ is a complete mapping. As $\phi(P)$ is of odd order, it admits a complete mapping by Corollary 3.6 and hence a normalized complete mapping. It follows that each of $P_1, \ldots, P_{q+1}$ is admissible. The proof that $\bigcup_{i=1}^{q(q+1)/2} M_i$ is admissible is very similar to the proof that H is admissible in Theorem 5.24.

By Lemma 5.31, $\bigcup_{i=1}^{q(q-1)/2} F_i^{\#}$ is admissible: for each i, $F_i^{\#} = F_i \setminus \{f_i\}$, where f_i is the unique involution of F_i; each f_i is a conjugate of $\phi(E)$. We leave it as an exercise to prove that $\{f_1, \ldots, f_{q(q-1)/2}\}$ admits a complete mapping, thus completing the proof. □

To prove that GL(n,q) is admissible for $n \geq 2$ and q odd, we first prove the admissibility of GL$(2,q)$ using the triple transitivity of GL$(2,q)$ on the one-dimensional subspaces of V_2 followed by induction on n.

Theorem 5.33 (Dalla Volta and Gavioli, 1997). GL(n,q) *is admissible if* q *is odd and* $n \geq 2$.

Proof. We know that GL$(2,q)$ acts triply transitively on the set of one-dimensional subspaces of V_2, which we have denoted π_1. Let

$$N^2_{a,b,c} = \begin{pmatrix} a & b \\ 0 & c \end{pmatrix}.$$

The stabilizer of $\langle (1,0)^T \rangle$ is $G_2 = \{N^2_{a,b,c} \mid a,b,c \in \mathrm{GF}(q), a,c \neq 0\}]$. Now $N = \{N^2_{1,b,1} \mid b \in \mathrm{GF}(q)\} \cong \mathrm{GF}(q)^+$ is a normal subgroup of G_2 and, being of odd order, is admissible by Corollary 3.6. Further $G_2/N \cong \mathbb{Z}_{q-1} \times \mathbb{Z}_{q-1}$, an abelian group with a noncyclic Sylow 2-subgroup, and, hence, is admissible by Theorem 3.9. It follows from Corollary 1.38 that G_2 is admissible. Thus, as G_2 is a point stabilizer in a 3-transitive action of GL$(2,q)$ on π_1, GL$(2,q)$ is admissible by Lemma 4.24. The construction in Lemma 4.24 yields an admissible dual system of coset representatives, D_2, for G_2 in GL$(2,q)$.

Let

$$N^n_{a,B,A} = \begin{pmatrix} a & B \\ 0 & A \end{pmatrix},$$

$a \in \mathrm{GF}(q)$, B a $1 \times (n-1)$ matrix with entries from $\mathrm{GF}(q)$ and A an $(n-1) \times (n-1)$ matrix with entries from $\mathrm{GF}(q)$. Let

$$G_n = \{N^n_{a,B,A} \mid a \neq 0, A \in \mathrm{GL}(n-1, q)\}.$$

We will show inductively that, for all $n \geq 2$, G_n is admissible, and there exist an admissible dual system of coset representatives, D_n, for G_n in $\mathrm{GL}(n, q)$. The admissibility of $\mathrm{GL}(n, q)$ for all $n \geq 2$ will then follow from Theorem 4.1. As we have already shown this to be true for $n = 2$, let us assume that $n > 2$.

Let

$$K_n = \{N^n_{a,0,A} \mid a \neq 0, A \in \mathrm{GL}(n-1, q)\}, \; M_n = \{N^n_{a,0,I} \mid a \neq 0\},$$

$$N_n = \{N^n_{1,0,A} \mid A \in \mathrm{GL}(n-1, q)\}, \; H_n = \{N^n_{a,0,A} \mid a \neq 0, A \in G_{n-1}\},$$

and

$$L_n = \{N^n_{1,0,A} \mid A \in G_{n-1}\}.$$

By the inductive hypothesis, G_{n-1} is admissible, and there is an admissible dual system of coset representatives, D_{n-1}, for G_{n-1} in $\mathrm{GL}(n-1, q)$. It follows that L_n is admissible and the set $T_n = \{N^n_{1,0,A} \mid A \in D_{n-1}\}$ is an admissible dual system of coset representatives for L_n in N_n.

Let

$$U_n = \{N^n_{a,0,A} \mid a \neq 0, A \in K_{n-1}\}, \quad \text{and} \quad Y_n = \{N^n_{1,0,A} \mid a \neq 0, A = N^{n-1}_{1,B,I}\}.$$

Now $U_n \cong \mathbb{Z}_{q-1} \times \mathbb{Z}_{q-1} \times G_{n-1}$ and $Y_n \cong (\mathrm{GF}(q)^+)^{n-2}$. If $n = 3$, then $U_n \cong \mathbb{Z}_{q-1} \times \mathbb{Z}_{q-1} \times \mathbb{Z}_{q-1}$ which is admissible by Theorem 3.9. If $n > 3$, then, as $\mathbb{Z}_{q-1} \times \mathbb{Z}_{q-1}$ is admissible by Theorem 3.9 and G_{n-2} is admissible by the inductive hypothesis, U_n is admissible by Corollary 1.38. As $|Y_n|$ is odd, Y_n is admissible by Corollary 3.6. Now $U_n \cap Y_n = \{I\}$ and $H_n = U_n Y_n$. Hence, H_n is admissible by Theorem 4.8. Further, as $\{I\} \times T_n$ is an admissible dual system of coset representatives for H_n in K_n, K_n is admissible by Theorem 4.1. Next, let

$$W_n = \{N^n_{1,B,I} \mid B \text{ a } 1 \times (n-1) \text{ matrix over } \mathrm{GF}(q)\}.$$

Now $G_n = K_n W_n$, W_n is of odd order and hence admissible by Corollary 3.6, and $K_n \cap W_n = \{I\}$, G_n is admissible by Theorem 4.8.

It remains to show that there exists an admissible dual system of coset representatives for G_n in GL(n, q). Now G_n is the stabilizer of the one-dimensional subspace $\langle(1, 0, \ldots, 0)^T\rangle$ of V_n. Let $X_1, \ldots, X_m$, $m = (q^{n-1} - 1)/(q - 1)$, be the two-dimensional subspaces of V_n that contain $\langle(1, 0, \ldots, 0)^T\rangle$, and let G_{X_i} be the stabilizer in GL(n, q) of X_i. As G_{X_i} acts three-transitively on the one-dimensional subspaces of X_i, there exists an admissible dual system of coset representatives, D_i, for $G_n \cap G_{X_i}$ in G_{X_i} by Lemma 4.24. Let $D = \bigcup_{i=0}^{m} D_i$. We claim that D is a dual system of coset representatives for G_n in GL(n, q). Let $d \in D_i \setminus \{I\}$ and $d' \in D_j \setminus \{I\}$. If $i = j$ and $d \neq d'$, then, as D_i is a dual system of coset representatives for $G_n \cap G_{X_i}$ in G_{X_i}, neither $d^{-1}d'$ nor $d'd^{-1}$ is in G_n. If $i \neq j$, then, as $X_i \cap X_j = \langle(1, 0, \ldots, 0)^T\rangle$, neither $d^{-1}d'$ nor $d'd^{-1}$ can stabilize $\langle(1, 0, \ldots, 0)^T\rangle$, and so neither $d^{-1}d'$ nor $d'd^{-1}$ is in G_n. It follows that D is a partial dual system of coset representatives for G_n in GL(n, q). By the construction in Lemma 4.24, $I \in D_i$ for each $i = 1, \ldots, m$. Hence $\{I\}, D_1 \setminus \{I\}, \ldots, D_1 \setminus \{I\}$ partitions D. Now $|D_i| = |G_{X_i}|/|G_{X_i} \cap G_n| = q + 1$. Thus

$$|D| = qm + 1 = \frac{q(q^{n-1} - 1)}{q - 1} + 1 = \frac{q^n - 1}{q - 1} = |\mathrm{GL}(n, q)|/|G_n|$$

and, hence, D is a dual system of coset representatives for G_n in GL(n, q). By the construction in Lemma 4.24, each D_i admits a complete mapping that fixes I. Further the mapping $I \mapsto I$ is a complete mapping of $\{I\}$. It follows that D is admissible by Lemma 5.2 and, hence, that GL(n, q) is admissible by Theorem 4.1. □

The proof that PGL(n, q) is admissible for q odd and $n \geq 3$ is similar to the proof of Theorem 5.33.

Theorem 5.34 (Dalla Volta and Gavioli, 1997). PGL(n, q) *is admissible if* q *is odd and* $n \geq 3$.

Proof. Let $Z = \{aI \mid a \in \mathrm{GF}(q), a \neq 0\}$, and let $\phi\colon \mathrm{GL}(n, q) \to \mathrm{GL}(n, q)/Z \cong \mathrm{PGL}(n, q)$ be the canonical homomorphism. Let G_n, K_n, W_n, N_n, and T_n be as in the proof in Theorem 5.33. Now $\phi(G_n) = \phi(K_n)\phi(W_n)$, $\phi(K_n) \cap \phi(W_n) = \{I\}$, and $\phi(K_n) \cong N_n$. As L_n is admissible and T_n is an admissible dual system of coset representatives for L_n in N_n, N_n, and hence $\phi(K_n)$ is admissible by Theorem 4.1. Further, as $\phi(W_n)$ is of odd order, $\phi(W_n)$ is admissible by Corollary 3.6, and hence $\phi(G_n)$ is admissible by Theorem 4.1. The proof that there exists an admissible dual system of coset representatives for $\phi(G_n)$ in PGL(n, q) is similar to the proof that there exists an admissible dual system of coset representatives for G_n in GL(n, q) given in the proof of Theorem 5.33. □

Chapter 6
Minimal Counterexamples to the Hall-Paige Conjecture

If the Hall-Paige conjecture is false, what might a minimal counterexample look like? This question was first raised by Aschbacher [21] in 1990. He proved that any minimal counterexample must be "close" to being a simple group. Note that, we will use the term minimal counterexample to mean a counterexample of least order. In 1992 Evans [120] tried to improve on this by extending complete mappings of H, a subgroup of G of index 2, to complete mappings of G, and complete mappings of G/H, $H \cong \mathbb{Z}_2$, to complete mappings of G. Evans' proofs require that certain technical conditions hold for the extensions to work. In 2001 Dalla Volta and Gavioli [85] improved Aschbacher's reduction, and in 2009, using elementary techniques, Wilcox [329] showed that any minimal counterexample to the Hall-Paige conjecture must be a nonabelian finite simple group. We will present these reductions in this chapter.

6.1 Aschbacher's Reduction

In 1990 Aschbacher [21] tackled the Hall-Paige conjecture. He considered the problem of what a minimal counterexample to the Hall-Paige conjecture might look like and presented his results in lectures at NSA in the summer of 1990. Aschbacher proved that any minimal counterexample to the Hall-Paige conjecture must be "close" to being simple.

6.1.1 Some Needed Group Theory

Let us introduce some group theory that will play an important role in Aschbacher's reduction. By Theorem 4.22, finite solvable groups satisfy the Hall-Paige conjecture. Thus any minimal counterexample to the Hall-Paige conjecture must be

A. B. Evans, *Orthogonal Latin Squares Based on Groups*, Developments in Mathematics 57, https://doi.org/10.1007/978-3-319-94430-2_6

nonsolvable. We begin by showing that any finite nonsolvable group must have a noncyclic Sylow 2-subgroup. This is a consequence of the celebrated Feit-Thompson Theorem.

Theorem 6.1 (Feit and Thompson, 1963). *If $|G|$ is odd, then G is solvable.*

Proof. See [141]. □

The structure of the Sylow 2-subgroup of a finite nonsolvable group can now be deduced.

Corollary 6.2. *If G is finite and nonsolvable, then the Sylow 2-subgroup of G is noncyclic.*

Proof. If G is finite and nonsolvable, then it cannot be of odd order as groups of odd order are solvable by the Feit-Thompson Theorem. Thus, the Sylow 2-subgroup of G is nontrivial. If the Sylow 2-subgroup S of G is cyclic, then by Lemma 3.33, G has a normal subgroup H of odd order for which $G/H \cong S$. By the Feit-Thompson Theorem, H is solvable and so G is solvable, a contradiction from which the result follows. □

Thus the Hall-Paige conjecture is equivalent to Conjecture 3.40 that all nonsolvable groups are admissible, a conjecture made by Dénes and Keedwell [92] in 1989.

If X and Y are subsets of a group G, then $[X, Y] = \{x^{-1}y^{-1}xy \mid x \in X, y \in Y\}$. The *commutator subgroup* of G, also called the *derived subgroup* of G, is $G' = [G, G]$. A group G is *quasisimple* if $G' = G$ and $G/Z(G)$ is simple. Aschbacher proved that minimal counterexamples to the Hall-Paige conjecture are "almost" quasisimple. A subgroup H is a *subnormal subgroup* of G if there exists a normal series $H = G_0 \trianglelefteq G_1 \trianglelefteq \cdots \trianglelefteq G_n = G$. A *component* of G is a subnormal quasisimple subgroup of G. We will use $E(G)$ to denote the subgroup of G generated by the components of G. Define L_n recursively by $L_1 = G$ and $L_n = [L_{n-1}, G]$. A group G is *nilpotent* if $L_n = \{1\}$ for some positive integer n. Note that all nilpotent groups are solvable. The *Fitting subgroup* of G, $F(G)$ is the largest nilpotent normal subgroup of G. The *generalized Fitting subgroup* of G is $F^*(G) = F(G)E(G)$. An easy argument proves that $F^*(G)$ is a normal subgroup of G.

Lemma 6.3. *The generalized Fitting subgroup, $F^*(G)$, is a normal subgroup of G.*

Proof. If $f_1, f_2 \in F(G)$ and $e_1, e_2 \in E(G)$, then

$$f_1e_1f_2e_2 = f_1(e_1f_2e_1^{-1})e_1e_2 \in F(G)E(G) = F^*(G).$$

Thus $F^*(G)$ is a subgroup of G. $F(G)$ is a normal subgroup of G by definition, and, as the conjugate of a component of G is a component of G, $E(G)$ is also a normal subgroup of G. The result follows. □

It should be noted that $F^*(G)$ is in fact a characteristic subgroup of G. We use $O(G)$ to denote the largest normal subgroup of G of odd order and $O_2(G)$ the largest normal 2-subgroup of G. More generally, if $\pi(G) = \{p \text{ prime} \mid p||G|\}$, then, for $p \in \pi(G)$, $O_p(G)$ is the largest normal p-subgroup of G.

Lemma 6.4. *The Fitting subgroup, $F(G)$, is the direct product of the groups $O_p(G)$, $p \in \pi(G)$.*

Proof. See result 31.8 in [23]. □

Lemma 6.5. *If $F^*(G)$ is abelian, then $G/F^*(G)$ is isomorphic to a subgroup of* Aut$(F^*(G))$.

Proof. If we define $\phi_x : F^*(G) \to F^*(G)$, $x \in G$, by $\phi_x(y) = xyx^{-1}$, then the mapping $x \mapsto \phi_x$ defines a homomorphism $G \to \mathrm{Aut}(F^*(G))$. By 31.13 in [23], $C_G(F^*(G)) \subseteq F^*(G)$. Hence, as $F^*(G)$ is abelian, $C_G(F^*(G)) = F^*(G)$. It follows that $G/F^*(G)$ is isomorphic to a subgroup of $\mathrm{Aut}(F^*(G))$. □

For more information on the generalized Fitting subgroup, consult [23]. There are two further results that we will need, the Frattini argument and the fact that the automorphism group of a cyclic 2-group is a 2-group.

Theorem 6.6 (The Frattini Argument). *If H is a normal subgroup of G, p is a prime dividing $|G|$, and P is a Sylow p-subgroup of H, then $G = HN_G(P)$.*

Proof. See result 6.2 in [23]. □

Theorem 6.7. *The automorphism group of a cyclic* 2*-group is a* 2*-group.*

Proof. Let G be a cyclic group of order 2^n with generator g. If α is an automorphism of G, then $\alpha(g) = g^i$ for some integer $i \in \{1, \ldots, n-1\}$, and $\alpha(x) = x^i$ for all $x \in G$. Further, the mapping $x \mapsto x^i$ is an automorphism of G if and only if $\gcd(i, n) = 1$. Hence $|\mathrm{Aut}(G)| = \phi(n) = 2^{n-1}$. It follows that Aut$(G)$ is a 2-group.
□

6.1.2 More on HP-Systems

Recall that an HP-system (H, D, θ, η) for a group G consists of a subgroup H of G: a dual system of coset representatives, D, for H in G; and two bijections $\theta, \eta : D \to D$ satisfying $x\theta(x)H = \eta(x)H$ for all $x \in D$. We showed in Theorem 4.1 that, if an HP-system, (H, D, θ, η), for a group G exists and H is admissible, then G is admissible. Aschbacher made use of HP-systems in his reduction using a reverse quotient group construction of HP-systems.

Lemma 6.8 (Aschbacher, 1990). *Let K, H, G be finite groups, $K \leq H \leq G$, K normal in G. If there exists an HP-system for H/K in G/K, then there exists an HP-system for H in G.*

Proof. Let $D^* = \{d_1K, \ldots, d_mK\}$ be a dual system of coset representatives for $H^* = H/K$ in $G^* = G/K$, and set $D = \{d_1, \ldots, d_k\}$. Let $(H^*, D^*, \theta^*, \eta^*)$ be an

HP-system for H^* in G^*, and define $\theta, \eta\colon D \to D$ by $\theta(d_i) = d_j$ if $\theta^*(d_i K) = d_j K$ and $\eta(d_i) = d_j$ if $\eta^*(d_i K) = d_j K$.

We claim that (H, D, θ, η) is an HP-system for H in G. Let $\pi\colon G \to G/K$ be the canonical homomorphism defined by $\pi(g) = gK$. If $d_i H = d_j H$, then $\pi(d_i H) = d_i H^* = d_j H^* = \pi(d_j H)$, which implies that $i = j$. Similarly, if $Hd_i = Hd_j$, then $i = j$, which implies, as $|D| = |G/H|$, that D is a dual system of coset representatives for H in G. Clearly θ and η are bijections. Thus, for $i \in \{1, \ldots, m\}$, $d_i\theta(d_i)H = \eta(d_j)H$ for some j. But then $\pi(d_i\theta(d_i)H) = d_i K\theta^*(d_i K)H^* = \eta^*(d_i K)H^*$ and $\pi(\eta(d_j)H) = \eta^*(d_j K)H^*$, which implies that $j = i$, completing the proof. □

6.1.3 *Minimal Counterexamples*

Aschbacher established the structure of a possible minimal counterexample to the Hall-Paige conjecture through a sequence of reductions, the first of which are described in Lemmas 6.9 through 6.13.

Lemma 6.9. *If G is a minimal counterexample to the Hall-Paige conjecture and H is a proper, nontrivial, normal subgroup of G, then either H has a nontrivial, cyclic Sylow 2-subgroup or G/H has a nontrivial, cyclic Sylow 2-subgroup.*

Proof. If neither H nor G/H has a nontrivial, cyclic Sylow 2-subgroup, then, by the minimality of G, both are admissible. But then G is admissible by Corollary 1.38, a contradiction. □

Lemma 6.10. *If G is a minimal counterexample to the Hall-Paige conjecture, then $O(G) = \{1\}$.*

Proof. If $O(G) \neq \{1\}$, then $H = O(G)$ has a trivial Sylow 2-subgroup, and thus, by Lemma 6.9, G/H must have a nontrivial, cyclic Sylow 2-subgroup. This implies that the Sylow 2-subgroup of G/H, and hence of G, is nontrivial and cyclic and, hence, that G cannot be a counterexample to the Hall-Paige conjecture. □

Lemma 6.11. *If G is a minimal counterexample to the Hall-Paige conjecture and H has a noncyclic Sylow 2-subgroup and is a proper, normal subgroup of G, then G/H is a cyclic 2-group.*

Proof. Assume this to be false and let $\pi : G \to G/H$ be the canonical homomorphism. Let S be a Sylow 2-subgroup of G. By Lemma 6.9, $S^* = \pi(S)$ must be cyclic. Thus, by Lemma 3.33 and Theorem 6.1, $G^* = \pi(G)$ is solvable, and hence, by Lemma 4.21, $G^* = N^*S^*$, where N^* is an odd-order subgroup of G^*. Set $K = SH$. Then $K^* = \pi(K) = S^*$. Define mappings $\theta, \eta\colon N^* \to N^*$ by $\theta(x) = x$ and $\eta(x) = x^2$ for $x \in N^*$. Then $\{K^*, N^*, \theta, \eta\}$ is an HP-system for K^* in G^*. The result follows from Lemma 6.8. □

Lemma 6.12. *If G is a minimal counterexample to the Hall-Paige conjecture, then $F^*(G) \neq O_2(G)$.*

Proof. Let us first note that a minimal counterexample to the Hall-Paige conjecture cannot be a 2-group. Assume that $F^*(G) = O_2(G)$. If $F^*(G)$ is cyclic, then, by Theorem 6.7, $\mathrm{Aut}(F^*(G))$ is a 2-group, and, by Lemma 6.5, $G/F^*(G) \subseteq \mathrm{Aut}(F^*(G))$ is a 2-group. Thus G is a 2-group, a contradiction. If $F^*(G)$ is noncyclic, then $G/F^*(G)$ is a cyclic 2-group by Lemma 6.11, and so G is again a 2-group, a contradiction from which the result follows. □

Lemma 6.13. *If G is a minimal counterexample to the Hall-Paige conjecture, then there exists a component L of G for which $M = \langle g^{-1}Lg \mid g \in G\rangle$ is a normal subgroup of G satisfying $C_G(M)$ and G/M are cyclic 2-groups.*

Proof. Assume G to be a minimal counterexample to the Hall-Paige conjecture. Thus, in particular, the Sylow 2-subgroup of G is noncyclic.

If G has no component, then $E(G) = \{1\}$ and so $F^*(G) = F(G)$. As $F(G)$ is nilpotent, $F(G)$ is solvable, and so $G/F(G)$ is nonsolvable as otherwise; $F(G)$ and $G/F(G)$ being solvable would imply that G is solvable and satisfies the Hall-Paige conjecture by Theorem 4.22. By Corollary 6.2, the Sylow 2-subgroup of $G/F(G)$ is noncyclic. Hence, the Sylow 2-subgroup of $F(G)$ is cyclic by Lemma 6.9. By Lemma 6.12, $F(G) \neq O_2(G)$. Therefore, by Lemma 6.4, $O_p(G) \neq \{1\}$ for some odd $p \in \pi(G)$. But then $O(G) \neq \{1\}$, contradicting Lemma 6.10. Hence G has a component L.

Set $M = \langle g^{-1}Lg \mid g \in G\rangle$. Then M is normal in G and has a noncyclic Sylow 2-subgroup by Corollary 6.2. Thus, by Lemma 6.11, G/M is a cyclic 2-group. Set $K = C_G(M)$. It is a simple exercise to show that K is normal in G. Set $G^* = G/K$. By the second isomorphism theorem, as K is a normal subgroup of G and L is a subgroup of G, $L \cap K$ is a normal subgroup of L and $L/(L \cap K) \cong LK/K$. As K is a subgroup of $C_G(L)$, $L \cap K$ is a subgroup of $Z(L)$. It follows that $L/(L \cap K)$ and, hence, LK/K is nonsolvable, and so LK/K, and hence G^* has a noncyclic Sylow 2-subgroup by Corollary 6.2. By Lemma 6.9, as G^* has a noncyclic Sylow 2-subgroup, K has a cyclic Sylow 2-subgroup. Therefore, by a Lemma 3.33, K has a normal subgroup H of odd order such that K/H is a cyclic 2-group. Let the cosets of H in K be $H = H_0, \ldots, H_s$. If $g \in H_i$, $i \neq 0$, is of odd order, then H_i is of odd order in K/H, an impossibility. Therefore H is the set of odd order elements of K, and, hence, H is a normal subgroup of G. By Lemma 6.10, $O(G) = \{1\}$. Hence $H = \{1\}$ and so K is a cyclic 2-group. □

To complete Aschbacher's reduction, we need to show that the groups M and L are equal.

Theorem 6.14 (Aschbacher, 1990). *If G is a minimal counterexample to the Hall-Paige conjecture, then there exists a quasisimple normal subgroup L of G such that $C_G(L)$ and G/L are cyclic 2-groups.*

Proof. Let G, K, M, and L be as in the proof of Lemma 6.13. If $M = L$, then $G/M = G/L$ is a cyclic 2-group, and as $K = C_G(M) = C_G(L)$ is a cyclic 2-group, we are done. In what follows, we will outline Aschbacher's proof that $M = L$.

If $M \neq L$, then there exist subgroups $L_1, \ldots, L_r$ of M, $r > 1$, such that each $L_i^* = L_i/K$ is a conjugate of the simple group $L^* = L/K$ and $M^* = M/K = L_1^* \times \cdots \times L_r^*$.

As $G^* = G/K$ is of even order, G^* contains an involution $t^* = tK$. As a permutation, acting via conjugation on the L_i^*, t^* is a product of $r/2$ disjoint transpositions. If I denotes the set of involutions in t^*M^*, then I is nonempty, and the action of M on I by conjugation is transitive. Thus $t^* \in I$ and by a Frattini argument $G = MH$, where $H^* = H/K = C_{G^*}(t^*)$. Further $H^* \cap M^* = H_1^* \times \cdots \times H_{r/2}^*$, where H_i^* is a full diagonal subgroup of $L_i^* L_i^t$. In particular $D^* = L_1^* \times \cdots \times L_{r/2}^*$ is a complement to H^* in G^*.

Now $(H^*, D^*, \theta^*, \eta^*)$ is an HP-system for G^*, where θ^* is a complete mapping of D^*: H^* is admissible by minimality. Thus, by Lemma 6.8, G is not a minimal counterexample to the Hall-Paige conjecture, a contradiction, hence the result. □

6.1.4 The HP-System Conjecture

Aschbacher [21] made a conjecture, the HP-system conjecture, which he showed would imply the Hall-Paige conjecture. His conjecture applies to almost simple groups. A group G is an *almost simple group* if $H \leq G \leq \mathrm{Aut}(H)$ for some nonabelian simple group H.

Conjecture 6.15 (HP-System Conjecture). If G is an almost simple group such that $G/F^*(G)$ is a cyclic 2-group, then there exists an HP-system (H, D, θ, η) for which H is a proper subgroup of G with a noncyclic Sylow 2-subgroup.

Aschbacher showed that if the HP-system conjecture is true, then the Hall-Paige conjecture is true.

Theorem 6.16. *The HP-system conjecture implies the Hall-Paige conjecture.*

Proof. Let G be a minimal counterexample to the Hall-Paige conjecture. By Theorem 6.14, G has a quasisimple normal subgroup L for which $K = C_G(L)$ and G/L are cyclic 2-groups. $G^* = G/K$ is almost simple and G^*/L^* is a cyclic 2-group, where $L^* = L/K$. If the HP-system conjecture holds, then there exists an HP-system $(H^*, D^*, \theta^*, \eta^*)$ for G^* for which H^* has noncyclic Sylow 2-subgroups and $H^* \neq G^*$. If H is the preimage of H^* in G, then H has a noncyclic Sylow 2-subgroup and $H \neq G$. Thus H is admissible by minimality. The proof follows from Lemma 6.8. □

Aschbacher showed that the HP-system conjecture holds for groups with minimal normal subgroup a Mathieu group or minimal normal subgroup one of a number of finite simple groups of Lie type. These results are presented here without proof. Finite simple groups of Lie type will be described in Section 7.2.

Theorem 6.17 (Aschbacher, 1990). *If G is an almost simple group, with minimal normal subgroup a Mathieu group, then G satisfies the HP-system conjecture.*

Theorem 6.18 (Aschbacher, 1990). *Let G be an almost simple group with minimal normal subgroup L of Lie type over $GF(q)$, G/L a cyclic 2-group. If the HP-system conjecture does not hold for G, then one of the following must be true:*

1. *L is of Lie rank 1, q is odd, and G induces inner diagonal automorphisms on L.*
2. *$L = L_n(q)$, $n \geq 3$, G nontrivial on the Dynkin diagram of L.*
3. *$L = \mathrm{Sp}_4(q)$ or $F_4(q)$, q even, G nontrivial on the Dynkin diagram of L.*
4. *$L =^2 F_4(q)$, q even.*

6.2 Further Quotient Constructions

By Theorem 6.14, if G is a minimal counterexample to the Hall-Paige conjecture, then there exist subgroups H and K of G satisfying H is a cyclic 2-group, H is normal in K, K/H is a nonabelian simple group, K is a normal subgroup of G, and G/K is a cyclic 2-group. This suggest two questions for a group G with a normal subgroup H.

1. If $G/H \cong \mathbb{Z}_2$, under what conditions does the admissibility of H imply the admissibility of G?
2. If $H \cong \mathbb{Z}_2$, under what conditions does the admissibility of G/H imply the admissibility of G?

Partial answers to these questions were given by Evans [120] in 1992. We will show that, under certain conditions, if $H \cong \mathbb{Z}_2$ is normal in G and G/H is admissible, then G is admissible; and, under certain conditions, if H is an admissible, normal subgroup of G and $G/H \cong \mathbb{Z}_2$, then G is admissible. The proofs of these results require some graph theory.

6.2.1 Some Graph Theory

A *graph* Γ consists of a set $V(\Gamma)$, whose elements are called *vertices*, and a set $E(\Gamma)$, consisting of two element subsets of $V(\Gamma)$, whose elements are called *edges*. A *walk* in Γ is a sequence, $v_0, \ldots, v_m$, of vertices of Γ for which $\{v_i, v_{i+1}\} \in E(\Gamma)$ for $i = 0, \ldots, m-1$. The *length* of this walk is m and this walk is *closed* if $v_m = v_0$. Γ is *connected* if every pair of vertices of Γ is joined by a walk. It is easy to see that a graph can be decomposed into *connected components*: two vertices are joined by a walk if and only if they are in the same connected component. A graph Γ is *bipartite* with *bipartite classes* X and Y if X and Y partition $V(\Gamma)$ and, for any $\{u, v\} \in E(\Gamma)$, either $u \in X$ and $v \in Y$ or $v \in X$ and $u \in Y$. Note that if Γ is bipartite, then the choice of bipartite classes need not be unique as Γ need not be connected. The following is a well-known characterization of bipartite graphs.

Lemma 6.19. *A graph is bipartite if and only if every closed walk is of even length.*

Proof. Let Γ be a bipartite graph with bipartite classes X and Y, and let $v_0, \ldots, v_m = v_0$ be a closed walk in Γ. Without loss of generality $v_0 \in X$. It is easy to see that

$$v_i \in \begin{cases} Y & \text{if } i \text{ is odd,} \\ X & \text{if } i \text{ is even.} \end{cases}$$

It follows that m is even and, hence, that any closed walk in Γ is of even length.

Next suppose that any closed walk in Γ is of even length. We may assume without loss of generality that Γ is connected. Let x be a vertex of Γ, and set

$$X = \{v \in V(\Gamma) \mid v \text{ joined to } x \text{ by a walk of even length}\},$$

and

$$Y = \{v \in V(\Gamma) \mid v \text{ joined to } x \text{ by a walk of odd length}\}.$$

The sets X and Y are disjoint as were there to exist a vertex $u \in X \cap Y$, then u would be joined to x by a walk of even length and a walk of odd length: the concatenation of these two walks would be a closed walk of odd length. Hence, X and Y partition $V(\Gamma)$. If $\{u, v\} \in E(\Gamma)$ and $x = v_0, \ldots, v_{m-1} = u$ is a walk of length m in Γ joining x to u, then $x = v_0, \ldots, v_{m-1}, u, v$ is a walk of length $m + 1$ in Γ joining x to v. It follows that one of u, v is in X and one in Y and so Γ is bipartite. □

6.2.2 G/H Admissible, $H \cong \mathbb{Z}_2$

We are now in a position to prove that, under certain conditions, if $H \cong \mathbb{Z}_2$ is normal in G, and G/H is admissible, then G is admissible.

Theorem 6.20 (Evans, 1992). *Let $H = \langle a \rangle$, a an involution in $Z(G)$, and let G/H be admissible. If there exists a 2-element $b \in G \setminus \{a\}$, $b^i \neq a$ for any even integer i, then G is admissible.*

Proof. Let $\phi\colon G \to G/H$ be the canonical homomorphism, and let θ be a complete mapping of G/H.

Define a graph Γ, whose vertices are the elements of G and whose edge set is

$$E(\Gamma) = \{\{x, xa\} \mid x \in G\} \cup \{\{x, xb\} \mid x \in G\} \cup \{\{x, xb^{-1}\} \mid x \in G\}.$$

Let $v_0, v_1, \ldots, v_m = v_0$ be a closed walk in Γ. Set

$$i = |\{t \mid v_{t+1} = v_t a\}|, \quad j = |\{t \mid v_{t+1} = v_t b\}|, \quad \text{and} \quad k = |\{t \mid v_{t+1} = v_t b^{-1}\}|,$$

subscripts being added modulo m. Then $m = i + j + k$ and $a^i b^{j-k} = 1$, and so $b^{k-j} = a^i$. If i is even, then $b^{k-j} = 1$, and so $k - j$ is even, and hence $m = i+k+j$ is even. If i is odd, then $b^{k-j} = a$, and so $k - j$ is odd, and hence $k + j$ is odd, from which it follows that $m = i + j + k$ is even. As all closed walks in Γ are of even length, by Lemma 6.19, Γ is bipartite. Let X and Y be bipartite classes of Γ.

Define $\phi'\colon G/H \to X$ by $\phi'(y) = x$ if $\phi(x) = y$ and $x \in X$. The mapping ϕ' is one to one as $\phi(x_1) = \phi(x_2)$ if and only if $x_2 = x_1$ or $x_1 a$, and in the latter case, exactly one of x_1 or x_2 is in X. Further ϕ' is onto as, if $x \in X$, then $\phi(x) = y$ for some $y \in G/H$. Thus ϕ' is a bijection. Set

$$W = \{x \mid x\phi'\theta\phi(x) \in X\} \text{ and } Z = \{x \mid x\phi'\theta\phi(x) \in Y\}.$$

As X and Y partition G, W and Z also partition G. As $\{x, xa\} \in E(\Gamma)$ for all $x \in X$, the mapping $x \mapsto xa$ is a mapping from $X \to Y$: this mapping is onto as, if $y \in Y$, then $\{ay, y\} \in E(\Gamma)$ and $ay \in X$. Now

$$\begin{aligned} Wa &= \{xa \mid x\phi'\theta\phi(x) \in X\} \\ &= \{y \mid ya\phi'\theta\phi(ya) \in X\} \\ &= \{y \mid y\phi'\theta\phi(ya) \in Xa\} \\ &= \{y \mid y\phi'\theta\phi(y) \in Y\}. \end{aligned}$$

Hence $Z = Wa$.

We next form a new graph Γ', whose vertices are the elements of W and whose edge set is

$$E(\Gamma') = \{\{x, y\} \mid \phi'\theta\phi(x)ab = \phi'\theta\phi(y)\} \cup \{\{x, y\} \mid \phi'\theta\phi(y)ab = \phi'\theta\phi(x)\}.$$

Let $x_0, \ldots, x_m = x_0$ be a closed walk of length m in Γ' and set

$$i = |\{t \mid \phi'\theta\phi(x_{t+1}) = \phi'\theta\phi(x_t)ab\}|, \quad \text{and}$$

$$j = |\{t \mid \phi'\theta\phi(x_{t+1}) = \phi'\theta\phi(x_t)(ab)^{-1}\}|,$$

indices added modulo m. Now

$$\phi'\theta\phi(x_0) = \phi'\theta\phi(x_{m-1})(ab)^{\varepsilon_{m-1}} = \cdots = \phi'\theta\phi(x_0)(ab)^{i-j},$$

where ε_{m-1} is either 1 or -1. Hence $(ab)^{i-j} = 1$, and so $i - j$ is even, and hence $m = i + j$ is even. It follows that every closed walk in Γ' is of even length, and so, by Lemma 6.19, Γ' is a bipartite graph. Let A and B be bipartite classes of Γ'. Now A and B partition W, and so, as $Z = Wa$ and W and Z partition G, the sets A, B, Aa, and Ba also partition G.

Let us define $\theta' : G \to G$ by

$$\theta'(x) = \begin{cases} \phi'\theta\phi(x) & \text{if } x \in A, \\ \phi'\theta\phi(x)b & \text{if } x \in B, \\ \phi'\theta\phi(x)ab & \text{if } x \in Aa, \text{ and} \\ \phi'\theta\phi(x)a & \text{if } x \in Ba. \end{cases}$$

We claim that θ' is a complete mapping of G. To see that θ' is a bijection, first, note that $\phi'\theta\phi(x_1) = \phi'\theta\phi(x_2)$ if and only if $\theta\phi(x_1) = \theta\phi(x_2)$ if and only if $\phi(x_1) = \phi(x_2)$ if and only if $x_1 = x_2$ or $x_2 a$. Thus $\phi'\theta\phi(x_1) \neq \phi'\theta\phi(x_2)$ if x_1 and x_2 are two distinct elements of A, or two distinct elements of Aa, or two distinct elements of B, or two distinct elements of Ba. The sets $\{\phi'\theta\phi(x) \mid x \in A\}$ and $\{\phi'\theta\phi(x)ab \mid x \in Aa\}$ are disjoint as, if $y = \phi'\theta\phi(x_1) = \phi'\theta\phi(x_2)ab$, $x_1 \in A$, $x_2 \in Aa$, then $x_2 a \in A$ and $y = \phi'\theta\phi(x_1) = \phi'\theta\phi(x_2 a)ab$, which implies that $\{x_1, x_2 a\} \in E(\Gamma')$, an impossibility as A is a bipartite class of Γ'. Similarly, the sets $\{\phi'\theta\phi(x)b \mid x \in B\}$ and $\{\phi'\theta\phi(x)a \mid x \in Ba\}$ are disjoint. The sets $\{\phi'\theta\phi(x) \mid x \in A\}$ and $\{\phi'\theta\phi(x)b \mid x \in B\}$ are disjoint as $\phi'\theta\phi(x_1) \in X$ and $\phi'\theta\phi(x_2)b \in Y$. Similarly the sets $\{\phi'\theta\phi(x)ab \mid x \in Aa\}$ and $\{\phi'\theta\phi(x)a \mid x \in Ba\}$ are disjoint.

To see that θ' is actually a complete mapping, first, note that

$$\{x\phi'\theta\phi(x)ab \mid x \in Aa\} = \{(xa)\phi'\theta\phi(xa)b \mid xa \in A\} = \{x\phi'\theta\phi(x)b \mid x \in A\},$$

and

$$\{x\phi'\theta\phi(x)a \mid x \in Ba\} = \{(xa)\phi'\theta\phi(xa) \mid xa \in B\} = \{x\phi'\theta\phi(x) \mid x \in B\}.$$

Thus, the images of $x \mapsto x\theta'(x)$ are

$$\{x\phi'\theta\phi(x) \mid x \in W\} \cup \{x\phi'\theta\phi(x)b \mid x \in W\} = X \cup Y = G,$$

and so the mapping $x \mapsto x\theta'(x)$ is onto and, hence, is a bijection as needed. □

Theorem 6.20 can be used to give a shorter proof of Theorem 4.20: Hall and Paige's proof that noncyclic 2-groups are admissible.

Corollary 6.21. *Noncyclic 2-groups are admissible.*

Proof. Suppose that this is false and let G be a minimal counterexample. We know that G must be nonabelian, as noncyclic abelian 2-groups are admissible by

Theorem 3.9. Let u be an involution in $Z(G)$. If $G/\langle u\rangle$ is cyclic of order m generated by $g\langle u\rangle$, then $g^m \in \langle u\rangle$, implying that G is abelian. Hence $G/\langle u\rangle$ must be noncyclic. Further, G cannot contain another involution as then, by Theorem 6.20, G would be admissible. It follows that G must be a quaternion group by Theorem 4.18. Complete mappings for quaternion groups were constructed in Lemma 4.19: a contradiction from which the result follows. □

6.2.3 $|G/H| = 2$, H Admissible

It was shown in 1992 by Evans [120] that, if G has noncyclic Sylow 2-subgroups and H is a normal subgroup of G of index 2, then, under certain conditions, complete mappings of H can be extended to complete mappings of G.

Theorem 6.22 (Evans, 1992). *Let G be a finite group, and let H be a normal subgroup of G of index 2 in G, and suppose that there exist 2-elements $a, b \in G \setminus H$ satisfying $xa^ix^{-1} \neq b^j$ for any $x \in H$ and any i, j odd. Then, if H is admissible, G is admissible.*

Proof. Let θ be a complete mapping of H, and let Γ be the graph whose vertices are the elements of H and whose edge set is

$$\begin{aligned} E(\Gamma) = \quad & \{\{x, axb\} \mid x \in H\} \cup \{\{x, axb^{-1}\} \mid x \in H\} \\ & \cup\{\{x, a^{-1}xb\} \mid x \in H\} \cup \{\{x, a^{-1}xb^{-1}\} \mid x \in H\}. \end{aligned}$$

It is important to note that the edges of Γ are sets of order two as, if $x = a^{\varepsilon}xb^{\delta}$, for some $\varepsilon, \delta = \pm 1$, then $x^{-1}a^{\varepsilon}x = b^{-\delta}$, an impossibility as a^{ε} is an odd power of a and $b^{-\delta}$ is an odd power of b. Let $x = x_0, x_1, \ldots, x_m = x$ be a closed walk in Γ. Thus $x_i = a^{\varepsilon_i}x_{i-1}b^{\delta_i}$ for $i = 1, \ldots, m$ and some $\varepsilon_i, \delta_i = \pm 1$. Then

$$x = a^{\sum_{i=1}^m \varepsilon_i}xb^{\sum_{i=1}^m \delta_i}, \quad \text{that is,} \quad x^{-1}a^{\sum_{i=1}^m \varepsilon_i}x = b^{-\sum_{i=1}^m \delta_i}.$$

Let $s = |\{i \in \{1, \ldots, m\} \mid \varepsilon_i = 1\}|$ and $t = |\{i \in \{1, \ldots, m\} \mid \varepsilon_i = -1\}|$. Then $\sum_{i=1}^m \varepsilon_i = s - t \equiv s + t \pmod 2 \equiv m \pmod 2$. Similarly $-\sum_{i=1}^m \delta_i \equiv m \pmod 2$. Thus m must be even, and hence any closed walk of Γ has even length, and so Γ is a bipartite graph by Lemma 6.19. Let X and Y be bipartite classes of Γ and set

$$A = \{x \mid x \in H, \theta(x) \in X\} \text{ and } B = \{x \mid x \in H, \theta(x) \in Y\}.$$

As X and Y partition H and θ is a complete mapping of H, the sets A and B also partition H; and, as the sets H and Ha partition G, the sets A, B, Aa, and Ba also partition G.

Define $\theta' \colon G \to G$ by

$$\theta'(x) = \begin{cases} \theta(x) & \text{if } x \in A, \\ \theta(x)b & \text{if } x \in B, \\ a^{-1}\theta(xa^{-1})b & \text{if } x \in Aa, \text{ and} \\ a^{-1}\theta(xa^{-1}) & \text{if } x \in Ba. \end{cases}$$

We will show that θ' is a complete mapping of G. The image set of θ' consists of four sets:

$$S_1 = \{\theta(\alpha) \mid \alpha \in A\},$$

$$S_2 = \{\theta(\beta)b \mid \beta \in B\},$$

$$S_3 = \{a^{-1}\theta(\alpha)b \mid \alpha \in A\}, \text{ and}$$

$$S_4 = \{a^{-1}\theta(\beta) \mid \beta \in B\}.$$

As θ is a bijection, the mapping $\alpha \mapsto \theta(\alpha)$ is a bijection $A \to S_1$, the mapping $\beta \mapsto \theta(\beta)b$ is a bijection $B \to S_2$, the mapping $\alpha \mapsto a^{-1}\theta(\alpha)b$ is a bijection $A \to S_3$, and the mapping $\beta \mapsto a^{-1}\theta(\beta)$ is a bijection $B \to S_4$. Thus to show that θ' is a bijection, we need only show that the sets S_1, S_2, S_3, and S_4 are disjoint, thus implying that θ' is one to one and, hence, by the finiteness of G, is a bijection. The sets S_1 and S_3 are disjoint from S_2 and S_4, as S_1 and S_3 are subsets of H, whereas S_2 and S_4 are subsets of Hb. The sets S_1 and S_3 are disjoint as $S_1 \subseteq X$ and $S_3 \subseteq Y$, and S_2 and S_4 are disjoint as $S_2b^{-1} \subseteq Y$ and $S_4b^{-1} \subseteq X$. Hence θ' is a bijection.

To complete the proof that θ' is a complete mapping, we need to show that the mapping $\eta' : G \to G$, defined by $\eta'(x) = x\theta'(x)$, is a bijection. This proof is similar to the proof that θ' is a bijection. The image set of η' consists of four sets:

$$T_1 = \{\alpha\theta(\alpha) \mid \alpha \in A\},$$

$$T_2 = \{\beta\theta(\beta)b \mid \beta \in B\},$$

$$T_3 = \{\alpha\theta(\alpha)b \mid \alpha \in A\}, \text{ and}$$

$$T_4 = \{\beta\theta(\beta) \mid \beta \in B\}.$$

As A and B partition H and θ is a complete mapping of H, $T_1 \cup T_4 = H$, and $T_2 \cup T_3 = Hb$. Thus η' is onto and hence, by the finiteness of G, is a bijection, completing the proof. □

6.3 Dalla Volta and Gavioli's Reduction

In 2001 Dalla Volta and Gavioli [85] improved Aschbacher's reduction. We proved in Corollary 1.38 that if H is a normal subgroup of G, and H and G/H are both admissible, then G is admissible. This fact is used repeatedly in Dalla Volta and Gavioli's reduction to show that certain groups cannot be minimal counterexamples to the Hall-Paige conjecture. In the following lemma, we collect together some of the special cases that were used.

Lemma 6.23. *If G is nonsolvable and H is a normal subgroup of G, then G is not a minimal counterexample to the Hall-Paige conjecture if any of the following hold.*

1. *H is an elementary abelian 2-group and $H \ncong \mathbb{Z}_2$.*
2. *G/H and H are both nonsolvable.*
3. *G/H is solvable and not a cyclic 2-group.*

Proof. Suppose that G is a minimal counterexample to the Hall-Paige conjecture, that G is nonsolvable, and that H is a normal subgroup of G.

If H is an elementary 2-group and $H \ncong \mathbb{Z}_2$, then H is admissible by Theorem 4.22. As G is nonsolvable, G/H must be nonsolvable, and hence, by Corollary 6.2, the Sylow 2-subgroup of G/H must be noncyclic. Now, as G is a minimal counterexample to the Hall-Paige conjecture, G/H must be admissible from which it follows, by Corollary 1.38, that G is admissible: a contradiction.

If G/H and H are both nonsolvable, then both have noncyclic Sylow 2-subgroups, and hence, as G is a minimal counterexample to the Hall-Paige conjecture, both are admissible. It follows from Corollary 1.38 that G is admissible: a contradiction.

Next suppose that G/H is solvable and not a cyclic 2-group. We will show that G/G' is a cyclic 2-group. If $G = G'$, then G/G' is trivially a cyclic 2-group. If $G \neq G'$, then, as G/G' is abelian, there exists a normal subgroup K of G containing G' for which K/G' is a 2-group and G/K is of odd order. Now $G = K$ as otherwise G/K would be admissible by Corollary 3.6 and, by the minimality of G, K would have to be admissible making G admissible by Corollary 1.38. Similarly, K/G' cannot be a noncyclic 2-group as K/G' would be admissible by Theorem 4.20 making G admissible by Corollary 1.38. It follows that G/G' is a cyclic 2-group. □

Similar results hold for factorable groups. However we will only need one such result dealing with wreath products. Let G and H be groups, and let H act on $\{1, \ldots, n\}$: we will use i^h to denote the image of i under h. The *wreath product* of G with H is the group $G \wr H = \{(g_1, \ldots, g_n, h) \mid g_1, \ldots, g_n \in G, h \in H\}$ with the product

$$(g_1, \ldots, g_n, h)(g'_1, \ldots, g'_n, h') = (g_{1^h} g'_1, \ldots, g_{n^h} g'_n, hh').$$

Many wreath products cannot be minimal counterexamples to the Hall-Paige conjecture.

Lemma 6.24. *If G and H are nontrivial finite groups, and the Sylow 2-subgroup of G is noncyclic, then $G \wr H$ is not a minimal counterexample to the Hall-Paige conjecture.*

Proof. Let $|H| = n$. As H is nontrivial, $n > 1$. Let

$$A = \{(g_1, \ldots, g_{n-1}, 1, 1) \mid g_1, \ldots, g_{n-1} \in G\}$$

and

$$B = \{(1, \ldots, 1, g, h) \mid g \in G, h \in H\}.$$

As $G \wr H = AB$ and $A \cap B = \{(1, \ldots, 1)\}$, $G \wr H$ is factorable with factors A and B. If $G \wr H$ were a minimal counterexample to the Hall-Paige conjecture, then, as A and B both have noncyclic Sylow 2-subgroups, A and B would both be admissible from which it would follow from Theorem 4.8 that $G \wr H$ is admissible. Hence $G \wr H$ cannot be a minimal counterexample to the Hall-Paige conjecture □

A subgroup H of G is a *characteristic subgroup* if $\alpha(H) = H$ for all $\alpha \in \mathrm{Aut}(G)$. The subgroups $\{1\}$ and G are characteristic subgroups of G, and G is *characteristically simple* if these are the only characteristic subgroups of G. Finite characteristically simple subgroups have been characterized.

Theorem 6.25. *If G is a finite characteristically simple group, then G is a direct product of isomorphic simple groups.*

Proof. See Satz 9.12 in [191]. □

Recall that $O(G)$ denotes the largest odd-order normal subgroup of G. The *socle* of G, $\mathrm{Soc}(G)$, is the subgroup of G generated by the minimal subgroups of G.

Theorem 6.26 (Dalla Volta and Gavioli, 2001). *If G is a minimal counterexample to the Hall-Paige conjecture, then G has a unique minimal normal subgroup, and one of the following holds.*

1. *G is an almost simple group, the socle S of G is simple, and G/S is a cyclic 2-group.*
2. *G has only one involution, the Sylow 2-subgroup of G is a quaternion group, $|G/G'| \leq 2$, $G' \cong \mathrm{SL}(2, q)$ for some $q > 5$ odd, and $G/Z(G')$ is isomorphic to a subgroup of $\mathrm{PGL}(2, q)$ containing $\mathrm{PSL}(2, q)$.*

Proof. Let G be a minimal counterexample to the Hall-Paige conjecture. By Theorem 4.22, the Hall-Paige conjecture is true for solvable groups, and so G must be nonsolvable. Hence, as G/G' is abelian, G' must be nonsolvable, and so, by Corollary 6.2, the Sylow 2-subgroup of G' must be noncyclic. By Lemma 6.23, G/G' is a cyclic 2-group.

The possible structures of $\mathrm{Soc}(G)$ are determined in part by the possible structures of minimal normal subgroups of G. Let H be a minimal normal subgroup of G. By Lemma 6.23, H cannot be of odd order. Hence H is of even order. Let K be a characteristic subgroup of H. As H is a normal subgroup of G, any inner automorphism of G is an automorphism of H and hence fixes K. It follows that H is characteristically simple, and thus, by Theorem 6.25, H is an elementary abelian 2-group or a direct product of copies of a nonabelian simple group. If H is an elementary 2-group, then $H \cong \mathbb{Z}_2$ by Theorem 4.22. Hence H is either $\mathbb{Z}_2$ or the direct product of copies of a nonabelian simple group. Further, G can have only one minimal normal subgroup. To see this, let H_1 and H_2 be two distinct minimal normal subgroups of G. If both H_1 and H_2 are isomorphic to $\mathbb{Z}_2$, then it is an exercise to show that the subgroup, K, of G generated by H_1 and H_2 is a normal subgroup of G isomorphic to $\mathbb{Z}_2 \times \mathbb{Z}_2$. But then, by Lemma 6.23, G cannot be a minimal counterexample to the Hall-Paige conjecture. It follows that at most one minimal normal subgroup of G can be isomorphic to $\mathbb{Z}_2$. Next, suppose that H_1 and H_2 are both nonsolvable. Now $H_2 \cong S \times \cdots \times S$ for some nonabelian simple group S. Hence G/H_1 must contain a subgroup isomorphic to S, and so both G/H_1 and H_1 are nonsolvable. Thus, by Lemma 6.23, G cannot be a minimal counterexample to the Hall-Paige conjecture. It follows that at most one minimal normal subgroup of G can be nonsolvable. Finally, let us suppose that $H_1 \cong \mathbb{Z}_2$ and H_2 is nonsolvable. It is now an exercise to show that $\mathrm{Soc}(G) = H_1 \times H_2$. Let a be a generator of H_1 and b an involution in H_2. As a and b satisfy the conditions of Theorem 6.20, it follows that G cannot be a minimal counterexample to the Hall-Paige conjecture.

We have shown that $\mathrm{Soc}(G)$ is the only minimal normal subgroup of G and is isomorphic to $\mathbb{Z}_2$ or a nonsolvable group that is the a direct product of copies of a nonabelian simple group. Thus there are two cases to consider.

Case 1. $\mathrm{Soc}(G)$ is nonabelian.

The minimal normal subgroup of G is $H = \mathrm{Soc}(G)$, which, by Theorem 6.25, must be isomorphic to a direct product of n copies of a nonabelian simple group S. The action of G on H by conjugation defines a homomorphism from G into $\mathrm{Aut}(H)$, the kernel K of which is the centralizer of H in G. As $Z(H)$ is trivial, $H \cap K = \{1\}$. As G can only have one minimal normal subgroup, K must be trivial. It follows that G is a subgroup of $\mathrm{Aut}(H)$. It is an exercise to show that $\mathrm{Aut}(H) \cong \mathrm{Aut}(S) \wr S_n$. G/H is a cyclic 2-group as otherwise, by Lemma 6.23, G cannot be a minimal counterexample to the Hall-Paige conjecture. If $H \neq S$, then G is the wreath product of S and a nontrivial cyclic 2-group. But then, by Lemma 6.24, G cannot be a minimal counterexample to the Hall-Paige conjecture. Hence $H = S$ and G is almost simple.

Case 2. $\mathrm{Soc}(G) \cong \mathbb{Z}_2$.

The minimal normal subgroup of G is $H = \mathrm{Soc}(G)$.

See [85] for the proof that, in this case, G has only one involution, the Sylow 2-subgroup of G is a quaternion group, $|G/G'| \leq 2$, $G' \cong \mathrm{SL}(2, q)$ for some $q > 5$ odd, and $G/Z(G')$ is isomorphic to a subgroup of $\mathrm{PGL}(2, q)$ containing $\mathrm{PSL}(2, q)$.

□

6.4 Wilcox's Reduction

In Theorem 6.14, we saw Aschbacher's [21] proof that any minimal counterexample to the Hall-Paige conjecture must be close to being simple, and in Theorem 6.26 we saw Dalla Volta and Gavioli's [85] improvement of Aschbacher's reduction. In 2009 Wilcox [329] used more elementary arguments than Aschbacher's to show that, in fact, any minimal counterexample to the Hall-Paige conjecture must be simple. He did this by showing that, if the Sylow 2-subgroup of G is noncyclic and G has a normal subgroup H of order 2, then from any complete mapping of G/H, we can construct a complete mapping of G; or if G has a normal subgroup H of index 2, then any complete mapping of H can be extended to a complete mapping of G: this significantly improved on the reductions of Theorems 6.20 and 6.22. From these two results, we will be able to derive our reduction.

6.4.1 *The Case G/H Admissible and $H \cong \mathbb{Z}_2$*

Recall that, under special conditions, a construction of complete mappings of G, a group with noncyclic Sylow 2-subgroups, from complete mappings of G/H, H a normal subgroup of order 2, was given in Theorem 6.20. Wilcox [329] removed the special conditions, thus showing that this construction is always possible. First we need a combinatorial result.

Lemma 6.27. *If I is a finite set acted on by involutions σ and ϕ that have no fixed points, then I can be partitioned into subsets J and K for which $\sigma(J) = \phi(J) = K$.*

Proof. Let us form a graph Γ whose vertices are the elements of I, $\{a, b\}$ being an edge of Γ if and only if either $\sigma(a) = b$ or $\phi(a) = b$. We will call an edge $\{a, b\}$ a σ-edge if $\sigma(a) = b$ and a ϕ-edge if $\phi(a) = b$: it may well be that some edges are both σ-edges and ϕ-edges. As two σ-edges cannot share a common vertex and two ϕ-edges cannot share a common vertex, the edges in any cycle in Γ must alternate, σ-edge, ϕ-edge. This implies that any cycle in Γ is of even length and hence, by Lemma 6.19, that Γ is bipartite. The result is then proved by choosing J and K to be bipartite classes in Γ. □

Note that the partition in Lemma 6.27 need not be unique as the graph Γ need not be connected.

The existence of a complete mapping of a group G is equivalent to the existence of three bijections $\mathrm{a}, \mathrm{b}, \mathrm{c}\colon I \to G$, I an index set, satisfying $\mathrm{a}(i)\mathrm{b}(i) = \mathrm{c}(i)$. To see this, if $\mathrm{a}, \mathrm{b}, \mathrm{c}\colon I \to G$ are bijections satisfying $\mathrm{a}(i)\mathrm{b}(i) = \mathrm{c}(i)$, then $\theta\colon G \to G$, defined by $\theta(\mathrm{a}(i)) = \mathrm{b}(i)$, is a complete mapping of G. On the other hand, if θ is a complete mapping of G and I is an index set satisfying $|I| = |G|$, then let $\mathrm{a}\colon I \to G$ be any bijection and define $\mathrm{b}, \mathrm{c}\colon I \to G$ by $b(i) = \theta(\mathrm{a}(i))$ and $\mathrm{c}(i) = \mathrm{a}(i)\mathrm{b}(i)$: $\mathrm{a}, \mathrm{b}, \mathrm{c}\colon I \to G$ are bijections satisfying $\mathrm{a}(i)\mathrm{b}(i) = \mathrm{c}(i)$.

Theorem 6.28 (Wilcox, 2009). *Let G be a finite group with a noncyclic Sylow 2-subgroup, and let $H \cong \mathbb{Z}_2$ be a normal subgroup of G. If G/H is admissible, then G is admissible.*

Proof. Let $H = \langle x \rangle$, and let $\pi\colon G \to G/H$ be the canonical homomorphism. $|G/H|$ must be even, as otherwise the Sylow 2-subgroup of G would be cyclic, and G/H would not be admissible. Thus G/H contains an involution y^*. Let $\mathrm{r}_y\colon G/H \to G/H$ be defined by $\mathrm{r}_y(z^*) = z^*y^*$. As G/H is admissible, there exist bijections $\mathrm{a}^*, \mathrm{b}^*, \mathrm{c}^*\colon I \to G/H$, I an index set, satisfying $\mathrm{a}^*(i)\mathrm{b}^*(i) = \mathrm{c}^*(i)$. The mappings $\sigma = \mathrm{b}^{*-1}\mathrm{r}_y\mathrm{b}^*$ and $\phi = \mathrm{c}^{*-1}\mathrm{r}_y\mathrm{c}^*$ are bijections, in fact involutions, from I to I with no fixed points. Thus, by Lemma 6.27, I can be partitioned into subsets J and K for which $\sigma(J) = \phi(J) = K$.

Next, we pick $y \in \pi^{-1}(y^*)$ and mappings $\mathrm{b}, \mathrm{c}\colon J \to G$ to satisfy $\pi\mathrm{b} = \mathrm{b}^*$ and $\pi\mathrm{c} = \mathrm{c}^*$. As σ and ϕ are bijections from J to K, we will extend b and c to mappings from I to G by setting $\mathrm{b}(\sigma(j)) = \mathrm{b}(j)y$ and $\mathrm{c}(\phi(j)) = \mathrm{c}(j)y$ for all $j \in J$. We define $\mathrm{a}\colon I \to G$ by $\mathrm{a}(i) = \mathrm{c}(i)\mathrm{b}(i)^{-1}$. With the extended definition, b and c still satisfy $\pi\mathrm{b} = \mathrm{b}^*$ and $\pi\mathrm{c} = \mathrm{c}^*$ as

$$\begin{aligned}\pi(\mathrm{b}(\sigma(j))) &= \pi(\mathrm{b}(j)y) = \mathrm{b}^*(j)y^* = \mathrm{b}^*(\sigma(j)), \quad \text{and}\\ \pi(\mathrm{c}(\phi(j))) &= \pi(\mathrm{c}(j)y) = \mathrm{c}^*(j)y^* = \mathrm{c}^*(\phi(j));\end{aligned}$$

and $\pi a = a^*$ as

$$\pi(\mathrm{a}(i)) = \pi(\mathrm{c}(i))\pi(\mathrm{b}(i))^{-1} = \mathrm{c}^*(i)\mathrm{b}^*(i)^{-1} = \mathrm{a}^*(i).$$

We will use the maps a, b, and c to define bijections $\alpha, \beta, \gamma\colon I \times H \to G$ that satisfy $\alpha(i, z)\beta(i, z) = \gamma(i, z)$ for all $(i, z) \in I \times H$. The mapping $\alpha(i, z) \mapsto \beta(i, z)$ will then be a complete mapping of G. We set

$$\alpha(i, z) = \begin{cases} \mathrm{a}(i) & \text{if } z = 1;\\ \mathrm{a}(i)x & \text{if } z = x;\end{cases}$$

$$\beta(i, z) = \begin{cases} \mathrm{b}(i) & \text{if } i \in J \text{ and } z = 1,\\ & \text{or } i \in K \text{ and } z = x;\\ \mathrm{b}(i)yx & \text{if } i \in J \text{ and } z = x;\\ \mathrm{b}(i)y^{-1}x & \text{if } i \in K \text{ and } z = 1;\end{cases}$$

$$\gamma(i, z) = \begin{cases} \mathrm{c}(i) & \text{if } i \in J \text{ and } z = 1;\\ \mathrm{c}(i)y & \text{if } i \in J \text{ and } z = x;\\ \mathrm{c}(i)y^{-1}x & \text{if } i \in K \text{ and } z = 1;\\ \mathrm{c}(i)x & \text{if } i \in K \text{ and } z = x.\end{cases}$$

The mapping α is onto, and hence is a bijection, as

$$\alpha(I \times H) = \alpha(I) \cup \alpha(I)x = G,$$

the mapping β is onto, and hence is a bijection, as

$$\begin{aligned}\beta(I \times H) &= \beta(J \times \{1\}) \cup \beta(K \times \{x\}) \cup \beta(J \times \{x\}) \cup \beta(K \times \{1\}),\\ &= \mathrm{b}(J) \cup \mathrm{b}(K) \cup \mathrm{b}(J)yx \cup \mathrm{b}(K)y^{-1}x\\ &= \mathrm{b}(J) \cup \mathrm{b}(K) \cup \mathrm{b}(K)x \cup \mathrm{b}(J)x\\ &= \mathrm{b}(I) \cup \mathrm{b}(I)x = G,\end{aligned}$$

and γ is onto, and hence is a bijection, as

$$\begin{aligned}\gamma(I \times H) &= \gamma(J \times \{1\}) \cup \gamma(J \times \{x\}) \cup \gamma(K \times \{1\}) \cup \gamma(K \times \{x\}),\\ &= \mathrm{c}(J) \cup \mathrm{c}(J)y \cup \mathrm{c}(K)y^{-1}x \cup \mathrm{c}(K)x\\ &= \mathrm{c}(J) \cup \mathrm{c}(K) \cup \mathrm{c}(J)x \cup \mathrm{c}(K)x\\ &= \mathrm{c}(I) \cup \mathrm{c}(I)x = G.\end{aligned}$$

If $i \in J$ and $z = 1$, then $\alpha(i, z)\beta(i, z) = \mathrm{a}(i)\mathrm{b}(i) = \mathrm{c}(i) = \gamma(i, z)$. If $i \in J$ and $z = x$, then $\alpha(i, z)\beta(i, z) = \mathrm{a}(i)x\mathrm{b}(i)yx = \mathrm{a}(i)\mathrm{b}(i)y = \mathrm{c}(i)y = \gamma(i, z)$. If $i \in K$ and $z = 1$, then $\alpha(i, z)\beta(i, z) = \mathrm{a}(i)\mathrm{b}(i)y^{-1}x = \mathrm{c}(i)y^{-1}x = \gamma(i, z)$. If $i \in K$ and $z = x$, then $\alpha(i, z)\beta(i, z) = \mathrm{a}(i)x\mathrm{b}(i) = \mathrm{c}(i)x = \gamma(i, z)$. Hence $\alpha(i, z)\beta(i, z) = \gamma(i, z)$ for all $(i, z) \in I \times H$. □

6.4.2 *The Case H Admissible and* $G/H \cong \mathbb{Z}_2$

We showed in Theorem 6.22 that if G has noncyclic Sylow 2-subgroups and H is a normal subgroup of G of index 2, then if there exist 2-elements $a, b \in G \setminus H$ satisfying $xa^ix^{-1} \neq b^j$ for any $x \in H$ and any i, j odd, complete mappings of H can be extended to complete mappings of G. Wilcox [329] proved that the condition, $xa^ix^{-1} \neq b^j$ for any $x \in H$ and any i, j odd, in Theorem 6.22 always holds. However, before we can prove Wilcox's result, we need more information about 2-elements and subgroups generated by 2-elements. The following is a result of Frobenius: a proof can be found in Chapter 1, Section 10 of [338].

Lemma 6.29. *Let $a \in G$, a group of order m, and let the conjugacy class of a be of order t. Then the number of solutions to $x^n = a$ in G is divisible by* $\gcd(tn, m)$.

We will only need the following special case.

Corollary 6.30. *If G is a finite group and S is a Sylow 2-subgroup of G, then the number of 2-elements of G is divisible by $|S|$.*

Proof. Apply Lemma 6.29 with $a = 1$, hence $t = 1$, and $n = |S|$. □

The next result is well-known: see Theorem 1.6 in [306], for instance.

Lemma 6.31. *If G is a finite 2-group and H a proper subgroup of G, then $N_G(H) \neq H$.*

The next lemma is key to showing that the conditions of Theorem 6.22 always hold.

Lemma 6.32. *Let G be a finite group with noncyclic Sylow 2-groups, and let H be a subgroup of G of index 2 in G. Then there exist 2-elements $u, v \in G \setminus H$ for which $\langle u \rangle$ and $\langle v \rangle$ are not conjugate.*

Proof. Let X denote the set of 2-elements in $G \setminus H$ and Y the set of cyclic subgroups generated by elements of X. If $x \in X$, then the even powers of x are contained in H, the odd powers in $G \setminus H$, and any odd power of x is a generator of $\langle x \rangle$.

Let us assume the lemma to be false, in which case the elements of Y are conjugate to each other. Let $K \in Y$ and let S be a Sylow 2-subgroup of G containing K. Then, as G is assumed to act transitively on Y by conjugation,

$$|Y| = \frac{|G|}{|N_G(K)|}.$$

As, if x is a generator of an element $K' \in Y$, any odd power of x also generates K', the sets of odd powers of generators of elements of Y partition $\left(\bigcup_{L \in Y} L\right) \setminus H$. Thus

$$|X| = \sum_{L \in Y} |L \setminus H| = \left(\frac{|K|}{2}\right) |Y| = \frac{|K||G|}{2\,|N_G(K)|}.$$

As $|S|$ is the order of a Sylow 2-subgroup of G and $|S|/2$ is the order of a Sylow 2-subgroup of $|H|$, by Corollary 6.30 the numbers of 2-elements of both G and H are divisible by $|S|/2$. Thus the number of 2-elements of G not in H, $|X|$, is divisible by $|S|/2$. Note that $|G|/|S|$ is odd. Further, as K is cyclic and S is not, by Lemma 6.31, K is a proper subgroup of $N_S(K)$. Thus, $|N_S(K) : K|$ is even. It follows that $|G|/(|S||N_S(K) : K|)$ is not an integer. But, noting that $N_S(K)$ is a subgroup of $N_G(K)$,

$$\frac{|G|}{|S||N_S(K) : K|} = \frac{|K||G|\,|N_G(K)|}{|N_G(K)|\,|S|\,|N_S(K)|} = \frac{|X|}{(|S|/2)} |N_G(K) : N_S(K)|,$$

an integer, a contradiction from which the result follows. □

We are now ready to prove that complete mappings of H extend to complete mappings of G, whenever H is of index 2 in G, and the Sylow 2-subgroups of G are noncyclic.

Theorem 6.33 (Wilcox, 2009). *Let G be a finite group whose Sylow 2-subgroups are noncyclic, and suppose that H is a subgroup of G of index 2 in G. Then, if H is admissible, G is admissible.*

Proof. By Lemma 6.32, there exist 2-elements $a, b \in G \setminus H$ for which $\langle a \rangle$ is not conjugate to $\langle b \rangle$. Then, as any odd power of a generates $\langle a \rangle$ and any odd power of b generates $\langle b \rangle$, $xa^i x^{-1} \neq b^j$ for any $x \in H$ and any i, j odd. The admissibility of G follows from Theorem 6.22. □

6.4.3 Possible Minimal Counterexamples

We now turn to the question of what a minimal counterexample to the Hall-Paige conjecture might be. We know that a counterexample cannot be solvable as Hall and Paige [171] proved their conjecture true for solvable groups in 1955: see Theorem 4.22. From Theorems 6.28 and 6.33, we will deduce that any minimal counterexample must, in fact, be simple. To prove this, we will need a simple fact about groups with nontrivial, cyclic Sylow 2-subgroups.

Lemma 6.34. *If G is a finite group with nontrivial cyclic Sylow 2-subgroups, then G has a normal subgroup of index 2.*

Proof. By Lemma 3.33, there is a homomorphism from G onto a nontrivial cyclic 2-group K. As K is cyclic of even order, K has a unique subgroup of index 2. Thus, by the Correspondence Theorem, G must have a normal subgroup of index 2. □

Wilcox used Lemma 6.34, Theorem 6.28, and Theorem 6.33, to prove that the only possible minimal counterexamples to the Hall-Paige conjecture are simple groups: this makes a proof of the conjecture possible either constructively or, as in Chapter 7, recursively.

Theorem 6.35 (Wilcox, 2009). *If G is a minimal counterexample to the Hall-Paige conjecture, then G is simple.*

Proof. Let G be a minimal counterexample to the Hall-Paige conjecture, and let us suppose that G is not simple. Thus G has a nontrivial, proper normal subgroup H which we will assume to contain no smaller nontrivial normal subgroup of G.

By the minimality of G, at least one of H or G/H has noncyclic Sylow 2-subgroups as otherwise both are solvable, and hence G is solvable, and thus admissible by Theorem 4.22.

Neither H nor G/H can be of odd order as groups of odd order are admissible by Corollary 3.6; the other group would have to have noncyclic Sylow 2-subgroups and hence be admissible by the minimality of G; and G would then be admissible by Corollary 1.38.

It cannot be that both H and G/H have noncyclic Sylow 2-subgroups as then both H and G/H would be admissible by the minimality of G and, thus, G would be admissible by Corollary 1.38. Thus both H and G/H have nontrivial Sylow 2-subgroups: one has cyclic Sylow 2-subgroups and the other noncyclic Sylow 2-subgroups.

If H has noncyclic Sylow 2-subgroups and G/H has cyclic Sylow 2-subgroups, then G/H has a normal subgroup of index 2 by Lemma 6.34. Thus, by the Correspondence Theorem, G has a normal subgroup of index 2, which has noncyclic Sylow 2-subgroups, and so, by the minimality of G, is admissible: the admissibility of G follows from Theorem 6.33.

The last possibility to consider is G/H having noncyclic Sylow 2-subgroups and H having cyclic Sylow 2-subgroups. We will show that H must have order 2. By Lemma 3.33, H has a normal subgroup K of odd order for which H/K has order a power of 2. By Theorem 1.3 in [157], as K is normal in H and $|K|$ and $|H/K|$ are relatively prime, K is a characteristic subgroup of H and so is a normal subgroup of G. Thus, as H contains no nontrivial normal subgroup of G other than itself, K is trivial, and H is a cyclic 2-group. But then H has a characteristic subgroup of index 2 in H, which is thus normal in G and hence is trivial. Therefore H has order 2, and, thus, G is admissible by Theorem 6.28, a contradiction from which the result follows. □

We already know that the Hall-Paige conjecture is true for cyclic groups of prime order (Corollary 3.2) and alternating groups (Theorem 4.5). This leaves the simple groups of Lie type and the sporadic simple groups as possible minimal counterexamples to the Hall-Paige conjecture.

Chapter 7
A Proof of the Hall-Paige Conjecture

In 2009 Wilcox [329] proved that any minimal counterexample to the Hall-Paige conjecture must be a finite nonabelian simple group: his reduction was covered in Section 6.4. He further proved that no finite simple group of Lie type, with the possible exception of ${}^2F_4(2)'$, the Tits group, could be a minimal counterexample to this conjecture. As the alternating groups were proved to be admissible in 1955 by Hall and Paige [171], and the Mathieu groups were proved admissible in 1993 by Dalla Volta and Gavioli [83], this left 22 possible minimal counterexamples to the Hall-Paige conjecture. Building on Wilcox's work, in 2009 Evans [128] reduced the number of possible minimal counterexamples to the Hall-Paige conjecture to just one: Janko's fourth group, J_4. This last group was shown not to be a minimal counterexample by Bray [50], thus completing a proof of the conjecture. It should be noted that Bray has yet to publish his work.

The proof that no finite nonabelian simple group can be a minimal counterexample to the Hall-Paige conjecture is in two parts.

- A proof that no group of Lie type can be a minimal counterexample, and
- a proof that no sporadic simple group can be a minimal counterexample.

We have already proved a number of nonabelian simple groups to be admissible. The alternating groups, A_n, $n \geq 5$, were proved admissible in Theorem 4.5; the Mathieu groups in Theorem 4.25; the Suzuki groups, $\mathrm{Sz}\left(2^{2n+1}\right) ={}^2 B_2\left(2^{2n+1}\right)$, $n \geq 1$, in Theorem 4.27; and the projective special unitary groups, $\mathrm{PSU}(3, q) = U_3(q) ={}^2 A_2(q)$, q even, $q > 2$, in Theorem 4.35. For q odd, $\mathrm{PSL}(2, q) = L_2(q) = A_1(q)$ was proved admissible for $q \equiv 3 \pmod 4$ in Theorem 5.21 and for $q \equiv 1 \pmod 4$ in Theorem 5.24. For q even and q and n not both 2, $\mathrm{PSL}(n, q) = L_n(q) = A_{n-1}(q)$ was proved admissible in Theorem 5.28. New proofs that each of these groups cannot be a minimal counterexample to the Hall-Paige conjecture will be described in this chapter.

In Section 7.1 we will describe W-systems, the essential tool in proving the Hall-Paige conjecture. In Section 7.2 we will use W-systems to prove that no finite simple

A. B. Evans, *Orthogonal Latin Squares Based on Groups*, Developments in Mathematics 57, https://doi.org/10.1007/978-3-319-94430-2_7

group of Lie type can be a counterexample to the Hall-Paige conjecture and in Section 7.3 to prove that no sporadic simple group can be a counterexample to the Hall-Paige conjecture.

7.1 Double Cosets and W-Systems

Just as Hall and Paige [171] used permutations of a dual system of coset representatives of a subgroup to extend a complete mapping of the subgroup to a complete mapping of the whole group (see Theorem 4.1), Wilcox [329] used permutations of the set of double cosets of a subgroup to extend a complete mapping of the subgroup to a complete mapping of the whole group. These W-systems, as we will call them, are the essential tool that we will need to prove the Hall-Paige conjecture.

7.1.1 Double Cosets

Let G be a finite group and H a subgroup of G. The set $HgH = \{h_1gh_2 \mid h_1, h_2 \in H\}$ is a *double coset* of H in G. We will use $\mathscr{D} = \{HgH \mid g \in G\}$ to denote the set of double cosets of H in G. It is clear from the definition that each double coset $D \in \mathscr{D}$ is a union of right cosets of H in G and also a union of left cosets of H in G.

Theorem 7.1. *The double cosets of H in G partition the element set of G.*

Proof. Clearly, any $g \in G$ is an element of the double coset HgH of H in G. It remains to show that two distinct double cosets of H in G are disjoint. Let $k \in HgH \cap Hg'H$. Then $k = h_1gh_2 = h_1'g'h_2'$ for some $h_1, h_2, h_1', h_2' \in H$, and, if k' is any other element of HgH, then $k' = h_3gh_4$ for some $h_3, h_4 \in H$. But then

$$k' = h_3gh_4 = (h_3h_1^{-1}h_1')g'(h_2'h_2^{-1}h_4) \in Hg'H.$$

Thus $HgH \subseteq Hg'H$. Similarly $Hg'H \subseteq HgH$, and hence $HgH = Hg'H$, completing the proof. □

As any double coset of H in G is a disjoint union of left cosets of H in G, if D is a double coset of H in G, then the *index* of H in D, written $|D : H|$, is the number of distinct left cosets of H contained in D, namely, $|D|/|H|$: these indices are also called subdegrees for reasons that will become apparent in Section 7.3. The following is an immediate consequence of Theorem 7.1.

Corollary 7.2. *If $\mathscr{D}$ is the set of double cosets of H in G, then*

$$\sum_{D \in \mathscr{D}} |D : H| = |G : H|.$$

The index of D, a double coset of H in G, can be calculated as the index of a subgroup of H in H, as in the following.

Theorem 7.3. *If $D = HgH$ is a double coset of H in G, then*

$$|D : H| = |H : H \cap gHg^{-1}|.$$

Proof. Any left coset of H in D is of the form hgH for some $h \in H$. Now, for $h, k \in H, hgH = kgH$ if and only if $g^{-1}h^{-1}kg \in H$, if and only if $h^{-1}k \in gHg^{-1}$. But $h^{-1}k \in H$, hence the result. □

7.1.2 W-Systems

In Section 4.1 of Chapter 4, we described Hall and Paige's [171] approach to using permutations of a dual system of coset representatives of a subgroup H in G to extend complete mappings of H to complete mappings of G. In a similar manner, Wilcox [329] used permutations of the set of double cosets of H in G to extend complete mappings of H to complete mappings of G. Wilcox's approach turns out to be more powerful than Hall and Paige's approach and provides us with the tool that we will need to prove the Hall-Paige conjecture. To describe Wilcox's approach, we need some lemmas, starting with the well-known Hall's marriage theorem.

Lemma 7.4 (Hall's Marriage Theorem). *Let $\mathscr{A} = \{A_1, \dots, A_n\}$ be a collection of finite sets. There exists a set $X = \{x_1, \dots, x_n\}$, $x_i \in A_i$, if and only if*

$$\left| \bigcup_{A_i \in \mathscr{B}} A_i \right| \geq k$$

for all $\mathscr{B} \subseteq \mathscr{A}$ satisfying $|\mathscr{B}| = k$.

Given a subgroup H of a finite group G, Hall's marriage theorem can be used to prove the existence of systems of left coset representatives of H in G that are also systems of distinct representatives for other partitions of the element set of G.

Lemma 7.5. *If H is a subgroup of a finite group G and $\mathscr{A} = \{A_1, \dots, A_n\}$, $|A_i| = |H|$ for all i, is a partition of G, then there exists a system of distinct left coset representatives $L = \{x_1, \dots, x_n\}$ of H in G with the property that $x_i \in A_i$.*

Proof. Let G/H be the set of left cosets of H in G, and define a mapping $\Phi \colon \mathscr{A} \to \mathscr{P}(G/H)$, the set of subsets of G/H, by

$$\Phi(A_i) = \{B \in G/H \mid B \cap A_i \neq \emptyset\}.$$

For $\mathscr{B}$ a subset of $\mathscr{A}$, let $\cup\mathscr{B}$ denote $\left(\bigcup_{A_i \in \mathscr{B}} A_i\right)$. Then

$$\bigcup_{A_i \in \mathscr{B}} \Phi(A_i) = \{B \in G/H \mid B \cap A_i \neq \emptyset \text{ for some } A_i \in \mathscr{B}\}$$

$$= \{B \in G/H \mid B \cap (\cup\mathscr{B}) \neq \emptyset\},$$

and

$$(\cup\mathscr{B}) \subseteq (\cup\mathscr{B})\, H = \bigcup\{B \in G/H \mid B \cap (\cup\mathscr{B}) \neq \emptyset\}.$$

Now

$$|(\cup\mathscr{B})| = |\mathscr{B}||H| \leq |\{B \in G/H \mid B \cap (\cup\mathscr{B}) \neq \emptyset\}||H|,$$

and thus

$$|\mathscr{B}| \leq |\{B \in G/H \mid B \cap (\cup\mathscr{B}) \neq \emptyset\}| = \left|\bigcup_{A_i \in \mathscr{B}} \Phi(A_i)\right|.$$

Thus, by Lemma 7.4 there exist distinct left cosets $B_1, \ldots, B_n$ of H in G (actually an ordering of the cosets) satisfying $B_i \subseteq \Phi(A_i)$ for each i. Now, for each i, $A_i \cap B_i \neq \emptyset$ and so we may choose $x_i \in A_i \cap B_i$. Then $L = \{x_1, \ldots, x_n\}$ is a system of left coset representatives of H in G with the property that $x_i \in A_i$. □

The following is an immediate corollary:

Corollary 7.6. *If H and K are subgroups of G, $|H| = |K|$, then there exists a system of distinct left coset representatives for H in G that is also a system of distinct left coset representatives for K in G.*

Lemma 7.5 can also be used to prove the existence of a system of distinct left coset representatives for H in G that is also a system of distinct right coset representatives for K in G, and, by setting $K = H$, the existence of a dual system coset of representatives for H in G.

In Corollary 1.38, we saw that if H is a normal subgroup of G, then from complete mappings of H and G/H, we can construct complete mappings of G. The proof relies on the fact that the product of two left cosets of H is again a left coset of H. This is not the case if H is not normal in G, though the product of two left cosets of H will be a union of left cosets of H. In this case, from complete mappings of H and permutations of G/H, satisfying a certain subset relation, we can construct complete mappings of G.

Lemma 7.7. *Let H be a subgroup of G and let I be a set of order $|G : H|$. Suppose that there exist bijections $\theta, \phi, \eta \colon I \to G/H$, G/H the set of left cosets of H in G, satisfying $\theta(i)\phi(i) \supseteq \eta(i)$ for all i. Then, if H is admissible, G is admissible.*

Proof. Let $Y = \{y_i \mid i \in I\}$ be a dual system of distinct coset representatives for H in G, ordered so that $y_i \in \phi(i)$ for each i. Now

$$\eta(i) \subseteq \theta(i)\phi(i) = \bigcup_{x \in \theta(i)} x\phi(i).$$

Thus, for each i, choose $x_i \in \theta(i)$ to satisfy $\eta(i) = x_i\phi(i)$. This determines, for each i, a unique $z_i \in \eta(i)$ that satisfies $z_i = x_i y_i$. Now $X = \{x_i \mid i \in I\}$ and $Z = \{z_i \mid i \in I\}$ are both systems of distinct left coset representatives for H in G.

The set of equalities $x_i y_i = z_i$ resembles an HP-system and our proof that this set of equalities enables us to construct a complete mapping of G from a complete mapping of H is similar to the proof of Theorem 4.1. Let ν be a complete mapping of H, and let $\eta_\nu(h) = h\nu(h)$. As Y is a dual system of distinct coset representatives for H in G,

$$y_i h = k_{i,h} y_{f(i,h)}$$

for some $k_{i,h} \in H$ and some $y_{f(i,h)} \in Y$. Thus

$$z_i \eta_\nu(h) = x_i y_i h\nu(h) = x_i k_{i,h} y_{f(i,h)} \nu(h).$$

Let us define $\theta', \eta' \colon G \to G$ by

$$\theta'(x_i k_{i,h}) = y_{f(i,h)}\nu(h) \text{ and } \eta'(x_i k_{i,h}) = (x_i k_{i,h})\theta'(x_i k_{i,h}) = z_i \eta_\nu(h).$$

We need to show that θ' and η' are well-defined bijections. To prove this, it suffices to show that every element of G can be uniquely expressed in the form $x_i k_{i,h}$, $i \in I$ and $h \in H$; in the form $y_{f(i,h)}\nu(h)$, $i \in I$ and $h \in H$; and in the form $z_i\eta_\nu(h)$, $i \in I$ and $h \in H$. Suppose that $x_i k_{i,h} = x_{i'} k_{i',h'}$. As $k_{i,h}, k_{i',h'} \in H$ and X is a system of distinct left coset representatives for H in G, $i = i'$, and thus $k_{i,h} = k_{i',h'}$ and

$$y_{f(i,h)} h^{-1} = k_{i,h}^{-1} y_i = k_{i',h'}^{-1} y_{i'} = y_{f(i',h')} h'^{-1}.$$

As Y is a system of distinct left coset representatives for H in G, $y_{f(i,h)} = y_{f(i',h')}$, from which it follows that $h^{-1} = h'^{-1}$ and so $h = h'$. Hence every element of G can be uniquely expressed in the form $x_i k_{i,h}$, $i \in I$ and $h \in H$. If $y_{f(i,h)}\nu(h) = y_{f(i',h')}\nu(h')$, as Y is a system of distinct left coset representatives for H in G, $y_{f(i,h)} = y_{f(i',h')}$ and $\nu(h) = \nu(h')$. Thus $h = h'$ and

$$k_{i,h}^{-1} y_i = y_{f(i,h)} h^{-1} = y_{f(i',h')} h'^{-1} = k_{i',h'}^{-1} y_{i'}.$$

As Y is a system of distinct right coset representatives for H in G, $i = i'$. Hence every element of G can be uniquely expressed in the form $y_{f(i,h)}\nu(h)$, $i \in I$ and $h \in H$. If $z_i\eta_\nu(h) = z_{i'}\eta_\nu(h')$, then $i = i'$ as Z is a system of distinct left coset representatives for H in G, and so $h = h'$ as ν is a bijection. Hence every element of G can be uniquely expressed in the form $z_i\eta_\nu(h)$, $i \in I$ and $h \in H$, completing the proof. □

In Lemma 7.7, permutations of G/H satisfying a certain subset relation were used to extend complete mappings of H to complete mappings of G. Similarly permutations of the set of double cosets of H in G, satisfying a certain subset relation, can also be used to extend complete mappings of H to complete mappings of G.

Theorem 7.8 (Wilcox, 2009). *Suppose that H is an admissible subgroup of a finite group G and that $\mathscr{D}$ is the set of double cosets of H in G. If there exist bijections $\phi, \psi \colon \mathscr{D} \to \mathscr{D}$ satisfying $|D| = |\psi(D)| = |\phi(D)|$ and $\psi(D) \subseteq D\phi(D)$ for all $D \in \mathscr{D}$, then G is admissible.*

Proof. Let $\mathscr{D} = \{D_1, \ldots, D_n\}$ and set $n_i = |D_i : H|$. For each i we pick $z_i \in \psi(D_i)$. By the conditions of the theorem, for each i we can pick $x_i \in D_i$ and $y_i \in \phi(D_i)$ satisfying $z_i = x_i y_i$. Now

$$|D_i| = \frac{|H|^2}{|H \cap x_i H x_i^{-1}|} = |\phi(D_i)| = \frac{|H|^2}{|H \cap y_i H y_i^{-1}|} = |\psi(D_i)| = \frac{|H|^2}{|H \cap z_i H z_i^{-1}|}.$$

Let X_i be a system of distinct left coset representatives for $H \cap x_i H x_i^{-1}$ in H, as well as for $H \cap z_i H z_i^{-1}$ in H: such an X_i exists by Corollary 7.6. Also let Y_i be a system of distinct left coset representatives for $H \cap y_i H y_i^{-1}$ in H. Thus

$$D_i = \coprod_{h \in X_i} h x_i H, \quad \psi(D_i) = \coprod_{h \in X_i} h z_i H, \quad \text{and } \phi(D_i) = \coprod_{h \in Y_i} h y_i H.$$

Let $\mu_i : X_i \to Y_i$ be a bijection, and set $I = \{(i,h) \mid i = 1, \ldots, n, h \in X_i\}$. Define mappings $\theta, \phi, \eta \colon I \to G/H$ by

$$\theta(i,h) = h x_i H, \quad , \phi(i,h) = \mu_i(h) y_i H, \quad \text{and } \eta(i,h) = h z_i H.$$

Now

$$\eta(i,h) = h z_i H = h x_i y_i H \subseteq h x_i H \mu_i(h) y_i H = \theta(i,h)\phi(i,h).$$

It remains to show that η, θ and ϕ are bijections.

If $\eta(i,h) = \eta(j,k)$, then $h z_i H = k z_j H$, and, as $h, k \in H$, $D_i = D_j$. Hence $i = j$, $h z_i H z_i^{-1} = k z_i H z_i^{-1}$, and $h, k \in X_i$, and so $h = k$. It follows that η is injective and hence is a bijection.

If $\theta(i,h) = \theta(j,k)$, then $hx_iH = kx_jH$, and, as $h,k \in H$, $D_i = D_j$. Hence $i = j$, $hx_iHx_i^{-1} = kx_iHx_i^{-1}$, $h,k \in X_i$, and so $h = k$. It follows that θ is injective and hence is a bijection.

If $\phi(i,h) = \phi(j,k)$, then $\mu_i(h)y_iH = \mu_j(k)y_jH$, and, as $\mu_i(h), \mu_j(k) \in H$, $D_i = D_j$. Hence $i = j$, $\mu_i(h)y_iHy_i^{-1} = \mu_i(k)y_iHy_i^{-1}$, and $\mu_i(h), \mu_i(k) \in Y_i$, and so $\mu_i(h) = \mu_i(k)$, and, as μ_i is a bijection, $h = k$. It follows that ϕ is injective and hence is a bijection.

The result then follows from Lemma 7.7. □

We will call $(\mathscr{D}, \phi, \psi)$ in Theorem 7.8 a *W-system*, and we will say that a W-system $(\mathscr{D}, \phi, \psi)$ is a *simple W-system* if ϕ and ψ are equal to the identity mapping as in the following lemma.

Corollary 7.9. *Suppose that H is an admissible subgroup of a finite group G and that $\mathscr{D}$ is the set of double cosets of H in G. If $D \subseteq D^2$ for all $D \in \mathscr{D}$, then G is admissible.*

Proof. Let ϕ and ψ be the identity mapping in Theorem 7.8. □

7.1.3 The Tits Group

A number of the W-systems that we will construct will be W-simple. We will say that a double coset D is *W-simple* if $D \subseteq D^2$. Thus Corollary 7.9 implies that, if every double coset of H in G is W-simple and H is admissible, then G is admissible. The following lemma gives a simple test to determine whether a double coset is W-simple or not.

Lemma 7.10. *Suppose that H is a subgroup of a group G and that $\mathscr{D}$ is the set of double cosets of H in G. Then $D \in \mathscr{D}$ is W-simple if and only if $g^2 \in D$ for some $g \in D$.*

Proof. Let $D \in \mathscr{D}$. If, for some $g \in D$, $g^2 \in D$, then

$$D^2 = HgHHgH \supseteq Hg^2H = D,$$

and so D is W-simple.

If D is W-simple and $h \in D$, then $h \in D^2$ and $h = m_1hm_2hm_3$ for some $m_1, m_2, m_3 \in H$. Setting $g = hm_2$, we see that $g \in D$ and

$$\begin{aligned} g^2 &= hm_2hm_2 = (m_1^{-1}m_1)hm_2h(m_3m_3^{-1})m_2 \\ &= m_1^{-1}(m_1hm_2hm_3)m_3^{-1}m_2 = m_1^{-1}hm_3^{-1}m_2 \end{aligned}$$

and so $g^2 \in D$. □

As an example, we will show that the group ${}^2F_4(2)'$, the Tits group, is not a minimal counterexample to the Hall-Paige conjecture. The Tits group, ${}^2F_4(2)'$, is an exceptional group of Lie type that some count among the sporadic simple groups.

Theorem 7.11 (Evans, 2009). *$^2F_4(2)'$ is not a minimal counterexample to the Hall-Paige conjecture.*

Proof. Using the permutation representation of degree 1,600 given in [330], we verified the following using magma. In the description given in [330], $^2F_4(2)'$ is generated by two elements, x and y, that satisfy the relations

$$x^2 = y^3 = (xy)^{13} = [x, y]^5 = [x, yxy]^4 = ((xy)^4xy^2)^6 = 1,$$

where $[a, b]$ denotes the commutator $a^{-1}b^{-1}ab$. Let H be the subgroup of $^2F_4(2)'$ generated by y and $(xy)^3(xy^{-1})^4xy(xy^{-1})^2(xy)^2x$. Then H is a maximal subgroup of $^2F_4(2)'$ of index 1,600 and has a noncyclic Sylow 2-subgroup. The set of double cosets of H in $^2F_4(2)'$ is $\mathscr{D} = \{D_i = Hd_iH \mid i = 1, \ldots, 4\}$, where

$$\begin{aligned} d_1 &= 1,\\ d_2 &= (xy^2)^4,\\ d_3 &= xy^{-1}xy(xy^{-1})^2(xy)^2x, \text{ and}\\ d_4 &= xyx. \end{aligned}$$

Now

$$\begin{aligned} |D_1 : H| &= 1,\\ |D_2 : H| &= 312,\\ |D_3 : H| &= 351, \text{ and}\\ |D_4 : H| &= 936. \end{aligned}$$

Further $d_i^2 \in D_i$ for $i = 1, \ldots, 4$. Hence, by Lemma 7.10, each of $D_1, \ldots, D_4$ is W-simple, and thus, by Corollary 7.9, $^2F_4(2)'$ is not a minimal counterexample to the Hall-Paige conjecture. □

7.2 Simple Groups of Lie Type

In 2009 Wilcox [329] proved that any finite simple group of Lie type, with the possible exception of the Tits group, cannot be a minimal counterexample to the Hall-Paige conjecture. In this section we will outline Wilcox's proof.

7.2.1 The Chevalley and Twisted Chevalley Groups

Simple groups of Lie type over the field of complex numbers have been known for some time. In 1955 Chevalley [72] constructed analogues of these groups over finite fields, the *adjoint Chevalley groups*: these groups are $A_n(q)$, $B_n(q)$, $C_n(q)$, $D_n(q)$, $E_6(q)$, $E_7(q)$, $E_8(q)$, $F_4(q)$, and $G_2(q)$. In 1959 Steinberg [299] modified Chevalley's construction to obtain the *twisted Chevalley groups*, $^2A_n(q)$, $^2B_n(q)$, $^2D_n(q)$, $^3D_4(q)$, $^2E_6(q)$, $^2F_4(q)$, and $^2G_2(q)$. The simple Chevalley and twisted Chevalley groups are the simple groups of *Lie type*. The first column of Table 7.1 lists the adjoint Chevalley groups and the twisted Chevalley groups, the second column lists the orders of the corresponding universal Chevalley groups and the universal twisted Chevalley groups, and the third column lists the ratios. In this table n is a positive integer, q a prime power, and p a prime. Most of the adjoint Chevalley groups and twisted Chevalley groups in this table are simple. The nonsimple exceptions involve small values of n and/or q. As examples both $A_1(2)$ and $A_1(3)$ are solvable, but $A_n(q)$ is simple for all other choices of n and q. Also ${}^2F_4(2)$ is not a simple group, but ${}^2F_4(2)'$, the Tits group, is.

We have already shown some of these groups to be admissible. The groups $A_n(q) = \mathrm{PSL}(n+1, q)$ were considered in Chapter 5, where we proved $\mathrm{PSL}(n, q)$ to be admissible for $n = 2$ if $q \neq 2$ and for q even and $n > 2$. The groups ${}^2B_2\left(2^{2k+1}\right) = \mathrm{Sz}\left(2^{2k+1}\right)$ were shown to be admissible for $k \geq 1$ in Section 4.3, and the groups ${}^2A_2(q) = \mathrm{PSU}(3, q)$ were shown to be admissible for q even in Section 4.4.

We give here a brief overview of the structure of simple groups of Lie type. For a detailed description of these groups, see the 1972 book by Carter [65]. We will find it useful to view simple groups of Lie type as groups with (B, N)-pairs. A group G has a *(B, N)-pair* if the following hold.

- B and N are subgroups of G and $G = \langle B, N\rangle$.
- $H = B \cap N$ is a normal subgroup of N.
- $W = N/H$ is generated by a set S of involutions. Let $\pi\colon N \to W$ be the natural homomorphism.
- $(BmB)(BnB) \subseteq BnB \cup BmnB$ for $m \in N$, $\pi(m) \in S$, and $n \in N$.

Table 7.1 The adjoint Chevalley and adjoint twisted Chevalley groups

H	|G|	|G|/|H|
$A_n(q)$	$q^{n(n+1)/2}\prod_{i=1}^{n}\left(q^{i+1}-1\right)$	$\gcd(n+1, q-1)$
$B_n(q)$, n odd	$q^{n^2}\prod_{i=1}^{n}\left(q^{2i}-1\right)$	$\gcd(2, q-1)$
$C_n(q)$	$q^{n^2}\prod_{i=1}^{n}\left(q^{2i}-1\right)$	$\gcd(2, q-1)$
$D_n(q)$, n even	$q^{n(n-1)}\left(q^n-1\right)\prod_{i=1}^{n-1}\left(q^{2i}-1\right)$	$\gcd(4, q^n-1)$
$E_6(q)$	$q^{36}\left(q^2-1\right)\left(q^5-1\right)\left(q^6-1\right)\left(q^8-1\right)$ $\left(q^9-1\right)\left(q^{12}-1\right)$	$\gcd(3, q-1)$
$E_7(q)$	$q^{63}\left(q^2-1\right)\left(q^6-1\right)\left(q^8-1\right)\left(q^{10}-1\right)$ $\times\left(q^{12}-1\right)\left(q^{14}-1\right)\left(q^{18}-1\right)$	$\gcd(2, q-1)$
$E_8(q)$	$q^{120}\left(q^2-1\right)\left(q^8-1\right)\left(q^{12}-1\right)\left(q^{14}-1\right)$ $\times\left(q^{18}-1\right)\left(q^{20}-1\right)\left(q^{24}-1\right)\left(q^{30}-1\right)$	1
$F_4(q)$	$q^{24}\left(q^2-1\right)\left(q^6-1\right)\left(q^8-1\right)\left(q^{12}-1\right)$	1
$G_2(q)$	$q^6\left(q^2-1\right)\left(q^6-1\right)$	1
${}^2A_n(q)$, $q = p^{2k}$	$q^{n(n+1)/2}\prod_{i=1}^{n}\left(q^{i+1}-(-1)^{i+1}\right)$	$\gcd(n+1, q+1)$
${}^2B_2(q)$, $q = 2^{2k+1}$	$q^2\left(q^2+1\right)(q-1)$	1
${}^2D_n(q)$, $q = p^{2k}$	$q^{n(n-1)}\left(q^n+1\right)\prod_{i=1}^{n-1}\left(q^{2i}-1\right)$	$\gcd(4, q^n+1)$
${}^3D_4(q)$, $q = p^{3k}$	$q^{12}\left(q^2-1\right)\left(q^6-1\right)\left(q^8+q^4+1\right)$	1
${}^2E_6(q)$, $q = p^{2k}$	$q^{36}\left(q^2-1\right)\left(q^5+1\right)\left(q^6-1\right)\left(q^8-1\right)$ $\left(q^9+1\right)\left(q^{12}-1\right)$	$\gcd(3, q+1)$
${}^2F_4(q)$, $q = 2^{2k+1}$	$q^{12}(q-1)\left(q^3+1\right)\left(q^4-1\right)\left(q^6+1\right)$	1
${}^2G_2(q)$, $q = 3^{2k+1}$	$q^3\left(q^3+1\right)(q-1)$	1

- $nBn \neq B$, for $n \in N, n \neq 1$.

The 4-tuple (G, B, N, S) is called a *Tits system*, B is the *Borel subgroup* of G, $H = B \cap N$ is the *Cartan subgroup* of G, W is the *Weyl group* of G, and the elements of S are the *roots* of W.

Theorem 7.12. *Any Chevalley or twisted Chevalley group has a (B, N)-pair.*

Proof. See Propositions 8.2.1 and 13.5.4 in [65]. □

Theorem 7.13. *If G is a group with a (B, N)-pair, then $G = BNB$.*

Proof. See Proposition 8.2.2 in [65]. □

Theorem 7.14. *Let G be a group with a (B, N)-pair, let W be the Weyl group of G, let S be the set of involutions that generate W, and let $\pi : N \to W$ be the natural homomorphism. For $T \subseteq S$, let W_T denote the subgroup of W generated by T. Then $P_T = B\pi^{-1}(W_T)B$ is a subgroup of G.*

Proof. See Proposition 8.2.2 in [65]. □

The groups P_T in Theorem 7.14 and their conjugates are the *parabolic subgroups* of G.

Theorem 7.15. *Let G be a group with a (B, N)-pair, let W be the Weyl group of G, and let S be the set of roots of W. For $U, V \subseteq S$ the following hold:*

1. $P_U \subseteq P_V$ *if and only if* $U \subseteq V$.
2. $P_U = P_V$ *if and only if* $U = V$.
3. $P_U \cap P_V = P_{U \cap V}$.

Proof. See Propositions 8.3.3 and 8.3.4 in [65]. □

For r a root, we will set $P_r = P_{S \setminus \{r\}}$: P_r is the *maximal parabolic subgroup associated with* r. These maximal parabolic subgroups will play an important role in Wilcox's proof that any finite simple group of Lie type, with the possible exception of the Tits group, cannot be a minimal counterexample to the Hall-Paige conjecture.

7.2.2 Wilcox's Proof

Let G be a simple group of Lie type, excluding the Tits group, r a root, and let P_r be the maximal parabolic subgroup of G associated with r. For most simple groups G of Lie type, Wilcox proved the existence of maximal parabolic subgroups P_r of G that have trivial or noncyclic Sylow 2-subgroups and for which $D \subseteq D^2$ for all double cosets D of P_r. Thus these groups cannot be minimal counterexamples to the Hall-Paige conjecture by Corollary 7.9. For the remaining simple groups of Lie type, for which this simple W-system approach fails, Wilcox constructed complete mappings.

Theorem 7.16 (Wilcox, 2009). *If G is a simple group of Lie type, not the Tits group, then G is not a minimal counterexample to the Hall-Paige conjecture.*

Proof. Here we will outline Wilcox's proof: for the details see [329].

Case 1. G is not one of $A_1(q)$, ${}^2A_2(q)$, ${}^2G_2(q)$, ${}^2A_3(q)$, or $B_2(q)$. In this case, for all roots r, the Sylow 2-subgroup of P_r, the maximal parabolic subgroup of G associated with r, is trivial or noncyclic and, for some root r, $D \subseteq D^2$ for all double cosets D of P_r. Hence, by Corollary 7.9, G is not a minimal counterexample to the Hall-Paige conjecture.

Case 2. G is one of ${}^2A_3(q)$ or $B_2(q)$. In this case, for some root r, the Sylow 2-subgroup of P_r, the maximal parabolic subgroup of G associated with r, is trivial or noncyclic, and, for all roots r, $D \subseteq D^2$ for all double cosets D of P_r. Hence, by Corollary 7.9, G is not a minimal counterexample to the Hall-Paige conjecture.

Case 3. G is one of $A_1(q)$, ${}^2A_2(q)$ or ${}^2G_2(q)$, and either q is even or $|H|$, H the Cartan subgroup of G, is odd. For some root r, the Sylow 2-subgroup of P_r, the maximal parabolic subgroup of G associated with r, is trivial or noncyclic and, for all roots r, $D \subseteq D^2$ for all double cosets D of P_r. Hence, by Corollary 7.9, G is not a minimal counterexample to the Hall-Paige conjecture.

Case 4. G is one of $A_1(q)$, ${}^2A_2(q)$, or ${}^2G_2(q)$, and q is odd and $|H|$, H the Cartan subgroup of G, is even. Constructions of complete mappings show G to be admissible. □

7.2.3 Aschbacher's Work Revisited

Let us revisit Aschbacher's work: we covered this in Chapter 6. In 1990 Aschbacher [21] proposed the HP-system conjecture—Conjecture 6.15—and, in Theorem 6.16, we proved that the HP-system conjecture implies the Hall-Paige conjecture. Theorem 6.18 described his list of finite almost simple groups G with minimal normal subgroup L of Lie type over $GF(q)$, G/L a cyclic 2-group, for which the HP-system conjecture might not hold. In view of Wilcox's 2009 proof [329] that any minimal counterexample to the Hall-Paige conjecture must be simple (see Theorem 6.35), we can rewrite Theorem 6.18 as follows.

Theorem 7.17. *Any finite simple group of Lie type that is a minimal counterexample to the Hall-Paige conjecture must be one of the following.*

1. $A_1(q)$, *q odd*, $q > 3$.
2. $A_n(q)$, $n \geq 2$.
3. $F_4(q)$, *q even*.
4. ${}^2A_2(q)$, *q odd*, $q \neq 4$.
5. ${}^2F_4(q)$, $q \neq 2$.
6. ${}^2F_4'(2)$.

7. $^2G_2(q)$, *q odd*, $q > 3$.

Of the groups in Theorem 7.17, in Chapter 5, the group $A_1(q)$ was proved to be admissible for $q \neq 2$, and the group $A_n(q)$ was proved to be admissible for q even and $n > 2$.

7.3 Sporadic Simple Groups

Having shown that any minimal counterexample to the Hall-Paige conjecture must be a sporadic simple group, we will show that these groups also cannot be minimal counterexamples, thus completing a proof of the conjecture. As the Mathieu groups have already been shown to be admissible, we are left with a list of 21 possible minimal counterexamples to the Hall-Paige conjecture. Our proof that none of these remaining groups can be a minimal counterexample to the conjecture is outlined in Table 7.2. For each group in this list, we will construct a W-system, usually a simple W-system, from which a proof of nonminimality will follow. In Table 7.2, for a group G and chosen maximal subgroup H, we list the isomorphism type of H, the index of H in G, the rank of the permutation representation of G acting on the right cosets of H, and the theorem in which a construction of a W-system on the double cosets of H in G is given. The constructions for groups other than J_4 are by Evans [128], and the construction for J_4 is by Bray [50]. In our proof that none of these 21 groups can be a minimal counterexample to the Hall-Paige conjecture, we will treat doubly transitive simple groups and rank-3 simple groups separately. The nonzero entries in collapsed adjacency matrices of orbital digraphs will be used to construct W-systems. Our proof for the Janko groups J_1 and J_3 uses the distribution of elements of order 3. All of our constructions of W-systems will be of simple W-systems, except for the Janko groups J_1 and J_3, the Held group He, the Thompson group Th, and possibly the Janko group J_4.

The notation for simple groups and their subgroups that we are using is the notation used in the ATLAS of Finite Groups [81]. There are other notations in the literature for the sporadic simple groups. In Aschbacher's book Sporadic Groups, [22] the monster, M, is denoted F_1; the baby monster, B, is denoted F_2; the Thompson group, Th, is denoted F_3; the Harada-Norton group, HN, is denoted F_5; the Fischer group, Fi_{22}, is denoted $M(22)$; the Fischer group, Fi_{23}, is denoted $M(23)$; and the Fischer group, Fi_{24}', is denoted $M(24)'$. In the notation for subgroups, n represents the cyclic group $\mathbb{Z}_n$ and 2^n the elementary abelian group of order 2^n. The symbol 2_+^{1+2n} represents the extraspecial 2-group that is isomorphic to the central product of n copies of D_8, the dihedral group of order 8. Extraspecial 2-groups play an important role in the study of sporadic simple groups: the theory of these groups is described in [22] and [241]. For products of groups, $K \times L$ represents the direct product of K and L, and the notation $K.L$ is used more generally to represent a group with a normal subgroup isomorphic to K and corresponding quotient group isomorphic to L. As a special case of $K.L$, $K : L$

represents a semi-direct product: given a homomorphism $\phi\colon L \to \mathrm{Aut}(K)$ that maps l to ϕ_l, the elements of $K : L$ are $\{(l,k) \mid k \in K, l \in L\}$, and the product of (l_1,k_1) and (l_2,k_2) is $(l_1l_2, \phi_{l_2}(k_1)k_2)$. The notation $K{\cdot}L$ may be used for an instance of $K.L$ which is not a direct or semi-direct product. Products are to be read from left to right, and so $K \times L{\cdot}M$ is to be read as $(K \times L){\cdot}M$. For brevity we use $3M_{22}$ in place of $(3.M_{22})$ in $2_+^{1+12}{\cdot}3M_{22} : 2$.

Table 7.2 The last 21 groups

G	H	Index	Rank	Proof
J_1	$A_5 \times 2$	1,463	22	Theorem 7.40
J_2	$U_3(3)$	100	3	Theorem 7.33
HS	$U_3(5) : 2$	176	2	Theorem 7.21
J_3	$L_2(19)$	14,688	14	Theorem 7.40
McL	$U_4(3)$	275	3	Theorem 7.33
He	$S_4(4) : 2$	2,058	5	Theorem 7.34
Ru	${}^2F_4(2)$	4,060	3	Theorem 7.33
Suz	$G_2(4)$	1,782	3	Theorem 7.33
O′N	$L_3(7) : 2$	122,760	5	Theorem 7.34
Co_3	McL : 2	276	2	Theorem 7.21
Co_2	$U_6(2) : 2$	2,300	3	Theorem 7.33
Fi_{22}	$2{\cdot}U_6(2)$	3,510	3	Theorem 7.33
	$O_7(3)$	14,080	3	Theorem 7.33
HN	2.HS.2	1,539,000	9	Theorem 7.35
Ly	$G_2(5)$	8,835,156	5	Theorem 7.34
Th	$2^5.L_5(2)$	283,599,225	11	Theorem 7.35
Fi_{23}	$2{\cdot}\mathrm{Fi}_{22}$	31,671	3	Theorem 7.33
	$O_8^+(3) : S_3$	137,632	3	Theorem 7.33
Co_1	Co_2	98,280	4	Theorem 7.34
J_4	$2_+^{1+12}{\cdot}3M_{22} : 2$	3,980,549,947		Bray [50]
Fi'_{24}	Fi_{23}	306,936	3	Theorem 7.33
B	$2_+^{1+22}.\mathrm{Co}_2$	11,707,448,673,375	10	Theorem 7.35
M	$2.B$	97,239,461,142,009,186,000	9	Theorem 7.35

Let G be a finite group, H a subgroup of G, and let $\mathscr{D} = \{HgH \mid g \in G\}$ be the set of double cosets of H in G. Recall that D is W-simple if $D \subseteq D^2$, and thus, by Corollary 7.9, if H is admissible and every $D \in \mathscr{D}$ is W-simple, then G is admissible. Recall also that, for $D \in \mathscr{D}$, the *index* of H in D, denoted $|D : H|$, is the number of distinct left cosets of H contained in D, and if $D = HgH$, then $|D : H| = |D|/|H| = |H : H \cap H^g|$, where $H^g = g^{-1}Hg$. We will call the double coset $HeH = H$ the *trivial* double coset. We will say that $D \in \mathscr{D}$ is *solitary* if $D' \in \mathscr{D}$ and $|D' : H| = |D : H|$ implies that $D' = D$. If all the double cosets of H in G are solitary, then the double coset that an element $g \in G$ is contained in can be determined by computing $|H : H \cap H^g|$. It is important to note that, if $\mathscr{D}$ contains a solitary double coset that is not W-simple, then a W-system cannot exist.

7.3.1 Doubly Transitive Groups

Suppose that G acts on a set X and $x \in X$. It is well-known that G acts doubly transitively on X if and only if there are only two double cosets of G_x in G: see Theorem 7.3 in [157]. There is a particularly simple test that can be used to prove a doubly transitive simple group to be admissible.

Lemma 7.18. *If H is a point-stabilizer in a doubly transitive permutation representation of a finite simple group G and H is admissible, then G is admissible.*

Proof. If G acts doubly transitively on a set X and $H = G_x$, for some $x \in X$, then the set of double cosets of H in G is $\mathscr{D} = \{H, D\}$. Clearly H is W-simple. To prove that D is W-simple, we will assume the contrary, and thus, by Lemma 7.10, that $g^2 \in H$ for all $g \in G$. In particular, if $|g| = m$ is odd, then $g = (g^{(m+1)/2})^2 \in H$. Let K be the subgroup of G generated by the set of odd-order elements of G. K is a nontrivial characteristic subgroup of G and is contained in H, contradicting the simplicity of G. Thus D is W-simple and the result then follows from Corollary 7.9. □

For simple groups this strengthens a 1993 result of Dalla Volta and Gavioli [83]: they proved that the admissibility of a point-stabilizer in a 3-transitive permutation representation of a finite group implies that the group is admissible: see Lemma 4.24. In 1955 Hall and Paige [171] used HP-systems to prove that the alternating groups are admissible: see Theorem 4.5. As the alternating groups of degree 5 and greater are multiply transitive, we can use Lemma 7.18 to give an alternative proof that these groups are admissible.

Corollary 7.19. *The alternating groups A_n, $n \geq 5$, are admissible.*

Proof. In the natural permutation representation on $\{1, 2, \ldots, n\}$, A_n is doubly transitive with each point-stabilizer being isomorphic to A_{n-1}. The group A_4 is solvable and has noncyclic Sylow 2-subgroups and hence is admissible by Theorem 4.22. The result then follows by induction. □

In 1993 Dalla Volta and Gavioli [83] proved the Mathieu groups to be admissible by expressing them as one-point extensions of admissible doubly transitive groups: see Theorem 4.25. Lemma 7.18 yields another proof of the admissibility of the Mathieu groups.

Corollary 7.20. *The Mathieu groups M_{11}, M_{12}, M_{22}, M_{23}, and M_{24} are admissible.*

Proof. We will use the data on doubly transitive permutation representations of the Mathieu groups given in [271] and [330].

The smallest Mathieu group, M_{11}, has a doubly transitive permutation representation of degree 11 with point-stabilizer isomorphic to $A_6.2$. As A_6 is admissible by Corollary 7.19 (equivalently Theorem 4.5), $A_6.2$ is admissible by Theorem 6.33, and thus M_{11} is admissible by Lemma 7.18.

M_{12} has a doubly transitive permutation representation of degree 12 with point-stabilizer isomorphic to M_{11}. The admissibility of M_{12} follows from the admissibility of M_{11} and Lemma 7.18.

M_{22} has a doubly transitive permutation representation of degree 22 with point-stabilizer isomorphic to $L_3(4)$. As $L_3(4)$ is admissible by Theorem 5.28, M_{22} is admissible by Lemma 7.18.

M_{23} has a doubly transitive permutation representation of degree 23 with point-stabilizer isomorphic to M_{22}. The admissibility of M_{22} implies the admissibility of M_{23} by Lemma 7.18.

Lastly, M_{24} has a doubly transitive permutation representation of degree 24 with point-stabilizer isomorphic to M_{23}. As M_{23} is admissible, M_{24} is admissible by Lemma 7.18. □

Two other sporadic simple groups have doubly transitive permutation representations: the Higman-Sims group, HS, and the third Conway group, Co_3.

Theorem 7.21 (Evans, 2009). *The groups* HS *and* Co_3 *are not minimal counterexamples to the Hall-Paige conjecture.*

Proof. From data in [271] and [330], we see that the group HS has a doubly transitive permutation representation of degree 176 with point-stabilizer isomorphic to $U_3(5) : 2$, and the group Co_3 has a doubly transitive permutation representation of degree 276 with point-stabilizer isomorphic to $\mathrm{McL} : 2$. In each case, the point-stabilizer has noncyclic Sylow 2-subgroups. The result follows from Lemma 7.18. □

7.3.2 *Orbitals, Orbits, and Double Cosets*

As Theorem 7.21 suggests, the action of a group G on a set X can be used to construct W-systems. For group actions that are transitive, but not doubly transitive, we will make use of the action of G on $X \times X$ to construct double cosets from orbitals. These constructions will prove useful in constructing W-systems for sporadic simple groups.

Let G be a finite group acting transitively on a set X. We will use x^g to denote the image of x under the action of the group element g. As the exponential notation suggests, the action of products of group elements should be read from left to right; thus $x^{gh} = (x^g)^h$. The action of G on X extends naturally to an action of G on $X \times X$: the image of (x, y) under the action of the group element g is $(x, y)^g = (x^g, y^g)$. The orbits of the action of the group G on $X \times X$ are called *orbitals*, and the number of orbitals is called the *rank* of the permutation representation of G on

X. The simplest orbital, $\{(x, x) \mid x \in X\}$, is called the *trivial* or *diagonal* orbital. From the orbitals of the action of G on X, we can construct the orbits of G_x on X and the double cosets of G_x in G.

Theorem 7.22. *Let G be a finite group acting transitively on a set X, let $x \in X$, and let $E_1, \ldots, E_r$ be the corresponding set of orbitals. Let the sets $O_1, \ldots, O_r$ be defined by*

$$O_i = \{y \mid (x, y) \in E_i\},$$

and the sets $D_1, \ldots, D_r$ by

$$D_i = \{g \in G \mid x^g \in O_i\}.$$

Then $O_1, \ldots, O_r$ are the distinct orbits of G_x on X, and $D_1, \ldots, D_r$ are the distinct double cosets of G_x in G.

Proof. Let $z \in O_i$ be given. $y \in O_i$ if and only if $(x, y) \in E_i$, if and only if $(x, y) = (x, z)^g$ for some $g \in G$, if and only if $x^g = x$ and $z^g = y$ for some $g \in G$. But this is true if and only if $g \in G_x$ and y and z are in the same orbit of G_x. Thus $O_1, \ldots, O_r$ are distinct orbits of G_x. These are all the orbits of G_x as if y is any element of X, then $(x, y) \in E_i$ for some i and, hence, $y \in O_i$ for some i.

Let $h, k \in G_x$, $g \in D_i$, and $y = x^g$. Then $x^{hgk} = x^{gk} = y^k$. But, as $y \in O_i$, $(x, y^k) = (x, y)^k \in E_i$ and so $y^k \in O_i$. Thus $hgk \in D_i$ and so $G_x g G_x \subseteq D_i$. Let $g' \in D_i$ and let $y' = x^{g'}$. As $y, y' \in O_i$ there exists $k \in G_x$ for which $y^k = y'$. But then $x^{gk} = x^{g'}$, from which it follows that $h = g'(gk)^{-1} \in G_x$. Hence $g' = hgk$ and $D_i \subseteq G_x g G_x$. Thus $D_i = G_x g G_x$ and each D_i is a double coset of G_x in G. If $i \neq j$, then $D_i \cap D_j = \emptyset$ as if $g \in D_i \cap D_j$, then $y = x^g \in O_i \cap O_j = \emptyset$, an impossibility. Thus $D_1, \ldots, D_r$ are distinct double cosets of G_x in G and constitute all the double cosets by the transitivity of G. □

Theorem 7.22 establishes one-to-one correspondences between the set of orbitals, the set of orbits of G_x on X, and the set of double cosets of G_x in G. For an orbital E, $O = \{y \mid (x, y) \in E\}$ and $D = \{g \in G \mid x^g = y \text{ for some } y \in O\}$ are the *corresponding orbit* of G_x and the *corresponding double coset* of G_x in G, respectively. For the trivial orbital, the corresponding double coset is the trivial double coset, G_x, of G_x in G. The action of G on X induces a homomorphism from G into the symmetric group on X: $|X|$ is called the *degree*, and $|O_i|$, $i = 1, \ldots, r$, are called the *subdegrees* of this permutation representation. The subdegrees $|O_i|$ and the indices $|D_i : G_x|$ are, in fact, equal.

Corollary 7.23. *If G_x is a point-stabilizer of a point x in the transitive action of G on a set X, O is an orbit of G_x on X, and D is the corresponding double coset of G_x in G, then $|O| = |D : G_x|$.*

Proof. Let $G_x g_1, \ldots, G_x g_m$ be the distinct right cosets of G_x in D. By Theorem 7.22, $O = \{y \in X \mid y = x^{hg_i}, \text{ some } hg_i \in D\}$. But $x^{hg_i} = x^{g_i}$ and $x^{g_i} = x^{g_j}$ if and only if $g_i g_j^{-1} \in G_x$, if and only if $i = j$. Thus $|O| = m = |D : G_x|$. □

For E an orbital, we define E^p by $E^p = \{(y,x) \mid (x,y) \in E\}$, and for D a double coset of G_x in G, we define $D^{(-1)}$ by $D^{(-1)} = \{g^{-1} \mid g \in D\}$. In fact E^p is an orbital and $D^{(-1)}$ is a double coset. Here are some elementary facts about E^p and $D^{(-1)}$.

Theorem 7.24. *Let G be a finite group acting transitively on a set X, let E be an orbital of G, and let D be a double coset of G_x in G for some $x \in X$. Then the following hold.*

1. *E^p is an orbital of G, $(E^p)^p = E$, and $|E^p| = |E|$.*
2. *$D^{(-1)}$ is a double coset of G_x in G, $\left(D^{(-1)}\right)^{(-1)} = D$, and $\left|D^{(-1)}\right| = |D|$.*
3. *If D is the double coset corresponding to E, then $D^{(-1)}$ is the double coset corresponding to E^p.*
4. *If D is the double coset corresponding to E, then $D^{(-1)} = D$ if and only if $E^p = E$.*
5. *$|D : G_x| = |D^{(-1)} : G_x|$.*

Proof.

1. If $(y,z) \in E$, then

$$\begin{aligned} E = &\ \{(y,z)^g \mid g \in G\} = \{(y^g, z^g) \mid g \in G\}, \quad \text{and} \\ E^p = &\ \{(z^g, y^g) \mid g \in G\} = \{(z,y)^g \mid g \in G\}, \end{aligned}$$

an orbital of G.
2. If $g \in D$, then $D = G_x g G_x$ and $D^{(-1)} = G_x g^{-1} G_x$, a double coset of G_x in G.
3. If D is the double coset corresponding to the orbital E, then $D^{(-1)}$ is the double coset corresponding to the orbital E^p, as $(x, x^g) \in E$ if and only if $(x^g, x) \in E^p$ if and only if

$$\left(x, x^{g^{-1}}\right) = (x^g, x)^{g^{-1}} \in E^p.$$

4. This follows from 3.
5. By Theorem 7.3, if $D = G_x g G_x$, then, as

$$D^{(-1)} = G_x g^{-1} G_x \quad \text{and} \quad g^{-1}(G_x \cap g G_x g^{-1}) g = G_x \cap g^{-1} G_x g,$$

$$\begin{aligned} |D : G_x| &= |G_x : G_x \cap g G_x g^{-1}| \\ &= \frac{|G_x|}{|G_x \cap g G_x g^{-1}|} = \frac{|G_x|}{|G_x \cap g^{-1} G_x g|} \\ &= |G_x : G_x \cap g^{-1} G_x g| = |D^{(-1)} : G_x|. \end{aligned}$$

□

For an orbital E, the orbital E^p is its *paired orbital*: E is *self-paired* if $E^p = E$. For a double coset D, the double coset $D^{(-1)}$ is the *inverse* of D: D is *self-inverse* if $D^{(-1)} = D$. As a consequence of Theorem 7.24, we can easily show that any solitary double coset must also be self-inverse.

Corollary 7.25. *If D is solitary, then D is self-inverse.*

Proof. By Theorem 7.24(5), $|D : H| = |D^{(-1)} : H|$, and so $D^{(-1)} = D$. □

For each orbital E of the action of G on $X \times X$, we can construct a graph or digraph $\mathcal{G}_E$ depending on whether E is self-paired or not. The vertices of $\mathcal{G}_E$ are the elements of X: if E is self-paired, then $\mathcal{G}_E$ is a graph and x and y are adjacent if and only if $(x, y) \in E$; if E is not self-paired, then $\mathcal{G}_E$ is a digraph and there is a directed edge from x to y if and only if $(x, y) \in E$. $\mathcal{G}_E$ is called an *orbital (di)graph.* If $E_1, \ldots, E_r$ are the orbitals of G ordered so that $|E_1| \leq |E_2| \leq \cdots \leq |E_r|$, E_1 the trivial orbital, $O_1, \ldots, O_r$ are the corresponding orbits of a given point-stabilizer H on X, $D_1, \ldots, D_r$ are the corresponding double cosets of H in G, and $x = a_1, \ldots, a_r$ are representatives of $O_1, \ldots, O_r$, respectively, then the *collapsed adjacency matrix* for $\mathcal{G}_{E_k}$ is the $r \times r$ matrix A^k with ijth entry $A^k_{ij} = |\{y \in O_j \mid (a_i, y) \in E_k\}|$. As an example, consider a finite group with a rank-2 permutation representation.

Example 7.26. Suppose that a finite group G acts doubly transitively on a set X and that $x \in X$. Then this is a rank-2 permutation representation of G. The orbitals are

$$E_1 = \{(y, y) \mid y \in X\} \quad \text{and} \quad E_2 = \{(y, z) \mid y, z \in X, y \neq z\},$$

each of which is self-paired, the corresponding orbits of G_x on X are $O_1 = \{x\}$ and $O_2 = X \setminus \{x\}$, the corresponding double cosets of G_x in G are $D_1 = G_x$ and $D_2 = G \setminus G_x$, and the corresponding subdegrees are 1 and $|X| - 1$. The orbital graph $\mathcal{G}_{E_1}$ is obtained from the edgeless graph on $|X|$ vertices by joining a loop to each vertex, and the orbital graph $\mathcal{G}_{E_2}$ is the complete graph on $|X|$ vertices. The collapsed adjacency matrices are

$$A^1 = \begin{array}{c|cc} & 1 & 2 \\ \hline 1 & 1 & 0 \\ 2 & 0 & 1 \end{array}, \quad \text{and} \quad A^2 = \begin{array}{c|cc} & 1 & 2 \\ \hline 1 & 0 & |X|-1 \\ 2 & 1 & |X|-2 \end{array}.$$

Collapsed adjacency matrices are useful in constructing W-systems.

Lemma 7.27. *$A^k_{ij} > 0$ if and only if $D_k D_i \supseteq D_j$.*

Proof. If $A^k_{ij} > 0$, then there exists $y \in O_j$ and $z \in O_i$ for which $(z, y) \in E_k$. But $z = x^g$ for some $g \in D_i$ and $y = x^{hg}$ for some h satisfying $hg \in D_j$. It follows that

$$(x, x^h) = (x^g, x^{hg})^{g^{-1}} = (z, y)^{g^{-1}} \in E_k,$$

and so $h \in D_k$. Thus

$$D_k D_i = HhHHgH \supseteq HhgH = D_j.$$

If $D_k D_i \supseteq D_j$, then there exists $h \in D_k$ and $g \in D_i$ for which $hg \in D_j$. Thus $(x, x^h) \in E_k$. and $(x^g, x^{hg}) = (x, x^h)^g \in E_k$, and, as $x^g \in O_i$ and $x^{hg} \in O_j$, $A^k_{ij} > 0$. □

It follows that, given a subgroup H of G, the collapsed adjacency matrices completely determine the possible W-systems for H in G. In particular, it is easy to determine from the collapsed adjacency matrices whether a double coset is W-simple or not.

Corollary 7.28. *D_k is W-simple if and only if $A^k_{kk} > 0$.*

We will use Corollary 7.28 and the following special case of Lemma 7.27 to construct our W-systems.

Corollary 7.29. *If $A^k_{kj} > 0$, then $D_k^2 \supseteq D_j$.*

From Corollary 7.28, we can derive a stronger version of Lemma 7.18.

Corollary 7.30. *If a finite group G acts doubly transitively on X, $|X| > 2$, $x \in X$, and G_x is admissible, then G is admissible.*

Proof. By Example 7.26, $A^2_{22} = |X| - 2 > 0$. Thus by Corollary 7.28, the nontrivial double coset D_2 is W-simple. As the trivial double coset is also W-simple, the result follows from Corollary 7.9. □

In Corollary 7.30, we can give an alternative proof that the nontrivial double coset D_2 is W-simple: D_2^2 is a union of double cosets of G_x in G and $D_2^2 \neq G_x$ as

$$|D_2^2|/|G_x| \geq |D_2 : G_x| > 1 = |G_x|/|G_x|.$$

Thus D_2^2 is either D_2 or $D_2 \cup G_x$: in either case D_2 is W-simple.

7.3.3 Rank-3 Groups

Having seen that simple groups with rank-2 permutation representations, which have admissible point-stabilizers, are admissible, the next question to ask is, is the same true of simple groups with rank-3 permutation representations? If the conditions are right, then a rank-3 permutation representation can be used to construct simple W-systems and, hence, if the point-stabilizers are admissible, proofs of admissibility. As several sporadic simple groups have rank-3 permutation representations, this will enable us to prove that several more sporadic simple groups cannot be minimal counterexamples to the Hall-Paige conjecture.

Let E_1, E_2, E_3 be the orbitals in a rank-3 permutation representation of a finite group G on X, E_1 the trivial orbital. Let $H = G_x$ for some $x \in X$, and let O_1, O_2, O_3 be the orbits of H on X corresponding to the orbitals E_1, E_2, E_3, respectively. Note that either E_2 and E_3 are paired or both are self-paired. In fact, if G_X is trivial and $|G|$ is even, then E_2 and E_3 are self-paired. In this case there exists an involution $g \in G$ and $y, z \in X$, $y \neq z$, for which $y^g = z$ and $z^g = y$. Now $(y, z) \in E_i$, for some $i = 1, 2$, and $(z, y) = (y, z)^g \in E_i$. Thus E_i is self-paired and so E_2 and E_3 are both self-paired. A similar argument shows that E_2 and E_3 are self-paired if G/G_X is of even order. Thus, if G/G_X is of even order, then $\mathcal{G}_{E_2}$ and $\mathcal{G}_{E_3}$ are both graphs. We then, as in [23], set $n = |X|$; $k = |O_2|$; $l = |O_3|$; for a fixed $w \in O_2$, $\lambda = |\{y \in O_2 \mid (w, y) \in E_2\}|$; and, for a fixed $z \in O_3$, $\mu = |\{y \in O_2 \mid (z, y) \in E_2\}|$. The integers (n, k, l, λ, μ) are the *parameters* of the rank-3 graph $\mathcal{G}_{E_2}$. These parameters uniquely determine the collapsed adjacency matrices.

Lemma 7.31. *Let E_1, E_2, E_3 be the orbitals in a rank-3 permutation representation of a finite group G on X, $|G/G_X|$ even, E_1 the trivial orbital, and let the parameters for $\mathcal{G}_{E_2}$ be (n, k, l, λ, μ). Then the collapsed adjacency matrices are*

$$A^1 = \begin{array}{c|ccc} & 1 & 2 & 3 \\ \hline 1 & 1 & 0 & 0 \\ 2 & 0 & 1 & 0 \\ 3 & 0 & 0 & 1 \end{array}, \quad A^2 = \begin{array}{c|ccc} & 1 & 2 & 3 \\ \hline 1 & 0 & k & 0 \\ 2 & 1 & \lambda & k-\lambda-1 \\ 3 & 0 & \mu & k-\mu \end{array}, \quad \textit{and}$$

$$A^3 = \begin{array}{c|ccc} & 1 & 2 & 3 \\ \hline 1 & 0 & 0 & l \\ 2 & 0 & k-\lambda-1 & l-k+\lambda+1 \\ 3 & 1 & k-\mu & l-k+\mu-1 \end{array}$$

Proof. The entries of A^1, the first row and column of A^2, and the first row and column of A^3 are easily determined from the definitions. It is also easy to show that

$$\sum_t A^t_{ij} = |O_j|,$$

and, as $\mathcal{G}_{E_2}$ is a regular graph of degree k, and $\mathcal{G}_{E_3}$ is a regular graph of degree l,

$$\sum_j A^2_{ij} = k \quad \text{and} \quad \sum_j A^3_{ij} = l.$$

By definition $A^2_{22} = \lambda$ and $A^2_{32} = \mu$, from which the remaining entries can be computed. □

From Lemma 7.31 and Corollary 7.28, we obtain a simple test for the existence of W-systems for groups with rank-3 permutation representations.

Lemma 7.32. *Let E_1, E_2, E_3 be the orbitals in a rank-3 permutation representation of a finite group G acting on X, $|G/G_X|$ even, E_1 the trivial orbital, and let the parameters for $\mathcal{G}_{E_2}$ be (n, k, l, λ, μ). If H, a point-stabilizer, is admissible, $\lambda > 0$, and $l - k + \mu - 1 > 0$, then G is admissible.*

Proof. By Lemma 7.31, $A_{22}^2 = \lambda$ and $A_{33}^3 = l - k + \mu - 1$. Thus, if $\lambda > 0$, then, by Corollary 7.28, the double coset of H in G corresponding to the orbital E_2 is W-simple, and if $l - k + \mu - 1 > 0$, then, by Corollary 7.28, the double coset of H in G corresponding to the orbital E_3 is W-simple. As H is also W-simple, the result follows from Corollary 7.9. □

Several sporadic simple groups have rank-3 permutation representations, HS, J_2, McL, Ru, Suz, Co_2, Fi_{22}, Fi_{23}, and Fi_{24}'. Lemma 7.32 yields proofs that eight of these groups cannot be minimal counterexamples to the Hall-Paige conjecture.

Theorem 7.33 (Evans, 2009). *J_2, McL, Ru, Suz, Co_2, Fi_{22}, Fi_{23}, and Fi_{24}' are not minimal counterexamples to the Hall-Paige conjecture.*

Proof. From the collapsed adjacency matrices in [271], we can read off the parameters for $\mathcal{G}_{E_2}$ in the rank-3 permutation representations of these groups. These parameters are displayed in Table 7.3. In each case the point-stabilizer, G_x, has noncyclic Sylow 2-subgroups, $\lambda > 0$, and $l - k + \mu - 1 > 0$. The result then follows from Lemma 7.32. □

Note that the parameters for $\mathcal{G}_{E_2}$ in the rank-3 permutation representation of HS, from [271], are $(100, 22, 77, 0, 6)$. Thus $\lambda = 0$, and no W-system can be constructed from this representation.

7.3.4 Rank-4 and Rank-5 Sporadic Simple Groups

Praeger and Soicher's book [271] is a particularly useful resource, containing collapsed adjacency matrices for permutation representations of sporadic simple

Table 7.3 Some rank-3 permutation representations

G	G_x	n	k	l	λ	μ
J_2	$U_3(3)$	100	36	63	14	12
McL	$U_4(3)$	275	112	162	30	56
Ru	${}^2F_4(2)$	4,060	1,755	2,304	730	780
Suz	$G_2(4)$	1,782	416	1,365	100	96
Co_2	$U_6(2):2$	2,300	891	1,408	378	324
Fi_{22}	$2^{\cdot}U_6(2)$	3,510	693	2,816	180	126
	$O_7(3)$	14,080	3,159	10,920	918	648
Fi_{23}	$2^{\cdot}\mathrm{Fi}_{22}$	31,671	3,510	28,160	693	351
	$O_8^+(3):S_3$	137,632	28,431	109,200	6,030	5,832
Fi_{24}'	Fi_{23}	306,936	31,671	275,264	3,510	3,240

groups of rank 5 or less. A number of proofs that a given sporadic simple group is not a minimal counterexample to the Hall-Paige conjecture can be deduced from these matrices. We have already used data from their book in the proofs of Theorems 7.21 and 7.33 and will make further use of their data to establish the existence of W-systems for He, O′N, Ly, and Co_1.

Theorem 7.34 (Evans, 2009). He, O′N, Ly, *and* Co_1 *are not minimal counterexamples to the Hall-Paige conjecture.*

Proof. The Held group He*:* From [271] we see that He has a rank-5 permutation representation of degree 2,058 with point-stabilizer H isomorphic to $S_4(4):2$. The indices are

$|D_1 : H| = 1,$
$|D_2 : H| = 136,$
$|D_3 : H| = 136,$
$|D_4 : H| = 425,$ and
$|D_5 : H| = 1{,}360.$

From the collapsed adjacency matrices in [271], we see that

$A^1_{11} = 1,$
$A^2_{23} = 36,$
$A^3_{32} = 36,$
$A^4_{44} = 136,$ and
$A^5_{55} = 894.$

Thus, by Corollary 7.28, D_1, D_4, and D_5 are W-simple; and, by Corollary 7.29, $D_2^2 \supseteq D_3$ and $D_3^2 \supseteq D_2$. If we define $\phi, \psi\colon \mathscr{D} \to \mathscr{D}$ by $\phi(D_i) = D_i$ for all i, and $\psi(D_i) = D_i$ if $i \neq 2, 3$, $\psi(D_2) = D_3$, and $\psi(D_3) = D_2$, then $\{\mathscr{D}, \phi, \psi\}$ is a W-system. As H has noncyclic Sylow 2-subgroups, by Theorem 7.8, He is not a minimal counterexample to the Hall-Paige conjecture.

The O'Nan group O′N*:* From [271] we see that O′N has a rank-5 permutation representation of degree 122,760 with point-stabilizer H isomorphic to $L_3(7):2$. The indices are

$|D_1 : H| = 1,$
$|D_2 : H| = 5{,}586,$
$|D_3 : H| = 6{,}384,$
$|D_4 : H| = 52{,}136,$ and
$|D_5 : H| = 58{,}653.$

From the collapsed adjacency matrices in [271], we see that

$A^1_{11} = 1,$
$A^2_{22} = 364,$
$A^3_{33} = 349,$
$A^4_{44} = 22{,}057,$ and

$A_{55}^5 = 27{,}972.$

Thus, by Corollary 7.28, each double coset of H in O′N is W-simple. As H has noncyclic Sylow 2-subgroups, by Corollary 7.9, O′N is not a minimal counterexample to the Hall-Paige conjecture.

The Lyons group Ly*:* From [271] we see that Ly has a rank-5 permutation representation of degree 8,835,156 with point-stabilizer H isomorphic to $G_2(5)$. The indices are

$|D_1 : H| = 1,$
$|D_2 : H| = 19{,}530,$
$|D_3 : H| = 968{,}750,$
$|D_4 : H| = 2{,}034{,}375,$ and
$|D_5 : H| = 5{,}812{,}500.$

From the collapsed adjacency matrices in [271], we see that

$A_{11}^1 = 1,$
$A_{22}^2 = 154,$
$A_{33}^3 = 114{,}013,$
$A_{44}^4 = 468{,}670,$ and
$A_{55}^5 = 3{,}821{,}006.$

Thus, by Corollary 7.28, each double coset of H in Ly is W-simple. As H has noncyclic Sylow 2-subgroups, by Corollary 7.9, Ly is not a minimal counterexample to the Hall-Paige conjecture.

The Conway group Co_1*:* From [271] we see that Co_1 has a rank-4 permutation representation of degree 98,280 with point-stabilizer H isomorphic to Co_2. The indices are

$|D_1 : H| = 1,$
$|D_2 : H| = 4{,}600,$
$|D_3 : H| = 46{,}575,$ and
$|D_4 : H| = 47{,}104.$

From the collapsed adjacency matrices in [271], we see that

$A_{11}^1 = 1,$
$A_{22}^2 = 892,$
$A_{33}^3 = 21{,}582,$ and
$A_{44}^4 = 22{,}528.$

Thus, by Corollary 7.28, each double coset of H in Co_1 is W-simple. As H has noncyclic Sylow 2-subgroups, by Corollary 7.9, Co_1 is not a minimal counterexample to the Hall-Paige conjecture. □

In [271] there are no permutation representations of rank 5 or less for the groups J_3, HN, Th, J_4, and M. There is a rank-5 permutation representation for J_1: each double coset for this representation is solitary, but, as two of these double cosets are not W-simple, this representation does not yield a W-system. There is also a rank-5 permutation representation for B: and again each double coset is solitary, but, as all but one of the double cosets for this representation are W-simple, this representation does not yield a W-system.

7.3.5 *The Groups* HN, Th, *B, and M*

For permutation representations of rank greater than 5, there is a GAP database available in [51]. In this database a different definition of collapsed adjacency matrices is used. The ijth entry of the kth collapsed adjacency matrix in [51] is defined to be $P^k_{ij} = |\{y \in O_i \mid (y, a_j) \in E_k\}|$. We can easily translate collapsed adjacency matrices as defined in [51] into collapsed adjacency matrices as defined here, in [271] and in [128]. If we let k^p be defined by $E_{k^p} = E^p_k$, then

$$P^k_{ij} = |\{y \in O_i \mid (y, a_j) \in E_k\}| = |\{y \in O_i \mid (a_j, y) \in E_{k^p}\}| = A^{k^p}_{ji},$$

by the definition used here, in [271] and in [128]. It follows that if the kth orbital is self-paired, then the kkth entry of the kth collapsed adjacency matrix will be the same in either definition. The collapsed adjacency matrices in this database allow us to construct W-systems for four more sporadic simple groups, HN, Th, B, and M.

Theorem 7.35 (Evans, 2009). HN, Th, *B, and M are not minimal counterexamples to the Hall-Paige conjecture.*

Proof. The Harada-Norton group HN*:* The GAP database in [51] contains four representations of HN. The third representation is a rank-9 permutation representation of degree 1,140,000 with point-stabilizer H isomorphic to 2.HS.2, a group of order 319,979,520. The indices are

$|D_1 : H| = 1$,
$|D_2 : H| = 1{,}408$,
$|D_3 : H| = 2{,}200$,
$|D_4 : H| = 5{,}775$,
$|D_5 : H| = 35{,}200$,
$|D_6 : H| = 123{,}200$,
$|D_7 : H| = 277{,}200$,
$|D_8 : H| = 354{,}816$, and
$|D_9 : H| = 739{,}200$.

We used the GAP database in [51] and the GAP program in this database to compute the collapsed adjacency matrices for this representation from the character table in GAP for HN. Translating from the definition of collapsed adjacency matrices used in the database to the definition of collapsed adjacency matrices used here, in [271] and in [128], we found the following.

$A_{11}^{1} = 1$,
$A_{22}^{2} = 53$,
$A_{33}^{3} = 106$,
$A_{44}^{4} = 30$,
$A_{55}^{5} = 1{,}856$,
$A_{66}^{6} = 9{,}516$,
$A_{77}^{7} = 48{,}960$,
$A_{88}^{8} = 81{,}684$, and
$A_{99}^{9} = 354{,}576$,

It follows from Corollary 7.28 that each double coset of H in HN is W-simple. As H has noncyclic Sylow 2-subgroups, by Corollary 7.9, HN is not a minimal counterexample to the Hall-Paige conjecture.

The Thompson group Th*:* The GAP database in [51] contains two representations of Th. The second representation is a rank-11 permutation representation of degree 283,599,225 with point-stabilizer H isomorphic to $2^5.L_5(2)$, a group of order 319,979,520. The indices are

$|D_1 : H| = 1$,
$|D_2 : H| = 248$,
$|D_3 : H| = 59{,}520$,
$|D_4 : H| = 2{,}064{,}384$,
$|D_5 : H| = 2{,}064{,}384$,
$|D_6 : H| = 2{,}539{,}520$,
$|D_7 : H| = 6{,}666{,}240$,
$|D_8 : H| = 35{,}553{,}280$,
$|D_9 : H| = 63{,}995{,}904$,
$|D_{10} : H| = 63{,}995{,}904$, and
$|D_{11} : H| = 106{,}659{,}840$.

We used the GAP database in [51] and the GAP program in this database to compute the collapsed adjacency matrices for this representation from the character table in GAP for Th. Translating from the definition of collapsed adjacency matrices used in the database to the definition of collapsed adjacency matrices used here, in [271] and in [128], we found the following.

$A_{11}^{1} = 1$,
$A_{22}^{2} = 7$,
$A_{33}^{3} = 488$,
$A_{45}^{4} = 13{,}517$,
$A_{54}^{5} = 13{,}517$,
$A_{66}^{6} = 30{,}856$,

$A_{77}^{7} = 149{,}968$,
$A_{88}^{8} = 4{,}511{,}008$,
$A_{99}^{9} = 14{,}435{,}003$,
$A_{10,10}^{10} = 14{,}451{,}643$, and
$A_{11,11}^{11} = 40{,}032{,}879$.

Thus, by Corollary 7.28, D_i is W-simple for $i \in \{1,\ldots,11\} \setminus \{4,5\}$, and, by Corollary 7.29, $D_4^2 \supseteq D_5$ and $D_5^2 \supseteq D_4$. If we define $\phi, \psi \colon \mathscr{D} \to \mathscr{D}$ by $\phi(D_i) = D_i$ for all i, and $\psi(D_i) = D_i$ for all $i \neq 4, 5$, $\psi(D_4) = D_5$, and $\psi(D_5) = D_4$, then $\{\mathscr{D}, \phi, \psi\}$ is a W-system. As H has noncyclic Sylow 2-subgroups, by Theorem 7.8, Th is not a minimal counterexample to the Hall-Paige conjecture.

The Baby Monster B: The GAP database in [51] contains four representations of B. The third representation is a rank-10 permutation representation of degree 11,707,448,673,375 with point-stabilizer H isomorphic to $2^{1+22}.Co_2$, a group of order 354,883,595,661,213,696,000. The indices are

$|D_1 : H| = 1$,
$|D_2 : H| = 93{,}150$,
$|D_3 : H| = 7{,}286{,}400$,
$|D_4 : H| = 262{,}310{,}400$,
$|D_5 : H| = 4{,}196{,}966{,}400$,
$|D_6 : H| = 9{,}646{,}899{,}200$,
$|D_7 : H| = 470{,}060{,}236{,}800$,
$|D_8 : H| = 537{,}211{,}699{,}200$,
$|D_9 : H| = 4{,}000{,}762{,}036{,}224$, and
$|D_{10} : H| = 6{,}685{,}301{,}145{,}600$.

We used the GAP database in [51] and the GAP program in this database to compute the collapsed adjacency matrices for this representation from the character table in GAP for B. Translating from the definition of collapsed adjacency matrices used in the database to the definition of collapsed adjacency matrices used here, in [271] and in [128], we found the following.

$A_{11}^{1} = 1$,
$A_{22}^{2} = 925$,
$A_{33}^{3} = 20{,}224$,
$A_{44}^{4} = 254{,}160$,
$A_{55}^{5} = 4{,}277{,}088$,
$A_{66}^{6} = 35{,}507{,}968$,
$A_{77}^{7} = 18{,}907{,}609{,}600$,
$A_{88}^{8} = 27{,}166{,}947{,}840$,
$A_{99}^{9} = 1{,}368{,}536{,}219{,}648$, and
$A_{10,10}^{10} = 3{,}815{,}388{,}193{,}024$.

Thus, by Corollary 7.28, each double coset of H in B is W-simple. As H has noncyclic Sylow 2-subgroups, by Corollary 7.9, B is not a minimal counterexample to the Hall-Paige conjecture.

The Monster M: The GAP database in [51] contains exactly one representation of M. This representation is a rank-9 permutation representation of degree 97,239,461,142,009,186,000 with point-stabilizer H isomorphic to $2.B$, a group of order 8,309,562,962,452,852,382,355,161,088,000,000. The indices are

$|D_1 : H| = 1$,
$|D_2 : H| = 27{,}143{,}910{,}000$,
$|D_3 : H| = 11{,}707{,}448{,}673{,}375$,
$|D_4 : H| = 2{,}031{,}941{,}058{,}560{,}000$,
$|D_5 : H| = 91{,}569{,}524{,}834{,}304{,}000$,
$|D_6 : H| = 1{,}102{,}935{,}324{,}621{,}312{,}000$,
$|D_7 : H| = 1{,}254{,}793{,}905{,}192{,}960{,}000$,
$|D_8 : H| = 30{,}434{,}513{,}446{,}055{,}706{,}624$, and
$|D_9 : H| = 64{,}353{,}605{,}265{,}653{,}760{,}000$.

We used the GAP database in [51] and the GAP program in this database to compute the collapsed adjacency matrices for this representation from the character table in GAP for M. Translating from the definition of collapsed adjacency matrices used in the database to the definition of collapsed adjacency matrices used here, in [271] and in [128], we found the following.

$A^1_{11} = 1$,
$A^2_{22} = 3{,}968{,}056$,
$A^3_{33} = 262{,}403{,}550$,
$A^4_{44} = 595{,}638{,}739{,}456$,
$A^5_{55} = 86{,}562{,}946{,}252{,}800$,
$A^6_{66} = 15{,}549{,}421{,}463{,}884{,}800$,
$A^7_{77} = 15{,}863{,}482{,}471{,}113{,}216$,
$A^8_{88} = 9{,}560{,}800{,}836{,}253{,}237{,}248$, and
$A^9_{99} = 42{,}569{,}854{,}610{,}071{,}384{,}576$.

Thus, by Corollary 7.28, each double coset of H in M is W-simple. As H has noncyclic Sylow 2-subgroups, by Corollary 7.9, M is not a minimal counterexample to the Hall-Paige conjecture. □

For J_1, the database in [51] contains a rank-11 permutation representation: three of the double cosets are solitary, but only two of these solitary double cosets are W-simple, and so this representation does not yield W-systems. For J_3, the database in [51] contains a rank-8 permutation representation: six of the double cosets are solitary, but only five of these solitary double cosets are W-simple, and so this representation does not yield W-systems. For J_4, the database in [51] contains a rank-7 and a rank-11 permutation representation: in each case, all of the double cosets are solitary, and in each case, all but one double coset is W-simple, and so these representations do not yield W-systems.

7.3.6 The Janko Groups

The second Janko group, J_2, was proved not to be a minimal counterexample to the Hall-Paige conjecture in Theorem 7.33, as it had a rank-3 permutation representation with the right parameters. This leaves the first, third, and fourth Janko groups, J_1, J_3, and J_4, as the only possible minimal counterexamples to the Hall-Paige conjecture. For J_1 we have to use a rank-22 permutation representation to construct a W-system and for J_3 a rank-14 permutation representation. While we could establish the existence of W-systems for these groups by computing entries in collapsed adjacency matrices, we will show that the distribution of the elements of order 3 can be used to indirectly prove the existence of W-systems. For the last group, J_4, Bray [50] has proved this not to be a minimal counterexample to the Hall-Paige conjecture: his proof has not been published as yet.

Lemma 7.36. *Suppose that H is a subgroup of a finite group G and that $\mathscr{D}$ is the set of double cosets of H in G. If $D \in \mathscr{D}$ contains an element of order* 3*, then*

$$D \subseteq (D^{(-1)})^2 \quad \textit{and} \quad D^{(-1)} \subseteq D^2.$$

Proof. Let $g \in D$ be of order 3. Then

$$D^2 = HgHHgH \supseteq Hg^2H = Hg^{-1}H = D^{(-1)}.$$

Similarly $(D^{(-1)})^2 \supseteq D^2$. □

The following is an easy corollary:

Corollary 7.37. *Suppose that H is a subgroup of a finite group G and that $\mathscr{D}$ is the set of double cosets of H in G. If $D \in \mathscr{D}$ is solitary and contains an element of order* 3*, then D is W-simple.*

Proof. If D is solitary, then $D^{(-1)} = D$, and so D is W-simple by Lemma 7.36. □

From Lemma 7.36 we obtain a simple criterion for establishing the admissibility of a group.

Lemma 7.38. *If H is an admissible subgroup of a finite group G and every nontrivial double coset of H in G contains an element of order* 3*, then G is admissible.*

Proof. Let $\mathscr{D}$ be the set of double cosets of H in G. Define $\phi, \psi \colon \mathscr{D} \to \mathscr{D}$ by $\phi(D) = D$ and $\psi(D) = D^{(-1)}$. Clearly ϕ and ψ are bijections. Further, by Lemma 7.36,

$$D\phi(D) = D^2 \supseteq D^{(-1)} = \psi(D).$$

Hence $\{\mathscr{D}, \phi, \psi\}$ is a W-system by Lemma 7.36, and so G is admissible by Theorem 7.8. □

To determine that a double coset D of H in a finite group G contains an element of order 3 or not, it is sufficient to check any right coset of H contained in D.

Lemma 7.39. *If H is a subgroup of a finite group G and D is a double coset of H in G, then a right coset of H in D contains an element of order* 3 *if and only if every right coset of H in D contains an element of order* 3.

Proof. If Hg is a right coset of H in D, where $|g| = 3$, then any other right coset of H in D is of the form $H(gm)$ for some $m \in H$, and $H(gm)$ contains the element $m^{-1}gm$ of order 3. □

Using the rank-22 permutation representation of J_1 and the rank-14 permutation representation of J_3, we can prove that neither of these groups can be a minimal counterexample to the Hall-Paige conjecture.

Theorem 7.40 (Evans, 2009). *J_1 and J_3 are not minimal counterexamples to the Hall-Paige conjecture.*

Proof. The first Janko group J_1: Using the permutation representation of degree 266 for J_1 given in [330], we verified the following using magma. In the description given in [330], J_1 is generated by two elements, x and y, that satisfy the relations

$$x^2 = y^3 = (xy)^7 = (xy(xyxy^2)^3)^5 = (xy(xyxy^2)^6xyxy(xy^2)^2)^2 = 1.$$

Let H be the subgroup of J_1 generated by

$$x \quad \text{and} \quad (xyxy^2)^{-4}(xy)^{-1}(xy(xyxy^2)^2)^3xy(xyxy^2)^4.$$

Then H is a maximal subgroup of J_1 of index 1,463, and is isomorphic to $A_5 \times 2$, and hence has noncyclic Sylow 2-subgroups. The set of double cosets of H in J_1 is $\mathscr{D} = \{D_i = Hd_iH \mid i = 1, \ldots, 22\}$, where the indices $|D_i : H|$ are

1,
12,
15 twice,
20 twice,
60 nine times, and
120 seven times.

Using magma we verified that, for $i = 1, \ldots, 22$, the right coset Hd_i contains an element of order 3, and hence, each $D \in \mathscr{D}$ contains an element of order 3. Thus, by Lemma 7.38, J_1 is not a minimal counterexample to the Hall-Paige conjecture.

The third Janko group J_3*:* Using the permutation representation of degree 6,156 for J_3 given in [330], we verified the following using magma. In the description given in [330], J_3 is generated by two elements, x and y, that satisfy the relations

$$x^2 = y^3 = (xy)^{19} = [x, y]^9 = ((xy)^6(xy^2)^5)^2 = ((xyxyxy^2)^2xyxy^2xy^2xyxy^2)^2$$

$$= xyxy(xyxy^2)^3xyxy(xyxy^2)^4xy^2(xyxy^2)^3 = (xyxyxyxyxy^2xyxy^2)^4 = 1,$$

where $[x, y]$ is the commutator $x^{-1}y^{-1}xy$. Let H be the subgroup of J_3 generated by

$$y^2xy \quad \text{and} \quad (xy^2)^{-4}(xyxy^2)^3(xy^2)^4.$$

Then H is a maximal subgroup of J_3 of index 14,688, and is isomorphic to $L_2(19)$, and hence has noncyclic Sylow 2-subgroups. The set of double cosets of H in J_3 is $\mathscr{D} = \{D_i = Hd_iH \mid i = 1, \ldots, 14\}$, where the indices $|D_i : H|$ are

1,
285,
342,
380,
570 twice,
855 twice,
1,140 twice,
1,710 three times, and
3,420.

Using magma we verified that, for $i = 1, \ldots, 14$, the right coset Hd_i contains an element of order 3, and hence, each $D \in \mathscr{D}$ contains an element of order 3. Thus, by Lemma 7.38, J_3 is not a minimal counterexample to the Hall-Paige conjecture. □

This leaves us with only one possible minimal counterexample to the Hall-Paige conjecture, Janko's fourth group, J_4. Bray [50] has computed the collapsed adjacency matrices for the permutation representation of J_4 of degree 3,980,549,947 with point-stabilizer isomorphic to $2_+^{1+12}{\cdot}3M_{22} : 2$ and from this has determined that J_4 is not a minimal counterexample to the Hall-Paige conjecture. Thus, the Hall-Paige conjecture has been proved.

The proof of the Hall-Paige conjecture has implications for the existence of duplexes, a generalization of transversals. A *duplex* in a Latin square L of order n is a set of $2n$ cells of L, two cells from each column and two cells from each row, each symbol of L appearing twice. It has been conjectured that all nontrivial Latin squares contain duplexes.

Conjecture 7.41 (Rodney's Conjecture). Every Latin square of order $n \geq 2$ contains a duplex.

In 2003 Vaughan and Wanless [316] proved that the Hall-Paige conjecture implied that the Cayley table of any nontrivial finite group contains a duplex, a special case of Rodney's conjecture.

Corollary 7.42. *The Cayley table of a nontrivial finite group contains a duplex.*

Proof. See Corollary 3.1 in [316]. □

There are two other papers that we should mention concerning the Hall-Paige conjecture. In 2009, the same year that most of the proof of the Hall-Paige was published, Moghaddamfar and Zokayi [244] proved that nonsolvable groups with dihedral Sylow 2-subgroups are admissible: they also proved that, under certain conditions, if a finite group G is factorable with two simple groups as factors, then G is admissible. Also in 2009 Pula [272] considered analogues of the Hall-Paige conjecture for loops.

Part III
Orthomorphism Graphs of Groups

Chapter 8
Orthomorphism Graphs of Groups

In Section 1.3 we defined orthomorphisms and orthogonality of orthomorphisms. As orthogonality is a symmetric relation, we can define and study orthomorphism graphs: graphs whose vertices are orthomorphisms, and in which adjacency implies orthogonality. In this chapter we introduce orthomorphism graphs of groups. In Section 8.1 we describe the main problems of interest in the study of orthomorphism graphs, and we describe automorphisms and congruences of orthomorphism graphs. In subsequent sections we describe some classes of orthomorphism graphs. In Section 8.2 we study the orthomorphism graph $\mathscr{P}(G)$, composed of orthomorphisms of the form $x \mapsto x^r$. In Section 8.3 we study orthomorphism graphs derived from difference sets. In Section 8.4 we study orthomorphism graphs obtained from automorphisms, and in Section 8.5 we study orthomorphism graphs whose vertices are strong complete mappings.

8.1 Orthomorphisms Graphs of Groups

Let us recall some basic definitions and results from Section 1.3. We defined an *orthomorphism* of a group G to be a bijection $\theta\colon G \to G$ for which the mapping $x \mapsto x^{-1}\theta(x)$ is also a bijection, two orthomorphisms θ and ϕ of G were said to be *orthogonal*, written $\theta \perp \phi$, if the mapping $x \mapsto \theta(x)^{-1}\phi(x)$ is a bijection, and we established that $\theta \perp \phi$ if and only if $\phi \perp \theta$. We were interested in orthogonal sets of orthomorphisms because from r pairwise orthogonal orthomorphisms of a group of order n, we can construct a set of $r + 1$ MOLS of order n; from a maximal set of r pairwise orthogonal orthomorphisms of a group of order n, we can construct a maximal set of $r + 1$ MOLS of order n; and from a complete set orthomorphisms of a group of order n, we can construct a complete set of MOLS of order n.

A. B. Evans, *Orthogonal Latin Squares Based on Groups*, Developments in Mathematics 57, https://doi.org/10.1007/978-3-319-94430-2_8

8.1.1 Orthomorphism Graphs

In Section 6.2 we defined a graph Γ to consist of a set $V(\Gamma)$, whose elements are called the *vertices* of Γ and a set $E(\Gamma)$ of 2-element subsets of $V(\Gamma)$ called *edges*. The *order* of Γ is $|V(\Gamma)|$. Two vertices are *adjacent* if they are contained in a common edge. The *degree* of a vertex v is the number of vertices adjacent to v, and the *neighbors* of v are the vertices adjacent to v. A *subgraph* of Γ consists of a subset of $V(\Gamma)$ and a subset of $E(\Gamma)$ that together form a graph. An *induced subgraph* Γ' of Γ is a subgraph of Γ with the property that any two vertices u, v of Γ' are adjacent in Γ' if and only if they are adjacent in Γ. The *neighborhood* $N(v)$ of a vertex v of Γ is the subgraph induced by the neighbors of v. A *complete subgraph* of Γ is an induced subgraph in which every vertex is adjacent to every other. An r-*clique* of Γ is a complete subgraph of order r, and the *clique number* of Γ, denoted $\omega(\Gamma)$, is the order of the largest complete subgraph of Γ.

The *orthomorphism graph* of a group G is the graph with vertices the normalized orthomorphisms of G, adjacency being synonymous with orthogonality. We will use $\text{Orth}(G)$ to denote the orthomorphism graph of G as well as the set of normalized orthomorphisms of G. An *orthomorphism graph* of G is a subgraph of $\text{Orth}(G)$, usually an induced subgraph. The clique number of $\text{Orth}(G)$ will be abbreviated to $\omega(G)$. Orthomorphism graphs were first studied by Johnson, Dulmage, and Mendelsohn [196] in 1961.

Given a group G and an orthomorphism graph $\mathscr{H}$ of G, the problems that concern us are the following:

1. Determine $\omega(\mathscr{H})$ or find good bounds for $\omega(\mathscr{H})$.
2. Find maximal cliques in $\mathscr{H}$. Are any of these maximal in $\text{Orth}(G)$?
3. Does $\mathscr{H}$ contain a complete set of orthomorphisms? If the answer is yes, then classify these complete sets.
4. What is the structure of $\mathscr{H}$?

In the special case $\mathscr{H} = \text{Orth}(G)$, these problems have been solved for small groups only: the data will be presented in Chapter 13.

A standard way to define orthomorphism graphs is to begin with a class of mappings and ask which of these are orthomorphisms. We shall see that, in many cases, the answer to this question also determines the adjacency rules for the orthomorphism graph. In this chapter we will consider some general constructions of orthomorphism graphs. In Section 8.2 we will study the class of orthomorphisms of the form $x \mapsto x^r$, from which we will be able to deduce a lower bound for $\omega(G)$. Orthomorphism graphs constructed from difference sets and difference families will be studied in Section 8.3. Group automorphisms that are also orthomorphisms will be studied in Section 8.4, and the orthomorphism graph consisting of strong complete mappings will be studied in Section 8.5. The elementary abelian groups, being additive groups of finite fields, are in a class by themselves and will be studied in Chapters 9 and 10. In Chapter 11 we will consider extensions of orthomorphism graphs, in Chapter 12 pairwise orthogonal orthomorphisms of dihedral groups and general linear groups of even characteristic, and in Chapter 13 we will present data for groups of small order.

8.1.2 Products of Orthomorphism Graphs

We now give a direct product construction for orthomorphisms. This construction was given in difference matrix form by Jungnickel [200] in 1978. This yields lower bounds for the clique numbers of orthomorphism graphs. These bounds, though weak, are the best lower bounds presently available for many groups. Let G and H be groups and let $\theta\colon G \to G$ and $\phi\colon H \to H$ be mappings. We define the *direct product* $\theta \times \phi\colon G \times H \to G \times H$ of θ and ϕ by $\theta \times \phi(x, y) = (\theta(x), \phi(y))$.

Theorem 8.1. *If* $\theta \in \mathrm{Orth}(G)$ *and* $\phi \in \mathrm{Orth}(H)$, *then* $\theta \times \phi \in \mathrm{Orth}(G \times H)$. *Further, if* $\theta, \theta' \in \mathrm{Orth}(G)$ *and* $\phi, \phi' \in \mathrm{Orth}(H)$, *then* $\theta \times \phi \perp \theta' \times \phi'$ *if and only if* $\theta \perp \theta'$ *and* $\phi \perp \phi'$.

Proof. We will make use of the following easy observation. Let S and T be sets. For $\mathrm{f}\colon S \to S$ and $\mathrm{h}\colon T \to T$, let $\mathrm{f} \times \mathrm{h}\colon S \times T \to S \times T$ be defined by $\mathrm{f} \times \mathrm{h}(s, t) = (\mathrm{f}(s), \mathrm{h}(t))$. Then $\mathrm{f} \times \mathrm{h}$ is a bijection if and only if f and h are bijections.

Now $\theta \times \phi$ is a bijection if and only if θ and ϕ are bijections, and the mapping

$$(x, y) \mapsto (x, y)^{-1}\theta \times \phi(x, y) = (x^{-1}\theta(x), y^{-1}\phi(y))$$

is a bijection if and only if the mappings $x \mapsto x^{-1}\theta(x)$ and $y \mapsto y^{-1}\phi(y)$ are both bijections. Hence, if $\theta \in \mathrm{Orth}(G)$ and $\phi \in \mathrm{Orth}(H)$, then $\theta \times \phi$ is an orthomorphism of $G \times H$.

Now $\theta \times \phi \perp \theta' \times \phi'$ if and only if the mapping

$$(x, y) \mapsto \theta \times \phi(x, y)^{-1}\theta' \times \phi'(x, y) = (\theta(x)^{-1}\theta'(x), \phi(y)^{-1}\phi'(y))$$

is a bijection, if and only if the mappings $x \mapsto \theta(x)^{-1}\theta'(x)$ and $y \mapsto \phi(y)^{-1}\phi'(y)$ are both bijections, if and only if $\theta \perp \theta'$ and $\phi \perp \phi'$. □

In Theorem 8.1 we have actually proved a stronger result, namely that, for any pair of mappings $\theta\colon G \to G$ and $\phi\colon H \to H, \theta \times \phi$ is an orthomorphism of $G \times H$ if and only if θ is an orthomorphism of G and ϕ is an orthomorphism of H. The following consequences of Corollary 1.38 are also corollaries of Theorem 8.1.

Corollary 8.2. $\omega(G \times H) \geq \min\{\omega(G), \omega(H)\}$.

More generally:

Corollary 8.3. $\omega(G_1 \times \cdots \times G_n) \geq \min\{\omega(G_1), \ldots, \omega(G_n)\}$.

If $\mathscr{H}$ is an orthomorphism graph of G and $\mathscr{K}$ an orthomorphism graph of H, then we define the *direct product* of $\mathscr{H}$ and $\mathscr{K}$ to be the orthomorphism graph of

$G \times H$ defined by $\mathscr{H} \times \mathscr{K} = \{\theta \times \phi \mid \theta \in \mathscr{H}, \phi \in \mathscr{K}\}$. The following are immediate corollaries of Theorem 8.1.

Corollary 8.4. *If $\mathscr{H}$ is an orthomorphism graph of G and $\mathscr{K}$ is an orthomorphism graph of H, then $\omega(\mathscr{H} \times \mathscr{K}) = \min\{\omega(\mathscr{H}), \omega(\mathscr{K})\}$.*

More generally:

Corollary 8.5. *If $\mathscr{H}_1, \ldots, \mathscr{H}_n$ are orthomorphism graphs of $G_1, \ldots, G_n$, respectively, then $\omega(\mathscr{H}_1 \times \cdots \times \mathscr{H}_n) = \min\{\omega(\mathscr{H}_1), \ldots, \omega(\mathscr{H}_n)\}$.*

8.1.3 Automorphisms and Congruences of Orth(G)

For T: Orth(G) $\to$ Orth(G), we shall use T[θ] to denote the image of θ under T. A bijection T: Orth(G) $\to$ Orth(G) is an *automorphism* of Orth(G) if T[θ] $\perp$ T[ϕ] if and only if $\theta \perp \phi$, and a *congruence* of Orth(G) if the neighborhood of T[θ] is isomorphic to the neighborhood of θ for all $\theta \in$ Orth(G). Let us define the following mappings from Orth(G) into Orth(G). For $\alpha \in$ Aut(G) the *homology* H_α is defined by $\mathrm{H}_\alpha[\theta] = \alpha\theta\alpha^{-1}$, for $g \in G$ the *translation* T_g is defined by $\mathrm{T}_g[\theta](x) = \theta(xg)\theta(g)^{-1}$, the *reflection* R is defined by $\mathrm{R}[\theta](x) = x\theta(x^{-1})$, and the *inversion* I is defined by $\mathrm{I}[\theta](x) = \theta^{-1}(x)$. These mappings and their relations were described implicitly by Singer [295] in 1960, for cyclic groups. There is one automorphism in Singer's paper that does not appear in our list. For a cyclic group, written additively, Singer defined $\overline{\mathrm{C}}$ by $\overline{\mathrm{C}}[\theta](x) = x - \theta(x)$. In fact, if G is an additively written abelian group, then $\alpha\colon x \mapsto -x$ is an automorphism of G and $\overline{\mathrm{C}} = \mathrm{RH}_\alpha$. Translations, homologies, and inversions were used by Johnson, Dulmage, and Mendelsohn [196] in 1961; though they used different notations and terminologies. Chang, Hsiang, and Tai [69] in 1964 defined reflections for nonabelian groups.

Theorem 8.6. H_α, T_g, *and* R *are automorphisms of* Orth(G).

Proof. Let $\theta, \phi \in$ Orth(G), $\theta \perp \phi$, and let $\alpha \in$ Aut(G). Define mappings ι, r_a, l_a, η_θ, η_ϕ, $\eta_{\theta,\phi}\colon G \to G$ by $\iota(x) = x^{-1}$, $\mathrm{r}_a(x) = xa$, $\mathrm{l}_a(x) = ax$, $\eta_\theta(x) = x^{-1}\theta(x)$, $\eta_\phi(x) = x^{-1}\phi(x)$, and $\eta_{\theta,\phi}(x) = \theta(x)^{-1}\phi(x)$. Each of θ, ϕ, α, ι, r_a, l_a, η_θ, η_ϕ, $\eta_{\theta,\phi}$ is a bijection. The proof of this theorem depends on writing various mappings as compositions of these mappings and/or their inverses, thus demonstrating that they are also bijections.

As $\mathrm{H}_\alpha[\theta] = \alpha\theta\alpha^{-1}$, $\mathrm{T}_g[\theta] = \mathrm{r}_{\theta(g)^{-1}}\theta\mathrm{r}_g$, and $\mathrm{R} = \eta_\theta\iota$, each of $\mathrm{H}_\alpha[\theta]$, $\mathrm{T}_g[\theta]$, and R[θ] is a bijection. As $x^{-1}\mathrm{H}_\alpha[\theta](x) = \alpha\eta_\theta\alpha^{-1}(x)$, $\mathrm{H}_\alpha[\theta]$ is an orthomorphism; as $x^{-1}\mathrm{T}_g[\theta](x) = \mathrm{l}_g\mathrm{r}_{\theta(g)^{-1}}\eta_\theta\mathrm{r}_g(x)$, $\mathrm{T}_g[\theta]$ is an orthomorphism; and as $x^{-1}\mathrm{R}[\theta](x) = \theta\iota(x)$, R[$\theta$] is an orthomorphism. It is easily shown that each of $\mathrm{H}_\alpha[\theta]$, $\mathrm{T}_g[\theta]$, and R[θ] fixes the identity and, thus, is in Orth(G). That each of H_α, T_g, and R are actually bijections Orth(G) $\to$ Orth(G) follows from the fact that each is invertible: this will be an easy consequence of Theorem 8.8. To see that H_α is an automorphism of Orth(G), note that $\mathrm{H}_\alpha[\theta] \perp \mathrm{H}_\alpha[\phi]$ if and only if the mapping $x \mapsto H_\alpha[\theta](x)^{-1}H_\alpha[\phi](x) = \alpha\eta_{\theta,\phi}\alpha^{-1}(x)$ is a bijection, if and only if the mapping $x \mapsto \eta_{\theta,\phi}(x)$ is a bijection, if and only if $\theta \perp \phi$. To see that T_g is an automorphism of $\mathrm{Orth}(G)$, note that $T_g[\theta] \perp T_g[\phi]$ if and only if the mapping $x \mapsto T_g[\theta](x)^{-1}T_g[\phi](x) = l_{\theta(g)}r_{\phi(g)^{-1}}\eta_{\theta,\phi}r_g(x)$ is a bijection, if and only if the mapping $x \mapsto \eta_{\theta,\phi}(x)$ is a bijection, if and only if $\theta \perp \phi$. Finally $R[\theta] \perp R[\phi]$ if and only if the mapping $x \mapsto R[\theta](x)^{-1}R[\phi](x) = \eta_{\theta,\phi}\iota(x)$ is a bijection, if and only if the mapping $x \mapsto \eta_{\theta,\phi}(x)$ is a bijection, if and only if $\theta \perp \phi$. □

For I, we can prove only that it is a congruence.

Theorem 8.7. *I is a congruence of* $\mathrm{Orth}(G)$.

Proof. Let $\theta \in \mathrm{Orth}(G)$ and let ι and η_θ be defined as in the proof of Theorem 8.6. Clearly $I[\theta]$ is a bijection that fixes the identity. Further the mapping $x \mapsto x^{-1}I[\theta](x) = \iota\eta_\theta\theta^{-1}(x)$ is a bijection. Hence $I[\theta] \in \mathrm{Orth}(G)$. Further I is a bijection $\mathrm{Orth}(G) \to \mathrm{Orth}(G)$ as it is invertible, being its own inverse.

To show that I is a congruence, note that if $\phi_1, \ldots, \phi_r$ are the neighbors of θ then, by Corollary 1.31, $\phi_1\theta^{-1}, \ldots, \phi_r\theta^{-1}$ are orthomorphisms, and they are also neighbors of θ^{-1} as $(\phi_i\theta^{-1})(\theta^{-1})^{-1} = \phi_i$ is an orthomorphism for $i = 1, \ldots, r$. Further, as $(\phi_i\theta^{-1})(\phi_j\theta^{-1})^{-1} = \phi_i\phi_j^{-1}$, it follows that $\phi_i\theta^{-1} \perp \phi_j\theta^{-1}$ if and only if $\phi_i \perp \phi_j$. Thus the neighborhood of θ is isomorphic to an induced subgraph of the neighborhood of $I[\theta]$. Similarly the neighborhood of $I[\theta]$ is isomorphic to an induced subgraph of the neighborhood of $\theta = I[I[\theta]]$, and so the neighborhood of θ is isomorphic to the neighborhood of $I[\theta]$. □

Note that in general I is not an automorphism of $\mathrm{Orth}(G)$. As an example, Figure 8.1 describes an orthogonal pair of orthomorphisms θ and ϕ of $\mathbb{Z}_2 \times \mathbb{Z}_4$. In this description, the element $(i, j), i \in \mathbb{Z}_2 = \{0, 1\}, j \in \mathbb{Z}_4 = \{0, 1, 2, 3\}$, is written as ij for short. It is easy to check that $\theta, \phi \in \mathrm{Orth}(\mathbb{Z}_2 \times \mathbb{Z}_4)$ and that $\theta \perp \phi$. However $\theta^{-1} \not\perp \phi^{-1}$ as $\theta^{-1}(10) - \phi^{-1}(10) = 12 - 13 = 03 = 11 - 12 = \theta^{-1}(02)\phi^{-1}(02)$.

Fig. 8.1 I is not an automorphism of $\mathrm{Orth}(\mathbb{Z}_2 \times \mathbb{Z}_4)$

x	00	10	01	11	02	12	03	13
$\theta(x)$	00	13	11	02	03	10	12	01
$\phi(x)$	00	03	12	13	01	02	11	10
$\theta^{-1}(x)$	00	12	13	01	11	03	02	10
$\phi^{-1}(x)$	00	13	02	03	12	01	10	11

The mapping R together with mappings of the form H_α, T_g generates a group of automorphisms of $\mathrm{Orth}(G)$, and these together with I generate a group of congruences of $\mathrm{Orth}(G)$ called the group of congruent mappings in [69]. Products of congruences are described in the next theorem. It is clear that $T_1 = 1$, and, if α is the identity automorphism, that $H_\alpha = 1$.

Theorem 8.8. *Let* $g, h \in G$, $\alpha, \beta \in \mathrm{Aut}(G)$, *and let* τ_g *denote the inner automorphism* $\tau_g(x) = gxg^{-1}$. *Then the following hold.*

1. $H_\alpha H_\beta = H_{\alpha\beta}$.
2. $T_g T_h = T_{gh}$.
3. $R^2 = I^2 = 1$.
4. $H_\alpha R = RH_\alpha$.
5. $H_\alpha T_g = T_{\alpha(g)} H_\alpha$.
6. $T_g R = RT_{g^{-1}} H_{\tau_g}$.
7. $H_\alpha I = IH_\alpha$.
8. $(IR)^3 = 1$.
9. $T_g I[\theta] = IT_{\theta^{-1}(g)}[\theta]$.

Proof. (1), ..., (7) and (9) follow from routine calculations. For (8) we will prove the equivalent equality IRI $=$ RIR. Let $\iota, \eta_\theta, \eta_{\theta^{-1}}: G \to G$ be defined by $\iota(x) = x^{-1}$, $\eta_\theta(x) = x^{-1}\theta(x)$, and $\eta_{\theta^{-1}}(x) = x^{-1}\theta^{-1}(x)$. Then $R[\theta](x) = x\theta(x^{-1}) = \eta_\theta(x^{-1}) = \eta_\theta\iota(x)$, $IR[\theta](x) = R[\theta]^{-1}(x) = \iota\eta_\theta^{-1}(x) = \eta_\theta^{-1}(x)^{-1}$, and $RIR[\theta](x) = xIR[\theta](x^{-1}) = x\eta_\theta^{-1}(x^{-1})^{-1}$. Further,

$$RI[\theta](x) = xI[\theta](x^{-1}) = x\theta^{-1}(x^{-1}) = \eta_{\theta^{-1}}(x^{-1}) = \eta_{\theta^{-1}}\iota(x),$$

and $IRI[\theta](x) = RI[\theta]^{-1}(x) = \iota\eta_{\theta^{-1}}^{-1}(x) = \eta_{\theta^{-1}}^{-1}(x)^{-1}$.

Thus we need to show that

$$x\eta_\theta^{-1}(x^{-1})^{-1} = \eta_{\theta^{-1}}^{-1}(x)^{-1}.$$

Let us first express $\eta_{\theta^{-1}}$ in terms of η_θ and θ^{-1}.

$$\eta_{\theta^{-1}}(x) = x^{-1}\theta^{-1}(x) = (\theta^{-1}(x)^{-1}\theta(\theta^{-1}(x)))^{-1} = \eta_\theta(\theta^{-1}(x))^{-1} = \iota\eta_\theta\theta^{-1}(x),$$

and so $\eta_{\theta^{-1}}^{-1}(x) = \theta(\eta_\theta^{-1}(x^{-1}))$. Hence, by setting $w = \eta_\theta^{-1}(x^{-1})$, our equation reduces to $xw^{-1} = \theta(w)^{-1}$, which is true if and only if $x^{-1} = w^{-1}\theta(w) = \eta_\theta(w) = x^{-1}$. The result follows. □

Groups of congruences can be used to classify orthomorphisms and orthomorphism graphs. Some examples of this will be seen in later chapters. In particular, in Section 8.4 we will study orthomorphisms fixed by translations. Cyclotomic orthomorphisms will be studied in Sections 9.3 and 10.1: these can be characterized by the homologies that fix them. In Section 15.2 we will see how subgroups of $\langle R, I\rangle$ can be used to classify neofields.

8.2 Powers

For a finite group G, the simplest class of mappings that we might consider are the mappings $\phi_r: G \to G$ defined by $\phi_r(x) = x^r$. In this section we will study the orthomorphism graph consisting of mappings of this type. First we need a simple lemma that tells us when ϕ_r is a bijection.

Lemma 8.9. *If $|G| = n$, then ϕ_r is a bijection if and only if* $\gcd(r, n) = 1$.

Proof. If $\gcd(r, n) \neq 1$, then there exists a prime p that divides both r and n, and there exists $g \in G$ of order p. Then $\phi_r(g) = 1 = \phi_r(1)$ and so ϕ_r cannot be a bijection.

If $\gcd(r, n) = 1$, then there exists r' satisfying $r'r \equiv 1 \pmod n$ and $\phi_r(x^{r'}) = x$ for all $x \in G$. Hence ϕ_r is onto and hence, by finiteness, is a bijection. □

We can now give criteria for a mapping ϕ_r to be an orthomorphism and criteria for orthogonality.

Theorem 8.10. *If G is a group of order n, then ϕ_r is an orthomorphism if and only if* $\gcd(r, n) = \gcd(r - 1, n) = 1$. *Further $\phi_r \perp \phi_s$ if and only if* $\gcd(s - r, n) = 1$.

Proof. By Lemma 8.9, ϕ_r is a bijection if and only if $\gcd(r, n) = 1$, and $x \mapsto x^{-1}\phi_r(x) = x^{r-1} = \phi_{r-1}(x)$ is a bijection if and only if $\gcd(r - 1, n) = 1$.

Further $\phi_r \perp \phi_s$ if and only if the mapping $x \mapsto \phi_r(x)^{-1}\phi_s(x) = x^{s-r} = \phi_{s-r}(x)$ is a bijection, if and only if $\gcd(s - r, n) = 1$. □

Using a similar proof, we can establish criteria for ϕ_r to be a complete mapping.

Corollary 8.11. *If G is a group of order n, then ϕ_r is a complete mapping if and only if* $\gcd(r, n) = \gcd(r + 1, n) = 1$.

It follows from Corollary 8.11 that, if $|G|$ is odd, then ϕ_1 is a complete mapping of G.

8.2.1 The Orthomorphism Graph $\mathscr{P}(G)$

We will define the orthomorphism graph $\mathscr{P}(G)$ to be the subgraph of Orth(G) induced by mappings of the form ϕ_r. By Theorem 8.10, $\mathscr{P}(G) = \{\phi_r \mid \gcd(r, n) = \gcd(r - 1, n) = 1\}$, where $n = |G|$. The following are immediate corollaries of Theorem 8.10.

Corollary 8.12. *$\mathscr{P}(G)$ is nonempty if and only if $|G|$ is odd.*

Corollary 8.13. *If G_1 and G_2 are groups for which $|G_1| = |G_2|$, then $\mathscr{P}(G_1) \cong \mathscr{P}(G_2)$.*

Thus the structure of $\mathscr{P}(G)$ depends not on the structure of G but on the order of G. We can determine the exact value of $\omega(\mathscr{P}(G))$ and the size of all maximal cliques of $\mathscr{P}(G)$.

Theorem 8.14. *If G is a group of order n and p is the smallest prime dividing n, then $\omega(\mathscr{P}(G)) = p - 2$, and any maximal clique of $\mathscr{P}(G)$ is a $(p - 2)$-clique.*

Proof. Clearly $S = \{\phi_2, \cdots, \phi_{p-1}\} \subseteq \mathscr{P}(G)$, as $\gcd(k, n) = \gcd(k - 1, n) = 1$ for $k = 2, \ldots, p - 1$. Further S is a clique of $\mathscr{P}(G)$ as $\gcd(i - j, n) = 1$ for all $i, j \in \{2, \ldots, p - 1\}$, $i \neq j$. Hence $\omega(\mathscr{P}(G)) \geq p - 2$.

Next let $\{\phi_{k_1}, \ldots, \phi_{k_r}\}$ be a clique of $\mathscr{P}(G)$. Then $k_i \not\equiv 0, 1 \pmod{p}$ and $k_i \not\equiv k_j \pmod{p}$ if $i \neq j$. Hence $r \leq p-2$ and so $\omega(\mathscr{P}(G)) = p-2$.

Next we will prove that any r-clique of $\mathscr{P}(G)$ can be extended to an $(r+1)$-clique of $\mathscr{P}(G)$ if $r < p-2$. Let $\phi_{a_1}, \ldots, \phi_{a_r}$ be an r-clique of $\mathscr{P}(G)$, $r < p-2$, and let $p = p_1 < \cdots < p_m$ be the prime divisors of n. Now for each j, there exists c_j such that $c_j \not\equiv 0, 1 \pmod{p_j}$ and $c_j \not\equiv a_i \pmod{p_j}$ for $i = 1, \ldots, r$. By the Chinese remainder theorem, there exists $a \equiv c_j \pmod{p_j}$ for all j, $1 < a < n$. Clearly $\phi_a \in \mathscr{P}(G)$ and $\phi_a \perp \phi_{a_i}$ for all i, thus extending our clique. Hence an r-clique of $\omega(\mathscr{P}(G))$ is maximal if and only if $r = p-2$. □

In Theorem 8.14 we have shown certain cliques to be maximal in $\mathscr{P}(G)$. This does not imply their maximality in Orth(G). We will establish in Corollary 8.24 that for some groups, in particular cyclic groups, maximal cliques of $\mathscr{P}(G)$ are also maximal in Orth(G).

As a corollary to Theorem 8.14, we obtain bounds on $\omega(G)$ for any group.

Corollary 8.15. *If G is a nontrivial group and p is the smallest prime divisor of $|G|$, then $p-2 \leq \omega(G) \leq |G|-2$.*

For many classes of groups, these are the best bounds that we have for $\omega(G)$.

The next theorem summarizes the actions of automorphisms and congruences of Orth(G) on $\mathscr{P}(G)$.

Theorem 8.16. *Let G be a group of order n.*

1. $\mathrm{H}_\alpha[\phi_r] = \phi_r$ *for all* $\alpha \in \mathrm{Aut}(G)$.
2. $\mathrm{T}_g[\phi_r] = \phi_r$ *for all* $g \in Z(G)$.
3. $\mathrm{R}[\phi_r] = \phi_{1-r}$.
4. $\mathrm{I}[\phi_r] = \phi_{r'}$, *where r' is the unique integer between* 1 *and* $n-1$ *that satisfies* $r'r \equiv 1 \pmod{n}$.

Proof. If $\alpha \in \mathrm{Aut}(G)$, then $\mathrm{H}_\alpha[\phi_r](x) = \alpha(\alpha^{-1}(x)^r) = x^r = \phi_r(x)$. If $g \in Z(G)$, then $\mathrm{T}_g[\phi_r](x) = (xg)^r(g^r)^{-1} = \phi_r(x)$. Further, $\mathrm{R}[\phi_r](x) = x(x^{-1})^r = x^{1-r} = \phi_{1-r}(x)$, and $\mathrm{I}[\phi_r](x) = \phi_r^{-1}(x) = \phi_{r'}(x)$. □

8.2.2 Maximal Sets of MOLS

We know from Theorem 1.32 that the existence of a maximal set of r orthomorphisms of a group G of order n implies the existence of a maximal set of $r+1$ mutually orthogonal Latin squares of order n and also, by Theorem 1.35, the existence of a maximal $(n, r; 1)$-difference matrix over G. If D is a $(n, r; 1)$-difference matrix over G and m a positive integer, then mD is the difference matrix $(D \cdots D)$, consisting of m consecutive copies of D. It was discovered by Evans [117] in 1991 that the maximality of mD can also be used to construct maximal sets of mutually orthogonal Latin squares.

Theorem 8.17 (Evans, 1991). *Let D be an $(n, r; 1)$-difference matrix over G for which mD is maximal. If there exist $r - 1$ mutually orthogonal Latin squares of order m then, there exists a maximal set of $r - 1$ mutually orthogonal Latin squares of order nm.*

Proof. If D' is obtained from D by permuting rows and columns, or by multiplying all the elements of any given row of D on the right by a constant or all the elements of any given column of D on the left by a constant, then D' will still be a difference matrix, and mD' will still be maximal. Thus we are free to assume that $D = (d_{ij})$, $d_{1j} = 1$ for all j. Let $L_k = (a^k_{ij})$, $k = 1, \ldots, r - 1$, be a set of mutually orthogonal Latin squares of order m, on the symbol set $\{0, \ldots, m - 1\}$. Define $L'_k = (b^k_{ij})$ by $b^k_{ij} = d_{2i}d_{k+1,j}$. Then $L'_1, \ldots, L'_{r-1}$ is a set of mutually orthogonal Latin squares of order n.

Form the Kronecker products $L_1 \times L'_1, \ldots, L_{r-1} \times L'_{r-1}$. This is a set of mutually orthogonal Latin squares of order mn with entries from the symbol set $\{(i, g) \mid g \in G, i = 0, \ldots, m - 1\}$. Suppose that there exists a Latin square M orthogonal to each of $L_1 \times L'_1, \ldots, L_{r-1} \times L'_{r-1}$. The symbol $(0, 1)$ will occur exactly once as an entry in each row and column of M. Let the corresponding cells of $L_1 \times L'_1$ contain $(a^1_{i_s j_s}, b^1_{u_s v_s})$, $s = 1, \ldots, mn$. Define an $(r + 1) \times mn$ matrix $D' = (d'_{ij})$, $d'_{1s} = 1$ for $s = 1, \ldots, mn$, $d'_{k+1,s} = d_{k+1,v_s}$ for $s = 1, \ldots, mn$ and $k = 1, \ldots, r - 1$, and $d'_{r+1,s} = d^{-1}_{2u_s}$ for $s = 1, \ldots, mn$. The sequence $\{(d'_{1s})^{-1}d'_{r+1,s} \mid s = 1, \ldots, mn\} = \{d^{-1}_{2u_s} \mid s = 1, \ldots, mn\}$ contains each element of G exactly m times as the sequence $\{u_s \mid s = 1, \ldots, mn\}$ contains m copies of each integer $i = 1, \ldots, n$. The sequence

$$\begin{aligned}\{(d_{k+1,s})^{-1}d_{r+1,s} \mid s = 1, \ldots, mn\} =& \{(d_{2u_s}d_{k+1,v_s})^{-1} \mid s = 1, \ldots, mn\}\\ =& \{b_{u_s v_s}k^{-1} \mid s = 1, \ldots, mn\}\end{aligned}$$

contains each element of G exactly m times as each element of the symbol set occurs exactly once in the sequence $\{(a^k_{i_s j_s}, b^k_{u_s v_s}) \mid s = 1, \ldots, mn\}$. Thus D' is an $(n, r + 1; m)$-difference matrix over G. This contradicts the maximality of mD. □

Thus we are led to ask for which $(n, r; 1)$-difference matrices D over which groups G, and positive integers m, is mD a maximal difference matrix? Obvious necessary conditions are that D be maximal and that n does not divide m. These are not sufficient; however, as if $G = \mathbb{Z}_n$, n even, and D is any $(n, 2; 1)$-difference matrix over G, then D will be maximal, but mD will be maximal if and only if m is odd.

One source of $(n, r; 1)$-difference matrices D, and positive integers m, for which mD is maximal is a result by Drake [105]: in 1977 he proved that if n does not divide m, then the existence of a projective plane of order n and the existence of $n - 1$ mutually orthogonal Latin squares of order m imply the existence of a maximal set of $n - 1$ mutually orthogonal Latin squares of order mn. A difference matrix result is obtained by a straightforward modification of Drake's proof.

Theorem 8.18. *If D is an $(n, n; 1, G)$-difference matrix, then mD is maximal if and only if n does not divide m.*

Theorem 8.18 gives rise to difference matrix proofs of some known results. Corollary 8.19, a 1977 result of Drake [105], and Corollary 8.20, a 1963 result of Bruck [57], follow immediately from Theorems 8.17 and 8.18.

Corollary 8.19 (Drake, 1977). *If q is a prime power, q does not divide m, and there exist $q-1$ mutually orthogonal Latin squares of order m, then there exists a maximal set of $q-1$ mutually orthogonal Latin squares of order qm.*

Corollary 8.20 (Bruck, 1963). *If $n = q_1 \ldots q_k$, $q_1 < \cdots < q_k$ are powers of distinct primes, then there exists a maximal set of $q_1 - 1$ mutually orthogonal Latin squares of order n.*

To determine new parameter pairs for the existence of maximal sets of mutually orthogonal Latin squares, we will construct, for p a prime, maximal sets of $p-1$ mutually orthogonal Latin squares of order p^r, and maximal sets of $p-1$ mutually orthogonal Latin squares of order mp^r, for certain values of m. First we need two lemmas.

Lemma 8.21. *Let $n = p^r$, p a prime, and let $\phi\colon \{0, \ldots, p-1\} \to \{0, \ldots, n-1\}$ be any function satisfying $\phi(i) \equiv i \pmod{p}$, and let $d_1, \ldots, d_{p-1}$ be integers. Then the system of equations*

$$\sum_{i=1}^{p-1} x_i \phi(i)^j \equiv d_j \pmod{n},\ j = 1, \ldots, p-1,$$

has a unique solution.

Proof. For $r = 1$ this is known to be true. Suppose this to be true for $r = k$, and consider the case $r = k+1$. Set $x_i = y_i + z_i p^k$, where $0 \le y_i < p^k$ and $0 \le z_i < p$.

$$\sum_{i=1}^{p-1} x_i \phi(i)^j \equiv \sum_{i=1}^{p-1} y_i \phi(i)^j + \Big(\sum_{i=1}^{p-1} z_i \phi(i)^j\Big) p^k \equiv d_j \pmod{n}.$$

Modulo p^k the y_is are uniquely determined:

$$\sum_{i=1}^{p-1} y_i \phi(i)^j \equiv c_j \pmod{n}, \quad \text{and} \quad c_j \equiv d_j \pmod{p^k}.$$

Thus

$$\Big(\sum_{i=1}^{p-1} z_i \phi(i)^j\Big) p^k \equiv d_j - c_j \pmod{n}$$

or

$$\sum_{i=1}^{p-1} z_i \phi(i)^j \equiv (d_j - c_j)/p_k \pmod{p}.$$

Thus the z_is are also uniquely determined. Hence the result. □

Lemma 8.22. *If n is a power of a prime p, then*

$$\sum_{i=0}^{n-1} i^{p-1} \equiv (n/p)(p-1) \pmod{n}.$$

Proof. We know this formula to be correct if n is a prime or $p = 2$, so let us assume that $n = p^r$, p an odd prime, $r > 1$, and that we have proved the formula correct for all smaller powers of p. Let g be primitive modulo n. Then

$$\begin{aligned}\sum_{i=0}^{n-1} i^{p-1} &\equiv \sum_{j=1}^{n-(n/p)} (g^j)^{p-1} + \sum_{j=0}^{(n/p)-1} (jp)^{p-1} \\ &\equiv \sum_{j=1}^{n-(n/p)} g^{j(p-1)} + p^{p-1} \sum_{j=0}^{(n/p)-1} j^{p-1} \pmod{n}.\end{aligned}$$

If $r = 2$ then $p^{p-1} \equiv 0 \pmod{n}$. If $r > 2$ then, by the inductive hypothesis,

$$\sum_{j=0}^{(n/p)-1} j^{p-1} \equiv 0 \pmod{n/p^2} \text{ and so } p^{p-1} \sum_{j=0}^{(n/p)-1} j^{p-1} \equiv 0 \pmod{n}.$$

Thus

$$\begin{aligned}\sum_{i=0}^{n-1} i^{p-1} \equiv \sum_{j=1}^{n-(n/p)} g^{j(p-1)} &\equiv (p-1) \sum_{k=0}^{(n/p)-1} (kp+1) \\ &\equiv (p-1) \sum_{k=0}^{(n/p)-1} kp + (p-1)(n/p) \\ &\equiv (p-1)(n/p) \pmod{n}.\end{aligned}$$

□

In 1991, Evans [117] established the existence of a class of difference matrices D for which mD is maximal.

Theorem 8.23 (Evans, 1991). *Let G be the group $\mathbb{Z}_n$, $n = p^r$, p a prime, let*

$$\phi\colon \{0, \ldots, p-1\} \to \{0, \ldots, n-1\}$$

be any function satisfying $\phi(i) \equiv i \pmod{p}$, *and suppose that* p *does not divide* m. *Let* D *be the* $(n, p; 1)$*-difference matrix over* G *with* ij*th entry* $\phi(i-1)j \pmod{n}$, $i = 1, \ldots, p$, $j = 0, \ldots, n-1$. *Then* mD *is a maximal difference matrix.*

Proof. Let d_{ij} denote the ijth entry of mD, $i = 1, \ldots, p$ and $j = 1, \ldots, mn$, and suppose that mD is not maximal. Then we may add an extra row whose entry in the jth column is d_j, $j = 1, \ldots, mn$.

By Lemma 8.21, we may choose $a_1, \ldots, a_{p-1}$ to satisfy

$$\sum_{i=1}^{p-1} a_i \phi(i)^j \equiv \begin{cases} 0 \pmod{n} & \text{if } j < p-1, \\ 1 \pmod{n} & \text{if } j = p-1. \end{cases}$$

Then

$$\begin{aligned}
&m(a_1 + \cdots + a_{p-1}) \sum_{x=0}^{n-1} x^{p-1} \\
&\equiv \sum_{i=2}^{p} a_{i-1} \sum_{j=1}^{mn} (d_j - d_{ij})^{p-1} \\
&\equiv \sum_{i=2}^{p} a_{i-1} \sum_{j=1}^{mn} (d_j - \phi(i-1) d_{2j}/\phi(1))^{p-1} \\
&\equiv \sum_{j=1}^{mn} \sum_{k=0}^{p-1} \binom{p-1}{k} (-1)^{p-1-k} d_j^k (d_{2j}/\phi(1))^{p-1-k} \sum_{i=2}^{p} a_{i-1} \phi(i-1)^{p-1-k} \\
&\equiv \sum_{j=1}^{mn} d_j^{p-1} \sum_{i=2}^{p} a_{i-1} + (-1)^{p-1} \sum_{j=1}^{mn} (d_{2j}/\phi(1))^{p-1} \\
&\equiv m(a_1 + \cdots + a_{p-1} + \epsilon) \sum_{x=0}^{n-1} x^{p-1} \pmod{n},
\end{aligned}$$

where

$$\epsilon = \begin{cases} 1 & \text{if } p > 2, \\ -1 & \text{if } p = 2. \end{cases}$$

Thus $m \sum_{x=0}^{n-1} x^{p-1} \equiv 0 \pmod{n}$, contradicting Lemma 8.22, hence the result. □

As a corollary to Theorem 8.23, we obtain a proof that, in some circumstances, maximal cliques in $\mathscr{P}(G)$ are maximal sets of orthomorphisms in $\mathrm{Orth}(G)$.

Corollary 8.24. *If the Sylow p-subgroup of a group G is cyclic and p is the smallest prime divisor of* $|G|$, *then any* $(p-2)$*-clique of* $\mathscr{P}(G)$ *is a maximal set of orthomorphisms of G.*

Proof. Let $G = \{g_1, \ldots, g_n\}$, $n = p^r m$, $\gcd(m, p) = 1$, let $\phi_{k_1}, \ldots, \phi_{k_{p-2}}$ be a clique in $\mathscr{P}(G)$, and let $S = \mathbb{Z}_{p^r} = \{h_1, \ldots, h_{p^r}\}$. Let $D_G = \{d_{ij}\}$ be the $(n, p; 1, G)$-difference matrix defined by

$$d_{ij} = \begin{cases} 1 & \text{if } i = 1, \\ g_j & \text{if} i = 2, \\ g_j^{k_{i-2}} & \text{if } i > 2. \end{cases}$$

Let $D_p = \{d'_{ij}\}$ be the $(p^r, p; 1, S)$-difference matrix defined by

$$d'_{ij} = \begin{cases} 1 & \text{if } i = 1, \\ h_j & \text{if } i = 2, \\ h_j^{k_{i-2}} & \text{if } i > 2. \end{cases}$$

As p is the smallest prime dividing $|G|$ and the Sylow p-subgroup of G is cyclic, by, for example, Corollary 1.4.18 in [241], there is a homomorphism f from G onto S. Each column of D_p appears exactly m times in f(D_G). Thus f(D_G) is mD_p with the columns permuted, and hence, by Theorem 8.23, f(D_G) is maximal, from which it follows that D_G is maximal. □

It is easy to see that Theorem 8.23 generalizes the Hall-Paige theorem (Theorem 3.34), on the nonexistence of complete mappings, by setting $p = 2$ in Corollary 8.24. The following is a special case of Corollary 8.24:

Corollary 8.25. *If p is a prime and* $r \geq 1$, *then there exists a maximal set of* $p-1$ *mutually orthogonal Latin squares of order* p^r.

Corollary 8.26. *Let* $n = p^r m$, *p a prime,* $r \geq 1$, *and p and m relatively prime. If there exist* $p-1$ *mutually orthogonal Latin squares of order m, then there exists a maximal set of* $p-1$ *mutually orthogonal Latin squares of order n.*

Proof. This follows from Theorems 8.17 and 8.23. □

We define n_k to be the largest number for which the existence of a mutually orthogonal set of r Latin squares of order n is unknown. The following is a natural consequence of Corollary 8.26. The values of n_k, for $k \leq 15$ and $k = 30$, are given in [79].

Corollary 8.27. *Let* $n = p^r m$, *p a prime. If* $m > n_{p-1}$ *and* $\gcd(m, p) = 1$, *then there exists a maximal set of* $p-1$ *mutually orthogonal Latin squares of order n.*

There are several special cases of Corollaries 8.26 and 8.27 that deserve mentioning. If $n = 2^s m$, m odd, then there exists a Latin square of order n that has no orthogonal mate: this was proved by Euler [110] in 1779 (see Theorem 3.1). If $n = 3^s m$, m not divisible by 3, then there exists a maximal set of 2 mutually orthogonal Latin squares of order n whenever $m \neq 2, 6$: the special case $m = 1$, $s = 2$, can be found in Jungnickel and Grams [207], where it is proved using a computer search. If $n = 5^s m$, m not divisible by 5, then there exists a maximal set of 4 mutually orthogonal Latin squares of order n whenever $m > n_4 = 22$. More generally, if $n = p^s m$, m not divisible by p, p a prime, then there exists a maximal set of $p - 1$ mutually orthogonal Latin squares of order n whenever m is sufficiently large.

8.2.3 Graph Representations

We will see that the structure of $\mathscr{P}(G)$ is actually fairly complex as any finite graph can be regarded as an induced subgraph of $\mathscr{P}(G)$ for some group G. We will first establish a more general result. Let Γ be a graph with vertices $v_1, \cdots, v_r$ and let n be a natural number. We say that Γ is *representable modulo n* if there exist distinct integers $a_1, \cdots, a_r$, $0 \leq a_i < n$, satisfying $\gcd(a_i - a_j, n) = 1$ if and only if v_i is adjacent to v_j. We call $\{a_1, \cdots, a_r\}$ a *representation* of Γ modulo n, and we call n the *order* of the representation.

In 1989 Erdős and Evans [108] proved that any graph is representable modulo some positive integer. The proof will require a lemma about sets of prime numbers.

Lemma 8.28. *For any positive integer m, there exist distinct primes $p_1, \cdots, p_m$ such that for all pairs A, B of disjoint, nonempty subsets of $\{p_1, \cdots, p_m\}$,*

$$\gcd\left(\prod\{p_i \mid p_i \in A\} - \prod\{p_j \mid p_j \in B\}, p_1 p_2 \cdots p_m\right) = 1.$$

Proof. Any set of two primes satisfies the conditions of the lemma. We will show that any set of primes $\{p_1, \ldots, p_s\}$, for which $s < m$ and $p_i > 3^m$ for all i, that satisfies the conditions of the lemma can be extended to a larger set of primes, all greater than 3^m, that satisfies the conditions of the lemma.

We will fix i, and for A, B disjoint subsets of $\{p_1, \cdots, p_m\} \setminus \{p_i\}$ we will use a to denote $\prod\{p_j \mid p_j \in A\}$ and b to denote $\prod\{p_j \mid p_j \in B\}$. There exists a $c_i \neq p_i$ satisfying $a - bc_i \not\equiv 0 \pmod{p_i}$ for all choices of A and B, as there are at most $3^{s-1} < 3^m < p_i$ such inequivalences, because for each j there are three choices, p_j in A, in B, or in neither.

By the Chinese remainder theorem, there exists a unique c satisfying $0 \leq c < p_1 \cdots p_s$ and $c \equiv c_i \pmod{p_i}$ for each i. As $\gcd(c, p_1 \cdots p_s) = 1$, by a theorem of Dirichlet, there exist infinitely many primes in the arithmetic sequence $\{c + np_1 \cdots p_s\}_{n=1}^{\infty}$. Let p_{s+1} be a prime in this sequence that is not equal to any of $p_1, \ldots, p_s$ and that does not divide $\prod\{p_i \mid p_i \in A\} - \prod\{p_j \mid p_j \in B\}$ for any pairs A, B of disjoint, nonempty subsets of $\{p_1, \cdots, p_s\}$. The set $\{p_1, \ldots, p_{s+1}\}$ satisfies the conditions of the lemma. □

We are now in a position to prove the result of Erdős and Evans [108] on the representability of graphs.

Theorem 8.29 (Erdős and Evans, 1989). *Any finite graph can be represented modulo some positive integer.*

Proof. Let Γ be a graph with vertices $v_1, \cdots, v_r$. Form a new graph Γ' by adjoining an isolated vertex v_0 to Γ. Let $e_1, \cdots, e_m$ be the edges of the complement of Γ', and let $p_1, \cdots, p_m$ be primes satisfying the conditions of Lemma 8.28. Then, for $i = 1, \cdots, r$, set $a_i = \prod\{p_j \mid e_j$ incident with v_i in the complement of $\Gamma'\}$, and set $n = p_1 p_2 \cdots p_m$. Then $\{a_1, \cdots, a_r\}$ is a representation of Γ modulo n. □

The proof of Theorem 8.29 gives representations of very large orders. If we let n_k denote the smallest number that is the product of k primes satisfying the conditions of Lemma 8.28, then it was proved in [109] that the sequence $\{\log_2 n_k/k^2\}$ is bounded and all its limit points lie in the interval $[1, \log_2 3]$. For a graph with $n \geq 3$ vertices and m edges, Theorem 8.29 gives a representation of order n_k, where $k = n + (n! - m)$. We can do much better: in 2003 Narayan [250] proved that any graph with $r \geq 3$ vertices can be represented modulo the product of the first $r - 1$ primes greater than or equal to $r - 1$. The central problem in the theory of graph representations is the determination of the smallest order of a representation of Γ for a given graph Γ.

Readers interested in graph representations should consult Evans, Fricke, Maneri, McKee, and Perkel [136] for the initial development of the theory; Evans, Isaak, and Narayan [137] for a survey and further results; and Evans, Narayan, and Urick [139] for a list of problems.

Before we can prove that any graph is isomorphic to an induced subgraph of some $\mathscr{P}(G)$, we need a lemma.

Lemma 8.30. *If $\{a_1, \cdots, a_r\}$ is a representation of Γ modulo n, then so is $\{ba_1 + c, \cdots, ba_r + c\}$, where $\gcd(b, n) = 1$ and addition and multiplication are performed modulo n.*

Proof. $\gcd((ba_i + c) - (ba_j + c), n) = \gcd(b(a_i - a_j), n) = 1$ if and only if $\gcd(a_i - a_j, n) = 1$. □

Lemma 8.30 enables us to construct representations in which any given adjacent pair of vertices is represented by 0 and 1: this is the key to proving that any finite graph can be embedded in $\mathscr{P}(G)$ for some G.

Theorem 8.31. *Any finite graph can be realized as an induced subgraph of $\mathscr{P}(G)$ for some finite group G.*

Proof. Let Γ be a finite graph with vertices $v_1, \cdots, v_r$, and let Γ' be the graph obtained from Γ by adjoining vertices v_{r+1} and v_{r+2} adjacent to each of $v_1, \cdots, v_r$ and to each other. By Theorem 8.29, Γ' has a representation $\{a_1, \ldots, a_{r+2}\}$ modulo n for some n. By Lemma 8.30, Γ' has a representation $\{a_1 - a_{r+1}, \ldots, a_r - a_{r+1}, 0, a_{r+2} - a_{r+1}\}$ modulo n. As v_{r+1} and v_{r+2} are adjacent, $a_{r+2} - a_{r+1}$ must be relatively prime to n, and so there exists $c \in \{1, \ldots, n-1\}$ satisfying $(a_{r+2} -$

$a_{r+1})c \equiv 1 \pmod{n}$. Setting $b_i = (a_i - a_{r+1})c$ and applying Lemma 8.30, we see that $\{b_1, \ldots, b_r, 0, 1\}$ is a representation of Γ' modulo n. Let G be any group of order n. Then the subgraph of $\mathscr{P}(G)$ induced by the orthomorphisms $\phi_{b_1}, \cdots, \phi_{b_r}$ is isomorphic to Γ. Thus Γ is realized as an induced subgraph of $\mathscr{P}(G)$. □

Of particular interest to us is the graph X_n, the *unitary Cayley graph*, whose representation is $\{0, 1, \ldots, n-1\}$ modulo n. If G has order n, then $\mathscr{P}(G)$ is isomorphic to the common neighborhood in X_n of 0 and 1. Many properties of the unitary Cayley graph are determined in [221]. The structure of X_n was described in [136] as a product of complete multipartite graphs. The *product* of graphs $\Gamma_1, \ldots, \Gamma_r$, denoted $\Gamma_1 \times \cdots \times \Gamma_r$, is the graph with vertices $v_1v_2\cdots v_r$, v_i is a vertex of Γ_i for each i, $u_1u_2\cdots u_r$ and $v_1v_2\cdots v_r$ being adjacent if and only if u_i is adjacent to v_i for each i. In the literature products are also called conjunctions, weak products, categorical products, and Kronecker products: the latter because the adjacency matrix of $\Gamma_1 \times \Gamma_2$ is the Kronecker product of the adjacency matrices of Γ_1 and Γ_2. A complete multipartite graph $K_{m_1,\ldots,m_r}$ is a graph with $m_1 + \cdots + m_r$ vertices partitioned into r partite classes, the ith partite class containing m_i vertices, such that two vertices are adjacent if and only if they are in different partite classes. We will use $K_{r(k)}$ to denote the complete multipartite graph with r partite classes, each containing k vertices.

Theorem 8.32. *If $n = p_1^{k_1}\cdots p_r^{k_r}$, where $p_1, \ldots, p_r$ are distinct primes, then the unitary Cayley graph X_n is isomorphic to $K_{p_1\left(p_1^{k_1-1}\right)} \times \cdots \times K_{p_r\left(p_r^{k_r-1}\right)}$.*

Proof. Let us define $\mathrm{f}_i : X_n \to X_{p_i^{k_i}}$ for $i = 1, \ldots, r$, by $\mathrm{f}_i(m) = m \pmod{p_i^{k_i}}$. Then the mapping $\mathrm{f}\colon X_n \to X_{p_1^{k_1}} \times \cdots \times X_{p_r^{k_r}}$, defined by $\mathrm{f}(m) = \mathrm{f}_1(m)\cdots \mathrm{f}_r(m)$, is an isomorphism, as $\gcd(m - t, n) = 1$ if and only if p_i does not divide $m - t$ for all i, if and only if $\gcd(\mathrm{f}_i(m) - \mathrm{f}_i(t), p_i^{k_i}) = 1$ for all i. It is easily seen that each $X_{p_i^{k_i}}$ is a complete multipartite graph with partite classes $\{0, p_i, 2p_i, \ldots, p_i^{k_i} - p_i\}$, $\{1, p_i+1, 2p_i+1, \ldots, p_i^{k_i} - p_i + 1\}, \ldots, \{p_i - 1, 2p_i - 1, 3p_i - 1, \ldots, p_i^{k_i} - 1\}$. □

In the course of proving Theorem 8.32, we implicitly established the following.

Corollary 8.33. *If $\gcd(m, n) = 1$, then $X_{mn} \cong X_m \times X_n$.*

Lindner, Mendelsohn, Mendelsohn, and Wolk [230] defined an orthogonal Latin square graph to be a graph, all of whose vertices are Latin squares of the same order, adjacency being orthogonality. In 1970, they proved that any finite graph is realizable as an orthogonal Latin square graph. Their theorem is an immediate consequence of Theorem 8.31.

Corollary 8.34. *Any finite graph can be realized as an orthogonal Latin square graph.*

Lindner, Mendelsohn, Mendelsohn, and Wolk actually proved that any finite graph can be realized as an orthogonal Latin square graph using idempotent Latin squares. This extra condition is a by-product of the construction used in their proof.

It should be noted that for a graph to be representable modulo n, it is necessary though not sufficient for the smallest prime divisor of n to be at least as large as the clique number of the graph. This is in marked contrast to the situation for orthogonal Latin square realizations, in which Lindner, Mendelsohn, Mendelsohn, and Wolk proved that only finitely many Latin square orders are ruled out for a given graph.

8.3 Difference Sets

In 1980 and 1981, Jungnickel [201, 202] gave constructions of difference matrices using planar difference sets, affine difference sets, difference families, and partial difference families. From these we can derive new orthomorphism graphs which will be described in this section. Jungnickel's constructions yield some of the largest theoretically determined lower bounds for clique numbers of orthomorphism graphs. Other large lower bounds have been obtained for small groups, using computers: these bounds will be described in Chapter 13.

8.3.1 The Orthomorphism Graph $\mathscr{D}_D(G)$

If G is a group of order v, then a (v, k, λ)-*difference set* in G is a k-element subset D of G for which each nonidentity element of G is expressible in exactly λ ways in the form $d^{-1}d'$, $d, d' \in D$: the *order* of the difference set is $n = k - \lambda$. A difference set in G is *cyclic* if G is cyclic, *abelian* if G is abelian, and *nonabelian* if G is nonabelian.

We are interested in the special case $\lambda = 1$: a $(v, k, 1)$-difference set is called a *planar difference set*. If q is a prime power and g is a generator of $GF(q^3)^*$, the multiplicative group of $GF(q^3)$, then the set $\{i \mid 0 \le i < q^2 + q + 1, g^i + (g^i)^q + (g^i)^{q^2} = 0\}$ is a $(q^2 + q + 1, q + 1, 1)$-difference set: these are *Singer difference sets*, and the sum $g^i + (g^i)^q + (g^i)^{q^2}$ is the *trace* of g^i over $GF(q)$. For the smallest example, if $q = 2$, and g is a root of $x^3 + x + 1$, then g is a generator of $GF(8)^*$, and $g^i + g^{2i} + g^{4i} = 0$, for $i = 1, 2, 4$: thus $\{1, 2, 4\}$ is a $(7, 3, 1)$-difference set in $\mathbb{Z}_7$, a fact which can be verified directly. An easy counting argument shows that a planar difference set of order n is an $(n^2 + n + 1, n + 1, 1)$-difference set. If D is a planar difference set of order n in G, then we can construct a projective plane π of order n as follows. The points of π are the elements of G, and the lines of π are the sets $Dg = \{dg \mid d \in D\}$, $g \in G$, incidence being inclusion. We leave it as an exercise to show that π does indeed satisfy the axioms of a projective plane of order n and that G acts regularly on the points and lines of π. A good survey of difference sets and planar difference sets can be found in [41]: a more recent survey is given in [337]. Algebraic methods for studying difference sets are described in [270], and many examples of difference sets were given in 2007 by Jungnickel, Pott, and Smith [209].

In 1980 Jungnickel [201] used planar difference sets over $G = \mathbb{Z}_{q^2+q+1}$, q a prime power, and MOLS of order $q + 1$ to construct difference matrices over G.

Theorem 8.35 (Jungnickel, 1980). *If q is a prime power and their exists a set of r MOLS of order $q+1$, then there exists a $(q^2+q+1, r+1; 1)$-difference matrix over $\mathbb{Z}_{q^2+q+1}$.*

Proof. If q is a prime power, then there exists a $(q^2+q+1, r+1; 1)$-difference set in $\mathbb{Z}_{q^2+q+1}$: the Singer difference set, for example. Let $S = \{s_0, \ldots, s_q\}$ be a planar difference set in $\mathbb{Z}_{q^2+q+1}$. Let $L_1, \ldots, L_r$ be a set of r MOLS of order $q+1$ on the symbol set S and with the 0th row of each equal to $\begin{pmatrix} s_0 \cdots s_q \end{pmatrix}$, and let L_0 be a $(q+1)\times(q+1)$ square with each row equal to $\begin{pmatrix} s_0 \cdots s_q \end{pmatrix}$. Note that while L_0 is not a Latin square, the squares $L_0, \ldots, L_r$ are pairwise orthogonal. Let l^i_j denote the jth row of L_i, $j = 0, \ldots, q$, and define an $(r+1)\times(q^2+q+1)$ matrix D as follows.

$$D = \{d_{ij}\} = \begin{bmatrix} 1 \; l_1^0 \; l_2^0 \cdots l_q^0 \\ 1 \; l_1^1 \; l_2^1 \cdots l_q^1 \\ \vdots \;\; \vdots \;\; \vdots \qquad \vdots \\ 1 \; l_1^r \; l_2^r \ldots l_q^r \end{bmatrix}$$

Let $i, k \in \{0, \ldots, r\}$, $i \neq k$. As L_i and L_k are orthogonal $\{d_{ij}^{-1} d_{kj} \mid j = 0, \ldots, q^2+q+1\} = \mathbb{Z}_{q^2+q+1}$, hence D is a difference matrix over $\mathbb{Z}_{q^2+q+1}$. □

Let us illustrate the proof of Theorem 8.35 for the special case $q = 4$: this case is also described in Jungnickel's paper. It is easy to verify that $S = \{3, 6, 7, 12, 14\}$ is a planar difference set of order 4 in $\mathbb{Z}_{21}$. A set of 5 pairwise orthogonal squares of order $q+1 = 5$ is shown in Figure 8.2. The squares $L_1, \ldots, L_4$ form a set of MOLS of order 5 on the symbol set S, each having the same 0th row, $\begin{pmatrix} 3 & 6 & 7 & 12 & 14 \end{pmatrix}$. To form the matrix D in Figure 8.3, strip away the 0th rows of $L_0, \ldots, L_4$, form the ith row of D by starting with 0, and then concatenate the remaining rows of L_i. It is routine, but tedious, to verify that D is a $(21, 5; 1)$-difference matrix over $\mathbb{Z}_{21}$. From this we obtain a set of 4 MOLS of order 21 based on the group $\mathbb{Z}_{21}$, a result that was first obtained by Parker [264] in 1959, through a computer search. This has since been improved to 5 MOLS (see [6]).

$$L_0 = \begin{pmatrix} 3 & 6 & 7 & 12 & 14 \\ 3 & 6 & 7 & 12 & 14 \\ 3 & 6 & 7 & 12 & 14 \\ 3 & 6 & 7 & 12 & 14 \\ 3 & 6 & 7 & 12 & 14 \end{pmatrix}, \quad L_1 = \begin{pmatrix} 3 & 6 & 7 & 12 & 14 \\ 6 & 7 & 12 & 14 & 3 \\ 7 & 12 & 14 & 3 & 6 \\ 12 & 14 & 3 & 6 & 7 \\ 14 & 3 & 6 & 7 & 12 \end{pmatrix}, \quad L_2 = \begin{pmatrix} 3 & 6 & 7 & 12 & 14 \\ 7 & 12 & 14 & 3 & 6 \\ 14 & 3 & 6 & 7 & 12 \\ 6 & 7 & 12 & 14 & 3 \\ 12 & 14 & 3 & 6 & 7 \end{pmatrix},$$

$$L_3 = \begin{pmatrix} 3 & 6 & 7 & 12 & 14 \\ 12 & 14 & 3 & 6 & 7 \\ 6 & 7 & 12 & 14 & 3 \\ 14 & 3 & 6 & 7 & 12 \\ 7 & 12 & 14 & 3 & 6 \end{pmatrix}, \quad \text{and} \quad L_4 = \begin{pmatrix} 3 & 6 & 7 & 12 & 14 \\ 14 & 3 & 6 & 7 & 12 \\ 12 & 14 & 3 & 6 & 7 \\ 7 & 12 & 14 & 3 & 6 \\ 6 & 7 & 12 & 14 & 3 \end{pmatrix}$$

Fig. 8.2 5 pairwise orthogonal squares of order 5

$$D = \left(\begin{array}{c|cccccc|cccccc|cccccc|cccccc}
0 & 3 & 6 & 7 & 12 & 14 & & 3 & 6 & 7 & 12 & 14 & & 3 & 6 & 7 & 12 & 14 & & 3 & 6 & 7 & 12 & 14 \\
\hline
0 & 6 & 7 & 12 & 14 & 3 & & 7 & 12 & 14 & 3 & 6 & & 12 & 14 & 3 & 6 & 7 & & 14 & 3 & 6 & 7 & 12 \\
0 & 7 & 12 & 14 & 3 & 6 & & 14 & 3 & 6 & 7 & 12 & & 6 & 7 & 12 & 14 & 3 & & 12 & 14 & 3 & 6 & 7 \\
0 & 12 & 14 & 3 & 6 & 7 & & 6 & 7 & 12 & 14 & 3 & & 14 & 3 & 6 & 7 & 12 & & 7 & 12 & 14 & 3 & 6 \\
0 & 14 & 3 & 6 & 7 & 12 & & 12 & 14 & 3 & 6 & 7 & & 7 & 12 & 14 & 3 & 6 & & 6 & 7 & 12 & 14 & 3
\end{array}\right)$$

Fig. 8.3 A $(21, 5; 1)$-difference matrix over $\mathbb{Z}_{21}$

Theorem 8.35 yields lower bounds on $N(q^2 + q + 1)$ whenever q is a prime power.

Corollary 8.36. *Let q be a prime power.*

1. $\omega(\mathbb{Z}_{q^2+q+1}) \geq N(q + 1) - 1$.
2. *If $q + 1$ is a prime power, then* $\omega(\mathbb{Z}_{q^2+q+1}) \geq q - 1$.

Proof. By Theorem 1.21, the existence of a $(q^2+q+1, r+1; 1)$-difference matrix over $\mathbb{Z}_{q^2+q+1}$ is equivalent to the existence of an $r - 1$ clique in $\mathrm{Orth}(\mathbb{Z}_{q^2+q+1})$. Hence $\omega(\mathbb{Z}_{q^2+q+1}) \geq N(q + 1) - 1$ by Theorem 8.35. If $q + 1$ is also a prime power, then $N(q + 1) = q$, in which case $\omega(\mathbb{Z}_{q^2+q+1}) \geq q - 1$. □

In Table 8.1 we list some lower bounds for $N(q^2 + q + 1)$ and $\omega(\mathbb{Z}_{q^2+q+1})$, q a prime power. The lower bounds for $N(q+1)$ and $N(q^2+q+1)$ are from [6]. Lower bounds for $N(q^2+q+1)$ that match lower bounds derived from Corollary 8.36 are marked with a prime, as are the lower bounds for $\omega(\mathbb{Z}_{q^2+q+1})$ that match those obtained from Corollary 8.36. Table 8.1 shows that the lower bounds for $N(q^2 + q + 1)$, from Corollary 8.36, have mostly been exceeded: however, this is not true for the lower bounds for $\omega\left(\mathbb{Z}_{q^2+q+1}\right)$.

Table 8.1 Some lower bounds for $N(q^2 + q + 1)$ and $\omega(\mathbb{Z}_{q^2+q+1})$

q	$q+1$	$N(q+1) \geq$	q^2+q+1	$N(q^2+q+1) \geq$	$\omega(\mathbb{Z}_{q^2+q+1}) \geq$
7	8	7	57	7′	6′
16	17	16	273	16′	15′
23	24	7	553	7′	6′
31	32	31	993	31′	30′
43	44	5	1,893	15	4′
61	62	5	3,783	30	4′
64	65	7	4,161	30	6′
67	68	5	4,557	15	4′
73	74	5	5,403	30	4′
79	80	9	6,321	15	8′
81	82	8	6,643	15	7′
97	98	6	9,507	15	5′

Jungnickel's proof of Theorem 8.35 is easily generalized to yield difference matrices of index 1, and hence orthogonal sets of orthomorphisms, for any group that contains a planar difference set. We will adapt Jungnickel's proof to obtain classes of orthomorphisms and corresponding orthomorphism graphs. Let $\{a_1, \ldots, a_k\}$ be a $(v, k, 1)$-difference set in a group G, and let $M = \{m_{ij}\}$ be a square matrix of order k with entries $\{1, \ldots, k\}$. Define $\theta_M \colon G \to G$ by $\theta_M(a_j^{-1}a_i) = a_j^{-1}a_{m_{ij}}$. We set $\mathscr{D}_D(G) = \{\theta_M \mid \theta_M \in \mathrm{Orth}(G)\}$. The following theorem characterizes elements of $\mathscr{D}_D(G)$ and orthogonality in $\mathscr{D}_D(G)$.

Theorem 8.37. *$\theta_M \in \mathscr{D}_D(G)$ if and only if M is a Latin square for which $m_{ii} = i$ for all i. Further, if $\theta_M, \theta_K \in \mathscr{D}_D(G)$, then $\theta_M \perp \theta_K$ if and only if M is orthogonal to K.*

Proof. $\theta_M(1) = \theta_M(a_i^{-1}a_i) = a_i^{-1}a_{m_{ii}}$. Thus θ_M is well-defined and fixes 1 if and only if $m_{ii} = i$ for all i. θ_M is a bijection if and only if, for all a, the equation $a_j^{-1}a_{m_{ij}} = a$ has a unique solution for i and j. But $a = a_j^{-1}a_r$ for unique values of j and r. Thus the equation $a_j^{-1}a_{m_{ij}} = a$ uniquely determines j and m_{ij}. It follows that θ_M is a bijection if and only if each column of M contains each symbol of $\{1, \cdots, k\}$ exactly once. Similarly the mapping $x \to x^{-1}\theta_M(x)$ is a bijection if and only if, for all a, the equation $a_i^{-1}a_{m_{ij}} = a$ has a unique solution for i and j and if and only if each row of M contains each symbol of $\{1, \cdots, k\}$ exactly once. Hence $\theta_M \in \mathscr{D}_D(G)$ if and only if M is a Latin square for which $m_{ii} = i$ for all i.

If $\theta_M, \theta_K \in \mathscr{D}_D(G)$, then $\theta_M \perp \theta_K$ if and only if, for all a, the equation $a_{k_{ij}}^{-1}a_{m_{ij}} = a$ has a unique solution for i and j; if and only if, for all r and s, the equations $a_{m_{ij}} = a_r$ and $a_{k_{ij}} = a_s$ have a unique solution for i and j; and if and only if M and K are orthogonal. □

An immediate corollary:

Corollary 8.38. *If D is a planar difference set of order n in G, then $\mathscr{D}_D(G)$ is isomorphic to the induced subgraph of the orthogonal Latin square graph, with vertices Latin squares of order $n+1$ and adjacency synonymous with orthogonality, whose squares all have the symbols in standard order on the main diagonal.*

This fact enables us to bound $\omega(\mathscr{D}_D(G))$.

Corollary 8.39. *If there exists a planar difference set of order n in G, then*

$$N(n+1) - 1 \leq \omega(\mathscr{D}_D(G)) \leq N(n+1).$$

Proof. The upper bound is an immediate consequence of Corollary 8.38. To obtain the lower bound, let $r = N(n+1)$ and let $L_1, \ldots, L_r$ be a set of MOLS of order $n+1$. By permuting the columns and symbols, we may, without loss of generality, assume that the same symbol appears in every cell of the main diagonal of L_1 and that the symbols appear in standard order on the main diagonal of each L_i for $i = 2, \ldots, r$. $\theta_{L_2}, \ldots, \theta_{L_r}$ is an $(r-1)$-clique of $\mathscr{D}_D(G)$, from which the result follows. □

The question arises as to which groups contain planar difference sets. This has been most studied for abelian groups. The principal conjecture is the following.

Conjecture 8.40 (Prime Power Conjecture). Any abelian planar difference set is of prime power order.

As the existence of a planar difference set of order n implies the existence of a projective plane of order n, the prime power conjecture for abelian planar difference sets is a special case of the prime power conjecture for projective planes. Conjecture 8.40 was proved true for orders $n < 2{,}000{,}000$ by Gordon [156] in 1994. For cyclic groups this was extended to $n < 2 \times 10^9$ by Baumert and Gordon [30] in 2004. For the structure of groups that contain planar difference sets, we have the following conjecture.

Conjecture 8.41. Any abelian planar difference set is cyclic.

No example of a noncyclic abelian planar difference set is known. While much less effort has been devoted to the study of nonabelian planar difference sets, some examples are known, all of prime power order. In 1955 Bruck [55] discovered a class of nonabelian planar difference sets of order q, q a prime power, $q \equiv 1 \pmod 3$.

Theorem 8.42 (Bruck, 1955). *If $v = q^2 + q + 1$, q is a prime power, $q \equiv 1 \pmod 3$, and $G = \langle a, b \mid a^3 = b^{v/3} = 1, a^{-1}ba = b^q \rangle$, then G contains a planar difference set.*

A construction of these difference sets from Singer difference sets can be found in [41].

Corollary 8.43. *If $v = q^2 + q + 1$, q a prime power, $q \equiv 1 \pmod 3$, and $G = \langle a, b \mid a^3 = b^{v/3} = 1, a^{-1}ba = b^q \rangle$, then*

$$N(n+1) - 1 \leq \omega(\mathscr{D}_D(G)) \leq N(n+1).$$

Proof. This follows from Corollary 8.39 and Theorem 8.42. □

Note that the group $G = \langle a, b \mid a^3 = b^{v/3} = 1, a^{-1}ba = b^q \rangle \cong \mathbb{Z}_3 \ltimes \mathbb{Z}_{v/3}$ is the semidirect product of $\langle a \rangle \cong \mathbb{Z}_3$ with $\langle b \rangle \cong \mathbb{Z}_{v/3}$. Corollary 8.43 yields lower bounds for $\omega(\mathbb{Z}_3 \ltimes \mathbb{Z}_{v/3})$. Some examples follow.

Example 8.44.

1. $\omega(\mathbb{Z}_3 \ltimes \mathbb{Z}_7) \geq N(5) - 1 = 4$.
2. $\omega(\mathbb{Z}_3 \ltimes \mathbb{Z}_{19}) \geq N(8) - 1 = 6$.
3. $\omega(\mathbb{Z}_3 \ltimes \mathbb{Z}_{91}) \geq N(17) - 1 = 15$.
4. $\omega(\mathbb{Z}_3 \ltimes \mathbb{Z}_{331}) \geq N(32) - 1 = 30$.

Generalizing Bruck's construction, Gao and Wei [152] found more classes of nonabelian groups that contain planar difference sets.

Theorem 8.45 (Gao and Wei, 1993). *If p and q are primes, $p \neq 2, 3$, u is a positive integer, and p divides $q^{2u} + q^u + 1$, then the group*

$$G = \langle a, b \mid a^{pm} = b^{v/(pm)} = 1, a^{-1}ba = b^{q^{3u}} \rangle$$

contains a planar difference set of order $n = q^{pu}$, *where*

$$v = q^{2pu} + q^{pu} + 1 \quad \textit{and} \quad m = \frac{v}{\gcd\{v, 1 + q^{3u} + \cdots + (q^{3u})^{p-1}\}}.$$

Gao and Wei found a number of values of p, q, u, and n, which satisfy Theorem 8.45: these are listed in Table 8.2.

Corollary 8.46. *If* p *and* q *are primes,* $p \neq 2, 3$, u *is a positive integer, and* p *divides* $q^{2u} + q^u + 1$, *and*

Table 8.2 Parameters for Gao and Wei's construction

p	q	u	n
7	2 or 4 (mod 7)	1 or 2 (mod 3)	q^{7u}
7	3 or 5 (mod 7)	2 or 4 (mod 6)	q^{7u}
13	2, 6, 7 or 11 (mod 13)	4 or 8 (mod 12)	q^{13u}
13	4 or 10 (mod 13)	2 or 4 (mod 6)	q^{13u}
13	3 or 9 (mod 13)	1 or 2 (mod 3)	q^{13u}

$$G = \langle a, b \mid a^{pm} = b^{v/(pm)} = 1, a^{-1}ba = b^{q^{3u}} \rangle,$$

where

$$v = q^{2pu} + q^{pu} + 1 \quad \textit{and} \quad m = \frac{v}{\gcd\{v, 1 + q^{3u} + \cdots + (q^{3u})^{p-1}\}},$$

then

$$N(q^{pu} + 1) - 1 \leq \omega(\mathscr{D}_D(G)) \leq N(q^{pu} + 1).$$

Proof. This follows from Corollary 8.39 and Theorem 8.45. □

For the smallest example of Corollary 8.46, $p = 7$, $q = 2$, and $u = 1$, $|G| = 16{,}513$ and $n + 1 = 129$: thus $\omega(G) \geq \omega(\mathscr{D}_D(G)) \geq N(129) - 1 \geq 6$.

8.3.2 *The Orthomorphism Graph* $\mathscr{D}_{D,H}(G)$

If G is a group of order mn and N is a normal subgroup of G of order n, then a (m, n, k, λ)-*difference set* in G relative to N is a k-element subset D of G for which each element of $G \setminus N$ is expressible in exactly λ ways in the form $d^{-1}d'$, $d, d' \in D$, $d \neq d'$, and no element of N can be so expressed: D is called a *relative difference set*, and N is the *forbidden subgroup*. As with difference sets, a relative difference set in G is *cyclic* if G is cyclic, *abelian* if G is abelian, and *nonabelian* if G is nonabelian.

We are interested in the special case $\lambda = 1$. An $(n+1, n-1, n, 1)$-difference set is an *affine difference set*. The reason for this name is that, if D is an $(n+1, n-1, n, 1)$-difference set in G relative to N, then the incidence structure with points $G \cup \{\infty\}$, lines $\{Dg \mid g \in G\} \cup \{Ng \cup \{\infty\} \mid g \in G\}$ is an affine plane of order n. If q is a prime power, then an affine difference set exists in $\mathbb{Z}_{q^2-1}$. To construct such a difference set, let g be a generator of $\mathrm{GF}(q^2)$. Then $D = \{i \mid 0 \le i < q^2-1, (g^i)+(g^i)^q = 1\}$ is a $(q+1, q-1, q, 1)$-difference set in $\mathbb{Z}_{q^2-1}$ relative to $\mathbb{Z}_{q-1}$. As an example, if $q = 3$ and g is a root of $x^2 + x + 2$, then $(g^i) + (g^i)^q = 1$ for $i = 4, 5, 7$, and so $\{4, 5, 7\}$ is a $(4, 2, 3, 1)$-difference set in $\mathbb{Z}_8$ relative to $\mathbb{Z}_2$, a result that is easily verified. A good survey of affine difference sets can be found in [270].

In 1980 Jungnickel [201] used affine difference sets in $\mathbb{Z}_{q^2-1}$, q a prime power, and difference matrices over $\mathbb{Z}_{q-1}$ to construct difference matrices over $\mathbb{Z}_{q^2-1}$.

Theorem 8.47 (Jungnickel, 1980). *If q is a prime power and there exists a $(q-1, r+1; 1)$-difference matrix over $\mathbb{Z}_{q-1}$, then there exists a $(q^2-1, r+1; 1)$-difference matrix over $\mathbb{Z}_{q^2-1}$.*

Proof. If q is a prime power, then there exists a $(q+1, q-1, q, 1)$-difference set in $\mathbb{Z}_{q^2-1}$ relative to $\mathbb{Z}_{q-1}$. Let $S = \{s_1, \ldots, s_q\}$ be a $(q+1, q-1, q, 1)$-difference set in $\mathbb{Z}_{q^2-1}$. Let $L_1, \ldots, L_r$ be a set of r MOLS of order q, without loss of generality on the symbol set S, and with the 0th row of each equal to $\begin{pmatrix} s_1 & \cdots & s_q \end{pmatrix}$. Let l^i_j denote the jth row of L_i, $j = 0, \ldots, q-1$, and define an $(r+1) \times (q^2-q)$ matrix A as follows.

$$A = \{a_{ij}\} = \begin{bmatrix} l^1_1 & l^1_2 & \cdots & l^1_{q-1} \\ l^2_1 & l^2_2 & \cdots & l^2_{q-1} \\ \vdots & \vdots & & \vdots \\ l^r_1 & l^r_2 & \ldots & l^r_{q-1} \end{bmatrix}$$

Let $H \cong \mathbb{Z}_{q-1}$ be the unique subgroup of $\mathbb{Z}_{q^2-1}$ of order $q-1$, and let B be a $(q-1, r+1; 1)$-difference matrix over H. We claim that $D = (\, A \;\, B \,)$ is a difference matrix over $\mathbb{Z}_{q^2-1}$. □

The result of Theorem 8.47 can only be of interest if q is even, as if q is odd, then r cannot be greater than one. In Table 8.3 we list some of the consequences of this result.

Table 8.3 Some lower bounds for $\omega(\mathbb{Z}_{q^2-1})$

q	$q-1$	$\omega(\mathbb{Z}_{q-1}) \ge$	q^2-1	$\omega(\mathbb{Z}_{q^2-1}) \ge$
4	3	1	15	1
8	7	5	63	5
16	15	3	255	3
32	31	29	1,023	29
64	63	5	4,095	5

Jungnickel's proof of Theorem 8.47 can easily be adapted to yield orthomorphisms of $\mathbb{Z}_{q^2-1}$. Let $\{a_1, \ldots, a_k\}$ be a $(v, k, 1; H)$-difference set in a group G, let $M = \{m_{ij}\}$ be a square matrix of order k with entries $\{1, \ldots, k\}$, and let ϕ be a mapping $H \to H$. Define $\theta_{M,\phi}: G \to G$ by $\theta_{M,\phi}(a_j^{-1}a_i) = a_j^{-1}a_{M_{ij}}$, and $\theta_{M,\phi}(h) = \phi(h)$ for $h \in H$. Set $\mathscr{D}_{D,H}(G) = \{\theta_{M,\phi} \mid \theta_{M,\phi} \in \text{Orth}(G)\}$.

Theorem 8.48. *$\theta_{M,\phi} \in \mathscr{D}_{D,H}(G)$ if and only if M is a Latin square with $m_{ii} = i$ for all i and $\phi \in Orth(H)$. Further, if $\theta_{M,\phi}, \theta_{K,\psi} \in \mathscr{D}_{D,H}(G)$, then $\theta_{M,\phi} \perp \theta_{K,\psi}$ if and only if M is orthogonal to K and $\phi \perp \psi$.*

Proof.

$$\theta_{M,\phi}(1) = \theta_{M,\phi}(a_i^{-1}a_i) = a_i^{-1}a_{m_{ii}}.$$

Thus $\theta_{M,\phi}$ will be well defined and will fix 1 if and only if $m_{ii} = i$ for all i, and ϕ fixes 1. $\theta_{M,\phi}$ will be a bijection if and only if ϕ is a bijection and, for all $a \notin H$, the equation $a_j^{-1}a_{m_{ij}} = a$ can be solved uniquely for i and j. But $a = a_j^{-1}a_r$ for unique values of j and r. Thus the equation $a_j^{-1}a_{m_{ij}} = a$ uniquely determines j and m_{ij}. It follows that $\theta_{M,\phi}$ can be a bijection if and only if ϕ is a bijection and each column of M contains each symbol of $\{1, \cdots, k\}$ exactly once. Similarly the mapping $x \mapsto x^{-1}\theta_{M,\phi}(x)$ will be a bijection if and only if the mapping $\eta: h \mapsto h^{-1}\phi(h)$ is a bijection $H \to H$ and, for all $a \notin H$, the equation $a_i^{-1}a_{m_{ij}} = a$ has a unique solution for i and j. The equation $a_i^{-1}a_{m_{ij}} = a$ has a unique solution for i and j if and only if each row of M contains each symbol of $\{1, \cdots, k\}$ exactly once. Hence $\theta_{M,\phi} \in \mathscr{D}_D(G)$ if and only if $\phi \in \text{Orth}(H)$ and M is a Latin square for which $m_{ii} = i$ for all i.

If $\theta_{M,\phi}, \theta_{K,\psi} \in \mathscr{D}_D(G)$, then $\theta_{M,\phi} \perp \theta_{K,\psi}$ if and only if, for all a, the equation $a_{k_{ij}}^{-1}a_{m_{ij}} = a$ can be uniquely solved for i and j and $\phi \perp \psi$, if and only if, for all r and s, the equations $a_{m_{ij}} = a_r$ and $a_{k_{ij}} = a_s$ can be uniquely solved for i and j and $\phi \perp \psi$, if and only if M and K are orthogonal and $\phi \perp \psi$. □

Recall that we defined the product of two graphs Γ_1 and Γ_2, denoted $\Gamma_1 \times \Gamma_2$, to be the graph with vertex set $V(\Gamma_1) \times V(\Gamma_2)$, (x_1, x_2) being adjacent to (y_1, y_2) if x_1 is adjacent to y_1 in Γ_1 and x_2 is adjacent to y_2 in Γ_2. The structure of $\mathscr{D}_{D,H}(G)$ can be deduced immediately from Theorem 8.48.

Corollary 8.49. *Let $\mathscr{M}_k$ denote the orthogonal Latin square graph with vertices Latin squares $M = (m_{ij})$ of order k that satisfy $m_{ii} = i$. If there exists a $(v, k, 1; H)$-difference set in a group G, then $\mathscr{D}_{D,H}(G) \cong \mathscr{M}_k \times \text{Orth}(H)$.*

Corollary 8.50. *If there exists a $(n + 1, n - 1, n, 1)$-difference set in a group G relative to H, then*

$$\min\{N(n) - 1, \omega(H)\} \leq \omega(\mathscr{D}_{D,H}(G)) \leq \min\{N(n), \omega(H)\}.$$

A special case of Corollary 8.50:

Corollary 8.51. *If n is a prime power, then $\omega(\mathscr{D}_{D,H}(G)) = \omega(H)$.*

Proof. If n is a prime power, then $N(n) = n-1$, whereas, as $|H| = n-1$, $\omega(H) \leq n-3$. Thus $\min\{N(n)-1, \omega(H)\} = \min\{N(n), \omega(H)\} = \omega(H)$. □

The question remains as to which groups Theorem 8.48 might apply to. As with planar difference sets, there is a version of the prime power conjecture for affine difference sets.

Conjecture 8.52. Any abelian affine difference set is of prime power order.

This conjecture has been proved true for all orders less than or equal to 10,000 by Jungnickel and Pott [208] in 1989. For nonabelian groups, a class of affine difference sets was constructed by Ganley and Spence [151] in 1975: their difference sets are of prime power order.

To describe Ganley and Spence's construction, let $n = p^r$, p a prime, and let $(G, +) = \mathbb{Z}_{n^2-1}$. We will modify addition in G to obtain a nonabelian group $(G, \oplus)$. Let $q = p^h$, where $hv = 2r$, v an integer, all prime factors of v divide $q-1$, and, if $q \equiv 3 \pmod 4$, then $v \not\equiv 0 \pmod 4$. Thus $n^2 = q^v$. Let $r(j)$ be the unique i (mod v) satisfying $q^i \equiv 1+j(q-1) \pmod{v(q-1)}$. Define $\oplus$ by $i \oplus j \equiv iq^{r(j)}+j \pmod{q^v-1}$. Then $(G, \oplus)$ is a nonabelian group if $v > 1$.

Theorem 8.53 (Ganley and Spence, 1975). *Let A be an affine difference set in the group $(G, +) = \mathbb{Z}_{n^2-1}$ with forbidden subgroup $(H, +)$ and suppose that $n = p^r$, p a prime. If $(H, \oplus)$ is a normal subgroup of $(G, \oplus)$ and A is fixed by $x \mapsto px$, then A is an affine difference set in $(G, \oplus)$ with forbidden subgroup $(H, \oplus)$.*

Proof. See Theorem 4.1 in [151]. □

If the conditions of Theorem 8.53 hold for $(G, +) = \mathbb{Z}_{n^2-1}$, then, by Corollary 8.51, $\omega(\mathscr{D}_{D,(H,\oplus)}((G, \oplus))) \geq \min\{N(n)-1, \omega((H, \oplus))\}$. As an example, suppose that $p = 2$ and $r = 3$. Thus $n = 8$ and $|G| = 63$. Now $hv = 6$, and the only choice of h and $v > 1$ that satisfies the condition that every prime divisor of v is a divisor of $2^h - 1$ is $h = 2$ and $v = 3$. Thus $q = 4$. To compute $r(j)$ note that $4^0 \equiv 1 + 0 \times 3 \pmod 9$, $4^1 \equiv 1 + 1 \times 3 \pmod 9$, and $4^2 \equiv 1 + 2 \times 3 \pmod 9$. Hence $r(0) = 0$, $r(1) = 1$, and $r(2) = 2$. Thus

$$i \oplus j = \begin{cases} i+j & \text{if } j \equiv 0 \pmod 3, \\ 4i+j & \text{if } j \equiv 1 \pmod 3, \\ 16i+j & \text{if } j \equiv 2 \pmod 3, \end{cases}$$

where the additions are performed modulo 63. The forbidden subgroup $(H, \oplus) = \{0, 9, 18, 27, 36, 45, 54\}$ is a normal subgroup of $(G, \oplus)$ and is isomorphic to $\mathbb{Z}_7$. The set $A = \{21, 31, 42, 47, 55, 59, 61, 62\}$ is an affine difference set in $(G, +)$ relative to $(H, +)$, and A is fixed by $x \mapsto 2x$. It follows from Theorem 8.53 that

$$\omega(\mathscr{D}_{D,(H,\oplus)}((G, \oplus))) \geq \min\{N(8)-1, \omega(\mathbb{Z}_7)\} = \min\{6, 5\} = 5.$$

8.3.3 The Orthomorphism Graphs $\mathscr{F}_D(G)$ and $\mathscr{F}_{D,H}(G)$

The concept of a difference set generalizes to the concept of a difference family. If G is a group of order v, then a $(v, \{k_1, \ldots, k_s\}, \lambda)$-*difference family* in G is a set $D = \{D_1, \ldots, D_s\}$ of subsets of the element set of G, $|D_i| = k_i$ for $i = 1, \ldots, s$, for which each nonidentity element of G is expressible in exactly λ ways in the form $d^{-1}d'$, where $d, d' \in D_i$ for some $i \in \{1, \ldots, s\}$. If $k_1 = \cdots = k_s = k$, then we may use the term (v, k, λ)-difference family in place of the term $(v, \{k, \ldots, k\}, \lambda)$-difference family: if $s = 1$ this is a (v, k, λ)-difference set.

The difference families with $\lambda = 1$ are the ones that can be used to construct orthomorphisms. Let $D = \{D_1, \ldots, D_s\}$, $D_i = \{a_{ij} \mid j = 1, \ldots, k_i\}$, be a $(v, \{k_1, \ldots, k_s\}, 1)$-difference family in a group G, and let $\mathscr{M} = \{M_1, \ldots, M_s\}$ be a family of square matrices, $M_i = (m^i_{jr})$ of order k_i with entries $\{1, \ldots, k_i\}$. Define $\theta_{\mathscr{M}} : G \to G$ by $\theta_{\mathscr{M}}(a_{ij}^{-1}a_{ir}) = a_{ij}^{-1}a_{im^i_{jr}}$. We set $\mathscr{F}_D(G) = \{\theta_{\mathscr{M}} \mid \theta_{\mathscr{M}} \in \text{Orth}(G)\}$. The following characterizes membership in $\mathscr{F}_D(G)$ and orthogonality in $\mathscr{F}_D(G)$.

Theorem 8.54. *Let D be a $(v, \{k_1, \ldots, k_s\}, 1)$-difference family in G, and let $\mathscr{M} = \{M_1, \ldots, M_s\}$, M_i a $k_i \times k_i$ matrix with entries in $\{1, \ldots, k_i\}$ for $i = 1, \ldots, s$. Then $\theta_{\mathscr{M}} \in \mathscr{F}_D(G)$ if and only if, for $i = 1, \ldots, s$, the matrix M_i is a Latin square satisfying $m^i_{jj} = j$ for $j = 1, \ldots, k_i$. Further, if $\mathscr{M} = \{M_1, \ldots, M_s\}$ and $\mathscr{K} = \{K_1, \ldots, K_s\}$ and $\theta_{\mathscr{M}}, \theta_{\mathscr{K}} \in \mathscr{F}_D(G)$, then $\theta_{\mathscr{M}} \perp \theta_{\mathscr{K}}$ if and only if M_i is orthogonal to K_i for $i = 1, \ldots, s$.*

Proof. This is a straightforward modification of the proof of Theorem 8.37. □

An immediate corollary:

Corollary 8.55. *If D is a $(v, \{k_1, \ldots, k_s\}, 1)$-difference family in G, then*

$$\min\{N(k_i) - 1 \mid i = 1, \ldots, s\} \leq \omega(\mathscr{F}_D(G)) \leq \min\{N(k_i) \mid i = 1, \ldots, s\}.$$

From Theorem 8.54 we can derive a 1981 result of Jungnickel [202] which generalizes Theorem 8.35. Note that Jungnickel's original proof is geometric.

Corollary 8.56 (Jungnickel, 1981). *If q is a prime power, $v = q^n + \cdots + q + 1$, n an even integer, and their exists a set of r MOLS of order $q + 1$, then there exists a $(v, r + 1; 1)$-difference matrix over $\mathbb{Z}_v$.*

Proof. Jungnickel [202] constructed a $(v, q + 1, 1)$-difference family in $\mathbb{Z}_v$. The result then follows from Theorem 8.54. □

We obtain a stronger result when $q + 1$ is a prime power.

Corollary 8.57 (Jungnickel, 1981). *If q and $q + 1$ are prime powers and $v = q^n + \cdots + q + 1$, n an even integer, then there exists a $(v, q + 1; 1)$-difference matrix over $\mathbb{Z}_v$.*

Proof. As $N(q+1) = q$ when $q+1$ is a prime power, the result follows from Corollary 8.56. □

In orthomorphism form, Corollaries 8.56 and 8.57 become the following:

Corollary 8.58. *If q is a prime power and $v = q^n + \cdots + q + 1$, n even, then $\omega(\mathbb{Z}_v) \geq N(q+1) - 1$. Further, if $q+1$ is a prime power, then $\omega(\mathbb{Z}_v) \geq q - 1$.*

The concept of a relative difference set generalizes to the concept of a relative difference family. Let $D = \{D_1, \ldots, D_s\}$, $|D_i| = k_i$ for $i = 1, \ldots, s$, be a family of subsets of the element set of a group G, $|G| = v$, and let H be a subgroup of G. We call D a $(v, \{k_1, \ldots, k_s\}, \lambda; H)$*-difference family* in G if each nonidentity element of $G \setminus H$ is expressible in exactly λ ways in the form $d^{-1}d'$, $d, d' \in D_i$, for some $i \in \{1, \ldots, s\}$, and no element of H can be so expressed. If $k_1 = \cdots = k_s = k$, then we use the term $(v, k, \lambda; H)$-difference family instead of the term $(v, \{k, \ldots, k\}, \lambda; H)$-difference family.

Relative difference families with $\lambda = 1$ can be used to construct of orthomorphisms. Let $D = \{D_1, \ldots, D_s\}$, $D_i = \{a_{ij} \mid j = 1, \ldots, k_i\}$, be a $(v, \{k_1, \ldots, k_s\}, 1; H)$-relative difference family in a group G, and let $\mathcal{M} = \{M_1, \ldots, M_s\}$ be a family of square matrices, $M_i = (m^i_{jr})$ of order k_i for $i = 1, \ldots, s$ with entries $\{1, \ldots, k_i\}$, and let ϕ be a mapping $H \to H$. Define $\theta_{\mathcal{M},\phi}\colon G \to G$ by $\theta_{\mathcal{M},\phi}(a_{ij}^{-1}a_{ir}) = a_{ij}^{-1}a_{im^i_{jr}}$ and $\theta_{\mathcal{M},\phi}(h) = \phi(h)$ for $h \in H$. We set $\mathcal{F}_{D,H}(G) = \{\theta_{\mathcal{M},\phi} \mid \theta_{\mathcal{M},\phi} \in \mathrm{Orth}(G)\}$. The following theorem characterizes membership in $\mathcal{F}_{D,H}(G)$ and orthogonality in $\mathcal{F}_{D,H}(G)$.

Theorem 8.59. *Let D be a $(v, \{k_1, \ldots, k_s\}, 1; H)$-difference family in G relative to H; let $\mathcal{M} = \{M_1, \ldots, M_s\}$ and $\mathcal{K} = \{K_1, \ldots, K_s\}$, where M_i and K_i are $k_i \times k_i$ matrices with entries in $\{1, \ldots, k_i\}$ for $i = 1, \ldots, s$; and let ϕ be a mapping $H \to H$. Then $\theta_{\mathcal{M},\phi} \in \mathcal{F}_{D,H}(G)$ if and only if for all i, the matrix M_i is a Latin square for which $m^i_{jj} = j$, $j = 1, \ldots, k_i$ and $\phi \in \mathrm{Orth}(H)$. Further, if $\theta_{\mathcal{M},\phi}, \theta_{\mathcal{K},\psi} \in \mathcal{F}_{D,H}(G)$, then $\theta_{\mathcal{M},\phi} \perp \theta_{\mathcal{K},\psi}$ if and only if M_i is orthogonal to K_i for $i = 1, \ldots, s$, and $\phi \perp \psi$.*

Proof. This is a straightforward modification of the proof of Theorem 8.48. □

An immediate corollary:

Corollary 8.60. *If D is a $(v, \{k_1, \ldots, k_s\}, 1; H)$-difference family in G relative to H, $|H| = m$, then*

$$\min(\{N(k_i) - 1 \mid i = 1, \ldots, s\} \cup \{\omega(\mathbb{Z}_m)\}) \leq \omega(\mathcal{F}_{D,H}(G)) \leq \min(\{N(k_i) \mid i = 1, \ldots, s\} \cup \{\omega(\mathbb{Z}_m)\}).$$

A special case is described next.

Corollary 8.61. *If q is a prime power and $v = q^n - 1$, then $\omega(\mathbb{Z}_v) \geq \omega(\mathbb{Z}_{q-1})$, and, if $q-1$ is also a prime power, then $\omega(\mathbb{Z}_v) \geq q - 3$.*

Proof. Let $G = \mathbb{Z}_v$ and let H be the unique subgroup of G of order $q - 1$. In 1981 Jungnickel [202] constructed a $(v, q, 1; H)$-difference family in G relative to H. Now $N(q) = q - 1$ and $\omega(\mathbb{Z}_{q-1}) \leq q - 3$. Thus, by Corollary 8.60, $\omega(\mathscr{F}_{D,H}(G)) = \omega(\mathbb{Z}_{q-1})$.

If $q - 1$ is a prime power, then $\omega(\mathbb{Z}_{q-1}) = q - 3$. As $\omega(\mathbb{Z}_v) \geq \omega(\mathscr{F}_{D,H}(G))$, the result follows. □

From Corollary 8.61, a 1981 result of Jungnickel [202], which generalizes Theorem 8.47, follows immediately.

Corollary 8.62 (Jungnickel, 1981). *If q is a prime power, $v = q^n - 1$, and their exists a $(q-1, r+1; 1)$-difference matrix over $\mathbb{Z}_{q-1}$, then there exists a $(v, r+1; 1)$-difference matrix over $\mathbb{Z}_v$.*

Another special case.

Corollary 8.63. *If q is a prime power and $v = q^n + \cdots + q + 1$, n odd, then $\omega(\mathbb{Z}_v) \geq \omega(\mathbb{Z}_{q+1})$, and, if $q + 1$ is also a prime power, then $\omega(\mathbb{Z}_v) \geq q - 1$.*

Proof. Let $G = \mathbb{Z}_v$ and let H be the unique subgroup of G of order $q + 1$. In 1981 Jungnickel [202] constructed a $(v, q + 1, 1; H)$-difference family in G relative to H. Now $N(q + 1) > \omega(\mathbb{Z}_{q+1})$. Thus, by Corollary 8.60, $\omega(\mathscr{F}_{D,H}(G)) = \omega(\mathbb{Z}_{q+1})$.

If $q + 1$ is a prime power, then $\omega(\mathbb{Z}_{q+1}) = q - 1$. As $\omega(\mathbb{Z}_v) \geq \omega(\mathscr{F}_{D,H}(G))$, the result follows. □

From Corollary 8.63, a 1981 result of Jungnickel [202] follows immediately.

Corollary 8.64 (Jungnickel, 1981). *If q is a prime power, $v = q^n + \cdots + q + 1$, n an odd integer, and their exists a $(q + 1, r + 1; 1)$-difference matrix over $\mathbb{Z}_{q+1}$, then there exists a $(v, r + 1; 1)$-difference matrix over $\mathbb{Z}_v$.*

8.4 Orthomorphism Graphs from Automorphisms

We now turn our attention to the construction of orthomorphisms from automorphisms of groups. We will describe two classes of orthomorphisms obtained from automorphisms of a group. The first of these classes, $\mathscr{A}(G)$, consists of those automorphisms of G that are also orthomorphisms of G: the second, $\mathscr{A}^-(G)$, consists of mappings of the form $x \mapsto \alpha(x)^{-1}$, $\alpha \in \mathrm{Aut}(G)$. The construction of MOLS based on a group G from automorphisms of G was pioneered by Mann [237] in 1942: this is known as *Mann's automorphism method*.

8.4.1 Fixed-Point-Free Automorphisms

An automorphism α of a group G is *fixed-point-free* if $\alpha(x) = x$ implies that $x = 1$. As an example, let $G = \mathrm{GF}(q)^+$, $q = p^n$, p prime. Viewed as an n-dimensional vector space over $\mathrm{GF}(p)$, the automorphisms of G are the nonsingular linear transformations, and the fixed-point-free automorphisms of G are the nonsingular linear transformations for which 1 is not an eigenvalue. If $\alpha \in \mathrm{Aut}(G)$ is fixed-point-free, then clearly $x^{-1}\alpha(x) = 1$ implies that $x = 1$ and $\alpha(x)x^{-1} = 1$ implies that $x = 1$, but a stronger result holds.

Theorem 8.65. *If $\alpha \in \mathrm{Aut}(G)$ is fixed-point-free, then the mappings $x \mapsto x^{-1}\alpha(x)$ and $x \mapsto \alpha(x)x^{-1}$ are bijections $G \to G$.*

Proof. Let α be a fixed-point-free automorphism of G. As α is an automorphism, $x^{-1}\alpha(x) = y^{-1}\alpha(y)$ if and only if $\alpha(xy^{-1}) = xy^{-1}$. But then, as α is fixed-point-free, $\alpha(xy^{-1}) = xy^{-1}$ if and only if $xy^{-1} = 1$, which is true if and only if $x = y$. Hence the mapping $x \mapsto x^{-1}\alpha(x)$ is one to one and so, by finiteness, is a bijection.

The proof that the mapping $x \mapsto \alpha(x)x^{-1}$ is a bijection is similar. □

The characterization of automorphisms that are orthomorphisms follows.

Corollary 8.66. *If $\alpha \in \mathrm{Aut}(G)$, then $\alpha \in \mathrm{Orth}(G)$ if and only if α is fixed-point-free.*

Proof. If $\alpha \in \mathrm{Aut}(G)$ is an orthomorphism, then it must be fixed-point-free, as otherwise, for some $a \neq 1$, $a^{-1}\alpha(a) = a^{-1}a = 1 = 1^{-1}\alpha(1)$.

If $\alpha \in \mathrm{Aut}(G)$ is fixed-point-free, then, by Theorem 8.65, the mapping $x \mapsto x^{-1}\alpha(x)$ is a bijection. As α is a bijection, it follows that α is also an orthomorphism. □

Other properties of fixed-point-free automorphisms follow from Theorem 8.65.

Corollary 8.67. *If $\alpha \in \mathrm{Aut}(G)$ is fixed-point-free and $\alpha(x) = g^{-1}xg$ for some g, then $x = 1$.*

Proof. Let $\alpha \in \mathrm{Aut}(G)$ be fixed-point-free and suppose that $\alpha(x) = g^{-1}xg$ for some g. By Theorem 8.65, there exists $h \in G$ satisfying $h^{-1}\alpha(h) = g$. It follows that $\alpha(x) = g^{-1}xg = \alpha(h)^{-1}hxh^{-1}\alpha(h)$. Hence $\alpha(hxh^{-1}) = hxh^{-1}$, and, as α is fixed-point-free, we conclude that $hxh^{-1} = 1$, from which it follows that $x = 1$. □

Corollary 8.68. *Let G be a group of order n. If $\alpha \in \mathrm{Aut}(G)$ is fixed-point-free, then*

$$x\alpha(x)\alpha^2(x)\cdots\alpha^{n-1}(x) = 1 \text{ and } \alpha(x)^{n-1}\cdots\alpha^2(x)\alpha^1(x)x = 1.$$

Proof. Let α be a fixed-point-free automorphism of G and let $x \in G$. Note that, as $|G| = n$, $\alpha^n(y) = y$ for any $y \in G$. By Theorem 8.65, there exists $y \in G$ satisfying $y^{-1}\alpha(y) = x$. Thus

$$
\begin{aligned}
x\alpha(x)\alpha^2(x)\cdots\alpha^{n-1}(x) &= (y^{-1}\alpha(y))\alpha(y^{-1}\alpha(y))\alpha^2(y^{-1}\alpha(y))\cdots\alpha^{n-1}(y^{-1}\alpha(y))\\
&= y^{-1}\alpha(y)\alpha(y)^{-1}\alpha^2(y)\cdots\alpha^{n-1}(y)^{-1}\alpha^n(y)\\
&= y^{-1}\alpha^n(y)\\
&= 1.
\end{aligned}
$$

Similarly $\alpha(x)^{n-1}\cdots\alpha^2(x)\alpha^1(x)x = 1$. □

Corollary 8.68 should be compared with Theorem 3.25: in this theorem, the existence of a complete mapping is used to prove that the product of all of the elements of the group, in some order, is the identity. If α is fixed-point-free, then, by Corollary 8.66, α is an orthomorphism, and so the mapping $x \mapsto x^{-1}\alpha(x)$ is a complete mapping. While the product in Corollary 8.68 need not contain all the elements of G, the proof of Corollary 8.68 is very similar to the proof of Theorem 3.25.

If H is a normal subgroup of G, a fixed-point-free automorphism of G that leaves H invariant induces a fixed-point-free automorphism of G/H.

Theorem 8.69. *Let $\alpha \in \mathrm{Aut}(G)$ be fixed-point-free. If H is a normal subgroup of G for which $\alpha(H) = H$, then the mapping $\alpha'\colon gH \mapsto \alpha(g)H$ is a fixed-point-free automorphism of G/H.*

Proof. α' is well-defined as if $gH = kH$, then $k = gh$ for some $h \in H$ and $\alpha'(kH) = \alpha(g)\alpha(h)H = \alpha(g)H = \alpha'(gH)$. As α is onto, α' must also be onto, and hence, by finiteness, α' must be a bijection. Further, $\alpha'(gHkH) = \alpha'(gkH) = \alpha(g)\alpha(k)H = \alpha(g)H\alpha(k)H = \alpha'(gH)\alpha'(kH)$, and hence $\alpha' \in \mathrm{Aut}(G/H)$.

To see that α' is fixed-point-free, assume that $\alpha'(gH) = gH$ for some $g \in G$. Thus $\alpha(g)H = gH$, from which it follows that $g^{-1}\alpha(g) \in H$. By Theorem 8.65, the mapping $\eta\colon x \mapsto x^{-1}\alpha(x)$ is a permutation of G, and as the restriction of α to H is a fixed-point-free automorphism of H, by Theorem 8.65, the restriction of η to H is a permutation of H. Similarly, the restriction of η to $G \setminus H$ is a permutation of $G \setminus H$. Hence $g \in H$ and so $gH = H$. We thus conclude that α' is a fixed-point-free automorphism of G/H. □

The mapping $\alpha'\colon G/H \to G/H$, described in Theorem 8.69, is the fixed-point-free automorphism of G/H *induced* by α. As a corollary:

Corollary 8.70. *If H is a characteristic subgroup of G and G admits fixed-point-free automorphisms, then H and G/H also admit fixed-point-free automorphisms.*

Proof. Let $\alpha \in \mathrm{Aut}(G)$ be fixed-point-free. If H is a characteristic subgroup of G, then H is a normal subgroup of G and $\alpha(H) = H$. Hence, by Theorem 8.69, α induces a fixed-point-free automorphism of G/H. Further, $\alpha|_H$ is a fixed-point-free automorphism of H. □

Much work has gone into the study of groups that admit fixed-point-free automorphisms. In a celebrated result, Thompson [310] proved that a group that admits a fixed-point-free automorphism of prime order must be nilpotent, thus settling a long-standing conjecture. More generally, it was conjectured that a group that admits a fixed-point-free automorphism must be solvable. As a consequence of Corollary 8.70, to prove the conjecture that any group admitting fixed-point-free automorphisms must be solvable, it is sufficient to prove that the only characteristically simple groups that admit fixed-point-free automorphisms are elementary abelian groups. The characteristically simple groups are characterized as direct products of isomorphic simple groups: see Theorem 10.2.1 in [157] for a proof. With the completion of the classification of finite simple groups, all characteristically simple groups became known, yielding a proof of the conjecture.

Theorem 8.71. *If G admits fixed-point-free automorphisms, then G is solvable.*

For more information on properties of fixed-point-free automorphisms, consult [157] or [191].

8.4.2 The Orthomorphism Graph $\mathscr{A}(G)$

Let us define $\mathscr{A}(G)$ to be the orthomorphism graph of G consisting of the fixed-point-free automorphisms of G: thus $\mathscr{A}(G) = \mathrm{Aut}(G) \cap \mathrm{Orth}(G)$. In Corollary 8.66, we characterized the elements of $\mathscr{A}(G)$. In the next theorem, we characterize orthogonality in $\mathscr{A}(G)$.

Theorem 8.72. *If $\alpha, \beta \in \mathscr{A}(G)$, then $\alpha \perp \beta$ if and only if $\beta\alpha^{-1}$ is fixed-point-free if and only if $\beta^{-1}\alpha$ is fixed-point-free.*

Proof. By Theorem 1.20, $\alpha \perp \beta$ if and only if $\beta\alpha^{-1}$ is an orthomorphism of G, and by Corollary 8.66, this is true if and only if $\beta\alpha^{-1}$ is fixed-point-free.

$\beta\alpha^{-1}$ is fixed-point-free if and only if $\beta\alpha^{-1}(x) = x$ implies that $x = 1$, if and only if $\alpha^{-1}(x) = \beta^{-1}(x)$ implies that $x = 1$, if and only if $\alpha^{-1}(\beta(x)) = \beta^{-1}(\beta(x))$ implies that $\beta(x) = 1$, if and only if $\alpha^{-1}\beta(x) = x$ implies that $x = 1$, if and only if $\beta\alpha^{-1}$ is fixed-point-free. □

A subgroup $\mathscr{H}$ of $\mathrm{Aut}(G)$ is a *group of fixed-point-free automorphisms* of G if each nonidentity element of $\mathscr{H}$ is fixed-point-free. Note that what we have called a group of fixed-point-free automorphisms is called a fixed-point-free automorphism group in [191]: this latter term is used in [157] to denote a subgroup $\mathscr{H}$ of $\mathrm{Aut}(G)$ for which $C_{\mathscr{H}}(G) = \{1\}$. Groups of fixed-point-free automorphisms provide cliques in $\mathscr{A}(G)$.

Corollary 8.73. *If $\mathscr{H}$ is a group of fixed-point-free automorphisms of G, then the nonidentity elements of $\mathscr{H}$ form a clique in $\mathscr{A}(G)$.*

Proof. If α and β are distinct nonidentity elements of $\mathscr{H}$, then $\alpha\beta^{-1}$ is a non-identity element of $\mathscr{H}$. By Theorem 8.72, $\alpha \perp \beta$, hence the result. □

Elementary abelian groups form a special case.

Corollary 8.74. *If q is a prime power, then $\omega(\mathscr{A}(\mathrm{GF}(q)^+)) = q-2$.*

Proof. The mappings $x \mapsto ax$, $a \neq 0$, form a group of fixed-point-free automorphisms of $\mathrm{GF}(q)^+$ of order $q-1$. Hence, by Corollary 8.73, $\omega(\mathscr{A}(\mathrm{GF}(q)^+)) \geq q-2$, and by Corollary 1.23, $\omega(\mathscr{A}(\mathrm{GF}(q)^+)) \leq q-2$, hence the result. □

To determine bounds for $\omega(\mathscr{A}(G))$, we need an analogue of Theorem 8.69 that deals with orthogonality.

Theorem 8.75. *Let H be a normal subgroup of G, and let $\alpha, \beta \in \mathscr{A}(G)$, satisfy $\alpha(H) = \beta(H) = H$. If $\alpha \perp \beta$ and α' and β' are the corresponding induced fixed-point-free automorphisms of G/H, then $\alpha' \perp \beta'$.*

Proof. If $\alpha \perp \beta$, then $\alpha\beta^{-1}$ is fixed-point-free by Theorem 8.72. Hence $\alpha\beta^{-1} \in \mathscr{A}(G)$ by Corollary 8.66. Hence $\alpha\beta^{-1}$ induces a fixed-point-free automorphism γ of G/H by Theorem 8.69. Now

$$\gamma(gH) = \alpha\beta^{-1}(g)H = \alpha'(\beta^{-1}(g)H) = \alpha'(\beta^{-1})'(gH),$$

from which it follows that $\gamma = \alpha'(\beta^{-1})'$. But $(\beta^{-1})' = \beta'^{-1}$ as $\beta'(\beta^{-1})'(gH) = \beta\beta^{-1}(g)H = gH$. Hence $\gamma = \alpha'\beta'^{-1}$ and thus $\alpha' \perp \beta'$ by Theorem 8.72. □

Theorem 8.75 yields bounds on $\omega(\mathscr{A}(G))$.

Corollary 8.76. *If H is a characteristic subgroup of G, then*

$$\omega(\mathscr{A}(G)) \leq \min\{\omega(\mathscr{A}(H)), \omega(\mathscr{A}(G/H))\}.$$

Proof. As H is a characteristic subgroup of G, $\alpha(H) = H$ for all $\alpha \in \mathrm{Aut}(G)$, any r-clique in $\mathscr{A}(G)$ induces an r-clique in $\mathscr{A}(H)$, and so $\omega(\mathscr{A}(G)) \leq \omega(\mathscr{A}(H))$. Further, by Theorem 8.75, from any r-clique in $\mathscr{A}(G)$, we can construct an r-clique in $\mathscr{A}(G/H)$, and so $\omega(\mathscr{A}(G)) \leq \omega(\mathscr{A}(G/H))$, hence the result. □

We can improve on Corollary 8.76 when the group is a direct product.

Theorem 8.77. $\omega(\mathscr{A}(G \times H)) \geq \min\{\omega(\mathscr{A}(H)), \omega(\mathscr{A}(G))\}$. *If $|G|$ and $|H|$ are relatively prime, then $\omega(\mathscr{A}(G \times H)) = \min\{\omega(\mathscr{A}(H)), \omega(\mathscr{A}(G))\}$.*

Proof. Let $r = \min\{\omega(\mathscr{A}(H)), \omega(\mathscr{A}(G))\}$, and let $\alpha_1, \ldots, \alpha_r$ be an r-clique of $\mathscr{A}(G)$ and $\beta_1, \ldots, \beta_r$ an r-clique of $\mathscr{A}(H)$. The mappings $(\alpha_i, \beta_i)\colon G \times H \to G \times H$, $i = 1, \ldots, r$, defined by $(\alpha_i, \beta_i)(g, h) = (\alpha_i(g), \beta_i(h))$, are pairwise orthogonal fixed-point-free automorphisms of $G \times H$. Hence

$$\omega(\mathscr{A}(G \times H)) \geq \min\{\omega(\mathscr{A}(H)), \omega(\mathscr{A}(G))\} = r.$$

If $|G|$ and $|H|$ are relatively prime, then G and H are characteristic subgroups of $G \times H$ and

$$\omega(\mathscr{A}(G \times H)) \leq \min\{\omega(\mathscr{A}(H)), \omega(\mathscr{A}(G))\} = r$$

by Corollary 8.76. The result follows. □

Recall the following definitions from Chapter 6. If $g, h \in G$, then the *commutator* of g and h is $[g, h] = ghg^{-1}h^{-1}$, and if H and K are subgroups of G, then $[H, K]$ is the subgroup of G generated by the commutators $[h, k]$, $h \in H$, $k \in K$. Let us define a sequence of subgroups $G_1 = G$ and $G_i = [G_{i-1}, G]$. A group G is *nilpotent* if $G_i = \{1\}$ for some i. The main properties of nilpotent groups are described in a number of group theory books (see, e.g., [157]). The most important characterization of nilpotent groups for our purposes is as direct products.

Lemma 8.78. *A group is nilpotent if and only if it is a direct product of its Sylow subgroups.*

Proof. See Theorem 2.3.5 in [157], for instance. □

Corollary 8.79. *If G is nilpotent, then*

$$\omega(\mathscr{A}(G)) = \min\{\omega(\mathscr{A}(H)) \mid H \text{ a Sylow } p\text{-subgroup of } G,\ p||G|\}.$$

Proof. By Lemma 8.78, if G is a nilpotent group, then G is a direct product of its Sylow subgroups. The result follows from Theorem 8.77. □

For G abelian, the exact value of $\omega(\mathscr{A}(G))$ was determined by Bailey and Jungnickel [25] in 1990. By Theorem 8.77, to determine $\omega(\mathscr{A}(G))$ for abelian groups, we need only to determine $\omega(\mathscr{A}(G))$ for abelian p-groups. We will use $n\mathbb{Z}_q$ to denote the direct product of n copies of $\mathbb{Z}_q$. Any abelian p-group, p a prime, can be uniquely written in the form $m_k\mathbb{Z}_{q_k} \times \cdots \times m_1\mathbb{Z}_{q_1}$, where $q_1 < \cdots < q_k$ are powers of p. Thus a first step toward determining $\omega(\mathscr{A}(G))$, for G abelian, is to determine $\omega(\mathscr{A}(m\mathbb{Z}_q))$ when q is a prime power: this was determined by Jungnickel [205] in 1989.

Lemma 8.80 (Jungnickel, 1989). *If q is a power of a prime p, then*

$$\omega(\mathscr{A}(m\mathbb{Z}_q)) = p^m - 2.$$

Proof. Let q be a power of a prime p. If $G = m\mathbb{Z}_q$ is written additively, then $H = \{g \mid g \in G, pg = 0\}$ is a characteristic subgroup of G, and $H \cong m\mathbb{Z}_p \cong \mathrm{GF}(p^m)^+$. By Corollarys 8.76 and 8.74,

$$\omega(\mathscr{A}(G)) \leq \omega(\mathscr{A}(H)) = p^m - 2.$$

Let us view G as a set of m-vectors with entries in $\mathbb{Z}_q$ and H as a field of order p^m. Let $\mathbf{0}$ and $\mathbf{1}$ denote the additive and multiplicative identities of H, respectively. For $\mathbf{a} \in H$, the mapping $\alpha_{\mathbf{a}}\colon \mathbf{x} \mapsto \mathbf{ax}$ is a linear transformation of H, regarded as a vector space over $\mathbb{Z}_p$: let $A_{\mathbf{a}}$ denote the matrix of $\alpha_{\mathbf{a}}$ with respect to the standard basis. For each $\mathbf{a} \in H$, the entries of $A_{\mathbf{a}}$ are integers, and if $\mathbf{a} \neq \mathbf{0}$, then $A_{\mathbf{a}}$ is nonsingular. Similarly, if $\mathbf{a} \neq \mathbf{b}$, then $A_{\mathbf{a}} - A_{\mathbf{b}}$ is nonsingular. It follows that $\det(A_{\mathbf{a}})$ is not divisible by p if $\mathbf{a} \neq \mathbf{0}$ and $\det(A_{\mathbf{a}} - A_{\mathbf{b}})$ is not divisible by p if $\mathbf{a} \neq \mathbf{b}$. Thus $\det(A_{\mathbf{a}})$ is a unit in $\mathbb{Z}_q$ if $\mathbf{a} \neq \mathbf{0}$ and $\det(A_{\mathbf{a}} - A_{\mathbf{b}})$ is a unit in $\mathbb{Z}_q$ if $\mathbf{a} \neq \mathbf{b}$. For

$\mathbf{a} \in H$, let $\beta_{\mathbf{a}}$ be the linear transformation of $m\mathbb{Z}_q$ with matrix $A_{\mathbf{a}}$ with respect to the standard basis. If $\mathbf{a} \neq \mathbf{0}, \mathbf{1}$, then $\beta_{\mathbf{a}}$ is an automorphism of $\mathbb{Z}_q$, and as $\beta_{\mathbf{a}}(\mathbf{x}) = \beta_{\mathbf{1}}(\mathbf{x}) = \mathbf{x}$ if and only if $\mathbf{x} = \mathbf{0}$, $\beta_{\mathbf{a}}$ is fixed-point-free. Similarly, if $\mathbf{a} \neq \mathbf{b}$ and $\mathbf{a}, \mathbf{b} \neq \mathbf{0}, \mathbf{1}$, then $\beta_{\mathbf{a}}(\mathbf{x}) = \beta_{\mathbf{b}}(\mathbf{x})$ if and only if $\mathbf{x} = \mathbf{0}$. Thus $\beta_{\mathbf{b}}^{-1}\beta_{\mathbf{a}}$ is fixed-point-free and hence, by Theorem 8.72, $\beta_{\mathbf{b}} \perp \beta_{\mathbf{a}}$. It follows that $\{\beta_{\mathbf{a}} \mid \mathbf{a} \in H \setminus \{\mathbf{0}, \mathbf{1}\}\}$ is a $(p^m - 2)$-clique of $\mathscr{A}(m\mathbb{Z}_q)$. Hence $\omega(\mathscr{A}(G)) \geq p^m - 2$, and the result follows. □

We are now in a position to determine $\omega(\mathscr{A}(G))$ for G an abelian p-group.

Theorem 8.81 (Bailey and Jungnickel, 1990). *If $m = \min\{m_i \mid i = 1, \ldots, k\}$, $G = m_k\mathbb{Z}_{q_k} \times \cdots \times m_1\mathbb{Z}_{q_1}$, and $q_1 < \cdots < q_k$ are powers of a prime p, then $\omega(\mathscr{A}(G)) = p^m - 2$.*

Proof. By Theorem 8.77 and Lemma 8.80,

$$\omega(\mathscr{A}(G)) \geq \min\{\omega(\mathscr{A}(m_k\mathbb{Z}_{q_k})), \ldots, \omega(\mathscr{A}(m_1\mathbb{Z}_{q_1}))\} = p^m - 2.$$

To establish equality, set $H_1 = \{g \mid g \in G, g^p = 1\}$, $H_i = \{h \mid h \in H_1, g^{q_{i-1}} = h \text{ for some } g \in G\}$, $i = 2, \ldots, k$, and $H_{k+1} = \{1\}$. Then $H_{i+1} \subseteq H_i$ for $i = 1, \ldots, k$, H_i is a characteristic subgroup of G for all i, and $|H_i| = p^{m_k + \cdots + m_i}$ for $i = 1, \ldots, k$. Thus $|H_i/H_{i+1}| = p^{m_i}$ for $i = 1, \ldots, k$. Thus, by Corollary 8.76, $\omega(\mathscr{A}(G)) \leq p^m - 2$, hence the result. □

We are now able to determine $\omega(\mathscr{A}(G))$ for any abelian group G.

Theorem 8.82 (Bailey and Jungnickel, 1990). *Let G be an abelian group, p a prime divisor of $|G|$, and S_p the Sylow p-subgroup of G. Suppose that $S_p = m_k\mathbb{Z}_{q_k} \times \cdots \times m_1\mathbb{Z}_{q_1}$, $q_1 < \cdots < q_k$ powers of p, and $m_p = \min\{m_i \mid i = 1, \ldots, k\}$. Then $\omega(\mathscr{A}(G)) = \min\{p^{m_p} - 2 \mid p \text{ divides } |G|\}$.*

Proof. By Theorem 8.81, $\omega(\mathscr{A}(S_p)) = p^{m_p} - 2$. As G is a direct product of its Sylow p-subgroups, the result follows from Theorem 8.77. □

8.4.3 The Orthomorphism Graphs $\mathscr{A}^-(G)$ and $\mathscr{A}(G) \cup \mathscr{A}^-(G)$

The fixed-point-free automorphisms are not the only classes of orthomorphisms that we can construct from automorphisms. For $\alpha \in \mathrm{Aut}(G)$, let us define $\alpha^- : G \to G$ by $\alpha^-(x) = \alpha(x)^{-1}$, and set $\mathscr{A}^-(G) = \{\alpha^- \mid \alpha \in \mathrm{Aut}(G), \alpha^- \in \mathrm{Orth}(G)\}$. If G is abelian, then $\mathscr{A}^-(G) = \mathscr{A}(G)$. Thus the orthomorphism graph $\mathscr{A}^-(G)$ is only of interest if G is nonabelian.

Theorem 8.83. *If $\alpha \in \mathrm{Aut}(G)$, then α^- is an orthomorphism of G if and only if $\alpha(x) = y^{-1}x^{-1}y$, for some y, implies that $x = 1$, if and only if α is a complete mapping of G.*

Proof. α^- is an orthomorphism if and only if $x^{-1}\alpha^-(x) = y^{-1}\alpha^-(y)$ implies that $x = y$, if and only if $x^{-1}\alpha(x^{-1}) = y^{-1}\alpha(y^{-1})$ implies that $x = y$, if and only if α is a complete mapping. Now $x\alpha(x) = y\alpha(y)$ if and only if $\alpha(xy^{-1}) = x^{-1}y$, if and only if $\alpha(z) = y^{-1}z^{-1}y$, where $z = xy^{-1}$. Thus $\alpha \in \text{Aut}(G)$ is a complete mapping if and only if $\alpha(x) = y^{-1}x^{-1}y$, for some y, implies that $x = 1$. □

By analogy with Corollary 8.66, we might expect that the condition, $\alpha(x) = y^{-1}x^{-1}y$ for some y implies that $x = 1$, in Theorem 8.83, could be replaced by the condition, $\alpha(x) = x^{-1}$ implies that $x = 1$. That this is not the case can be illustrated by a small example: let $G = \langle a, b \mid a^3 = b^2 = 1, ab = ba^2\rangle$, the dihedral group of order 6, and define $\alpha\colon G \to G$ by

$$\alpha(a^i b^j) = \begin{cases} a^i & \text{if } j = 0, \\ a^{i+2}b & \text{if } j = 1. \end{cases}$$

It is easily verified that α is an automorphism and $\alpha(x) = x^{-1}$ implies that $x = 1$. However α^- cannot be an orthomorphism as G admits no orthomorphisms by Theorem 3.34: the Sylow 2-subgroup of G is nontrivial and cyclic. For G the condition, $\alpha(x) = y^{-1}x^{-1}y$ for some y implies that $x = 1$ is not satisfied as $\alpha(a) = b^{-1}a^{-1}b$.

Determining orthogonality between elements of $\mathscr{A}^-(G) \cup \mathscr{A}(G)$ is straightforward.

Theorem 8.84. *If $\alpha, \beta \in \text{Aut}(G)$, then $\alpha^- \perp \beta^-$ if and only if $\alpha \perp \beta$ and $\alpha^- \perp \beta$ if and only if $\alpha\beta^{-1}$ is a complete mapping.*

Proof. $\alpha^- \perp \beta^-$ if and only if the mapping $x \mapsto \alpha^-(x)^{-1}\beta^-(x) = \alpha(x^{-1})^{-1}\beta(x^{-1})$ is a bijection, if and only if $\alpha \perp \beta$.

$\alpha^- \perp \beta$ if and only if the mapping

$$x \mapsto \beta(x)^{-1}\alpha^-(x) = \beta(x^{-1})\alpha(x^{-1}) = \beta(x^{-1})\alpha\beta^{-1}(\beta(x^{-1}))$$

is a bijection, if and only if $\alpha\beta^{-1}$ is a complete mapping. □

In what follows, we will use τ_g to denote the inner automorphism $x \mapsto gxg^{-1}$. Note that $(\tau_g\alpha)^-(x) = (\tau_g\alpha)(x^{-1}) = \tau_g\alpha^-(x)$: thus $(\tau_g\alpha)^- = \tau_g\alpha^-$. Similarly $(\alpha\tau_g)^- = \alpha^-\tau_g$.

Lemma 8.85. *If $\alpha, \beta \in \text{Aut}(G)$ and $g, h \in G$, then the following hold.*

1. $\tau_g\alpha \in \text{Orth}(G)$ *if and only if* $\alpha \in \text{Orth}(G)$.
2. $\alpha\tau_g \in \text{Orth}(G)$ *if and only if* $\alpha \in \text{Orth}(G)$.
3. $\tau_g\alpha^- \in \text{Orth}(G)$ *if and only if* $\alpha^- \in \text{Orth}(G)$.
4. $\alpha^-\tau_g \in \text{Orth}(G)$ *if and only if* $\alpha^- \in \text{Orth}(G)$.
5. $\tau_g\alpha \perp \tau_h\beta$ *if and only if* $\alpha \perp \beta$.
6. $\tau_g\alpha^- \perp \tau_h\beta^-$ *if and only if* $\alpha^- \perp \beta^-$.
7. $\tau_g\alpha^- \perp \tau_h\beta$ *if and only if* $\alpha^- \perp \beta$.

8. $\alpha \not\perp \tau_g\alpha$.
9. $\alpha^- \not\perp \tau_g\alpha^-$.

Proof.

1. Let us assume that $\alpha \in \mathrm{Orth}(G)$. By Corollary 8.66, α is fixed-point-free. Now, if $\tau_g\alpha(x) = x$, then $\alpha(x) = g^{-1}xg$, which, by Corollary 8.67, implies that $x = 1$. Hence $\tau_g\alpha$ is fixed-point-free, and so $\tau_g\alpha \in \mathrm{Orth}(G)$ by Corollary 8.66.
   To prove the converse, assume that $\tau_g\alpha \in \mathrm{Orth}(G)$. It follows, by the above, that $\tau_{g^{-1}}\tau_g\alpha = \alpha \in \mathrm{Orth}(G)$.
2. Let us assume that $\alpha \in \mathrm{Orth}(G)$. By Corollary 8.66, α is fixed-point-free. Now, if $\alpha(x)\tau_g = x$, then $\alpha(g^{-1}xg) = x = g(g^{-1}xg)g^{-1}$, which, by Corollary 8.67, implies that $g^{-1}xg = 1$, and hence $x = 1$. Hence $\alpha\tau_g$ is fixed-point-free, and so $\alpha\tau_g \in \mathrm{Orth}(G)$ by Corollary 8.66.
   To prove the converse, assume that $\alpha\tau_g \in \mathrm{Orth}(G)$. It follows, by the above, that $\alpha\tau_g\tau_{g^{-1}} = \alpha \in \mathrm{Orth}(G)$.
3. Let us assume that $\alpha^- \in \mathrm{Orth}(G)$. Thus, by Theorem 8.83, $\alpha(x) = y^{-1}x^{-1}y$ implies that $x = 1$. If $\tau_g\alpha(x) = y^{-1}x^{-1}y$, then $\alpha(x) = (yg)^{-1}x^{-1}(yg)$, and so $x = 1$. Hence $(\tau_g\alpha)^- = \tau_g\alpha^- \in \mathrm{Orth}(G)$ by Theorem 8.83.
   To prove the converse, assume that $\tau_g\alpha^- \in \mathrm{Orth}(G)$. It follows, by the above, that $\tau_{g^{-1}}\tau_g\alpha^- = \alpha^- \in \mathrm{Orth}(G)$.
4. Let us assume that $\alpha^- \in \mathrm{Orth}(G)$. Thus, by Theorem 8.83, $\alpha(x) = y^{-1}x^{-1}y$ implies that $x = 1$. If $\alpha\tau_g(x) = y^{-1}x^{-1}y$, then $\alpha(gxg^{-1}) = (gy)^{-1}(gx^{-1}g^{-1})(gy)$, and so $x = 1$. Hence $(\alpha\tau_g)^- = \alpha^-\tau_g \in \mathrm{Orth}(G)$ by Theorem 8.83.
   To prove the converse, assume that $\alpha^{-\tau_g} \in \mathrm{Orth}(G)$. It follows, by the above, that $\alpha^-\tau_g\tau_{g^{-1}} = \alpha^- \in \mathrm{Orth}(G)$.
5. By Theorem 8.72, $\tau_g\alpha \perp \tau_h\beta$ if and only if $\tau_h\beta(\tau_g\alpha)^{-1} = \tau_h(\beta\alpha^{-1})\tau_{g^{-1}}$ is fixed-point-free, if and only if, by Corollary 8.66, $\tau_h(\beta\alpha^{-1})\tau_{g^{-1}} \in \mathrm{Orth}(G)$, if and only if, by (2), $\tau_h(\beta\alpha^{-1})\tau_{g^{-1}}\tau_{g^{-1}} = \tau_h(\beta\alpha^{-1}) \in \mathrm{Orth}(G)$, if and only if, by (1), $\tau_h^{-1}\tau_h(\beta\alpha^{-1}) = \beta\alpha^{-1} \in \mathrm{Orth}(G)$, if and only if, by Corollary 8.66, $\beta\alpha^{-1}$ is fixed-point-free, if and only if, by Theorem 8.72, $\alpha \perp \beta$.
6. This is an immediate consequence of (5) and Theorem 8.84.
7. By Theorem 8.84, $\tau_g\alpha^- \perp \tau_h\beta$ if and only if $\tau_g\alpha^-(\tau_h\beta)^{-1}$ is a complete mapping of G. By Theorem 8.83, $\tau_g\alpha^-(\tau_h\beta)^{-1}$ is a complete mapping of G if and only if $\tau_g\alpha^-\beta^{-1}\tau_{h^{-1}}(x) = y^{-1}x^{-1}y$ implies $x = 1$, if and only if $\alpha^-\beta^{-1}(h^{-1}xh) = (h^{-1}yg)^{-1}h^{-1}x^{-1}h(h^{-1}yg)$ implies $x = 1$, if and only if, by Theorem 8.83, $\alpha\beta^{-1}$ is a complete mapping of G, if and only if, by Theorem 8.84, $\alpha^- \perp \beta$.
8. $\alpha \perp \tau_g\alpha$ if and only if $\tau_g\alpha\alpha^{-1} = \tau_g$ is a fixed-point-free automorphism of G, which is impossible as $\tau_g(g) = g$.
9. This is an immediate consequence of (8) and Theorem 8.84. □

Theorem 8.86 (Evans, 1989). *Let H be a nontrivial subgroup of G, and let $\mathscr{H} = \{gHg^{-1} \mid g \in G\}$. Let $A_{\mathscr{H}} = \{\alpha \mid \alpha$ or $\alpha^- \in \mathrm{Aut}(G), \alpha(\mathscr{H}) = \mathscr{H}\}$, and let*

$A_H = \{\alpha|_H \mid \alpha$ *or* $\alpha^- \in \mathrm{Aut}(G), \alpha(H) = H\}$. *Then*

$$\omega(A_{\mathscr{H}} \cap \mathrm{Orth}(G)) \leq \omega(A_H \cap \mathrm{Orth}(H)).$$

Proof. If $\alpha_1, \ldots, \alpha_n$ is a clique of $A_{\mathscr{H}} \cap \mathrm{Orth}(G)$, then $\alpha_i(H) = g_i^{-1}Hg_i$, for some $g_i \in G$. Set $\beta_i = \tau_{g_i}\alpha_i$. Then, by Lemma 8.85, $\beta_1, \ldots, \beta_n$ is a clique of $A_{\mathscr{H}} \cap \mathrm{Orth}(G)$. Further $\beta_i(H) = g_i g_i^{-1} H g_i g_i^{-1} = H$, and hence $\beta_1|_H, \ldots, \beta_n|_H$ is a clique of $A_H \cap \mathrm{Orth}(H)$, from which the result follows. □

Several upper bounds on $\omega(\mathscr{A}(G) \cup \mathscr{A}^-(G))$ can be derived from Theorem 8.86 by choosing the right subgroup H.

Corollary 8.87. *If* $p||G|$ *and* H *is a Sylow* p*-subgroup of* G*, then*

$$\omega(\mathscr{A}(G) \cup \mathscr{A}^-(G)) \leq \omega(\mathscr{A}(H) \cup \mathscr{A}^-(H)).$$

Proof. Let $p||G|$ and let H be a Sylow p-subgroup of G. Then $\mathscr{H}$ is the set of Sylow p-subgroups of G. If α or $\alpha^- \in \mathrm{Aut}(G)$, then $\alpha(\mathscr{H}) = \mathscr{H}$. Thus $A_{\mathscr{H}} \cap \mathrm{Orth}(G) = \mathscr{A}(G) \cup \mathscr{A}^-(G)$. As $A_H \cap \mathrm{Orth}(G) = \mathscr{A}(H) \cup \mathscr{A}^-(H)$, the result follows from Theorem 8.86. □

Corollary 8.88. *If* $Z(G)$ *is nontrivial, then*

$$\omega(\mathscr{A}(G) \cup \mathscr{A}^-(G)) \leq \omega(\mathscr{A}(Z(G)) \cup \mathscr{A}^-(Z(G))).$$

Proof. Let $H = Z(G)$. Then $\mathscr{H} = \{Z(G)\}$, and if α or $\alpha^- \in \mathrm{Aut}(G)$, then $\alpha(H) = H$ and $\alpha(\mathscr{H}) = \mathscr{H}$. Thus $A_{\mathscr{H}} \cap \mathrm{Orth}(G) = \mathscr{A}(G) \cup \mathscr{A}^-(G)$ and $A_H \cap \mathrm{Orth}(G) = \mathscr{A}(H) \cup \mathscr{A}^-(H)$. The result follows from Theorem 8.86. □

Corollary 8.89. *If* $Z(G)$ *is nontrivial,* $p||G|$*, and* H *is a Sylow* p*-subgroup of* $Z(G)$*, then*

$$\omega(\mathscr{A}(G) \cup \mathscr{A}^-(G)) \leq \omega(\mathscr{A}(H) \cup \mathscr{A}^-(H)).$$

Proof. By Corollary 8.88, $\omega(\mathscr{A}(G) \cup \mathscr{A}^-(G)) \leq \omega(\mathscr{A}(Z(G)) \cup \mathscr{A}^-(Z(G)))$, and, by Corollary 8.87, $\omega(\mathscr{A}(Z(G)) \cup \mathscr{A}^-(Z(G))) \leq \omega(\mathscr{A}(H) \cup \mathscr{A}^-(H))$. □

Corollary 8.90. *If* $p||G|$ *and* H *is a Sylow* p*-subgroup of* G*, then*

$$\omega(\mathscr{A}(G) \cup \mathscr{A}^-(G)) \leq \omega(\mathscr{A}(Z(H)) \cup \mathscr{A}^-(Z(H))).$$

Proof. By Corollary 8.87, $\omega(\mathscr{A}(G) \cup \mathscr{A}^-(G)) \leq \omega(\mathscr{A}(H) \cup \mathscr{A}^-(H))$, and, by Corollary 8.88, $\omega(\mathscr{A}(H) \cup \mathscr{A}^-(H)) \leq \omega(\mathscr{A}(Z(H)) \cup \mathscr{A}^-(Z(H)))$. □

Corollary 8.91. *Let* H *be a nontrivial characteristic subgroup of* G*. Then*

$$\omega(\mathscr{A}(G) \cup \mathscr{A}^-(G)) \leq \omega(\mathscr{A}(H) \cup \mathscr{A}^-(H)).$$

Proof. As H is a characteristic subgroup of G, $\mathscr{H} = \{H\}$ and if α or $\alpha^- \in \mathrm{Aut}(G)$, then $\alpha(H) = H$ and $\alpha(\mathscr{H}) = \mathscr{H}$. Thus $A_{\mathscr{H}} \cap \mathrm{Orth}(G) = \mathscr{A}(G) \cup \mathscr{A}^-(G)$ and $A_H \cap \mathrm{Orth}(G) = \mathscr{A}(H) \cup \mathscr{A}^-(H)$. The result follows from Theorem 8.86. □

We know that, if G is nontrivial, then $\omega(\mathscr{A}(G) \cup \mathscr{A}^-(G)) \leq |G| - 2$ as this upper bound holds for all orthomorphism graphs. The natural question to ask is, when does equality hold? This question was answered by Evans [115] in 1989.

Theorem 8.92 (Evans, 1989). *If G is nontrivial, then $\omega(\mathscr{A}(G) \cup \mathscr{A}^-(G)) = |G| - 2$ if and only if G is elementary abelian.*

Proof. If G is elementary abelian, then $\omega(\mathscr{A}(G) \cup \mathscr{A}^-(G)) = |G| - 2$ by Corollary 8.74.

Let us assume that $\omega(\mathscr{A}(G) \cup \mathscr{A}^-(G)) = |G| - 2$. If $p \mid |G|$ and H is a Sylow p-subgroup of G, then

$$|G| - 2 = \omega(\mathscr{A}(G) \cup \mathscr{A}^-(G)) \leq \omega(\mathscr{A}(H) \cup \mathscr{A}^-(H)) \leq |H| - 2 \leq |G| - 2,$$

by Corollary 8.87. It follows that $G = H$, a p-group. Further G must be characteristically simple as, if K is a nontrivial, proper, characteristic subgroup of G, then

$$|G| - 2 = \omega(\mathscr{A}(G) \cup \mathscr{A}^-(G)) \leq \omega(\mathscr{A}(K) \cup \mathscr{A}^-(K)) \leq |K| - 2 < |G| - 2,$$

by Corollary 8.91. As characteristically simple groups are characterized as direct products of isomorphic simple groups (see Theorem 10.2.1 in [157]) and the only simple p-groups are cyclic, G must be elementary abelian. □

Restricted to $\mathscr{A}(G)$, Theorem 8.92 yields a special case of a result obtained by Jungnickel [204] in 1989. Jungnickel's proof used partial congruence partitions: these will be described when we discuss translation nets.

Corollary 8.93 (Jungnickel, 1989). *If G is nontrivial, then $\omega(\mathscr{A}(G)) = |G| - 2$ if and only if G is elementary abelian.*

8.4.4 Translation Nets

We studied nets in Section 1.4. A net $\mathscr{N}$ of order n and degree k has n^2 points and nk lines, each line contains n points, the lines are divided into k parallel classes of n lines each, the lines in each parallel class partition the points of $\mathscr{N}$, each point is on k lines, and nonparallel lines intersect in exactly one point. $\mathscr{N}$ is a *translation net* if it has a collineation group that acts regularly on the points and preserves parallel classes. A collineation group of a translation net $\mathscr{N}$ that acts regularly on the points and preserves parallel classes of $\mathscr{N}$ is called a *translation group* of $\mathscr{N}$. Clearly, if $\mathscr{N}$ has order n, then any translation group of $\mathscr{N}$ has order n^2.

A *partial congruence partition* of a group G of order n^2 is a set U of k subgroups of G of order n, such that $H \cap H' = \{1\}$ for all distinct $H, H' \in U$. The elements of U are called the *components* of the partial congruence partition, and k is called its *degree*. From a partial congruence partition U of a group G, we can form an incidence structure $\mathcal{N}(U)$ whose points are the elements of G and whose lines are $\{Hg \mid g \in G, H \in U\}$. Note that G acts on the points and lines of $\mathcal{N}(U)$ by right multiplication. In 1982 Sprague [298] showed that translation nets and partial congruence partitions are equivalent objects.

Theorem 8.94 (Sprague, 1982). *Let G be a group of order n^2 and let U be a partial congruence partition of G of degree k. Then $\mathcal{N}(U)$ is a translation net of order n and degree k with translation group G. Further, if $\mathcal{N}$ is a translation net of order n with translation group G, then $\mathcal{N} = \mathcal{N}(U)$ for some partial congruence partition U of G.*

Proof. Let $\mathcal{N}$ be a translation net of order n and degree k with translation group G, and let P be a point of $\mathcal{N}$. Let $L_1, \ldots, L_k$ be the lines through P, and set $H_i = \{g \in G \mid Pg \in L_i\}$ for $i = 1, \ldots, k$. As G acts regularly on the points of $\mathcal{N}$, there is one element of H_i for each point of L_i, and hence, $|H_i| = n$. If $h \in H_i$, then, as $1 \in H_i$, $h \in H_i h$. Thus, as h preserves parallel classes, $H_i = H_i h$, H_i is closed under multiplication and, hence, is a subgroup of G. As L_i intersects L_j in precisely one point if $i \neq j$, $|H_i \cap H_j| = 1$ if $i \neq j$. It follows that $U = \{H_1, \ldots, H_k\}$ is a partial congruence partition of G.

If G is a group of order n^2 and U is a partial congruence partition of G of degree k, then it is routine to show that $\mathcal{N}(U)$ is a translation net of order n and degree k with translation group G. □

An immediate corollary:

Corollary 8.95. *If G is a translation group of a finite translation net $\mathcal{N}$, $g \in G$, and P is a point of $\mathcal{N}$, then $Pg = P$ if and only if $g = 1$.*

By Theorem 8.94 in our discussion of finite translation nets, we may restrict ourselves to partial congruence partitions U of a finite group G and the corresponding net $\mathcal{N}(U)$. We note that, if $H \in U$, then the right cosets of H form a parallel class in $\mathcal{N}(U)$. For $g \in G$ let us define the mapping $\tau_g\colon G \to G$ by $\tau_g(x) = xg$. For each $g \in G$, τ_g is a collineation of $\mathcal{N}(U)$ that preserves parallel classes. Recall that in Section 1.4, we defined a strict translation of a net to be a collineation that fixes no points and fixes all lines of a given parallel class: strict translations are called central translations in [164]. By Lemma 1.47, strict translations fix parallel classes. For which $g \in G$, if any, is τ_g a strict translation of $\mathcal{N}(U)$?. By Corollary 8.95, if $g \neq 1$, then τ_g fixes no points of $\mathcal{N}(U)$. Thus, if $g \neq 1$, then τ_g is a strict translation of $\mathcal{N}(U)$ if and only if τ_g fixes all lines of some parallel class of $\mathcal{N}(U)$. If τ_g fixes some parallel class of $\mathcal{N}(U)$, the right cosets of $H \in U$ say, then $Hg = H$ and so $g \in H$.

Lemma 8.96. *Let G be a group of order n^2, let U be a partial congruence partition of G of degree k, and let $H \in U$. If $h \in H \setminus \{1\}$, then τ_h is a strict translation of $\mathcal{N}(U)$ fixing all right cosets of H if and only if $ghg^{-1} \in H$ for all $g \in G$.*

Proof. τ_h fixes all left cosets of H if and only if $Hgh = Hg$ for all $g \in G$, if and only if $ghg^{-1} \in H$ for all $g \in G$. □

Lemma 8.97. *Let G be a group of order n^2, let U be a partial congruence partition of G of degree k, and let $H \in U$. H is normal in G if and only if τ_h is a strict translation of $\mathcal{N}(U)$ for all $h \in H \setminus \{1\}$.*

Proof. This follows from Lemma 8.96. □

We are now in a position to establish a relationship between partial congruence partitions with a normal component and pairwise orthogonal sets of orthomorphisms. A more general version of the following result was proved by Hachenberger and Jungnickel [164] in 1990.

Theorem 8.98 (Hachenberger and Jungnickel, 1990). *Let G be a group of order n^2, let U be a partial congruence partition of G of degree $k > 3$, and let $H \in U$. If H is normal in G, then there exists a pairwise orthogonal set of $k-3$ orthomorphisms of H.*

Proof. This follows from Lemma 8.97 and Theorem 1.48. □

In the special case in which G is elementary abelian of order p^{2m}, p a prime, G can be viewed as the additive group of a vector space of dimension $2m$ over GF(p). If U is a partial congruence partition of G of degree k, all of whose components are subspaces of dimension m, then we call U a *partial spread*. If $|U| = p^m + 1$, then $\mathcal{N}(U)$ is a *translation plane*: translation planes will be discussed in Section 14.2.

We are particularly interested in partial congruence partitions U of G in which at least two components are normal subgroups of G. The corresponding translation nets are called *splitting translation nets*. Sprague [298] showed that, for such partial congruence partitions, any two normal components will be isomorphic, and $G \cong H \times H$, where $H \in U$ is a normal subgroup of G.

Theorem 8.99. *Let G be a group of order n^2, let U be a partial congruence partition of G of degree $k > 3$, and let $H, K \in U$, $H \neq K$. If H and K are normal in G, then $H \cong K$ and $G \cong H \times H$.*

Proof. In the proof of Lemma 2.9 in [298], it is shown that $H \cong K$, and, by Lemma 2.8 of [298], $G = HK$. Hence, as H and K are normal subgroups of G that intersect trivially, $G \cong H \times K \cong H \times H$. □

Let us revisit the construction of nets from pairwise orthogonal sets of orthomorphisms. Let H be a group of order n, let $G = H \times H$, and let $\theta_1, \ldots, \theta_{k-3}$ be a pairwise orthogonal set of normalized orthomorphisms of H. In the proof of Theorem 1.48, we constructed a net of order n and degree k from a set of pairwise orthogonal orthomorphisms of a group of order n. Let us use $\mathcal{N}(\theta_1, \ldots, \theta_{k-3})$ to denote the net constructed from $\theta_1, \ldots, \theta_{k-3}$. The points of $\mathcal{N}(\theta_1, \ldots, \theta_{k-3})$ are the elements of G, and the lines of $\mathcal{N}(\theta_1, \ldots, \theta_{k-3})$ belong to k parallel classes A, B, C, and D_i, $i = 1, \ldots, k-3$. The lines of A are $\{(x, y) \mid y \in H\}, x \in H$; the lines of B are $\{(x, y) \mid x \in H\}, y \in H$; the lines of C are $\{(x, xh) \mid x \in H\}, h \in H$; and the lines of D_i are $\{(x, \theta_i(x)h) \mid x \in H\}, h \in H, i = 1, \ldots, k-3$. The following connection between splitting translation nets and fixed-point-free automorphisms was proved by Hachenberger and Jungnickel [164] in 1990.

Theorem 8.100 (Hachenberger and Jungnickel, 1990). *Let H be a group of order n, and let $\theta_1, \ldots, \theta_{k-3}$ be a pairwise orthogonal set of normalized orthomorphisms of H. Then $\mathcal{N}(\theta_1, \ldots, \theta_{k-3})$ is a translation net if and only if $\theta_1, \ldots, \theta_{k-3} \in \mathcal{A}(H)$.*

Proof. If $\mathcal{N}(\theta_1, \ldots, \theta_{k-3})$ is a translation net, then the mapping $\tau_{(g,1)}\colon (x, y) \mapsto (xg, y)$ is a strict translation of $\mathcal{N}(\theta_1, \ldots, \theta_{k-3})$. Thus

$$\tau_{(g,1)}(x, \theta_i(x)h) = (xg, \theta_i(x)h) = (xg, \theta_i(xg)k)$$

for some $k \in H$ and all $x \in H$. Hence $\theta_i(x)h = \theta_i(xg)k$ for some $k \in H$ and all $x \in H$. Setting $x = 1$ yields $k = \theta_i(g)^{-1}h$ and so $\theta_i(x) = \theta_i(xg)\theta_i(g)^{-1}$. Thus $\theta_i(x)\theta_i(g) = \theta_i(xg)$ for all $x, g \in H$. It follows that $\theta_1, \ldots, \theta_{k-3} \in \mathcal{A}(H)$.

The converse is routine. □

Thus all splitting translation nets can be constructed from orthomorphisms that are also automorphisms, i.e., fixed-point-free automorphisms. Many of our results for $\mathcal{A}(G)$ were in fact originally proved in the context of translation nets. The corresponding construction of mutually orthogonal Latin squares from group automorphisms dates back to Mann [237] in 1942 and is usually referred to as Mann's automorphism method. Most constructions in the literature of translation nets are of translation nets with elementary abelian translation groups. There are a few exceptions. In 1991 Hachenberger [162] constructed an infinite series of translation nets with nonabelian translation groups, and in 1992 he [163] constructed a number of translation nets of prime power order whose translation groups are not elementary abelian. The translation nets that we have discussed are of index one: for information on more general translation nets, see the 1992 survey by Hachenberger and Jungnickel [165].

8.5 Strong Complete Mappings

A *strong complete mapping* is a complete mapping that is also an orthomorphism. The term strong complete mapping was first used by Hsu and Keedwell [187] in 1985: these were called strong orthomorphisms in [16] and strong permutations in [181]. We will use $\mathcal{S}(G)$ to denote the set of normalized strong complete mappings of G and the corresponding orthomorphism graph. Note that, for $|G|$

odd, the existence of a strong complete mapping of G implies the existence of a 2-clique of Orth(G), as any strong complete mapping of G will be orthogonal to $\phi_{-1}\colon x \mapsto x^{-1}$, which, by Theorem 8.10, is an orthomorphism of G. In particular, for $|G|$ odd, $\omega(\mathcal{S}(G)) \le \omega(G)-1$. Any automorphism or congruence of Orth(G) that fixes ϕ_{-1} must also map strong complete mappings to strong complete mappings. In particular $\mathcal{S}(G)$ is fixed by H_α, $\alpha \in \mathrm{Aut}(G)$, T_g, $g \in Z(G)$, and I.

8.5.1 Some Classes of Strong Complete Mappings

We ask ourselves, among the classes of orthomorphisms that we have already studied, which are elements of $\mathcal{S}(G)$? Recall that in Section 8.2, we defined the mapping $\phi_r\colon G \to G$ by $\phi_r(x) = x^r$ and set $\mathcal{P}(G) = \{\phi_r \mid \phi_r \in \mathrm{Orth}(G)\}$. Facts about the intersection of $\mathcal{S}(G)$ with $\mathcal{P}(G)$ can be gleaned from Section 8.2.

Theorem 8.101.

1. *If* $\phi_r \in \mathcal{P}(G)$, *then* $\phi_r \in \mathcal{S}(G)$ *if and only if* $\gcd(r,n) = \gcd(r-1,n) = \gcd(r+1,n) = 1$.
2. *If* $|G|$ *is even, then* $\mathcal{P}(G) \cap \mathcal{S}(G)$ *is empty.*
3. *If* $|G|$ *is odd and* p *is the smallest prime divisor of* $|G|$, *then*

$$\omega(\mathcal{P}(G) \cap \mathcal{S}(G)) = p-3.$$

4. $\mathcal{P}(G) \cap \mathcal{S}(G)$ *is nonempty if and only if* $\gcd(6,|G|) = 1$.

Proof.

1. This follows from Theorem 8.10 and the definition of a strong complete mapping.
2. By Corollary 8.12, $\mathcal{P}(G)$ is empty if $|G|$ is even.
3. As $\phi_2,\dots,\phi_{p-2},\phi_{-1}$ is a clique of $\mathcal{P}(G)$, $\phi_2,\dots,\phi_{p-2}$ is a clique of $\mathcal{P}(G) \cap \mathcal{S}(G)$. Thus $\omega(\mathcal{P}(G) \cap \mathcal{S}(G)) \ge p-3$. Equality follows from the fact that an r-clique of $\mathcal{P}(G) \cap \mathcal{S}(G)$ extends to an $(r+1)$-clique of $\mathcal{P}(G)$ by adding ϕ_{-1}, and, hence, by Theorem 8.14, $r+1 \le p-2$.
4. This follows from (2) and (3). □

In Section 8.4, we answered the question, which automorphisms are orthomorphisms? It is natural to also ask, which automorphisms are strong complete mappings?

Theorem 8.102. *If* $\alpha \in \mathcal{A}(G)$, *then* $\alpha \in \mathcal{S}(G)$ *if and only if* $\alpha^- \in \mathcal{A}^-(G)$.

Proof. If $\alpha \in \mathcal{A}(G)$, then α is an orthomorphism by definition, and by Theorem 8.83, α is a complete mapping if and only if α^- is an orthomorphism of G, i.e., $\alpha^- \in \mathcal{A}^-(G)$. □

The following is an alternative formulation of Theorem 8.102.

Theorem 8.103. *If* $\alpha \in \mathrm{Aut}(G)$*, then* $\alpha \in \mathscr{S}(G)$ *if and only if* $\alpha(x) = x$ *implies that* $x = 1$ *and* $\alpha(x) = y^{-1}x^{-1}y$*, for any* y*, implies that* $x = 1$.

Proof. By Corollary 8.66, α is an orthomorphism if and only if $\alpha(x) = x$ implies that $x = 1$, and by Theorem 8.83, α is a complete mapping if and only if $\alpha(x) = y^{-1}x^{-1}y$, for any y, implies that $x = 1$. □

In the abelian case, the reformulation of Theorem 8.102 is a little simpler.

Corollary 8.104. *If* G *is abelian and* $\alpha \in \mathrm{Aut}(G)$*, then* $\alpha \in \mathscr{S}(G)$ *if and only if* $\alpha(x) = x$ *implies that* $x = 1$ *and* $\alpha(x) = x^{-1}$ *implies that* $x = 1$.

For $a \in \mathrm{GF}(q)$, we will use $[a]$ to denote the mapping $x \mapsto ax$ and set $\mathscr{C}_1(q) = \{[a] \mid [a] \in \mathrm{Orth}(\mathrm{GF}(q)^+)\}$. The elements of $\mathscr{C}_1(q)$ are *linear orthomorphisms* and form the simplest class of cyclotomic orthomorphisms: cyclotomic orthomorphisms will be studied in Chapter 10.

Theorem 8.105. *If* q *is a prime power, then the following hold.*

1. *If* $a \in \mathrm{GF}(q)$*, then* $[a] \in \mathscr{S}(\mathrm{GF}(q)^+)$ *if and only if* $a \neq 0, \pm 1$.
2. $\mathscr{C}_1(q) \cap \mathscr{S}(\mathrm{GF}(q)^+)$ *is nonempty if and only if* $q > 3$.
3. *If* $q > 3$*, then* $\omega(\mathscr{C}_1(q) \cap \mathscr{S}(\mathrm{GF}(q)^+)) = q - 3$.

Proof.

1. It is easy to prove that $[a]$ is an orthomorphism of $\mathrm{GF}(q)^+$ if and only if $a \neq 0, 1$, and a complete mapping of $\mathrm{GF}(q)^+$ if and only if $a \neq 0, -1$. The result follows.
2. $\mathscr{C}_1(q) \cap \mathscr{S}(\mathrm{GF}(q)^+)$ is nonempty if and only if there exists $a \in GF(q)$, $a \neq 0, \pm 1$, if and only if $q > 3$.
3. By (1), $|\mathscr{C}_1(q) \cap \mathscr{S}(\mathrm{GF}(q)^+)| = q - 3$. Further it is easy to prove that $[a] \perp [b]$ if and only if $a \neq b$, from which it follows that the elements of $\mathscr{C}_1(q) \cap \mathscr{S}(\mathrm{GF}(q)^+)$ form a pairwise orthogonal set of strong complete mappings of $\mathrm{GF}(q)^+$. □

8.5.2 The Existence of Strong Complete Mappings

A problem of interest is: which finite groups admit strong complete mappings? The proof of the Hall-Paige conjecture (see Chapter 7) tells us that a finite group admits orthomorphisms, equivalently complete mappings, if and only if its Sylow 2-subgroup is either trivial or noncyclic. Results proved in 1990 by Horton [181] and Evans [116] indicates that the existence of strong complete mappings may depend, not just on the structure of the Sylow 2-subgroup, but also on the structure of the Sylow 3-subgroup.

Theorem 8.106 (Evans, 1990). *If* G *has a nontrivial, cyclic Sylow* 3*-subgroup* S *and* H *is a normal subgroup of* G *for which* $G/H \cong S$*, then* G *does not admit strong complete mappings.*

Proof. Let $|G| = mn$, $(n,m) = 1$, m a power of 3, and let $\phi\colon G \to \mathbb{Z}_m \cong G/H$ be the canonical homomorphism. If G admits a strong complete mapping θ, then the following holds.

$$\begin{aligned}
2\sum_{g\in G}\phi(g)^2 &= \sum_{g\in G}\phi(g\theta(g))^2 + \sum_{g\in G}\phi(g^{-1}\theta(g))^2\\
&= \sum_{g\in G}(\phi(\theta(g)) + \phi(g))^2 + \sum_{g\in G}(\phi(\theta(g)) - \phi(g))^2\\
&=2\sum_{g\in G}\phi(\theta(g))^2 + 2\sum_{g\in G}\phi(g)^2\\
&=4\sum_{g\in G}\phi(g)^2.
\end{aligned}$$

This implies that $\sum_{g\in G}\phi(g)^2 = 0$. But by direct computation, we see that

$$\sum_{g\in G}\phi(g)^2 = n(0^2+1^2+\cdots+(m-1)^2) = n(m-1)m(2m-1)/6 \not\equiv 0 \pmod m,$$

a contradiction from which the result follows. □

The special case for groups of odd order was proved by Horton [181].

Corollary 8.107 (Horton, 1990). *If a finite group G of odd order has a nontrivial, cyclic Sylow 3-subgroup, then G does not admit strong complete mappings.*

Proof. If 3 is the smallest prime divisor of $|G|$ and the Sylow 3-subgroup S is nontrivial and cyclic, then there is a normal subgroup H of G for which $G/H \cong S$ (See Corollary 1.4.18 in [241] for a proof). The result then follows from Theorem 8.106. □

Theorem 8.106 has implications for abelian groups.

Corollary 8.108. *An abelian group with a nontrivial, cyclic Sylow 2-subgroup or a nontrivial, cyclic Sylow 3-subgroup does not admit strong complete mappings.*

Corollary 8.109. *If $\gcd(n,6) \neq 1$, then $\mathbb{Z}_n$ does not admit strong complete mappings.*

By Theorem 8.105, we know that elementary abelian groups of order at least 4 admit strong complete mappings, and by Theorem 8.101, that groups whose order is not divisible by either 2 or 3 admit strong complete mappings. What other classes of groups admit strong complete mappings? Several classes of abelian groups were shown to admit strong complete mappings by Horton [181] and Evans [116] in 1990. These existence results depend on a quotient group construction.

Theorem 8.110. *If H is a subgroup of an abelian group G and G/H and H both admit strong complete mappings, then G admits strong complete mappings.*

Proof. Let $\theta\colon H \to H$ and $\phi\colon G/H \to G/H$ be strong complete mappings of H and G/H, respectively, and let $g_1, \ldots, g_n$ be a set of coset representatives of H in G. Let $\alpha\colon G/H \to G$ be defined by $\alpha(g_i H) = g_i$, and let $\beta\colon G \to G$ be defined by $\beta(g_i h) = \alpha\phi(g_i H)\theta(h)$, $h \in H$.

We need to show that β is a strong complete mapping of G. β is a bijection as, if $\beta(g_i h) = \beta(g_j h')$, then $\theta(h) = \theta(h')$, from which it follows that $h = h'$, and $\alpha\phi(g_i H) = \alpha\phi(g_j H)$, from which it follows that $\phi(g_i H) = \phi(g_j H)$, from which it follows that $i = j$.

If $g_i h\beta(g_i h) = g_j h'\beta(g_j h')$, then $g_i h\alpha\phi(g_i H)\theta(h) = g_j h'\alpha\phi(g_j H)\theta(h')$. It follows that $g_i H\phi(g_i H) = g_j H\phi(g_j H)$. As ϕ is a complete mapping of G/H, $i = j$. Thus $h\theta(h) = h'\theta(h')$. As θ is a complete mapping of H, $h = h'$. Hence the mapping $g_i h \mapsto g_i h\beta(g_i h)$ is a bijection, and so β is a complete mapping.

Similarly the mapping $g_i h \mapsto (g_i h)^{-1}\beta(g_i h)$ is a bijection, and so β is an orthomorphism and, thus, a strong complete mapping. □

A simple induction argument establishes the existence of strong complete mappings for direct products.

Corollary 8.111. *If $G \cong H_1 \times \cdots \times H_k$ is a finite abelian group and each of H_i admits strong complete mappings, then G admits strong complete mappings.*

We are interested in the question, which groups admit strong complete mappings? A special case of this question is, which abelian groups admit strong complete mappings? In 1990 Evans [116] conjectured that a finite abelian group admits strong complete mappings if and only if its Sylow 2-subgroup is either trivial or noncyclic, and its Sylow 3-subgroup is also either trivial or noncyclic. In what follows we will prove this conjecture. We begin by proving that all noncyclic abelian 2-groups admit strong complete mappings. To do so we need the following lemma.

Lemma 8.112. *If $G \cong \mathbb{Z}_m \times \mathbb{Z}_2$, $m = 2^n$, then G admits a strong complete mapping.*

Proof. Let $\mathbb{Z}_2 = \langle h\rangle$ and $\mathbb{Z}_m = \langle g\rangle$. Set $S = \{1, \cdots, 2^{n-1}\}$. Let $\theta\colon G \to G$ be defined by

$$\theta(x) = \begin{cases} g^{2i} & \text{if } x = hg^i \text{ and } i \in S,\\ g^{2i-1} & \text{if } x = g^i \text{ and } i \in S,\\ hg^{2i-1} & \text{if } x = hg^i \text{ and } i \notin S,\\ hg^{2i} & \text{if } x = g^i \text{ and } i \notin S.\end{cases}$$

It was proved in Theorem 3.20 that the mapping $x \mapsto x^{-1}\theta(x)$ is a complete mapping of G. Thus the mappings $x \mapsto x^{-1}\theta(x)$ and $x \mapsto x(x^{-1}\theta(x)) = \theta(x)$ are bijections. The mapping $x \mapsto x\theta(x)$ is also a bijection as $\{x\theta(x) \mid x \in G\} = \{g^{3i-1} \mid i = 1, \ldots, m\} \cup \{hg^{3i} \mid i = 1, \ldots, m\} = G$. Hence θ is a strong complete mapping. □

Theorem 8.113. *Any noncyclic abelian 2-group admits strong complete mappings.*

Proof. Let us assume that this result is false and that G is a nonabelian 2-group of least order that does not admit strong complete mappings. Then $G \equiv \mathbb{Z}_{m_1} \times \cdots \times \mathbb{Z}_{m_k}$, where $\mathbb{Z}_{m_i} = \langle g_i \rangle$, $m_i = 2^{r_I}$, $r_i \geq 1$, for $i = 1, \ldots, k$, and $k \geq 2$.

If $k \geq 4$, then $H = \langle g_1 \rangle \times \langle g_2 \rangle \times \langle 1 \rangle \times \cdots \times \langle 1 \rangle$ is a noncyclic subgroup of G and G/H is also noncyclic. By the minimality of $|G|$, each of H and G/H admits strong complete mappings. Hence, by Theorem 8.110, G admits strong complete mappings. Thus $k = 2$ or 3.

If $k = 3$, then some $m_i > 2$ as otherwise G would admit strong complete mappings by Theorem 8.105. Without loss of generality $m_1 > 2$. Then $H = \langle g_1^2 \rangle \times \langle g_2 \rangle \times \langle 1 \rangle$ is a noncyclic subgroup of G and G/H is also noncyclic. By the minimality of $|G|$, each of H and G/H admits strong complete mappings. Hence, by Theorem 8.110, G admits strong complete mappings. Thus $k = 2$.

If $k = 2$ and $m_1, m_2 > 2$, then $H = \langle g_1^2 \rangle \times \langle g_2^2 \rangle$ is a noncyclic subgroup of G and G/H is also noncyclic. By the minimality of $|G|$, each of H and G/H admits strong complete mappings. Hence, by Theorem 8.110, G admits strong complete mappings. Thus either $m_1 = 2$ or $m_2 = 2$. But then G admits strong complete mappings by Lemma 8.112, a contradiction from which the result follows. □

For noncyclic abelian 3-groups, in 1990 Evans [116] conjectured an analogous result to Lemma 8.112: this result was proved by Evans [129] in 2012. If $m = 3$, then $\mathbb{Z}_3 \times \mathbb{Z}_m$ admits strong complete mappings by Theorem 8.105. Horton [181] found a strong complete mapping when $m = 9$. Horton's strong complete mapping θ is described in Figure 8.4. We use ij, $i \in \mathbb{Z}_3 = \{0, 1, 2\}$, $j \in \mathbb{Z}_9 = \{0, 1, \ldots, 8\}$, to denote an element of $\mathbb{Z}_3 \times \mathbb{Z}_9$. It is easy to check that θ is a strong complete mapping of $\mathbb{Z}_3 \times \mathbb{Z}_9$.

Let $G = \mathbb{Z}_3 \times \mathbb{Z}_{3m}$, $m \geq 3$ a power of 3, and let K be the subgroup of G generated by $(1, 0)$ and $(0, m)$. Thus K is isomorphic to the vector space of dimension 2 over GF(3). We will use ij to denote the element (i, jm) of K and, if $k = ij \in K$ and $l \in \{0, \ldots, m-1\}$, we will use $[k, l]$ or $[ij, l]$ to denote the element $(i, jm + l)$ of G. For $ij \in K$ we will call i the *first component* and j the *second component* of ij.

x	$\theta(x)$	x	$\theta(x)$	x	$\theta(x)$	x	$\theta(x)$	x	$\theta(x)$	x	$\theta(x)$	x	$\theta(x)$	x	$\theta(x)$	x	$\theta(x)$
00	00	11	08	24	02	03	10	10	11	20	12	05	22	14	26	22	25
01	05	12	03	27	06	04	14	13	15	21	16	07	27	15	20	23	28
02	01	18	07	28	04	06	17	16	13	25	18	08	23	17	21	26	24

Fig. 8.4 A strong complete mapping of $\mathbb{Z}_3 \times \mathbb{Z}_9$

For $i = 0, 1, 2$ and $j = 0, 1, \ldots, m-1$, let $\delta_{i,j}, \epsilon_{i,j} \in K$ be defined by

$$\delta_{i,j} = \begin{cases} 01 & \text{if } (3j + i \mod m) < j, \\ 00 & \text{otherwise,} \end{cases}$$

and

$$\epsilon_{i,j} = \begin{cases} 01 & \text{if } (3j + i \mod m) \geq m - j, \\ 00 & \text{otherwise.} \end{cases}$$

For $a \in \{1, \ldots, m-1\}$, $\gcd(a, m) = 1$, let a' denote the unique integer in $\{1, \ldots, m-1\}$ satisfying $aa' \equiv 1 \pmod{m}$.

Lemma 8.114. *$\delta_{i,j} = 01$ if and only if $j \in \{m/3, \ldots, (m-3)/2\} \cup \{2m/3, \ldots, m-2\}$ and $i = 0, 1, 2$, or $j = (m-1)/2$ and $i = 0$, or $j = m-1$ and $i = 0, 1$.*

Proof. By definition $\delta_{i,j} = 01$ if and only if $(3j + i \mod m) < j$, if and only if $0 \leq 3j + i - km < j$ for some integer k, equivalently $2j + i < km \leq 3j + i$. As $0 \leq j \leq m-1$ and $0 \leq i \leq 2$, this inequality can hold true for $k = 1$ or 2 only. If $k = 1$, then, solving for j, the inequality holds if and only if

$$\frac{m}{3} \leq j \leq \begin{cases} \frac{m-1}{2} & \text{if } i = 0, \\ \frac{m-3}{2} & \text{if } i = 1, 2. \end{cases}$$

If $k = 2$, then, solving for j, the inequality holds if and only if

$$\frac{2m}{3} \leq j \leq \begin{cases} m-1 & \text{if } i = 0, 1, \\ m-2 & \text{if } i = 2. \end{cases}$$

To see the left-hand side of these equalities, it must be remembered that $m \geq 3$ is a power of 3. The result follows. □

Lemma 8.115. *The group $G = \mathbb{Z}_3 \times \mathbb{Z}_{3m}$, $m \geq 3$ a power of 3, admits strong complete mappings if there exist $\alpha_{i,j} \in K$, $i = 0, 1, 2$ and $j = 0, 1, \ldots, m-1$, and $x_j \in \{01, 02\}$, $j = 0, 1, \ldots, m-1$, for which the following three conditions hold.*

1. *For $i = 0, 1, 2$ and $j = 0, 1, \ldots, \frac{m}{3} - 1$, the set*

$$\{\alpha_{i,j}, \alpha_{i,j+m/3}, \alpha_{i,j+2m/3}\}$$

is a system of distinct coset representatives for $\langle 11 \rangle$ in K.

2. *For $j = 0, 1, \ldots, m-1$, the set*

$$\{\alpha_{0,j} - \delta_{0,j}, \alpha_{1,j-2'} - x_{j-2'} - \delta_{1,j-2'}, \alpha_{2,j-1} + x_{j-1} - \delta_{2,j-1}\}$$

is a system of distinct coset representatives for $\langle 10 \rangle$ in K, where the indices are computed modulo m.

3. *For $j = 0, 1, \ldots, m-1$, the set*

$$\{\alpha_{0,j}, \alpha_{1,j-4'}, \alpha_{2,j-2'}\}$$

is a system of distinct coset representatives for $\langle 01 \rangle$ *in* K*, where the second indices are computed modulo* m*.*

Proof. Let H be the subgroup of K generated by 12, and let $A = \begin{pmatrix} 0 & 0 \\ 1 & 1 \end{pmatrix}$. Then $HA = \langle 11 \rangle$, $H(A-I) = \langle 10 \rangle$, and $H(A+I) = \langle 01 \rangle$. Let $\alpha_{i,j} \in K$, $i = 0, 1, 2$ and $j = 0, 1, \ldots, m-1$, and $x_j \in \{01, 02\}$, $j = 0, 1, \ldots, m-1$, satisfy the conditions of the lemma and let $\theta\colon G \to G$ be defined by

$$\theta([h + ix_j, j]) = [hA + \alpha_{i,j}, (3j + i \mod m)],$$

where $h \in H$, $i \in \{0, 1, 2\}$, and $j \in \{0, 1, \ldots, m-1\}$.

θ is well-defined as, if $[h_1 + i_1 x_{j_1}, j_1] = [h_2 + i_2 x_{j_2}, j_2]$, then $j_1 = j_2$ and so $x_{j_1} = x_{j_2}$. Now ix_{j_1}, $i = 0, 1, 2$, is a system of distinct representatives for H in K, from which it follows that $i_1 = i_2$ and hence that $h_1 = h_2$.

We will show that (1) implies that θ is a bijection, (2) that $x \mapsto \theta(x) - x$ is a bijection, and (3) that $x \mapsto \theta(x) + x$ is a bijection.

If $\theta([h_1 + i_1 x_{j_1}, j_1]) = \theta([h_2 + i_2 x_{j_2}, j_2])$, then $3j_1 + i_1 \equiv 3j_2 + i_2 \pmod{m}$. It follows that, as $i_1, i_2 \in \{0, 1, 2\}$, $i_1 = i_2$ and hence that $3j_1 \equiv 3j_2 \pmod{m}$. Thus j_1 is congruent to one of j_2, $j_2 + (m/3)$, or $j_2 + (2m/3)$ modulo m. By (2), the set $\{\alpha_{i,j}, \alpha_{i,j+m/3}, \alpha_{i,j+2m/3}\}$, $i = 0, 1, 2$ and $j = 0, 1, \ldots, \frac{m}{3} - 1$, is a system of distinct coset representatives for $\langle 11 \rangle = HA$ in K. Thus $j_1 = j_2$ and so $h_1 A = h_2 A$. It follows that $h_1 = h_2$ and hence that θ is a bijection.

If $\theta([h_1 + i_1 x_{j_1}, j_1]) - [h_1 + i_1 x_{j_1}, j_1] = \theta([h_2 + i_2 x_{j_2}, j_2]) - [h_2 + i_2 x_{j_2}, j_2]$, then $h_1(A - I) - i_1 x_{j_1} + \alpha_{i_1, j_1} - \delta_{i_1, j_1} = h_2(A - I) - i_2 x_{j_2} + \alpha_{i_2, j_2} - \delta_{i_2, j_2}$ and $2j_1 + i_1 \equiv 2j_2 + i_2 \pmod{m}$. Without loss of generality $i_2 - i_1 \in \{0, 1, 2\}$. If $i_1 = i_2$, then as $\gcd(2, m) = 1$, $j_1 = j_2$. It follows that $h_1(A - I) = h_2(A - I)$. If $i_2 - i_1 = 1$, then $j_2 = j_1 - 2'$ and so, by (2), $\alpha_{i_1, j_1} - i_1 x_{j_1} - \delta_{i_1, j_1}$ and $\alpha_{i_2, j_2} - i_2 x_{j_2} - \delta_{i_2, j_2}$ are in different cosets of $H(A - I)$. Hence $h_1(A - I) - i_1 x_{j_1} + \alpha_{i_1, j_1} - \delta_{i_1, j_1} \neq h_2(A - I) - i_2 x_{j_2} + \alpha_{i_2, j_2} - \delta_{i_2, j_2}$, a contradiction. Similarly, the case $i_2 = i_1 + 2$ leads to a contradiction. Hence θ is an orthomorphism.

If $\theta([h_1 + i_1 x_{j_1}, j_1]) + [h_1 + i_1 x_{j_1}, j_1] = \theta([h_2 + i_2 x_{j_2}, j_2]) + [h_2 + i_2 x_{j_2}, j_2]$, then $h_1(A + I) + i_1 x_{j_1} + \alpha_{i_1, j_1} + \epsilon_{i_1, j_1} = h_2(A + I) + i_2 x_{j_2} + \alpha_{i_2, j_2} + \epsilon_{i_2, j_2}$ and $4j_1 + i_1 \equiv 4j_2 + i_2 \pmod{m}$. Without loss of generality $i_2 - i_1 \in \{0, 1, 2\}$. If $i_1 = i_2$, then, as $\gcd(4, m) = 1$, $j_1 = j_2$. It follows that $h_1(A+I) = h_2(A+I)$. If $i_2 - i_1 = 1$, then $j_2 = j_1 - 4'$ and so, by (3), α_{i_1, j_1} and α_{i_2, j_2} are in different cosets of $H(A+I)$. Hence, as $x_j, \epsilon_{i,j} \in H(A + I)$ for all i, j, $h_1(A + I) + i_1 x_{j_1} + \alpha_{i_1, j_1} + \epsilon_{i_1, j_1} \neq h_2(A + I) + i_2 x_{j_2} + \alpha_{i_2, j_2} + \epsilon_{i_2, j_2}$, a contradiction. Similarly, the case $i_2 = i_1 + 2$ leads to a contradiction. It follows that θ is a strong complete mapping. □

Lemma 8.116. *If* $m = 3^n$, $n \geq 2$*, then* $\mathbb{Z}_3 \times \mathbb{Z}_{3m}$ *admits strong complete mappings.*

Proof. We will choose $\alpha_{i,j}$ and x_j, $i = 0, 1, 2$ and $j = 0, \ldots, m-1$, so that the conditions of Lemma 8.115 are satisfied.

We set

$$x_j = \begin{cases} 01 & \text{if } j = 2m/3, \ldots, (5m-9)/6, \\ 02 & \text{otherwise,} \end{cases}$$

$\alpha_{0,j} = 0k$, where

$$k = \begin{cases} 0 & \text{if } 0 < j < m/3, \text{ or } j = 2m/3, \\ 1 & \text{if } m/3 \le j < 2m/3, \\ 2 & \text{if } 2m/3 < j \le m-1, \text{ or } j = 0, \end{cases}$$

$\alpha_{1,j} = 1k$, where

$$k = \begin{cases} 0 & \text{if } 0 \le (j+2' \mod m) < m/3, \\ 1 & \text{if } m/3 \le (j+2' \mod m) < 2m/3, \\ 2 & \text{if } 2m/3 \le (j+2' \mod m) \le m-1. \end{cases}$$

and $\alpha_{2,j} = 2k$, where

$$k = \begin{cases} 0 & \text{if } 0 < (j+1 \mod m) \le m/3, \\ 1 & \text{if } m/3 < (j+1 \mod m) < 2m/3, \text{ or } (j+1 \mod m) = 0, \\ 2 & \text{if } 2m/3 \le (j+1 \mod m) \le m-1. \end{cases}$$

By our construction

$$\{\alpha_{0,0}, \alpha_{0,m/3}, \alpha_{0,2m/3}\} = \{02, 01, 00\}$$

and

$$\{\alpha_{2,m-1}, \alpha_{2,(m-3)/3}, \alpha_{2,(2m-3)/3}\} = \{21, 20, 22\}.$$

For any $j = 0, \ldots, m-1$ the set

$$\{\alpha_{i,j}, \alpha_{i,j+m/3}, \alpha_{i,j+2m/3}\} = \{i0, i1, i2\}$$

in some order. It follows that each of these sets is a system of distinct coset representatives for $\langle 11 \rangle$, and so Lemma 8.115(1) is satisfied.

Simple computation shows that the set

$$\{-\delta_{0,j}, -x_{j-2'} - \delta_{1,j-2'}, x_{j-1} - \delta_{2,j-1}\}$$

$$= \begin{cases} \{00, 01, 02\} & \text{if } j \in \{1, \ldots, (m-3)/3\} \cup \{(m+1)/2, \ldots, (2m-3)/3\}, \\ \{02, 00, 01\} & \text{if } j \in \{(m+3)/3, \ldots, (m-1)/2\} \cup \{(5m+3)/6, \ldots, m-1\}, \\ \{02, 01, 00\} & \text{if } j \in \{(2m+3)/3, \ldots, (5m-3)/6\}. \end{cases}$$

For each of these sets, the second components form a permutation of $\{0, 1, 2\}$, and hence each is a system of distinct coset representatives for $\langle 10 \rangle$. As $\alpha_{0,j}$, $\alpha_{0,j-2'}$, and $\alpha_{0,j-1}$ are the same if $j \neq 0, m/3$ or $2m/3$, it follows that $\{\alpha_{0,j} - \delta_{0,j}, \alpha_{1,j-2'} - x_{j-2'} - \delta_{1,j-2'}, \alpha_{2,j-1} + x_{j-1} - \delta_{2,j-1}\}$ is a system of distinct coset representatives for $\langle 10 \rangle$, if $j \neq 0, m/3$ or $2m/3$. For $j = 0, m/3$ or $2m/3$, the set $\{\alpha_{0,j} - \delta_{0,j}, \alpha_{1,j-2'} - x_{j-2'} - \delta_{1,j-2'}, \alpha_{2,j-1} + x_{j-1} - \delta_{2,j-1}\} = \{02, 11, 20\}$ if $j = 0$, $\{00, 11, 22\}$ if $j = m/3$, and $\{02, 10, 21\}$ if $j = 2m/3$: as, for each of these sets the second components form a permutation of $\{0, 1, 2\}$, each of these sets is a system of distinct coset representatives for $\langle 10 \rangle$. Hence Lemma 8.115(2) is satisfied.

For each $j = 0, \ldots, m-1$, the first components of $\{\alpha_{0,j}, \alpha_{1,j-4'}, \alpha_{2,j-2'}\}$ are $\{0, 1, 2\}$. Hence, for each $j = 0, \ldots, m-1$, the set $\{\alpha_{0,j}, \alpha_{1,j-4'}, \alpha_{2,j-2'}\}$ is a system of distinct coset representatives for $\langle 01 \rangle$, and so Lemma 8.115(3) is satisfied. Hence Lemma 8.115(1) is satisfied, and $\mathbb{Z}_3 \times \mathbb{Z}_{3m}$ admits strong complete mappings by Lemma 8.115. □

We can now prove the existence of strong complete mappings for noncyclic, abelian 3-groups.

Lemma 8.117. *Every noncyclic, abelian 3-group admits strong complete mappings.*

Proof. If G is a noncyclic, abelian 3-group then it is an exercise to show that G admits a subnormal series $\{1\} = K_1 < \cdots < K_s = G$, in which $H_i = K_i/K_{i-1} \cong \mathrm{GF}(q_i)^+$, $q_i > 3$, if $i < s$, and $H_s = K_s/K_{s-1} \cong \mathrm{GF}(q)^+$, $q > 3$, or $\mathbb{Z}_{3m} \times \mathbb{Z}_3$, $m = 3^n$. By Theorem 8.105, H_i admits strong complete mappings for $i < s$. If $H_s \cong \mathrm{GF}(q)^+, q > 3$, then H_s admits strong complete mappings by Theorem 8.105. If $H_s \cong \mathbb{Z}_{3m} \times \mathbb{Z}_3$, then H_s admits strong complete mappings by Theorem 8.105 if $m = 1$, by Figure 8.4 if $m = 3$, or by Lemma 8.116 if $m > 3$. Hence, by repeated applications of Theorem 8.110, G admits strong complete mappings. □

Lemma 8.117 yields a characterization of finite abelian groups that admit strong complete mappings.

Theorem 8.118 (Evans, 2012). *A finite abelian group admits strong complete mappings if and only if its Sylow 2-subgroup is trivial or noncyclic and its Sylow 3-subgroup is also trivial or noncyclic.*

Proof. Let G be a finite abelian group. For $p||G|$, p prime, we will use H_p to denote the Sylow p-subgroup of G. Then $G \cong \bigotimes_{p||G|} H_p$.

If $2||G|$ and H_2 is cyclic, then, by Corollary 3.2, G does not admit complete mappings and, hence, does not admit strong complete mappings. If $3||G|$ and H_3 is cyclic, then, by Theorem 8.106, G does not admit strong complete mappings.

Let us assume that, if $2||G|$, then H_2 is noncyclic and, if $3||G|$, then H_3 is noncyclic. By Theorem 8.113, if $2||G|$, then H_2 admits strong complete mappings. If $3||G|$, then we can prove that H_3 admits strong complete mappings by a simple adaptation of the proof of Theorem 8.113. If prime $p||G|$, $p \neq 2, 3$, then H_p admits strong complete mappings by Theorem 8.101. The result then follows from Corollary 8.111. □

For finite groups in general, it might seem reasonable to conjecture that a finite group is strongly admissible if and only if its Sylow 2-subgroup is trivial or noncyclic and its Sylow 3-subgroup is also trivial or noncyclic. However, in a preprint, Shieh, Hsiang, and Hsu [294] described a strong complete mapping for the dihedral group of order 12, a group with a nontrivial, cyclic Sylow 3-subgroup. In 2013 Evans [130] conjectured that a finite group G is strongly admissible if and only if its Sylow 2-subgroup is trivial or noncyclic and its Sylow 3-subgroup is trivial, noncyclic, or cyclic and not a homomorphic image of G. This conjecture was short-lived as the dihedral and quaternion groups of order 8 are not strongly admissible: this can easily be shown via a computer search. Embarassingly, these are the smallest groups that could have been counterexamples to this conjecture.

8.5.3 Knut Vic Designs

In a Latin square of order n, the cells $\{(i, k - i) \mid i = 0, \ldots, n - 1\}$, subtraction modulo n, form the kth *right diagonal*, and the cells $\{(i, k + i) \mid i = 0, \ldots, n - 1\}$, subtraction modulo n, form the kth *left diagonal*. Note that the main diagonal is the 0th left diagonal. A *Knut Vic design* of order n is a Latin square of order n in which each left diagonal is a transversal and each right diagonal is a transversal; thus the entries on each left diagonal are distinct and the entries on each right diagonal are distinct. There is a close connection between Knut Vic designs of order n and strong complete mappings of $\mathbb{Z}_n$.

Theorem 8.119. *A Knut Vic design of order n exists if and only if $\mathbb{Z}_n$ admits a strong complete mapping.*

Proof. Let K be a Knut Vic design of order n and let $a \in \mathbb{Z}_n$. Define a mapping $\theta \colon \mathbb{Z}_n \to \mathbb{Z}_n$ by $\theta(j) = a - i$ if the ijth entry of K is a. If $a - i = \theta(j) = \theta(j') = a - i'$, then $i = i'$ and so, as K is a Latin square, $j = j'$. Thus θ is a bijection. If $j + a - i = j + \theta(j) = j' + \theta(j') = j' + a - i'$, then $j - i = j' - i'$. Thus the ijth and $i'j'$th entries are on the same left diagonal and so $i = i'$ and $j = j'$. Thus θ is a complete mapping. If $a - i - j = \theta(j) - j = \theta(j') - j' = a - i' - j'$, then $j + i = j' + i'$. Thus the ijth and $i'j'$th entries are on the same right diagonal diagonal and so $i = i'$ and $j = j'$. Thus θ is an orthomorphism.

Conversely let θ be a strong complete mapping of $\mathbb{Z}_n$ and define K to be the square with ijth entry $i + \theta(j)$. As θ is a bijection, K is a Latin square. The entries on the kth right diagonal, $i + \theta(k - i) = k + (\theta(k - i) - (k - i))$, $i = 0, \ldots, n - 1$, are distinct as θ is an orthomorphism, and the entries on the kth left diagonal, $i + \theta(k+i) = -k+(\theta(k+i)+(k+i)), i = 0, \ldots, n-1$, are distinct as θ is a complete mapping. It follows that K is a Knut Vic design. □

Note that the Knut Vic design constructed from θ in Theorem 8.119 is in fact the Latin square that we denoted by L_θ in Section 1.3. In additive notation, if $G = \{g_1, \ldots, g_n\}$ is an abelian group and $\theta\colon G \to G$ is a bijection, then L_θ is the Latin square with ijth entry $g_i + \theta(g_j)$. In the proof of Theorem 8.119, we also proved the following.

Corollary 8.120. *If $\theta\colon \mathbb{Z}_n \to \mathbb{Z}_n$ is a strong complete mapping of $\mathbb{Z}_n$, then L_θ is a Knut Vic design of order n.*

In light of Theorem 8.119, known existence results for Knut Vic designs follow from existence results for strong complete mappings. In 1975 Hedayat and Federer [176] established the existence of Knut Vic designs of order n when neither 2 nor 3 divide n.

Theorem 8.121 (Hedayat and Federer, 1975). *If $\gcd(n, 6) = 1$, then Knut Vic designs of order n exist.*

Proof. By Theorem 8.101, if $\gcd(n, 6) = 1$, then $\mathbb{Z}_n$ admits strong complete mappings. This implies the existence of a Knut Vic design of order n by Theorem 8.119. □

Hedayat and Federer proved Theorem 8.121 by first using implicitly the mapping $x \mapsto -2x$ to construct Knut Vic designs of order p^r, p a prime other than 2 or 3. They then used a Kronecker product construction (this is described in Section 1.2) to show that Knut Vic designs existed for all orders not divisible by 2 or 3.

This left open the question, does a Knut Vic design of order n exist if 2 or 3 divides n? This question was answered in the negative by Hedayat [174] in 1977.

Theorem 8.122 (Hedayat, 1977). *A Knut Vic design of order n cannot exist if either 2 or 3 divides n.*

Proof. If either 2 or 3 divides n, then $\gcd(n, 6) \neq 1$. It follows then, from Corollary 8.109, that $\mathbb{Z}_n$ does not admit strong complete mappings, and so, by Theorem 8.119, no Knut Vic design of order n can exist. □

We say that two Knut Vic designs are *orthogonal* if they are orthogonal as Latin squares. Known results on the existence of mutually orthogonal Knut Vic designs follow from our results on strong complete mappings.

Lemma 8.123. *If there exists a set of k pairwise orthogonal strong complete mappings of $\mathbb{Z}_n$, then there exists a set of pairwise orthogonal Knut Vic designs of order n.*

Proof. If $\theta_1, \ldots, \theta_k$ is a set of pairwise orthogonal strong complete mappings of $\mathbb{Z}_n$, then $L_{\theta_1}, \ldots, L_{\theta_k}$ is a set of Knut Vic designs of order n by Corollary 8.120, which are pairwise orthogonal by Theorem 1.20. □

In 1977 Hedayat [174] proved the existence of $n-3$ mutually orthogonal Knut Vic designs of order n when n is prime.

Theorem 8.124 (Hedayat, 1977). *If n is a prime, then there exists a set of $n-3$ mutually orthogonal Knut Vic designs of order n.*

Proof. By Theorem 8.105, there exists a set of $n-3$ pairwise orthogonal strong complete mappings of $\mathbb{Z}_n$. Thus, by Lemma 8.123, there exists a set of $n-3$ mutually orthogonal Knut Vic designs of order n. □

In 1987 Afsarinejad [13] proved the existence of $p-3$ mutually orthogonal Knut Vic designs of order n, where p is the smallest prime divisor of n.

Theorem 8.125 (Afsarinejad, 1987). *If p is the smallest prime divisor of n, then there exist at least $p-3$ mutually orthogonal Knut Vic designs of order n.*

Proof. By Theorem 8.101, there exists a set of $p-3$ pairwise orthogonal strong complete mappings of $\mathbb{Z}_n$. Thus, by Lemma 8.123, there exists a set of $p-3$ mutually orthogonal Knut Vic designs of order n. □

Chapter 9
Elementary Abelian Groups. I

Elementary abelian groups can be thought of as additive groups of finite fields. As such, all of the tools of field theory are available to us in the study of orthomorphism graphs of these groups. In particular, any function from a finite field to itself, and thus any orthomorphism of the additive group of the field, can be realized as a polynomial function. Several interesting classes of orthomorphisms can be described as sets of orthomorphism polynomials by placing restrictions on the polynomials.

Classes of normalized orthomorphisms of additive groups of finite fields can also be defined using multiplication. The simplest such class is the class of linear orthomorphisms: these are of the form $x \mapsto ax$. It is clear that such a mapping is an orthomorphism if and only if $a \neq 0, 1$. In this chapter we will also study quadratic orthomorphisms: these map x to ax if x is a square and to bx if x is a nonsquare. In Chapter 10 we will generalize linear and quadratic orthomorphisms by partitioning the elements of a finite field into sets $\{C_i\}$ and then defining and studying classes of orthomorphisms consisting of mappings of the form $x \mapsto a_i x$, if $x \in C_i$. This leads to an important class of orthomorphisms, the class of cyclotomic orthomorphisms, when the partition consists of cyclotomic classes.

Combined, polynomials and cyclotomy provide us with some of our most powerful tools for defining and studying orthomorphism graphs of elementary abelian groups.

9.1 Permutation Polynomials

To any polynomial p over the finite field GF(q), we can associate a function $a \mapsto$ p(a) from GF(q) into itself. Polynomials whose associated functions are permutations—permutation polynomials—play an important role in the study of

A. B. Evans, *Orthogonal Latin Squares Based on Groups*, Developments in Mathematics 57, https://doi.org/10.1007/978-3-319-94430-2_9

orthomorphisms and complete mappings of elementary abelian groups as every orthomorphism and complete mapping of an elementary abelian group, viewed as the additive group of a finite field, is induced by a permutation polynomial.

9.1.1 Polynomials and Functions

There is a very close relationship between polynomials over $\mathrm{GF}(q)$ and functions from $\mathrm{GF}(q)$ to itself. Any polynomial over $\mathrm{GF}(q)$ induces a function from $\mathrm{GF}(q)$ to $\mathrm{GF}(q)$, and any function from $\mathrm{GF}(q)$ into itself is induced by a unique polynomial of degree at most $q - 1$: this was proved by Hermite [178] in 1863 for the case q a prime and by Dickson [97] in 1897 for the general case.

Theorem 9.1. *For any* $\mathrm{f}: \mathrm{GF}(q) \to \mathrm{GF}(q)$, *there exists a uniquely determined polynomial* $\mathrm{p}(x)$ *over* $\mathrm{GF}(q)$ *of degree at most* $q - 1$, *for which* $\mathrm{f}(a) = \mathrm{p}(a)$ *for all* $a \in \mathrm{GF}(q)$.

Proof. Let p and p′ be two polynomials of degree at most $q - 1$ over $\mathrm{GF}(q)$, satisfying $\mathrm{p}(a) = \mathrm{p}'(a)$ for all $a \in \mathrm{GF}(q)$. Now $\mathrm{p} = \mathrm{p}'$ as, otherwise, $\mathrm{d}(x) = \mathrm{p}(x) - \mathrm{p}'(x)$ has all q elements of $\mathrm{GF}(q)$ as roots, whereas $\mathrm{d}(x)$, being a nonzero polynomial of degree at most $q - 1$ over $\mathrm{GF}(q)$, can have at most $q - 1$ roots.

Thus the mapping from the set P, of polynomials of degree at most $q - 1$ over $\mathrm{GF}(q)$, to the set of associated functions $\mathrm{GF}(q) \to \mathrm{GF}(q)$ is an injection and is, in fact, a bijection as the number of functions $\mathrm{GF}(q) \to \mathrm{GF}(q)$ is q^q, which is the same as the number of polynomials over $\mathrm{GF}(q)$ of degree at most $q - 1$. □

As a consequence of Theorem 9.1, any permutation of the elements of $\mathrm{GF}(q)$ is induced by a polynomial over $\mathrm{GF}(q)$. A *permutation polynomial* of $\mathrm{GF}(q)$ is a polynomial over $\mathrm{GF}(q)$ whose associated function from $\mathrm{GF}(q)$ into $\mathrm{GF}(q)$ is a permutation of the elements of $\mathrm{GF}(q)$. Polynomials that induce the same function are very closely related.

Theorem 9.2. *Polynomials* f *and* g *over* $\mathrm{GF}(q)$ *induce the same function* $\mathrm{GF}(q) \to \mathrm{GF}(q)$ *if and only if* $\mathrm{f}(x) \equiv \mathrm{g}(x) \pmod{x^q - x}$.

Proof. If $\mathrm{f}(x) \equiv \mathrm{g}(x) \pmod{x^q - x}$, then $\mathrm{f}(x) = \mathrm{g}(x) + \mathrm{h}(x)(x^q - x)$ for some polynomial h over $\mathrm{GF}(q)$, and, as every element of $\mathrm{GF}(q)$ is a root of $x^q - x$, for any $a \in \mathrm{GF}(q)$, $\mathrm{f}(a) = \mathrm{g}(a) + \mathrm{h}(a)(a^q - a) = \mathrm{g}(a)$, and f and g represent the same function $\mathrm{GF}(q) \to \mathrm{GF}(q)$.

If, on the other hand, polynomials f and g represent the same function $\mathrm{GF}(q) \to \mathrm{GF}(q)$, that is, $\mathrm{f}(a) = \mathrm{g}(a)$ for all $a \in \mathrm{GF}(q)$, then there exist unique polynomials f′ and g′ of degree at most $q-1$ over $\mathrm{GF}(q)$ that satisfy $\mathrm{f}(x) \equiv \mathrm{f}'(x) \pmod{x^q - x}$ and $\mathrm{g}(x) \equiv \mathrm{g}'(x) \pmod{x^q - x}$. But then, f′ induces the same function $\mathrm{GF}(q) \to \mathrm{GF}(q)$ as f, which induces the same function as g, which induces the same function as g′. Thus $\mathrm{f}' = \mathrm{g}'$ by Theorem 9.1. Hence $\mathrm{f}(x) \equiv \mathrm{f}'(x) \equiv \mathrm{g}'(x) \equiv \mathrm{g}(x) \pmod{x^q - x}$, as claimed. □

Theorem 9.1 establishes a one-to-one correspondence between functions from GF(q) to GF(q) and polynomials of degree at most $q-1$ over GF(q). The Lagrange interpolation formula (see [229]) is used to determine polynomials that pass through a given set of points. The version of the Lagrange interpolation formula that we will give is from [64]: this gives an explicit formula for the unique polynomial of degree at most $q-1$ over GF(q) that induces a given function from GF(q) to GF(q).

Theorem 9.3 (Lagrange Interpolation Formula). *Let* f *be a function from* GF(q) *to* GF(q). *Then* f *is induced by the polynomial*

$$\begin{aligned} \mathrm{p}(x) &= -\sum_{a\in \mathrm{GF}(q)} \mathrm{f}(a)\left(\frac{x^q-x}{x-a}\right) \\ &= \sum_{a\in \mathrm{GF}(q)} \mathrm{f}(a)\left(1-(x-a)^{q-1}\right). \end{aligned}$$

Proof. The two versions of this interpolation formula are equivalent as

$$\begin{aligned} -(x-a)\left(1-(x-a)^{q-1}\right) &= -(x-a)+(x-a)^q \\ &= -\left(x-x^q\right)+\left(a-a^q\right) \\ &= x^q-x, \end{aligned}$$

for all $a \in \mathrm{GF}(q)$.

As

$$\left(1-(x-a)^{q-1}\right) = \begin{cases} 0 & \text{if } x \neq a, \\ 1 & \text{if } x = a, \end{cases}$$

$$\mathrm{p}(b) = \sum_{a\in \mathrm{GF}(q)} \mathrm{f}(a)\left(1-(b-a)^{q-1}\right) = \mathrm{f}(b).$$

Hence $\mathrm{p}(a) = \mathrm{f}(a)$ for all $a \in \mathrm{GF}(q)$, as claimed. □

9.1.2 Hermite's Criterion

Given a polynomial f over GF(q), we would like to determine whether f is a permutation polynomial of GF(q) or not. We will describe a test, first proved by Hermite [178] in 1863 for the case q a prime and subsequently generalized to arbitrary q by Dickson [97] in 1897.

In characterizing permutation polynomials, we are led to ask what algebraic properties a sequence $\{x_1, \ldots, x_q\}$ of elements of GF(q) might have for it to be a permutation of the elements of GF(q). The values of sums of powers of the x_is yield a characterization of sequences that are permutations.

Lemma 9.4. *The sequence* $\{x_1, \ldots, x_q\}$ *is a permutation of the elements of* GF(q) *if and only if*

$$\sum_{i=1}^{q} x_i^t = \begin{cases} 0 & \text{if } t = 1, \ldots, q-2, \\ q-1 & \text{if } t = q-1. \end{cases}$$

Proof. Set $S_t = \sum_{i=1}^{q} x_i^t$ and let g be a primitive element of GF(q).
If the sequence $\{x_1, \ldots, x_q\}$ is a permutation of the elements of GF(q), then

$$g^t S_t = \sum_{i=1}^{q} (gx_i)^t = \sum_{i=1}^{q} x_i^t = S_t.$$

It follows that $\left(g^t - 1\right) S_t = 0$ and thus $S_t = 0$, for $t = 1, \ldots, q-2$. For $t = q-1$,

$$S_{q-1} = \sum_{x_i \neq 0} 1 = q - 1.$$

Next assume that the sequence $\{x_1, \ldots, x_q\}$ of elements of GF(q) satisfies $S_t = 0$ for $t = 1, \ldots, q-2$, and $S_{q-1} = q-1$. By Theorem 9.1, for each i there exists a unique polynomial f_i of degree at most $q-1$ over GF(q) that satisfies $\mathrm{f}_i(x) = 0$ if $x \neq x_i$ and $\mathrm{f}_i(x_i) = 1$. By Theorem 9.3, $\mathrm{f}_i(x) = 1 - (x - x_i)^{q-1}$. Set $\mathrm{f}(x) = \sum_{i=1}^{q} \mathrm{f}_i(x)$. Then f is a polynomial of degree at most $q-1$ over GF(q) and, as $\mathrm{f}(a) = |\{i \mid x_i = a\}| \cdot 1$, $\{x_1, \ldots, x_q\}$ is a permutation of the elements of GF(q) if $\mathrm{f}(x) = 1$. As direct computation yields

$$\mathrm{f}(x) = \sum_{i=1}^{q} \left(1 - (x - x_i)^{q-1}\right) = \sum_{j=0}^{q-2} -(-1)^{q-1-j} \binom{q-1}{j} S_{q-1-j} x^j = 1,$$

the result follows. □

In Lemma 9.4, to show that $\{x_1, \ldots, x_q\}$ was a permutation of the elements of GF(q), we did not need S_{q-1} to be equal to $q-1$; it is sufficient for S_{q-1} to be nonzero.

Corollary 9.5. *The sequence* $\{x_1, \ldots, x_q\}$ *is a permutation of the elements of* GF(q) *if and only if* $\sum_{i=1}^{q} x_i^t = 0$ *for* $t = 1, \ldots, q-2$ *and is nonzero for* $t = q-1$.

Proof. This is essentially the same as the proof of Lemma 9.4, except that, in the second part of the proof, $S_{q-1} \neq 0$. It follows that the polynomial

$f(x) = -(-1)^{q-1}S_{q-1}$ is a nonzero constant. Thus $f(a) = |\{i \mid x_i = a\}| \cdot 1 \neq 0$, and hence $|\{i \mid x_i = a\}| \neq 0$ for all $a \in GF(q)$. Thus, for each $a \in GF(q)$, there exists an i for which $a = x_i$, and hence $\{x_1, \ldots, x_q\}$ is a permutation of the elements of $GF(q)$. □

We are now in a position to prove Hermite's criterion for a polynomial to be a permutation polynomial.

Theorem 9.6 (Hermite's Criterion). *A polynomial* f *over* $GF(q)$ *is a permutation polynomial of* $GF(q)$ *if and only if the coefficient of* x^{q-1} *in the reduction of* $f(x)^t$ *modulo* $x^q - x$ *is zero for* $t = 1, \ldots, q-2$, t *relatively prime to* q, *and nonzero for* $t = q - 1$.

Proof. Let f be a polynomial over $GF(q)$. We know that $f(x)^t = \sum_{i=0}^{q-1} a_{i,t}x^i + h_t(x)(x^q - x)$ for some polynomial h_t over $GF(q)$ and some $a_{i,t} \in GF(q)$. Set $S_t = \sum_{a \in GF(q)} a^t$. Then, by Lemma 9.4,

$$\sum_{a \in GF(q)} f(a)^t = \sum_{i=1}^{q-1} a_{i,t}S_i = (q-1)a_{q-1,t}.$$

Thus, by Lemma 9.4, it follows that f is a permutation polynomial of $GF(q)$ if and only if

$$a_{q-1,t} = \begin{cases} 0 & \text{if } t = 1, \ldots, q-2, \\ 1 & \text{if } t = q-1. \end{cases}$$

Hence, if f is a permutation polynomial of $GF(q)$, then the coefficient of x^{q-1} in the reduction of $f(x)^t$ modulo $x^q - x$ is zero for $t = 1, \ldots, q-2$, t relatively prime to q, and nonzero for $t = q - 1$.

Next suppose that the coefficient of x^{q-1} in the reduction of $f(x)^t$ modulo $x^q - x$ is zero for $t = 1, \ldots, q-2$, t relatively prime to q, and nonzero for $t = q-1$. If t is not relatively prime to q, then $t = mq^r$ for some m relatively prime to q and some positive integer r. Then $a_{q-1,t} = a_{q-1,m}^{q^r}$, and so $a_{q-1,t} = 0$ for $t = 1, \ldots, q-2$. The result then follows from Corollary 9.5. □

An alternative version of Hermite's criterion follows.

Corollary 9.7. *A polynomial* f *over* $GF(q)$ *is a permutation polynomial of* $GF(q)$ *if and only if the coefficient of* x^{q-1} *in the reduction of* $f(x)^t$ *modulo* $x^q - x$ *is zero for* $t = 1, \ldots, q-2$, t *relatively prime to* q, *and the equation* $f(x) = 0$ *has exactly one solution.*

Proof. If $f(x) = 0$ has exactly one solution in $GF(q)$, then $f(a)^{q-1} = 1$ for all but one $a \in GF(q)$, in which case $\sum_{a \in GF(q)} f(a)^{q-1} = q - 1$. The result then follows from Theorem 9.6. □

Setting $t = 1$ in Theorem 9.6 yields the following restriction on the degree of a permutation polynomial.

Corollary 9.8. *If $q > 2$, then the degree of a permutation polynomial over* GF(q), *reduced modulo $x^q - x$, is at most $q - 2$.*

More generally:

Corollary 9.9. *If* f *is a permutation polynomial of* GF(q) *of degree $n > 1$, then n cannot divide $q - 1$.*

Proof. If $\mathrm{f}(x) = a_n x^n + \cdots + a_0$ and n divides $q - 1$, then

$$\mathrm{f}(x)^{(q-1)/n} = a_n^{(q-1)/n} x^{q-1} + ((q-1)/n) a_0 a_n^{(q-n-1)/n} x^{q-2} + \cdots + a_0^{(q-1)/n}.$$

Hence the coefficient of x^{q-1} in the reduction of $\mathrm{f}(x)^{(q-1)/n}$ modulo $x^q - x$ is $a_n^{(q-1)/n} \neq 0$, and so f cannot be a permutation polynomial of GF(q) by Hermite's Criterion. □

In general, applying Hermite's criterion can be complex and tedious. For monomials, however, it yields a simple characterization of those that are permutation polynomials.

Corollary 9.10. *The monomial x^n is a permutation polynomial of* GF(q) *if and only if* $\gcd(n, q-1) = 1$.

Proof. If $\gcd(n, q-1) = d > 1$, then

$$(x^n)^{(q-1)/d} \pmod{x^q - x} = x^{q-1}.$$

Thus x^n is not a permutation polynomial of GF(q) by Hermite's criterion.

On the other hand, if $\gcd(n, q-1) = 1$, then $(x^n)^{q-1} \equiv x^{q-1} \pmod{x^q - x}$, and $(x^n)^t \not\equiv x^{q-1} \pmod{x^q - x}$ if $t = 1, \ldots, q-2$. Thus x^n is a permutation polynomial of GF(q) by Hermite's Criterion. □

9.2 More on Permutation Polynomials

As any permutation of the elements of GF$(q)^+$ can be represented by a permutation polynomial, any orthomorphism or complete mapping of GF$(q)^+$ can be represented by an orthomorphism or complete mapping polynomial. In this section we will describe some classes of permutation polynomials and orthomorphism/complete mapping polynomials. In particular we will describe all permutation and orthomorphism polynomials of degree at most six.

9.2.1 *Some Classes of Permutation Polynomials*

Permutations of the elements of GF(q) with simple cycle structures need not have simple associated polynomials. As an example, the permutations with the simplest structures are the transpositions. The transposition (a, b) has, using the Lagrange interpolation formula, the associated polynomial

$$x + (b - a)\left(1 - (x - a)^{q-1}\right) + (a - b)\left(1 - (x - b)^{q-1}\right),$$

which clearly has a more complex structure. An important problem in the study of permutation polynomials is the determination of classes of permutation polynomials that have simple polynomial structures. Several such classes of permutation polynomials have been discovered. Many permutation polynomials can be obtained from any given permutation polynomial using elementary polynomial arithmetic.

Lemma 9.11. *If* f *is a permutation polynomial of* GF(q)*, then so is* $\mathrm{h}(x) = \alpha\mathrm{f}(x + \beta) + \gamma$ *for any* $\alpha, \beta, \gamma \in \mathrm{GF}(q)$*,* $\alpha \neq 0$*.*

Proof. The polynomial h is the composition of four bijections, $a \mapsto a + \beta$, $a \mapsto f(a)$, $a \mapsto \alpha a$, and $a \mapsto a + \gamma$. □

We will say that two permutation polynomials f and h of GF(q) are *equivalent* if there exist $\alpha, \beta, \gamma \in \mathrm{GF}(q)$, $\alpha \neq 0$ for which $\mathrm{h}(x) = \alpha\mathrm{f}(x + \beta) + \gamma$: this is clearly an equivalence relation. In Lemma 9.11, we can choose α, β, and γ so that h is monic and has constant term 0. Further, should n, the degree of f, be relatively prime to q, then we may choose β so that the coefficient of x^{n-1} is zero. We shall call a degree n permutation polynomial of GF(q) a *normalized permutation polynomial* if it is monic, has zero constant term, and, should $\gcd(n, q) = 1$, has its coefficient of x^{n-1} equal to zero. As any permutation polynomial is equivalent to at least one normalized permutation polynomial, to determine all permutation polynomials of degree n, it is sufficient to determine all normalized permutation polynomials of degree n. A complete list of the normalized permutation polynomials of degree at most 5 was found by Dickson [97] in 1897: this list can also be found in [229] and in [252]. All normalized permutation polynomials of degree 5 or less are given in Theorem 9.12. Dickson also gave a list of normalized permutation polynomials of degree 6: this list will be given in Theorem 9.13. There was some question as to whether Dickson's list of permutation polynomials was complete or not. In 2010 Li, Chandler, and Xiang [227] gave a complete classification of permutation polynomials of degree 6 for even characteristic, and in 2013 Shallue and Wanless [291] classified all permutation polynomials of degree 6: these lists will be described in Theorems 9.14 and 9.15.

Theorem 9.12 (Dickson, 1897). *The following are the only normalized permutation polynomials of degree less than* 6*. Here* $\mathrm{GF}(9) = \{\alpha + \beta i \mid \alpha, \beta \in \mathrm{GF}(3), i^2 + 1 = 0\}$.

1. Normalized permutation polynomials of degree 1.

a. For all q, x.

2. Normalized permutation polynomials of degree 2.

a. For q even, x^2.

3. Normalized permutation polynomials of degree 3.

a. For $q \not\equiv 1 \pmod 3$, x^3.
b. For $q \equiv 0 \pmod 3$, $x^3 - ax$, *a a nonsquare.*

4. Normalized permutation polynomials of degree 4.

a. For q even, $x^4 + ax^2 + bx$, 0 *the only root in* $\mathrm{GF}(q)$.
b. For $q = 7$, $x^4 \pm 3x$.

5. Normalized permutation polynomials of degree 5.

a. For $q \not\equiv 1 \pmod 5$, x^5.
b. For $q \equiv 0 \pmod 5$, $x^5 - ax$, *a not a fourth power in* $\mathrm{GF}(q)$.
c. For $q \equiv 0 \pmod 5$, $x^5 - 2ax^3 + a^2x$, *a a nonsquare.*
d. For $q \equiv \pm 2 \pmod 5$, $x^5 + ax^3 + 5^{-1}a^2x$.
e. For $q = 7$, $x^5 \pm 2x^2$.
f. For $q = 7$, $x^5 + ax^3 \pm x^2 + 3a^2x$, *$a$ a nonsquare.*
g. For $q = 9$, $x^5 \pm ix$.
h. For $q = 13$, $x^5 + ax^3 + 3a^2x$, *a a nonsquare.*

Proof. The only possible normalized permutation polynomial of degree 1 is x, which is a permutation polynomial of $\mathrm{GF}(q)$ for all q.

By Corollary 9.9, a permutation polynomial of degree 2 cannot be a polynomial of $\mathrm{GF}(q)$ if q is odd, and if q is even, then the only possible normalized permutation polynomials of degree 2 are x^2 and x^2+ax, $a \neq 0$: the first of these is a permutation polynomial of $\mathrm{GF}(q)$ by Corollary 9.10, whereas the second has roots 0 and a and so cannot be a permutation polynomial.

The normalized permutation polynomials of degrees 3, 4, and 5 are determined through a tedious and routine application of Hermite's criterion. □

Computing all normalized permutation polynomials of degree n clearly becomes impractical as n increases. The following is Dickson's list of normalized permutation polynomials of degree 6 for the special case, q odd.

Theorem 9.13 (Dickson, 1897). *For q odd the following are normalized permutation polynomials of degree* 6. *For* $q = 9$, $\mathrm{GF}(9) = \{\alpha + \beta i \mid \alpha, \beta \in \mathrm{GF}(3), i^2 + 1 = 0\}$, $w^2 - w - 1 = 0$, *and* $w^2 = i$.

1. For $q = 9$:

a. $x^6 + ax^5 + a^3x^3 - a^4x^2 - a^5x$, *$a$ nonzero.*
b. $x^6 + ax^5 + ba^3x^3 - ba^4x^2 - ca^5x$, *$a$ nonzero,* $b = \pm(1 + c)$, $c = \pm i$.
c. $x^6 + ax^5 - a^3x^3 + a^4x^2 + (\pm i - 1)a^5x$, *$a$ nonzero.*

d. $x^6 + a^2x^4 + bx^3 + a^4x^2 + (2a^2b + a^5)x$, $b \in \{0, ia^3, 2ia^3, a^3\}$, $a \in \{1, \ldots, w^7\}$; *or* $b = 1 + i$, $a \in \{1, w^3, w^5, w^6, w^7\}$; *or* $b = 1 - i$, $a \in \{1, w, w^2, w^5, w^7\}$.
e. $x^6 + a^2x^4 + bx^3 + a^4x^2 + (2a^2b - a^5)x$, $b \in \{0, ia^3, 2ia^3, 2a^3\}$, $a \in \{1, \ldots, w^7\}$; *or* $b = 1 + i$, $a \in \{w, w^2, w^3, w^4, w^7\}$; *or* $b = 1 - i$, $a \in \{w, w^3, w^4, w^5, w^6\}$.

2. *For* $q = 11$*:*

a. $x^6 \pm 2x$.
b. $x^6 \pm 4x$.
c. $x^6 + a^2x^3 + ax^2 + 5x$, *a a nonzero square.*
d. $x^6 - a^2x^3 + ax^2 - 5x$, *a a nonzero square.*
e. $x^6 + 4a^2x^3 + ax^2 + 4x$, *a a nonsquare.*
f. $x^6 - 4a^2x^3 + ax^2 - 4x$, *a a nonsquare.*

3. *For* $q = 27$*:*

a. $x^6 + ax^5 - a^4x^2$, *a nonzero.*

Proof. See [97]. Dickson's list included the polynomials $x^6 + ax^4 + bx^3 + a^2x^2 + (2ab \pm a^{5/2})x$, $b \in \{0, \pm\sqrt{2}a^{3/2}, \pm a^{3/2}, \pm\sqrt{2} + 1\}$, a a nonzero square, $q = 9$. Here we replace a by a^2. Some of these polynomials are permutation polynomials and some are not, depending on the choice of sign and the choice of $a^{1/2}$ as was pointed out in [97]. We used magma to determine which of these are permutation polynomials. □

In 2010, Li, Chandler, and Xiang [227] gave a complete classification of permutation polynomials of degree 6 when q is even.

Theorem 9.14 (Li, Chandler, Xiang, 2010). *Any permutation polynomial of degree 6 over* GF(q)*,* q *even, is equivalent to one of the following permutation polynomials:*

1. *For* $q = 8$*:*

a. $x^6 + x^5 + x^3 + x^2 + x$.
b. $x^6 + x^5 + x^3 + ax^2 + ax$, $a^3 + a + 1 = 0$.
c. $x^6 + x^5 + ax^3$, $a^3 + a + 1 = 0$.
d. $x^6 + x^5 + x^4 + x^3 + x^2$.
e. $x^6 + x^5 + x^4 + x^3 + x$.
f. $x^6 + x^5 + x^4 + a^3x^3 + a^4x^2 + a^6x$, $a^3 + a + 1 = 0$.
g. $x^6 + x^5 + x^4$.
h. $x^6 + x^3 + x^2$.

2. *For* $q = 16$*:*

a. $x^6 + x^5 + x^3 + b^3x^2 + b^5x$, $b^4 + b + 1 = 0$.
b. $x^6 + x^5 + b^3x^4 + x^3 + bx^2 + b^6x$, $b^4 + b + 1 = 0$.
c. $x^6 + x^5 + b^3x^4 + x^3 + b^8x^2 + b^{13}x$, $b^4 + b + 1 = 0$.

3. *For* $q = 32$*:*

   a. $x^6 + x^5 + x^2$.

4. *For* $q = 2^n$*,* $n \geq 3$*,* n *odd:*

   a. x^6.

In 2013, Shallue and Wanless [291] gave a complete classification of permutation polynomials of degree 6.

Theorem 9.15 (Shallue and Wanless, 2013). *The following are the normalized permutation polynomials of degree* 6*. For* $q = 9$, $\mathrm{GF}(9) = \{\alpha + \beta i \mid \alpha, \beta \in \mathrm{GF}(3), i^2 + 1 = 0\}$

1. *For* $q = 8$*:*

   a. $x^6 + a^5b^2x^4 + a^4x^3 + (a^3 + a^3b + a^3b^4)x^2 + a^2b^2x$, *a nonzero, b arbitrary.*
   b. $x^6 + ax^5 + a^2(b^2 + b)x^4 + a^3x^3 + a^4(b^2 + c)x^2 + a^5(b + c)x$, *a nonzero,* $b^4 + b^2 + b = 0$, $c^4 + c^3 + c^2 + 1 = 0$.
   c. $x^6 + ax^5 + a^2(b^2 + b)x^4 + a^3cx^3 + a^4(b^4 + bc)x^2 + a^5(b^4 + b^2c)x$, *a nonzero, b arbitrary,* $c^3 + c + 1 = 0$.
   d. $x^6 + ax^5 + a^2(b^2 + b + 1)x^4 + a^3x^3 + a^4(b^2 + 1)x^2 + a^5bx$, *a nonzero, b arbitrary.*
   e. $x^6 + ax^5 + a^2(b^2 + b + 1)x^4 + a^3c^3x^3 + a^4(b^4 + bc^3 + c^4)x^2 + a^5(b^4 + b^2c^3 + c^3 + c^4)x$, *a nonzero, b arbitrary,* $c^4 + c^2 + c = 0$.

2. *For* $q = 9$*:*

   a. $x^6 + a^2x^4 + a^7bx^3 + a^4x^2 + a(2b + 1)x$, *a nonzero,* $b^6 - b^4 + b^3 + b^2 + b = 0$.
   b. $x^6 + ax^5 + 2abx^4 + (a^3 + ab^2 + 2b^3)x^3 + (2a^4 + ab^3)x^2 + (2a^5 + a^4b + 2ab^4)x$, *a nonzero, b arbitrary.*
   c. $x^6 + ax^5 + 2abx^4 + (ab^2 + 2b^3 + a^3c)x^3 + (ab^3 + 2a^4c)x^2 + (a^5c^2 + 2ab^4 + a^4bc)x$, *a nonzero, b arbitrary,* $c^4 + 1 = 0$.
   d. $x^6 + ax^5 + 2abx^4 + (2a^3 + ab^2 + 2b^3)x^3 + (a^4 + ab^3)x^2 + (2a^5 + a^5c + 2a^4b + 2ab^4)x$, *a nonzero, b arbitrary,* $c = \pm i$.

3. *For* $q = 11$*:*

   a. $x^6 \pm 2x$.
   b. $x^6 + a^2bx^3 + ax^2 + 5bx$, *a a nonzero square,* $b = \pm 1$.
   c. $x^6 + 4a^2bx^3 + ax^2 + 4bx$, $a = 0$ *or a nonsquare,* $b = \pm 1$.

4. *For* $q = 16$*:*

   a. $x^6 + ax^5 + a^2(b^6 + b^5 + c^4 + c^2 + c)x^4 + a^3x^3 + a^4(b^6 + b^5 + c^4 + c^2)x^2 + a^5(b + c)x$, *a nonzero,* $b^7 + b^4 + b^3 + b^2 + b = 0$, $c^8 + c^4 + c^2 + c + 1 = 0$.

5. *For* $q = 27$*:*

   a. $x^6 + ax^5 + 2abx^4 + (ab^2 + 2b^3)x^3 + (2a^4 + ab^3)x^2 + (a^4b + 2ab^4)x$, *a nonzero, b arbitrary.*

6. *For* $= 32$*:*

   a. $x^6 + ax^5 + a^2(b^2 + b)x^4 + a^4(b^4 + 1)x^2 + a^5b^4x$, *a nonzero, b arbitrary.*

7. *For* $q = 2^t$, $t \geq 3$ *odd:*

   a. $x^6 + a^2x^4 + a^4x^2$, *a arbitrary.*

One class of permutation polynomials discovered by Carlitz [63] in 1962 will prove useful in Section 9.3 of this chapter in our study of quadratic orthomorphisms.

Theorem 9.16. *If* q *is odd, then* $x^{(q+1)/2}+ax$ *is a permutation polynomial of* $\mathrm{GF}(q)$ *if and only if* $a^2 - 1$ *is a nonzero square.*

Proof. We will make use of the fact that $\mathrm{f}(x) = x^{(q+1)/2} + ax = \left(x^{(q-1)/2} + a\right)x$ and $b^{(q-1)/2}$ is 0 if $b = 0$, 1 if b is a nonzero square, and -1 if b is a nonsquare. The image of $\mathrm{GF}(q)$ under f is

$$\{0\} \cup \{(a+1)b \mid b \text{ a nonzero square}\} \cup \{(a-1)b \mid b \text{ a nonsquare}\}.$$

Let us set $S = \{(a+1)b \mid b \text{ a nonzero square}\}$ and $N = \{(a-1)b \mid b \text{ a nonsquare}\}$. As neither S nor N can contain 0, $a \neq \pm 1$. S is the set of nonzero squares if $a + 1$ is a nonzero square and the set of nonsquares if $a + 1$ is a nonsquare. Similarly N is the set of nonzero squares if $a - 1$ is a nonsquare and the set of nonsquares if $a - 1$ is a nonzero square. It follows that f is a surjection, and hence a bijection, if and only if $\mathrm{GF}(q) = \{0\} \cup S \cup N$, if and only if $a + 1$ and $a - 1$ are both nonzero squares or both nonsquares, if and only if $a^2 - 1$ is a nonzero square. □

For more information on permutation polynomials, see the 2013 handbook by Mullen and Panario [247], the 1993 survey article by Mullen [245], or the 2015 survey by Hou [182].

9.2.2 *Complete Mapping and Orthomorphism Polynomials*

As any function $\mathrm{GF}(q) \to \mathrm{GF}(q)$ can be realized as a polynomial function, so can any orthomorphism or complete mapping of $\mathrm{GF}(q)^+$. We will say that a polynomial f is an *orthomorphism polynomial* of $\mathrm{GF}(q)$ if both $\mathrm{f}(x)$ and $\mathrm{f}(x)-x$ are permutation polynomials of $\mathrm{GF}(q)$ and a *complete mapping polynomial* of $\mathrm{GF}(q)$ if both $\mathrm{f}(x)$ and $\mathrm{f}(x) + x$ are permutation polynomials of $\mathrm{GF}(q)$. Clearly a polynomial f over $\mathrm{GF}(q)$ is an orthomorphism polynomial of $\mathrm{GF}(q)$ if and only if the induced mapping $a \mapsto \mathrm{f}(a)$ is an orthomorphism of $\mathrm{GF}(q)^+$, and a complete mapping polynomial of $\mathrm{GF}(q)$ if and only if the induced mapping $a \mapsto \mathrm{f}(a)$ is a complete mapping of $\mathrm{GF}(q)^+$. The same connection between orthomorphisms and complete mappings holds for orthomorphism polynomials and complete mapping polynomials: $\mathrm{f}(x)$ is an orthomorphism polynomial of $\mathrm{GF}(q)$ if and only if $\mathrm{f}(x)-x$ is a complete mapping polynomial of GF(q) and a complete mapping polynomial of GF(q) if and only if f(x) + x is an orthomorphism polynomial of GF(q).

In Corollary 9.8 we saw that the degree of a permutation polynomial of GF(q), reduced modulo $x^q - x$, must be less than $q - 1$. In fact for orthomorphism polynomials and complete mapping polynomials, we can do better. The following improvement was obtained by Niederreiter and Robinson [252] in 1982 for q odd and by Wan [320] in 1986 for q even.

Theorem 9.17. *For $q > 2$ the degree of an orthomorphism polynomial or complete mapping polynomial of* GF(q)*, reduced modulo $x^q - x$, is at most $q - 3$.*

Proof. For q odd let f be an orthomorphism polynomial of GF(q). We may assume without loss of generality that the degree of f is less than q and as f is a permutation polynomial, by Corollary 9.8, that the degree of f is at most $q - 2$. As $\mathrm{f}(x) - x$ is a permutation polynomial, the degree of $(\mathrm{f}(x) - x)^2 = \mathrm{f}(x)^2 - 2x\mathrm{f}(x) - x^2$, reduced modulo $x^q - x$, will be less than $q - 1$. As the degree of $\mathrm{f}(x)^2$, reduced modulo $x^q - x$, is less than $q - 1$, the degree of $2x\mathrm{f}(x)$ must be less than $q - 1$, and hence the degree of f must be less than $q - 2$.

For the case $q = 2^n$, let p be an irreducible polynomial of degree n over GF(2), and let ζ be a root of p. Let $\eta\colon \mathbb{Z} \to \mathrm{GF}(2)$ be the canonical homomorphism which we extend to a homomorphism from the ring of polynomials over $\mathbb{Z}$ to the ring of polynomials over GF(2) by setting $\eta(x) = x$. Let P be a monic polynomial over $\mathbb{Z}$ satisfying $\eta(\mathrm{P}(x)) = \mathrm{p}(x)$ and let ν be a root of P. By setting $\eta(\nu) = \zeta$, η becomes a surjective homomorphism from the ring $E = \mathbb{Z}(\nu)$ of algebraic integers to GF(q) as well as a surjective homomorphism from the ring of polynomials over E to the ring of polynomials over GF(q). Let g be a primitive element of GF(q) and let $h \in E$ satisfy $\eta(h) = g$. Then $h^{q-1} \equiv 1 \pmod 2$ and $h^i \not\equiv 1 \pmod 2$ for $1 \le i < q - 1$. Now $\eta(h^q) = g$ and, if $h^{q-1} \not\equiv 1 \pmod 4$, then $h^{q-1} \equiv 3 \pmod 4$, $q - 1 \equiv 3 \pmod 4$ as $q > 2$, and

$$\begin{aligned} (h^q)^{q-1} &= \left(h\left(1 + 2\left(\frac{h^{q-1} - 1}{2}\right)\right)\right)^{q-1} \\ &\equiv h^{q-1} + 2h^{q-1}(q-1)\left(\frac{h^{q-1} - 1}{2}\right) \pmod 4 \\ &\equiv 1 \pmod 4. \end{aligned}$$

Thus we may assume, without loss of generality, that $h^{q-1} \equiv 1 \pmod 4$.

Set $S = \{h^i \mid 1 \le i \le q - 1\} \cup \{0\}$, let f be any permutation polynomial of GF(q), and let F be a polynomial over E for which $\eta(\mathrm{F}(x)) = \mathrm{f}(x)$. Then

$$\{\eta(s) \mid s \in S\} = \{\eta(\mathrm{F}(s)) \mid s \in S\} = \mathrm{GF}(q)$$

and so $\sum_{s\in S} \mathrm{F}(s)^2 = \sum_{s\in S} (s + 2\mathrm{G}(s))^2$ for some function $\mathrm{G}\colon E \to E$. Hence

$$\begin{aligned}\sum_{s\in S}\mathrm{F}(s)^2 &\equiv s^2 \pmod 4\\ &\equiv h^2\left(\frac{\left(h^{q-1}\right)^2-1}{h^2-1}\right) \pmod 4\\ &\equiv 0 \pmod 4.\end{aligned}$$

Thus, if f is an orthomorphism polynomial of GF(q) and $\eta(\mathrm{F}(x)) = \mathrm{f}(x)$, then $\sum_{s\in S}\mathrm{F}(s)^2 \equiv 0 \pmod 4$ and $\sum_{s\in S}(\mathrm{F}(s)-s)^2 \equiv 0 \pmod 4$. As $\sum_{s\in S} s^2 \equiv 0 \pmod 4$,

$$\begin{aligned}\sum_{s\in S}(\mathrm{F}(s)-s)^2 &\equiv \sum_{s\in S}\mathrm{F}(s)^2 - 2\sum_{s\in S} s\mathrm{F}(s) - \sum_{s\in S} s^2 \pmod 4\\ &\equiv -2\sum_{s\in S} s\mathrm{F}(s) \pmod 4\end{aligned}$$

Thus $2\sum_{s\in S} s\mathrm{F}(s) \equiv 0 \pmod 4$. Suppose that

$$\begin{aligned}\mathrm{f}(x) &= a_{q-2}x^{q-2} + a_{q-3}x^{q-3} + \cdots + a_0, \quad \text{and}\\ \mathrm{F}(x) &= b_{q-2}x^{q-2} + b_{q-3}x^{q-3} + \cdots + b_0.\end{aligned}$$

Then, as $\sum_{s\in S} s^i \equiv 0 \pmod 2$ for $i = 1, \cdots, q-2$, and $\sum_{s\in S} s^{q-1} \equiv 1 \pmod 2$, $2b_{q-2} \equiv 0 \pmod 4$ and so $a_{q-2} = \eta(b_{q-2}) = 0$ proving that the degree of f is at most $q-3$.

If f is a complete mapping polynomial, then $\mathrm{f}(x) + x$ is an orthomorphism polynomial. Hence, as the degree of $\mathrm{f}(x) + x$ is at most $q-3$, the degree of f is at most $q-3$. □

A different proof of Theorem 9.17 for the case q even was given by Sun and Zhang [304] in 1998. Other restrictions on the possible degrees of orthomorphism polynomials were given by Li [228] in 2008 for the special case q even. Note that Li uses the term orthomorphic permutation polynomials for what we are calling orthomorphism polynomials.

Given a list of normalized permutation polynomials, we wish to determine a list of orthomorphism polynomials.

Lemma 9.18. *If* θ *is an orthomorphism polynomial of* GF(q)*, then there exists a normalized permutation polynomial* f *of* GF(q)*, and* $a, b, c \in$ GF(q)*,* $a \neq 0$*, such that* $\theta(x) = a\mathrm{f}(x+b) + c$ *and* f *is orthogonal to the mapping* $x \mapsto a^{-1}x$.

If f *is a normalized permutation polynomial of* GF(q) *orthogonal to the mapping* $x \mapsto a^{-1}x$*,* $a \neq 0$*, then for all* $b, c \in$ GF(q)*, the polynomial* $a\mathrm{f}(x+b) + c$ *is an orthomorphism polynomial of* GF(q).

Proof. If θ is an orthomorphism polynomial of GF(q), then $\theta(x)$ and $\theta(x) - x$ are both permutation polynomials of GF(q), and by Lemma 9.11, so are $f(x) = \alpha\theta(x+\beta)+\gamma$ and $f(x) - \alpha x = \alpha\theta(x+\beta) - \alpha x + \gamma$ for any $\alpha, \beta, \gamma \in GF(q)$, $\alpha \neq 0$. Let us choose α, β, and γ so that f is a normalized permutation polynomial of GF(q). Then, setting $a = \alpha^{-1}$, $b = -\beta$, and $c = -\gamma$, $\theta(x) = af(x+b)+c$ and $f(x) \perp a^{-1}x$.

If f is a normalized permutation polynomial of GF(q) orthogonal to the mapping $x \mapsto a^{-1}x$, then $f(x)$ and $f(x) - a^{-1}x$ are permutation polynomials of GF(q), and, as $a \neq 0$, $af(x)$ and $af(x) - x$ are permutation polynomials of GF(q). Thus $af(x)$ is an orthomorphism polynomial of GF(q). It is then easy to show that $af(x+b)+c$ is an orthomorphism polynomial of GF(q) for all $b, c \in GF(q)$. □

If f is a normalized permutation polynomial of GF(q) orthogonal to $a^{-1}x$, then $g(x) = f(x) - a^{-1}x$ is also a permutation polynomial of GF(q). Further, if the degree of f is $n \geq 3$, then g is easily seen to be a normalized permutation polynomial of GF(q) of degree n. Thus, as we know all permutation polynomials of degrees 3, 4, and 5, we can determine all orthomorphism polynomials of GF(q) of degrees 3, 4, and 5. This approach was used by Niederreiter and Robinson [252] in 1982 to compute all complete mapping polynomials of degree at most 5. Adding x to each of these complete mapping polynomials yields the list of all orthomorphism polynomials of degree at most 5.

Theorem 9.19. *The following are the orthomorphism polynomials of degree less than* 6. *Here* $GF(9) = \{\alpha + \beta i \mid \alpha, \beta \in GF(3), i^2 + 1 = 0\}$.

1. *Orthomorphism polynomials of degree* 1.
   a. *For all* $q > 2$, $a(x+b)+c$, $a \neq 0, 1$, *b and c arbitrary.*
2. *There are no orthomorphism polynomials of degree* 2.
3. *Orthomorphism polynomials of degree* 3.
   a. *For* $q \equiv 0 \pmod 3$, $-ax^3 + x + c$, *a a nonsquare, c arbitrary.*
   b. *For* $q \equiv 0 \pmod 3$, $ax^3 + c$, *a a nonsquare, c arbitrary.*
   c. *For* $q \equiv 0 \pmod 3$, $(b-a)^{-1}x^3 + \left(1 - b(b-a)^{-1}\right)x + c$, *a and b distinct nonsquares, c arbitrary.*
4. *Orthomorphism polynomials of degree* 4.
   a. *For q even,* $q > 2$, $a^{-1}(x^4 + bx^2 + (a+c)x) + d$, $a \neq 0$, *d arbitrary, and* $x^4 + bx^2 + cx = 0$ *and* $x^4 + bx^2 + (a+c)x = 0$ *each have only* $x = 0$ *as a solution in* GF(q).
   b. *For* $q = 7$, $(x+a)^4 + 4x + b$, *a and b arbitrary.*
   c. *For* $q = 7$, $6(x+a)^4 + 4x + b$, *a and b arbitrary.*
5. *Orthomorphism polynomials of degree* 5.
   a. *For* $q \equiv 0 \pmod 5$, $-ax^5 + x + b$, *a not a fourth power, b arbitrary.*
   b. *For* $q \equiv 0 \pmod 5$, $ax^5 + b$, *a not a fourth power, b arbitrary.*

c. *For* $q \equiv 0 \pmod 5$, $(a-b)^{-1}x^5 + \left(1 - a(a-b)^{-1}\right)x + c$, $a \neq b$ *not fourth powers,* c *arbitrary.*
d. *For* $q = 9$, $\pm i(x+b)^5 + c$, b *and* c *arbitrary.*
e. *For* $q = 9$, $\pm i(x+b)^5 + x + c$, b *and* c *arbitrary.*
f. *For* $q = 9$, $\pm i(x+b)^5 - x + c$, b *and* c *arbitrary.*
g. *For* $q = 13$, $5a^{-2}(x+b)^5 + 5a^{-1}(x+b)^3 + 2x + c$, a *a nonsquare,* b *and* c *arbitrary.*
h. *For* $q = 13$, $8a^{-2}(x+b)^5 + 8a^{-1}(x+b)^3 + 12x + c$, a *a nonsquare,* b *and* c *arbitrary.*

Proof. Let θ be an orthomorphism polynomial of GF(q) of degree n. The case $n = 1$ is trivial. For the case $n = 2$, let us first note that if $f(x) = ax^2 + bx + c$, $a \neq 0$ is a permutation polynomial of GF(q), then $b = 0$ as otherwise $f(0) = c = f(-a^{-1}b)$: hence $f(x)$ and $f(x) - x$ cannot both be permutation polynomials of GF(q), and so no orthomorphism polynomials of degree 2 exist.

If $n = 3$, then $q \equiv 0$ or $2 \pmod 3$. If $q \equiv 2 \pmod 3$, then the only normalized permutation polynomial of GF(q) of degree 3 is x^3. But then $x^3 - a^{-1}x$ cannot be a permutation polynomial of GF(q) for any $a \neq 0$, and so no orthomorphism polynomials of GF(q) of degree 3 can exist if $q \equiv 2 \pmod 3$. If $q \equiv 0 \pmod 3$, then the only normalized permutation polynomials of GF(q) of degree 3 are x^3 and $x^3 - bx$, b a nonsquare. Now $x^3 \perp a^{-1}x$ if and only if a is a nonsquare, and, if b is a nonsquare, then $x^3 - bx \perp a^{-1}x$ if and only if $a^{-1} = -b$ or $a^{-1} + b$ is a nonsquare. The corresponding orthomorphism polynomials can be computed using Lemma 9.18. The cases $n = 4$ and $n = 5$ are dealt with similarly. □

Niederreiter and Robinson also used Dickson's list of normalized permutation polynomials to determine all degree-6 complete mapping polynomials of GF(q) when $\gcd(q, 6) = 1$. Adding x to each of their polynomials yields all degree-6 orthomorphism polynomials of GF(q) when $\gcd(q, 6) = 1$.

Theorem 9.20 (Niederreiter and Robinson, 1982). *If* f *is an orthomorphism polynomial of degree* 6 *and* q *is relatively prime to* 6*, then* $q = 11$*, and* $f(x)$ *is one of the following:*

1. $2(x+b)^6 + 4x + c$, b *and* c *arbitrary.*
2. $2(x+b)^6 + 8x + c$, b *and* c *arbitrary.*
3. $3(x+b)^6 + 6x + c$, b *and* c *arbitrary.*
4. $4(x+b)^6 + 6x + c$, b *and* c *arbitrary.*
5. $5(x+b)^6 + 2x + c$, b *and* c *arbitrary.*
6. $5(x+b)^6 + 10x + c$, b *and* c *arbitrary.*
7. $6(x+b)^6 + 2x + c$, b *and* c *arbitrary.*
8. $6(x+b)^6 + 10x + c$, b *and* c *arbitrary.*
9. $7(x+b)^6 + 6x + c$, b *and* c *arbitrary.*
10. $8(x+b)^6 + 6x + c$, b *and* c *arbitrary.*
11. $9(x+b)^6 + 4x + c$, b *and* c *arbitrary.*
12. $9(x+b)^6 + 8x + c$, b *and* c *arbitrary.*

Proof. Similar to the proof of Theorem 9.19. □

In 2013 Shallue and Wanless [291] extended Niederreiter and Robinson's list to include orthomorphism polynomials of GF(q) when $\gcd(q, 6) \neq 1$, i.e., when q is either even or a power of 3. They found that there were no orthomorphism polynomials of degree 6 unless $q = 9$, and they determined all degree-6 orthomorphism polynomials of GF(9).

Theorem 9.21 (Shallue and Wanless, 2013). *There are no orthomorphism polynomials of* GF(q) *of degree* 6 *if* q *is even or if* $q = 3^k$, $k \geq 3$.

Theorem 9.22 (Shallue and Wanless, 2013). *The degree-*6 *orthomorphism polynomials of* GF(9) *are the following. Here* $\mathrm{GF}(9) = \{\alpha + \beta i \mid \alpha, \beta \in \mathrm{GF}(3), i^2 + 1 = 0\}$.

1. $a(x+c)^6 + a^7(x+c)^4 + a^5(x+c)^2 + 2x + d$, *$a$ nonzero, c and d arbitrary.*
2. $a(x+c)^6 + a^7(x+c)^4 - a^2b(x+c)^3 + a^5(x+c)^2 + (b+2)x + d$, *$a$ nonzero, $b = \pm i$, c and d arbitrary.*
3. $a^5(x+c)^6 + a^4b(x+c)^5 + a^2b(x+c)^3 + a(x+c)^2 + 2(b+1)x + d$, *$a$ nonzero, $b = \pm i$, c and d arbitrary.*

In 1987 Mullen and Niederreiter [246] studied complete mapping polynomials constructed from Dickson polynomials. The *Dickson polynomials* are defined by

$$D_n(x, a) = \sum_{i=0}^{\lfloor n/2 \rfloor} \frac{n}{n-i} \binom{n-i}{i} (-a)^i x^{n-2i}.$$

Mullen and Niederreiter found necessary conditions for $\alpha D_n(x, a) + \beta$, $\alpha \neq 0$, to be a complete mapping polynomial. There are a number of constructions of complete mapping polynomials in the literature: for recent work on complete mapping polynomials, see the 2013 paper by Muratović-Ribić and Pasalic [249]; the 2014 paper by Wu, Li, Helleseth, and Zhang [336]; and the 2015 papers by Baofeng and Dongdai [27], by Guangkui and Cao [160], and by Zha, Hu, and Cao [340].

9.3 Linear and Quadratic Orthomorphisms

One approach to constructing normalized orthomorphisms of $\mathrm{GF}(q)^+$ is to partition the nonzero elements of GF(q) into sets $A_1, \ldots, A_n$ and then study mappings of the form

$$\theta(x) = \begin{cases} 0 & \text{if } x = 0, \\ a_i x & \text{if } x \in A_i. \end{cases}$$

Among the simplest classes of such orthomorphisms are the linear orthomorphisms, obtained when $n = 1$, and the quadratic orthomorphisms, obtained when $n = 2$, A_1 is the set of nonzero squares, and A_2 is the set of nonsquares.

9.3.1 Linear Orthomorphisms

The simplest partition of the nonzero elements of $\mathrm{GF}(q)$ consists of only one set. This partition yields mappings of the form $x \mapsto ax$: we shall use the symbol $[a]$ to denote the mapping $x \mapsto ax$, that is, $[a]$ is the function $\mathrm{GF}(q) \to \mathrm{GF}(q)$ induced by the polynomial ax. We know from Example 1.25 that $[a]$ is a bijection if and only if $a \neq 0$ and an orthomorphism if and only if $a \neq 0, 1$. We will use $\mathscr{C}_1(q)$, or $\mathscr{C}_1$ if there is no ambiguity, to denote the set of orthomorphisms $\{[a] \mid a \in \mathrm{GF}(q), a \neq 0, 1\}$ as well as the corresponding induced orthomorphism graph. The elements of $\mathscr{C}_1$ will be called *linear orthomorphisms*. As $[a] \perp [b]$ if and only if $a \neq b$, the $q - 2$ elements of $\mathscr{C}_1$ form a complete set of orthomorphisms of $\mathrm{GF}(q)^+$.

The action of congruences on $\mathscr{C}_1$ is easily determined.

Theorem 9.23.

1. *If* $\theta \in \mathscr{C}_1$ *then* $\mathrm{T}_g[\theta] = \theta$ *for all* g.
2. *If* q *is prime, then* $\theta \in \mathscr{C}_1$ *if and only if* $\mathrm{T}_g[\theta] = \theta$ *for all* g.
3. $\mathrm{H}_\alpha[\theta] = \theta$ *for all* $\alpha \in \mathscr{C}_1$ *if and only if* $\theta \in \mathscr{C}_1$.
4. $\mathrm{R}[[a]] = [1 - a]$.
5. $\mathrm{I}[[a]] = [1/a]$.

Proof.

1. If $\theta = [a]$, $a \neq 0, 1$, then $\mathrm{T}_g[\theta](x) = a(x + g) - ag = ax = \theta(x)$ for all $x, g \in \mathrm{GF}(q)$.
2. By (1), if q is prime and $\theta \in \mathscr{C}_1$, then $\mathrm{T}_g[\theta] = \theta$ for all g. If, on the other hand, $\mathrm{T}_g[\theta] = \theta$ for all g, then $\mathrm{T}_g[\theta](x) = \theta(x + g) - \theta(g) = \theta(x)$, from which it follows that $\theta(x + g) = \theta(x) + \theta(g)$ for all $x, g \in \mathrm{GF}(q)$. Thus θ is an automorphism of $\mathrm{GF}(q)^+$, and, when q is prime, all automorphisms of $\mathrm{GF}(q)^+$ are of the form $x \mapsto ax$.
3. If $\theta \in \mathscr{C}_1$, then $\theta = [a]$ for some $a \neq 0, 1$, and $\mathrm{H}_{[b]}[\theta](x) = b\theta(b^{-1}x) = bab^{-1}x = ax = [a](x)$ for all $x \in \mathrm{GF}(q)$, as claimed.

   If $\mathrm{H}_\alpha[\theta] = \theta$ for all $\alpha \in \mathscr{C}_1$, then $b\theta(b^{-1}x) = \theta(x)$ for all $x \in GF(q)$ and all $b \neq 0$. Set $a = \theta(1)$. Then $a \neq 0, 1$ as $a = \theta(1) \neq \theta(0) = 0$ and $a - 1 = \theta(1) - 1 \neq \theta(0) - 0 = 0$. For all $b \neq 0, 1$, $\theta(b) = b\theta(b^{-1}b) = b\theta(1) = ab$. Hence $\theta = [a]$.
4. $\mathrm{R}[[a]](x) = x + a(-x) = (1 - a)x = [1 - a](x)$.
5. $[1/a][a](x) = (1/a)(ax) = x = a((1/a)x) = [a][1/a](x)$ for all $x \in \mathrm{GF}(q)$. Thus $[1/a] = [a]^{-1} = \mathrm{I}[[a]]$. □

Theorem 9.23 contains two characterizations of the elements of $\mathscr{C}_1$ in terms of the congruences that fix them: for general q, $\theta \in \mathscr{C}_1$ if and only if $H_\alpha[\theta] = \theta$ for all $\alpha \in \mathscr{C}_1$; and for q prime, $\theta \in \mathscr{C}_1$ if and only if $T_g[\theta] = \theta$ for all g. The class $\mathscr{C}_1$ is only one of several classes of orthomorphisms that can be characterized by the congruences that fix them. The action of I and R on $\mathscr{C}_1$, described in Theorem 9.23, is summarized in Figure 9.1.

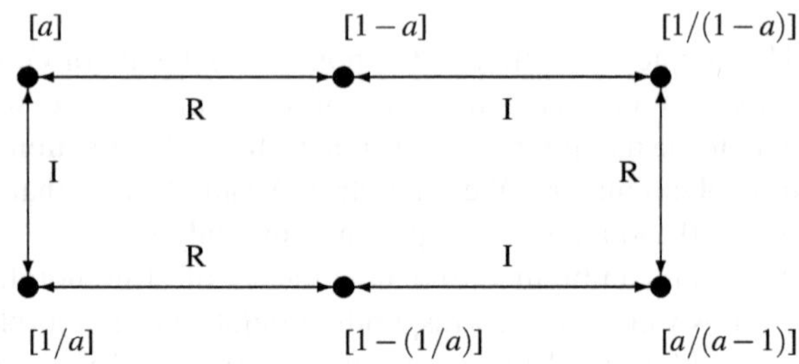

Fig. 9.1 The action of I and R on $\mathscr{C}_1$

9.3.2 *Quadratic Orthomorphisms*

For q odd, the nonzero elements of GF(q) can be partitioned into two sets, the nonzero squares and the nonsquares. If a and b are nonzero, then the mapping $[a,b]\colon \mathrm{GF}(q) \to \mathrm{GF}(q)$ is defined by

$$[a,b](x) = \begin{cases} 0 & \text{if } x = 0, \\ ax & \text{if } x \text{ is a nonzero square,} \\ bx & \text{if } x \text{ is a nonsquare.} \end{cases}$$

We will use $\mathscr{C}_2(q)$, or $\mathscr{C}_2$ if there is no ambiguity, to denote the set $\{[a,b] \mid [a,b]$ an orthomorphism of $\mathrm{GF}(q)^+\}$ as well as the induced orthomorphism graph. The elements of $\mathscr{C}_2$ will be called *quadratic orthomorphisms.* Clearly $\mathscr{C}_1(q) \subseteq \mathscr{C}_2(q)$ as $[a,a] = [a]$, and so linear orthomorphisms are also quadratic orthomorphisms. Quadratic orthomorphisms were first defined by Mendelsohn and Wolk [240] in 1985 and subsequently studied by Evans [111] in 1987. The basic properties of mappings of the form $[a,b]$ are given in the following theorem.

Theorem 9.24.

1. *The mapping* $[a,b]\colon \mathrm{GF}(q) \to \mathrm{GF}(q)$ *is induced by the polynomial*

$$((a-b)/2)x^{(q+1)/2} + ((a+b)/2)x.$$

2. $[a,b](x) - [c,d](x) = [a-c, b-d](x)$.

3. $[a, b](x) + [c, d](x) = [a + c, b + d](x)$.
4. *If c is a nonzero square, then* $[a, b](cx) = [ac, bc](x) = c[a, b](x)$.
5. *If c is a nonsquare, then* $[a, b](cx) = [bc, ac](x) = c[b, a](x)$.
6. *If ab is a nonzero square, then*

$$[c, d] \circ [a, b] = \begin{cases} [ca, db] & \text{if } a \text{ and } b \text{ are both nonzero squares,} \\ [da, cb] & \text{if } a \text{ and } b \text{ are both nonsquares.} \end{cases}$$

7. $[a, b]$ *is a bijection if and only if* a/b *is a nonzero square, if and only if ab is a nonzero square.*
8. $[a, b]$ *is an orthomorphism if and only if* a/b *and* $(a - 1)/(b - 1)$ *are both nonzero squares, if and only if ab and* $(a-1)(b-1)$ *are both nonzero squares.*
9. $[a, b]$ *is a complete mapping if and only if* a/b *and* $(a + 1)/(b + 1)$ *are both nonzero squares, if and only if ab and* $(a+1)(b+1)$ *are both nonzero squares.*
10. $[a, b] \perp [c, d]$ *if and only if* $(a - c)/(b - d)$ *is a nonzero square, if and only if* $(a - c)(b - d)$ *is a nonzero square.*

Proof.

1.

$$\left(\frac{a-b}{2}\right) x^{(q+1)/2} + \left(\frac{a+b}{2}\right) x = \left(\left(\frac{a-b}{2}\right) x^{(q-1)/2} + \left(\frac{a+b}{2}\right)\right) x$$

$$= \begin{cases} \left(\left(\frac{a+b}{2}\right) + \left(\frac{a-b}{2}\right)\right) x & \text{if } x \text{is a square,} \\ \left(\left(\frac{a+b}{2}\right) - \left(\frac{a-b}{2}\right)\right) x & \text{if } x \text{is a nonsquare.} \end{cases}$$

$$= \begin{cases} ax & \text{if } x \text{ is a square,} \\ bx & \text{if } x \text{ is a nonsquare.} \end{cases}$$

2. If x is a nonzero square, then $[a, b](x) - [c, d](x) = ax - cx = (a - c)x = [a - c, b - d](x)$. Similarly if x is a nonsquare.
3. Similar to (2).
4. Suppose that c is a nonzero square. If x is a nonzero square, then cx is also a nonzero square and $[a, b](cx) = acx = [ac, bc](x) = c[a, b](x)$. The case x a nonsquare is similar.
5. Similar to (4)
6. If ab is a nonzero square, then either a and b are both nonzero squares or both are nonsquares. If both a and b are nonzero squares and x is a nonzero square, then $[c, d] \circ [a, b](x) = [c, d](ax) = cax = [ca, db](x)$. The other cases are similar.
7. Suppose that $[a, b]$ is a bijection. If a is a nonzero square, then $[a, b]$ maps nonzero squares to nonzero squares and hence nonsquares to nonsquares, and so b must also be a nonzero square. Similarly if a is a nonsquare, then $[a, b]$

maps nonzero squares to nonsquares and hence nonsquares to nonzero squares, and so b must also be a nonsquare. In each case a/b is a nonzero square.
If a/b is a nonzero square, then either a and b are both nonzero squares, in which case $[a, b]$ maps the set of nonzero squares to the set of nonzero squares and the set of nonsquares to the set of nonsquares, or a and b are both nonsquares, in which case $[a, b]$ maps the set of nonzero squares to the set of nonsquares and the set of nonsquares to the set of nonzero squares. In either case $[a, b]$ is a bijection. Further a/b is a nonzero square if and only if $ab = (a/b)b^2$ is a nonzero square.

8. $[a, b]$ is an orthomorphism if and only if the mappings $x \mapsto [a, b](x)$ and $x \mapsto [a, b](x) - x = [a, b](x) - [1, 1](x) = [a - 1, b - 1](x)$ are bijections, if and only if a/b and $(a - 1)/(b - 1)$ are both nonzero squares, if and only if ab and $(a - 1)(b - 1)$ are both nonzero squares.
9. $[a, b]$ is a complete mapping if and only if the mappings $x \mapsto [a, b](x)$ and $x \mapsto [a, b](x) + x = [a, b](x) + [1, 1](x) = [a + 1, b + 1](x)$ are bijections, if and only if a/b and $(a + 1)/(b + 1)$ are both nonzero squares, if and only if ab and $(a + 1)(b + 1)$ are both nonzero squares.
10. $[a, b] \perp [c, d]$ if and only if the mapping $x \mapsto [a, b](x) - [c, d](x) = [a - c, b - d](x)$ is a bijection, if and only if $(a - c)/(b - d)$ is a nonzero square, if and only if $(a - c)(b - d)$ is a nonzero square. □

An immediate corollary:

Corollary 9.25.

1. *$[a, b]$ is an orthomorphism if and only if $[b, a]$ is an orthomorphism.*
2. *$[a, b] \perp [c, d]$ if and only if $[b, a] \perp [d, c]$.*
3. *If $a \neq b$ then $[a, b] \perp [b, a]$ if and only if -1 is a square.*

Proof.

1. By Theorem 9.24, $[a, b]$ is an orthomorphism if and only if ab and $(a-1)(b-1)$ are both nonzero squares, if and only if ba and $(b - 1)(a - 1)$ are both nonzero squares, if and only if $[b, a]$ is an orthomorphism.
2. $[a, b] \perp [c, d]$ if and only if $(a - c)(b - d)$ is a nonzero square, if and only if $(b - d)(a - c)$ is a nonzero square, if and only if $[b, a] \perp [d, c]$.
3. If $a \neq b$, then $[a, b] \perp [b, a]$ if and only if $(a - b)(b - a) = -(a - b)^2$ is a nonzero square, if and only if -1 is a square. □

The actions of the congruences I, R, T_g, and $H_{[c]}$ on $\mathscr{C}_2$ are described in the next theorem.

Theorem 9.26. *If $[a, b] \in \mathscr{C}_2$, then, the following hold.*

1.

$$I[[a, b]] = \begin{cases} \left[\frac{1}{a}, \frac{1}{b}\right] & \text{if } a \text{ and } b \text{are nonzero squares,} \\ \left[\frac{1}{b}, \frac{1}{a}\right] & \text{if } a \text{ and } b \text{ are nonsquares.} \end{cases}$$

2.

$$\mathrm{R}[[a,b]] = \begin{cases} [1-a, 1-b] & \textit{if } -1 \textit{ is a square,} \\ [1-b, 1-a] & \textit{if } -1 \textit{ is a nonsquare.} \end{cases}$$

3. *If $c \neq 0$, then*

$$\mathrm{H}_{[c]}[a,b] = \begin{cases} [a,b] & \textit{if } c \textit{ is a square,} \\ [b,a] & \textit{if } c \textit{ is a nonsquare.} \end{cases}$$

4. *If $a \neq b$ and $g \neq 0$ then $\mathrm{T}_g[[a,b]] \notin \mathscr{C}_2$.*

Proof.

1. As $[a,b] \in \mathscr{C}_2$, by Theorem 9.24, a and b are either both nonzero squares or both nonsquares. If a and b are both nonzero squares, then $1/a$ and $1/b$ are both nonzero squares and, by Theorem 9.24(6), $[a,b] \circ [1/a, 1/b] = [1,1]$ and so $\mathrm{I}[[a,b]] = [1/a, 1/b]$. If a and b are both nonsquares, then $1/a$ and $1/b$ are both nonsquares and, by Theorem 9.24(6), $[a,b] \circ [1/b, 1/a] = [1,1]$ and so $\mathrm{I}[[a,b]] = [1/b, 1/a]$.
2. If -1 is a square, then

$$\mathrm{R}[[a,b]](x) = x + [a,b](-x) = [1,1](x) - [a,b](x) = [1-a, 1-b](x).$$

If -1 is a nonsquare, then

$$\mathrm{R}[[a,b]](x) = x + [a,b](-x) = [1,1](x) - [b,a](x) = [1-b, 1-a](x).$$

3. If c is a nonzero square, then $\mathrm{H}_{[c]}[[a,b]](x) = c[a,b](c^{-1}x) = [a,b](x)$ by Theorem 9.24(4). If c is a nonsquare, then $\mathrm{H}_{[c]}[[a,b]](x) = c[a,b](c^{-1}x) = [b,a](x)$ by Theorem 9.24(5).
4. Let $[a,b] \in \mathscr{C}_2(q)$, $a \neq b$. As $\mathscr{C}_2(q) = \mathscr{C}_1(q)$ for $q \leq 5$, it must be that $q \geq 7$. Suppose, for $g \neq 0$, that $\mathrm{T}_g[[a,b]] = [c,d]$.
If g is a square, then there exists a nonzero square x for which $x+g$ is a nonzero square and a nonsquare y for which $y+g$ is a nonzero square. Then

$$cx = [c,d](x) = \mathrm{T}_g[[a,b]](x) = a(x+g) - ag = ax$$

and

$$dy = [c,d](y) = \mathrm{T}_g[[a,b]](y) = a(y+g) - ag = ay.$$

Hence $c = d = a$ and so $[a,b](x) = \mathrm{T}_{-g}\mathrm{T}_g[[a,b]](x) = \mathrm{T}_{-g}[[a]](x) = [a](x)$. A contradiction.

If g is a nonsquare, then there exists a nonzero square x for which $x + g$ is a nonsquare and a nonsquare y for which $y + g$ is a nonsquare. Then

$$cx = [c, d](x) = \mathrm{T}_g[[a, b]](x) = b(x + g) - bg = bx$$

and

$$dy = [c, d](y) = \mathrm{T}_g[[a, b]](y) = b(y + g) - bg = by.$$

Hence $c = d = b$ and so $[a, b](x) = \mathrm{T}_{-g}\mathrm{T}_g[[a, b]](x) = \mathrm{T}_{-g}[[b]](x) = [b](x)$. A contradiction. □

As a corollary we obtain a characterization of quadratic orthomorphisms.

Corollary 9.27. *If* $\theta \in \mathrm{Orth}(\mathrm{GF}(q)^+)$, *then*

$$\mathrm{H}_\alpha[\theta] = \theta \textit{ for all } \alpha \in \{[c] \mid c \textit{ a nonzero square}\},$$

if and only if $\theta \in \mathscr{C}_2(q)$.

Proof. By Theorem 9.26(3), if $[a, b] \in \mathscr{C}_2$ and c is a nonzero square, then

$$\mathrm{H}_{[c]}[[a, b]] = [a, b].$$

Next suppose that $\mathrm{H}_{[c]}[\theta] = \theta$ for all nonzero squares c. Let g be a primitive element of $\mathrm{GF}(q)$, let $a = \theta(1)$, and let $b = g^{-1}\theta(g)$. If c is a nonzero square, then

$$a = \theta(1) = \mathrm{H}_{[c^{-1}]}[\theta](1) = c^{-1}\theta(c),$$

and

$$bg = \theta(g) = \mathrm{H}_{[c^{-1}]}[\theta](g) = c^{-1}\theta(cg).$$

Thus, $\theta(c) = ac$ and $\theta(gc) = b(gc)$ for all nonzero squares c. It follows that $u\theta = [a, b]$. □

We saw in Theorem 9.24(8) that $[a, b]$ is an orthomorphism of $\mathrm{GF}(q)^+$ if and only if both a/b and $(a-1)/(b-1)$ are nonzero squares. This fact was used in [111] to provide a simple method for generating all nonlinear, quadratic orthomorphisms.

Theorem 9.28. *Let r and s be distinct nonzero squares different from* 1, *and set*

$$a = s\left(\frac{1-r}{s-r}\right) \quad \textit{and} \quad b = \frac{1-r}{s-r}.$$

Then $[a, b]$ is an orthomorphism and all nonlinear, quadratic orthomorphisms are of this form.

Proof. Simple computation shows that $a/b = s$ and $(a-1)/(b-1) = r$. Thus, as r and s are nonzero squares, $[a, b]$ is an orthomorphism by Theorem 9.24(8).

If $[a, b] \in \mathscr{C}_2$, $a \neq b$, then a/b and $(a-1)/(b-1)$ are both nonzero squares by Theorem 9.24(8). If we set

$$\frac{a}{b} = s \quad \text{and} \quad \frac{a-1}{b-1} = r,$$

then r and s are nonzero squares different from 1 and $r \neq s$ as otherwise $a(b-1) = b(a-1)$, which implies that $a = b$. Solving for a and b yields

$$a = s\left(\frac{1-r}{s-r}\right) \quad \text{and} \quad b = \frac{1-r}{s-r}.$$

□

This yields a formula for the number of quadratic orthomorphisms of $\mathrm{GF}(q)^+$.

Corollary 9.29. $|\mathscr{C}_2(q)| = (q-3)(q-5)/4 + (q-2)$.

Proof. The number of linear orthomorphisms of $\mathrm{GF}(q)^+$ is $q-2$. By Theorem 9.28, the number of nonlinear quadratic orthomorphisms is equal to the number of ordered pairs (r, s), r and s distinct nonzero squares different from 1. As the number of nonzero squares in $\mathrm{GF}(q)$ is $(q-1)/2$, it follows that the number of nonlinear quadratic orthomorphisms of $GF(q)^+$ is

$$\left(\frac{q-1}{2} - 1\right)\left(\frac{q-1}{2} - 2\right) = \frac{(q-3)(q-5)}{4}.$$

□

A different proof of this, using cyclotomic numbers, is given in [240]. Let C_0 be the set of nonzero squares and C_1 the set of nonsquares. We know that $[a, b] \in \mathscr{C}_2$ if and only if a and b are either both nonzero squares or both nonsquares, and $a-1$ and $b-1$ are either both nonzero squares or both nonsquares. It follows that

$$|\mathscr{C}_2| = \sum_{i,j=0}^{1} |\{x \mid x \in C_i, x-1 \in C_j\}|^2.$$

The numbers $|\{x \mid x \in C_i, x-1 \in C_j\}|$ are cyclotomic numbers: these will be studied in Section 10.1. These particular cyclotomic numbers will be computed in Theorem 10.5. For $q \equiv 1 \pmod 4$, this sum is

$$\left(\frac{q-5}{4}\right)^2 + 3\left(\frac{q-1}{4}\right)^2 = \frac{q^2-4q+7}{4} = \frac{(q-3)(q-5)}{4} + q - 2,$$

and for $q \equiv 3 \pmod 4$, this sum is

$$\left(\frac{q+1}{4}\right)^2 + 3\left(\frac{q-3}{4}\right)^2 = \frac{q^2-4q+7}{4} = \frac{(q-3)(q-5)}{4} + q - 2.$$

When q is a prime, we can read off the nonlinear quadratic orthomorphisms of $GF(q)^+$ by inspection. First we write down the nonzero elements of $GF(q)$ in order and mark the squares by, for instance, enclosing them in brackets. Proceeding from left to right, if, for instance, a is marked and $a-1$ is unmarked, find those b to the right of a for which b is marked and $b-1$ is unmarked. Then $[a, b]$ and $[b, a]$ are both nonlinear quadratic orthomorphisms of $GF(q)^+$. The following exemplifies this procedure for $q = 11$.

Example 9.30. We first list the nonzero elements of GF(11) in order with the squares in brackets.

[1] 2 [3] [4] [5] 6 7 8 [9] 10

For $a = 2$, as 2 is unmarked and $2 - 1 = 1$ is marked, our choices for b are 6 and 10. Thus [2, 6], [2, 10], [6, 2], and [10, 2] are nonlinear quadratic orthomorphisms of $GF(11)^+$. For $a = 3$, as 3 is marked and $3 - 1 = 2$ is unmarked, our only choice for b is 9. Thus [3, 9] and [9, 3] are nonlinear quadratic orthomorphisms of $GF(11)^+$. We present the 12 nonlinear quadratic orthomorphisms of $GF(11)^+$ below in a compact table form.

a	2	2	3	4	6	7	b
b	6	10	9	5	10	8	a

In the following examples, we describe the nonlinear quadratic orthomorphisms of $GF(7)^+$, $GF(13)^+$, and $GF(17)^+$.

Example 9.31. We list the nonzero elements of GF(7) in order with the squares in brackets.

[1] [2] 3 [4] 5 6

The nonlinear quadratic orthomorphisms of $GF(7)^+$ are [3, 5] and [5, 3].

Example 9.32. We list the elements of GF(13)* in order with the squares marked.

[1] 2 [3] [4] 5 6 7 8 [9] [10] 11 [12]

The 20 nonlinear quadratic orthomorphisms of $GF(13)^+$ can be read off from the following table.

a	2	2	3	3	4	5	6	6	7	9	b
b	5	11	9	12	10	11	7	8	8	12	a

Example 9.33. We list the elements of GF(17)* in order with the squares marked.

[1] [2] 3 [4] 5 6 7 [8] [9] 10 11 12 [13] 14 [15] [16]

The 42 nonlinear quadratic orthomorphisms of $\mathrm{GF}(17)^+$ can be read off from the following table.

a	2	2	3	3	3	4	4	4	5	5	6	6	6	7	7	8	8	9	10	11	13	b
b	9	16	5	10	14	8	13	15	10	14	7	11	12	11	12	13	15	16	14	12	15	a

While we are not in a position to completely determine the structure of $\mathscr{C}_2$, we can still determine some orthogonalities. Orthogonalities between nonlinear quadratic orthomorphisms and linear orthomorphisms were determined in [111] using a similar approach to that used in Theorem 9.28.

Theorem 9.34. *Let $[a, b]$ be a nonlinear quadratic orthomorphism and l a nonzero square, $l \neq 1, a/b, (a-1)/(b-1)$. If $c = (lb-a)/(l-1)$, then $[c]$ is a linear orthomorphism orthogonal to $[a, b]$. Further, all linear orthomorphisms orthogonal to $[a, b]$ can be generated in this way.*

Proof. $c \neq 0$ as then $l = a/b$, and $c \neq 1$ as then $l = (a-1)/(b-1)$. Thus $[c]$ is a linear orthomorphism. Further $(a-c)/(b-c) = l$, a nonzero square, and so $[a, b] \perp [c]$.

If $[c]$ is a linear orthomorphism orthogonal to $[a, b]$, then, setting $l = (a-c)/(b-c)$, and solving for c yields $c = (lb-a)/(l-1)$. □

Theorem 9.34 yields a simple formula for the number of linear orthomorphisms orthogonal to a given nonlinear quadratic orthomorphism.

Corollary 9.35. *If $[a, b]$ is a nonlinear quadratic orthomorphism of $\mathrm{GF}(q)^+$, then $[a, b]$ is orthogonal to precisely $(q-7)/2$ linear orthomorphisms of $\mathrm{GF}(q)^+$.*

Proof. By Theorem 9.34, the number of linear orthomorphisms orthogonal to $[a, b]$ is equal to the number of nonzero squares l different from $1, a/b$, and $(a-1)/(b-1)$. As $1, a/b$, and $(a-1)/(b-1)$ are distinct nonzero squares, the number of choices for l is

$$\frac{q-1}{2} - 3 = \frac{q-7}{2}.$$

□

As an example, in Figure 9.2 we list, for each nonlinear quadratic orthomorphism of $\mathrm{GF}(11)^+$, the linear neighbors.

Fig. 9.2 The linear neighbors of quadratic orthomorphisms in $\mathscr{C}_2(11)$

quadratic orthomorphisms	linear neighbors
$[2,6]$ and $[6,2]$	$[7]$ and $[8]$
$[2,10]$ and $[10,2]$	$[3]$ and $[9]$
$[3,9]$ and $[9,3]$	$[5]$ and $[7]$
$[4,5]$ and $[5,4]$	$[8]$ and $[9]$
$[6,10]$ and $[10,6]$	$[4]$ and $[5]$
$[7,8]$ and $[8,7]$	$[3]$ and $[4]$

Few other rules for orthogonality in $\mathscr{C}_2$ are known. The following from [111] yields many orthogonalities.

Theorem 9.36. *Let* $[a,b]$ *and* $[c,d]$ *be distinct nonlinear, quadratic orthomorphisms. If either*

$$\frac{c}{d}=\frac{a}{b} \quad or \quad \frac{(c-1)}{(d-1)}=\frac{(a-1)}{(b-1)},$$

then $[a,b]\perp[c,d]$.

Proof. If

$$s=\frac{a}{b}=\frac{c}{d}, \quad r=\frac{a-1}{b-1}, \quad \text{and} \quad r'=\frac{c-1}{d-1},$$

then s, r, and r' are distinct nonzero squares different from 1,

$$a=s\left(\frac{1-r}{s-r}\right), \qquad b=\frac{1-r}{s-r},$$

$$c=s\left(\frac{1-r'}{s-r'}\right), \quad \text{and} \quad d=\frac{1-r'}{s-r'}.$$

Thus

$$\frac{a-c}{b-d}=s\left(\frac{(1-r)(s-r')-(1-r')(s-r)}{(1-r)(s-r')-(1-r')(s-r)}\right)=s,$$

a nonzero square, and so $[a,b]\perp[c,d]$.

If

$$r=\frac{a-1}{b-1}=\frac{c-1}{d-1}, \quad s=\frac{a}{b}, \quad \text{and} \quad s'=\frac{c}{d},$$

then s, r, and r' are distinct nonzero squares different from 1,

$$a = s\left(\frac{1-r}{s-r}\right), \qquad b = \frac{1-r}{s-r},$$

$$c = s'\left(\frac{1-r}{s'-r}\right), \quad \text{and} \quad d = \frac{1-r}{s'-r}.$$

Thus

$$\frac{a-c}{b-d} = \left(\frac{s(1-r)(s'-r) - s'(1-r)(s-r)}{(1-r)(s'-r) - (1-r)(s-r)}\right) = r,$$

a nonzero square, and so $[a, b] \perp [c, d]$. □

From this we obtain a lower bound on the number of quadratic orthomorphisms orthogonal to a given nonlinear, quadratic orthomorphism.

Corollary 9.37. *A nonlinear quadratic orthomorphism of* $\mathrm{GF}(q)^+$ *is orthogonal to at least* $(3q-21)/2$ *quadratic orthomorphisms of* $\mathrm{GF}(q)^+$.

Proof. Let $[a, b]$ be a nonlinear quadratic orthomorphism of $\mathrm{GF}(q)^+$. By Corollary 9.35, $[a, b]$ is orthogonal to $(q-7)/2$ linear orthomorphisms of $\mathrm{GF}(q)^+$.

Let $[c, d] \in \mathscr{C}_2(q)$, and set

$$s = \frac{a}{b}, \qquad r = \frac{a-1}{b-1},$$

$$s' = \frac{c}{d}, \quad \text{and} \quad r' = \frac{c-1}{d-1}.$$

Note that $s' = s$ and $r' = r$ if and only if $[c, d] = [a, b]$. By Theorem 9.36, if $s' = s$ and $r' \neq r$ or $r' = r$ and $s' \neq s$, then $[c, d] \perp [a, b]$.

Let us count those $[c, d]$ for which $s' = s$ and $r' \neq r$. As $r' \neq 1$, s or r, which are distinct, there are $(q-1)/2 - 3 = (q-7)/2$ choices for r'. This yields $(q-7)/2$ nonlinear quadratic neighbors of $[a, b]$.

Similarly, counting those $[c, d]$, for which $r' = r$ and $s' \neq s$, yields a further $(q-7)/2$ nonlinear quadratic neighbors of $[a, b]$. The result follows. □

For $q = 7$, the bound in Corollary 9.37 is best possible. In Example 9.31, we saw that the only nonlinear quadratic orthomorphisms of $\mathrm{GF}(7)^+$ are $[3, 5]$ and $[5, 3]$. By Corollary 9.35, neither of these is orthogonal to any linear orthomorphism, and by Corollary 9.25(3), $[3, 5]$ and $[5, 3]$ are not orthogonal.

For $q = 11$, by Corollary 9.37, each nonlinear quadratic orthomorphism is orthogonal to at least 6 quadratic orthomorphism: 2 linear orthomorphisms and 4 nonlinear quadratic orthomorphisms. In fact, for $q = 11$, each nonlinear quadratic orthomorphism is orthogonal to 8 quadratic orthomorphisms.

For $q = 13$, by Corollary 9.37, each nonlinear quadratic orthomorphism is orthogonal to at least 9 quadratic orthomorphism: 3 linear orthomorphisms and 6 nonlinear quadratic orthomorphisms. In fact 12 of the nonlinear quadratic orthomorphisms of $GF(13)^+$ have 12 neighbors in $\mathscr{C}_2$, 6 have 14 neighbors in $\mathscr{C}_2$, and 2 have 16 neighbors in $\mathscr{C}_2$. Data on the orthomorphism graphs of $GF(11)^+$ and $GF(13)^+$ will be given in Chapter 13.

9.4 Maximal and Complete Sets of Orthomorphisms in $\mathscr{C}_2$

In this section we will use quadratic orthomorphisms to construct maximal sets of $(p-1)/2$ mutually orthogonal Latin squares of prime order p when $p \equiv 1 \pmod 4$ and maximal sets of $(p-3)/2$ mutually orthogonal Latin squares of prime order p when $p \equiv 3 \pmod 4$. We will also describe all complete sets of orthomorphisms in $\mathscr{C}_2(q)$.

9.4.1 Maximal Sets of Orthomorphisms in $\mathscr{C}_2$

One of the central problems in the study of orthomorphism graphs is, given an orthomorphism graph $\mathscr{H}$ of G, find maximal cliques in $\mathscr{H}$ and determine if these are maximal in $\mathrm{Orth}(G)$. The following maximal cliques in $\mathscr{C}_2(p)$, which are also maximal in $\mathrm{Orth}(GF(p)^+)$, p prime, were discovered by Evans [118] in 1992.

Theorem 9.38 (Evans, 1992). *Let $p \geq 7$ be a prime.*

1. *If $p \equiv 3 \pmod 4$, then $\mathscr{C}_2(p)$ contains a maximal set of $(p-5)/2$ orthomorphisms of $GF(p)^+$.*
2. *If $p \equiv 1 \pmod 4$, then $\mathscr{C}_2(p)$ contains a maximal set of $(p-3)/2$ orthomorphisms of $GF(p)^+$.*

Proof. Let $m = (p-7)/2$ and let $[a, b] \in \mathscr{C}_2(p) - \mathscr{C}_1(p)$. By Corollary 9.35, $[a, b]$ is orthogonal to precisely m elements of $\mathscr{C}_1(p)$: label these elements $[c_1], \ldots, [c_m]$. Now $[a, b]$ and $[b, a]$ are both orthogonal to $[c_1], \ldots, [c_m]$, and $[a, b]$ is orthogonal to $[b, a]$ if and only if $p \equiv 1 \pmod 4$. Thus $[a, b], [c_1], \ldots, [c_m]$ form a $(p-5)/2$-clique of $\mathscr{C}_2(p)$ and, if $p \equiv 1 \pmod 4$, then we can extend this set using $[b, a]$ to a $(p-3)/2$-clique of $\mathscr{C}_2(p)$. We will prove these cliques of orthomorphisms to be maximal in $\mathrm{Orth}(GF(p)^+)$.

The orthomorphism $[c_i]$ is represented by the orthomorphism polynomial $c_i x$, the orthomorphism $[a, b]$ by the orthomorphism polynomial

$$\frac{(a-b)}{2}x^{(p+1)/2} + \frac{(a+b)}{2}x,$$

and the orthomorphism $[b, a]$ by the orthomorphism polynomial

$$\frac{(b-a)}{2}x^{(p+1)/2} + \frac{(b+a)}{2}x.$$

First we will prove that the only orthomorphisms that can be orthogonal to $[c_1], \ldots, [c_m]$ are linear orthomorphisms, quadratic orthomorphisms, and their translates. Let

$$\mathrm{f}(x) = a_{1,1}x + a_{2,1}x^2 + \cdots + a_{p-1,1}x^{p-1}$$

be an orthomorphism polynomial orthogonal to $c_1x, \ldots, c_mx$. Thus $\mathrm{f}(x) - yx$ is a permutation polynomial of GF(p) for $y = 0, 1, c_1, \ldots, c_m$.

Let

$$\mathrm{f}(x)^t \equiv a_{1,t}x + a_{2,t}x^2 + \cdots + a_{p-1,t}x^{p-1} \pmod{x^p - x}.$$

Then, for $t < p-1$, the coefficient of x^{p-1} in the reduction of $(\mathrm{f}(x) - yx)^t$ modulo $x^p - x$ is

$$\sum_{i=0}^{t-1}(-1)^i \binom{t}{i} a_{p-1-i,t-i}\, y^i.$$

By Theorem 9.6 this coefficient must be congruent to 0 modulo $y(y-1)(y-c_1)\cdots(y-c_m)$ and so $a_{p-1-i,t-i} = 0$ for $0 \le i \le t-1 < m+2 = (p-3)/2$. This can be rewritten, by setting $k = t - i$, as

$$a_{(p-1)-i,k} = 0 \text{ for } i \ge 0,\ k \ge 1,\ \text{and } k + i \le (p-3)/2:$$

we will refer to these equations as the coefficient equations.

By setting $k = 1$ in the coefficient equations we find that

$$a_{p-1,1} = a_{p-2,1} = \cdots = a_{(p+3)/2,1} = 0.$$

There are now two cases to consider.

Case 1. $a_{(p+1)/2,1} \neq 0$. We may assume that $a_{(p-1)/2,1} = 0$, as this will certainly be true for some translate of $\mathrm{f}(x)$. Suppose that $a_{n,1} \neq 0$ for some $n < (p-1)/2$ and that $a_{n',1} = 0$ whenever $n < n' \le (p-1)/2$.

Considering $\mathrm{f}(x)^2$, one sees that

$$a_{n,1} = \frac{a_{n+(p+1)/2,2}}{2a_{(p+1)/2,1}}.$$

But this must be zero if $n + (p+1)/2 \ge p - 1 + 2 - (p-3)/2$, i.e., if $n \ge 2$. Thus $n = 1$ and $\mathrm{f}(x)$ represents a quadratic orthomorphism or a translate of a quadratic orthomorphism.

Case 2. $a_{(p+1)/2,1} = 0$. Suppose that $a_{n,1} \neq 0$ for some $n < (p+1)/2$ and that $a_{n',1} = 0$ whenever $n < n' \leq (p+1)/2$. Assume $n \neq 1$. Now n cannot divide $p-1$ and so $p-1 = qn + r$, $1 \leq r < n$. Thus $a_{qn,q} = (a_{n1})q \neq 0$. Therefore, by the coefficient equations, $(p-1) - qn + q \geq (p-1)/2$ and so $qn + r - qn + q \geq (qn+r)/2$ or $n \leq 2 + r/q < 2 + (n-1)/2$, as $q \geq 2$ and $r \leq n-1$ and we cannot have both $q = 2$ and $r = n-1$ (this would imply that $p = 3n$), and so $n < 3$, a contradiction. Thus $n = 1$ and $f(x)$ represents a linear orthomorphism.

We have thus shown that any orthomorphism orthogonal to $[c_1], \ldots, [c_m]$ must be either a linear orthomorphism, a quadratic orthomorphism, or a translate of a quadratic orthomorphism. Further, Theorem 8.6 implies that we may, without loss of generality, restrict ourselves to linear and quadratic orthomorphisms only.

To complete the proof, we need to show that only two nonlinear quadratic orthomorphisms can be orthogonal to $[c_1], \ldots, [c_m]$. Suppose that

$$f(x) = Ax^{(p+1)/2} + Bx \text{ and } f'(x) = A'x^{(p+1)/2} + B'x$$

are both nonlinear quadratic orthomorphisms orthogonal to $[c_1], \ldots, [c_m]$. The coefficients of x^{p-1} in the reductions of $(f(x) - yx)^{(p-1)/2}$ and $(f'(x) - yx)^{(p-1)/2}$ modulo $x^p - x$ are both polynomials in y of degree $m+2$, and so must be constant multiples of each other, as each is congruent to 0 modulo $y(y-1)(y-c_1)\cdots(y-c_m)$.

The coefficients of x^{p-1} in the reductions of $(f(x) - yx)^{(p-1)/2}$ and $(f'(x) - yx)^{(p-1)/2}$ modulo $x^p - x$ are

$$((p-1)/2)(B-y)^{(p-3)/2}A + \binom{\frac{p-1}{2}}{3}(B-y)^{(p-7)/2}A^3 + \text{ terms in } y^k,$$

$k < (p-7)/2$, and

$$((p-1)/2)(B'-y)^{(p-3)/2}A' + \binom{\frac{p-1}{2}}{3}(B'-y)^{(p-7)/2}A'^3 + \text{ terms in } y^k,$$

$k < (p-7)/2$, respectively. Comparing coefficients we see that $A' = \alpha A$, $A'B' = \alpha AB$, for some α, which implies that $B' = B$. Further $A'B'^2 = \alpha AB^2$, and so $A'^3 = \alpha A^3$. Thus $\alpha = \pm 1$ and so $A' = \pm A$. Thus the only orthomorphisms orthogonal to $[c_1], \ldots, [c_m]$ are the quadratic orthomorphisms $f(x) = Ax^{(p+1)/2} + Bx$ and $f'(x) = -Ax^{(p+1)/2} + Bx$, their translates, and linear orthomorphisms, hence the result. □

An immediate corollary.

Corollary 9.39. *Let $p \geq 7$ be a prime.*

1. *If $p \equiv 3 \pmod 4$, then there exists a maximal set of $(p-3)/2$ mutually orthogonal Latin squares of order p.*

2. *If* $p \equiv 1 \pmod 4$, *then there exists a maximal set of* $(p-1)/2$ *mutually orthogonal Latin squares of order* p.

As examples, Corollary 9.39 establishes the existence of maximal sets of r mutually orthogonal Latin squares of order n for $(n, r) = (7, 2), (11, 4), (13, 6), (17, 8), (19, 8), (23, 10)$, and $(29, 14)$. Of these only the first two parameter pairs were already known: the second pair was found through a computer search by Evans and Mcfarland [138] in 1984.

In 1993 Pott [269] gave a different proof of Theorem 9.38 using earlier results of Rédei [275] and Lovász and Schrijver [232] on the set of slopes of a function. For $\mathrm{f}\colon GF(q) \to GF(q)$, we define the *set of slopes*, S_{f}, of f by

$$S_{\mathrm{f}} = \left\{ \frac{\mathrm{f}(x) - \mathrm{f}(y)}{x - y} \;\middle|\; x, y \in \mathrm{GF}(q), x \neq y \right\}.$$

Note that $S_{\mathrm{f}} = \{m\}$ if and only if $\mathrm{f}(x) = mx + b$. The importance of the set of slopes should be made clear by the following result.

Lemma 9.40. *For* $\mathrm{f}\colon \mathrm{GF}(q) \to \mathrm{GF}(q)$, $\mathrm{f}(x) \perp mx$ *if and only if* $m \notin S_f$.

Proof. $\mathrm{f}(x) \perp mx$ if and only if $\mathrm{f}(x) - mx \neq \mathrm{f}(y) - my$ whenever $x \neq y$, if and only if $\mathrm{f}(x) - \mathrm{f}(y) \neq m(x - y)$ whenever $x \neq y$. □

It follows from Lemma 9.40 that a function $\mathrm{f}\colon \mathrm{GF}(q) \to \mathrm{GF}(q)$ is a bijection if and only if $0 \notin S_{\mathrm{f}}$, an orthomorphism if and only if $0, 1 \notin S_{\mathrm{f}}$, and a complete mapping if and only if $0, -1 \notin S_{\mathrm{f}}$. For the functions $[a, b]$,

$$S_{[a,b]} = \{a, b\} \cup \left\{ \frac{az - b}{z - 1} \;\middle|\; z \text{ a nonsquare} \right\},$$

from which an alternative proof to Corollary 9.35 can be obtained.

For nonlinear functions, in 1973 Rédei [275] established a lower bound on $|S_{\mathrm{f}}|$ in the case q prime.

Lemma 9.41 (Rédei, 1973). *If* p *is a prime and* $\mathrm{f}\colon \mathrm{GF}(p) \to GF(p)$ *is nonlinear, then* $|S_{\mathrm{f}}| \geq (p+3)/2$.

In 1981 Lovász and Schrijver [232] characterized those functions for which equality holds in Lemma 9.41.

Lemma 9.42 (Lovász, Schrijver, 1981). *If* p *is a prime and* $\mathrm{f}\colon \mathrm{GF}(p) \to \mathrm{GF}(p)$ *is nonlinear, then* $|S_{\mathrm{f}}| = (p+3)/2$ *if and only if*

$$\mathrm{f}(x) = a(x+b)^{(p+1)/2} + cx + d,$$

for some $a, b, c, d \in \mathrm{GF}(p)$, $a \neq 0$.

Note that proofs of Lemmas 9.41 and 9.42 are implicit in the proof of Theorem 9.38. From Lemmas 9.41 and 9.42, Pott [269] gave a shorter proof of Theorem 9.38.

Theorem 9.43. *If $p \geq 7$ is a prime, then there exists a maximal $(p, (p-1)/2; 1)$-difference matrix over* $\mathrm{GF}(p)^+$ *if* $p \equiv 3 \pmod 4$ *and a maximal $(p, (p+1)/2; 1)$-difference matrix over* $\mathrm{GF}(p)^+$ *if* $p \equiv 1 \pmod 4$.

Proof. Let $\mathrm{f}(x) = x^{(p+1)/2} = [1, -1](x)$. Then, by Lemma 9.42, $|S_\mathrm{f}| = (p+3)/2$. Thus, by Lemma 9.40, that $\mathrm{f}(x) \perp mx$ for $t = (p-3)/2$ values of m: let these values be denoted $m_1, \ldots, m_t$. Form a $(p, (p-1)/2; 1)$-difference matrix $D = (D_{ij})$ by setting $\mathrm{GF}(p) = \{g_1, \ldots, g_p\}$ and

$$d_{ij} = \begin{cases} m_i g_j & \text{for } i = 1, \ldots, t, \\ \mathrm{f}(g_j) & \text{for } i = t+1. \end{cases}$$

As was shown in the proof of Theorem 9.38, if D is not maximal, then any row added to D must have jth entry $\pm(g_j+b)^{(p+1)/2}+d$, where we may assume, without loss of generality, that $d = 0$. Set $\mathrm{h}(x) = (x+b)^{(p+1)/2}$ and $\mathrm{k}(x) = -(x+b)^{(p+1)/2}$. Now, if h is orthogonal to f, then $b \neq 0$ and

$$\frac{(x+b)^{(p+1)/2} - x^{(p+1)/2}}{b} = \frac{(x+b)^{(p+1)/2} - x^{(p+1)/2}}{(x+b) - x} \in S_{\mathrm{h}(x)-\mathrm{f}(x)},$$

which would imply that $|S_{\mathrm{h}(x)-\mathrm{f}(x)}| > (p+3)/2$, violating Lemma 9.41. Thus $\mathrm{h} \not\perp \mathrm{f}$.

If $\mathrm{k} \perp \mathrm{f}$ and $p \equiv 3 \pmod 4$, then $b = 0$ as

$$\mathrm{k}(0) - \mathrm{f}(0) = -b^{(p+1)/2} = \mathrm{k}(-b) - \mathrm{f}(-b).$$

But then $\mathrm{k} = [-1, 1]$ and, by Corollary 9.25(3), $\mathrm{k} \not\perp \mathrm{f}$, a contradiction.

If $p \equiv 1 \pmod 4$, then $[1, -1] \perp [-1, 1]$ by Corollary 9.25(3), and so we can extend D by adding the row with jth entry $-g_j^{(p+1)/2}$. This difference matrix is maximal as we can show, using the same argument we used to prove that $\mathrm{h} \not\perp \mathrm{f}$, that $\mathrm{k}(x) \not\perp -x^{(p+1)/2}$ for any value of b. □

From Theorems 1.35 and 9.43, we derive Theorem 9.38 as a corollary.

The proof of Theorem 9.38 rested on showing that the only orthomorphisms of $\mathrm{GF}(p)^+$, p an odd prime, that could be orthogonal to $(p-7)/2$ linear orthomorphisms are quadratic orthomorphisms and their translates. This was essentially proved by Stothers [302] in 1990.

Theorem 9.44 (Stothers, 1990). *Let p be a prime. If* $\mathrm{f}(x) + cx$ *is a permutation polynomial of* $\mathrm{GF}(p)$ *for more than $(p-3)/2$ values of c, then* f *is a linear polynomial.*

If $\mathrm{f}(x) + cx$ *is a permutation polynomial of* $\mathrm{GF}(p)$ *for exactly $(p-3)/2$ values of c, then* f *is a equivalent to a polynomial of the form* $x^{(p+1)/2} + ax$.

Proof. See [302]. □

Theorem 9.38 can easily be derived from Theorem 9.44. Does the construction in Theorem 9.38 still yield a maximal set of orthomorphisms in $\mathscr{C}_2(q)$ if q is odd but not prime? For the smallest case, $q = 9$, the answer is no as all 6 nonlinear quadratic orthomorphisms of $\mathrm{GF}(9)^+$ are orthogonal to the same linear orthomorphism as well as each other. For the next cases, $q = 25$, 27, and 49, Pott [269] used a computer search to show that the construction in Theorem 9.43 still yields maximal difference matrices: thus, there exists a maximal set of 11 MOLS of order 25, a maximal set of 11 MOLS of order 27, and a maximal set of 23 MOLS of order 49. Whether this result holds or not for all odd $q > 9$ remains unsolved. Some progress toward resolving this problem was made by Wan, Mullen, and Shiue [321, 322] in 1995. When q is an odd prime, we showed in the proof of Theorem 9.38 that any nonlinear orthomorphism orthogonal to $(q-7)/2$ linear orthomorphisms must be a quadratic orthomorphism or a translate of a quadratic orthomorphism. In the more general case, Wan, Mullen, and Shiue established a bound on the degree of the corresponding orthomorphism polynomial.

Theorem 9.45 (Wan, Mullen, and Shiue, 1995). *For q odd, let θ be an orthomorphism of $\mathrm{GF}(q)^+$. If θ is orthogonal to $(q-7)/2$ linear orthomorphisms of $\mathrm{GF}(q)^+$, then θ can be represented by a permutation polynomial of $\mathrm{GF}(q)$ of degree at most $(q+1)/2$.*

Proof. See [321]. □

Further, in the proof of Theorem 9.38, we showed that, for q an odd prime, if a nonlinear quadratic orthomorphism is orthogonal to $(q-7)/2$ linear orthomorphisms of $\mathrm{GF}(q)^+$, then only one other nonlinear quadratic orthomorphism would be orthogonal to this set of $(q-7)/2$ linear orthomorphisms of $\mathrm{GF}(q)^+$. Wan, Mullen, and Shiue showed that this also holds in the general case.

Theorem 9.46 (Wan, Mullen, and Shiue, 1995). *For q odd, let $[a,b], [c,d] \in \mathscr{C}_2(q)$, $a \neq b$ and $c \neq d$. If $[a,b]$ and $[c,d]$ are orthogonal to the same set of $(q-7)/2$ linear orthomorphisms of $\mathrm{GF}(q)^+$, then $\{a,b\} = \{c,d\}$.*

Proof. See [321]. □

If $\mathrm{f}(x)$ is a polynomial of $\mathrm{GF}(q)$, for how many values of $c \in \mathrm{GF}(q)$ is $\mathrm{f}(x)+cx$ a permutation polynomial of $\mathrm{GF}(q)$? There are a number of articles in the literature on this problem: see [321]. Relevant to us is a 1993 result of Chou [74].

Theorem 9.47 (Chou, 1993). *Let $\mathrm{f}(x)$ be a polynomial over $\mathrm{GF}(q)$ of degree $d \leq q-1$. If $\mathrm{f}(x)+cx$ is a permutation polynomial of $\mathrm{GF}(q)$ for m values of $c \in \mathrm{GF}(q)$, then $m \leq q-1-d$.*

Proof. See [74]. □

If in Theorem 9.47, $\mathrm{f}(x)$ is the orthomorphism polynomial corresponding to a nonlinear quadratic orthomorphism of $\mathrm{GF}(q)^+$, then $d = (q+1)/2$ and $m = (q-3)/2$, and equality holds.

9.4.2 Complete Sets of Orthomorphisms in $\mathscr{C}_2$

We know that $\mathscr{C}_1(q)$ is a complete set of orthomorphisms in $\mathscr{C}_2(q)$. A natural question is, how many $(q-2)$-cliques can we find in $\mathscr{C}_2(q)$? Let us first consider the special case, q prime. By Corollary 9.29, $\mathscr{C}_2(3) = \mathscr{C}_1(3)$ and $\mathscr{C}_2(5) = \mathscr{C}_1(5)$. By Example 9.31, $\mathscr{C}_2(7) \setminus \mathscr{C}_1(7) = \{[3, 5], [5, 3]\}$, and, by Corollaries 9.25(3) and 9.35, neither of these can be contained in a 5-clique in $\mathscr{C}_2(7)$. It was shown that the only $(q-2)$-clique in $\mathscr{C}_2(q)$ is $\mathscr{C}_1(q)$ for $q = 13$ and 17 by Mendelsohn and Wolk [240] in 1985 via a computer search.

For $q \le 47$, q prime, Evans [111] proved in 1987 that the only $(q-2)$-clique in $\mathscr{C}_2(q)$ is $\mathscr{C}_1(q)$, using simple hand calculations. Partition $\mathrm{GF}(q)\setminus\{0, 1\}$ into four sets

$$C_{i,j} = \{x \ne 0, 1 \mid x \in C_i, x-1 \in C_j\}, i, j = 0, 1,$$

where C_0 is the set of nonzero squares and C_1 is the set of nonsquares. Form a digraph Γ_q whose vertices are the elements of $\mathrm{GF}(q)\setminus\{0, 1\}$, there being an arc from x to y if $y - x \in C_0$: if $q \equiv 1 \pmod 4$, then Γ_q is a graph. The existence of $(q-2)$-cliques in $\mathscr{C}_2(q)$ is partly determined by isomorphisms of Γ_q that preserve each $C_{i,j}$.

Lemma 9.48. *If $\{[a_k, b_k] \mid k = 1, \ldots q-2\}$ is a $(q-2)$-clique in $\mathscr{C}_2(q)$, then the mapping $\mathrm{f}\colon a_k \mapsto b_k$ is an isomorphism of Γ_q that preserves $C_{i,j}$ for each i, j.*

Proof. f is clearly a bijection on the vertices of Γ_q, and if there is an arc from x to y in Γ_q, then $y - x \in C_0$, $x = a_u$ for some u, and $y = a_v$ for some v. By Theorem 9.24(10), as $[a_u, b_u] \perp [a_v, b_v]$, $\mathrm{f}(a_v) - \mathrm{f}(a_u) = b_v - b_u \in C_0$, and so f maps arcs to arcs, and, by Theorem 9.24(8), if $a_k \in C_{i,j}$, then $b_k = \mathrm{f}(a_k) \in C_{i,j}$. □

For small primes q the isomorphisms of Γ_q, preserving each $C_{i,j}$, can be determined, and no $(q-2)$-clique other than $\mathscr{C}_1(q)$ can be found in $\mathscr{C}_2(q)$.

Theorem 9.49 (Evans, 1987). *For $q \le 47$, q prime, the only $(q-2)$-clique in $\mathscr{C}_2(q)$ is $\mathscr{C}_1(q)$.*

Proof. We need only to consider the case $q \ge 11$.

For $q = 11$ the digraph Γ_{11} is shown in Figure 9.3. We see that any isomorphism of Γ_{11}, which preserves each $C_{i,j}$, must fix 3, 4, 5, 7, 8, and 9. If $b_u \ne a_u$ for some u, then $[a_u, b_u]$ is orthogonal to $6 > 2 = (11-7)/2$ linear orthomorphisms, contradicting Corollary 9.35. Thus the only 9-clique in $\mathscr{C}_2(11)$ is $\mathscr{C}_1(11)$.

For $q = 13$ the graph Γ_{13} is shown in Figure 9.4. We see that any isomorphism of Γ_{13}, which preserves each $C_{i,j}$, must fix 2, 7, and 12. As, by Corollary 9.35, any nonlinear, quadratic orthomorphism of $\mathrm{GF}(13)^+$ must be orthogonal to 3 linear orthomorphisms, the only possibility for a 11-clique in $\mathscr{C}_2(13)$ is [2], [7], [12], [4, 10], [10, 4], [3, 9], [9, 3], [6, 8], [8, 6], [5, 11], and [11, 5]. However $[3, 9] \not\perp [2]$ as $(3-2)(9-2) = 7$, a nonsquare, contradicting Theorem 9.24(10). Thus the only 11-clique in $\mathscr{C}_2(13)$ is $\mathscr{C}_1(13)$.

The other cases are similar but tedious. □

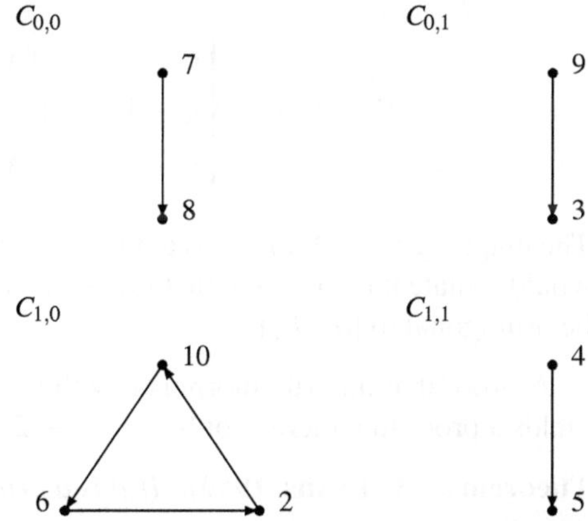

Fig. 9.3 The digraph Γ_{11}

In 1989 Evans [114] extended Theorem 9.49 to all primes by determining the existence of a nonlinear orthomorphism, orthogonal to many linear orthomorphisms, from a complete set of orthomorphisms in $\mathscr{C}_2$ other than $\mathscr{C}_1$, and proving that such an orthomorphism cannot exist.

Lemma 9.50. *Let p be an odd prime. If there exists a $(p-2)$-clique in $\mathscr{C}_2(p)$ other than $\mathscr{C}_1(p)$, then there exists $\theta \in Orth(\mathrm{GF}(p)^+)$ that is orthogonal to $[a]$ for all squares a.*

Proof. Let g be a nonsquare in $\mathrm{GF}(p)$, and let $[a_1, b_1], \ldots, [a_{p-2}, b_{p-2}]$ be a $(p-2)$-clique in $\mathscr{C}_2(p)$. Define $\theta\colon \mathrm{GF}(p) \to \mathrm{GF}(p)$ by

$$\theta(x) = \begin{cases} 0 & \text{if } x = 0, \\ g & \text{if } x = 1, \\ gb_i & \text{if } x = a_i, i = 1, \ldots, p-2. \end{cases}$$

Clearly, θ is a bijection, and

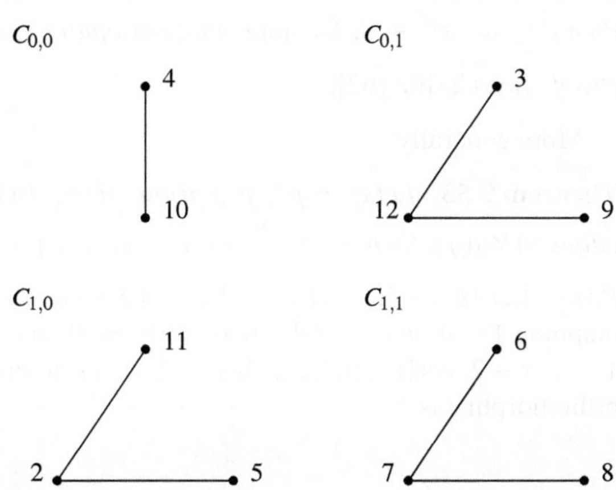

Fig. 9.4 The graph Γ_{13}

$$\theta(x) - x = \begin{cases} 0 & \text{if } x = 0, \\ g - 1 & \text{if } x = 1, \\ gb_i - a_i & \text{if } x = a_i, i = 1, \ldots, p-2. \end{cases}$$

The mapping $x \mapsto \theta(x) - x$ is a bijection as, for $i \neq j$, $gb_i - a_i \neq gb_j - a_j$, as this would violate the condition that $(a_i - a_j)/(b_i - b_j)$ must be a square for $[a_i, b_i]$ to be orthogonal to $[a_j, b_j]$. □

A proof that an orthomorphism with the properties in Lemma 9.50 cannot exist yields a proof that there is only one $(p-2)$-clique in $\mathscr{C}_2(p)$ when p is prime.

Theorem 9.51 (Evans, 1989). *If p is a prime, then only $(p-2)$-clique in $\mathscr{C}_2(p)$ is $\mathscr{C}_1(p)$.*

Proof. Similar to the proof of Theorem 9.38. □

An alternative proof of Theorem 9.51 can be obtained from Lemma 9.41. By Lemma 9.42, any orthomorphism orthogonal to $(p-7)/2$ linear orthomorphisms must be a translate of a quadratic orthomorphism, and by Corollary 9.35, no nonlinear quadratic orthomorphism can be orthogonal to $(p-5)/2$ linear orthomorphisms. The result then follows from Lemma 9.50.

If q is not a prime, then $\mathscr{C}_2(q)$ contains more than one $(q-2)$-clique. To see this suppose $q = p^r$, p a prime, $r \neq 1$, and let $t|q$. Then the orthomorphisms $[a, a^t]$, $a \neq 0, 1$, form a $(q-2)$-clique of $\mathscr{C}_2(q)$. In fact, each $t = 1, p, \cdots, p^{r-1}$ yields a distinct $(q-2)$-clique. This has immediate consequences for the case $q = 9$. By Corollary 9.29, $\mathscr{C}_2(9) - \mathscr{C}_1(9)$ has 6 elements. By Corollary 9.35 each of these orthomorphisms is orthogonal to exactly one element of $\mathscr{C}_1(9)$. Thus, a second 7-clique can only exist in $\mathscr{C}_2(9)$ if each orthomorphism of $\mathscr{C}_2(9) \setminus \mathscr{C}_1(9)$ is orthogonal to the same orthomorphism of $\mathscr{C}_1(9)$ as well as to each other. The complete determination of the $(q-2)$-cliques in $\mathscr{C}_2(q)$, for general q, follows from a 1960 result of Carlitz [62].

Lemma 9.52 (Carlitz, 1960). *If* f *is a permutation polynomial of* GF(p^n)*, p a prime, for which* $(\mathrm{f}(a) - \mathrm{f}(b))/(a - b)$ *is a nonzero square for all a, b, $a \neq b$, then* $\mathrm{f}(x) = cx^{p^i} + d$*, for some nonzero square c, and some i, $0 \leq i < n$.*

Proof. See Carlitz [62]. □

More generally.

Theorem 9.53. *Let $q = p^n$, p a prime. If $[a_1, b_1], \ldots, [a_{q-2}, b_{q-2}]$ is a $(q-2)$-clique in $\mathscr{C}_2(q)$, then $b_i = a_i^{p^j}$, for $i = 1, \ldots, q-2$, and some j, $1 \leq j < n$.*

Proof. Let $[a_1, b_1], \ldots, [a_{p-2}, b_{p-2}]$ be a $(p-2)$-clique in $\mathscr{C}_2(p)$, and define a mapping f: GF$(p) \rightarrow$ GF(p) by f(0) $= 0$, f(1) $= 1$, and f$(a_i) = b_i$ for $i = 1, \ldots, p-2$. As $[a_1, b_1], \ldots, [a_{p-2}, b_{p-2}]$ is a set of pairwise orthogonal quadratic orthomorphisms

$$\frac{f(a_i)-0}{a_i-0}=\frac{b_i}{a_i},\quad \frac{f(a_i)-f(1)}{a_i-1}=\frac{b_i-1}{a_i-1},\quad \text{and} \quad \frac{f(a_i)-f(a_j)}{a_i-a_j}=\frac{b_i-b_j}{a_i-a_j},$$

are all nonzero squares for $i, j = 1, \ldots, p-2, i \neq j$. Further $(f(1)-f(0))/(1-0) = 1$, a nonzero square. Thus f satisfies the conditions of Lemma 9.52 and so $b_i = cb_i^{p^j}$ for $i = 1, \ldots, p-2$. As $f(1) = 1$, $c = 1$ and the result follows. □

Theorem 9.51 is the special case of Theorem 9.53 with $n = 1$.

The smallest non-prime case of interest is $q = 9$. GF(9) can be thought of as consisting of elements of the form $a+bi, a, b \in GF(3)$, i a solution of $x^2+1=0$, and addition and subtraction modulo 3. The squares are 1, 2, i, and $2i$, and the nonsquares are $1+i$, $1+2i$, $2+i$, and $2+2i$.

When we list the quadratic orthomorphisms $[a, b]$, we obtain 7 linear orthomorphisms for which $a = b$; and, for $a \neq b$, we find 6 nonlinear, quadratic orthomorphisms corresponding to the following values of a and b.

a	i	1 + i	2 + i	b
b	2i	1 + 2i	2 + 2i	a

Each of these orthomorphisms is orthogonal to exactly one element of $\mathscr{C}_1$. In fact they are all orthogonal to [2], and the elements of $\mathscr{C}_2 \setminus \mathscr{C}_1$ together with [2] form a 7-clique in $\mathscr{C}_2(9)$. By Theorem 9.53, for each quadratic orthomorphism $[a, b]$ in this clique, $b = a^3$. In fact, it is easy to verify that $2^3 = 2$, $i^3 = 2i$, $(2i)^3 = i$, $(1+i)^3 = 1+2i$, $(1+2i)^3 = 1+i$, $(2+i)^3 = 2+2i$, and $(2+2i)^3 = 2+i$.

Chapter 10
Elementary Abelian Groups. II

In Chapter 9 we used permutation polynomials in the study of orthomorphisms and orthomorphism graphs of elementary abelian groups. We also studied linear orthomorphisms, orthomorphisms of the form $x \mapsto ax$; and quadratic orthomorphisms, orthomorphisms of the form x maps to ax if x is a square and bx if x is a nonsquare.

In this chapter we generalize linear and quadratic orthomorphisms by partitioning the elements of a finite field into cyclotomic classes $\{C_i\}$ and then defining and studying classes of orthomorphisms consisting of mappings of the form $x \mapsto a_i x$, if $x \in C_i$: this is the class of cyclotomic orthomorphisms.

We will also show how classes of orthomorphisms of elementary abelian groups can be found by solving systems of linear equations called (σ, ε)-systems.

10.1 Cyclotomy

Subgroups of the multiplicative group of GF(q) and their cosets, called cyclotomic classes, are useful in several combinatorial constructions, in particular in the construction of difference sets: see [301], for instance. They will also prove useful in the construction of classes of orthomorphisms of GF$(q)^+$. Solutions to $1+x = y$ with x and y from given cyclotomic classes will prove invaluable.

10.1.1 Cyclotomic Classes

For q a prime power and e a divisor of $q - 1$, let $q = ef + 1$, and let g denote a primitive element of GF(q). Then $w = g^f$ is a primitive eth root of unity and $v = g^e$ is a primitive fth root of unity. Set $C_i = \{g^{je+i} \mid j = 0, \ldots, f - 1\}$, $i = 0, \ldots, e - 1$. The set C_i is the *ith cyclotomic class*. It is clear that C_0 is the unique

A. B. Evans, *Orthogonal Latin Squares Based on Groups*, Developments in Mathematics 57, https://doi.org/10.1007/978-3-319-94430-2_10

subgroup of index e of the multiplicative subgroup of GF(q) and that $C_0, \ldots, C_{e-1}$ are its cosets. It should be noted that the ordering of the cyclotomic classes depends on the choice of primitive element as, if $\gcd(m, q-1) = 1$, then g^m is another primitive element of GF(q).

Lemma 10.1. *If* $\gcd(m, q-1) = 1$ *and we use* C_i' *to denote the* i*th cyclotomic class computed from the primitive element* g^m*, then* $C_i' = C_{mi}$*, where* mi *is computed modulo* e*.*

Proof. If we use C_i' to denote the ith cyclotomic class computed from the primitive element g^m, then

$$C_i' = \{g^{m(je+i)} \mid j=0, \ldots, f-1\} = \{g^{(mj)e+mi} \mid mj \equiv 0, \ldots, f-1 \pmod{f}\} = C_{mi},$$

where mi is computed modulo e. □

As an example, suppose that $q = 13$. It is easily verified that $g = 2$ is a primitive element of GF(13) and that the other primitive elements are $g^5 = 6$, $g^7 = 11$, and $g^{11} = 7$. Choosing $e = 3$, the cyclotomic classes, computed using $g = 2$, are

$$C_0 = \{1, 5, 8, 12\}, \quad C_1 = \{2, 3, 10, 11\}, \quad C_2 = \{4, 6, 7, 9\}.$$

These are the same cyclotomic classes that we would have computed had we used $g^7 = 11$ as the primitive element. Had we used $g^5 = 6$ or $g^{11} = 7$ instead, then C_0 would remain the same, but C_1 and C_2 would be swapped.

Lemma 10.2.

1. $C_i = g^i C_0 = g^i \langle v \rangle$*, where* $\langle v \rangle$ *is the subgroup, of the multiplicative group of* $GF(q)$*, generated by* $v = g^e$*.*
2. C_0 *consists of the nonzero* e*th powers.*
3. *If* q *is even or* f *is even, then* $-1 \in C_0$*.*
4. *If* e *is even and* f *is odd, then* $-1 \in C_{e/2}$*.*

Proof.

$$C_i = \{g^{je+i} \mid j = 0, \ldots, f-1\} = g^i \{g^{je} \mid j = 0, \ldots, f-1\} = g^i C_0$$

and

$$\begin{aligned} C_0 &= \{g^{je} \mid j = 0, \ldots, f-1\} \\ &= \{(g^e)^j \mid j = 0, \ldots, f-1\} \\ &= \{v^j \mid j = 0, \ldots, f-1\} \\ &= \langle v \rangle. \end{aligned}$$

If $x \in C_0$, then $x = g^{ej} = (g^j)^e$, an eth power. If $x = y^e \neq 0$, then $y = g^m$ for some m and $x = g^{me} \in C_0$.

If q is even, then $-1 = 1 \in C_0$.

If q is odd, then the multiplicative subgroup of GF(q), being cyclic, contains a unique subgroup, $\{1, -1\}$, of order 2. Thus $-1 \in C_0$ if and only if $|C_0| = f$ is even. If f is odd, then e is even and

$$-1 = g^{(q-1)/2} = g^{(ef)/2} = g^{(e(f-1))/2+e/2} \in C_{e/2}.$$

□

As a result of Lemma 10.2, we obtain the following consequences.

Corollary 10.3.

1. *If* $q \equiv 1 \pmod 4$, *then* -1 *is a square.*
2. *If* $q \equiv 3 \pmod 4$, *then* -1 *is a nonsquare.*
3. *If* e *is odd, then* $-1 \in C_0$.
4. *If* $q \equiv 1 \pmod 8$ *and* $e = 4$, *then* $-1 \in C_0$.
5. *If* $q \equiv 5 \pmod 8$ *and* $e = 4$, *then* $-1 \in C_2$.

Proof.

1. If $q \equiv 1 \pmod 4$ and $e = 2$, then f is even and so $-1 \in C_0$ by Lemma 10.2, and C_0 is the set of nonzero squares.
2. If $q \equiv 3 \pmod 4$ and $e = 2$, then f is odd and so $-1 \in C_1$ by Lemma 10.2, and C_1 is the set of nonsquares.
3. While we may derive this from Lemma 10.2, it is simpler to note that C_0 is the set of nonzero eth powers and, if e is odd, then $-1 = (-1)^e$ is an eth power and hence an element of C_0.
4. If $q \equiv 1 \pmod 8$ and $e = 4$, then f is even and so $-1 \in C_0$.
5. If $q \equiv 5 \pmod 8$ and $e = 4$, then f is odd and so $-1 \in C_2$. □

10.1.2 Cyclotomic Numbers

We define the *cyclotomic number of order* e, $\text{cycl}_e(i, j)$, to be the number of solutions to the equation $x_i + 1 = x_j$, $x_i \in C_i$ and $x_j \in C_j$. The cyclotomic numbers can be conveniently displayed in matrix form: the *cyclotomic matrix* is the $e \times e$ matrix with ijth entry $\text{cycl}_e(i, j)$. Note that the cyclotomic matrix and the cyclotomic numbers are not independent of the choice of primitive element of GF(q). If, instead of g, we choose g' as the primitive element of GF(q), then $g' = g^m$ for some m relatively prime to $q - 1$; and if we use $\text{cycl}'_e(i, j)$ to denote the new cyclotomic numbers, then $\text{cycl}'_e(i, j) = \text{cycl}_e(mi, mj)$.

Now let us consider some elementary properties of cyclotomic numbers.

Lemma 10.4.

1. $\mathrm{cycl}_e(i+me, j+ne) = \mathrm{cycl}_e(i, j)$.

2.

$$\mathrm{cycl}_e(i, j) = \begin{cases} \mathrm{cycl}_e(j, i) & \textit{if } f \textit{ is even or } q \textit{ is even,} \\ \mathrm{cycl}_e(j+e/2, i+e/2) & \textit{if } f \textit{ is odd and } e \textit{ is even.} \end{cases}$$

3. $\mathrm{cycl}_e(i, j) = \mathrm{cycl}_e(e-i, j-i)$.

4.

$$\sum_{j=0}^{e-1} \mathrm{cycl}_e(i, j) = \begin{cases} f & \textit{if } -1 \notin C_i, \\ f-1 & \textit{if } -1 \in C_i. \end{cases}$$

5.

$$\sum_{i=0}^{e-1} \mathrm{cycl}_e(i, j) = \begin{cases} f & \textit{if } 1 \notin C_j, \\ f-1 & \textit{if } 1 \in C_j. \end{cases}$$

6.

$$\mathrm{cycl}_e(0, j) \textit{ is } \begin{cases} \textit{odd} & \textit{if } 2 \in C_j, \\ \textit{even} & \textit{if } 2 \notin C_j. \end{cases}$$

Proof.

1. This follows from the fact that $C_{i+me} = C_i$ and $C_{j+ne} = C_j$.
2. If $x_i + 1 = x_j$, then $(-x_j) + 1 = (-x_i)$. If either f or e is even, then, by Lemma 10.2, $-1 \in C_0$ and so $-x_i \in C_i$ and $-x_j \in C_j$. If, on the other hand, f is odd and e is even, then, by Lemma 10.2, $-1 \in C_{e/2}$ and so $-x_i \in C_{i+e/2}$ and $-x_j \in C_{j+e/2}$.
3. If $x_i + 1 = x_j$, then $(1/x_i) + 1 = (x_j/x_i)$. Now $1/x_i \in C_{-i} = C_{e-i}$ and $x_j/x_i \in C_{j-i}$, the indices being subtracted modulo e.
4. $\sum_{j=0}^{e-1} \mathrm{cycl}_e(i, j) = |\{x_i \mid x_i + 1 \neq 0\}$ and this is f or $f-1$ according as $-1 \in C_i$ or not.
5. $\sum_{i=0}^{e-1} \mathrm{cycl}_e(i, j) = |\{x_j \mid x_j - 1 \neq 0\}$ and this is f or $f-1$ according as $1 \in C_j$ or not.
6. Let $S = \{(a, b) \mid a + 1 = b, a \in C_0, b \in C_j\}$. Now $(1, 2) \in S$ if and only if $2 \in C_j$ and if $(a, b) \in S$, $a \neq 1$, then $(1/a, 1/b) \in S$. Thus the elements of S other than $(1, 2)$ can be paired and so $|S \setminus \{(1, 2)\}|$ is even, hence the result. □

10.1.3 Cyclotomic Numbers of Order 2

In the special case $e = 2$, $q = 2f + 1$ is odd, C_0 is the set of nonzero squares, and C_1 is the set of nonsquares. In this case the cyclotomic numbers $\mathrm{cycl}_2(i, j)$ can be computed using only the properties in Lemma 10.4. For $e > 2$, the properties of cyclotomic numbers, given in Lemma 10.4, are not sufficient to compute the cyclotomic numbers.

Theorem 10.5. *If* $q \equiv 3 \pmod 4$, *then the cyclotomic matrix, for* $e = 2$, *is*

$$\begin{array}{c|cc} & 0 & 1 \\ \hline 0 & A & B \\ 1 & A & A \end{array}, \quad A = \frac{q-3}{4}, \quad B = \frac{q+1}{4}.$$

If $q \equiv 1 \pmod 4$, *then the cyclotomic matrix, for* $e = 2$, *is*

$$\begin{array}{c|cc} & 0 & 1 \\ \hline 0 & A & B \\ 1 & B & B \end{array}, \quad A = \frac{q-5}{4}, \quad B = \frac{q-1}{4}.$$

Proof. If $q \equiv 3 \pmod 4$, then f is odd and, as $e = 2$ is even, by Lemma 10.4(2),

$$\mathrm{cycl}_2(0, 0) = \mathrm{cycl}_2(0 + e/2, 0 + e/2) = \mathrm{cycl}_2(1, 1).$$

Further, by Lemma 10.4(3),

$$\mathrm{cycl}_2(1, 1) = \mathrm{cycl}_2(2 - 1, 1 - 1) = \mathrm{cycl}_2(1, 0).$$

Thus $\mathrm{cycl}_2(0, 0) = \mathrm{cycl}_2(1, 0) = \mathrm{cycl}_2(1, 1) = A$ and $\mathrm{cycl}_2(0, 1) = B$. As $q \equiv 3 \pmod 4$, -1 is a square. Thus, by Lemma 10.4(4 or 5), $A + A = f - 1 = (q-3)/2$ and $A + B = f = (q - 1)/2$, from which it follows that $A = (q - 3)/4$ and $B = (q + 1)/4$.

If $q \equiv 1 \pmod 4$, then f is even and so by Lemma 10.4(2),

$$\mathrm{cycl}_2(1, 0) = \mathrm{cycl}_2(0, 1).$$

Further, by Lemma 10.4(3),

$$\mathrm{cycl}_2(1, 1) = \mathrm{cycl}_2(2 - 1, 1 - 1) = \mathrm{cycl}_2(1, 0).$$

Thus $\mathrm{cycl}_2(0, 0) = A$ and $\mathrm{cycl}_2(0, 1) = \mathrm{cycl}_2(1, 0) = \mathrm{cycl}_2(1, 1) = B$. As $q \equiv 1 \pmod 4$, -1 is a nonsquare. Thus, by Lemma 10.4(4 or 5), $A + B = f - 1 = (q - 3)/2$ and $B + B = f = (q - 1)/2$, from which it follows that $A = (q - 5)/4$ and $B = (q - 1)/4$. □

We list here the cyclotomic classes and numbers for $e = 2$ and small values of q. We do not need to give a primitive element as, for $e = 2$, the cyclotomic classes and numbers are independent of the choice of primitive element.

Example 10.6. If $q = 3 \equiv 3 \pmod 4$, then $C_0 = \{1\}$, $C_1 = \{2\}$, and

$$\mathrm{cycl}_2(0,0) = \mathrm{cycl}_2(1,0) = \mathrm{cycl}_2(1,1) = 0, \quad \mathrm{cycl}_2(0,1) = 1.$$

Example 10.7. If $q = 5 \equiv 1 \pmod 4$, then $C_0 = \{1,4\}$, $C_1 = \{2,3\}$, and

$$\mathrm{cycl}_2(0,1) = \mathrm{cycl}_2(1,0) = \mathrm{cycl}_2(1,1) = 1, \quad \mathrm{cycl}_2(0,0) = 0.$$

Example 10.8. If $q = 7 \equiv 3 \pmod 4$, then $C_0 = \{1,2,4\}$, $C_1 = \{3,5,6\}$, and

$$\mathrm{cycl}_2(0,0) = \mathrm{cycl}_2(1,0) = \mathrm{cycl}_2(1,1) = 1, \quad \mathrm{cycl}_2(0,1) = 2.$$

Example 10.9. If $q = 9 \equiv 1 \pmod 4$, then, setting $\mathrm{GF}(9) = \{a + bi \mid i^2 + 1 = 0, a, b \in \mathrm{GF}(3)\}$, $C_0 = \{1, 2, i, 2i\}$, $C_1 = \{1+i, 1+2i, 2+i, 2+2i\}$, and

$$\mathrm{cycl}_2(0,1) = \mathrm{cycl}_2(1,0) = \mathrm{cycl}_2(1,1) = 2, \quad \mathrm{cycl}_2(0,0) = 1.$$

Example 10.10. If $q = 11 \equiv 3 \pmod 4$, then $C_0 = \{1,3,4,5,9\}$, $C_1 = \{2,6,7,8,10\}$, and

$$\mathrm{cycl}_2(0,0) = \mathrm{cycl}_2(1,0) = \mathrm{cycl}_2(1,1) = 2, \quad \mathrm{cycl}_2(0,1) = 3.$$

Example 10.11. If $q = 13 \equiv 1 \pmod 4$, then $C_0 = \{1,3,4,9,10,12\}$, $C_1 = \{2,5,6,7,8,11\}$, and

$$\mathrm{cycl}_2(0,1) = \mathrm{cycl}_2(1,0) = \mathrm{cycl}_2(1,1) = 3, \quad \mathrm{cycl}_2(0,0) = 2.$$

Example 10.12. If $q = 17 \equiv 1 \pmod 4$, then $C_0 = \{1,2,4,8,9,13,15,16\}$, $C_1 = \{3,5,6,7,10,11,12,14\}$, and

$$\mathrm{cycl}_2(0,1) = \mathrm{cycl}_2(1,0) = \mathrm{cycl}_2(1,1) = 4, \quad \mathrm{cycl}_2(0,0) = 3.$$

Example 10.13. If $q = 19 \equiv 3 \pmod 4$, then $C_0 = \{1,4,5,6,7,9,11,16,17\}$, $C_1 = \{2,3,8,10,12,13,14,15,18\}$, and

$$\mathrm{cycl}_2(0,0) = \mathrm{cycl}_2(1,0) = \mathrm{cycl}_2(1,1) = 4, \quad \mathrm{cycl}_2(0,1) = 5.$$

10.1.4 Cyclotomic Numbers of Higher Order

For order 2, the properties of cyclotomic numbers given in Lemma 10.4 are sufficient for us to determine the cyclotomic numbers. This is not true for orders 3 and higher. For order 2, C_0 is the set of nonzero squares and C_1 the set of nonsquares: this does not depend on the choice of primitive element. But for order 3, while C_0 is the set of nonzero cubes and C_1 and C_2 are the other cosets of C_0 in the multiplicative group of GF(q), which coset is which does depend on the choice of primitive element. Thus this choice of primitive element also affects the values of the cyclotomic numbers. This is not the whole story, as will become clear when we compute the cyclotomic numbers of order 3. To compute higher order cyclotomic numbers, we need more relations between these numbers, such as in the following lemma.

Lemma 10.14. *If $e \geq 3$, $-1 \in C_0$, and i_1, i_2, i_3 are distinct, then the number of solutions to*

$$1 + x_{i_1} + x_{i_2} + x_{i_3} = 0, \quad x_{i_j} \in C_{i_j}, \quad \textit{is} \quad \sum_{k=0}^{e-1} \text{cycl}_e(i_1, k)\text{cycl}_e(i_2 - k, i_3 - k).$$

Proof. First note none of $x_{i_1}, x_{i_2}, x_{i_3}$ can equal -1. If this could occur, then we may assume that $i_1 = 0$ and then, if $x_{i_1} = -1$, $x_{i_2} + x_{i_3} = 0$, from which it follows that $C_{i_2} \ni x_{i_2} = -x_{i_3} \in C_{i_3}$, an impossibility.

To count the solutions to $1 + x_{i_1} + x_{i_2} + x_{i_3} = 0$, for each $k \in \{0, \ldots, e-1\}$, $1 + x_{i_1} = x_k$, $x_k \in C_k$, exactly $\text{cycl}_e(i_1, k)$ times, and for each such solution $1 + x_{i_1} + x_{i_2} + x_{i_3} = x_k + x_{i_2} + x_{i_3} = x_k(1 + x_{i_2-k} + x_{i_3-k}) = 0$ has $\text{cycl}_e(i_2 - k, i_3 - k)$ solutions as this is the number of solutions to $1 + x_{i_2-k} = -x_{i_3-k} \in C_{i_3-k}$. The result then follows. □

The result of Lemma 10.14 can be generalized to counting the number of solutions to $1 + x_{i_1} + x_{i_2} + \cdots + x_{i_m} = 0$, $x_{i_j} \in C_{i_j}$, for $m \geq 3$ and -1 not necessarily in C_0. The results of Lemmas 10.4 and 10.14 enable us to determine the possible forms of the cyclotomic numbers of order 3.

Theorem 10.15. *If $q = 3f + 1$, then the cyclotomic matrix for $e = 3$ is*

	0	1	2
0	A	B	C
1	B	C	D
2	C	D	B

where $A = \dfrac{q - 8 + c}{9}$, $B = \dfrac{2q - 4 - c - 9d}{18}$,

$$C = \frac{2q - 4 - c + 9d}{18}, \quad D = \frac{q + 1 + c}{9},$$

and $q = (c^2 + 27d^2)/4$, for some integers c, d, $c \equiv 1 \pmod 3$.

Proof. Set $A = \mathrm{cycl}_3(0,0)$. By Lemma 10.4(2 and 3)

$$\mathrm{cycl}_3(1,0) = \mathrm{cycl}_3(0,1) = \mathrm{cycl}_3(2,2) = B,$$

$$\mathrm{cycl}_3(2,0) = \mathrm{cycl}_3(1,1) = \mathrm{cycl}_3(0,2) = C, \text{ and}$$

$$\mathrm{cycl}_3(2,1) = \mathrm{cycl}_3(1,2) = D.$$

By Lemma 10.4(4 or 5),

$$A + B + C = f - 1 \quad \text{and} \quad B + C + D = f.$$

By Corollary 10.3(3), $-1 \in C_0$, and so by Lemma 10.14, the number of solutions to the equation $1 + x_0 + x_1 + x_2 = 0$, $x_i \in C_i$, is

$$\mathrm{cycl}_3(0,0)\mathrm{cycl}_3(1,2) + \mathrm{cycl}_3(0,1)^2 + \mathrm{cycl}_3(0,2)\mathrm{cycl}_3(2,0) = AD + B^2 + C^2.$$

This is the same as the number of solutions to $1 + x_1 + x_2 + x_0 = 0$, which, by Lemma 10.14, equals

$$\mathrm{cycl}_3(1,0)\mathrm{cycl}_3(2,0) + \mathrm{cycl}_3(1,1)\mathrm{cycl}_3(1,2) + \mathrm{cycl}_3(1,2)\mathrm{cycl}_3(0,1)$$
$$= BC + CD + BD.$$

Thus $AD + B^2 + C^2 = BC + CD + BD$. Setting

$$c = 9A + 8 - q \text{ and } d = (3q - 12 - 9A - 18B)/9,$$

it follows that c and d are integers,

$$c \equiv 1 \pmod 3), \quad A = (q - 8 + c)/9, \quad \text{and } B = (2q - 4 - c - 9d)/18.$$

Further,

$$C = f - 1 - A - B = f - 1 - \frac{q - 8 + c}{9} - \frac{2q - 4 - c - 9d}{18} = \frac{2q - 4 - c + 9d}{18},$$

and

$$D = f - B - C = f - \frac{2q - 4 - c - 9d}{18} - \frac{2q - 4 - c + 9d}{18} = \frac{q + c + 1}{9}.$$

Substituting for A, B, C, and D in $AD + B^2 + C^2 = BC + CD + BD$ and solving for q yields $q = (c^2 + 27d^2)/4$. □

To complete the determination of the cyclotomic numbers of order 3, we need to know more about possible solutions to the equation $4q = c^2 + 27d^2$, where $q = p^m \equiv 1 \pmod 3$, p is prime, and $c \equiv 1 \pmod 3$. As $c \equiv 1 \pmod 3$, c^2 determines c, but d^2 only determines the absolute value of d: changing the sign of d has the effect of swapping B and C, which is exactly what happens if the cyclotomic classes C_1 and C_2 are interchanged.

Sometimes c^2 and d^2 are uniquely determined. If $n = 1$, then c^2 and d^2 are uniquely determined. Similarly, if n is even and $p \equiv 2 \pmod 3$, then c^2 and d^2 are uniquely determined: in fact, $d^2 = 0$ in this case. However, for $n > 1$ and $p \equiv 1 \pmod 3$, c^2 and d^2 need not be uniquely determined. In this latter case, we can force uniqueness by imposing the added restriction that q and c be relatively prime: in [301] it is shown that this condition yields the correct cyclotomic numbers.

Here are the smallest examples.

Example 10.16. If $q = 7$ and $e = 3$, then $c = 1$, $d = 1$; and $A = 0$, $B = 0$, $C = 1$, and $D = 1$. The cyclotomic matrix, with primitive element $g = 3$, is

	0	1	2
0	0	0	1
1	0	1	1
2	1	1	0

The cyclotomic classes are

$$C_0 = \{1, 6\}, \quad C_1 = \{3, 4\}, \quad \text{and } C_2 = \{2, 5\}.$$

Example 10.17. If $q = 13$ and $e = 3$, then $c = -5$, $d = 1$; and $A = 0$, $B = 1$, $C = 2$, and $D = 1$. The cyclotomic matrix, with primitive element $g = 2$ or 11, is

	0	1	2
0	0	1	2
1	1	2	1
2	2	1	1

The cyclotomic classes are

$$C_0 = \{1, 5, 8, 12\}, \quad C_1 = \{2, 3, 10, 11\}, \quad \text{and } C_2 = \{4, 6, 7, 9\}.$$

Example 10.18. If $q = 16$ and $e = 3$, then $c = -8$, $d = 0$; and $A = 0$, $B = 2$, $C = 2$, and $D = 1$. Let α be a root of the primitive polynomial $x^4 + x^3 + 1$ over GF(2). Then the cyclotomic matrix, with primitive element $g = \alpha$, or $\alpha^2 + \alpha$, or $\alpha^2 + \alpha + 1$, or $\alpha^3 + 1$, is

	0	1	2
0	0	2	2
1	2	2	1
2	2	1	2

The cyclotomic classes are

$$\begin{aligned} C_0 =&\{1, \alpha+1, \alpha^2+1, \alpha^3, \alpha^3+\alpha^2+\alpha+1\},\\ C_1 =&\{\alpha, \alpha^2+\alpha, \alpha^2+\alpha+1, \alpha^3+1, \alpha^3+\alpha\}, \text{ and}\\ C_2 =&\{\alpha^2, \alpha^3+\alpha+1, \alpha^3+\alpha^2, \alpha^3+\alpha^2+1, \alpha^3+\alpha^2+\alpha\}. \end{aligned}$$

Example 10.19. If $q = 19$ and $e = 3$, then $c = 7$, $d = 1$; and $A = 2$, $B = 1$, $C = 2$, and $D = 3$. The cyclotomic matrix, with primitive element $g = 2, 3$, or 14, is

	0	1	2
0	2	1	2
1	1	2	3
2	2	3	1

The cyclotomic classes are $C_0 = \{1, 7, 8, 11, 12, 18\}$, $C_1 = \{2, 3, 5, 14, 16, 17\}$, and $C_2 = \{4, 6, 9, 10, 13, 15\}$.

Cyclotomic numbers of small orders have been studied: see [301] or [225]. For q prime, some cyclotomic matrices of orders 3, 5, and 7 are given in [79] and in [104].

10.2 Cyclotomic Orthomorphisms

The linear and quadratic orthomorphisms studied in Section 9.3 generalize to cyclotomic orthomorphisms. If $q = ef + 1$, and $C_0, \ldots, C_{e-1}$ are the cyclotomic classes of order e, then we define the mapping $[a_0, \ldots, a_{e-1}]\colon \mathrm{GF}(q) \to \mathrm{GF}(q)$ by

$$[a_0, \ldots, a_{e-1}](x) = \begin{cases} 0 & \text{if } x = 0 \\ a_i x & \text{if } x \in C_i. \end{cases}$$

We use $\mathscr{C}_e(q)$, or $\mathscr{C}_e$ if there is no ambiguity, to denote both the set

$$\{[a_0, \ldots, a_{e-1}] \mid [a_0, \ldots, a_{e-1}] \in \mathrm{Orth}(\mathrm{GF}(q)^+)\}$$

as well as the corresponding induced orthomorphism graph of $\mathrm{GF}(q)^+$. The elements of $\mathscr{C}_e(q)$ are called *cyclotomic orthomorphisms* of index e.

10.2.1 The Orthomorphism Graph $\mathscr{C}_e(q)$

To describe elements of $\mathscr{C}_e(q)$, $e > 2$, just as with our description of elements of $\mathscr{C}_1(q)$ and $\mathscr{C}_2(q)$, we first need to determine when the mapping $[a_0, \ldots, a_{e-1}]$ is a bijection.

Theorem 10.20.

1. *$[a_0, \cdots, a_{e-1}]$ is a permutation if and only if the mapping $C_i \mapsto a_i C_i$ permutes the cyclotomic classes.*
2. *$[a_0, \cdots, a_{e-1}] \perp [b_0, \cdots, b_{e-1}]$ if and only if the mapping $C_i \mapsto (a_i - b_i)C_i$ permutes the cyclotomic classes.*
3. *$[a_0, \cdots, a_{e-1}]$ is an orthomorphism if and only if the mappings $C_i \mapsto a_i C_i$ and $C_i \mapsto (a_i - 1)C_i$ both permute the cyclotomic classes.*

Proof. If $a_i = 0$ for any i, then $[a_0, \ldots, a_{e-1}]$ cannot be a bijection, and if $a_i \neq 0$, then $a_i C_i$ is a cyclotomic class. It follows that $[a_0, \ldots, a_{e-1}]$ is a bijection if and only if $a_0 C_0, \ldots, a_{e-1} C_{e-1}$ are all of the cyclotomic classes.

Thus $[a_0, \cdots, a_{e-1}] \perp [b_0, \cdots, b_{e-1}]$ if and only if the mapping

$$x \mapsto [a_0, \cdots, a_{e-1}](x) - [b_0, \cdots, b_{e-1}](x) = [a_0 - b_0, \cdots, a_{e-1} - b_{e-1}](x)$$

is a bijection, if and only if the mapping $C_i \mapsto (a_i - b_i)C_i$ permutes the cyclotomic classes.

It also follows that $[a_0, \cdots, a_{e-1}]$ is an orthomorphism if and only if $[a_0, \cdots, a_{e-1}]$ and $[a_0 - 1, \cdots, a_{e-1} - 1]$ are bijections, if and only if the mappings $C_i \mapsto a_i C_i$ and $C_i \mapsto (a_i - 1)C_i$ both permute the cyclotomic classes. □

The linear and quadratic orthomorphisms, studied in Section 9.3, are the special cases of Theorem 10.20 with $e = 1$ and $e = 2$. At the other extreme, suppose $e = q - 1$. Then there are precisely $q - 1$ cyclotomic classes each consisting of one element only. Thus $\mathscr{C}_{q-1}(q) = \text{Orth}(\text{GF}(q)^+)$.

The effects of many of the congruences on elements of $\mathscr{C}_e(q)$ are described in the next result.

Theorem 10.21. *Let $[a_0, \cdots, a_{e-1}] \in \mathscr{C}_e(q)$.*

1. *If $\alpha(x) = ax$, $a \in C_k(q)$, then $\text{H}_\alpha[[a_0, \cdots, a_{e-1}]] = [a_{0-k}, \cdots, a_{e-1-k}] \in \mathscr{C}_e(q)$, where the indices are subtracted modulo e.*
2. *$\text{R}[[a_0, \cdots, a_{e-1}]] \in \mathscr{C}_e(q)$ and*

$$\text{R}[[a_0, \cdots, a_{e-1}]](x) = \begin{cases} [1 - a_0, \ldots, 1 - a_{e-1}] & \text{if } q \text{ or } f \text{ is even,} \\ [1 - a_{0+e/2}, \ldots, 1 - a_{3e/2-1}] & \text{if } e \text{ is even and } f \text{ is odd,} \end{cases}$$

where the indices are added modulo e.

3. $\mathrm{I}[[a_0, \cdots, a_{e-1}]] \in \mathscr{C}_e(q)$ *and, if* $a_i C_i = C_{\sigma(i)}$, *then*

$$\mathrm{I}[[a_0, \cdots, a_{e-1}]](x) = [1/a_{\sigma^{-1}(0)}, \ldots, 1/a_{\sigma^{-1}(e-1)}].$$

Proof.

1. If $a \in C_k$ and $x \in C_i$, then $a^{-1} \in C_{i-k}$ and

$$\mathrm{H}_\alpha[[a_0, \ldots, a_{e-1}]](x) = a[a_0, \ldots, a_{e-1}](a^{-1}x) = aa_{i-k}a^{-1}x = a_{i-k}x.$$

Thus $\mathrm{H}_\alpha[[a_0, \cdots, a_{e-1}]] = [a_{0-k}, \cdots, a_{e-1-k}] \in \mathscr{C}_e(q)$.

2. If q or f is even, then $-1 \in C_0$ by Lemma 10.2. In this case, if $x \in C_i$, then

$$\mathrm{R}[[a_0, \cdots, a_{e-1}]](x) = x + \mathrm{R}[[a_0, \cdots, a_{e-1}]](-x) = (1 - a_i)x.$$

If e is even and f is odd, then $-1 \in C_{e/2}$ by Lemma 10.2. In this case, if $x \in C_i$, then

$$\mathrm{R}[[a_0, \cdots, a_{e-1}]](x) = x + \mathrm{R}[[a_0, \cdots, a_{e-1}]](-x) = (1 - a_{i+e/2})x,$$

hence the result.

3. If $x \in C_i$, then $[a_0, \cdots, a_{e-1}](x) = a_i x \in C_{\sigma(i)}$ and

$$[1/a_{\sigma^{-1}(0)}, \ldots, 1/a_{\sigma^{-1}(e-1)}](a_i x) = \frac{1}{a_{\sigma^{-1}(\sigma(i))}} a_i x = x,$$

hence the result. □

We will see later, in Theorem 10.27, that, if $\theta \in \mathscr{C}_e(q) \setminus \mathscr{C}_1(q)$ and $g \neq 0$, then $\mathrm{T}_g[\theta] \notin \mathscr{C}_e(q)$. From Theorem 10.21 we can derive a simple characterization of the elements of $\mathscr{C}_e(q)$.

Theorem 10.22. $\theta \in \mathscr{C}_e(q)$ *if and only if* $\mathrm{H}_\alpha[\theta] = \theta$ *for all* $\alpha \in \{[a^e] \mid a \neq 0\}$.

Proof. If $a^e \in C_0$, $\alpha = [a^e]$, and $\theta \in \mathscr{C}_e(q)$, then $\mathrm{H}_\alpha[\theta] = \theta$ by Theorem 10.21.

Next assume that $\mathrm{H}_\alpha[\theta] = \theta$ for all $\alpha \in \{[a^e] \mid a \neq 0\}$. If $x, y \in C_i$, then $y = c^e x$ for some c and then $\alpha(x) = y = c^e x$. Thus, if $\theta(x) = bx$, then

$$\mathrm{H}_\alpha[\theta](y) = \mathrm{H}_\alpha[\theta](c^e x) = c^e b c^{-e} c^e x = by.$$

Hence $\theta \in \mathscr{C}_e(q)$ by definition. □

The characterization in Theorem 10.22 enables us to partially determine the lattice structure of the set of cyclotomic orthomorphisms.

Theorem 10.23. *Let* $a, b | (q-1)$. *If* $a | b$, *then* $\mathscr{C}_a(q) \subseteq \mathscr{C}_b(q)$.

Proof. If $\theta \in \mathscr{C}_a(q)$, then $\mathrm{H}_\alpha[\theta] = \theta$ for all $\alpha \in \{[c^a] \mid c \neq 0\}$ by Theorem 10.22. But $\{[c^b] \mid c \neq 0\} \subseteq \{[c^a] \mid c \neq 0\}$ and so $\mathrm{H}_\beta[\theta] = \theta$ for all $\beta \in \{[c^b] \mid c \neq 0\}$. Hence $\theta \in \mathscr{C}_b(q)$ by Theorem 10.22. □

It follows from Theorem 10.23 that the sets $\mathscr{C}_e(q)$ form a lattice which is isomorphic to the subgroup lattice of $\mathbb{Z}_{q-1}$, whenever the sets of cyclotomic orthomorphisms are distinct. That these two lattices need not always be isomorphic can be seen from small examples. If $q = 5$, then $\mathscr{C}_4 = \mathscr{C}_2 = \mathscr{C}_1$, whereas $\mathbb{Z}_4$, $\mathbb{Z}_2$, and $\mathbb{Z}_1$ are distinct. Similarly, if $q = 7$, then $\mathscr{C}_3 = \mathscr{C}_1$, whereas $\mathbb{Z}_3$ and $\mathbb{Z}_1$ are distinct.

10.2.2 The Number of Cyclotomic Orthomorphisms

It is natural to ask how many cyclotomic orthomorphisms there exist of any given index. This question was answered by Evans [113] in 1989.

Theorem 10.24 (Evans, 1989). *Suppose $q = ef + 1$. Then*

$$|\mathscr{C}_e(q)| = \sum \prod_{i=0}^{e-1} \mathrm{cycl}_e(\alpha(i) - i)(\beta(i) - i),$$

where subtraction is modulo e and the summation runs over all ordered pairs of permutations α, β of the set $\{0, \cdots, e-1\}$.

Proof. First, note that, if $[a_0, \cdots, a_{e-1}]$ is an orthomorphism, then

$$a_i C_i = C_{\beta(i)} \text{ and } (a_i - 1)C_i = C_{\alpha(i)},$$

for some pair of permutations α, β of the set $\{0, \cdots, e-1\}$. Thus

$$a_i \in C_{\beta(i)-i} \cap (C_{\alpha(i)-i} + 1),$$

where the indices are subtracted modulo e. Thus, for given permutations α and β of $\{0, \ldots, e-1\}$, the number of orthomorphisms $[a_0, \cdots, a_{e-1}]$ satisfying $a_i C_i = C_{\beta(i)}$ and $(a_i - 1)C_i = C_{\alpha(i)}$ is equal to

$$\prod_{i=0}^{e-1} \mathrm{cycl}_e(\alpha(i) - i)(\beta(i) - i).$$

Summing over all such pairs of permutations yields the number of elements of $\mathscr{C}_e$. □

The formula in Theorem 10.24 is only practical for relatively small values of e: as an example, if $e = q-1$, then this formula contains $(q-1)!^2$ terms, each of which is either 0 or 1. The formulae for $|\mathscr{C}_1(q)|$ and $|\mathscr{C}_2(q)|$ were given in Section 9.3. A formula for $|\mathscr{C}_3(q)|$ follows.

Corollary 10.25. *If* $e = 3$, *then*

$$|\mathscr{C}_3(q)| = 3\left(\sum_{j=0}^{2}\prod_{i=0}^{2}\mathrm{cycl}_3(i,j) + \sum_{i=0}^{2}\prod_{j=0}^{2}\mathrm{cycl}_3(i,j) + \sum_{k=0}^{2}\prod_{i=0}^{2}\mathrm{cycl}_3(i,i+k)\right) + \sum_{i,j=0}^{2}\mathrm{cycl}_3(i,j)^3.$$

Thus we can easily calculate $|\mathscr{C}_3(q)|$ from the cyclotomic matrix. These matrices were described in Examples 10.17, 10.18, and 10.19 for $q = 13, 16$, and 19, yielding $|\mathscr{C}_3(13)| = 65$, $|\mathscr{C}_3(16)| = 122$, and $|\mathscr{C}_3(19)| = 233$.

Terms in the formula for $|\mathscr{C}_e(q)|$, given in Theorem 10.24, can actually be used to construct elements of $\mathscr{C}_e(q)$. As an example, suppose $q = 13$, $e = 3$, and $\alpha = \beta = (1, 2, 3)$. This yields the term

$$\prod_{i=0}^{e-1}\mathrm{cycl}_3(\alpha(i)-i)(\beta(i)-i) = \mathrm{cycl}_3(1,1)^3 = 8.$$

Let $[a_0, a_1, a_2]$ be an orthomorphism counted by this term. Then $a_i \in C_1$ and $a_i + 1 \in C_1$ for $i = 0, 1, 2$. It follows that $a_i \in \{2, 10\}$ for $i = 0, 1, 2$, and each of the 8 choices for a_0, a_1, and a_2 yields an orthomorphism. Formulae for the number of cyclotomic orthomorphisms of indexes 3, 4, 5, and 6 were given by Bell [40] in 2013 with some restrictions on the values of q.

10.2.3 Maximal Sets of Orthomorphisms in $\mathscr{C}_e(q)$

In 1993 Pott [269], modifying the construction used to give an alternate proof of Theorem 9.38, constructed potential maximal difference matrices over $\mathrm{GF}(p)^+$, p prime.

Let $p = ef + 1$ and let g be a primitive element of $\mathrm{GF}(p)$. Pott considered the mapping $\phi\colon \mathrm{GF}(p) \to \mathrm{GF}(p)$ defined by

$$\phi(x) = \begin{cases} 0 & \text{if } x = 0, \\ ax & \text{if } x \in C_0, \\ bx & \text{if } x \in C_i,\ i \neq 0, \end{cases}$$

where $a \neq b$. We see that ϕ is really the cyclotomic mapping $[a, b, \ldots, b]$. Recall that, for $\mathrm{f}\colon \mathrm{GF}(q) \to \mathrm{GF}(q)$, we defined the *set of slopes*, S_{f}, of f by

$$S_{\mathrm{f}} = \left\{ \frac{\mathrm{f}(x) - \mathrm{f}(y)}{x - y} \;\middle|\; x, y \in \mathrm{GF}(q), x \neq y \right\}.$$

For $[a, b, \ldots, b]$,

$$S_{[a,b,\ldots,b]} = \{a, b\} \cup \left\{ \frac{az - b}{z - 1} \;\middle|\; z \notin C_0 \right\}.$$

It is easy to see that $|S_\phi| = p - f + 1$, and so, by Lemma 9.40, $\phi \perp [c]$ for $f - 1$ values of c: let these values be $c_1, \ldots, c_{f-1}$. Form a $(p, f; 1)$-difference matrix over $\mathrm{GF}(p)^+$, the columns indexed by the elements of $\mathrm{GF}(p)$, the ixth entry being $c_i x$, and the fxth entry being $\phi(x)$. Via a computer search, Pott found this difference matrix to be maximal when $p = 13$ and $e = 4$. However, this difference matrix is usually not maximal, and so extra rows must be added to obtain maximal difference matrices. Using a computer, Pott was able to establish the existence of maximal $(p, r; 1)$-difference matrices over $\mathrm{GF}(p)^+$ for $(p, r) = (13, 3), (13, 5), (17, 4), (17, 5), (19, 4)$, and $(19, 7)$.

10.2.4 Permutation Polynomials

Elements of $\mathscr{C}_1(q)$ are represented by orthomorphism polynomials of the form ax, and elements of $\mathscr{C}_2(q)$ by orthomorphism polynomials of the form $ax^{(q+1)/2} + bx$. There is a more general orthomorphism polynomial characterization of elements of $\mathscr{C}_e(q)$.

Theorem 10.26. *Let θ be an orthomorphism polynomial of* $\mathrm{GF}(q)^+$, $q = ef + 1$. *Then $\theta \in \mathscr{C}_e(q)$ if and only if*

$$\theta(x) \equiv \sum_{i=0}^{e-1} a_i x^{1+if} \pmod{x^q - x},$$

for some $a_i \in \mathrm{GF}(q)$.

Proof. Without loss of generality, we may assume that $\theta(x) = \sum_{j=0}^{q-1} a_j x^j$. Then, by Theorem 10.22, $\theta(x) \in \mathscr{C}_e(q)$ if and only if $\theta(ax) = a\theta(x)$ for all $a \in \mathrm{GF}(q)$ for which $a^f = 1$, if and only if $a^j = a$ whenever $a_j \neq 0$, if and only if $j - 1$ is divisible by f whenever $a_j \neq 0$. The result follows. □

The polynomial characterization of cyclotomic orthomorphisms allows us to say more about the effect of translations on cyclotomic orthomorphisms.

Theorem 10.27. *Let q be a prime. If $a, b | q-1$, $a, b \neq q-1$, and $g, h \in \mathrm{GF}(q)$, $g \neq h$, then*

$$\mathrm{T}_g[\mathscr{C}_a(q)] \cap \mathrm{T}_h[\mathscr{C}_b(q)] = \mathscr{C}_1(q).$$

Proof. Without loss of generality we may assume that $h = 0$ and $a, b \neq 1$. Clearly $\mathscr{C}_1(q) \subseteq \mathrm{T}_g[\mathscr{C}_a(q)] \cap \mathscr{C}_b(q)$. Next assume $\theta \in \mathrm{T}_g[\mathscr{C}_a(q)] \cap \mathscr{C}_b(q)$. Then there exists $\phi \in \mathscr{C}_a(q)$ such that $\theta = \mathrm{T}_g[\phi]$. Let $q = af + 1 = bf' + 1$. Then by Theorem 10.26

$$\theta(x) = \sum_{i=0}^{a-1} a_i (x+g)^{1+if} - \sum_{i=0}^{a-1} a_i g^{1+if} = \sum_{j=0}^{b-1} b_j x^{1+jf'},$$

for some $a_0, \cdots, a_{a-1}, b_0, \cdots, b_{b-1} \in \mathrm{GF}(q)$. Let $s = \sup\{i \mid a_i \neq 0\}$. If $s \neq 0$, then the coefficients of x^{1+sf} and x^{sf} are a_s and $a_s(1+sf)g$ respectively, both of which are nonzero since q is a prime. But, as $b \neq q-1$, $f' \geq 2$, and it follows that we cannot have consecutive powers of x with nonzero coefficients. Hence $s = 0$ and $\theta \in \mathscr{C}_1(q)$. □

Theorem 10.27 cannot be extended to non-prime q. To see why, let $q = p^n$, p prime, $n > 1$, and set $e = (q-1)/(p-1)$. Let $\theta(x) = cx^p + dx$ be an orthomorphism polynomial of GF(q). Then $\theta \in \mathscr{C}_e(q)$ and

$$\mathrm{T}_k[\theta](x) = \big(c(x+k)^p + d(x+k)\big) - \big(ck^p + dk\big) = cx^p + dx.$$

Thus

$$\theta \in (\mathrm{T}_g[\mathscr{C}_e(q)] \cap \mathrm{T}_h[\mathscr{C}_e(q)]) - \mathscr{C}_1(q) \text{ for all } g, h \in \mathrm{GF}(q).$$

For $a = b = 2$, however, Theorem 10.27 still holds for non-prime q. Further, there are no orthogonalities between nonlinear orthomorphisms in $\mathrm{T}_g[\mathscr{C}_2(q)]$ and nonlinear orthomorphisms in $\mathrm{T}_h[\mathscr{C}_2(q)]$ if $g \neq h$.

Theorem 10.28.

1. *Let q be odd and let $g, h \in \mathrm{GF}(q)$, $g \neq h$. Then $\mathrm{T}_g[\mathscr{C}_2(q)] \cap \mathrm{T}_h[\mathscr{C}_2(q)] = \mathscr{C}_1(q)$.*
2. *Let q be odd and let $g, h \in \mathrm{GF}(q)$, $g \neq h$. If $\theta, \phi \in \mathscr{C}_2(q) - \mathscr{C}_1(q)$ then $\mathrm{T}_g[\theta]$ cannot be orthogonal to $\mathrm{T}_h[\phi]$.*

Proof.

1. The result is clear if $q = 3$ or 5 as then $\mathscr{C}_{q-1}(q) = \mathscr{C}_1(q)$. If $q > 5$ then $\mathscr{C}_1(q) \subset \mathrm{T}_g[\mathscr{C}_2(q)] \cap \mathrm{T}_h[\mathscr{C}_2(q)]$. Without loss of generality we will assume that $h = 0$. Let $\theta \in \mathrm{T}_g[\mathscr{C}_2(q)] \cap \mathscr{C}_2(q)$. Then there exists $\phi \in \mathscr{C}_2(q)$ such that $\theta = \mathrm{T}_g[\phi]$. By Theorem 9.24(1), $\theta(x) = ax^{(q+1)/2} + bx$ and $\phi(x) = cx^{(q+1)/2} + dx$. Thus

$$a(x+g)^{(q+1)/2} + b(x+g) - ag^{(q+1)/2} - bg = cx^{(q+1)/2} + dx.$$

Comparing coefficients yields $a = c = 0$ and so $\theta \in \mathscr{C}_1(q)$.

2. Without loss of generality we may assume that $h = 0$. Let $\theta, \phi \in \mathscr{C}_2(q)$, $T_g[\theta] \perp \phi$. By Theorem 9.24(1), $\theta(x) = ax^{(q+1)/2} + bx$ and $\phi(x) = cx^{(q+1)/2} + dx$ and by Theorem 9.6 the coefficient of x^{q-1} in the reduction of $(T_g[\theta](x) - \phi(x))^2$ modulo $x^q - x$ must be zero. Hence the coefficient of x^{q-1} in the reduction of $T_g[\theta](x)\phi(x)$ modulo $x^q - x$ must be zero. But this coefficient is $acg^2(q^2 - 1)/8 \neq 0$, a contradiction from which the result follows. □

10.2.5 Niederreiter and Robinson's Construction

In 1982 Niederreiter and Robinson [252] studied the class of polynomials $ax^{1+f} + bx, a \neq 0$ over GF(q), $q = ef + 1$. By Theorem 10.26, each of these polynomials represents a cyclotomic mapping: specifically

$$ax^{1+f} + bx = [a + b, ag^f + b, ag^{2f} + b, \ldots, ag^{(e-1)f} + b].$$

Niederreiter and Robinson first determined when monic polynomials of this form were permutation polynomials.

Lemma 10.29. *Let $q = ef + 1$, $e \geq 2$, and let w be a primitive eth root of unity in* GF(q). *Then $x^{f+1} + bx$ is a permutation polynomial of* GF(q) *if and only if the following conditions hold.*

1. $(-b)^e \neq 1$.
2. $((b + w^i)(b + w^j)^{-1})^f \neq w^{j-i}$ *for all* $0 \leq i < j < e$.

Proof. Let w be a primitive eth root of unity in GF(q), $q = ef + 1$, and let $h(x) = x^{f+1} + bx$.

If $h(x)$ is a permutation polynomial then condition 1 holds as otherwise $-b = a^f$ for some $a \neq 0$ and so $h(0) = h(a) = 0$. If condition 2 does not hold, then for some $i, j, 0 \leq i < j < e$,

$$((b + w^i)(b + w^j)^{-1})^f = w^{j-i},$$

that is

$$(b + w^i)(b + w^j)^{-1} = g^{j-i},$$

for some primitive element g of GF(q) satisfying $g^f = w$, and so $h(g^j) = h(g^i)$, a contradiction.

Next suppose that conditions 1 and 2 hold. If $h(x) = 0$, then either $x = 0$ or $-b = x^f$, in which case $(-b)^e = 1$. Thus $h(x) = 0$ if and only if $x = 0$. If $h(a) = h(c)$ for $a, c \neq 0$, then for some i, j, $a^f = w^i$ and $c^f = w^j$. But then $(w^i + b)a = (w^j + b)c$ from which it follows that

$$((b + w^i)(b + w^j)^{-1})^f = w^{j-i}.$$

Thus $i = j$ and so $a = c$ and $h(x)$ is a permutation polynomial. □

From Lemma 10.29, Niederreiter and Robinson [252] constructed complete mapping polynomials. Clearly this also yields constructions of orthomorphism polynomials. Lemma 10.29 also enables us to determine orthogonalities within this class of orthomorphism polynomials.

Theorem 10.30. *Let $q = ef + 1$, $e \geq 2$, and let w be a primitive eth root of unity in* GF(q).

1. *If $\phi(x) = ax^{f+1} + bx$, $a \neq 0$, then $\phi \in \mathscr{C}_e(q)$ if and only if the following conditions hold for $c = b/a, (b-1)/a$:*
   a. $(-c)^e \neq 1$.
   b. $((c + w^i)(c + w^j)^{-1})^f \neq w^{j-i}$ *for all* $0 \leq i < j < e$.
2. *If $\phi(x) = ax^{f+1} + bx$, $a \neq 0$, then $\phi \perp [d]$ if and only if the following conditions hold for $c = (b-d)/a$:*
   a. $(-c)^e \neq 1$.
   b. $((c + w^i)(c + w^j)^{-1})^f \neq w^{j-i}$ *for all* $0 \leq i < j < e$.
3. *If $\phi(x) = ax^{f+1} + bx$ and $\phi'(x) = a'x^{f+1} + b'x$, $a, a' \neq 0$, then $\phi \perp \phi'$ if and only if $a = a'$ and $b \neq b'$ or the following conditions hold for $c = (b-b')/(a-a')$:*
   a. $(-c)^e \neq 1$.
   b. $((c + w^i)(c + w^j)^{-1})^f \neq w^{j-i}$ *for all* $0 \leq i < j < e$.

Proof. This is a routine application of Lemmas 9.18 and 10.29. □

10.3 A Linear Algebra Approach

In this section we will describe a method for generating orthorphisms of GF$(q)^+$ that is based on a simple observation. If θ is an orthomorphism of GF$(q)^+$ and $a \in$ GF$(q) \setminus \{0, 1\}$, then $x \mapsto a^{-1}\theta(x)$ and $x \mapsto (a-1)^{-1}\theta(x)$ are both permutations of the elements of GF(q). From this observation we will derive a method for generating orthomorphisms of GF$(q)^+$ by solving systems of linear equations over GF(q).

10.3.1 (σ, ε)-Systems

Let the nonzero elements of GF(q) be $x_1, \ldots, x_{q-1}$ and let $a \in$ GF(q), $a \neq 0, 1$. If θ is a normalized orthomorphism of GF$(q)^+$, then

$$\theta(x_i) = ax_{\sigma(i)}$$

for some permutation σ of $\{1, \ldots, q-1\}$, and

$$\theta(x_i) - x_i = ax_{\sigma(i)} - x_i = (a-1)x_{\varepsilon(i)}$$

for some permutation ε of $\{1, \ldots, q-1\}$. It is easy to see that $\sigma(i) = i$ if and only if $\varepsilon(i) = i$ if and only if $\sigma(i) = \varepsilon(i)$. If σ and ε are permutations of $\{1, \ldots, q-1\}$ satisfying $\sigma(i) = i$ if and only if $\varepsilon(i) = i$ if and only if $\sigma(i) = \varepsilon(i)$, then we call the system of equations $ax_{\sigma(i)} - x_i = (a-1)x_{\varepsilon(i)}$ a (σ, ε)-*system over* GF(q) *(defined by* θ*)* and we call $|\{i \mid \sigma(i) \neq i\}|$ the *order* of the system. It is a simple exercise to show that (σ, ε)-systems of orders 1 and 2 do not exist. (σ, ε)-systems were introduced and studied by Evans [112] in 1987.

A normalized orthomorphism θ of GF$(q)^+$ defines several (σ, ε)-systems: each choice of a may yield a different system. However we can always find a (σ, ε)-system, defined by θ, that has order at most $q-3$.

Theorem 10.31 (Evans, 1987). *If* θ *is a normalized orthomorphism of* $GF(q)^+$, *then there exists a* (σ, ε)*-system, defined by* θ*, of order at most* $q-3$.

Proof. Let $x_1, \ldots, x_{q-1}$ be the nonzero elements of GF(q) and let a_i, $i = 1, \ldots, q-1$, be defined by $\theta(x_i) = a_i x_i$. As the mappings $x_i \mapsto \theta(x_i) = a_i x_i$ and $x_i \mapsto \theta(x_i) - x_i = (a_i - 1)x_i$ are both bijections, $a_i \neq 0, 1$, for $i = 1, \ldots, q-1$. But then, as there are only $q-2$ choices for each a_i, for some i, j, $i \neq j$, we have $a_i = a_j = a$. The permutations σ and ε, defined by $\theta(x_i) = ax_{\sigma(i)}$ and $\theta(x_i) - x_i = (a-1)x_{\varepsilon(i)}$, form a (σ, ε)-system, defined by θ, of order at most $q-3$. □

To generate normalized orthomorphisms of GF$(q)^+$, we first pick permutations σ and ε of $\{1, \ldots, q-1\}$, satisfying $\sigma(i) = i$ if and only if $\varepsilon(i) = i$ if and only if $\sigma(i) = \varepsilon(i)$. We then solve the corresponding (σ, ε)-system for $a, x_1, \ldots, x_{q-1}$. If $a, x_1, \ldots, x_{q-1}$ is a solution satisfying $a \neq 0, 1$, $x_i \neq 0$ for all i, and $x_i = x_j$ if and only if $i = j$, then we define $\theta\colon$ GF(q) $\to$ GF(q) by

$$\theta(x) = \begin{cases} 0 & \text{if } x = 0, \\ ax_{\sigma(i)} & \text{if } x = x_i. \end{cases}$$

It is clear that θ is an orthomorphism of GF$(q)^+$. We have thus proved the following.

Theorem 10.32 (Evans, 1987). *Let* $a, x_1, \ldots, x_{q-1}$ *be a solution to a* (σ, ε)*-system over* GF(q). *If* $a \neq 0, 1$, *and* $x_1, \ldots, x_{q-1}$ *are distinct and nonzero, then the mapping* θ, *defined by*

$$\theta(x) = \begin{cases} 0 & \textit{if } x = 0 \\ ax_{\sigma(i)} & \textit{if } x = x_i, \end{cases}$$

is an orthomorphism of GF$(q)^+$.

We will say that the orthomorphism θ, described in Theorem 10.32, is *generated by the* (σ, ε)*-system*, and that θ is the orthomorphism *defined by the solution* a, $x_1, \ldots, x_n$ to the (σ, ε)-system.

Two (σ, ε)-systems are *equivalent* if one can be obtained from the other by reindexing. Formally, a (σ, ε)-system is equivalent to a (σ', ε')-system if there exists a permutation γ of $\{1, \ldots, q-1\}$ for which $\gamma^{-1}\sigma\gamma = \sigma'$ and $\gamma^{-1}\varepsilon\gamma = \varepsilon'$. Thus, we may, without loss of generality, assume that, in a (σ, ε)-system of order n, $\sigma(i) = i$ if $i > n$: any equation $ax_{\sigma(i)} - x_i = (a-1)x_{\varepsilon(i)}$ can then be ignored if $i > n$ as it is $ax_i - x_i = (a-1)x_i$, which is always true.

The trivial (σ, ε)-system, of order 0, generates the linear orthomorphisms as if θ is an orthomorphism generated by this system, then $\theta(x_i) = ax_i$ for some $a \neq 0, 1$. The smallest nontrivial (σ, ε)-system is of order 3: this system is unique up to equivalence. We will use this system to illustrate our method for generating orthomorphisms of $GF(q)^+$.

Example 10.33. Let $\sigma = (1\ 2\ 3)$: ε must be $(1\ 3\ 2)$ and the system is

$$\begin{aligned} ax_2 - x_1 &= (a-1)x_3 \\ ax_3 - x_2 &= (a-1)x_1 \\ ax_1 - x_3 &= (a-1)x_2. \end{aligned}$$

The coefficient matrix for this system of equations is

$$A = \begin{pmatrix} -1 & a & 1-a \\ 1-a & -1 & a \\ a & 1-a & -1 \end{pmatrix}.$$

As we always have the solution $x_1 = x_2 = x_3$, for any choice of a, the rank of the solution space of this system is at least 1 and, to obtain solutions with x_1, x_2, x_3 distinct, the rank of the solution space must be at least 2. Hence the rank of A must be 1. Thus the second row of A is a scalar multiple of the first, from which it follows that $a^2 - a + 1 = 0$. Under this condition, solving the system yields $(x_1, x_2, x_3) = b(0, a, 1) + c(1, 1, 1)$, which defines orthomorphisms when $\{c, ab + c, b + c\}$ are distinct and nonzero, that is, when $c \neq 0$ and $b \neq 0, -c, -c/a$.

Our method for generating orthomorphisms of $GF(q)^+$ can easily be extended to generating orthomorphisms of additive groups of rings with identity. Let R be a commutative ring with identity and suppose that there exists a unit a in R with the property that $a - 1$ is also a unit. If θ is a normalized orthomorphism of R^+ and $R \setminus \{0\} = \{x_1, \ldots, x_m\}$, then $\theta(x_i) = ax_{\sigma(i)}$ and $\theta(x_i) - x = ax_{\sigma(i)} - x_i = (a-1)x_{\varepsilon(i)}$ for uniquely determined permutations σ, ε of $\{1, \ldots, m\}$. Very little needs to be changed for our methods to generalize to additive groups of rings with identity: Theorem 10.31 need not hold, and solutions to (σ, ε)-systems are quite likely to be different. As an example, let us revisit Example 10.33. Given a commutative ring R with identity, a necessary condition for this system to generate orthomorphisms of R^+ is that $a^2 - a + 1$ not be a unit. The solutions are then $(x_1, x_2, x_3) = b(0, a, 1) + c(1, 1, 1)$, the same as in the field case except that b must satisfy $(a^2 - a + 1)b = 0$. A solution, satisfying these conditions, defines an orthomorphism if and only if $b \neq 0$ and $c \neq 0, -b, -ab$. As a special case, let $R = \mathbb{Z}_{15}$. As a and $a-1$ must both be units, $a = 2$, 8, or 14: in each case $a^2 - a + 1 \equiv 0 \pmod 3$, and a solution then defines orthomorphisms if and only if $b = 5$ or 10, and $c \neq 0$, 5, or 10. Note that $a^2 - a + 1 \equiv 0 \pmod{15}$ has no solutions. Further examples of solutions to (σ, ε)-systems for additive groups of commutative rings with identity can be found in [119].

10.3.2 A Special Class of (σ, ε)*-Systems*

The (σ, ε)-system in Example 10.33 is a special case of a class of (σ, ε)-systems, which we will call *cyclic* (σ, ε)-systems. These are defined as follows. Let r and n be integers satisfying $2 \leq r < n \leq q-3$. Let σ and ε be permutations of $\{1, 2, \ldots, n\}$ satisfying $\sigma(i) \equiv i+1 \pmod n$ and $\varepsilon(i) \equiv i+r \pmod n$ for $i = 1, \ldots, n$. We will call r the *shift* of the cyclic (σ, ε)-system. Thus a cyclic (σ, ε)-system is uniquely determined by its order and its shift. Note that in a cyclic (σ, ε)-system of order n and shift r, $\sigma = (1\ 2\ \cdots\ n)$ and ε is the disjoint product of d cycles of order n/d, where $d = \gcd(r, n)$. A class of solutions to cyclic (σ, ε)-systems is described in the following theorem.

Theorem 10.34 (Evans, 1987). *Suppose that* $2 \leq r < n \leq q-3$ *and that* n *divides* $q-1$. *Let* g *be a primitive* n*th root of unity. Set* $a = (1-g^r)/(g-g^r)$ *and let* $b, c \in \mathrm{GF}(q)$, $b \neq 0$ *and* c *not an* n*th root of unity. Then the mapping* θ, *defined by*

$$\theta(x) = \begin{cases} ab(g^{i+1} - c) & \text{if } x = b(g^i - c), \\ ax & \text{otherwise,} \end{cases}$$

is an orthomorphism of $\mathrm{GF}(q)^+$.

Proof. For $i = 1, \ldots, n$, set $x_i = b(g^i - c)$. As g is a primitive nth root of unity, $a \neq 0, 1$ and $x_1, \ldots, x_n$ are distinct and, as $b \neq 0$ and c is not an nth root of unity, $x_i \neq 0$ for $i = 1, \ldots, n$. Let $\sigma, \varepsilon\colon \{1, \ldots, n\} \to \{1, \ldots, n\}$ be the permutations satisfying $\sigma(i) \equiv i+1 \pmod n$ and $\varepsilon(i) \equiv i+r \pmod n$.

Now, for $i \in \{1, \ldots, n\}$, $\theta(x_i) = ax_{\sigma(i)}$ by definition and computation yields

$$ax_{\sigma(i)} - x_i = ab(g^{i+1} - c) - b(g^i - c) = (a-1)b(g^{i+r} - c) = (a-1)x_{\varepsilon(i)}.$$

Hence $a, x_1, \ldots, x_n$ form a solution to a (σ, ε)-system that satisfies the conditions of Theorem 10.32. The result follows. □

As a special case of Theorem 10.34, if q is odd, $n = (q-1)/2$, and $c = 0$, then the orthomorphisms described in Theorem 10.34 are precisely the quadratic orthomorphisms studied in Section 9.3.

We can verify that, if 3 divides $q - 1$, then the orthomorphisms generated in Theorem 10.34 are the same as those generated in Example 10.33. In this example, to generate orthomorphims, it was necessary that $a^2 - a + 1 = 0$. If 3 divides q, then $a = 2$ is the only solution to this equation. If 3 does not divide q, then, in order for the system to generate orthomorphisms, 3 must divide $q - 1$. To see this multiply the quadratic equation by $a + 1$. This yields $a^3 + 1 = 0$. Thus a must be a cube root of -1. If q is even, then 3 must divide $q - 1$. If q is odd, then $a^6 = 1$ and 6 divides $q - 1$. In either case 3 divides $q - 1$. We are lead to ask, if $\gcd(n, q) = 1$, for a cyclic (σ, ε)-system of order n to generate orthomorphisms, must it be that n divides $q - 1$, and must the orthomorphisms generated be precisely those described in Theorem 10.34? We will give a partial answer to this question in Theorem 10.35. We define the *rank* of a (σ, ε)-system, for a given $a \neq 0, 1$, to be the rank of the solution space of the system of equations $ax_{\sigma(i)} - x_i = (a - 1)x_{\varepsilon(i)}$, $\sigma(i) \neq i$.

Theorem 10.35. *Suppose that $2 \leq r < n \leq q - 3$ and $\gcd(n, q) = 1$. If the rank of a cyclic (σ, ε)-system of order n and shift r is at most two for all choices of a, then this system generates orthomorphisms if and only if n divides $q - 1$: the orthomorphisms generated are those described in Theorem 10.34.*

Proof. The system of equations is $ax_{i+1} - x_i = (a - 1)x_{i+r}$, $i = 1, \ldots, n$, indices added modulo n, and the coefficient matrix for this system is $A = aP - I + (1 - a)P^r$, where I is the identity matrix of order n and P is the permutation matrix of order n,

$$P = \begin{pmatrix} 0 & 1 & 0 & \cdots & 0 \\ \vdots & \ddots & \ddots & \ddots & \vdots \\ \vdots & & \ddots & \ddots & 0 \\ 0 & \cdots & \cdots & 0 & 1 \\ 1 & 0 & \cdots & \cdots & 0 \end{pmatrix},$$

a matrix with 1s above the main diagonal and in the lower left-hand corner, and 0s elsewhere. Let F be the nth cyclotomic extension of $\mathrm{GF}(q)$ and let g be a primitive nth root of unity in F. The eigenvalues of P are g^i, $i = 1, \ldots, n$, and the eigenspace corresponding to the eigenvalue g^i is of dimension 1 with basis $\{(g^i, g^{2i}, \ldots, 1)^T \mid i = 1, \ldots, n\}$: hence, these eigenvectors form a basis for the vector space of dimension n over F.

The eigenvalues of A are $\lambda_i = ag^i - 1 + (1 - a)g^{ir}$, $i = 1, \ldots, n$, and the eigenvectors of A are the eigenvectors of P, and hence form a basis for the vector space of dimension n over F: the eigenvector $(g^i, g^{2i}, \ldots, 1)^T$ corresponds to the eigenvalue λ_i. Thus A is similar to $\mathrm{diag}(\lambda_1, \ldots, \lambda_n)$ and so the rank of A is equal to the number of nonzero λ_is. Now $\lambda_n = 0$ and so, by the hypotheses of the theorem, $n - 2 \leq \text{rank } A \leq n - 1$. If rank $A = n - 1$, then the only solutions to our system of equations are constant and, hence, do not define orthomorphisms. Thus to generate orthomorphisms, we must have rank $A = n - 2$. Hence $\lambda_i = 0$

for some $i \neq n$. for this i, $ag^i - 1 + (1-a)g^{ir} = 0$. Solving for a yields $a = (1-g^{ir})/(g^i - g^{ir})$, which is not equal to either 0 or 1. As rank A is then $n-2$, the solutions to this system are linear combinations of $(g^i, g^{2i}, \ldots, 1)^T$ and $(1, \ldots, 1)^T$, the eigenvectors corresponding to λ_n and λ_i. Thus $x_j = bg^{ij} + c$. To define orthomorphisms, the x_js must be distinct, and so $b \neq 0$ and g^i must be a primitive nth root of unity, and as the x_js must be nonzero, $-c/b$ cannot be an nth root of unity. The orthomorphisms generated are thus the orthomorphisms described in Theorem 10.34.

It remains to show that n divides $q-1$. As $x_j \in \mathrm{GF}(q)$ for all j, $b + c = x_n \in \mathrm{GF}(q)$, $bg^i + c = x_1 \in \mathrm{GF}(q)$, and $bg^{2i} + c = x_2 \in \mathrm{GF}(q)$. Hence

$$g^i = \left(\frac{x_2 - x_n}{x_1 - x_n}\right) - 1 \in \mathrm{GF}(q).$$

As $g^i \in \mathrm{GF}(q)$ is a primitive nth root of unity, n must divide $q-1$. □

Whether Theorem 10.35 holds for ranks greater than 2 or not is an unanswered question. However, if $r = 2$ or $n-1$, then the rank of the cyclic (σ, ε)-system of order n and shift r is at most 2 for all choices of a, and so Theorem 10.35 holds.

Corollary 10.36. *If $n > 2$ and $\gcd(n, q) = 1$, then a cyclic (σ, ε)-system of order n and shift 2 or $n-1$ only generates orthomorphisms if n divides $q-1$, in which case the orthomorphisms generated are those described in Theorem 10.34.*

Proof. As in Theorem 10.35, let $A = aP - I + (1-a)P^r$ be the coefficient matrix of a cyclic (σ, ε)-system of order n and shift r, g a primitive nth root of unity, and $\lambda_i = ag^i - 1 + (1-a)g^{ir}$, $i = 1, \ldots, n$, the eigenvalues of A. If rank $A < n-2$, then for some $i, j \neq n$, $i \neq j$, $\lambda_i = \lambda_j$. It follows that

$$a = \frac{1-g^{ir}}{g^i - g^{ir}} = \frac{1-g^{jr}}{g^j - g^{jr}},$$

and so $(g^i - g^j) - (g^{ir} - g^{jr}) + g^{i+j}(g^{i(r-1)} - g^{j(r-1)}) = 0$. If $r = 2$ this equation reduces to $(g^i - g^j)(1-g^i)(1-g^j) = 0$, a contradiction as $1, g^i, g^j$ are distinct. If $r = n-1$ this equation reduces to $(g^i - g^j)(1 - g^{(n-1)i})(1 - g^{(n-1)j}) = 0$, a contradiction as $1, g^{(n-1)i}, g^{(n-1)j}$ are distinct. Hence, if $r = 2$ or $n-1$, then the rank of the cyclic (σ, ε)-system of order n and shift r is at most 2 for all choices of a. The result then follows from Theorem 10.35. □

10.4 Examples of (σ, ε)-Systems

We have seen that (σ, ε)-systems of orders 1 and 2 do not exist, and in Example 10.33 we solved the (σ, ε)-system of order 3. In this section we will solve all the remaining (σ, ε)-systems of order 4 and 5, as well as (σ, ε)-systems for all $q \leq 8$.

10.4.1 (σ, ε)-Systems of Small Order

We have already dealt with (σ, ε)-systems of orders 0 and 3. We will now solve, up to equivalence, all other (σ, ε)-systems of order at most 5. We may thus assume, without loss of generality, that $\sigma(i) = i$ for $i > n$ if the order of the system is n. As $x_1 = \cdots = x_n$ is always a solution, the rank of the system is always at least 1 for any choice of a. For the system to generate orthomorphisms its rank must be at least 2 for some choice of $a \neq 0, 1$. If A is the coefficient matrix of a (σ, ε)-system and B is any submatrix of A, then rank $B \leq$ rank A as, if k of the rows of B are independent, then the corresponding k rows of A must also be independent. It follows that, if B is the submatrix of A, obtained by removing the last row and column, then a necessary condition for the system to generate orthomorphisms is that $\det(B)$, called the *leading principal* $(n-1) \times (n-1)$ *minor of* A, be 0 for some choice of $a \neq 0, 1$. Our approach will be to first compute this leading principal minor, a polynomial in a, and then, for $a \neq 0, 1$, a root of this polynomial, to solve the system of equations using standard methods. Results for (σ, ε)-systems of order 4 and 5 are given in the following examples. The solutions to (σ, ε)-systems of order at most 5 were given by Evans [112] in 1987. This list did not however contain cyclic (σ, ε)-systems that were not covered by Theorem 10.35: we will include these systems in our list. We will also correct an error in the solution to the (σ, ε)-system with $\sigma = (1\,2\,3\,4\,5)$ and $\varepsilon = (1\,3)(2\,5\,4)$.

Example 10.37. Let $\sigma = (1\,2\,3\,4)$ and $\varepsilon = (1\,3)(2\,4)$. This is a cyclic (σ, ε)-system of order 4 and shift 2. The corresponding system of equations is

$$\begin{aligned} ax_2 - x_1 &= (a-1)x_3 \\ ax_3 - x_2 &= (a-1)x_4 \\ ax_4 - x_3 &= (a-1)x_1 \\ ax_1 - x_4 &= (a-1)x_2. \end{aligned}$$

The coefficient matrix of this system is

$$\begin{pmatrix} -1 & a & 1-a & 0 \\ 0 & -1 & a & 1-a \\ 1-a & 0 & -1 & a \\ a & 1-a & 0 & -1 \end{pmatrix}.$$

As the first and second equations are independent of each other, the rank of this system is at most 2. If 2 does not divide q, then, by Corollary 10.36, the solution is given in Theorem 10.34. We will therefore assume that 2 divides q. But then the leading principal 3×3 minor is $a^3(a+1)$. Hence, as $a \neq 0, 1$, this system generates no orthomorphisms if q is a power of 2.

Example 10.38. Let $\sigma = (1\,2\,3\,4)$ and $\varepsilon = (1\,4\,3\,2)$. This is a cyclic (σ, ε)-system of order 4 and shift 3. The corresponding system of equations is

$$\begin{aligned} ax_2 - x_1 &= (a-1)x_4 \\ ax_3 - x_2 &= (a-1)x_1 \\ ax_4 - x_3 &= (a-1)x_2 \\ ax_1 - x_4 &= (a-1)x_3. \end{aligned}$$

The coefficient matrix of this system is

$$\begin{pmatrix} -1 & a & 0 & 1-a \\ 1-a & -1 & a & 0 \\ 0 & 1-a & -1 & a \\ a & 0 & 1-a & -1 \end{pmatrix}.$$

As the first and second equations are independent of each other, the rank of this system is at most 2. If 2 does not divide q, then, by Corollary 10.36, the solution is given in Theorem 10.34. We will therefore assume that 2 divides q. The leading principal 3×3 minor is $1 \neq 0$ and, hence, this system generates no orthomorphisms if q is a power of 2.

Example 10.39. Let $\sigma = (1\,2)(3\,4)$ and $\varepsilon = (1\,3)(2\,4)$. The corresponding system of equations is

$$\begin{aligned} ax_2 - x_1 &= (a-1)x_3 \\ ax_1 - x_2 &= (a-1)x_4 \end{aligned}$$

$$\begin{aligned} ax_4 - x_3 &= (a-1)x_1 \\ ax_3 - x_4 &= (a-1)x_2. \end{aligned}$$

The coefficient matrix of this system is

$$\begin{pmatrix} -1 & a & 1-a & 0 \\ a & -1 & 0 & 1-a \\ 1-a & 0 & -1 & a \\ 0 & 1-a & a & -1 \end{pmatrix}.$$

As the first and third equations are independent of each other, the rank of this system is at most 2. The leading principal 3×3 minor is $2a(a-1)$. As $a \neq 0, 1$, the principal minor can only be 0 if q is a power of 2. The solutions to this system are then $(x_1, x_2, x_3, x_4) = b(0, a+1, a, 1) + c(1, 1, 1, 1)$, and x_1, x_2, x_3, x_4 are distinct if and only if $b \neq 0$ and nonzero if and only if $c/b \neq 0, 1, a, a+1$.

Example 10.40. Let $\sigma = (1\ 2)(3\ 4)$ and $\varepsilon = (1\ 3\ 2\ 4)$. The corresponding system of equations is

$$ax_2 - x_1 = (a-1)x_3$$
$$ax_1 - x_2 = (a-1)x_4$$

$$ax_4 - x_3 = (a-1)x_2$$
$$ax_3 - x_4 = (a-1)x_1.$$

The coefficient matrix of this system is

$$\begin{pmatrix} -1 & a & 1-a & 0 \\ a & -1 & 0 & 1-a \\ 0 & 1-a & -1 & a \\ 1-a & 0 & a & -1 \end{pmatrix}.$$

As the first and third equations are independent of each other, the rank of this system is at most 2. The leading principal 3×3 minor is $a^3 - a^2 + a - 1 = (a-1)(a^2 + 1)$. Hence, this system can only generate orthomorphisms if $q \equiv 1 \pmod 4$. The solutions to this system are then $(x_1, x_2, x_3, x_4) = b(0, a-1, a, -1) + c(1, 1, 1, 1)$, and x_1, x_2, x_3, x_4 are distinct if and only if $b \neq 0$ and nonzero if and only if $c/b \neq 0, 1, -a, 1-a$.

Example 10.41. Let $\sigma = (1\ 2)(3\ 4\ 5)$ and $\varepsilon = (1\ 3\ 5\ 2\ 4)$. The corresponding system of equations is

$$ax_2 - x_1 = (a-1)x_3$$
$$ax_1 - x_2 = (a-1)x_4$$

$$ax_4 - x_3 = (a-1)x_5$$
$$ax_5 - x_4 = (a-1)x_1$$
$$ax_3 - x_5 = (a-1)x_2.$$

The coefficient matrix of this system is

$$\begin{pmatrix} -1 & a & 1-a & 0 & 0 \\ a & -1 & 0 & 1-a & 0 \\ 0 & 0 & -1 & a & 1-a \\ 1-a & 0 & 0 & -1 & a \\ 0 & 1-a & a & 0 & -1 \end{pmatrix}.$$

As the second, third, and fourth equations form an independent set, the rank of this system is at most 2. The leading principal 4×4 minor is $1 - 2a + 3a^2 - 2a^3 = (1-a)(1-a+2a^2)$. Hence, as $a \neq 1$, this system can only generate orthomorphisms if $1 - a + 2a^2 = 0$. The solutions to this system are then $(x_1, x_2, x_3, x_4, x_5) = b(0, 1-3a, -a-1, 2-2a, -4a)+c(1, 1, 1, 1, 1)$, and x_1, x_2, x_3, x_4, x_5 are distinct if and only if $b \neq 0$ and nonzero if and only if $c/b \neq 0, 3a-1, a+1, 2a-2, 4a$.

Example 10.42. Let $\sigma = (1\ 2)(3\ 4\ 5)$ and $\varepsilon = (1\ 3\ 5)(2\ 4)$. The corresponding system of equations is

$$ax_2 - x_1 = (a-1)x_3$$
$$ax_1 - x_2 = (a-1)x_4$$

$$ax_4 - x_3 = (a-1)x_5$$
$$ax_5 - x_4 = (a-1)x_2$$
$$ax_3 - x_5 = (a-1)x_1.$$

The coefficient matrix of this system is

$$\begin{pmatrix} -1 & a & 1-a & 0 & 0 \\ a & -1 & 0 & 1-a & 0 \\ 0 & 0 & -1 & a & 1-a \\ 0 & 1-a & 0 & -1 & a \\ 1-a & 0 & a & 0 & -1 \end{pmatrix}.$$

As the second, third, and fifth equations form an independent set, the rank of this system is at most 2. The leading principal 4×4 minor is $2a - 3a^2 + 2a^3 - a^4 = a(1-a)(2-a+a^2)$. Hence, as $a \neq 0, 1$, this system generates orthomorphisms only if $2 - a + a^2 = 0$. The solutions to this system are then $(x_1, x_2, x_3, x_4, x_5) = b(0, -a-1, -2, 1-a, -2a) + c(1, 1, 1, 1, 1)$, and x_1, x_2, x_3, x_4, x_5 are distinct if and only if q is odd and $b \neq 0$ and nonzero if and only if $c/b \neq 0, a+1, 2, a-1, 2a$.

Example 10.43. Let $\sigma = (1\ 2\ 3\ 4\ 5)$ and $\varepsilon = (1\ 3\ 5\ 2\ 4)$. This is a cyclic (σ, ε)-system of order 5 and shift 2. If 5 does not divide q, then, by Corollary 10.36, the solution is given in Theorem 10.34. We will therefore assume that 5 divides q. The corresponding system of equations is

$$ax_2 - x_1 = (a-1)x_3$$
$$ax_3 - x_2 = (a-1)x_4$$
$$ax_4 - x_3 = (a-1)x_5$$
$$ax_5 - x_4 = (a-1)x_1$$
$$ax_1 - x_5 = (a-1)x_2.$$

The coefficient matrix of this system is

$$\begin{pmatrix} -1 & a & 1-a & 0 & 0 \\ 0 & -1 & a & 1-a & 0 \\ 0 & 0 & -1 & a & 1-a \\ 1-a & 0 & 0 & -1 & a \\ a & 1-a & 0 & 0 & -1 \end{pmatrix}.$$

As the first, second, and third equations form an independent set, the rank of this system is at most 2. The leading principal 4×4 minor is $1+3a+4a^2+2a^3+a^4 = (a+3)^4$. Hence $a = 2$. The solutions to this system are then $(x_1, x_2, x_3, x_4, x_5) = b(4, 3, 2, 1, 0)+c(1, 1, 1, 1, 1)$, and x_1, x_2, x_3, x_4, x_5 are distinct if and only if $b \neq 0$ and nonzero if and only if c/b is not in the prime subfield of GF(q).

Example 10.44. Let $\sigma = (1\ 2\ 3\ 4\ 5)$ and $\varepsilon = (1\ 4\ 2\ 5\ 3)$. This is a cyclic (σ, ε)-system of order 5 and shift 3. The corresponding system of equations is

$$\begin{aligned} ax_2 - x_1 &= (a-1)x_4 \\ ax_3 - x_2 &= (a-1)x_5 \\ ax_4 - x_3 &= (a-1)x_1 \\ ax_5 - x_4 &= (a-1)x_2 \\ ax_1 - x_5 &= (a-1)x_3. \end{aligned}$$

The coefficient matrix of this system is

$$A = \begin{pmatrix} -1 & a & 0 & 1-a & 0 \\ 0 & -1 & a & 0 & 1-a \\ 1-a & 0 & -1 & a & 0 \\ 0 & 1-a & 0 & -1 & a \\ a & 0 & 1-a & 0 & -1 \end{pmatrix}.$$

The leading principal 4×4 minor is $M_4 = 1 - a + a^2 - a^3 + a^4$. Hence this system generates orthomorphisms only if $M_4 = 0$. The leading principal 3×3 minor, the determinant of the matrix obtained from A by deleting all but the first three rows and columns, of the coefficient matrix is $M_3 = -1 + a^2 - a^3$. Now M_3 and M_4 are polynomials in a and $\gcd(M_3, M_4) = 1$ for any q. Hence, M_3 and M_4 cannot both be zero. It follows that, if $M_4 \neq 0$, then the rank of the system is 1, and, if $M_4 = 0$, then $M_3 \neq 0$ and the rank of the system is 2. In either case the rank of the system is at most 2 for all choices of a, and so, if 5 does not divide q, then the solution is given in Theorem 10.34. We will therefore assume that 5 divides q. Then $M_4 = (a+1)^4$ and so $a = 4$. The solutions to this system are then $(x_1, x_2, x_3, x_4, x_5) = b(0, 1, 2, 3, 4) + c(1, 1, 1, 1, 1)$, and x_1, x_2, x_3, x_4, x_5

are distinct if and only if $b \neq 0$ and nonzero if and only if c/b is not in the prime subfield of GF(q).

Example 10.45. Let $\sigma = (1\ 2\ 3\ 4\ 5)$ and $\varepsilon = (1\ 5\ 4\ 3\ 2)$. This is a cyclic (σ, ε)-system of order 5 and shift 4. If 5 does not divide q, then, by Corollary 10.36, the solution is given in Theorem 10.34. We will therefore assume that 5 divides q. The corresponding system of equations is

$$\begin{aligned} ax_2 - x_1 &= (a-1)x_5 \\ ax_3 - x_2 &= (a-1)x_1 \\ ax_4 - x_3 &= (a-1)x_2 \\ ax_5 - x_4 &= (a-1)x_3 \\ ax_1 - x_5 &= (a-1)x_4. \end{aligned}$$

The coefficient matrix of this system is

$$\begin{pmatrix} -1 & a & 0 & 0 & 1-a \\ 1-a & -1 & a & 0 & 0 \\ 0 & 1-a & -1 & a & 0 \\ 0 & 0 & 1-a & -1 & a \\ a & 0 & 0 & 1-a & -1 \end{pmatrix}.$$

As the first, third, and fourth equations form an independent set, the rank of this system is at most 2. The leading principal 4×4 minor is $1+2a+4a^2+3a^3+a^4 = (a+2)^4$. Hence this system generates orthomorphisms only if $a = 3$. The solutions to this system are then $(x_1, x_2, x_3, x_4, x_5) = b(4, 3, 2, 1, 0) + c(1, 1, 1, 1, 1)$, and x_1, x_2, x_3, x_4, x_5 are distinct if and only if $b \neq 0$ and nonzero if and only if c/b is not in the prime subfield of GF(q).

Example 10.46. Let $\sigma = (1\ 2\ 3\ 4\ 5)$ and $\varepsilon = (1\ 3\ 5\ 4\ 2)$. This is a cyclic (σ, ε)-system of order 5 and shift 2. If 5 does not divide q, then the solution is given in Theorem 10.34. We will therefore assume that 5 divides q. The corresponding system of equations is

$$\begin{aligned} ax_2 - x_1 &= (a-1)x_3 \\ ax_3 - x_2 &= (a-1)x_1 \\ ax_4 - x_3 &= (a-1)x_5 \\ ax_5 - x_4 &= (a-1)x_2 \\ ax_1 - x_5 &= (a-1)x_4. \end{aligned}$$

The coefficient matrix of this system is

$$\begin{pmatrix} -1 & a & 1-a & 0 & 0 \\ 1-a & -1 & a & 0 & 0 \\ 0 & 0 & -1 & a & 1-a \\ 0 & 1-a & 0 & -1 & a \\ a & 0 & 0 & 1-a & -1 \end{pmatrix}.$$

As the first, third, and fourth equations form an independent set, the rank of this system is at most 2. The leading principal 4×4 minor is $1-2a+3a^2-2a^3+a^4 = (a^2-a+1)^2$. Hence this system generates orthomorphisms only if $a^2-a+1 = 0$. The solutions to this system are then $(x_1, x_2, x_3, x_4, x_5) = b(a, a, a, 1, 0) + c(1, 1, 1, 1, 1)$, and, as $x_1 = x_2 = x_3$ for all choices of b and c, this system generates no orthomorphisms.

Example 10.47. Let $\sigma = (1\ 2\ 3\ 4\ 5)$ and $\varepsilon = (1\ 3)(2\ 5\ 4)$. The corresponding system of equations is

$$\begin{aligned} ax_2 - x_1 &= (a-1)x_3 \\ ax_3 - x_2 &= (a-1)x_5 \\ ax_4 - x_3 &= (a-1)x_1 \\ ax_5 - x_4 &= (a-1)x_2 \\ ax_1 - x_5 &= (a-1)x_4. \end{aligned}$$

The coefficient matrix of this system is

$$\begin{pmatrix} -1 & a & 1-a & 0 & 0 \\ 0 & -1 & a & 0 & 1-a \\ 1-a & 0 & -1 & a & 0 \\ 0 & 1-a & 0 & -1 & a \\ a & 0 & 0 & 1-a & -1 \end{pmatrix}.$$

As the first, second, and fifth equations form an independent set, the rank of this system is at most 2. The leading principal 4×4 minor is $a(2-3a+2a^2)$. Hence, as $a \neq 0, 1$, this system generates orthomorphisms only if $2-3a+2a^2 = 0$. The solutions to this system are then $(x_1, x_2, x_3, x_4, x_5) = b(0, -a, a-2, 2a-2, a) + c(1, 1, 1, 1, 1)$, and x_1, x_2, x_3, x_4, x_5 are distinct if and only if q is odd and $b \neq 0$ and nonzero if and only if $c/b \neq 0, \pm a, 2-a, 2-2a$. Note that the solution given by Evans [112] is incorrect as it gives the solutions as $(x_1, x_2, x_3, x_4, x_5) = b(0, -a, a-2, 3a-4, a) + c(1, 1, 1, 1, 1)$.

10.4.2 (σ, ε)-*Systems for Small q*

By Theorem 10.31, knowing the solutions to all (σ, ε)-systems of order 5 or less enables us to generate all normalized orthomorphisms of $GF(q)^+$ for $q \leq 8$. For small orders, $q \leq 5$, all normalized orthomorphisms of $GF(q)^+$ are linear.

All the nonlinear orthomorphisms of $GF(7)^+$ are generated in Example 10.33: these are all the translates of the nonlinear quadratic orthomorphisms of $GF(7)^+$. That each of these orthomorphisms is not orthogonal to any other orthomorphism of $GF(7)^+$ follows from Corollaries 9.25 and 9.35, and Theorem 10.28.

All the nonlinear orthomorphisms of $GF(8)^+$ are generated in Example 10.39: in this example $\sigma = (1\ 2)(3\ 4)$ and $\varepsilon = (1\ 3)(2\ 4)$. Recall that this system generates orthomorphisms only if q is even, and then the solutions are $(x_1, x_2, x_3, x_4) = b(0, a+1, a, 1) + c(1, 1, 1, 1)$, $a \neq 0, 1$, $b \neq 0$, and $c/b \neq 0, 1, a, a+1$. The orthomorphism θ defined by this solution acts on $\{x_1, x_2, x_3, x_4\}$ as follows

$$\begin{aligned} c &\mapsto a(ab+b+c) \\ ab+b+c &\mapsto ac \\ ab+c &\mapsto a(b+c) \\ b+c &\mapsto a(ab+c). \end{aligned}$$

Thus, if $x \in \{c, b+c, ab+c, b+ab+c\}$, then $\theta(x) = a(x+b+ab)$; and if $x \notin \{c, b+c, ab+c, b+ab+c\}$, then $\theta(x) = ax$. If $q = 8$, then the set $H_{a,b} = \{0, b, ab, ab+b\}$ is a subgroup of $GF(8)^+$ of index 2, and $c \notin H_{a,b}$. Hence, for each $a \neq 0, 1$ and $b \neq 0$, the mapping, $\phi_{a,b}\colon GF(8)^+ \to GF(8)^+$, defined by

$$\phi_{a,b}(x) = \begin{cases} ax & \text{if } x \in H_{a,b}, \\ a(x+ab+b) & \text{if } x \notin H_{a,b}, \end{cases}$$

is an orthomorphism of $GF(8)^+$. The set $\{\phi_{a,b} \mid a \neq 0, 1, b \neq 0\}$ is the set of normalized orthomorphisms of $GF(8)^+$ generated by Example 10.39. Orthogonalities between these orthomorphisms can be determined by simple, but tedious computation. Two examples are given below.

Example 10.48. If $a^3 + a + 1 = 0$, then $\phi_{a,1} \perp \phi_{a+1,b}$ if and only if $b \in \{1, a^2, 1 + a + a^2\}$. Note that a list of orthogonalities given by Evans [112] in 1987 included the incorrect orthogonality $\phi_{a,1} \perp \phi_{a+1,1+a^2}$.

Example 10.49. If $a^3 + a + 1 = 0$, then $\phi_{a,a} \perp \phi_{a+1,b}$ if and only if $b \in \{a, 1 + a, 1 + a^2\}$.

We next present a method of determining which linear orthomorphisms are orthogonal to a given orthomorphism, generated by a (σ, ε)-system.

Theorem 10.50 (Evans, 1987). *Let* $\mathrm{GF}(q) = \{0, x_1, \ldots, x_{q-1}\}$, *let* $\theta(x_i) = ax_{\sigma(i)}$ *and* $\theta(0) = 0$ *define an orthomorphism of* $\mathrm{GF}(q)^+$, *and suppose* $b \neq a$. *Then* $\theta \perp [b]$ *if and only if the mapping*

$$x_i \mapsto \left(\frac{ax_{\sigma(i)} - bx_i}{a - b}\right)$$

is a permutation of the set $\{x_i \mid i \neq \sigma(i)\}$.

Proof. $\theta \perp [b]$ if and only if

$$\theta(x_i) - bx_i = ax_{\sigma(i)} - bx_i = (a - b)x_{\delta(i)}$$

for some permutation δ of $\{1, \ldots, q-1\}$. As, if $\sigma(i) = i$, then $\delta(i) = i$, the result follows. □

As a consequence of this we obtain.

Corollary 10.51. *Let* $\mathrm{GF}(q) = \{0, x_1, \ldots, x_{q-1}\}$. *Let* $\{a, y_1, \ldots, y_{q-1}\}$, *defining an orthomorphism* θ *of* $\mathrm{GF}(q)^+$, *and* $\{a, z_1, \ldots, z_{q-1}\}$, *defining an orthomorphism* ϕ *of* $\mathrm{GF}(q)^+$, *be solutions of the same* (σ, ε)*-system and suppose that this system has rank two for this choice of* a. *Then* $\theta \perp [b]$ *if and only if* $\phi \perp [b]$.

Proof. As the rank of the (σ, ε)-system is two, it must be that $z_i = cy_i + d$, $i = 1, \ldots, n$, for some $c, d \in \mathrm{GF}(q)$, $c \neq 0$. Thus

$$\frac{az_{\sigma(i)} - bz_i}{a - b} = c\left(\frac{ay_{\sigma(i)} - by_i}{a - b}\right) + d,$$

and the result then follows by Theorem 10.50. □

As an example, for $q = 8$ the orthomorphism $\phi_{a,b}$ is orthogonal to exactly one linear orthomorphism, $[a^2]$.

Corollary 10.52. *For* $q = 8$, *the orthomorphism* $\phi_{a,b}$ *is orthogonal to* $[d]$ *if and only if* $d = a^2$.

Proof. By Corollary 10.51, we may assume, without loss of generality, that $b = 1$. Then, as $d \neq 0$,

$$\frac{\phi_{a,1}(x) + dx}{a + d} = \begin{cases} x & \text{if } x \in H_{a,1}, \\ x + \left(\frac{a^2+a}{a+d}\right) & \text{otherwise.} \end{cases}$$

This is a permutation of GF(8) if and only if

$$\frac{a^2 + a}{a + d} \in H_{a,1} = \{0, 1, a, 1 + a\}.$$

But the only values of d for which this is true are $d = 0$, $d = 1$, or $d = a^2$. □

Chapter 11
Extensions of Orthomorphism Graphs

In Chapters 9 and 10, we considered orthomorphisms and orthomorphism graphs of elementary abelian groups. Other than these groups, the other classes of abelian groups that have received significant attention are the cyclic groups and direct products of elementary abelian groups. In this chapter we will define the extension of the orthomorphism graph of a group G by a group H: this is an orthomorphism graph of $G \times H$. We will discuss two special cases: $H = \mathrm{GF}(3)^+$ and $G = \mathrm{GF}(q)^+$ and $H = \mathbb{Z}_3$ and $G = \mathbb{Z}_m$, $\gcd(m, 3) = 1$.

In Section 11.1, we will use the 1978 construction of a set of four MOLS of order 15 by Schellenberg, Van Rees, and Vanstone [289] to introduce extensions of orthomorphism graphs. In Section 11.2, we will discuss extensions by $\mathrm{GF}(3)^+$, in Section 11.3 we will use difference equations to study the extension of $\mathrm{Orth}(\mathrm{GF}(q)^+)$ by $\mathrm{GF}(3)^+$, in Section 11.4 we will consider extensions of the orthomorphism graph of $\mathrm{GF}(q)^+$ consisting of linear orthomorphisms, and in Section 11.5 we will extend these methods to the study of extensions of cyclic groups by $\mathrm{GF}(3)^+$.

11.1 Introduction

In 1978 Schellenberg, Van Rees, and Vanstone [289] discovered, via a computer search, a set of four MOLS of order 15 based on the group $\mathbb{Z}_{15}$. We will see that the orthomorphisms corresponding to this set of MOLS have an interesting structure when viewed modulo 5. This will lead us naturally to extensions of orthomorphism graphs.

A. B. Evans, *Orthogonal Latin Squares Based on Groups*, Developments in Mathematics 57, https://doi.org/10.1007/978-3-319-94430-2_11

11.1.1 Perpendicular Difference Arrays

For G an abelian group of odd order n, Schellenberg, Van Rees, and Vanstone [289] called an $r \times (n-1)/2$ matrix $P = (p_{ij})$ a *perpendicular difference array of order n and depth r*, denoted PDA(n, r), if P is a matrix with entries from G having the property that if i and k are distinct, then the set $\{\pm(p_{ij} - p_{kj}) \mid j = 1, \ldots, (n-1)/2\}$ consists of the nonzero elements of G. It should be noted that perpendicular difference arrays generalize easily to nonabelian groups. We can easily construct difference matrices from perpendicular difference arrays.

Lemma 11.1. *If G is a finite abelian group and P is a* PDA(n, r) *over G, then $\begin{pmatrix} 0 & P & -P \end{pmatrix}$ is an $(n, r; 1)$-difference matrix over G.*

Proof. Let $D = (d_{ij}) = \begin{pmatrix} 0 & P & -P \end{pmatrix}$. If i and k are distinct, then the differences $\{d_{ij} - d_{kj} \mid j = 1, \ldots, n\}$ are

$$\{0\} \cup \{(p_{ij} - p_{kj}) \mid j = 1, \ldots, (n-1)/2\} \cup \{-(p_{ij} - p_{kj}) \mid j = 1, \ldots, (n-1)/2\}:$$

by the definition of perpendicular difference arrays, every element of G occurs in this set and, by cardinality, exactly once. □

Schellenberg, Van Rees, and Vanstone found the following four PDAs: the first three via computer searches and the last using a difference set.

Example 11.2. The matrix

$$P = \begin{pmatrix} 0 & 0 & 0 & 0 & 0 & 0 & 0 \\ 1 & 2 & 3 & 4 & 5 & 6 & 7 \\ 2 & 5 & 7 & 9 & 12 & 4 & 1 \\ 6 & 3 & 14 & 10 & 7 & 13 & 4 \\ 10 & 6 & 1 & 11 & 2 & 7 & 12 \end{pmatrix}$$

is a PDA$(15, 5)$ over $\mathbb{Z}_{15}$. Thus, by Lemma 11.1, we can construct a $(15, 5; 1)$-difference matrix over $\mathbb{Z}_{15}$, and from this, by Theorem 1.17, a set of 4 MOLS of order 15 based on the group $\mathbb{Z}_{15}$.

Example 11.3. The matrix

$$P = \begin{pmatrix} 0 & 0 & 0 & 0 & 0 & 0 & 0 & 0 & 0 & 0 & 0 & 0 & 0 & 0 & 0 & 0 \\ 1 & 2 & 3 & 4 & 5 & 6 & 7 & 8 & 9 & 10 & 11 & 12 & 13 & 14 & 15 & 16 \\ 10 & 9 & 8 & 7 & 6 & 16 & 15 & 14 & 13 & 12 & 22 & 32 & 31 & 30 & 29 & 28 \\ 2 & 4 & 6 & 8 & 17 & 26 & 21 & 18 & 20 & 28 & 3 & 19 & 22 & 9 & 32 & 10 \end{pmatrix}$$

is a PDA$(33, 4)$ over $\mathbb{Z}_{33}$ yielding a set of 3 MOLS of order 33 based on the group $\mathbb{Z}_{33}$.

Example 11.4. The matrix

$$P = \begin{pmatrix} 0 & 0 & 0 & 0 & 0 & 0 & 0 & 0 & 0 & 0 & 0 & 0 & 0 & 0 & 0 & 0 & 0 & 0 & 0 \\ 1 & 2 & 3 & 4 & 5 & 6 & 7 & 8 & 9 & 10 & 11 & 12 & 13 & 14 & 15 & 16 & 17 & 18 & 19 \\ 12 & 11 & 10 & 9 & 8 & 7 & 19 & 18 & 17 & 16 & 15 & 14 & 26 & 38 & 37 & 36 & 35 & 34 & 33 \\ 3 & 1 & 6 & 8 & 10 & 18 & 26 & 24 & 20 & 35 & 32 & 27 & 5 & 23 & 25 & 22 & 33 & 11 & 2 \end{pmatrix}$$

is a PDA(39, 4) over $\mathbb{Z}_{39}$ yielding a set of 3 MOLS of order 39 based on the group $\mathbb{Z}_{39}$.

Example 11.5. The matrix

$$P = \begin{pmatrix} 3 & 6 & 7 & 12 & 14 & 3 & 6 & 7 & 12 & 14 \\ 6 & 7 & 12 & 14 & 3 & 7 & 12 & 14 & 3 & 6 \\ 7 & 12 & 14 & 3 & 6 & 14 & 3 & 6 & 7 & 12 \\ 12 & 14 & 3 & 6 & 7 & 6 & 7 & 12 & 14 & 3 \\ 14 & 3 & 6 & 7 & 12 & 12 & 14 & 3 & 6 & 7 \end{pmatrix}$$

is a PDA(21, 5) over $\mathbb{Z}_{21}$ yielding a set of 4 MOLS of order 21 based on the group $\mathbb{Z}_{21}$. This PDA was constructed from the (21, 5, 1)-difference set $\{3, 6, 7, 12, 14\}$: see Section 8.3 for information on difference sets.

11.1.2 The Construction of Schellenberg, Van Rees, and Vanstone

Example 11.2 shows that $\omega(\mathbb{Z}_{15}) \geq 3$ and $N(15) \geq 4$. Neither of these results has been improved: the first is the best result possible as two computer searches have established that $\omega(\mathbb{Z}_{15}) = 3$, one unpublished by Roth and Wilson [283] and one by Bedford and Whitaker [39] in 2000.

We saw in Section 1.3 that from difference matrices of index 1, we can construct pairwise orthogonal orthomorphisms. As with difference matrices, there are some elementary operations that transform perpendicular difference arrays into other perpendicular difference arrays: permuting rows, permuting columns, adding the same group element to all entries in a given column, and multiplying all entries in a given column by -1. With these operations we can transform any perpendicular difference array into a perpendicular difference array in *standard form*, by which we mean that all the entries in the first row are zero and the entries in the second row are in some standard order, such as $1, 2, \ldots, (n-1)/2$ if the group is $\mathbb{Z}_n$. Let P be a perpendicular difference array over G in standard form and let $D = \begin{pmatrix} 0 & P & -P \end{pmatrix}$ be the corresponding difference matrix. Then D is normalized. Recall how we can construct pairwise orthogonal orthomorphisms from normalized difference matrices of index 1: each row other than the first and second represents a permutation of the second row; these permutations are the orthomorphisms. Each orthomorphism

θ constructed from a perpendicular difference array in standard form satisfies $\theta(-x) = -\theta(x)$ and so is fixed by the homology H_α, where $\alpha(x) = -x$ for all x: it is an easy exercise to show that any set of pairwise orthogonal orthomorphisms fixed by H_α can be obtained from a perpendicular difference array.

The orthomorphisms of $\mathbb{Z}_{15}$, constructed from the perpendicular difference array in Example 11.2, are described in Figure 11.1 in two ways: first with the elements of $\mathbb{Z}_{15}$ as the symbols $\{0, 1, \ldots, 14\}$. In the second way, the elements of $\mathbb{Z}_{15}$ are rewritten by replacing each $x \in \mathbb{Z}_{15}$ by ij, where $x = 6i + 10j$. Note that 6 generates the unique subgroup of $\mathbb{Z}_{15}$ oforder 5 and is congruent to 1 modulo 5 and 0 modulo 3, while 10 generates the unique subgroup of $\mathbb{Z}_{15}$ of order 3 and is congruent to 0 modulo 5 and 1 modulo 3. Thus $x \equiv i \pmod 5$ and $x \equiv j \pmod 3$. This corresponds to writing $\mathbb{Z}_{15}$ as $\mathbb{Z}_5 \times \mathbb{Z}_3$. Using the second description of $\mathbb{Z}_{15}$ and letting $\pi\colon \mathbb{Z}_{15} \to \mathbb{Z}_5$ denote the canonical homomorphism $ij \mapsto i$, we see that

$$\pi\theta(xy) = \begin{cases} 4x & \text{if } y = 0, \\ 4x + 3 & \text{if } y = 1, \\ 4x - 3 & \text{if } y = 2, \end{cases}$$

$$\pi\phi(xy) = \begin{cases} 3x & \text{if } y = 0 \\ 3x + 3 & \text{if } y = 1, \\ 3x - 3 & \text{if } y = 2, \end{cases}$$

and

$$\pi\gamma(xy) = \begin{cases} 2x & \text{if } y = 0, \\ 2x + 3 & \text{if } y = 1, \\ 2x - 3 & \text{if } y = 2. \end{cases}$$

We observe that the mappings $x \mapsto \pi\theta(x0)$, $x \mapsto \pi\phi(x0)$, and $x \mapsto \pi\gamma(x0)$ form a set of three orthogonal orthomorphisms of $GF(5)^+$, as do the mappings $x \mapsto \pi\theta(x1)$, $x \mapsto \pi\phi(x1)$, and $x \mapsto \pi\gamma(x1)$ and $x \mapsto \pi\theta(x2)$, $x \mapsto \pi\phi(x2)$, and $x \mapsto \pi\gamma(x2)$.

Fig. 11.1 4 MOLS of order 15

x	0	1	2	3	4	5	6	7	8	9	10	11	12	13	14
$\theta(x)$	0	2	5	7	9	12	4	1	14	11	3	6	8	10	13
$\phi(x)$	0	6	3	14	10	7	13	4	11	2	8	5	1	12	9
$\gamma(x)$	0	10	6	1	11	2	7	12	3	8	13	4	14	9	5

x	00	11	22	30	41	02	10	21	32	40	01	12	20	31	42
$\theta(x)$	00	22	02	21	40	20	41	11	42	12	30	10	32	01	31
$\phi(x)$	00	10	30	42	01	21	31	41	12	22	32	02	11	20	40
$\gamma(x)$	00	01	10	11	12	22	21	20	30	32	31	41	42	40	02

11.1.3 Extensions of Orthomorphism Graphs

Let G and H be groups written additively. Any mapping $\theta\colon G \times H \to G \times H$, and hence any orthomorphism of $G \times H$, can be written as $\theta(x, y) = (\theta_{1,y}(x), \theta_{2,x}(y))$, where for each $y \in H$, $\theta_{1,y}$ is a mapping $G \to G$, and for each $x \in G$, $\theta_{2,y}$ is a mapping $H \to H$. We will call $\theta_{1,y}$ the *yth first component mapping of* θ, and we will call $\theta_{2,x}$ the *xth second component mapping of* θ.

We are interested in the orthomorphism graph of $G \times H$, in which each first component mapping is an orthomorphism of G, and in which orthomorphisms θ and ϕ of $G \times H$ are adjacent if they are orthogonal and, for each $y \in H$, $\theta_{1,y}$ is orthogonal to $\phi_{1,y}$. We will refer to this set of orthomorphisms, and the corresponding orthomorphism graph, as the *extension of* Orth(G) *by* H, denoted $\mathrm{Ext}_H(\mathrm{Orth}(G))$, or $\mathrm{Ext}_H(G)$ for short. As the vertices of orthomorphism graphs are normalized orthomorphisms, if $\theta \in \mathrm{Ext}_H(G)$, then $\theta_{1,0}$ must be normalized: if $y \neq 0$, then $\theta_{1,y}$ need not be normalized. There is a simple bound on $\omega(\mathrm{Ext}_H(G))$.

Theorem 11.6. $\min\{\omega(H), \omega(G)\} \le \omega(\mathrm{Ext}_H(G)) \le \omega(G)$.

Proof. Let $r = \min\{\omega(H), \omega(G)\}$, let $\phi_1, \ldots, \phi_r$ be a pairwise orthogonal set of normalized orthomorphisms of H, and let $\psi_1, \ldots, \psi_r$ be a pairwise orthogonal set of normalized orthomorphisms of G. By Theorem 8.1, $\psi_1 \times \phi_1, \ldots, \psi_r \times \phi_r$ is a pairwise orthogonal set of normalized orthomorphisms in $\mathrm{Ext}_H(G)$. Hence $\min\{\omega(H), \omega(G)\} \le \omega(Ext_H(G))$. The upper bound follows from the definition of $\mathrm{Ext}_H(G)$. □

Our main concern will be with the case $H = \mathrm{GF}(3)^+$ and $G = \mathrm{GF}(q)^+$, $q \ge 3$, or $G = \mathbb{Z}_n$, n odd, in which case $\min\{\omega(H), \omega(G)\} = 1$. In Section 11.3, we will improve the lower bound in Theorem 11.6 when $H = \mathrm{GF}(3)^+$ and $G = \mathrm{GF}(q)^+$, and in Section 11.5, we will improve the lower bound in Theorem 11.6 when $H = \mathrm{GF}(3)^+$ and $G = \mathbb{Z}_m$, $\gcd(m, 6) = 1$.

Schellenberg, Van Rees, and Vanstone's construction, in particular, suggests trying to construct large sets of MOLS of order $3p$, $p > 3$ a prime, using orthomorphisms of $\mathbb{Z}_{3p} \cong \mathbb{Z}_p \times \mathbb{Z}_3$ whose first component mappings are of the form

$$\theta_{1,i}(x) \equiv \begin{cases} ax & \text{if } i = 0, \\ ax + b & \text{if } i = 1, \\ ax - b & \text{if } i = 2, \end{cases}$$

where b is fixed. In Section 11.5, we will use orthomorphisms of this kind to improve the lower bound of Theorem 11.6.

11.2 Extensions by GF(3)$^+$

In this section we will lay the groundwork for studying the orthomorphism graph $\mathrm{Ext}_{\mathrm{GF}(3)^+}(G)$, which we will abbreviate to $\mathrm{Ext}_3(G)$: if $G = \mathrm{GF}(q)^+$, then we will further abbreviate this to $\mathrm{Ext}_3(q)$. We will show that if we specify the first component mappings, then the possible second component mappings can be determined by solving systems of difference equations over GF(3)$^+$. We will see that adjacencies also arise from solutions to difference equations over GF(3)$^+$. Throughout this section the groups G and $G \times \mathrm{GF}(3)^+$ will be written additively. Our development will follow that given in a preprint by Evans [135]: in this preprint $G = \mathrm{GF}(q)^+$.

11.2.1 Second Component Mappings

The bijective first component mappings for a bijection impose restrictions on the possible second component mappings.

Lemma 11.7. *If θ is a mapping $G \times \mathrm{GF}(3) \to G \times \mathrm{GF}(3)$ and, for each $y \in \mathrm{GF}(3)$, $\theta_{1,y}$ is a bijection, then θ is a bijection if and only if, for each $x \in G$, $\theta_{2,x}$ can be written as*

$$\theta_{2,x}(y) = y\mathrm{def}_\theta\theta_{1,0}^{-1}\theta_{1,y}(x) + \mathrm{sum}_\theta\theta_{1,0}^{-1}\theta_{1,y}(x)$$

for some $\mathrm{def}_\theta, \mathrm{sum}_\theta\colon G \to \mathrm{GF}(3)$, $\mathrm{def}_\theta(x) \neq 0$ *for all* $x \in G$.

Proof. First assume that θ is a bijection. For any given $z \in G$, the mapping $y \mapsto \theta_{2,\theta_{1,y}^{-1}(z)}(y)$ is a bijection $\mathrm{GF}(3) \to \mathrm{GF}(3)$, as θ is a bijection and

$$\theta(\theta_{1,y}^{-1}(z), y) = (z, a) = \theta(\theta_{1,w}^{-1}(z), w)$$

whenever $\theta_{2,\theta_{1,y}^{-1}(z)}(y) = a = \theta_{2,\theta_{1,w}^{-1}(z)}(w)$. Thus, using the fact that any permutation of the elements of GF(3) can be represented as a nonconstant linear polynomial over GF(3),

$$\theta_{2,\theta_{1,y}^{-1}(z)}(y) = y\mathrm{k}(z) + \mathrm{h}(z)$$

for some maps $\mathrm{k}, \mathrm{h}\colon G \to \mathrm{GF}(3)$, $\mathrm{k}(z) \neq 0$ for all $z \in G$. If we set $x = \theta_{1,y}^{-1}(z)$, $\mathrm{def}_\theta = \mathrm{k}\theta_{1,0}$, and $\mathrm{sum}_\theta = \mathrm{h}\theta_{1,0}$, then the conditions of the lemma are satisfied.

Next assume that $\theta_{2,x}$ can be written as

$$\theta_{2,x}(y) = y\mathrm{def}_\theta\theta_{1,0}^{-1}\theta_{1,y}(x) + \mathrm{sum}_\theta\theta_{1,0}^{-1}\theta_{1,y}(x)$$

for some $\mathrm{def}_\theta, \mathrm{sum}_\theta\colon G \to \mathrm{GF}(3)$, $\mathrm{def}_\theta(x) \neq 0$ for all $x \in G$. If $\theta(x_1, y_1) = (a, b) = \theta(x_2, y_2)$, then $\theta_{1,y_1}(x_1) = a = \theta_{1,y_2}(x_2)$ and

$$\begin{aligned}\theta_{2,x_1}(y_1) &= y_1 \mathrm{def}_\theta \theta_{1,0}^{-1} \theta_{1,y_1}(x_1) + \mathrm{sum}_\theta \theta_{1,0}^{-1} \theta_{1,y_1}(x_1)\\ &= b\\ &= y_2 \mathrm{def}_\theta \theta_{1,0}^{-1} \theta_{1,y_2}(x_2) + \mathrm{sum}_\theta \theta_{1,0}^{-1} \theta_{1,y_2}(x_2)\\ &= \theta_{2,x_2}(y_2).\end{aligned}$$

Thus

$$y_1 \mathrm{def}_\theta \theta_{1,0}^{-1}(a) + \mathrm{sum}_\theta \theta_{1,0}^{-1}(a) = y_2 \mathrm{def}_\theta \theta_{1,0}^{-1}(a) + \mathrm{sum}_\theta \theta_{1,0}^{-1}(a).$$

Hence, as $\mathrm{def}_\theta \theta_{1,0}^{-1}(a) \neq 0$, $y_1 = y_2$ and so $x_1 = x_2$. Thus θ is one-to-one and hence is a bijection. □

We will call the functions def_θ and sum_θ in Lemma 11.7 the *defining function* and the *sum function* for θ, respectively. Any bijection $\theta\colon G \times \mathrm{GF}(3) \to G \times \mathrm{GF}(3)$ with bijective first component mappings can be specified by stating its first component mappings $\{\theta_{1,y} \mid y \in \mathrm{GF}(3)\}$, its sum function sum_θ, and its defining function def_θ.

Lemma 11.8. *If θ and ϕ are bijections $G \times \mathrm{GF}(3) \to G \times \mathrm{GF}(3)$ with bijective first component mappings and $\theta_{1,y} = \phi_{1,y}$ for each $y \in \mathrm{GF}(3)$, then $\theta = \phi$ if and only if $\mathrm{def}_\theta = \mathrm{def}_\phi$ and $\mathrm{sum}_\theta = \mathrm{sum}_\phi$.*

Proof. $\theta = \phi$ if and only if $\theta_{2,x} = \phi_{2,x}$ for each $x \in G$, i.e., if and only if

$$\begin{aligned}y \mathrm{def}_\theta \theta_{1,0}^{-1} \theta_{1,y}(x) + \mathrm{sum}_\theta \theta_{1,0}^{-1} \theta_{1,y}(x) &= y \mathrm{def}_\phi \phi_{1,0}^{-1} \phi_{1,y}(x) + \mathrm{sum}_\phi \phi_{1,0}^{-1} \phi_{1,y}(x)\\ &= y \mathrm{def}_\phi \theta_{1,0}^{-1} \theta_{1,y}(x) + \mathrm{sum}_\phi \theta_{1,0}^{-1} \theta_{1,y}(x)\end{aligned}$$

for all $y \in \mathrm{GF}(3)$ and $x \in G$. If this equation holds, then setting $y = 0$ yields $\mathrm{sum}_\theta = \mathrm{sum}_\phi$ from which it follows that $\mathrm{def}_\theta = \mathrm{def}_\phi$. On the other hand, if $\mathrm{def}_\theta = \mathrm{def}_\phi$ and $\mathrm{sum}_\theta = \mathrm{sum}_\phi$, then this equation holds as $\theta_{1,y} = \phi_{1,y}$ for all $y \in \mathrm{GF}(3)$. □

The identity mapping $\iota\colon G \times \mathrm{GF}(3) \to G \times \mathrm{GF}(3)$ is easily seen to have identity mappings as first component mappings, $\iota_{1,y}(x) = x$ for all $y \in \mathrm{GF}(3)$, and constant defining function and sum function, $\mathrm{def}_\iota(x) = 1$ and $\mathrm{sum}_\iota(x) = 0$ for all $x \in G$.

11.2.2 *Difference Equations for* $\mathrm{Ext}_3(G)$

Let $\theta_{1,y}$, $y \in \mathrm{GF}(3)$, and $\phi_{1,y}$, $y \in \mathrm{GF}(3)$, be bijections $G \to G$. If $\theta_{1,y}$ is orthogonal to $\phi_{1,y}$ for each $y \in \mathrm{GF}(3)$, then we define $\mathrm{diff}_{\theta,\phi,y}$, the *yth difference function* for θ by ϕ, by

$$\mathrm{diff}_{\theta,\phi,y}(x) = \theta_{1,y}(x) - \phi_{1,y}(x),$$

and $\mathrm{incr}_{\theta,\phi,y}$, the *yth increment function* for θ by ϕ, by

$$\mathrm{incr}_{\theta,\phi,y} = \theta_{1,0}^{-1}\theta_{1,y}\mathrm{diff}_{\theta,\phi,y}^{-1}\mathrm{diff}_{\theta,\phi,0}.$$

When ϕ is the identity mapping ι, we will use $\mathrm{diff}_{\theta,y}$ in place of $\mathrm{diff}_{\theta,\iota,y}$, and $\mathrm{incr}_{\theta,y}$ in place of $\mathrm{incr}_{\theta,\iota,y}$. The functions $\mathrm{diff}_{\theta,y}$, $y \in \mathrm{GF}(3)$, will be called the *difference functions* for θ, and the functions $\mathrm{incr}_{\theta,y}$, $y \in \mathrm{GF}(3)$, will be called the *increment functions* for θ.

Lemma 11.9. *If* $\theta_{1,y}$, $y \in \mathrm{GF}(3)$, *and* $\phi_{1,y}$, $y \in \mathrm{GF}(3)$, *are bijections* $G \to G$ *and* $\theta_{1,y}$ *is orthogonal to* $\phi_{1,y}$ *for all* $y \in \mathrm{GF}(3)$, *then*

$$\mathrm{diff}_{\theta,\phi,y}^{-1}\mathrm{diff}_{\theta,\phi,0} = \mathrm{diff}_{\phi,\theta,y}^{-1}\mathrm{diff}_{\phi,\theta,0} \text{ for all } y \in \mathrm{GF}(3).$$

Proof. Let λ: $G \to G$ be defined by $\lambda(x) = -x$. Then $\mathrm{diff}_{\phi,\theta,y} = \lambda\mathrm{diff}_{\theta,\phi,y}$ and

$$\mathrm{diff}_{\phi,\theta,y}^{-1}\mathrm{diff}_{\phi,\theta,0} = (\lambda\mathrm{diff}_{\theta,\phi,y})^{-1}(\lambda\mathrm{diff}_{\theta,\phi,0}) = \mathrm{diff}_{\theta,\phi,y}^{-1}\mathrm{diff}_{\theta,\phi,0}.$$

□

To determine adjacency in $\mathrm{Ext}_3(G)$ we need to determine when two bijections with orthogonal bijective first component mappings are orthogonal. Given mappings f: $G \to G$ and g: $G \to \mathrm{GF}(3)^+$, we will set

$$\Delta_{\mathrm{f}}[\mathrm{g}] = \mathrm{g}(\mathrm{f}(x)) - \mathrm{g}(x).$$

Theorem 11.10 (Adjacency Difference Equations). *Let* θ, ϕ: $G \times \mathrm{GF}(3) \to G \times \mathrm{GF}(3)$ *be bijections with bijective first component mappings. Assume further that* $\theta_{1,y}$ *is orthogonal to* $\phi_{1,y}$ *for all* $y \in \mathrm{GF}(3)$. *Then* θ *is orthogonal to* ϕ *if and only if*

$$\Delta_{\mathrm{incr}_{\theta,\phi,y}}[\mathrm{sum}_\theta](x) - \Delta_{\mathrm{incr}_{\phi,\theta,y}}[\mathrm{sum}_\phi](x) =$$
$$y(\mathrm{discr}_{\theta,\phi}(x) - \mathrm{def}_\theta\mathrm{incr}_{\theta,\phi,y}(x) + \mathrm{def}_\phi\mathrm{incr}_{\phi,\theta,y}(x))$$

for all $y \in \mathrm{GF}(3)$, *and some* $\mathrm{discr}_{\theta,\phi}$: $G \to \mathrm{GF}(3)$, $\mathrm{discr}_{\theta,\phi}(x) \neq 0$ *for all* $x \in G$.

Proof. Assume that θ is orthogonal to ϕ. For the mapping $x \mapsto \theta(x, y) - \phi(x, y)$, the three images with first component $\mathrm{diff}_{\theta,\phi,0}(x)$ have second components

$$\mathrm{sum}_\theta\mathrm{incr}_{\theta,\phi,y}(x) - \mathrm{sum}_\phi\mathrm{incr}_{\phi,\theta,y}(x) + y(\mathrm{def}_\theta\mathrm{incr}_{\theta,\phi,y}(x) - \mathrm{def}_\phi\mathrm{incr}_{\phi,\theta,y}(x)),$$

and so, as these must all be different and any permutation of the elements of GF(3) can be represented as a nonconstant linear polynomial over GF(3),

$$\mathrm{sum}_\theta\mathrm{incr}_{\theta,\phi,y}(x) - \mathrm{sum}_\phi\mathrm{incr}_{\phi,\theta,y}(x) + y(\mathrm{def}_\theta\mathrm{incr}_{\theta,\phi,y}(x) - \mathrm{def}_\phi\mathrm{incr}_{\phi,\theta,y}(x)) = y\mathrm{discr}_{\theta,\phi}(x) + c_{\theta,\phi}(x)$$

for some $\mathrm{discr}_{\theta,\phi}, c_{\theta,\phi}\colon G \to \mathrm{GF}(3)$, $\mathrm{discr}_{\theta,\phi}(x) \neq 0$ for all $x \in G$. Setting $y = 0$, and noting that both $\mathrm{incr}_{\theta,\phi,0}$ and $\mathrm{incr}_{\phi,\theta,0}$ are the identity mapping, we see that $c_{\theta,\phi}(x) = \mathrm{sum}_\theta(x) - \mathrm{sum}_\phi(x)$. Thus

$$\big(\mathrm{sum}_\theta(\mathrm{incr}_{\theta,\phi,y}(x)) - \mathrm{sum}_\theta(x)\big) - \big(\mathrm{sum}_\phi(\mathrm{incr}_{\phi,\theta,y}(x)) - \mathrm{sum}_\phi(x)\big) = y(\mathrm{discr}_{\theta,\phi}(x) - \mathrm{def}_\theta\mathrm{incr}_{\theta,\phi,y}(x) + \mathrm{def}_\phi\mathrm{incr}_{\phi,\theta,y}(x))$$

as claimed.

Next assume that the difference equations of the theorem hold for all $y \in \mathrm{GF}(3)$, and for some $\mathrm{discr}_{\theta,\phi}\colon G \to \mathrm{GF}(3)$ satisfying $\mathrm{discr}_{\theta,\phi}(x) \neq 0$ for all $x \in G$.

If $\theta(x_1, y_1) - \phi(x_1, y_1) = \theta(x_2, y_2) - \phi(x_2, y_2)$, then, comparing the first components, we see that $\mathrm{diff}_{\theta,\phi,y_1}(x_1) = \mathrm{diff}_{\theta,\phi,y_2}(x_2)$ which, as $\theta_{1,y} \perp \phi_{1,y}$ for all y, must equal $\mathrm{diff}_{\theta,\phi,0}(x)$ for some $x \in \mathrm{GF}(q)$. Thus $x_i = \mathrm{diff}^{-1}_{\theta,\phi,y_i}\mathrm{diff}_{\theta,\phi,0}(x)$ for $i = 1, 2$. By Lemma 11.9, we also have $x_i = \mathrm{diff}^{-1}_{\phi,\theta,y_i}\mathrm{diff}_{\phi,\theta,0}(x)$ for $i = 1, 2$. Substituting into the difference equations of the theorem yields

$$\Big(\mathrm{sum}_\theta\theta^{-1}_{1,0}\theta_{1,y_1}(x_1) - \mathrm{sum}_\theta(x)\Big) - \Big(\mathrm{sum}_\phi\phi^{-1}_{1,0}\phi_{1,y_1}(x_1) - \mathrm{sum}_\phi(x)\Big) = y_1(\mathrm{discr}_{\theta,\phi}(x) - \mathrm{def}_\theta\theta^{-1}_{1,0}\theta_{1,y_1}(x_1) + \mathrm{def}_\phi\phi^{-1}_{1,0}\phi_{1,y_1}(x_1))$$

and

$$\Big(\mathrm{sum}_\theta\theta^{-1}_{1,0}\theta_{1,y_2}(x_2) - \mathrm{sum}_\theta(x)\Big) - \Big(\mathrm{sum}_\phi\phi^{-1}_{1,0}\phi_{1,y_2}(x_2) - \mathrm{sum}_\phi(x)\Big) = y_2(\mathrm{discr}_{\theta,\phi}(x) - \mathrm{def}_\theta\theta^{-1}_{1,0}\theta_{1,y_2}(x_2) + \mathrm{def}_\phi\phi^{-1}_{1,0}\phi_{1,y_2}(x_2)).$$

These are equivalent to

$$y_1\mathrm{discr}_{\theta,\phi}(x) = \Big(\theta_{2,y_1}(x_1) - \phi_{2,y_1}(x_1)\Big) - \big(\mathrm{sum}_\theta(x) - \mathrm{sum}_\phi(x)\big)$$

and

$$y_2\mathrm{discr}_{\theta,\phi}(x) = \Big(\theta_{2,y_2}(x_2) - \phi_{2,y_2}(x_2)\Big) - \big(\mathrm{sum}_\theta(x) - \mathrm{sum}_\phi(x)\big).$$

Now $\theta(x_1, y_1) - \phi(x_1, y_1) = \theta(x_2, y_2) - \phi(x_2, y_2)$. Hence $y_1\text{discr}_{\theta,\phi}(x) = y_2\text{discr}_{\theta,\phi}(x)$, and so, as $\text{discr}_{\theta,\phi}(x) \neq 0$ for all x, $y_1 = y_2$. Now $\text{diff}_{\theta,\phi,y_1}(x_1) = \text{diff}_{\theta,\phi,y_2}(x_2) = \text{diff}_{\theta,\phi,y_1}(x_2)$, and so, as $\text{diff}_{\theta,\phi,y_1}$ is a bijection, $x_1 = x_2$. Therefore the mapping $(x, y) \mapsto \theta(x, y) - \phi(x, y)$ is one-to-one and hence a bijection. □

We will call the function $\text{discr}_{\theta,\phi}$ in Theorem 11.10 the *discriminant function* for θ by ϕ. If $\theta, \phi \in \text{Ext}_3(G)$, then Theorem 11.10 can be used to determine if θ is adjacent to ϕ in $\text{Ext}_3(G)$ or not. If ϕ is the identity mapping ι, then, as $\text{sum}_\iota(x) = 0$ and $\text{def}_\iota(x) = 1$ for all $x \in G$, we obtain, as a corollary to Theorem 11.10, the following characterization of orthomorphisms in $\text{Ext}_3(G)$.

Corollary 11.11 (Difference Equations). *If* $\theta\colon G \times \text{GF}(3) \to G \times \text{GF}(3)$ *is a mapping, whose first component mappings are orthomorphisms of* G, $\theta_{1,0}(0) = 0$, *and* $\text{sum}_\theta(0) = 0$, *then* $\theta \in \text{Ext}_3(G)$ *if and only if* $\text{def}_\theta(x) \neq 0$ *for all* $x \in G$ *and* def_θ *and* sum_θ *satisfy*

$$\Delta_{\text{incr}_{\theta,y}}[\text{sum}_\theta](x) = y(\text{discr}_\theta(x) - \text{def}_\theta\text{incr}_{\theta,y}(x) + 1)$$

for all $y \in \text{GF}(3)$ *and some* $\text{discr}_\theta\colon G \to \text{GF}(3)$, $\text{discr}_\theta(x) \neq 0$ *for all* $x \in G$.

We will call the function discr_θ in Corollary 11.11 the *discriminant function* for θ.

For the pairwise adjacent orthomorphisms $\theta, \phi, \gamma \in \text{Ext}_3(5)$ in Figure 11.1, the increment functions for θ, ϕ, and γ are

$$\text{incr}_{\theta,0}(x) = \text{incr}_{\phi,0}(x) = \text{incr}_{\gamma,0}(x) = x,$$

$$\text{incr}_{\theta,1}(x) = \text{incr}_{\gamma,1}(x) = x + 1, \text{ and } \text{incr}_{\phi,1}(x) = x + 2,$$

$$\text{incr}_{\theta,2}(x) = \text{incr}_{\gamma,2}(x) = x - 1, \text{ and } \text{incr}_{\phi,2}(x) = x - 2.$$

The sum, defining, and discriminant functions for θ, ϕ, and γ are shown in Figure 11.2.

The adjacency increment and discriminant functions for θ, ϕ, and γ are given in Figure 11.3.

x	0	1	2	3	4
$\text{sum}_\theta(x)$	0	1	2	1	2
$\text{sum}_\phi(x)$	0	1	1	2	2
$\text{sum}_\gamma(x)$	0	1	2	1	2

,

x	0	1	2	3	4
$\text{def}_\theta(x)$	1	2	1	1	2
$\text{def}_\phi(x)$	1	1	2	2	1
$\text{def}_\gamma(x)$	1	2	1	1	2

, and

x	0	1	2	3	4
$\text{discr}_\theta(x)$	2	1	2	2	1
$\text{discr}_\phi(x)$	2	2	1	1	2
$\text{discr}_\gamma(x)$	2	1	2	2	1

.

Fig. 11.2 The sum, defining, and discriminant functions for θ, ϕ, and γ

$$\begin{array}{c|ccc} y & 0 & 1 & 2 \\ \hline \mathrm{incr}_{\theta,\phi,y} & x & x+2 & x-2 \\ \mathrm{incr}_{\theta,\gamma} & x & x+2 & x-2 \\ \mathrm{incr}_{\phi,\gamma} & x & x+1 & x-1 \end{array}, \quad \begin{array}{c|ccc} y & 0 & 1 & 2 \\ \hline \mathrm{incr}_{\phi,\theta} & x & x+1 & x-1 \\ \mathrm{incr}_{\gamma,\theta} & x & x-1 & x+1 \\ \mathrm{incr}_{\gamma,\phi} & x & x-1 & x+1 \end{array}, \text{ and } \begin{array}{c|ccccc} x & 0 & 1 & 2 & 3 & 4 \\ \hline \mathrm{discr}_{\theta,\phi}(x) & 1 & 2 & 2 & 2 & 2 \\ \mathrm{discr}_{\theta,\gamma}(x) & 2 & 1 & 1 & 1 & 1 \\ \mathrm{discr}_{\phi,\gamma}(x) & 1 & 2 & 2 & 2 & 2 \end{array}.$$

Fig. 11.3 The adjacency increment and discriminant functions for θ, ϕ, and γ

11.3 The Orthomorphism Graph $\mathrm{Ext}_3(q)$

In this section we will restrict ourselves to the extension of $\mathrm{Orth}(\mathrm{GF}(q)^+)$ by $\mathrm{GF}(3)^+$, $\mathrm{Ext}_3(q)$. The special case in which the first component mappings are linear will be considered in Section 11.4.

11.3.1 Some Orthomorphisms in $\mathrm{Ext}_3(q)$

The difference equations in Theorem 11.10 will be called the *adjacency difference equations*, and difference equations in Corollary 11.11 the *difference equations* for $\mathrm{Ext}_3(q)$. For $G = \mathrm{GF}(q)^+$, these equations were derived in 2018 by Evans [134] when the first component mappings are linear: the more general version was given in a preprint by Evans [135]. If we define an orthomorphism graph $\mathscr{H}$, a subgraph of $\mathrm{Ext}_3(q)$, by specifying which orthomorphisms or combinations of orthomorphisms can be used as first component mappings, then the difference equations can, in principle, be used to construct every orthomorphism in $\mathscr{H}$. For any choice of first component mappings, we have the trivial solutions $\mathrm{def}_\theta(x) = -1$, $\mathrm{sum}_\theta(x) = 0$, and $\mathrm{discr}_\theta(x) = 1$ for all $x \in \mathrm{GF}(q)$, giving us the orthomorphism $\theta(x, y) = (\theta_{1,y}(x), -y)$. In this section we will exhibit some classes of solutions to the difference equations that appeared in [135].

To construct orthomorphisms in $\mathrm{Ext}_3(q)$, we need to solve the difference equations. As the difference equation with $y = 0$ is simply $0 = 0$, we need only consider the difference equations with $y = 1$ and $y = 2$.

Lemma 11.12. *If* $\theta_{1,y}$ *are orthomorphisms of* $\mathrm{GF}(q)^+$ *for all* $y \in \mathrm{GF}(3)$, *then* $\mathrm{incr}_{\theta,y} = \mathrm{incr}_{\theta,z}$ *if and only if* $\theta_{1,y} = \theta_{1,z}$.

Proof. Let $\varepsilon_{\theta,y} = \theta_{1,y}\mathrm{diff}_{\theta,y}^{-1}$. Then

$$\varepsilon_{\theta,y}^{-1}(x) = \mathrm{diff}_{\theta,y}\theta_{1,y}^{-1}(x) = x - \theta_{1,y}^{-1}(x).$$

Thus $\theta_{1,y} = (1-\varepsilon_{\theta,y}^{-1})^{-1}$, and so it follows that $\varepsilon_{\theta,y} = \varepsilon_{\theta,z}$ if and only if $\theta_{1,y} = \theta_{1,z}$. Now

$$\text{incr}_{\theta,\phi,y} = \theta_{1,0}^{-1}\varepsilon_{\theta,y}\text{diff}_{\theta,\phi,0} = \theta_{1,0}^{-1}\varepsilon_{\theta,z}\text{diff}_{\theta,\phi,0} = \text{incr}_{\theta,z}$$

if and only if $\varepsilon_{\theta,y} = \varepsilon_{\theta,z}$, hence the result. □

The difference equations for $\text{Ext}_3(q)$ have very simple solutions if all of the first component mappings are equal.

Theorem 11.13. *Let* θ *be a bijection* $\text{GF}(q)^+ \times \text{GF}(3)^+ \to \text{GF}(q)^+ \times \text{GF}(3)^+$. *If* $\theta_{1,0} = \theta_{1,1} = \theta_{1,2}$, $\theta_{1,0}(0) = 0$, *and* $\text{sum}_\theta(0) = 0$, *then* $\theta \in \text{Ext}_3(q)$ *if and only if* $\text{def}_\theta(x) = -1$ *and* $\text{discr}_\theta(x) = 1$ *for all* $x \in \text{GF}(q)$.

Proof. By Lemma 11.12,

$$\text{incr}_{\theta,0} = \text{incr}_{\theta,1} = \text{incr}_{\theta,2}.$$

As $\text{incr}_{\theta,y}$ is the identity mapping for all $y \in \text{GF}(3)$, the difference equations reduce to

$$\text{discr}_\theta(x) - \text{def}_\theta(x) + 1 = 0 \text{ for all } x \in \text{GF}(q).$$

As $\text{def}_\theta(x)$ and $\text{discr}_\theta(x)$ can never be 0, this can be true if and only if $\text{def}_\theta(x) = -1$ and $\text{discr}_\theta(x) = 1$ for all $x \in \text{GF}(q)$. □

Corollary 11.14. *Let* $\theta_{1,y}$, $y \in \text{GF}(3)$ *be orthomorphisms of* $\text{GF}(q)^+$. *If* $\theta_{1,0} = \theta_{1,1} = \theta_{1,2}$, $\theta_{1,0}(0) = 0$, *and* $\text{sum}_\theta(0) = 0$, *then the number of orthomorphisms in* $\text{Ext}_3(q)$ *with first component mappings* $\theta_{1,y}$, $y \in \text{GF}(3)$, *is* 3^{q-1}.

Proof. By Theorem 11.13, the number of orthomorphisms in $\text{Ext}_3(q)$ with first component mappings $\theta_{1,y}$, $y \in \text{GF}(3)$, is equal to the number of choices of sum_θ, i.e., 3^{q-1}. □

The difference equations for $\text{Ext}_3(q)$ have almost the same simple solutions if two of the first component mappings are equal.

Theorem 11.15. *Let* θ *be a bijection* $\text{GF}(q)^+ \times \text{GF}(3)^+ \to \text{GF}(q)^+ \times \text{GF}(3)^+$. *If* $|\{\theta_{1,0}, \theta_{1,1}, \theta_{1,2}\}| = 2$, $\theta_{1,0}(0) = 0$, *and* $\text{sum}_\theta(0) = 0$, *then* $\theta \in \text{Ext}_3(q)$ *if and only if* $\text{def}_\theta(x) = -1$, $\text{discr}_\theta(x) = 1$ *for all* $x \in \text{GF}(q)$, *and* sum_θ *is constant on orbits of* $\text{incr}_{\theta,z}$, *where* $\text{incr}_{\theta,z} \neq \text{incr}_{\theta,0}$.

Proof. Assume that $\theta \in \text{Ext}_3(q)$. By Lemma 11.12, $|\{\text{incr}_{\theta,0}, \text{incr}_{\theta,1}, \text{incr}_{\theta,2}\}| = 2$. If $\text{incr}_{\theta,1} = \text{incr}_{\theta,0}$ (respectively, $\text{incr}_{\theta,2} = \text{incr}_{\theta,0}$), then, as $\text{incr}_{\theta,0}$ is the identity mapping, the difference equation with $y = 1$ (respectively, with $y = 2$) reduces to $\text{discr}_\theta(x) - \text{def}_\theta(x) + 1 = 0$ and so $\text{def}_\theta(x) = -1$ and $\text{discr}_\theta(x) = 1$ for all $x \in \text{GF}(q)$. If $\text{incr}_{\theta,1} = \text{incr}_{\theta,2}$, then subtracting the difference equation with $y = 1$ from the difference equation with $y = 2$ yields $\text{discr}_\theta(x) - \text{def}_\theta\text{incr}_{\theta,1}(x) + 1 = 0$ and again $\text{def}_\theta(x) = -1$ and $\text{discr}_\theta(x) = 1$ for all $x \in \text{GF}(q)$. Now, for some $z \in \{1, 2\}$, $\text{incr}_{\theta,z} \neq \text{incr}_{\theta,0}$, the difference equation with $y = z$ reduces to $\text{sum}_\theta\text{incr}_{\theta,z}(x) - \text{sum}_\theta(x) = 0$: i.e., sum_θ is constant on orbits of $\text{incr}_{\theta,z}$. The converse is routine. □

Corollary 11.16. *Let $\theta_{1,y}$, $y \in \mathrm{GF}(3)$, be orthomorphisms of* $\mathrm{GF}(q)^+$, *and suppose that* $|\{\theta_{1,0}, \theta_{1,1}, \theta_{1,2}\}| = 2$, $\theta_{1,0}(0) = 0$. *Then the number of orthomorphisms in* $\mathrm{Ext3}(q)$ *with first component mappings* $\{\theta_{1,y} \mid y \in \mathrm{GF}(3)\}$ *is* 3^{m-1}, *where m is the number of orbits of* $\mathrm{incr}_{\theta,z}$, $\mathrm{incr}_{\theta,z} \neq \mathrm{incr}_{\theta,0}$.

Another class of orthomorphisms in $\mathrm{Ext3}(q)$ for which we can easily solve the difference equations is the class for which $\mathrm{incr}_{\theta,1} = \mathrm{incr}_{\theta,2}^{-1}$.

Theorem 11.17. *Let θ be a bijection* $\mathrm{GF}(q)^+ \times \mathrm{GF}(3)^+ \to \mathrm{GF}(q)^+ \times \mathrm{GF}(3)^+$ *with first component mappings which are orthomorphisms of* $\mathrm{GF}(q)^+$, $\theta_{1,0}(0) = 0$, *and* $\mathrm{sum}_\theta(0) = 0$.

We will assume that a set of representatives of the orbits of $\mathrm{incr}_{\theta,1}$ *is given, and we will use* $m_{\mathrm{incr}_{\theta,1}}$ *to denote the mapping* $\mathrm{GF}(q)^+ \to \mathbb{Z}$ *defined by*

$$m_{\mathrm{incr}_{\theta,1}}(x) = \min\{i \mid i \geq 0, \mathrm{incr}_{\theta,1}^{i}(a) = x\},$$

where a is the representative of the orbit of $\mathrm{incr}_{\theta,1}$ *containing x.*

If $\mathrm{incr}_{\theta,1} = \mathrm{incr}_{\theta,2}^{-1}$, $\mathrm{incr}_{\theta,1} \neq \mathrm{incr}_{\theta,0}, \mathrm{incr}_{\theta,2}$, *then* $\theta \in \mathrm{Ext3}(q)$ *if and only if for each orbit O of* $\mathrm{incr}_{\theta,1}$ *one of the following is true:*

1. $|O| \equiv 0 \pmod 3$, $\mathrm{def}_\theta(x) = \mathrm{discr}_\theta(x) = -1$ *for all* $x \in \mathrm{GF}(q)$, *and* $\mathrm{sum}_\theta(x) = m_{\mathrm{incr}_{\theta,1}}(x) + C_O$ *for some* $C_O \in \mathrm{GF}(3)$ *and all* $x \in O$.

   or
2. $|O| \equiv 0 \pmod 3$, $\mathrm{def}_\theta(x) = \mathrm{discr}_\theta(x) = 1$ *for all* $x \in \mathrm{GF}(q)$, *and* $\mathrm{sum}_\theta(x) = m_{\mathrm{incr}_{\theta,1}}(x) + d_O$ *for some* $d_O \in \mathrm{GF}(3)$ *and all* $x \in O$.

   or
3. $\sum_{x \in O} \mathrm{def}_\theta(x) = -|O|$, $\mathrm{discr}_\theta(x) = -\mathrm{def}_\theta(x)$, *and*

$$\mathrm{sum}_\theta(x) = -\sum_{i=1}^{m_{\mathrm{incr}_{\theta,1}}(x)} \mathrm{def}_\theta \mathrm{incr}_{\theta,1}^{i-1}(a) - \sum_{i=1}^{m_{\mathrm{incr}_{\theta,1}}(x)} \mathrm{def}_\theta \mathrm{incr}_{\theta,1}^{i}(a) + m_{\mathrm{incr}_{\theta,1}}(x) + C_O$$

for some $C_O \in \mathrm{GF}(3)$ *and all* $x \in O$, *where a is the representative of the orbit O.*

Proof. Assume that $\theta \in \mathrm{Ext3}(q)$. As $\mathrm{incr}_{\theta,2} = \mathrm{incr}_{\theta,1}^{-1}$,

$$\Delta_{\mathrm{incr}_{\theta,2}}[\mathrm{sum}_\theta](\mathrm{incr}_{\theta,1}(x)) = -\Delta_{\mathrm{incr}_{\theta,1}}[\mathrm{sum}_\theta](x)$$

and the difference equations reduce to

$$\begin{aligned} \Delta_{\mathrm{incr}_{\theta,1}}[\mathrm{sum}_\theta](x) &= \mathrm{discr}_\theta(x) - \mathrm{def}_\theta \mathrm{incr}_{\theta,1}(x) + 1 \\ &= \mathrm{discr}_\theta \mathrm{incr}_{\theta,1}(x) - \mathrm{def}_\theta(x) + 1 \end{aligned} \tag{11.1}$$

and so $\Delta_{\mathrm{incr}_{\theta,1}}[\mathrm{discr}_\theta + \mathrm{def}_\theta](x) = 0$. Hence $\mathrm{discr}_\theta(x) + \mathrm{def}_\theta(x) = \mathrm{c}(x)$ for some mapping $\mathrm{c}\colon \mathrm{GF}(q) \to \mathrm{GF}(3)$ that is constant on orbits of $\mathrm{incr}_{\theta,1}$: we will set $\mathrm{c}(x) = c_O$ for all $x \in O$. We will need to solve (11.1) for each orbit of $\mathrm{incr}_{\theta,1}$.

Let O be an orbit of $\mathrm{incr}_{\theta,1}$. Summing over all $x \in O$, we obtain

$$\sum_{x\in O} \mathrm{discr}_\theta(x) - \sum_{x\in O} \mathrm{def}_\theta(x) + \sum_{x\in O} 1 = 0.$$

Now $\mathrm{discr}_\theta(x) + \mathrm{def}_\theta(x) = c_O$ for all $x \in O$.

If $c_O = 1$, then $\mathrm{discr}_\theta(x) = \mathrm{def}_\theta(x) = -1$ for all $x \in O$, $|O| \equiv 0 \pmod 3$, and (11.1) becomes $\Delta_{\mathrm{incr}_{\theta,1}}[\mathrm{sum}_\theta](x) = 1$. The solution to this equation is $\mathrm{sum}_\theta(x) = m_{\mathrm{incr}_{\theta,1}}(x) + C_O$ for some $C_O \in \mathrm{GF}(3)$ and all $x \in O$.

If $c_O = -1$, then $\mathrm{discr}_\theta(x) = \mathrm{def}_\theta(x) = 1$ for all $x \in O$, $|O| \equiv 0 \pmod 3$, and (11.1) again becomes $\Delta_{\mathrm{incr}_{\theta,1}}[\mathrm{sum}_\theta](x) = 1$. The solution is $\mathrm{sum}_\theta(x) = m_{\mathrm{incr}_{\theta,1}}(x) + C_O$ for some $C_O \in \mathrm{GF}(3)$ and all $x \in O$.

If $c_O = 0$, then $\mathrm{discr}_\theta(x) = -\mathrm{def}_\theta(x)$, and $\sum_{x\in O} \mathrm{def}_\theta(x) = -|O|$. The equation ($\star$) becomes

$$\Delta_{\mathrm{incr}_{\theta,1}}[\mathrm{sum}_\theta](x) = -\mathrm{def}_\theta(x) - \mathrm{def}_\theta \mathrm{incr}_{\theta,1}(x) + 1.$$

The solution to this equation is

$$\mathrm{sum}_\theta(x) = -\sum_{i=1}^{m_{\mathrm{incr}_{\theta,1}}(x)} \mathrm{def}_\theta \mathrm{incr}_{\theta,1}^{i-1}(a) - \sum_{i=1}^{m_{\mathrm{incr}_{\theta,1}}(x)} \mathrm{def}_\theta \mathrm{incr}_{\theta,1}^{i}(a) + m_{\mathrm{incr}_{\theta,1}}(x) + C_O$$

for some $C_O \in \mathrm{GF}(3)$ and all $x \in O$.

The converse is routine. □

Corollary 11.18. *If* $\theta_{1,y}$, $y \in \mathrm{GF}(3)$, *are orthomorphisms of* $\mathrm{GF}(q)^+$ *for which* $\mathrm{incr}_{\theta,1} = \mathrm{incr}_{\theta,2}^{-1}$ *and* $\mathrm{incr}_{\theta,1} \neq \mathrm{incr}_{\theta,0}, \mathrm{incr}_{\theta,2}$, *then the number of orthomorphisms in* $\mathrm{Ext}_3(q)$ *with first component mappings* $\theta_{1,y}$, $y \in \mathrm{GF}(3)$ *is*

$$\frac{1}{3} \prod_{\substack{O \text{ orbit } \mathrm{incr}_{\theta,1},\\ |O| \not\equiv 0 \bmod 3}} \left(3 \sum_{i\geq 0} \binom{|O|}{3i}\right) \prod_{\substack{O \text{ orbit } \mathrm{incr}_{\theta,1},\\ |O| \equiv 0 \bmod 3}} \left(6 + 3 \sum_{i\geq 0} \binom{|O|}{3i}\right).$$

Proof. For each orbit O of $\mathrm{incr}_{\theta,1}$, once def_θ and discr_θ have been chosen, there are three choices for sum_θ: one if $0 \in O$. If $|O| \not\equiv 0 \pmod 3$, then $\sum_{x\in O} \mathrm{def}_\theta(x) = -|O|$, $\mathrm{discr}_\theta(x) = -\mathrm{def}_\theta(x)$, and there are $\sum_{i\geq 0} \binom{|O|}{3i}$ choices for def_θ and discr_θ. If $|O| \equiv 0 \pmod 3$, then there an are additional two choices for def_θ and discr_θ.

□

11.3.2 Some Upper Bounds on Clique Numbers

We will now give several conditions on $\mathscr{H} \subseteq \text{Ext}_3(q)$ that imply small clique numbers. These bounds are from [135].

Theorem 11.19. *Let $\mathscr{H}$ be a subgraph of* Ext$_3(q)$.

- *If for all $\theta, \phi \in \mathscr{H}$, for which $\theta_{1,y}$ is orthogonal to $\phi_{1,y}$ for all $y \in$* GF(3), $\text{diff}_{\theta,\phi,1} = \text{diff}_{\theta,\phi,0}$, *or*
- *for all $\theta, \phi \in \mathscr{H}$, for which $\theta_{1,y}$ is orthogonal to $\phi_{1,y}$ for all $y \in$* GF(3), $\text{diff}_{\theta,\phi,2} = \text{diff}_{\theta,\phi,0}$, *or*
- *for all $\theta, \phi \in \mathscr{H}$, for which $\theta_{1,y}$ is orthogonal to $\phi_{1,y}$ for all $y \in$* GF(3), $\text{diff}_{\theta,\phi,2} = \text{diff}_{\theta,\phi,1}$,

then $\omega(\mathscr{H}) \leq 3$.

Proof. Let $\theta_1, \ldots, \theta_4$ be a 4-clique of $\mathscr{H}$. We will use $\theta_{i,1,y}$ instead of $\theta_{i_{1,y}}$, and sum_i instead of sum_{θ_i}, etc.

If $\text{diff}_{i,j,1} = \text{diff}_{i,j,0}$ for all $i, j \in \{1, \ldots, 4\}$, $i \neq j$, then $\text{incr}_{i,j,1} = \theta_{i,1,0}^{-1}\theta_{i,1,1}$, which is independent of j, and if we set $G_i = \Delta_{\text{incr}_{i,j,1}}[\text{sum}_i](0) + \text{def}_i\text{incr}_{i,j,1}(0)$, then G_i is independent of j. If $i \neq j$, the adjacency difference equations with $y = 1$ and $x = 0$ yield $G_i - G_j = \text{discr}_{i,j}(0) \neq 0$. Thus $G_1, \ldots, G_4$ are four distinct elements of GF(3): a contradiction. It follows that if for all $\theta, \phi \in \mathscr{H}$, for which $\theta_{1,y}$ is orthogonal to $\phi_{1,y}$ for all $y \in$ GF(3), $\text{diff}_{\theta,\phi,1} = \text{diff}_{\theta,\phi,0}$, then $\omega(\mathscr{H}) \leq 3$.

If $\text{diff}_{i,j,2} = \text{diff}_{i,j,0}$ for all $i, j \in \{1, \ldots, 4\}$, $i \neq j$, then $\text{incr}_{i,j,2} = \theta_{i,1,0}^{-1}\theta_{i,1,2}$, which is independent of j, and if we set $G_i = \Delta_{\text{incr}_{i,j,2}}[\text{sum}_i](0) - \text{def}_i\text{incr}_{i,j,2}(0)$, then G_i is independent of j. If $i \neq j$, the adjacency difference equations with $y = 2$ and $x = 0$ yield $G_i - G_j = \text{discr}_{i,j}(0) \neq 0$. Thus $G_1, \ldots, G_4$ are four distinct elements of GF(3): a contradiction. It follows that, if for all $\theta, \phi \in \mathscr{H}$, for which $\theta_{1,y}$ is orthogonal to $\phi_{1,y}$ for all $y \in$ GF(3), $\text{diff}_{\theta,\phi,1} = \text{diff}_{\theta,\phi,0}$, then $\omega(\mathscr{H}) \leq 3$.

If $\text{diff}_{i,j,2} = \text{diff}_{i,j,1}$ for all $i, j \in \{1, \ldots, 4\}$, $i \neq j$, then we set

$$w = \text{diff}_{i,j,2}^{-1}\text{diff}_{i,j,0}(x) = \text{diff}_{i,j,1}^{-1}\text{diff}_{i,j,0}(x).$$

Noting that $\text{diff}_{i,j,y}^{-1}\text{diff}_{i,j,0} = \text{diff}_{j,i,y}^{-1}\text{diff}_{j,i,0}$ by Lemma 11.9, the adjacency difference equations for $y = 1$ and $y = 2$ then become

$$(\text{sum}_i\theta_{i,1,0}^{-1}\theta_{i,1,1}(w) - \text{sum}_i(x)) - (\text{sum}_j\theta_{j,1,0}^{-1}\theta_{j,1,1}(w) - \text{sum}_j(x))$$
$$= \text{discr}_{i,j}(x) - \text{def}_i\theta_{i,1,0}^{-1}\theta_{i,1,1}(w) + \text{def}_j\theta_{j,1,0}^{-1}\theta_{j,1,1}(w).$$

and

$$(\text{sum}_i\theta_{i,1,0}^{-1}\theta_{i,1,2}(w) - \text{sum}_i(x)) - (\text{sum}_j\theta_{j,1,0}^{-1}\theta_{j,1,2}(w) - \text{sum}_j(x))$$
$$= \text{discr}_{i,j}(x) - \text{def}_i\theta_{i,1,0}^{-1}\theta_{i,1,2}(w) + \text{def}_j\theta_{j,1,0}^{-1}\theta_{j,1,2}(w).$$

Subtracting these equations we obtain the equation

$$(\mathrm{sum}_i\theta_{i,1,0}^{-1}\theta_{i,1,1}(w)-\mathrm{sum}_i\theta_{i,1,0}^{-1}\theta_{i,1,2}(w)+\mathrm{def}_i\theta_{i,1,0}^{-1}\theta_{i,1,1}(w)-\mathrm{def}_i\theta_{i,1,0}^{-1}\theta_{i,1,2}(w))-$$
$$(\mathrm{sum}_j\theta_{j,1,0}^{-1}\theta_{j,1,1}(w)-\mathrm{sum}_j\theta_{j,1,0}^{-1}\theta_{j,1,2}(w)+\mathrm{def}_j\theta_{j,1,0}^{-1}\theta_{j,1,1}(w)-\mathrm{def}_j\theta_{j,1,0}^{-1}\theta_{j,1,2}(w))$$
$$=-\mathrm{discr}_{i,j}\mathrm{diff}_{i,j,0}^{-1}\mathrm{diff}_{i,j,2}(w).$$

Setting

$$G_i=\mathrm{sum}_i\theta_{i,1,0}^{-1}\theta_{i,1,1}(0)-\mathrm{sum}_i\theta_{i,1,0}^{-1}\theta_{i,1,2}(0)+\mathrm{def}_i\theta_{i,1,0}^{-1}\theta_{i,1,1}(0)-\mathrm{def}_i\theta_{i,1,0}^{-1}\theta_{i,1,2}(0),$$

which is independent of j, this equation, with $w=0$, becomes

$$G_i-G_j=-\mathrm{discr}_{i,j}\mathrm{diff}_{i,j,0}^{-1}\mathrm{diff}_{i,j,2}(0)\neq 0.$$

Thus $G_1,\ldots,G_4$ are four distinct elements of GF(3): a contradiction. It follows that, if for all $\theta,\phi\in\mathcal{H}$, for which $\theta_{1,y}$ is orthogonal to $\phi_{1,y}$ for all $y\in\mathrm{GF}(3)$, $\mathrm{diff}_{\theta,\phi,2}=\mathrm{diff}_{\theta,\phi,1}$, then $\omega(\mathcal{H})\leq 3$. □

Theorem 11.20. *Let $\mathcal{H}$ be a subgraph of* $\mathrm{Ext}_3(q)$. *If*

- $\theta_{1,0}=\theta_{1,1}$ *for all* $\theta\in\mathcal{H}$, *or*
- $\theta_{1,0}=\theta_{1,2}$ *for all* $\theta\in\mathcal{H}$, *or*
- $\theta_{1,1}=\theta_{1,2}$ *for all* $\theta\in\mathcal{H}$,

then $\omega(\mathcal{H})\leq 1$.

Proof. By Theorem 11.15 if $\theta\in\mathcal{H}$, then $\mathrm{def}_\theta(x)=-1$ for all $x\in\mathrm{GF}(q)$. Let θ, ϕ be an adjacent pair of orthomorphisms in $\mathcal{H}$. The adjacency difference equations reduce to

$$\Delta_{\mathrm{incr}_{\theta,\phi,y}}[\mathrm{sum}_\theta](x)-\Delta_{\mathrm{incr}_{\phi,\theta,y}}[\mathrm{sum}_\phi](x)=y\mathrm{discr}_{\theta,\phi}(x).$$

If $\theta_{1,0}=\theta_{1,1}$ for all $\theta\in\mathcal{H}$, then $\mathrm{incr}_{\theta,\phi,1}$ and $\mathrm{incr}_{\phi,\theta,1}$ are both the identity mapping, and the adjacency difference equation for $y=1$ becomes $\mathrm{discr}_{\theta,\phi}(x)=0$. But $\mathrm{discr}_{\theta,\phi}(x)\neq 0$ for all $x\in\mathrm{GF}(q)$, a contradiction. We arrive at a similar contradiction if $\theta_{1,0}=\theta_{1,2}$ for all $\theta\in\mathcal{H}$.

If $\theta_{1,2}=\theta_{1,1}$ for all $\theta\in\mathcal{H}$, then $\mathrm{incr}_{\theta,\phi,1}=\mathrm{incr}_{\theta,\phi,2}$ and $\mathrm{incr}_{\phi,\theta,1}=\mathrm{incr}_{\phi,\theta,2}$. Subtracting the adjacency difference equation with $y=1$ from the adjacency difference equation with $y=2$ again yields the contradiction $\mathrm{discr}_{\theta,\phi}(x)=0$, hence the result. □

Theorem 11.21. *Let $\mathcal{H}$ be an induced subgraph of* $\mathrm{Ext}_3(q)$.

1. *If for all $\theta,\phi\in\mathcal{H}$, with $\theta_{1,y}$ orthogonal to $\phi_{1,y}$ for all $y\in\mathrm{GF}(3)$, $\mathrm{incr}_{\theta,\phi,1}$ is independent of ϕ then $\omega(\mathcal{H})\leq 3$.*

2. *If for all* $\theta, \phi \in \mathscr{H}$, *with* $\theta_{1,y}$ *orthogonal to* $\phi_{1,y}$ *for all* $y \in \mathrm{GF}(3)$, $\mathrm{incr}_{\theta,\phi,2}$ *is independent of* ϕ *then* $\omega(\mathscr{H}) \leq 3$.
3. *If for all* $\theta, \phi \in \mathscr{H} \cup \{\iota\}$, *with* $\theta_{1,y}$ *orthogonal to* $\phi_{1,y}$ *for all* $y \in \mathrm{GF}(3)$, $\mathrm{incr}_{\theta,\phi,1}$ *is independent of* ϕ *then* $\omega(\mathscr{H}) \leq 2$.
4. *If for all* $\theta, \phi \in \mathscr{H} \cup \{\iota\}$, *with* $\theta_{1,y}$ *orthogonal to* $\phi_{1,y}$ *for all* $y \in \mathrm{GF}(3)$, $\mathrm{incr}_{\theta,\phi,2}$ *is independent of* ϕ *then* $\omega(\mathscr{H}) \leq 2$.

Proof. To prove (1), let $\theta_1, \ldots, \theta_4$ be a 4-clique of $\mathscr{H}$. We will use $\theta_{i,1,y}$ instead of $\theta_{i_{1,y}}$, sum_i instead of sum_{θ_i}, etc. If $\mathrm{incr}_{\theta_i,\theta_j,1}$, the increment function for θ_i by θ_j, is independent of θ_j, then we will use incr_i instead of $\mathrm{incr}_{\theta_i,\theta_j,1}$ and G_i to denote $\Delta_{\mathrm{incr}_i}[\mathrm{sum}_i](0)+\mathrm{def}_i\mathrm{incr}_i(0)$. Then, by the adjacency difference equations, G_1, G_2, G_3, and G_4 are distinct elements of GF(3): an impossibility. Hence $\omega(\mathscr{H}) \leq 3$.

The proofs of (2) through (4) are similar. □

Some examples will illustrate the bounds given in Theorems 11.19, 11.20., and 11.21.

Example 11.22. In 1999 Evans [122] defined $\mathscr{SVV}_b(3q)$, for a fixed $b \in \mathrm{GF}(q)$, to consist of all orthomorphisms of Ext3(q) with first component mappings of the form

$$(\theta_{1,0}(x), \theta_{1,1}(x), \theta_{1,2}(x)) = (ax, ax + b, ax - b).$$

If $\theta, \phi \in \mathscr{SVV}_b(3q)$ have first component mappings

$$(\theta_{1,0}(x), \theta_{1,1}(x), \theta_{1,2}(x)) = (ax, ax + b, ax - b)$$

and

$$(\phi_{1,0}(x), \phi_{1,1}(x), \phi_{1,2}(x)) = (Ax, Ax + b, Ax - b),$$

respectively, $A \neq a$, then the difference function for θ by ϕ is given by $\mathrm{diff}_{\theta,\phi,y}(x) = (a - A)x$ for all $y \in \mathrm{GF}(3)$, and so $\omega(\mathscr{SVV}_b(3q)) \leq 3$ by Theorem 11.19: this was proved by Evans [122] in 1999 for the special case q a prime. Schellenberg, Van Rees, and Vanstone's [289] construction of four MOLS of order 15 shows that this bound is attained when $q = 5$ and $b = 3$: see Figure 11.1.

It was shown in [122] that the value of $\omega(\mathscr{SVV}_b(3q))$ depends only on whether $b = 0$ or not. If $b = 0$, then $\theta_{1,y}$ is independent of y, and so $\omega(\mathscr{SVV}_0(3q)) \leq 1$ by Theorem 11.20.

Example 11.23. If we define $\mathscr{H}$ to be the subgraph of Ext3(q), consisting of all orthomorphisms with first component mappings of the form

$$(\theta_{1,0}(x), \theta_{1,1}(x), \theta_{1,2}(x)) = (ax, \left(\frac{a}{2}\right)x, \alpha(x)),$$

then the increment function for θ by ϕ satisfies $incr_{\theta,\phi,1}(x) = x$ for all $\theta, \phi \in \mathscr{H}$, with corresponding first component mappings orthogonal, and it follows from Theorem 11.21 that $\omega(\mathscr{H}) \leq 3$.

Similarly, if we define $\mathscr{H}$ to be the subgraph of $\mathrm{Ext}_3(q)$ consisting of all orthomorphisms with first component mappings of the form

$$(\theta_{1,0}(x), \theta_{1,1}(x), \theta_{1,2}(x)) = (ax, \alpha(x), \left(\frac{a}{2}\right) x).$$

then the increment function for θ by ϕ satisfies $incr_{\theta,\phi,2}(x) = x$ for all $\theta, \phi \in \mathscr{H}$, with corresponding first component mappings orthogonal, and it follows from Theorem 11.21 that $\omega(\mathscr{H}) \leq 3$.

Example 11.24. If we fix $b \in GF(3)$, $b \neq 0$, and define $\mathscr{H}$ to be the subgraph of $\mathrm{Ext}_3(q)$ consisting of all orthomorphisms with first component mappings of the form

$$(\theta_{1,0}(x), \theta_{1,1}(x), \theta_{1,2}(x)) = (ax, ax + (a-1)b, \alpha(x)),$$

then the increment function for θ by ϕ satisfies $incr_{\theta,\phi,1}(x) = x - b$ for all $\theta, \phi \in \mathscr{H} \cup \{\iota\}$, with corresponding first component mappings orthogonal, and we see, by Theorem 11.21, that $\omega(\mathscr{H}) \leq 2$.

Similarly, if we fix $c \in GF(3)$, $c \neq 0$, and define $\mathscr{H}$ to be the subgraph of $\mathrm{Ext}_3(q)$ consisting of all orthomorphisms with first component mappings of the form

$$(\theta_{1,0}(x), \theta_{1,1}(x), \theta_{1,2}(x)) = (ax, \alpha(x), ax + (a-1)c),$$

then the increment function for θ by ϕ satisfies $incr_{\theta,\phi,2}(x) = x - c$ for all $\theta, \phi \in \mathscr{H} \cup \{\iota\}$, with corresponding first component mappings orthogonal, and we see, by Theorem 11.21, that $\omega(\mathscr{H}) \leq 2$.

11.4 Linear First Component Mappings

Orthomorphism graphs that are subgraphs of $\mathrm{Ext}_3(q)$ can be defined by specifying the form of the first component mappings. The simplest mappings to use as first component mappings are the linear mappings $x \mapsto ax + b$. We define $\mathscr{L}_3(q)$ to be the orthomorphism graph consisting of orthomorphisms in $\mathrm{Ext}_3(q)$ with linear first component mappings of the form $ax + b_y$: as these orthomorphisms are normalized, $b_0 = 0$. The orthomorphism graph $\mathscr{L}_3(q)$ was studied in 2018 by Evans [135].

11.4.1 Introduction

The orthomorphism graph, $\mathcal{L}_3(q)$, has the interesting property that adjacency is synonymous with orthogonality.

Lemma 11.25. $\theta, \phi \in \mathcal{L}_3(q)$ *are adjacent in* $\mathcal{L}_3(q)$ *if and only if they are orthogonal orthomorphisms in* $\mathrm{Orth}(\mathrm{GF}(q)^+ \times \mathrm{GF}(3)^+)$.

Proof. Let $\theta, \phi \in \mathcal{L}_3(q)$. If θ and ϕ are adjacent in $\mathcal{L}_3(q)$, then they are orthogonal by definition. On the other hand, if they are orthogonal orthomorphisms, $\theta_{1,y}(x) = ax + b_y$, and $\phi_{1,y}(x) = Ax + B_y$, then $a \neq A$, as otherwise $\theta_{1,y}(x) - \phi_{1,y}(x) = b_y - B_y$ for all $x \in \mathrm{GF}(q)$, and so θ and ϕ must be adjacent in $\mathcal{L}_3(q)$. □

The increment functions for $\mathcal{L}_3(q)$ are easily computed.

Lemma 11.26. *Let* $\theta \in \mathcal{L}_3(q)$ *have first component mappings* $\theta_{1,y}(x) = ax + b_y$. *Then the increment functions for* θ *are*

$$\mathrm{incr}_{\theta,y}(x) = x + C_{\theta,y} = x - \frac{b_y}{a(a-1)}.$$

If $\phi \in \mathcal{L}_3(q)$, *with first component mappings* $\phi_{1,y}(x) = Ax + B_y$, *is adjacent to* θ, *then the adjacency increment functions for* θ *by* ϕ *are*

$$\mathrm{incr}_{\theta,\phi,y}(x) = x + C_{\theta,\phi,y} = x + \frac{aB_y - Ab_y}{a(a-A)}.$$

The increment functions $C_{\theta,y}$, $C_{\phi,y}$, $C_{\theta,\phi,y}$, *and* $C_{\phi,\theta,y}$ *are related by*

$$aC_{\theta,\phi,y} = aA\left(\frac{a-1}{a-A}\right)C_{\theta,y} + aA\left(\frac{A-1}{A-a}\right)C_{\phi,y} = AC_{\phi,\theta,y}.$$

Proof. Routine computation. □

For a group G, the automorphisms H_α, T_g, and R of $\mathrm{Orth}(G)$ were described in Section 8.1. Additively, these can be written as:

- $\mathrm{T}_g[\theta](x) = \theta(x+g) - \theta(g)$;
- $\mathrm{R}[\theta](x) = x + \theta(-x)$; and
- $\mathrm{H}_\alpha[\theta] = \alpha\theta\alpha^{-1}$.

Their actions on $\mathcal{L}_3(q)$ are as follows:

Lemma 11.27. *Let* $\theta \in \mathcal{L}_3(q)$ *have first component mappings* $\theta_{1,y}(x) = ax + b_y$, $b_1 = b$, *and* $b_2 = c$. *Then*

$$(\mathrm{R}[\theta]_{1,0}(x), \mathrm{R}[\theta]_{1,1}(x), \mathrm{R}[\theta]_{1,2}(x)) = ((1-a)x, (1-a)x + c, (1-a)x + b),$$

$$(\mathrm{T}_{(d,e)}[\theta]_{1,0}(x), \mathrm{T}_{(d,e)}[\theta]_{1,1}(x), \mathrm{T}_{(d,e)}[\theta]_{1,2}(x)) = (ax, ax + b, ax + c) \textit{ if } e = 0,$$

$$(\mathrm{T}_{(d,e)}[\theta]_{1,0}(x), \mathrm{T}_{(d,e)}[\theta]_{1,1}(x), \mathrm{T}_{(d,e)}[\theta]_{1,2}(x)) = (ax, ax+(c-b), ax-b) \text{ if } e = 1,$$

$$(\mathrm{T}_{(d,e)}[\theta]_{1,0}(x), \mathrm{T}_{(d,e)}[\theta]_{1,1}(x), \mathrm{T}_{(d,e)}[\theta]_{1,2}(x)) = (ax, ax-c, ax+(b-c)) \text{ if } e = 2,$$

$$(\mathrm{H}_\alpha[\theta]_{1,0}(x), \mathrm{H}_\alpha[\theta]_{1,1}(x), \mathrm{H}_\alpha[\theta]_{1,2}(x)) = (ax, ax+db, ax+dc) \text{ if } \alpha(x, y) = (dx, y),$$

and

$$(\mathrm{H}_\alpha[\theta]_{1,0}(x), \mathrm{H}_\alpha[\theta]_{1,1}(x), \mathrm{H}_\alpha[\theta]_{1,2}(x)) = (ax, ax+c, ax+b) \text{ if } \alpha(x, y) = (x, -y).$$

Proof. Routine computation. □

Note that each of H_α, T_g, and R preserve $\mathscr{L}_3(q)$.

11.4.2 The Number of Orthomorphisms in $\mathscr{L}_3(q)$

For $\theta \in \mathscr{L}_3(q)$ with first component mappings $\theta_{1,y}(x) = ax + b_y$, we will use $C_{\theta,y}$ to denote $-b_y/(a(a-1))$. By Lemma 11.26, the increment functions for θ are given by $\mathrm{incr}_{\theta,y}(x) = x + C_{\theta,y}$. If we regard GF$(q)$ as a vector space over its prime subfield, then the subspace spanned by $\{C_{\theta,1}, C_{\theta,2}\}$, equivalently $\{b_1, b_2\}$, has dimension at most two: the dimension of this subspace will be called the *dimension* of θ.

The description of all orthomorphisms in $\mathscr{L}_3(q)$ of dimension zero follows from Theorem 11.13, and the description of many orthomorphisms in $\mathscr{L}_3(q)$ of dimension one follows from Theorem 11.15 and Theorem 11.17. These results, combined with Lemma 11.27, yield descriptions of all orthomorphisms in $\mathscr{L}_3(q)$ of dimension at most one when the characteristic of GF(q) is either 2, 3, or 5.

Theorem 11.28.

- *If the characteristic of* GF(q) *is* 2, *then the number of orthomorphisms in* $\mathscr{L}_3(q)$ *of dimension at most one is*

$$(q-2)3^{q-1} + (q-1)(q-2)3^{q/2}.$$

- *If the characteristic of* GF(q) *is* 3, *then the number of orthomorphisms in* $\mathscr{L}_3(q)$ *of dimension at most one is*

$$(q-2)3^{q-1} + (q-1)(q-2)3^{q/3} + (q-1)(q-2)3^{(q/3)-1}4^{q/3}.$$

- *If the characteristic of* GF(q) *is* 5, *then the number of orthomorphisms in* $\mathscr{L}_3(q)$ *of dimension at most one is*

$$(q-2)3^{q-1} + (q-1)(q-2)3^{q/5} + (q-1)(q-2)3^{q/5}11^{q/5}.$$

Proof. For $a \in GF(q)$, $a \neq 0, 1$, the number of orthomorphisms in $Ext_3(q)$ with each first component mapping equal to ax is 3^{q-1} by Corollary 11.14. As there are $q-2$ choices for a, we have accounted for $(q-2)3^{q-1}$ elements of $\mathscr{L}_3(q)$. It remains to count the number of orthomorphisms in $\mathscr{L}_3(q)$ of dimension one.

Suppose that the characteristic of $GF(q)$ is p. For $a, b_y \in GF(q)$, $a \neq 0, 1$, $b_0 = b_2 = 0 \neq b_1$, the number of orthomorphisms in $Ext_3(q)$ with first component mappings $ax + b_y$ is $3^{(q/p)-1}$ by Corollary 11.16. As there are $q-2$ choices for a and $q-1$ choices for b_1, we have accounted for a further $(q-1)(q-2)3^{(q/p)-1}$ elements of $\mathscr{L}_3(q)$. The remaining cases, $b_0 = b_1 = 0 \neq b_2$ and $b_1 = b_2 \neq 0 = b_0$, are similar, yielding a further $2(q-1)(q-2)3^{(q/p)-1}$ elements of $\mathscr{L}_3(q)$.

So far we have counted the orthomorphisms in $\mathscr{L}_3(q)$ with first component mappings $\theta_{1,y}(x) = ax + b_y$, for which $|\{b_0, b_1, b_2\}| \leq 2$, yielding

$$(q-2)3^{q-1} + (q-1)(q-2)3^{q/p}$$

orthomorphisms in $\mathscr{L}_3(q)$.

It is easily seen that if $p = 2$, there are no other orthomorphisms in $\mathscr{L}_3(q)$ of dimension one or less.

If $p = 3$, for $a, b_y \in GF(q)$, $a \neq 0, 1$, $b_1 = -b_2 \neq 0 = b_0$, by Corollary 11.18, the number of orthomorphisms in $Ext_3(q)$ with first component mappings $ax + b_y$ is

$$\frac{1}{3}\left(6 + 3\sum_{i \geq 0}\binom{3}{3i}\right)^{q/3} = 4^{q/3}3^{(q/3)-1}.$$

As there are $q-2$ choices for a and $q-1$ choices for b_1, we have accounted for a further $(q-1)(q-2)4^{q/3}3^{(q/3)-1}$ elements of $\mathscr{L}_3(q)$.

As it is easily seen that these are the only possibilities when $p = 3$ and the dimension of θ is at most one, the result for $p = 3$ follows.

If $p = 5$, for $a, b_y \in GF(q)$, $a \neq 0, 1$, $b_1 = -b_2 \neq 0 = b_0$, by Corollary 11.18, the number of orthomorphisms in $Ext_3(q)$ with first component mappings $ax + b_y$ is

$$\frac{1}{3}\left(3\sum_{i \geq 0}\binom{5}{3i}\right)^{q/5} = 11^{q/5}3^{(q/5)-1}.$$

As there are $q-2$ choices for a and $q-1$ choices for b_1, we have accounted for a further $(q-1)(q-2)11^{q/5}3^{(q/5)-1}$ elements of $\mathscr{L}_3(q)$.

If $b_2 = 2b_1 \neq 0 = b_0$, then, by Lemma 11.27, $T_{d,1}[\theta]$ has first component mappings

$$(ax, ax + b_1, ax - b_1),$$

and if $b_2 = 3b_1 \neq 0 = b_0$, then, by Lemma 11.27, $H_\alpha[\theta]$, where $\alpha(x, y) = (x, -y)$, has first component mappings

$$(ax, ax + b_2, ax + 2b_2).$$

This accounts for a further

$$2(q-1)(q-2)11^{q/5}3^{(q/5)-1}$$

elements of $\mathscr{L}_3(q)$. The result for $p = 5$ follows. □

Theorems 11.13, 11.15, and 11.17 are sufficient to describe all orthomorphisms in $\mathscr{L}_3(q)$ of dimension at most one when the characteristic of GF(q) is at most 5. Any other orthomorphism in $\mathscr{L}_3(q)$ of dimension one must have first component mappings of the form $ax + b_y$, $y \in \mathrm{GF}(3)$, $b_2 = m.b_1 \neq 0 = b_0$ for some positive integer m, $1 < m < p - 1$, p the characteristic of GF(q). To extend the result of Theorem 11.28 to the case $p = 7$, we will need to describe more classes of orthomorphisms in $\mathscr{L}_3(q)$.

Lemma 11.29. *Let* $\theta_{1,y}(x) = ax + b_y$*,* $y \in \mathrm{GF}(3)$*, be orthomorphisms of* GF(q)*,* $b_2 = m \cdot b_1 \neq 0 = b_0$*,* $0 < m < p$*,* p *the characteristic of* GF(q)*, and let functions* g, d: GF(q) → GF(3) *satisfy* g(x), d(x) $\neq 0$ *for all* $x \in \mathrm{GF}(q)$*. Then there exists* $\theta \in \mathscr{L}_3(q)$ *with first component mappings* $\theta_{1,y}$*,* $y \in \mathrm{GF}(3)$*, defining function* g*, and discriminant function* d*, if and only if, for all* $x \in \mathrm{GF}(q)$*,*

$$\sum_{i=0}^{p-1} \mathrm{d}(x + iB) - \sum_{i=0}^{p-1} \mathrm{g}(x + iB) + p = 0 \tag{11.2}$$

and

$$\mathrm{d}(x) - \mathrm{g}(x + mB) + \sum_{i=0}^{m-1} \mathrm{d}(x + iB) - \sum_{i=1}^{m} \mathrm{g}(x + iB) + m + 1 = 0, \tag{11.3}$$

where $B = C_{\theta,1} = -b_1/(a(a-1))$*.*

Proof. Suppose that $\theta \in \mathscr{L}_3(q)$ has first component mappings $\theta_{1,y}(x) = ax + b_y$. By Lemma 11.26, $\mathrm{incr}_{\theta,1}(x) = x + B$ and $\mathrm{incr}_{\theta,2}(x) = x + mB$. The difference equations become

$$\mathrm{sum}_\theta(x + B) - \mathrm{sum}_\theta(x) = \mathrm{discr}_\theta(x) - \mathrm{def}_\theta(x + B) + 1,$$

and

$$\mathrm{sum}_\theta(x + mB) - \mathrm{sum}_\theta(x) = -\mathrm{discr}_\theta(x) + \mathrm{def}_\theta(x + mB) - 1.$$

From equation 11.2 we see that

$$\mathrm{sum}_\theta(x + pB) - \mathrm{sum}_\theta(x) = \sum_{i=0}^{p-1} \mathrm{discr}_\theta(x + iB) - \sum_{i=0}^{p-1} \mathrm{def}_\theta(x + iB) + p = 0,$$

and

$$\begin{aligned}\mathrm{sum}_\theta(x + mB) - \mathrm{sum}_\theta(x) &= \textstyle\sum_{i=0}^{m-1} \mathrm{discr}_\theta(x + iB) - \sum_{i=0}^{m-1} \mathrm{def}_\theta(x + iB) + m \\ &= -\mathrm{discr}_\theta(x) + \mathrm{def}_\theta(x + mB) - 1.\end{aligned}$$

We leave it to the reader to verify that if $\theta \in \mathscr{L}_3(q)$, then the defining and discriminant functions for θ satisfy equations 11.2 and 11.3. □

We will now apply Lemma 11.29 in the special case $m = 3$ and the characteristic of $\mathrm{GF}(q)$ is 7.

Theorem 11.30. *Let the characteristic of* $\mathrm{GF}(q)$ *be* 7*; let* $\theta_{1,y}(x) = ax + b_y$, $a \neq 0, 1$, $b_2 = 3b_1 \neq 0 = b_0$*; and let* $\mathrm{g}, \mathrm{d}\colon \mathrm{GF}(q) \to \mathrm{GF}(3)$ *satisfy* $\mathrm{g}(x), \mathrm{d}(x) \neq 0$ *for all* $x \in \mathrm{GF}(q)$*. Set* $B = C_{\theta,1} = -b_1/(a(a-1))$*. There exists an orthomorphism* $\theta \in \mathscr{L}_3(q)$ *with first component mappings* $\theta_{1,y}(x) = ax + b_y$, $a \neq 0, 1$, $b_2 = 3b_1 \neq 0 = b_0$*, defining function* g*, and discriminant function* d*, if and only if, for each orbit* O *of* $x \mapsto x + B$*,*

$$\mathrm{d}(x) = \mathrm{g}(x + 6B) - \mathrm{g}(x + 5B) - \mathrm{g}(x + 4B) - \mathrm{g}(x + 2B) - 1,$$

for all $x \in O$*, and either* $\mathrm{g}(x) = -1$ *for all* $x \in O$ *and* sum_θ *is constant on* O*, or* $(\mathrm{g}(x), \mathrm{g}(x+B), \ldots, \mathrm{g}(x+6B))$ *is a cyclic permutation of* $(1, 1, -1, -1, -1, 1, -1)$ *and* sum_θ *is given by*

$$\mathrm{sum}_\theta(x+mB) = \sum_{i=1}^{m} (\mathrm{g}(x+6iB) - \mathrm{g}(x+5iB) - \mathrm{g}(x+4iB) - \mathrm{g}(x+2iB) - \mathrm{g}(x+iB)) + c,$$

where c *is constant on* O*.*

Proof. By Lemma 11.29, the difference equations for θ have a solution only if

$$-\mathrm{d}(x) + \mathrm{d}(x + B) + \mathrm{d}(x + 2B) = \mathrm{g}(x + B) + \mathrm{g}(x + 2B) - \mathrm{g}(x + 3B) - 1,$$

for all $x \in \mathrm{GF}(q)$, a system of linear equations, which when solved for d yields

$$\mathrm{d}(x) = \mathrm{g}(x + 6B) - \mathrm{g}(x + 5B) - \mathrm{g}(x + 4B) - \mathrm{g}(x + 2B) - 1.$$

As this satisfies the equation

$$\sum_{i=0}^{6} \mathrm{d}(x + iB) - \sum_{i=0}^{6} \mathrm{g}(x + iB) + 1 = 0,$$

we see that g can be a defining function if and only if it satisfies the inequality

$$\mathrm{g}(x + 6B) - \mathrm{g}(x + 5B) - \mathrm{g}(x + 4B) - \mathrm{g}(x + 2B) \neq 1, \tag{11.4}$$

for all $x \in \mathrm{GF}(q)$.

Let O be an orbit of $x \mapsto x + B$ and set $N = |\{x \in O \mid \mathrm{g}(x) = 1\}|$. We will show that $N \in \{0, 3\}$.

It is easy to rule out the possibilities $N = 1, 6$, or 7.

If $N = 2$ then $\mathrm{g}(y) = \mathrm{g}(y + jB) = 1$ for some $y \in O$ and some $j \in \{1, 2, 3\}$. If $j = 1$ then setting $x = y - 4B$ in equation 11.4 leads to a contradiction: if $j = 2$ then setting $x = y - 2B$ in equation 11.4 leads to a contradiction; and if $j = 3$ then setting $x = y - 2B$ in equation 11.4 leads to a contradiction. Thus $N \neq 2$.

If $N = 5$ then $\mathrm{g}(y) = \mathrm{g}(y + jB) = -1$ for some $y \in O$ and some $j \in \{1, 2, 3\}$. If $j = 1$ then setting $x = y$ in equation 11.4 leads to a contradiction: if $j = 2$ then setting $x = y - B$ in equation 11.4 leads to a contradiction; and if $j = 3$ then setting $x = y$ in equation 11.4 leads to a contradiction. Thus $N \neq 5$.

If $N = 4$ then there are two possibilities. It could be that $\mathrm{g}(y) = \mathrm{g}(y + B) = \mathrm{g}(y + jB) = -1$ for some $y \in O$ and some $j \in \{2, 3, 4, 5\}$: in this case if $j = 2$ then setting $x = y - 5B$ in equation 11.4 yields a contradiction; if $j = 3$ then setting $x = y - 5B$ in equation 11.4 yields a contradiction; if $j = 4$ then setting $x = y - 2B$ in equation 11.4 yields a contradiction; and if $j = 5$ then setting $x = y - B$ in equation 11.4 yields a contradiction. The other possibility is that $(\mathrm{g}(y), \ldots, \mathrm{g}(y + 6B)) = (-1, 1, -1, 1, -1, 1, 1)$ for some $y \in O$. In this case setting $x = y + 3B$ in (I) yields a contradiction. Thus $N \neq 4$ and so $N \in \{0, 3\}$.

If $N = 0$ then $\mathrm{g}(x) = -1$ for all $x \in O$, equation 11.4 is satisfied, $\mathrm{d}(x) = 1$ for all $x \in O$, and the sum function is constant on O.

If $N = 3$ then there are two possibilities. It could be that $(\mathrm{g}(y), \ldots, \mathrm{g}(y + 6B)) = (1, -1, 1, -1, 1, -1, -1)$ for some $y \in O$, in which case setting $x = y$ in equation 11.4 yields a contradiction, or it could be that $\mathrm{g}(y) = \mathrm{g}(y + B) = \mathrm{g}(y + jB) = 1$ for some $y \in O$ and some $j \in \{2, 3, 4, 5\}$: in this case if $j = 2$ then setting $x = y - 2B$ in equation 11.4 yields a contradiction; if $j = 3$ then setting $x = y - B$ in equation 11.4 yields a contradiction; if $j = 4$ then setting $x = y - B$ in equation 11.4 yields a contradiction; and if $j = 5$ then equation 11.4 is satisfied. Once the defining function is determined, then the discriminant function is known, and the sum function can then be computed. □

Theorems 11.13, 11.15, 11.17, and 11.30 yield $|\mathscr{L}_3(q)|$ when the characteristic of $\mathrm{GF}(q)$ is 7.

Corollary 11.31. *If the characteristic of* GF(q) *is 7, then the number of orthomorphisms in* $\mathscr{L}_3(q)$ *of dimension at most one is*

$$(q-2)3^{q-1}+(q-1)(q-2)3^{q/7}$$
$$+(q-1)(q-2)3^{q/7}43^{q/7}+2(q-1)(q-2)8^{q/7}3^{(q/7)-1}.$$

Proof. The proof is similar to the proof of Theorem 11.28. □

In principle the techniques we have used can be employed to compute all orthomorphisms in $\mathscr{L}_3(q)$ of dimension one for any given characteristic of GF(q). It is clear though that the length of the proof must increase as the characteristic of GF(q) increases, making this approach impractical in general. This problem gets worse in the dimension two case. We will describe all orthomorphisms in $\mathscr{L}_3(q)$ when the characteristic of GF(q) is 2 only. First we need a dimension two analogue of Lemma 11.29.

Lemma 11.32. *Let* $\theta_{1,y}(x)=ax+b_y$, $y\in$ GF(3), $b_0=0$, *be orthomorphisms of* GF(q) *for which* $\{b_1,b_2\}$ *spans a two-dimensional subspace of* GF(q), *viewed as a vector space over its prime subfield, and let functions* g, d: GF$(q)\to$ GF(3) *satisfy* g(x), d$(x)\neq 0$ *for all* $x\in$ GF(q). *Then there exists* $\theta\in\mathscr{L}_3(q)$ *with first component mappings* $\theta_{1,y}$, $y\in$ GF(3), *defining function* g, *and discriminant function* d, *if and only if, for all* $x\in$ GF(q),

$$\sum_{i=0}^{p-1}\mathrm{d}(x+iB_y)-\sum_{i=0}^{p-1}\mathrm{g}(x+iB_y)+p=0$$

for $y=1,2$, *and*

$$\mathrm{d}(x)+\mathrm{d}(x+B_1)+\mathrm{d}(x+B_2)+\mathrm{g}(x+B_1)+\mathrm{g}(x+B_1+B_2)+\mathrm{g}(x+B_2)=0,$$

where p *is the characteristic of* GF(q), *and* $B_y=C_{\theta,y}=-b_y/(a(a-1))$.

Proof. Similar to the proof of Lemma 11.29. □

Using Lemma 11.32, we can now describe all orthomorphisms in $\mathscr{L}_3(q)$ when the characteristic of GF(q) is 2.

Theorem 11.33. *Let the characteristic of* GF(q) *be* 2, $\theta_{1,y}(x)=ax+b_y$, $a\neq 0,1$, $b_1,b_2\neq 0=b_0$, $b_2\neq b_1$, *and let* g, d: GF$(q)\to$ GF(3) *satisfy* g(x), d$(x)\neq 0$ *for all* $x\in$ GF(q). *Set* $B_y=C_{\theta,y}=b_y/(a(a+1))$. *There exists an orthomorphism* $\theta\in\mathscr{L}_3(q)$ *with first component mappings* $\theta_{1,y}(x)=ax+b_y$, $a\neq 0,1$, $b_1,b_2\neq 0=b_0$, $b_2\neq b_1$, *defining function* g, *and discriminant function* d, *if and only if, for each orbit* O *of* $\langle x\mapsto x+B_1, x\to x+B_2\rangle$, *either* g$(x)=-1$ *and* d$(x)=1$ *for all* $x\in O$ *in which case the sum function is constant on* O, *or, for some* $y\in O$, g$(y)=-1$, g$(x)=1$ *for all* $x\in O$, $x\neq y$, d$(y+B_1+B_2)=1$, *and* d$(x)=-1$

for all $x \in O$, $x \neq y + B_1 + B_2$, *and* sum_Θ *satisfies*

$$(\text{sum}_\Theta(y), \text{sum}_\Theta(y+B_1), \text{sum}_\Theta(y+B_2), \text{sum}_\theta(y+B_1+B_2)) = (c, -1+c, 1+c, c),$$

where c is a constant.

Proof. Let $\Theta \in \mathscr{L}_3(q)$ have first component mappings $\theta_{1,y}(x) = ax + b_y$, $a \neq 0, 1$, $b_1, b_2 \neq 0 = b_0$, defining function g, and discriminant function d. The equations of Lemma 11.32 reduce to

$$\mathrm{d}(x) + \mathrm{d}(x + B_y) = \mathrm{g}(x) + \mathrm{g}(x + B_y) + 1$$

and

$$\mathrm{d}(x) + \mathrm{d}(x + B_1) + \mathrm{d}(x + B_2) + \mathrm{g}(x + B_1) + \mathrm{g}(x + B_1 + B_2) + \mathrm{g}(x + B_2) = 0.$$

It follows that

$$\mathrm{d}(x) = \mathrm{g}(x + B_1 + B_2) - \mathrm{g}(x) - \mathrm{g}(x + B_1) - \mathrm{g}(x + B_2) - 1,$$

and so

$$\mathrm{g}(x + B_1 + B_2) - \mathrm{g}(x) - \mathrm{g}(x + B_1) - \mathrm{g}(x + B_2) \neq 1,$$

for all $x \in \mathrm{GF}(q)$.

For O an orbit of $\langle x \mapsto x + B_1, x \mapsto x + B_2\rangle$, let $N = |\{x \in O \mid \mathrm{g}(x) = 1\}|$. It is easy to prove that $N \in \{0, 3\}$.

If $N = 0$ then $\mathrm{g}(x) = -1$ and $\mathrm{d}(x) = 1$ for all $x \in O$, and so the sum function is constant on O.

If $N = 3$ then let y be the unique element in O for which $\mathrm{g}(y) = -1$. Then $\mathrm{g}(x) = 1$ for all $x \in O, x \neq y$, and simple computation shows that $\mathrm{d}(y+B_1+B_2) = 1$ and $\mathrm{d}(x) = -1$ for all $x \in O$, $x \neq y + B_1 + B_2$. The sum function can then be easily computed.

We leave the converse to the reader. □

When the characteristic of GF(q) is 2, by counting the number of orthomorphisms in $\mathscr{L}_3(q)$ of dimension two, as described in Theorem 11.33, and adding this to the number of orthomorphisms in $\mathscr{L}_3(q)$ of dimension at most one, as counted in Theorem 11.28, we can compute $|\mathscr{L}_3(q)|$.

Corollary 11.34. *If* GF(q) *is of characteristic two, then*

$$|\mathscr{L}_3(q)| = (q-2)3^{q-1} + (q-1)(q-2)3^{q/2} + (q-2)^2(q-1)3^{(q/4)-1}5^{q/4}.$$

As in the dimension one case, the amount of computation needed to describe all orthomorphisms in $\mathscr{L}_3(q)$ of dimension two for any given characteristic increases as the characteristic increases. For any $\theta \in \mathscr{L}_3(q)$, the defining function g and the discriminant function d satisfy $d(x) = \pm g(x)$ for any $x \in GF(q)$. In the special cases $d(x) = -g(x)$ for all $x \in GF(q)$ and $d(x) = g(x)$ for all $x \in GF(q)$, we can use Lemma 11.32 to determine all orthomorphisms of dimension two in $\mathscr{L}_3(q)$ that have defining function g and discriminant function d.

Theorem 11.35. *Let $\theta_{1,y}(x) = ax + b_y$, $y \in GF(3)$, $b_0 = 0$, be orthomorphisms of $GF(q)^+$ for which $\{b_1, b_2\}$ spans a two-dimensional subspace of GF(q), viewed as a vector space over its prime subfield, let the function $g\colon GF(q) \to GF(3)$ satisfy $g(x) \neq 0$ for all $x \in GF(q)$, and let p be the characteristic of GF(q) and $B_y = C_{\theta,y} = -b_y/(a(a-1))$. For O an orbit of $\langle x \mapsto x+B_1, x \mapsto x+B_2\rangle$, let G_O be the $p \times p$ matrix with ijth entry $g(a + (i-1)B_1 + (j-1)B_2)$, where $a \in O$ is fixed.*

1. *There exists $\theta \in \mathscr{L}_3(q)$ with first component mappings $\theta_{1,y}$, $y \in GF(3)$, defining function g, and discriminant function $-g$, if and only if, for each orbit O of $\langle x \mapsto x + B_1, x \mapsto x + B_2\rangle$, the matrix G_O is a circulant matrix with first row summing to $-p$. Then sum_θ is defined on O by*

$$\mathrm{sum}_\theta(a + iB_1 + jB_2) = \sum_{k=0}^{j-1} g(a + (i-k)B_1) + \sum_{k=1}^{j} g(a + (i-k)B_1)$$
$$- \sum_{k=0}^{i-1} g(a + kB_1) - \sum_{k=1}^{i} g(a + kB_1) + i - j + c,$$

where $a \in O$ is fixed and c is a constant.

2. *There exists $\theta \in \mathscr{L}_3(q)$ with first component mappings $\theta_{1,y}$, $y \in GF(3)$, defining function g, and discriminant function g, if and only if $p = 3$ and, for each orbit O of $\langle x \mapsto x + B_1, x \mapsto x + B_2\rangle$, each row of G_O is either $(1, 1, 1)$ or $(-1, -1, -1)$, or each column of G_O is either $(1, 1, 1)^T$ or $(-1, -1, -1)^T$. Then sum_θ is defined on O by*

$$\mathrm{sum}_\theta(a + iB_1 + jB_2) = \sum_{k=0}^{i-1} g(a + kB_1) - \sum_{k=1}^{i} g(a + kB_1) + i - j + c,$$

if each row of G_O is either $(1, 1, 1)$ or $(-1, -1, -1)$, and

$$\mathrm{sum}_\theta(a + iB_1 + jB_2) = \sum_{k=1}^{j} g(a + kB_2) - \sum_{k=0}^{j-1} g(a + kB_2) + i - j + c,$$

if each column of G_O is either $(1, 1, 1)^T$ or $(-1, -1, -1)^T$, where $a \in O$ is fixed and c is a constant.

Proof. 1. If d $= -$g, then the equations of Lemma 11.32 become

$$\sum_{i=0}^{p-1} \mathrm{g}(x + iB_y) = -p, \tag{11.5}$$

for $y = 1, 2$ and all $x \in \mathrm{GF}(q)$, and

$$\mathrm{g}(x + B_1 + B_2) = \mathrm{g}(x), \tag{11.6}$$

for all $x \in \mathrm{GF}(q)$.

Equation 11.6 implies that, for each orbit O of $\langle x \mapsto x + B_1, x \mapsto x + B_2\rangle$, G_O must be a circulant matrix, and so each row and each column sum must be the same, $-p$ by equation 11.5. Any circulant $p \times p$ matrix with row sum $-p$ satisfies equations 11.5 and 11.6.

By induction on equation 11.6, we see that $\mathrm{g}(x + iB_1 + jB_2) = \mathrm{g}(x + (i - j)B_1) = \mathrm{g}(x + (j - i)B_2)$. The sum function is then easily computed.

2. If d $=$ g then the equations of Lemma 11.32 become $p = 0$, i.e., $p = 3$, and $\mathrm{g}(x+B_1+B_2)+\mathrm{g}(x) = \mathrm{g}(x+B_1)+\mathrm{g}(x+B_2)$. It is an exercise to show that this is equivalent to either each row of G_O being either $(1, 1, 1)$ or $(-1, -1, -1)$, or each column of G_O being either $(1, 1, 1)^T$ or $(-1, -1, -1)^T$, for each orbit O of $\langle x \mapsto x + B_1, x \mapsto x + B_2\rangle$. The sum function can then easily be computed.

□

Corollary 11.36. *Let p be the characteristic of* GF(q). *The number of orthomorphisms in* $\mathscr{L}_3(q)$ *of dimension two is at least*

$$(q-2)(q-1)(q-p)3^{(q/p^2)-1}\left(\sum_{i\geq 0}\binom{p}{3i}\right)^{q/p^2} \quad \textit{if } p > 3,$$

and at least

$$(q-2)(q-1)(q-3)3^{(q/9)-1}4^{q/9} \textit{ if } p = 3.$$

Proof. Routine. □

Using Theorem 11.28, Corollary 11.31, and Corollary 11.36, we can now establish lower bounds for $|\mathscr{L}_3(q)|$ when the characteristic of GF(q) is 3, 5, or 7.

Corollary 11.37. *1. If the characteristic of* GF(q) *is* 3, *then*

$$|\mathscr{L}_3(q)| \geq (q-2)3^{q-1} + (q-1)(q-2)3^{q/3} + (q-1)(q-2)3^{(q/3)-1}4^{q/3} \\ +(q-2)(q-1)(q-3)3^{(q/9)-1}4^{q/9}.$$

2. *If the characteristic of* GF(q) *is* 5, *then*

$$|\mathscr{L}_3(q)| \geq (q-2)3^{q-1} + (q-1)(q-2)3^{q/5} + (q-1)(q-2)3^{q/5}11^{q/5} + (q-2)(q-1)(q-5)3^{(q/25)-1}11^{q/25}.$$

3. *If the characteristic of* GF(q) *is* 7, *then*

$$|\mathscr{L}_3(q)| \geq (q-2)3^{q-1} + (q-1)(q-2)3^{q/7} + (q-1)(q-2)3^{q/7}43^{q/7} + 2(q-1)(q-2)8^{q/7}3^{(q/7)-1} + (q-1)(q-2)(q-7)3^{(q/49)-1}43^{q/49}.$$

11.4.3 Bounds on $\omega(\mathscr{L}_3(q))$

We will now establish lower bounds on $\omega(\mathscr{L}_3(q))$ when q is not a power of 3. For $\theta, \phi \in \mathscr{L}_3(q)$, $\text{discr}_{\theta,\phi}$ will denote the discriminant function for θ by ϕ, $\text{discr}_{\phi,\theta}$ will denote the discriminant function for ϕ by θ, and, if the first component mappings for θ are $ax + b_y$, $y \in \text{GF}(3)$, and the first component mappings for ϕ are $Ax + B_y$, $y \in \text{GF}(3)$, then, as in Lemma 11.26, $C_{\theta,\phi,y} = (aB_y - Ab_y)/(a(a - A))$, and $C_{\phi,\theta,y} = (Ab_y - aB_y)/(A(A - a))$. First let us note an easily established upper bound.

Theorem 11.38. $\omega(\mathscr{L}_3(q)) \leq q - 2$.

Proof. Let $\theta_1, \ldots, \theta_n$ be a pairwise orthogonal set of orthomorphisms in $\mathscr{L}_3(q)$, and let the first component mappings of θ_i be $a_i x + b_{i,y}$ for $i = 1, \ldots, n$. Then $0, 1, a_1, \ldots, a_n$ are distinct elements of GF(q). The result follows. □

In [124] it was shown that $\omega(\mathbb{Z}_{3n}) \geq 2$ for all $n \geq 5$, n not divisible by 2 or 3. We will show that this result also holds for $\omega(\mathscr{L}_3(q))$ when the characteristic of GF(q) is not 3. First, let us establish this for characteristic greater than 3.

Theorem 11.39 (Evans, 2018). *If $q > 2$ is a power of a prime greater than* 3, *then* $\omega(\mathscr{L}_3(q)) \geq 2$.

Proof. Let p be the characteristic of GF(q). Let θ, ϕ: GF$(q) \times$ GF(3) be bijections; let θ have first component mappings $\theta_{1,0}(x) = 2x$, $\theta_{1,1}(x) = 2x - 2$, and $\theta_{1,2}(x) = 2x + 2$; and let ϕ have first component mappings $\phi_{1,0}(x) = -x$, $\phi_{1,1}(x) = -x - 2$, and $\phi_{1,2}(x) = -x + 2$. By Lemma 11.26, $C_{\theta,1} = 1 = C_{\phi,1}$ and $C_{\theta,2} = -1 = C_{\phi,2}$. Thus the difference equations for θ are identical to the difference equations for ϕ. Let us impose the conditions

$$\text{sum}_\theta = \text{sum}_\phi = \text{sum},$$
$$\text{def}_\theta = \text{def}_\phi = \text{def}, \quad \text{and}$$
$$\text{discr}_\theta = \text{discr}_\phi = \text{discr}.$$

By Theorem 11.17, discr $=$ $-$def. Then the difference equations for both θ and ϕ reduce to

$$\mathrm{sum}(x+1) - \mathrm{sum}(x) = -\mathrm{def}(x) - \mathrm{def}(x+1) + 1. \tag{11.7}$$

Equation (11.7) has a solution if and only if, for each orbit O of $x \mapsto x+1$,

$$\sum_{x\in O} \mathrm{def}(x) = -p.$$

Now $C_{\theta,\phi,1} = -1$, $C_{\theta,\phi,2} = 1$, $C_{\phi,\theta,1} = 2$ and $C_{\phi,\theta,2} = -2$, and the adjacency difference equations for θ and ϕ reduce to

$$\begin{aligned} &\mathrm{sum}(x-1) - \mathrm{sum}(x+2) \\ &= \mathrm{discr}_{\theta,\phi}(x) - \mathrm{def}(x-1) + \mathrm{def}(x+2) \\ &= \mathrm{discr}_{\theta,\phi}(x+1) - \mathrm{def}(x+2) + \mathrm{def}(x-1). \end{aligned} \tag{11.8}$$

Equation (11.7) can be used repeatedly to compute $\mathrm{sum}(x+m) - \mathrm{sum}(x+n)$ for all integers m and n. Doing so yields

$$\begin{aligned} \mathrm{sum}(x+2) - \mathrm{sum}(x-1) &= -\mathrm{discr}_{\theta,\phi}(x) + \mathrm{def}(x-1) - \mathrm{def}(x+2) \\ &= -\mathrm{discr}_{\theta,\phi}(x+1) + \mathrm{def}(x+2) - \mathrm{def}(x-1) \\ &= -\mathrm{def}(x+2) + \mathrm{def}(x+1) + \mathrm{def}(x) - \mathrm{def}(x-1). \end{aligned} \tag{11.9}$$

Thus

$$\mathrm{discr}_{\theta,\phi}(x) = -\mathrm{def}(x+1) - \mathrm{def}(x) - \mathrm{def}(x-1).$$

Hence θ can be orthogonal to ϕ if and only if $\sum_{x\in O} \mathrm{def}(x) = -p$, for each orbit O of $x \mapsto x+1$, i.e., $\theta, \phi \in \mathscr{L}_3(q)$, and

$$\mathrm{def}(x) + \mathrm{def}(x+1) + \mathrm{def}(x+2) \neq 0$$

for all $x \in GF(q)$. Such a function def exists. For example, set $\mathrm{def}(w), \mathrm{def}(w+1), \ldots, \mathrm{def}(w+p-1) = 1, -1, 1, -1, \ldots, 1, -1, \epsilon$, where $\epsilon \equiv -p \pmod 3$ and w is a given element of an orbit O of $x \mapsto x+1$. □

In Theorem 11.52 we will improve the bound of Theorem 11.39 to $\omega(\mathscr{L}_3(q)) \geq 3$ when the characteristic of GF(q) is not 2, 3, 7, or 17. When the characteristic of GF(q) is 2, we can find solutions to the difference equations and the adjacency difference equations that yield orthogonal pairs of orthomorphisms in $\mathscr{L}_3(q)$.

Theorem 11.40 (Evans, 2018). *If $q > 2$ is even, then $\omega(\mathscr{L}_3(q)) \geq 2$.*

Proof. There are two cases to consider, q an even power of 2 and q an odd power of 2.

Case 1. Let q be an even power of 2, let a be a solution to $x^2+x+1=0$ in GF(q), and let $A=a+1$. Let θ, ϕ: GF(q) $\times$ GF(3) be bijections; let θ have first component mappings $\theta_{1,0}(x)=ax$, $\theta_{1,1}(x)=ax+1$, and $\theta_{1,2}(x)=ax$; and let ϕ have first component mappings $\phi_{1,0}(x)=Ax$, $\phi_{1,1}(x)=Ax$, and $\phi_{1,2}(x)=Ax+1$. By Lemma 11.26, $C_{\theta,1}=1=C_{\phi,2}$, $C_{\theta,2}=0=C_{\phi,1}$, $C_{\theta,\phi,1}=a^2+1$, $C_{\theta,\phi,2}=1$, $C_{\phi,\theta,1}=1$, and $C_{\phi,\theta,2}=a^2$.

By Theorem 11.15, $\mathrm{def}_\phi(x)=\mathrm{def}_\theta(x)=-1$ for all $x\in$ GF(q), $\mathrm{discr}_\phi(x)=\mathrm{discr}_\theta(x)=1$ for all $x\in$ GF(q), and sum_ϕ and $\mathrm{sum}_\theta(x)$ are constant on orbits of $x\mapsto x+1$.

Let us impose the condition $\mathrm{sum}_\theta=\mathrm{sum}_\phi=\mathrm{sum}$. It is clear that $\theta,\phi\in\mathscr{L}_3(q)$ if and only if $\mathrm{sum}(0)=0$, sum is constant on orbits of $x\mapsto x+1$, and $\mathrm{def}_\theta(x)=\mathrm{def}_\phi(x)=-1$ for all $x\in$ GF(q).

With these conditions on θ and ϕ, the adjacency difference equations are

$$\mathrm{sum}(x+a^2+1)-\mathrm{sum}(x+1)=\mathrm{discr}_{\theta,\phi}(x),$$

and

$$\mathrm{sum}(x+1)-\mathrm{sum}(x+a^2)=-\mathrm{discr}_{\theta,\phi}(x).$$

These two equations are equivalent as sum is constant on orbits of $x\mapsto x+1$. It follows from these equations that θ and ϕ are orthogonal if $\mathrm{sum}(x+a^2)\neq\mathrm{sum}(x)$ for all $x\in$ GF(q). Set $H=\{0,1,a^2,1+a^2\}$. H is a subgroup of GF(q)$^+$. Let $0=w_0, w_1, \dots, w_{(q/4)-1}$ be a system of coset representatives for H in GF(q)$^+$, and set $\mathrm{sum}(w_i)=\mathrm{sum}(w_i+1)=0$ and $\mathrm{sum}(w_i+a^2)=\mathrm{sum}(w_i+a^2+1)=1$ for $i=0,\dots,(q/4)-1$. With this choice of sum, θ and ϕ are orthogonal orthomorphisms in $\mathscr{L}_3(q)$.

Case 2. Let q be an odd power of 2, let $a^2+a+1\neq 0$, and let $A=a+1$. Let θ, ϕ: GF(q) $\times$ GF(3) be bijections; let θ have first component mappings $\theta_{1,0}(x)=ax$, $\theta_{1,1}(x)=ax+a^2+a$, $\theta_{1,2}(x)=ax$; and let ϕ have first component mappings $\phi_{1,0}(x)=Ax$, $\phi_{1,1}(x)=Ax$, $\phi_{1,2}(x)=Ax+a^2+a$. By Lemma 11.26, $C_{\theta,1}=1=C_{\phi,2}$, $C_{\theta,2}=0=C_{\phi,1}$, $C_{\theta,\phi,1}=a^2+1$, $C_{\theta,\phi,2}=a^2+a$, $C_{\phi,\theta,1}=a^2+a$, and $C_{\phi,\theta,2}=a^2$. By Theorem 11.15, $\mathrm{def}_\phi(x)=\mathrm{def}_\theta(x)=-1$ for all $x\in$ GF(q), $\mathrm{discr}_\phi(x)=\mathrm{discr}_\theta(x)=1$ for all $x\in$ GF(q), and sum_ϕ and $\mathrm{sum}_\theta(x)$ are constant on orbits of $x\mapsto x+1$.

As in the case q an even power of 2, we will impose the condition $\mathrm{sum}_\theta=\mathrm{sum}_\phi=\mathrm{sum}$. It is clear that $\theta,\phi\in\mathscr{L}_3(q)$ if and only if $\mathrm{sum}(0)=0$, sum is constant on orbits of $x\mapsto x+1$, and $\mathrm{def}_\theta(x)=\mathrm{def}_\phi(x)=-1$ for all $x\in$ GF(q).

With these conditions on θ and ϕ, the adjacency difference equations are

$$\mathrm{sum}(x+a^2+1)-\mathrm{sum}(x+a^2+a)=\mathrm{discr}_{\theta,\phi}(x),$$

Fig. 11.4 Orthogonal orthomorphisms in $\mathscr{L}_3(4)$

x	000 010 100 110	001 011 101 111	002 012 102 112
$\theta(x)$	000 111 010 101	100 012 110 002	001 112 011 102
$\phi(x)$	000 101 110 011	002 100 112 010	102 001 012 111

and

$$\mathrm{sum}(x + a^2 + a) - \mathrm{sum}(x + a^2) = -\mathrm{discr}_{\theta,\phi}(x).$$

These two equations are equivalent as sum is constant on orbits of $x \mapsto x + 1$. It follows from these equations that θ and ϕ are orthogonal if $\mathrm{sum}(x + a) \neq \mathrm{sum}(x)$ for all $x \in \mathrm{GF}(q)$. Set $H = \{0, 1, a, 1 + a\}$. H is a subgroup of $\mathrm{GF}(q)^+$. Let $0 = w_0, w_1, \ldots, w_{(q/4)-1}$ be a system of coset representatives for H in $\mathrm{GF}(q)^+$, and set $\mathrm{sum}(w_i) = \mathrm{sum}(w_i+1) = 0$ and $\mathrm{sum}(w_i+a) = \mathrm{sum}(w_i+a+1) = 1$ for $i = 0, \ldots, (q/4) - 1$. With this choice of sum, θ and ϕ are orthogonal orthomorphisms in $\mathscr{L}_3(q)$. □

The smallest example, where $q = 4$, is given in Figure 11.4, where an element of $\mathbb{Z}_2 \times \mathbb{Z}_2 \times \mathbb{Z}_3$ is denoted ijk, $i, j = 0, 1$, and $k = 0, 1, 2$. A pair of orthogonal orthomorphisms of $\mathbb{Z}_2 \times \mathbb{Z}_2 \times \mathbb{Z}_3$ was first found via a computer search (see [106]): subsequent computer searches (see [43] or [196]) improved this to four pairwise orthogonal orthomorphisms.

11.5 A Lower Bound for $\omega(\mathrm{Ext}_3(\mathbb{Z}_n))$

In this section we will improve the lower bound on $\omega(\mathrm{Ext}_3(\mathbb{Z}_n))$ to 3, for $n > 1$, $n \neq 7, 17$, n not divisible by 2 or 3. Combined with results in the literature, which are summarized in [8], this reduces the problem of determining for which v there exist three pairwise orthogonal orthomorphisms of $\mathbb{Z}_v$ to the case $v = 9p$, p an odd prime: such constructions exist for primes $p = 3, 5, 7, 11, 13, 17, 23, 29, 31$, and 109 (See [8]), and these are the only primes less than 113 for which such constructions are known.

11.5.1 The Orthomorphisms

For the remainder of this section, we will assume that n is odd and not divisible by 3. Note that we will be using $\mathbb{Z}_n$ to denote both the additive group of integers modulo n and the ring of integers modulo n: the meaning should be clear from the context. If u is a unit of the ring $\mathbb{Z}_n$, we will use u' to denote the unique element of $\mathbb{Z}_n$ satisfying $uu' = 1$. We will construct three orthomorphisms $\theta, \phi, \gamma \in \mathrm{Ext}_3(\mathbb{Z}_n)$. The first component mappings will be as follows:

$$\theta_{1,y}(x) = \begin{cases} 2x & \text{if } y = 0, \\ 2x+2 & \text{if } y = 1, \\ 2x-2 & \text{if } y = 2; \end{cases}$$

$$\phi_{1,y}(x) = \begin{cases} -x & \text{if } y = 0, \\ -x+2 & \text{if } y = 1, \\ -x-2 & \text{if } y = 2; \end{cases}$$

and

$$\gamma_{1,y}(x) = \begin{cases} 2'x & \text{if } y = 0, \\ 2'x+2 & \text{if } y = 1, \\ 2'x-2 & \text{if } y = 2. \end{cases}$$

The increment functions for θ, ϕ, and γ are

$$\begin{aligned} \mathrm{incr}_{\theta,0}(x) = x = \mathrm{incr}_{\phi,0}(x) &= \mathrm{incr}_{\gamma,0}(x), \\ \mathrm{incr}_{\theta,1}(x) = x - 1 &= \mathrm{incr}_{\phi,1}(x), \\ \mathrm{incr}_{\theta,2}(x) = x + 1 &= \mathrm{incr}_{\phi,2}(x), \\ \mathrm{incr}_{\gamma,1}(x) = x + 8, \text{ and } \mathrm{incr}_{\gamma,2}(x) &= x - 8. \end{aligned}$$

As the increment functions for θ and ϕ are identical, so are the corresponding difference equations. This motivates us to impose a simplifying condition

$$\mathrm{def}_\theta = \mathrm{def}_\phi = \mathrm{def}, \quad \text{and} \quad \mathrm{sum}_\theta = \mathrm{sum}_\phi = \mathrm{sum}.$$

With these conditions on θ and ϕ, with slightly different first component mappings, in 2002 Evans [124] gave solutions to the difference and adjacency difference equations, thus establishing that $\omega(\mathbb{Z}_{3n}) \geq \omega(\mathrm{Ext}_3(\mathbb{Z}_n)) \geq 2$. In 2007 Evans [127] improved this to $\omega(\mathbb{Z}_{3n}) \geq \omega(\mathrm{Ext}_3(\mathbb{Z}_n)) \geq 3$ for $n \neq 7, 17$. The proof that we will give in this section will be the proof from [127].

We note, by Lemma 11.7, that the second component mappings of θ, ϕ, and γ are

$$\theta_{2,x}(y) = \begin{cases} \mathrm{sum}(x) & \text{if } y = 0, \\ \mathrm{def}(x+1) + \mathrm{sum}(x+1) & \text{if } y = 1, \\ -\mathrm{def}(x-1) + \mathrm{sum}(x-1) & \text{if } y = 2; \end{cases}$$

$$\phi_{2,x}(y) = \begin{cases} \mathrm{sum}(x) & \text{if } y = 0, \\ \mathrm{def}(x-2) + \mathrm{sum}(x-2) & \text{if } y = 1, \\ -\mathrm{def}(x+2) + \mathrm{sum}(x+2) & \text{if } y = 2; \end{cases}$$

and

$$\gamma_{2,x}(y) = \begin{cases} \mathrm{sum}_\gamma(x) & \text{if } y = 0, \\ \mathrm{def}_\gamma(x+4) + \mathrm{sum}_\gamma(x-4) & \text{if } y = 1, \\ -\mathrm{def}_\gamma(x-4) + \mathrm{sum}_\gamma(x+4) & \text{if } y = 2, \end{cases}$$

where $\mathrm{def}(x) \neq 0$ for all $x \in \mathbb{Z}_n$ and $\mathrm{def}_\gamma(x) \neq 0$ for all $x \in \mathbb{Z}_n$.

The following lemma gives sufficient conditions on def, sum, def_γ, and sum_γ for θ, ϕ, and γ to be pairwise orthogonal orthomorphisms in $\mathrm{Ext}_3(\mathbb{Z}_n)$.

Lemma 11.41. *If there exist functions* def *and* def_γ *that satisfy*

1. $\mathrm{def}(x), \mathrm{def}_\gamma(x) \neq 0$ *for all* $x \in \mathbb{Z}_n$,
2. $\sum_{x \in \mathbb{Z}_n} \mathrm{def}(x) = \sum_{x \in \mathbb{Z}_n} \mathrm{def}_\gamma(x) = -n$,
3. $\mathrm{def}(x) + \mathrm{def}(x+1) + \mathrm{def}(x+2) \neq 0$ *for all* $x \in \mathbb{Z}_n$, *and*
4. $\mathrm{def}_\gamma(x+4) - \mathrm{def}_\gamma(x) = \mathrm{def}(x+5) - \mathrm{def}(x+3) + \mathrm{def}(x+1) - \mathrm{def}(x-1)$ *for all* $x \in \mathbb{Z}_n$,

then there exist three pairwise orthogonal orthomorphisms of $\mathrm{Ext}_3(\mathbb{Z}_n)$.

Proof. Following the proof of Theorem 11.17, we can show that θ, ϕ, and γ are orthomorphisms in $\mathrm{Ext}_3(\mathbb{Z}_n)$ if and only if

$$\mathrm{def}(x), \mathrm{def}_\gamma(x) \neq 0 \text{ for all } x \in \mathbb{Z}_n,$$

$$\sum_{x \in \mathbb{Z}_n} \mathrm{def}(x) = \sum_{x \in \mathbb{Z}_n} \mathrm{def}_\gamma(x) = -n,$$

$$\mathrm{discr}_\theta(x) = \mathrm{discr}_\phi(x) = -\mathrm{def}(x) \text{ for all } x \in \mathbb{Z}_n,$$

$$\mathrm{discr}_\gamma(x) = -\mathrm{def}_\gamma(x) \text{ for all } x \in \mathbb{Z}_n,$$

$$\mathrm{sum}(x) = -\sum_{i=1}^{x} \mathrm{def}(1-i) - \sum_{i=1}^{x} \mathrm{def}(-i) + x,$$

and

$$\mathrm{sum}_\gamma(x) = -\sum_{i=1}^{x} \mathrm{def}_\gamma(8(i-1)) - \sum_{i=1}^{x} \mathrm{def}_\gamma(8i) + x,$$

where x is taken to be either an element of $\mathbb{Z}_n = \{0, 1, \ldots, n-1\}$ or an integer in $\{0, 1, \ldots, n-1\}$ depending on the context.

Substituting

$$\mathrm{discr}_{\theta,\phi}(x) = -\mathrm{def}(x-1) - \mathrm{def}(x) - \mathrm{def}(x+1),$$
$$\mathrm{discr}_{\theta,\gamma}(x) = \mathrm{def}(x-1) + \mathrm{def}(x) + \mathrm{def}(x+1),$$

and

$$\mathrm{discr}_{\phi,\gamma}(x) = -\mathrm{def}(x-1) - \mathrm{def}(x) - \mathrm{def}(x+1),$$

into the adjacency difference equations of Theorem 11.10 shows θ, ϕ, and γ to be pairwise adjacent orthomorphisms in $\mathrm{Ext}_3(\mathbb{Z}_n)$. □

11.5.2 Some Small Examples

We give below the smallest cases:

Example 11.42. For $n = 5$, we may define def and def_γ as follows:

x	0	1	2	3	4
$\mathrm{def}(x)$	1	1	2	1	2
$\mathrm{def}_\gamma(x)$	2	2	1	1	1

It is routine to check that def and def_γ satisfy the conditions of Lemma 11.41.

From this we can compute the three pairwise orthogonal orthomorphisms of $\mathbb{Z}_{15}$. Writing $10y + 6x \mod 15$ in place of (y, x), θ, ϕ, and γ are as follows:

i	0	1	2	3	4	5	6	7	8	9	10	11	12	13	14
$\theta(i)$	0	14	12	11	10	8	7	6	4	13	2	5	9	3	1
$\phi(i)$	0	6	11	2	8	13	4	10	5	1	12	7	3	14	9
$\gamma(i)$	0	5	9	4	14	3	8	13	2	12	7	1	11	6	10

Schellenberg, Van Rees, and Vanstone's 1978 construction [289] of three pairwise orthogonal orthomorphisms of $\mathbb{Z}_{15}$ was described in Figure 11.1. Other sets of three pairwise orthogonal orthomorphisms of $\mathbb{Z}_{15}$, obtained via computer searches, can be found in [281].

Example 11.43. For $n = 7$ there are no choices of def and def_γ that satisfy the conditions of Lemma 11.41.

By condition (1) of Lemma 11.41, the sequence $\mathrm{def}(0), \mathrm{def}(1), \ldots, \mathrm{def}(6)$ consists of 1s and 2s, and by condition (2) the number of 1s in this sequence is 0, 3, or 6. Condition (3) can only be satisfied if $\mathrm{def}(0), \mathrm{def}(1), \ldots, \mathrm{def}(6)$ contains precisely three 1s and is a cyclic shift of 1212122. Using conditions (4) and (2) to compute def_γ, we find that $\text{def}_\gamma(0), \text{def}_\gamma(1), \ldots, \text{def}_\gamma(6) = 1, 1, 0, 1, 1, 2, 2$, violating condition (1).

Example 11.44. For $n = 11$, we may define def and def_γ as follows:

x	0	1	2	3	4	5	6	7	8	9	10
$\text{def}(x)$	2	1	1	2	1	1	2	1	2	2	1
$\text{def}_\gamma(x)$	2	1	1	2	1	1	2	2	1	1	2

It is routine to check that def and def_γ satisfy the conditions of Lemma 11.41.

Example 11.45. For $n = 13$, we may define def and def_γ as follows:

x	0	1	2	3	4	5	6	7	8	9	10	11	12
$\text{def}(x)$	1	1	2	1	2	2	1	2	1	2	2	1	2
$\text{def}_\gamma(x)$	1	1	2	2	1	1	2	2	2	1	1	2	2

It is routine to check that def and def_γ satisfy the conditions of Lemma 11.41.

11.5.3 Generating Functions

For $\mathrm{k}\colon \mathbb{Z}_n \to \mathbb{Z}_3$, we define the *generating function* of k to be the element $K(z) = \sum_{x\in\mathbb{Z}_n} \mathrm{k}(x)z^x$ of the group ring $\mathbb{Z}_3[\mathbb{Z}_n]$. As a basis for this group ring, we will use $\{z^x \mid x \in \mathbb{Z}_n\} = \{1, z, z^2, \ldots, z^{n-1}\} \cong \mathbb{Z}_n$. Note that $z^i z^j = z^{i+j}$ with the powers being added modulo n. We will use $\mathrm{wt}\colon \mathbb{Z}_3[\mathbb{Z}_n] \to \mathbb{Z}_3$ to denote the ring homomorphism defined by $\mathrm{wt}(\sum_{x\in\mathbb{Z}_n} \mathrm{k}(x)z^x) = \sum_{x\in\mathbb{Z}_n} \mathrm{k}(x)$. $D(z)$ will denote the generating function of def, and $D_\gamma(z)$ the generating function of def_γ. Set $I = \sum_{x\in\mathbb{Z}_n} z^x$. Conditions (2) and (4) of Lemma 11.41 can be rewritten, in generating function form, as

$$\mathrm{wt}(D(z)) = \mathrm{wt}(D_\gamma(z)) = -n$$

and

$$(z^{-4} - 1)D_\gamma(z) = (z^{-5} - z^{-3} + z^{-1} - z)D(z),$$

which, by multiplying by $-z^4$, we can write as

$$(z^4 - 1)D_\gamma(z) = (z - z^{-1} + z^5 - z^3)D(z).$$

If $S \subseteq \mathbb{Z}_n$ then the *characteristic function* of S is the mapping $\chi_S\colon \mathbb{Z}_n \to \mathbb{Z}_3$ defined by

$$\chi_S(x) = \begin{cases} 1 & \text{if } x \in S \\ 0 & \text{if } x \notin S \end{cases},$$

and the corresponding generating function is denoted $X_S(z) = \sum_{x \in S} z^x$. Three elementary observations that we will use repeatedly are presented in the following lemmas:

Lemma 11.46. *If* $\text{k}\colon \mathbb{Z}_n \to \mathbb{Z}_3$ *satisfies* $\text{k}(x) \neq 0$ *for all* $x \in \mathbb{Z}_n$, *then* $\text{k}(a)+\text{k}(b)+\text{k}(c) = 0$ *if and only if* $\text{k}(a) = \text{k}(b) = \text{k}(c)$.

Proof. Routine. □

Lemma 11.47. *If* t *is relatively prime to* n, *then* $(z^t - 1)H(z) = (z^t - 1)K(z)$ *if and only if* $H(z) = K(z) + cI$ *for some* $c \in \mathbb{Z}_3$.

Proof. The mapping $H(z) \mapsto (z^t - 1)H(z)$ is a linear transformation $\mathbb{Z}_3[\mathbb{Z}_n] \to \mathbb{Z}_3[\mathbb{Z}_n]$ with kernel $\{cI \mid c \in Z_3\}$. □

Lemma 11.48. $\text{k}\colon \mathbb{Z}_n \to \mathbb{Z}_3$ *satisfies* $k(x) \neq 0$, *for all* $x \in \mathbb{Z}_n$, *if and only if* $K(z) = \sum_{i=1}^{t} \epsilon_i X_{S_i}$ *for some partition* $\{S_1, \ldots, S_t\}$ *of the element set of* $\mathbb{Z}_n$, *where* $\epsilon_i \in \{1, -1\}$ *for* $i = 1, \ldots, t$.

11.5.4 A Proof That $\omega(\text{Ext}_3(\mathbb{Z}_n)) \geq 3$

To prove that $\omega(\text{Ext}_3(\mathbb{Z}_n)) \geq 3$, $n > 1$, $\gcd(n, 6) = 1$, $n \neq 7, 17$, we will construct sequences $\text{def}(0), \ldots, \text{def}(n-1)$ and $\text{def}_\gamma(0), \ldots, \text{def}_\gamma(n-1)$ that satisfy Lemma 11.41. It will then follow that θ, ϕ, and γ are pairwise adjacent orthomorphisms in $\text{Ext}_3(\mathbb{Z}_n)$. We will make extensive use of the following subsequences:

- $L = 11212212$,
- $M = 11212$,
- $S = 112$,
- $\bar{L} = 22121121$,
- $\bar{M} = 22121$, and
- $\bar{S} = 221$.

To each of these subsequences, we can associate a generating function:

- $L(z) = 1 + z - z^2 + z^3 - z^4 - z^5 + z^6 - z^7$ to L,
- $M(z) = 1 + z - z^2 + z^3 - z^4$ to M,
- $S(z) = 1 + z - z^2$ to S,
- $\bar{L}(z) = -1 - z + z^2 - z^3 + z^4 + z^5 - z^6 + z^7$ to $\bar{L}$,
- $\bar{M}(z) = -1 - z + z^2 - z^3 + z^4$ to $\bar{M}$, and
- $\bar{S}(z) = -1 - z + z^2$ to $\bar{S}$.

Each of these polynomials can be described in terms of the polynomial $U(z) = 1 + z - z^2 + z^3$. We see that

- $L(z) = (1 - z^4)U(z)$,
- $M(z) = U(z) - z^4$,
- $S(z) = U(z) - z^3$,
- $\bar{L}(z) = -L(z) = (z^4 - 1)U(z)$,
- $\bar{M}(z) = -M(z) = z^4 - U(z)$, and
- $\bar{S}(z) = -S(z) = z^3 - U(z)$.

In Theorem 11.49 we will express $D(z)$ in terms of these polynomials and compute $D_\gamma(z)$ from $(z^4 - 1)D_\gamma(z) = (z^5 - z^3 + z - z^{-1})D(z)$, verifying that the mappings def and def_γ so described satisfy the conditions of Lemma 11.41.

Theorem 11.49 (Evans, 2007). *If $n > 1$ is neither 7 nor 17 and is not divisible by 2 or 3, then there exist three pairwise adjacent orthomorphisms in $\text{Ext}_3(\mathbb{Z}_n)$.*

Proof. There are eight cases to consider as n is congruent to 1, 3, 5, or 7 modulo 8 and 1 or 2 modulo 3.

Case 1. If $n \equiv 5 \pmod 8$ and $n \equiv 2 \pmod 3$, set

$$\text{def}(0), \text{def}(1), \ldots, \text{def}(n-1) = ML^m,$$

where $n = 8m + 5$. Note that $m \equiv 0 \pmod 3$. Clearly $\text{def}(x) \neq 0$ for all $x \in \mathbb{Z}_n$, and by Lemma 11.46, $\text{def}(x) + \text{def}(x+1) + \text{def}(x+2) \neq 0$ for all $x \in \mathbb{Z}_n$. The generating function for def is

$$D(z) = M(z) + z^5 \left(\sum_{i=0}^{m-1} z^{8i}\right) L(z) = M(z) + z^5 \left(\sum_{i=0}^{m-1} z^{8i}\right)(1 - z^4)U(z).$$

It follows from this that $\text{wt}(D(z)) = \text{wt}(M(z)) + m \times \text{wt}(L(z)) = 1 = -n$. Now

$$\begin{aligned}(z^4 - 1)D_\gamma(z) &= (z^4 + 1)(z - z^{-1})M(z) + (z - z^{-1})z^5 \left(\sum_{i=0}^{m-1} z^{8i}\right)(1 - z^8)U(z) \\ &= (z^4 + 1)(z - z^{-1})M(z) + (z - z^{-1})z^5(1 - z^{8m})U(z) \\ &= z^4 - z^6 = z^4\left(1 - (z^4)^{4m+3}\right).\end{aligned}$$

It follows from Lemma 11.47 that

$$D_\gamma(z) = -z^4 \sum_{i=0}^{4m+2} z^{4i} + cI.$$

As $\text{wt}(D_\gamma(z)) = cn$, to satisfy condition (2) of Lemma 11.41, it must be that $c = -1$. It is then clear that $\text{def}_\gamma(x) \neq 0$ for all $x \in \mathbb{Z}_n$, and so the conditions of Lemma 11.41 are satisfied.

Case 2. If $n \equiv 3 \pmod 8$ and $n \equiv 2 \pmod 3$, set

$$\text{def}(0), \text{def}(1), \ldots, \text{def}(n-1) = SL^m,$$

where $n = 8m + 3$. Note that $m \equiv 1 \pmod 3$. Clearly $\text{def}(x) \neq 0$ for all $x \in \mathbb{Z}_n$, and by Lemma 11.46, $\text{def}(x) + \text{def}(x+1) + \text{def}(x+2) \neq 0$ for all $x \in \mathbb{Z}_n$. The generating function for def is

$$D(z) = S(z) + z^3 \left(\sum_{i=0}^{m-1} z^{8i} \right) (1 - z^4) U(z).$$

It follows from this that $\text{wt}(D(z)) = 1 = -n$. Now

$$\begin{aligned}(z^4 - 1) D_\gamma(z) &= (z^4 + 1)(z - z^{-1}) S(z) + (z - z^{-1}) z^3 (1 - z^{8m}) U(z) \\ &= z^3 - z^5 \\ &= z^3 \left(1 - (z^4)^{4m+2}\right).\end{aligned}$$

It follows from Lemma 11.47 that

$$D_\gamma(z) = -z^3 \sum_{i=0}^{4m+1} z^{4i} + cI.$$

As $\text{wt}(D_\gamma(z)) = cn$, to satisfy condition (2) of Lemma 11.41, it must be that $c = -1$. It is then clear that $\text{def}_\gamma(x)$ 0 for all $x \in \mathbb{Z}_n$, and so the conditions of Lemma 11.41 are satisfied.

Case 3. If $n \equiv 1 \pmod 8$ and $n \equiv 1 \pmod 3$, then set

$$\text{def}(0), \text{def}(1), \ldots, \text{def}(n-1) = M^5 L^m,$$

where $n = 8m + 25$. Note that $m \equiv 0 \pmod 3$. Clearly $\text{def}(x) \neq 0$ for all $x \in \mathbb{Z}_n$, and by Lemma 11.46, $\text{def}(x) + \text{def}(x+1) + \text{def}(x+2) \neq 0$ for all $x \in \mathbb{Z}_n$. The generating function for def is

$$D(z) = (1 + z^5 + z^{10} + z^{15} + z^{20}) M(z) + z^{25} \left(\sum_{i=0}^{m-1} z^{8i} \right) (1 - z^4) U(z).$$

It follows from this that $\mathrm{wt}(D(z)) = 5 = -n$. Now

$$\begin{aligned}(z^4 - 1)D_\gamma(z) &= (z^4 + 1)(z - z^{-1})(1 + z^5 + z^{10} + z^{15} + z^{20})M(z) \\ &\quad + (z - z^{-1})z^{25}(1 - z^{8m})U(z) \\ &= z^4 - z^6 + z^9 - z^{11} + z^{14} - z^{16} + z^{19} - z^{21} + z^{24} - z^{26} \\ &= (z^4 + z^9 + z^{14} + z^{19} + z^{24})\left(1 - (z^4)^{4m+13}\right).\end{aligned}$$

It follows from Lemma 11.47 that

$$D_\gamma(z) = -(z^4 + z^9 + z^{14} + z^{19} + z^{24}) \sum_{i=0}^{4m+12} z^{4i} + cI.$$

As $\mathrm{wt}(D_\gamma(z)) = 1 + cn$, to satisfy condition (2) of Lemma 11.41, it must be that $c = 1$. To complete this case, we need to show that $\mathrm{def}_\gamma(x) \neq 0$ for all $x \in \mathbb{Z}_n$, thus satisfying the conditions of Lemma 11.41. Set

$$J = \{4i \mid i = 1, \ldots, 4m + 12\}$$

and

$$K = \{4i \mid i = 4m + 13, \ldots, 8m + 24\}.$$

Then J, K, $\{0\}$ partitions $\mathbb{Z}_n$, and inductively we can prove that

$$z^{2r} X_J(z) = \begin{cases} I - X_J(z) + \sum_{i=1}^{r} (-1)^i z^{2i} & \text{if } r \text{ is odd} \\ X_J(z) - \sum_{i=1}^{r} (-1)^i z^{2i} & \text{if } r \text{ is even.} \end{cases}$$

Now

$$\begin{aligned}D_\gamma(z) &= -(z^{2\times 2} + zz^{2\times 4} + z^{2\times 7} + zz^{2\times 9} + z^{2\times 12})(X_J(z) + 1) + I \\ &= -I - X_J(z) - (z^2 + z^9 + z^{13} + z^{14} + z^{17} + z^{18} + z^{22}) \\ &\quad + (z^{11} + z^{15} + z^{16} + z^{20})\end{aligned}$$

If we set

$$S = \{2, 9, 13, 14, 17, 18, 22\}$$

and

$$T = \{11, 15, 16, 20\},$$

then $T \subseteq J$ and $S \subseteq K$. Thus T, $J \setminus T$, S, $K \setminus S$, $\{0\}$ partitions $\mathbb{Z}_n$ and

$$D_\gamma(z) = X_{J\setminus T}(z) - X_T(z) + X_S(z) - X_{K\setminus S}(z) - X_{\{0\}}(z).$$

Thus $\mathrm{def}_\gamma \neq 0$ for all $x \in \mathbb{Z}_n$ by Lemma 11.48.

Case 4. If $n \equiv 7 \pmod 8$ and $n \equiv 1 \pmod 3$, set

$$\mathrm{def}(0), \mathrm{def}(1), \ldots, \mathrm{def}(n-1) = S^5 L^m,$$

where $n = 8m+15$. Note that $m \equiv 2 \pmod 3$ and that the case $n = 7$ is not covered by this construction. Clearly $\mathrm{def}(x) \neq 0$ for all $x \in \mathbb{Z}_n$, and by Lemma 11.46, $\mathrm{def}(x) + \mathrm{def}(x+1) + \mathrm{def}(x+2) \neq 0$ for all $x \in \mathbb{Z}_n$. The generating function for def is

$$D(z) = (1 + z^3 + z^6 + z^9 + z^{12})S(z) + z^{15}\left(\sum_{i=0}^{m-1} z^{8i}\right)(1 - z^4)U(z).$$

It follows from this that $\mathrm{wt}(D(z)) = 5 = -n$. Now

$$\begin{aligned}(z^4 - 1)D_\gamma(z) &= (z^4 + 1)(z - z^{-1})(1 + z^3 + z^6 + z^9 + z^{12})S(z) \\ &\quad + (z - z^{-1})z^{15}(1 - z^{8m})U(z) \\ &= z^3 - z^5 + z^6 - z^8 + z^9 - z^{11} + z^{12} - z^{14} + z^{15} - z^{17} \\ &= (z^3 + z^6 + z^9 + z^{12} + z^{15})\left(1 - (z^4)^{4m+8}\right).\end{aligned}$$

It follows from Lemma 11.47 that

$$D_\gamma(z) = -(z^3 + z^6 + z^9 + z^{12} + z^{15}) \sum_{i=0}^{4m+7} z^{4i} + cI.$$

As $\mathrm{wt}(D_\gamma(z)) = 1 + cn$, to satisfy condition (2) of Lemma 11.41, it must be that $c = 1$. To complete this case, we need to show that $\mathrm{def}_\gamma(x) \neq 0$ for all $x \in \mathbb{Z}_n$, thus satisfying the conditions of Lemma 11.41. Set

$$J = \{4i \mid i = 1, \ldots, 4m + 7\}$$

and

$$K = \{4i \mid i = 4m + 8, \ldots, 8m + 14\}.$$

Then J, K, $\{0\}$ partitions $\mathbb{Z}_n$, and inductively we can prove that

$$z^{2r}X_J(z) = \begin{cases} I - X_J(z) + \sum_{i=1}^{r}(-1)^i z^{2i} & \text{if } r \text{ is odd} \\ X_J(z) - \sum_{i=1}^{r}(-1)^i z^{2i} & \text{if } r \text{ is even.} \end{cases}$$

Now

$$\begin{aligned} D_\gamma(z) &= -(zz^{2\times 1} + z^{2\times 3} + zz^{2\times 4} + z^{2\times 6} + zz^{2\times 7})(X_J(z) + 1) + I \\ &= I + zX_J(z) + z(z^7 + z^{10}) - z(z^5 + z^8 + z^9 + z^{12}) \end{aligned}$$

If we set

$$S = \{7, 10\}$$

and

$$T = \{5, 8, 9, 12\},$$

then $T \subseteq J$ and $S \subseteq K$. Thus T, $J \setminus T$, S, $K \setminus S$, $\{0\}$ partitions $\mathbb{Z}_n$ and

$$D_\gamma(z) = z(X_{K\setminus S}(z) + X_{\{0\}}(z) - X_S(z) - X_{J\setminus T}(z) + X_T(z)).$$

Thus $\mathrm{def}_\gamma(x) \neq 0$ for all $x \in \mathbb{Z}_n$ by Lemma 11.48.

Case 5. If $n \equiv 5 \pmod 8$ and $n \equiv 1 \pmod 3$ then set

$$\mathrm{def}(0), \mathrm{def}(1), \ldots, \mathrm{def}(n-1) = \bar{M}\bar{L}^m,$$

where $n = 8m + 5$. Note that $m \equiv 1 \pmod 3$. Clearly $\mathrm{def}(x) \neq 0$ for all $x \in \mathbb{Z}_n$, and by Lemma 11.46, $\mathrm{def}(x) + \mathrm{def}(x+1) + \mathrm{def}(x+2) \neq 0$ for all $x \in \mathbb{Z}_n$. The generating function for def is

$$D(z) = -M(z) - z^5 \left(\sum_{i=0}^{m-1} z^{8i}\right) L(z) = -M(z) - z^5 \left(\sum_{i=0}^{m-1} z^{8i}\right) (1 - z^4)U(z).$$

It follows from this that $\mathrm{wt}(D(z)) = \mathrm{wt}(M(z)) + m \times \mathrm{wt}(L(z)) = -1 = -n$. Now

$$\begin{aligned} (z^4 - 1)D_\gamma(z) &= -(z^4 + 1)(z - z^{-1})M(z) - (z - z^{-1})z^5 \left(\sum_{i=0}^{m-1} z^{8i}\right) (1 - z^8)U(z) \\ &= -(z^4 + 1)(z - z^{-1})M(z) - (z - z^{-1})z^5(1 - z^{8m})U(z) \\ &= z^6 - z^4 = z^4\left((z^4)^{4m+3} - 1\right). \end{aligned}$$

It follows from Lemma 11.47 that

$$D_\gamma(z) = z^4 \sum_{i=0}^{4m+2} z^{4i} + cI.$$

As $\mathrm{wt}(D_\gamma(z)) = 1 + cn$, to satisfy condition (2) of Lemma 11.41, it must be that $c = 1$. It is then clear that $\mathrm{def}_\gamma(x) \neq 0$ for all $x \in \mathbb{Z}_n$, and so the conditions of Lemma 11.41 are satisfied.

Case 6. If $n \equiv 3 \pmod 8$ and $n \equiv 1 \pmod 3$, set

$$\mathrm{def}(0), \mathrm{def}(1), \ldots, \mathrm{def}(n-1) = \bar{S}\bar{L}^m,$$

where $n = 8m + 3$. Note that $m \equiv 2 \pmod 3$. Clearly $\mathrm{def}(x) \neq 0$ for all $x \in \mathbb{Z}_n$, and by Lemma 11.46, $\mathrm{def}(x) + \mathrm{def}(x+1) + \mathrm{def}(x+2) \neq 0$ for all $x \in \mathbb{Z}_n$. The generating function for def is

$$D(z) = -S(z) - z^3 \left(\sum_{i=0}^{m-1} z^{8i} \right) (1 - z^4) U(z).$$

It follows from this that $\mathrm{wt}(D(z)) = -1 = -n$. Now

$$\begin{aligned}
(z^4 - 1) D_\gamma(z) &= -(z^4 + 1)(z - z^{-1}) S(z) - (z - z^{-1}) z^3 (1 - z^{8m}) U(z) \\
&= z^5 - z^3 \\
&= z^3 \left((z^4)^{4m+2} - 1 \right).
\end{aligned}$$

It follows from Lemma 11.47 that

$$D_\gamma(z) = z^3 \sum_{i=0}^{4m+1} z^{4i} + cI.$$

As $\mathrm{wt}(D_\gamma(z)) = 1 + cn$, to satisfy condition (2) of Lemma 11.41, it must be that $c = 1$. It is then clear that $\mathrm{def}_\gamma(x) \neq 0$ for all $x \in \mathbb{Z}_n$, and so the conditions of Lemma 11.41 are satisfied.

Case 7. If $n \equiv 1 \pmod 8$ and $n \equiv 2 \pmod 3$, set

$$\mathrm{def}(0), \mathrm{def}(1), \ldots, \mathrm{def}(n-1) = \bar{M}^5 \bar{L}^m,$$

where $n = 8m + 25$. Note that $m \equiv 2 \pmod 3$ and that the case $n = 17$ is not covered by this construction. Clearly $\mathrm{def}(x) \neq 0$ for all $x \in \mathbb{Z}_n$, and by Lemma 11.46, $\mathrm{def}(x) + \mathrm{def}(x+1) + \mathrm{def}(x+2) \neq 0$ for all $x \in \mathbb{Z}_n$. The generating function for def is

$$D(z) = -(1 + z^5 + z^{10} + z^{15} + z^{20})M(z) - z^{25}\left(\sum_{i=0}^{m-1} z^{8i}\right)(1 - z^4)U(z).$$

It follows from this that $\mathrm{wt}(D(z)) = -5 = -n$. Now

$$\begin{aligned}(z^4 - 1)D_\gamma(z) &= -(z^4 + 1)(z - z^{-1})(1 + z^5 + z^{10} + z^{15} + z^{20})M(z)\\ &\quad -(z - z^{-1})z^{25}(1 - z^{8m})U(z)\\ &= -z^4 + z^6 - z^9 + z^{11} - z^{14} + z^{16} - z^{19} + z^{21} - z^{24} + z^{26}\\ &= (z^4 + z^9 + z^{14} + z^{19} + z^{24})\left((z^4)^{4m+13} - 1\right).\end{aligned}$$

It follows from Lemma 11.47 that

$$D_\gamma(z) = (z^4 + z^9 + z^{14} + z^{19} + z^{24})\sum_{i=0}^{4m+12} z^{4i} + cI.$$

As $\mathrm{wt}(D_\gamma(z)) = cn$, to satisfy condition (2) of Lemma 11.41, it must be that $c = -1$. To complete this case, we need to show that $\mathrm{def}_\gamma(x) \neq 0$ for all $x \in \mathbb{Z}_n$, thus satisfying the conditions of Lemma 11.41. Set

$$J = \{4i \mid i = 1, \ldots, 4m + 12\}$$

and

$$K = \{4i \mid i = 4m + 13, \ldots, 8m + 24\}.$$

Then J, K, $\{0\}$ partitions $\mathbb{Z}_n$, and inductively we can prove that

$$z^{2r}X_J(z) = \begin{cases} I - X_J(z) + \sum_{i=1}^{r}(-1)^i z^{2i} & \text{if } r \text{ is odd}\\ X_J(z) - \sum_{i=1}^{r}(-1)^i z^{2i} & \text{if } r \text{ is even.}\end{cases}$$

Now

$$\begin{aligned}D_\gamma(z) &= (z^{2\times 2} + zz^{2\times 4} + z^{2\times 7} + zz^{2\times 9} + z^{2\times 12})(X_J(z) + 1) - I\\ &= I + X_J(z) + (z^2 + z^9 + z^{13} + z^{14} + z^{17} + z^{18} + z^{22})\\ &\quad -(z^{11} + z^{15} + z^{16} + z^{20}).\end{aligned}$$

If we set

$$S = \{2, 9, 13, 14, 17, 18, 22\}$$

and

$$T = \{11, 15, 16, 20\},$$

then $T \subseteq J$ and $S \subseteq K$. Thus T, $J \setminus T$, S, $K \setminus S$, $\{0\}$ partitions $\mathbb{Z}_n$ and

$$D_\gamma(z) = X_{K\setminus S}(z) - X_S(z) - X_{J\setminus T}(z) + X_T(z) + X_{\{0\}}(z).$$

Thus $\mathrm{def}_\gamma(x) \neq 0$ for all $x \in \mathbb{Z}_n$ by Lemma 11.48.

Case 8. If $n \equiv 7 \pmod 8$ and $n \equiv 2 \pmod 3$, set

$$\mathrm{def}(0), \mathrm{def}(1), \ldots, \mathrm{def}(n-1) = \bar{S}^5 \bar{L}^m,$$

where $n = 8m + 15$. Note that $m \equiv 1 \pmod 3$. Clearly $\mathrm{def}(x) \neq 0$ for all $x \in \mathbb{Z}_n$, and by Lemma 11.46, $\mathrm{def}(x) + \mathrm{def}(x+1) + \mathrm{def}(x+2) \neq 0$ for all $x \in \mathbb{Z}_n$. The generating function for def is

$$D(z) = -(1 + z^3 + z^6 + z^9 + z^{12})S(z) - z^{15}\left(\sum_{i=0}^{m-1} z^{8i}\right)(1 - z^4)U(z).$$

It follows from this that $\mathrm{wt}(D(z)) = -5 = -n$. Now

$$\begin{aligned}(z^4 - 1)D_\gamma(z) &= -(z^4 + 1)(z - z^{-1})(1 + z^3 + z^6 + z^9 + z^{12})S(z)\\ &\quad -(z - z^{-1})z^{15}(1 - z^{8m})U(z)\\ &= -z^3 + z^5 - z^6 + z^8 - z^9 + z^{11} - z^{12} + z^{14} - z^{15} + z^{17}\\ &= (z^3 + z^6 + z^9 + z^{12} + z^{15})\left((z^4)^{4m+8} - 1\right).\end{aligned}$$

It follows from Lemma 11.47 that

$$D_\gamma(z) = (z^3 + z^6 + z^9 + z^{12} + z^{15})\sum_{i=0}^{4m+7} z^{4i} + cI.$$

As $\mathrm{wt}(D_\gamma(z)) = cn$, to satisfy condition (2) of Lemma 11.41, it must be that $c = -1$. To complete this case, we need to show that $\mathrm{def}_\gamma(x) \neq 0$ for all $x \in \mathbb{Z}_n$, thus satisfying the conditions of Lemma 11.41. Set

$$J = \{4i \mid i = 1, \ldots, 4m + 7\}$$

and

$$K = \{4i \mid i = 4m + 8, \ldots, 8m + 14\}.$$

Then J, K, $\{0\}$ partitions $\mathbb{Z}_n$, and inductively we can prove that

$$z^{2r}X_J(z) = \begin{cases} I - X_J(z) + \sum_{i=1}^{r}(-1)^i z^{2i} & \text{if } r \text{ is odd} \\ X_J(z) - \sum_{i=1}^{r}(-1)^i z^{2i} & \text{if } r \text{ is even.} \end{cases}$$

Now

$$\begin{aligned} D_\gamma(z) &= (zz^{2\times 1} + z^{2\times 3} + zz^{2\times 4} + z^{2\times 6} + zz^{2\times 7})(X_J(z) + 1) - I \\ &= -I - zX_J(z) - z(z^7 + z^{10}) + z(z^5 + z^8 + z^9 + z^{12}). \end{aligned}$$

If we set

$$S = \{7, 10\}$$

and

$$T = \{5, 8, 9, 12\},$$

then $T \subseteq J$ and $S \subseteq K$. Thus T, $J \setminus T$, S, $K \setminus S$, $\{0\}$ partitions $\mathbb{Z}_n$ and

$$D_\gamma(z) = z(-X_{K\setminus S}(z) + X_S(z) + X_{J\setminus T}(z) - X_T(z) - X_{\{0\}}).$$

Thus $\mathrm{def}_\gamma(x) \neq 0$ for all $x \in \mathbb{Z}_n$ by Lemma 11.48. □

An immediate corollary:

Corollary 11.50. *If $n > 1$ is neither 7 nor 17 and is not divisible by 2 or 3, then $\omega(\mathbb{Z}_{3n}) \geq 3$.*

A number of special cases of Corollary 11.50 had already been proved: these results were summarized by Abel, Finizio, Ge, and Greig [8] in 2006. In the following theorem, we give a partial answer to the question: For which v is $\omega(\mathbb{Z}_v) \geq 3$?

Theorem 11.51. *$\omega(\mathbb{Z}_v) \geq 3$ if and only if v is odd, $v \neq 3, 9$, with the possible exception of $v = 9n$, n not divisible by 2, 3, 5, 7, 11, 13, 17, 23, 29, 31, or 109.*

Proof. If v is even, then $\omega(\mathbb{Z}_v) = 0$ by Corollary 3.2.

If p is the smallest prime divisor of v, then $\omega(\mathbb{Z}_v) \geq p - 2$ by Theorem 8.14. Thus $\omega(\mathbb{Z}_v) \geq 3$ if v is divisible by neither 2 nor 3.

If $v = 3^m$, $m \geq 3$, then $\omega(\mathbb{Z}_v) \geq 3$: this was proved by Abel, Finizio, Ge, and Greig [8] in 2006. Thus, by Corollary 1.39, $\omega(\mathbb{Z}_v) \geq 3$, whenever v is odd and a multiple of 27.

The cases that remain are $v = 3n$ and $v = 9n$, n not divisible by 2 or 3. It is easy to establish that $\omega(\mathbb{Z}_3) = 1$ and $\omega(\mathbb{Z}_9) = 1$: see Chapter 13.

If $v = 3n$, $n > 3$ is not divisible by 2 or 3, and $n \neq 7, 17$, then $\omega(\mathbb{Z}_v) \geq 3$ by Theorem 11.49. In 1959 Parker [264] showed that $\omega(\mathbb{Z}_{21}) \geq 3$, and in 2005 Abel and Ge [9] showed that $\omega(\mathbb{Z}_{51}) \geq 4$.

For the case, $v = 9n$, in 2006 Abel, Finizio, Ge, and Greig [8] proved that $\omega(\mathbb{Z}_{9n}) \geq 3$ for $n = 5, 7, 11, 13, 17, 23, 29, 31$, and 109. Thus, by Corollary 1.39, if n, not divisible by 2 or 3, is divisible by at least one of $5, 7, 11, 13, 17, 23, 29, 31$, or 109, then $\omega(\mathbb{Z}_v) \geq 3$. □

The result of Theorem 11.49 enables us to improve the bound $\omega(\mathcal{L}_3(q)) \geq 2$ if the characteristic of GF(q) is not 3, given in Section 11.4, when the characteristic of GF(q) is not 2, 3, 7, or 17.

Theorem 11.52 (Evans, 2018). *If the characteristic of* GF(q) *is not* 2, 3, 7, *or* 17, *then*

$$\omega(\mathcal{L}_3(q)) \geq 3.$$

Proof. Let p be the characteristic of GF(q), $p \neq 2, 3, 7, 17$, and let GF(p) be the prime subfield of GF(q). By Theorem 11.49, $\omega(\mathcal{L}_3(p)) \geq 3$. With sum, def, sum_γ, and def_γ be as in Theorem 11.49, the orthomorphisms θ, ϕ, and γ in $\mathcal{L}_3(p)$ are pairwise orthogonal. We will extend θ, ϕ, and γ to θ', ϕ', and γ', three pairwise orthogonal orthomorphisms in $\mathcal{L}_3(q)$.

Let θ', ϕ', γ': GF(q) $\times$ GF(3) be bijections; let θ' have first component mappings $\theta'_{1,0}(x) = 2x$, $\theta'_{1,1}(x) = 2x + 2$, and $\theta'_{1,2}(x) = 2x - 2$; let ϕ' have first component mappings $\phi'_{1,0}(x) = -x$, $\phi'_{1,1}(x) = -x + 2$, and $\phi'_{1,2}(x) = -x - 2$; and let γ' have first component mappings $\gamma'_{1,0}(x) = ((p+1)/2)x$, $\gamma'_{1,1}(x) = ((p+1)/2)x+2$, and $\gamma'_{1,2}(x) = ((p+1)/2)x-2$. Restricted to GF($p$), the first component mappings of θ, ϕ, and γ are the same as the first component mappings of θ', ϕ', and γ', respectively.

Let $0 = w_0, w_1, \ldots, w_{(q/p)-1}$ be representatives of the orbits of $x \mapsto x + 1$ in GF(q), and let us define sum$'$, def$'$, sum'_γ, and def'_γ, by setting $\text{sum}'(w_i + x) = \text{sum}(x)$, $\text{def}'(w_i + x) = \text{def}(x)$, $\text{sum}'_\gamma(w_i + x) = \text{sum}_\gamma(x)$, and $\text{def}'_\gamma(w_i + x) = \text{def}_\gamma(x)$, for $i = 0, \ldots, (q/p) - 1$ and $x \in \text{GF}(p)$. We leave it to the reader to verify that these functions solve the difference equations for θ', ϕ', and γ' as well as the adjacency difference equations for any pair of θ', ϕ', and γ'. □

Chapter 12
$\omega(G)$ for Some Classes of Nonabelian Groups

Almost all attempts to improve the lower bounds for $\omega(G)$ have been for abelian groups, in particular direct products of elementary abelian groups. These lower bounds were described in Chapter 11 and, for small groups, will be described in Chapter 13. The only classes of nonabelian groups for which attempts have been made to improve the lower bound for $\omega(G)$ are the dihedral groups and some of the linear groups of even characteristic. We present these improvements in this chapter.

Improved lower bounds for $\omega(D_{2n})$, D_{2n} the dihedral group of order $2n$, will be given in Section 12.1. These improvements will be obtained from improved lower bounds for $\omega(\mathbb{Z}_n)$ given in Chapter 11 and improved lower bounds for $\omega(D_{2n})$ when n is a power of 2. While proofs of admissibility show that $\omega(G) \geq 1$ when G is nonsolvable, very little work has been done on improving this bound. In Section 12.2 we will show that $\omega(G) \geq \min\{\omega(\mathbb{Z}_{q-1}), \omega(\mathbb{Z}_{q+1})\}$, when $G = \mathrm{GL}(2, q)$ or $\mathrm{SL}(2, q)$, q even, $q \neq 2$. To establish this bound, we will use partitions of the element sets of the groups. For higher dimensions, in Section 12.3, we will establish lower bounds for $\omega(\mathrm{GL}(n, q))$, q even, $q \neq 2$, using even-odd decompositions and partitions of the sets of odd-order elements of the groups.

12.1 Dihedral Groups

Chapter 7 of [119] contains a list of problems dealing with orthomorphism graphs of groups. In problem 38 we are asked what can be said about $\omega(D_{2n})$, where D_{2n} is the dihedral group of order $2n$. Recall that

$$D_{2n} = \langle a, b \mid a^n = b^2 = 1, bab = a^{-1} \rangle.$$

We know that if n is odd, then $\omega(D_{2n}) = 0$ by Theorem 3.34, and, if n is even, then $\omega(D_{2n}) \geq 1$ by Theorem 4.22. We also know $\omega(D_{4m})$ for small values of m:

A. B. Evans, *Orthogonal Latin Squares Based on Groups*, Developments in Mathematics 57, https://doi.org/10.1007/978-3-319-94430-2_12

$\omega(D_4) = 2$, $\omega(D_8) = 1$, and $\omega(D_{12}) = 2$. Values of $\omega(G)$ for small groups will be given in Chapter 13. In 1997 Bowler [48] improved on the lower bound for $\omega(D_{4m})$ for infinitely many odd integers m.

Theorem 12.1 (Bowler, 1997). *If $\omega(\mathbb{Z}_m) \geq 2$, then $\omega(D_{4m}) \geq 2$.*

Proof. Let $n = 2m$ and $H = \langle a^2 \rangle$. As $H \cong \mathbb{Z}_m$; there exists an orthogonal pair of orthomorphisms θ and ϕ of H. Define $\theta', \phi' : D_{4m} \to D_{4m}$ by

$$\theta'(x) = \begin{cases} \theta(a^{2i}) & \text{if } x = a^{2i},\ i = 0, \dots, m-1, \\ \theta(a^{2i})b & \text{if } x = a^{2i+1},\ i = 0, \dots, m-1, \\ \theta(a^{2i})ab & \text{if } x = a^{2i}b,\ i = 0, \dots, m-1, \\ \theta(a^{2i})a & \text{if } x = a^{2i+1}b,\ i = 0, \dots, m-1, \end{cases}$$

and

$$\phi'(x) = \begin{cases} \phi(a^{2i}) & \text{if } x = a^{2i},\ i = 0, \dots, m-1, \\ \phi(a^{2i})ab & \text{if } x = a^{2i+1},\ i = 0, \dots, m-1, \\ \phi(a^{2i})a & \text{if } x = a^{2i}b,\ i = 0, \dots, m-1, \\ \phi(a^{2i})b & \text{if } x = a^{2i+1}b\, i = 0, \dots, m-1. \end{cases}$$

It is easily seen that D_{4m} is partitioned by H, Ha, Hb, and Hab. To see that θ' is a bijection, note that

$$\begin{aligned} \{\theta'(x) \mid x \in D_{4m}\} = & \quad \{\theta(a^{2i}) \mid x = a^{2i}, i = 0, \dots, m-1\} \\ & \cup \{\theta(a^{2i})b \mid x = a^{2i+1}, i = 0, \dots, m-1\} \\ & \cup \{\theta(a^{2i})ab \mid x = a^{2i}b, i = 0, \dots, m-1\} \\ & \cup \{\theta(a^{2i})a \mid x = a^{2i+1}b, i = 0, \dots, m-1\} \\ = & \quad H \cup Hb \cup Hab \cup Ha. \end{aligned}$$

It follows that θ' is onto and hence, by finiteness, is a bijection. Similarly

$$\{\phi'(x) \mid x \in D_{4m}\} = H \cup Hab \cup Ha \cup Hb$$

from which it follows that ϕ' is onto and hence, by finiteness, is a bijection.

To see that θ' and ϕ' are orthomorphisms, note that

$$x^{-1}\theta'(x) = \begin{cases} a^{-2i}\theta(a^{2i}) & \text{if } x = a^{2i},\ i = 0, \ldots, m-1, \\ a^{-2i}\theta(a^{2i})a^{-1}b & \text{if } x = a^{2i+1},\ i = 0, \ldots, m-1, \\ \left(a^{-2i}\theta(a^{2i})\right)^{-1} a^{-1} & \text{if } x = a^{2i}b,\ i = 0, \ldots, m-1, \\ \left(a^{-2i}\theta(a^{2i})\right)^{-1} b & \text{if } x = a^{2i+1}b,\ i = 0, \ldots, m-1, \end{cases}$$

and

$$x^{-1}\phi'(x) = \begin{cases} a^{-2i}\phi(a^{2i}) & \text{if } x = a^{2i},\ i = 0, \ldots, m-1, \\ a^{-2i}\phi(a^{2i})b & \text{if } x = a^{2i+1},\ i = 0, \ldots, m-1, \\ \left(a^{-2i}\phi(a^{2i})\right)^{-1} a^{-1}b & \text{if } x = a^{2i}b,\ i = 0, \ldots, m-1, \\ \left(a^{-2i}\phi(a^{2i})\right)^{-1} a & \text{if } x = a^{2i+1}b,\ i = 0, \ldots, m-1, \end{cases}$$

from which it follows that

$$\{x^{-1}\theta'(x) \mid x \in D_{4m}\} = H \cup Hab \cup Ha \cup Hb$$

and

$$\{x^{-1}\phi'(x) \mid x \in D_{4m}\} = H \cup Hb \cup Hab \cup Ha.$$

Hence, the mappings $x \mapsto x^{-1}\theta'(x)$ and $x \mapsto x^{-1}\phi'(x)$ are both onto and, hence, bijections.

To see that $\theta' \perp \phi'$, note that

$$\phi'(x)^{-1}\theta'(x) = \begin{cases} \phi(a^{-2i})^{-1}\theta(a^{2i}) & \text{if } x = a^{2i},\ i = 0, \ldots, m-1, \\ \theta(a^{2i})^{-1}\phi(a^{-2i})a & \text{if } x = a^{2i+1},\ i = 0, \ldots, m-1, \\ \phi(a^{-2i})^{-1}\theta(a^{2i})b & \text{if } x = a^{2i}b,\ i = 0, \ldots, m-1, \\ \theta(a^{2i})^{-1}\phi(a^{-2i})a^{-1}b & \text{if } x = a^{2i+1}b,\ i = 0, \ldots, m-1, \end{cases}$$

from which it follows that

$$\{\phi' i(x)^{-1}\theta'(x) \mid x \in D_{4m}\} = H \cup Ha \cup Hb \cup Hab.$$

Hence, the mapping $x \mapsto \phi'(x)^{-1}\theta'(x)$ is onto and, hence, is a bijection. □

In Chapter 11 we obtained a number of lower bounds for $\omega(\mathbb{Z}_m)$. These bounds, combined with the result of Theorem 12.1, yield lower bounds for $\omega(D_{4m})$ when m is odd.

Corollary 12.2. *If $m > 3$ is odd and not divisible by 9, then $\omega(D_{4m}) \geq 2$. Further, if $m = 9n$, $n > 1$, n not divisible by 2, 3, 5, 7, 11, 13, 17, 23, 29, 31, or 109, then $\omega(D_{4m}) \geq 2$.*

Table 12.1 Three pairwise orthogonal orthomorphisms of D_{16}

x	00	10	20	30	40	50	60	70	01	11	21	31	41	51	61	71
$\theta_1(x)$	00	01	61	60	41	40	20	21	30	71	10	51	31	70	11	50
$\theta_2(x)$	00	30	41	40	31	21	10	71	70	60	51	50	61	11	20	01
$\theta_3(x)$	00	31	30	71	51	01	40	20	11	41	61	60	21	50	70	10

For dihedral 2-groups, we know that $\omega(D_8) = 1$. For D_{16} the exact value of $\omega(D_{16})$ is unknown. In 1999 Quinn [273] improved the lower bound for $\omega(D_{16})$ from 1 to 3.

Theorem 12.3 (Quinn, 1999). $\omega(D_{16}) \geq 3$.

Proof. Three mappings $\theta_1, \theta_2, \theta_3 : D_{16} \to D_{16}$ are described in Table 12.1: ij is used to represent $a^i b^j$. It is routine, but tedious, to establish that θ_1, θ_2, and θ_3 are three pairwise orthogonal orthomorphisms of D_{16}. □

There are a number of quotient group constructions of difference matrices and pairwise orthogonal sets of orthomorphisms for abelian groups in the literature. In 1999 Quinn [273] gave a quotient group construction for pairwise orthogonal sets of orthomorphisms of nonabelian groups: see Corollary 1.39. This yields a lower bound for $\omega(D_{2n})$.

Theorem 12.4. *If m is odd, then $\omega(D_{2^t m}) \geq \min\{\omega(D_{2^t}), \omega(\mathbb{Z}_m)\}$.*

Proof. Let $H = \langle a^{2^{t-1}} \rangle$. Then H is a normal subgroup of $G = D_{2^t m}$, $H \cong \mathbb{Z}_m$ and $G/H \cong D_{2^t}$. The result then follows from Corollary 1.39. □

Note that, as $\omega(D_4) = 2$, Theorem 12.1 is an immediate corollary of Theorem 12.4. The results from Chapter 11, combined with the results of Theorems 12.3 and 12.4, yield lower bounds for $\omega(D_{16m})$ when m is odd.

Corollary 12.5. *If $m > 3$ is odd and not divisible by 9, then $\omega(D_{16m}) \geq 2$.*

Corollary 12.6. *If m is odd, $m \neq 3, 9$, and m is not of the form $9k$, where k is not divisible by 3, 5, 7, 11, 13, 17, 23, 29, 31, or 109, then $\omega(D_{16m}) \geq 3$.*

To obtain better lower bounds for $\omega(D_{2n})$ using Theorem 12.4, we need better lower bounds for $\omega(D_{2^t})$. If $t \geq 5$, our current lower bound is $\omega(D_{2^t}) \geq 1$. This can certainly be improved.

12.2 $\omega(\mathrm{GL}(2,q))$ and $\omega(\mathrm{SL}(2,q))$ for q Even

For most nonsolvable groups, the current lower bound for $\omega(G)$ is one, a consequence of the proof of the Hall-Paige conjecture, given in Chapter 7. There is only one class of nonsolvable groups for which we have better bounds, the linear groups of even characteristic. For $\mathrm{GL}(2, q)$ and $\mathrm{SL}(2, q)$, q even, $q \neq 2$, improved bounds were given by Evans [121] in 1993: these bounds are described in this section.

12.2.1 Partitions

In Chapter 5, we constructed complete mappings of groups piecewise by partitioning the elements of the group and constructing complete mappings for each set in the partition. This same idea can be applied to construct pairwise orthogonal sets of orthomorphisms. Let X be a proper subset of the set of elements of a finite group G. We will say that $\theta\colon X \to X$ is an *orthomorphism* of X if both θ and the mapping $x \mapsto x^{-1}\theta(x)$ are permutations of X, and, if $\theta, \phi\colon X \to X$, then we say that θ is *orthogonal* to ϕ, written $\theta \perp \phi$, if the mapping $x \mapsto \theta(x)^{-1}\phi(x)$ is a permutation of X. Note that, if θ and ϕ are orthomorphisms of X, then $\theta\perp\phi$ and $\phi\perp\theta$ are equivalent statements if and only if X is closed under inverses. If X is closed under inverses, then a set $\{\theta_1, \ldots, \theta_r\}$ of orthomorphisms of X is said to be *pairwise orthogonal* if $\theta_i\perp\theta_j$ for all i, j, $1 \leq i, j \leq r$, $i \neq j$. If X is closed under inverses, then we define $\mathrm{Orth}(X)$ to be both the set of orthomorphisms of X as well as the orthomorphism graph of X, whose vertices are the orthomorphisms of X, adjacency being synonymous with orthogonality. The clique number of $\mathrm{Orth}(X)$, the maximum number of pairwise orthogonal orthomorphisms of X, is denoted $\omega(\mathrm{Orth}(X))$, or $\omega(X)$ for short, and by convention $\omega(\{1\}) = \infty$. Clearly, if $\{X_i \mid i \in I\}$ is a partition of X, a subset of the element set of G, and each X_i is closed under inverses, then $\omega(X) \geq \min\{\omega(X_i) \mid i \in I\}$. From pairwise adjacent orthomorphisms of sets in a partition of the elements of a group G, we can construct pairwise orthogonal orthomorphisms of G.

Lemma 12.7. *If $X_1, \ldots\ldots, X_r$ is a partition of the elements of a group G, and each X_i is closed under inverses, then*

$$\omega(G) \geq \min\{\omega(X_i) \mid i = 1, \ldots, r\}.$$

Proof. Routine. □

We shall be interested in a particular class of partitions. Let Γ be a subgroup of $\mathrm{Aut}(G)$ and define an equivalence relation on the set of elements of G by setting x equivalent to y whenever

$$\Gamma_x = \{\alpha \mid \alpha \in \Gamma, \alpha(x) = x\} = \{\alpha \mid \alpha \in \Gamma, \alpha(y) = y\} = \Gamma_y.$$

The *Γ-partition* of G is the partition of the set of elements of G induced by this equivalence relation. It is not necessary to determine the clique number for each set in a Γ-partition, only for representatives of orbits of the action of the normalizer in $\mathrm{Aut}(G)$ of Γ on the sets in the Γ-partition.

Lemma 12.8. *If $X_1, \ldots, X_m$ is a Γ-partition of the set of elements of G, then each X_i is closed under inverses. Further, if $\alpha \in \mathrm{Aut}(G)$ normalizes Γ, then α permutes the classes of the partition, and, if $\alpha(X_j) = X_i$, then $\mathrm{Orth}(X_i) \cong \mathrm{Orth}(X_j)$.*

Proof. Each X_i is closed under inverses as, for each $x \in G$, $\Gamma_{x^{-1}} = \Gamma_x$ and so $x^{-1} \equiv x$.

Let $\alpha \in \mathrm{Aut}(G)$ normalize Γ. If $\beta \in \Gamma_x$, then $(\alpha\beta\alpha^{-1})\alpha(x)$ and, hence, $\alpha\Gamma_x\alpha^{-1} \subseteq \Gamma_{\alpha(x)}$ and $|\alpha\Gamma_x\alpha^{-1}| \leq |\Gamma_{\alpha(x)}|$. Similarly, if $\beta \in \Gamma_{\alpha(x)}$, then $\alpha^{-1}\beta\alpha(x) = \alpha^{-1}\alpha(x) = x$ and, hence, $\alpha^{-1}\Gamma_{\alpha(x)}\alpha \subseteq \Gamma_x$, $\Gamma_{\alpha(x)} \subseteq \alpha\Gamma_x\alpha^{-1}$, and $|\Gamma_{\alpha(x)}| \leq |\alpha\Gamma_x\alpha^{-1}|$. It follows that $\Gamma_{\alpha(x)} = \alpha\Gamma_x\alpha^{-1}$ and the mapping $X_i \mapsto \alpha(X_i)$ is a permutation of $X_1, \ldots, X_m$.

Let $\alpha \in \mathrm{Aut}(G)$ normalize Γ and suppose that $\alpha(X_j) = X_i$. Let $\theta_1, \ldots, \theta_n$ be the set of orthomorphisms of X_i. It is routine to show that $\alpha\theta_1\alpha^{-1}, \ldots, \alpha\theta_n\alpha^{-1}$ is the set of orthomorphisms of X_j and that $\alpha\theta_s\alpha^{-1} \perp \alpha\theta_t\alpha^{-1}$ if and only if $\theta_s \perp \theta_t$. □

12.2.2 $\omega(\mathrm{GL}(2, q))$, q Even

For $\mathrm{GL}(2, q)$, q even, Evans [121] proved in 1993 that $\omega(\mathrm{GL}(2, q))$ was bound below by $\omega(\mathbb{Z}_{q-1})$ and $\omega(\mathbb{Z}_{q+1})$. This proof was obtained via a Γ-partition of $\mathrm{GL}(2, q)$ with Γ being the inner automorphism group of $\mathrm{GL}(2, q)$.

Theorem 12.9. *For q even, $q \neq 2$, $\omega(\mathrm{GL}(2, q)) \geq \min\{\omega(\mathbb{Z}_{q-1}), \omega(\mathbb{Z}_{q+1})\}$.*

Proof. Let $G = \mathrm{GL}(2, q)$, q even, $q \neq 2$, and let $\Gamma = \mathrm{Inn}(G)$. By Lemma 12.8, to determine lower bounds on $\omega(X)$ for each set in the Γ-partition, it is sufficient to bound clique numbers for the Γ-equivalence classes of one element of each conjugacy class.

If $A \in \mathrm{GL}(2, q)$, then the characteristic polynomial $x^2 + bx + a$ of $A \in \mathrm{GL}(2, q)$ either has one root $\lambda \in \mathrm{GF}(q)$, in which case A is similar to either $\begin{pmatrix} \lambda & 0 \\ 0 & \lambda \end{pmatrix}$ or $\begin{pmatrix} \lambda & 1 \\ 0 & \lambda \end{pmatrix}$; or two distinct roots $\lambda_1, \lambda_2 \in \mathrm{GF}(q)$, in which case A is similar to $\begin{pmatrix} \lambda_1 & 0 \\ 0 & \lambda_2 \end{pmatrix}$; or is irreducible, in which case A is similar to $\begin{pmatrix} 0 & 1 \\ a & b \end{pmatrix}$. We thus have four cases to consider.

Case 1. A is similar to $B = \begin{pmatrix} \lambda & 0 \\ 0 & \lambda \end{pmatrix}$. Let X be the Γ-equivalence class containing B. Simple computation shows that

$$X = \left\{ \begin{pmatrix} \alpha & 0 \\ 0 & \alpha \end{pmatrix} \middle| \alpha \neq 0 \right\}.$$

Now X is isomorphic to the multiplicative group of $\mathrm{GF}(q)$, which is isomorphic to $\mathbb{Z}_{q-1}$ and so $\omega(X) = \omega(\mathbb{Z}_{q-1})$.

Case 2. A is similar to $C = \begin{pmatrix} \lambda & 1 \\ 0 & \lambda \end{pmatrix}$. Let Y be the Γ-equivalence class containing C. Simple computation shows that

$$Y = \left\{ \begin{pmatrix} \alpha & \beta \\ 0 & \alpha \end{pmatrix} \middle| \alpha, \beta \neq 0 \right\} = \left\{ \begin{pmatrix} \alpha & 0 \\ 0 & \alpha \end{pmatrix} \begin{pmatrix} 1 & \beta/\alpha \\ 0 & 1 \end{pmatrix} \middle| \alpha, \beta \neq 0 \right\}.$$

Thus Y is a direct product of two groups,

$$X = \left\{ \begin{pmatrix} \alpha & 0 \\ 0 & \alpha \end{pmatrix} \middle| \alpha \neq 0 \right\} \cong \mathbb{Z}_{q-1} \text{ and } \left\{ \begin{pmatrix} 1 & \gamma \\ 0 & 1 \end{pmatrix} \middle| \gamma \neq 0 \right\} \cong \mathrm{GF}(q)^+.$$

Hence, by Corollary 8.2,

$$\omega(Y) \geq \min\{\omega(\mathbb{Z}_{q-1}), \omega(\mathrm{GF}(q)^+)\} = \min\{\omega(\mathbb{Z}_{q-1}), q-2\} = \omega(\mathbb{Z}_{q-1}).$$

Case 3. A is similar to $D = \begin{pmatrix} \lambda_1 & 0 \\ 0 & \lambda_2 \end{pmatrix}$, $\lambda_1 \neq \lambda_2$. Let Z be the Γ-equivalence class containing D. Simple computation shows that

$$Z = \left\{ \begin{pmatrix} \alpha & 0 \\ 0 & \beta \end{pmatrix} \mid \alpha, \beta \neq 0, \alpha \neq \beta \right\}.$$

Now

$$X \cup Z = \left\{ \begin{pmatrix} \alpha & 0 \\ 0 & \alpha \end{pmatrix} \begin{pmatrix} 1 & 0 \\ 0 & \beta \end{pmatrix} \middle| \alpha, \beta \neq 0 \right\},$$

which is a subgroup of $\mathrm{GL}(2,q)$ isomorphic to $\mathbb{Z}_{q-1} \oplus \mathbb{Z}_{q-1}$. Set $Z^* = \{\mathrm{diag}(1,\beta) \mid \beta \neq 0\}$ and $r = \omega(\mathbb{Z}_{q-1})$. Now X and Z^* are both isomorphic to $\mathbb{Z}_{q-1}$ and $X \cup Z = X \times Z^*$. Let $\theta_1, \ldots, \theta_r$ be a set of r pairwise orthogonal normalized orthomorphisms of X and $\phi_1, \ldots, \phi_r$ a set of r pairwise orthogonal normalized orthomorphisms of Z^*. By Theorem 8.1, $\theta_1 \times \phi_1, \ldots, \theta_r \times \phi_r$ is a set of pairwise orthogonal orthomorphisms of $X \cup Z$. Further $\theta_1 \times \phi_1|_X, \ldots, \theta_r \times \phi_r|_X$ is a set of r pairwise orthogonal orthomorphisms of X, and hence, as $Z \cap X = \emptyset$, $\theta_1 \times \phi_1|_Z, \ldots, \theta_r \times \phi_r|_Z$ is a set of r pairwise orthogonal orthomorphisms of Z. It follows that

$$\omega(Z) \geq \omega(\mathbb{Z}_{q-1}).$$

Case 4. The characteristic polynomial $x^2 + bx + a$ of A is irreducible over $\mathrm{GF}(q)$, and A is similar to $E = \begin{pmatrix} 0 & 1 \\ a & b \end{pmatrix}$. Let W be the Γ-equivalence class containing E. Simple computation shows that

$$W = \left\{ \alpha \begin{pmatrix} 0 & 1 \\ a & b \end{pmatrix} + \begin{pmatrix} \beta & 0 \\ 0 & \beta \end{pmatrix} \middle| \alpha \neq 0 \right\}.$$

Now $X \cup \{0\}$ with the operations addition and multiplication is a field of order q, and hence is isomorphic to GF(q). As E is a root of an irreducible quadratic polynomial over GF(q), with the operations addition and multiplication, $X \cup \{0\} \cup W \cong \mathrm{GF}(q^2)$. In particular, under multiplication, $X \cup W \cong \mathrm{GF}(q^2)^*$, the multiplicative group of GF(q^2). As X is the unique subgroup of $\mathrm{GF}(q^2)^*$ of order $q-1$ and $\gcd(q-1, q+1) = 1$, $X \cup W \cong \mathbb{Z}_{q-1} \times \mathbb{Z}_{q+1}$. By Corollary 8.2, $\omega(\mathbb{Z}_{q-1} \times \mathbb{Z}_{q+1}) \geq \min\{\omega(\mathbb{Z}_{q-1}), \omega(\mathbb{Z}_{q+1})\}$. By a similar argument to that used in Case 3, it follows that

$$\omega(W) \geq \min\{\omega(\mathbb{Z}_{q-1}), \omega(\mathbb{Z}_{q+1})\}.$$

By Lemmas 12.7 and 12.8,

$$\omega(G) \geq \min\{\omega(X), \omega(Y), \omega(Z), \omega(W)\} = \min\{\omega(\mathbb{Z}_{q-1}), \omega(\mathbb{Z}_{q+1})\}.$$

□

12.2.3 ω(SL(2, q)), q Even

For SL$(2, q)$, q even, Evans [121] established in 1993 that the lower bounds that he had obtained for $\omega(\mathrm{GL}(2, q))$ also held for $\omega(\mathrm{SL}(2, q))$. The proof is similar to the proof of Theorem 12.9 and uses the inner automorphism group of GL$(2, q)$ as Γ instead of the inner automorphism group of SL$(2, q)$.

Theorem 12.10. *For q even, $q \neq 2$, $\omega(\mathrm{SL}(2, q)) \geq \min\{\omega(\mathbb{Z}_{q-1}), \omega(\mathbb{Z}_{q+1})\}$.*

Proof. Let $G = \mathrm{SL}(2, q)$, let $\Gamma = \mathrm{Inn}(\mathrm{GL}(2, q))$, and let X, Y, Z, and W be as in Theorem 12.9. For U one of X, Y, Z, or W, let U' denote $U \cap G$. Then the Γ-equivalence classes to be considered are

$$X' = \left\{ \begin{pmatrix} 1 & 0 \\ 0 & 1 \end{pmatrix} \right\}, Y' = \left\{ \begin{pmatrix} \alpha & 0 \\ 0 & \alpha^{-1} \end{pmatrix} \middle| \alpha \neq 0, 1 \right\}, Z' = \left\{ \begin{pmatrix} 1 & \beta \\ 0 & 1 \end{pmatrix} \middle| \beta \neq 0 \right\},$$

and, for $x^2 + bx + a$ irreducible over GF(q),

$$W' = \left\{ \alpha \begin{pmatrix} 0 & 1 \\ a & b \end{pmatrix} + \begin{pmatrix} \beta & 0 \\ 0 & \beta \end{pmatrix} \middle| \alpha \neq 0, a\alpha^2 + b\beta\alpha + \beta^2 = 1 \right\}.$$

Now $X' \cup Y' \cong \mathbb{Z}_{q-1}$, $X' \cup Z' \cong \mathrm{GF}(q)^+$, and $X' \cup W' \cong \mathrm{GF}(q^2)^*/\mathrm{GF}(q)^* \cong \mathbb{Z}_{q+1}$. Thus, by Lemmas 12.7 and 12.8,

$$\begin{aligned}\omega(\mathrm{SL}(2,q)) &\geq \min\{\omega(X'), \omega(Y'), \omega(Z'), \omega(W')\}\\ &= \min\{\infty, \omega(\mathbb{Z}_{q-1}), \omega(\mathrm{GF}(q)^{+}), \omega(\mathbb{Z}_{q+1})\}\\ &= \min\{\omega(\mathbb{Z}_{q-1}), \omega(\mathbb{Z}_{q+1})\}.\end{aligned}$$

□

To apply Theorems 12.9 and 12.10, we need values of, or bounds on, $\omega(\mathbb{Z}_n)$. Exact values of $\omega(\mathbb{Z}_n)$ for small n will be given in Chapter 13; for general n, bounds were given in Chapter 11; and some bounds were obtained in Section 8.3, using difference sets and difference families. As examples, if $q = 16$, then

$$\omega(\mathrm{GL}(2,16)), \omega(\mathrm{SL}(2,16)) \geq \min\{\omega(\mathbb{Z}_{15}), \omega(\mathbb{Z}_{17})\} \geq \min\{3, 15\} = 3;$$

if $q = 32$, then

$$\omega(\mathrm{GL}(32)), \omega(\mathrm{SL}(32)) \geq \min\{\omega(\mathbb{Z}_{31}), \omega(\mathbb{Z}_{33})\} \geq \min\{29, 4\} = 4;$$

and if $q = 64$, then

$$\omega(\mathrm{GL}(64)), \omega(\mathrm{SL}(64)) \geq \min\{\omega(\mathbb{Z}_{63}), \omega(\mathbb{Z}_{65})\} \geq \min\{5, 3\} = 3.$$

12.3 The Group GL(n, q), q Even, $q \neq 2$

In 2014 Evans [131] improved lower bounds for $\omega(\mathrm{GL}(n,q))$, for $n \geq 2$ and q even, $q \neq 2$. These improved lower bounds were obtained by "stitching" together pairwise orthogonal orthomorphisms of the set of 2-elements of a group with pairwise orthogonal orthomorphisms of the set of odd-order elements of a group.

12.3.1 The Even-Odd Decomposition

For G, a finite group, let U be the set of 2-elements of G and S the set of odd-order elements of G. Note that $U \cap S = \{1\}$. We will construct pairwise orthogonal sets of orthomorphisms of G from pairwise orthogonal sets of orthomorphisms of U and pairwise orthogonal sets of orthomorphisms of S. To do this we need to use a decomposition of elements of G into commuting products of elements of U and S.

Lemma 12.11 (Even-Odd Decomposition). *Let G be a finite group, U the set of 2-elements of G, and S the set of odd order elements of G. Then each $g \in G$ can be uniquely written as a product $g = us$, where $us = su$, $u \in U$, and $s \in S$.*

Proof. Let $g \in G$ be of order mn, m a power of 2, and n odd. As $\gcd(m, n) = 1$, there exist integers m' and n' satisfying $mm' + nn' = 1$. If we set $u = g^{nn'}$ and $s = g^{mm'}$, then $g = us = su$, $u \in U$, and $s \in S$.

To prove uniqueness, suppose that $g = us$, where $us = su$, $u \in U$, and $s \in S$. Let $k = |u|$ and $l = |s|$. As $\gcd(k, l) = 1$, there exist positive integers k' and s' satisfying $kk' \equiv 1 \pmod{l}$ and $ll' \equiv 1 \pmod{k}$. It follows that $s = g^{kk'}$ and $u = g^{ll'}$ are uniquely determined. □

We now present a method for constructing orthomorphisms of G from orthomorphisms of U and S, using even-odd decompositions. It is important to note that U and S are both closed under inverses. Given two mappings $\theta\colon U \to U$ and $\phi\colon S \to S$, we will define the *product of* θ *and* ϕ (with respect to U and S), written $\theta \times_J \phi$, by $\theta \times_J \phi(g) = \theta(u)\phi(s)$, where $g = us$ is the even-odd decomposition of g. If K is a subset of the element set of G and $x \in G$, then we will use $C_K(x) = \{k \in K \mid kx = xk\}$ to denote the *centralizer* of x in K. The centralizers that we will make use of are $C_U(s)$, where $s \in S$, and $C_S(u)$, where $u \in U$. A product construction of orthomorphisms is described in the following lemma.

Lemma 12.12. *Let θ be an orthomorphism of U that acts on $C_U(s)$ for each $s \in S$, and let ϕ be an orthomorphism of S that acts on $C_S(u)$ for each $u \in U$. Then $\theta \times_J \phi$ is an orthomorphism of G.*

Proof. If $u \in U$ and $s \in S$ satisfy $us = su$, then, as θ acts on $C_U(s) = C_U(s^{-1})$, $\theta(u)s^{-1} = s^{-1}\theta(u)$; as ϕ acts on $C_S(\theta(u))$, $\phi(s)\theta(u) = \theta(u)\phi(s)$; and, as ϕ acts on $C_S(u) = C_S(u^{-1})$, $u^{-1}\phi(s) = \phi(s)u^{-1}$. Thus, if $g = us$, $u \in U$ and $s \in S$, is the even-odd decomposition of g, then $\theta \times_J \phi(g) = \theta(u)\phi(s)$ is the even-odd decomposition of $\theta \times_J \phi(g)$, and $g^{-1}(\theta \times_J \phi(g)) = (u^{-1}\theta(u))(s^{-1}\phi(s))$ is the even-odd decomposition of $g^{-1}(\theta \times_J \phi(g))$. We will use these even-odd decompositions to prove that each of the maps $g \mapsto \theta \times_J \phi(g)$ and $g \mapsto g^{-1}\theta \times_J \phi(g)$ is a bijection and, hence, that $\theta \times_J \phi$ is an orthomorphism.

Let $g = us$ and $h = vt$, $u, v \in U$ and $s, t \in S$, be the even-odd decompositions of g and h in G. If $k = \theta \times_J \phi(g) = \theta \times_J \phi(h)$, then $k = \theta(u)\phi(s)$ and $k = \theta(v)\phi(t)$ are both the even-odd decomposition of k, and thus $\theta(u) = \theta(v)$ and $\phi(s) = \phi(t)$, which implies that $u = v$ and $s = t$. Hence $\theta \times_J \phi$ is one to one, and hence, by finiteness, is a bijection. Similarly, if $g^{-1}(\theta \times_J \phi(g)) = h^{-1}(\theta \times_J \phi(h))$, then $u^{-1}\theta(u) = v^{-1}\theta(v)$ and $s^{-1}\phi(s) = t^{-1}\phi(t)$, which implies that $u = v$ and $s = t$. Hence the mapping $g \mapsto g^{-1}\theta \times_J \phi(g)$ is one to one, and hence, by finiteness, is a bijection. It follows that $\theta \times_J \phi$ is an orthomorphism of G. □

We can adapt the proof of Lemma 12.12 to give a construction of sets of pairwise orthogonal orthomorphisms of G from sets of pairwise orthogonal othomorphisms of U and S.

Lemma 12.13. *Let $\{\theta_1, \ldots, \theta_r\}$ be a set of pairwise orthogonal orthomorphisms of U, θ_i acting on $C_U(s)$ for each $i = 1, \ldots, r$ and $s \in S$, and let $\{\phi_1, \ldots, \phi_r\}$ be a set of pairwise orthogonal orthomorphisms of S, ϕ_i acting on $C_S(u)$ for each*

$i = 1, \ldots, r$ and $u \in U$. Then $\{\theta_1 \times_J \phi_1, \ldots, \theta_r \times_J \phi_r\}$ is a set of pairwise orthogonal orthomorphisms of G.

Proof. By Lemma 12.12, each of $\theta_1 \times_J \phi_1, \ldots, \theta_r \times_J \phi_r$ is an orthomorphism of G. Let $g = us$, $u \in U$ and $s \in S$, be the even-odd decompositions of g in G, and let $i, j \in \{1, \ldots, r\}$, $i \neq j$. Then

$$(\theta_i \times_J \phi_i(g))^{-1}(\theta_j \times_J \phi_j(g)) = (\theta_i(u)^{-1}\theta_j(u))(\phi_i(s)^{-1}\phi_j(s)),$$

is the even-odd decomposition of $(\theta_i \times_J \phi_i(g))^{-1}(\theta_j \times_J \phi_j(g))$ by the same argument as in the proof of Lemma 12.12. Similarly, using the same argument as in the proof of Lemma 12.12, we can show that the mapping $g \mapsto (\theta_i \times_J \phi_i(g))^{-1}(\theta_j \times_J \phi_j(g))$ is one to one and, hence, by finiteness, is bijective. It follows that $\theta_i \times_J \phi_i(g) \perp \theta_j \times_J \phi_j(g)$. The result follows. □

Let X and Y be subsets of the element set of G and suppose that X is closed under inverses. We will use $\omega_Y(X)$ to denote the maximum possible number of orthomorphisms in a set of pairwise orthogonal orthomorphisms of X, each of which acts on $C_X(y)$ for each $y \in Y$.

Theorem 12.14. $\omega(G) \geq \min\{\omega_S(U), \omega_U(S)\}$.

Proof. Let $\min\{\omega_S(U), \omega_U(S)\} = r$, let $\theta_1, \ldots, \theta_r$ be a pairwise orthogonal set of orthomorphisms of U that acts on $C_U(s)$ for each $s \in S$, and let $\phi_1, \ldots, \phi_r$ be a pairwise orthogonal set of orthomorphisms of S that acts on $C_S(u)$ for each $u \in U$. By Lemma 12.12, $\theta_1 \times_J \phi_1, \ldots, \theta_r \times_J \phi_r$ are orthomorphisms of G, and, by Lemma 12.13, they are pairwise orthogonal. □

In the following we will establish a lower bound for ω(GL(n, q), q even, $q \neq 2$, by establishing lower bounds for $\omega_S(U)$ and $\omega_U(S)$.

12.3.2 A Lower Bound for $\omega_S(U)$

For the group GL(n, q), q a power of 2, $q \neq 2$, we will give explicit constructions of orthomorphisms of U that act on $C_U(B)$ for each $B \in S$. Let $M_n(q)$ denote the set of $n \times n$ matrices over GF(q), q a power of 2. For each $a \in$ GF$(q)^*$, the multiplicative group of GF(q), we define the mapping

$$\theta_a \colon M_n(q) \to M_n(q) \text{ by } \theta_a(A) = I + a(I + A).$$

If G is a subgroup of GL(n, q), q a power of 2, let

$$F_G = \{a \in \text{GF}(q)^* \mid \theta_a(U) \subseteq G\}.$$

Note that, if $A^m = I$, m a power of 2, then $\theta_a(A)^m = I$ and so θ_a maps 2-elements to 2-elements. Thus, if $\theta_a(U) \subseteq G$, then $\theta_a(U) \subseteq U$. The next lemma establishes the properties of this class of mappings.

Lemma 12.15. *If G is a subgroup of* GL(n, q), *q a power of* 2, *$q \neq 2$, then the following hold.*

1. *F_G is a subgroup of* GF$(q)^*$, *the multiplicative group of* GF(q).
2. *If $a \in F_G$, $a \neq 1$, then $\theta_a \mid_U$ is an orthomorphism of U which acts on $C_U(B)$ for each $B \in S$.*
3. *If $a, b \in F_G \setminus \{1\}$, $a \neq b$, then $\theta_a \mid_U \perp \theta_b \mid_U$.*

Proof.

1. As $\theta_1(A) = A$ for all A, $1 \in F_G$. Let $a, b \in F_G$. Then $\theta_b(U) \subseteq U \subseteq G$, $\theta_a(U) \subseteq U \subseteq G$ and
$$\theta_a \circ \theta_b(A) = \theta_a(I + b(I + A)) = I + ab(I + A) = \theta_{ab}(A).$$
Thus F_G is closed under multiplication, and hence, by finiteness, is a group.
2. Let $a \in F_G$, $a \neq 1$, and define η_a by $\eta_a(A) = A^{-1}\theta_a(A)$. As A commutes with $\theta_a(A)$ and $\theta_a(U) \subseteq U$, $\eta_a(U) \subseteq U$. As $\theta_a \mid_U$ is invertible, with inverse $\theta_{1/a} \mid_U$, it is a bijection $U \to U$. As
$$\eta_a(A) = A^{-1}(I + a(I + A)) = I + (1 + a)(I + A^{-1}) = \theta_{1+a}(A^{-1}),$$
$\eta_a \mid_U$ is a bijection $U \to U$ and thus $\theta_a \mid_U$ is an orthomorphism of U. To see that $\theta_a \mid_U$ acts on $C_U(B)$ for all $B \in S$, let $A \in C_U(B)$, $B \in S$. Then, as $\theta_a(A)$ is a polynomial in A, $\theta_a(A) \in C_U(B)$.
3. Let $a, b \in F_G \setminus \{1\}$, $a \neq b$, and let $A \in U$. Now
$$a\theta_b(A) + b\theta_a(A) = a(I + b(I + A)) + b(I + a(I + A)) = (a + b)I,$$
and so
$$\theta_a(A)^{-1}\theta_b(A) = (b/a)I + (1 + b/a)\theta_a(A)^{-1} = \theta_{1+b/a}(\theta_a(A)^{-1}).$$
It follows that the mapping $A \mapsto \theta_a(A)^{-1}\theta_b(A)$ is a bijection $U \to U$. Hence $\theta_a \mid_U \perp \theta_b \mid_U$. □

From Lemma 12.15, we obtain a lower bound for $\omega_S(U)$ when G is a subgroup of GL(n, q). Note that, if $q = 2$ and $G \subseteq$ GL(n, q), then $F_G = \{1\}$ and Lemma 12.15 yields no constructions of orthomorphisms of U.

Lemma 12.16. *If G is a subgroup of* GL(n, q), *q a power of* 2, *$q \neq 2$, then*
$$\omega_S(U) \geq |F_G| - 1.$$

Proof. By Lemma 12.15, $\{\theta_a|_U| \ a \in F_G, a \neq 1\}$ is a set of $|F_G| - 1$ pairwise orthogonal orthomorphisms of U, each of which acts on $C_U(B)$ for each $B \in S$, and so $\omega_S(U) \geq |F_G| - 1$. □

For the linear group GL(n, q), Lemma 12.16 yields a simple lower bound for $\omega_S(U)$.

Theorem 12.17. *For $G = \mathrm{GL}(n, q)$, q a power of 2, $q \neq 2$, $\omega_S(U) \geq q - 2$.*

Proof. If $G = \mathrm{GL}(n, q)$, then $F_G = \mathrm{GF}(q)^*$, and hence, by Lemma 12.16, $\omega_S(U) \geq q - 2$. □

12.3.3 A Lower Bound for $\omega_U(S)$

We will now derive lower bounds for $\omega_U(S)$ for the group GL(n, q), q a power of 2, $q \neq 2$: these bounds will be determined by the existence of pairwise orthogonal orthomorphisms of cyclic groups. Recall that, if F is an algebraic closure of GF(q), then any $n \times n$ matrix A over GF(q) is diagonalizable if and only if there exists a matrix $B \in \mathrm{GL}(n, F)$ for which BAB^{-1} is a diagonal matrix: the diagonal entries are the eigenvalues of A. Also for any $n \times n$ matrix A over GF(q), there exist a matrix $B \in \mathrm{GL}(n, F)$ for which $BAB^{-1} = \mathrm{diag}(J_1, \ldots, J_r)$, where, for $i = 1, \ldots, r$, J_i is the square matrix

$$\begin{pmatrix} \lambda_i & 1 & 0 & \ldots & 0 \\ 0 & \ddots & \ddots & \ddots & \vdots \\ \vdots & \ddots & \ddots & \ddots & 0 \\ \vdots & & \ddots & \ddots & 1 \\ 0 & \ldots & \ldots & 0 & \lambda_i \end{pmatrix},$$

with λ_is on the main diagonal, 1s just above the main diagonal, and 0s everywhere else. Recall that diag$(J_1, \ldots, J_r)$ is the *Jordan canonical form* for A, $J_1, \ldots, J_r$ are the *Jordan blocks* of A, and the Jordan canonical form for A is uniquely determined up to the order of the Jordan blocks. A is diagonalizable if and only if each of its Jordan blocks is of order 1. When q is even, the set of odd-order elements of GL(n, q) are precisely the set of diagonalizable matrices in GL(n, q).

Lemma 12.18. *If q is a power of 2 and S is the set of odd-order elements of GL(n, q), then $A \in S$ if and only if A is diagonalizable.*

Proof. Let F be an algebraic closure of GF(q), and let $W \in \mathrm{GL}(n, q)$. Then $W = BYB^{-1}$ for some $B, Y \in \mathrm{GL}(n, F)$, Y in Jordan canonical form. If Y is diagonal, then, as each nonzero element of F is of odd multiplicative order, Y, and hence W, is of odd multiplicative order. If Y is not diagonal, then Y contains, for some $k \geq 2$, a $k \times k$ Jordan block $\lambda I + N = (\lambda I)(I + (1/\lambda)N)$, where N has ones above the main diagonal and zeroes elsewhere. Now λI and $I + (1/\lambda)N$ commute, and it is easy to show that $N^r = O$ and $(I + (1/\lambda)N)^r = I$, where $r = 2^{k-1}$. It follows that $I + (1/\lambda)N$ is of multiplicative order a power of 2 and so Y, and hence W, is of even multiplicative order, hence the result. □

To determine lower bounds for $\omega_U(S)$, we will need to partition S. As with groups in Lemma 12.7, if $\{X_i \mid i \in I\}$ is a partition of S, and each X_i is closed under inverses, then $\omega(S) \geq \min\{\omega(X_i) \mid i \in I\}$. For $G = \mathrm{GL}(n, q)$, let us define an equivalence relation on the elements of S; we will say that $A, B \in S$ are *polynomially equivalent* over $\mathrm{GF}(q)$, written $A \equiv_p B$, if there exist polynomials p_1 and p_2 over $\mathrm{GF}(q)$, satisfying $\mathrm{p}_1(A) = B$ and $\mathrm{p}_2(B) = A$: it is routine to establish that this is indeed an equivalence relation. We will use $P_{n,q}(A)$ to denote the polynomial equivalence class in $\mathrm{GL}(n, q)$ of A over $\mathrm{GF}(q)$. If $A \in \mathrm{GL}(n, q)$ has multiplicative order m, then $A^{-1} = A^{m-1}$ and $\left(A^{-1}\right)^{m-1} = A$. It follows that each polynomial equivalence class is closed under inverses. In the partition of S, that we will use, each set in the partition will be a polynomial equivalence class over $\mathrm{GF}(q)$. The values of $\omega(X)$ for polynomial equivalence classes yields a bound for $\omega_U(S)$.

Lemma 12.19. *For $G = \mathrm{GL}(n, q)$, q a power of 2, $q \neq 2$,*

$$\omega_U(S) \geq \min\{\omega(P_{n,q}(A)) \mid A \in S\}.$$

Proof. Clearly $\omega(S) \geq \min\{\omega(P_{n,q}(A)) \mid A \in S\}$. To prove that this bound also holds for $\omega_U(S)$, we need to show that any orthomorphism of S, whose restriction to $P_{n,q}(A)$ is an orthomorphism, acts on $C_S(u)$ for each $u \in U$. This holds as, if $A \in S$ centralizes $B \in U$, and p is a polynomial over $\mathrm{GF}(q)$, then $\mathrm{p}(A)$ also centralizes B. □

Define $\mathrm{quot}_n(q)$ by

$$\mathrm{quot}_n(q) = \begin{cases} q - 1 & \text{if } n = 1, \\ (q^n - 1)/\mathrm{lcm}\{q^i - 1 \mid i \text{ divides } n,\ i \neq n\} & \text{if } n > 1, \end{cases}$$

and set $w_n(q) = \omega(\mathbb{Z}_{\mathrm{quot}_n(q)})$. Our lower bound for $\omega_U(S)$ will be the minimum of $w_1(q), \ldots, w_n(q)$.

The set of integers $\{q^k - 1 \mid k \text{ a positive integer}\}$, q a nontrivial prime power, plays an import role in our construction. This set forms a lattice under division which is isomorphic to the lattice of positive integers under division: the isomorphism is $q^k - 1 \mapsto k$. The greatest lower bound in each lattice is the greatest common divisor.

Lemma 12.20. *If q is a nontrivial prime power and π is a finite nonempty set of positive integers, then* $\gcd\{q^i - 1 \mid i \in \pi\} = q^{\gcd\{i \mid i \in \pi\}} - 1$.

Proof. Routine application of the Euclidean algorithm. □

However, the least upper bound in our lattice is not the lowest common multiple, as it is in the lattice of positive integers. As an example, $\text{lcm}\{2, 3\} = 6$, whereas $\text{lcm}\{q^2 - 1, q^3 - 1\} = (q^2 - 1)(q^3 - 1)/(q - 1)$, not $q^6 - 1$. The formula for the lowest common multiple is as follows.

Lemma 12.21. *If q is a nontrivial prime power, $n \geq 1$, and $k_1, \ldots, k_n \geq 1$, then*

$$\text{lcm}\{q^{k_1} - 1, \ldots, q^{k_n} - 1\} = \frac{\prod_{\pi \subseteq \{1,\ldots,n\}, |\pi|\ odd} \gcd\{q^{k_i} - 1 \mid i \in \pi\}}{\prod_{\pi \subseteq \{1,\ldots,n\}, |\pi|\ even, \pi \neq \emptyset} \gcd\{q^{k_i} - 1 \mid i \in \pi\}}.$$

Proof. Let p be a prime and let p^{s_i} be the largest power of p that divides $q^{k_i} - 1$. We will assume that $s_1 \leq s_2 \leq \cdots \leq s_n$. The largest power of p that divides the left-hand side of the equality is p^{s_n}, and the largest power of p dividing $\gcd\{q^{k_i} - 1 \mid i \in \pi\}$ is p^{s_j}, where $j = \min\{i \mid i \in \pi\}$. Thus the largest power of p dividing the right-hand side of the equality is

$$\prod_{i=1}^{n} \left(p^{s_i}\right)^{|\{\pi | \pi \subseteq \{i+1,\ldots,n\}, |\pi|\ \text{even}\}| - |\{\pi | \pi \subseteq \{i+1,\ldots,n\}, |\pi|\ \text{odd}\}|}.$$

Now

$$\left(p^{s_i}\right)^{|\{\pi | \pi \subseteq \{i+1,\ldots,n\}, |\pi|\ \text{even}\}| - |\{\pi | \pi \subseteq \{i+1,\ldots,n\}, |\pi|\ \text{odd}\}|} = \begin{cases} 1 & \text{if } i < n; \\ p^{s_n} & \text{if } i = n, \end{cases}$$

and so

$$\prod_{i=1}^{n} \left(p^{s_i}\right)^{|\{\pi | \pi \subseteq \{i+1,\ldots,n\}, |\pi|\ \text{even}\}| - |\{\pi | \pi \subseteq \{i+1,\ldots,n\}, |\pi|\ \text{odd}\}|} = p^{s_n}.$$

As p is arbitrary, the result follows. □

Lemma 12.22. *If $n \geq 1$, then*

$$\prod_{m|n} \text{quot}_m(q) = q^n - 1.$$

Proof. This is clearly true if $n = 1$. If $n = p^r$, p a prime, then, as $\text{quot}_{p^i}(q) = (q^{p^i} - 1)/(q^{p^{i-1}} - 1)$,

$$\prod_{m|n} \text{quot}_m(q) = (q - 1) \prod_{i=1}^{r} \frac{q^{p^i} - 1}{q^{p^{i-1}} - 1} = q^n - 1.$$

Let us assume that the result holds for all proper divisors of n. Let $p_1, \ldots, p_r$ be the distinct prime divisors of n and, for $\pi \subseteq \{1, \ldots, r\}$, set $t_\pi = n/\prod_{i\in\pi} p_i$.
Then

$$\begin{aligned}
\mathrm{quot}_n(q) &= \frac{(q^n-1)}{\mathrm{lcm}\{q^i-1 \mid i \neq n, i \mid n\}} \\
&= \frac{(q^n-1)}{\mathrm{lcm}\{q^{t\{i\}}-1 \mid i=1,\ldots,r\}} \\
&= \frac{(q^n-1)\prod_{\pi\subseteq\{1,\ldots,r\},|\pi|\text{ even},\pi\neq\emptyset}(q^{t_\pi}-1)}{\prod_{\pi\subseteq\{1,\ldots,r\},|\pi|\text{ odd}}(q^{t_\pi}-1)}.
\end{aligned}$$

Now

$$\begin{aligned}
\prod_{m|n}\mathrm{quot}_m(q) &= \mathrm{quot}_n(q)\frac{\prod_{\pi\subseteq\{1,\ldots,r\},|\pi|\text{ odd}}\prod_{m|t_\pi}\mathrm{quot}_m(q)}{\prod_{\pi\subseteq\{1,\ldots,r\},|\pi|\text{ even},\pi\neq\emptyset}\prod_{m|t_\pi}\mathrm{quot}_m(q)} \\
&= \mathrm{quot}_n(q)\frac{\prod_{\pi\subseteq\{1,\ldots,r\},|\pi|\text{ odd}}(q^{t_\pi}-1)}{\prod_{\pi\subseteq\{1,\ldots,r\},|\pi|\text{ even},\pi\neq\emptyset}(q^{t_\pi}-1)} \\
&= q^n-1.
\end{aligned}$$

The result then follows by induction. □

It should be noted that Lemmas 12.20, 12.21, and 12.22 hold for odd q, not just even q. We need to modify the quotient group construction given in Corollary 1.39 to construct pairwise orthogonal sets of orthomorphisms of subsets of groups. For this purpose we will need to use normalized orthomorphisms. Recall from Section 1.3 that an orthomorphism θ of a group G is normalized if $\theta(1) = 1$, that if θ is not normalized, then the mapping θ' defined by $\theta'(x) = \theta(x)\theta(1)^{-1}$ is a normalized orthomorphism of G, the normalization of θ, and that normalization preserves orthogonality.

Lemma 12.23. *If H is a subgroup of an abelian group G and $t = \min\{\omega(H), \omega(G/H)\}$, then $\omega(G \setminus H) \geq t$.*

Proof. Let $1 = x_1, \ldots, x_r$ be a system of distinct coset representatives for H in G. Let $\theta_1, \ldots, \theta_t$ be a set of pairwise orthogonal orthomorphisms of H and $\phi_1, \ldots, \phi_t$ a set of pairwise orthogonal normalized orthomorphisms of G/H. For $i = 1, \ldots, t$, define $\phi_i' : \{x_1, \ldots, x_r\} \to \{x_1, \ldots, x_r\}$ by $\phi_i'(x_u) = x_v$ if $\phi_i(x_u H) = x_v H$.

We claim that the mappings $\gamma_i : x_u h \mapsto \phi_i'(x_u)\theta_i(h)$, $u = 1, \ldots, r$, $h \in H$ form a set of t pairwise orthogonal orthomorphisms of G. If $\gamma_i(x_u h) = \gamma_i(x_v k)$, then $\phi_i'(x_u) = \phi_i'(x_v)$ which implies that $x_u = x_v$, which implies that $\theta_i(h) = \theta_i(k)$, which implies that $h = k$. Hence γ_i is one to one, which, by finiteness, implies that γ_i is a bijection. If $(x_u h)^{-1}\gamma_i(x_u h) = (x_v k)^{-1}\gamma_i(x_v k)$, then $(x_u^{-1}\phi_i'(x_u)(h^{-1}\theta_i(h)) = (x_v^{-1}\phi_i'(x_v)(k^{-1}\theta_i(k))$, which implies that $x_u^{-1}\phi_i'(x_u) = x_v^{-1}\phi_i'(x_v)$ and $h^{-1}\theta_i(h) = k^{-1}\theta_i(k)$, which implies that $x_u = x_v$ and $h = k$. Hence, the mapping $x_u h \mapsto (x_u h)^{-1}\gamma_i(x_u h)$ is one to one, which, by finiteness, implies that the mapping $x_u h \mapsto (x_u h)^{-1}\gamma_i(x_u h)$ is a bijection and, hence, that

γ_i is an orthomorphism. If $i \neq j$ and $\gamma_i(x_u h)^{-1}\gamma_j(x_u h) = \gamma_i(x_v k)^{-1}\gamma_j(x_v k)$, then $(\phi_i(x_u)^{-1}\phi_j(x_u))(\theta_i(h)^{-1}\theta_j(h)) = (\phi_i(x_v)^{-1}\phi_j(x_v))(\theta_i(k)^{-1}\theta_j(k))$, which implies that $\phi_i(x_u)^{-1}\phi_j(x_u) = \phi_i(x_v)^{-1}\phi_j(x_v)$ and $\theta_i(h)^{-1}\theta_j(h) = \theta_i(k)^{-1}\theta_j(k)$, which implies that $x_u = x_v$ and $h = k$. Hence, the mapping $x_u h \mapsto \gamma(x_u h)^{-1}\gamma_j(x_u h)$ is one to one, which, by finiteness, implies that the mapping $x_u h \mapsto \gamma_i(x_u h)^{-1}\gamma_j(x_u h)$ is a bijection and, hence, that $\gamma_i \perp \gamma_j$. This proves the claim.

As $\phi_i'(1) = 1$ for all $i = 1, \ldots, t$, $\gamma_1, \ldots, \gamma_r$ fix both H and $G \setminus H$, and the restriction of $\gamma_1, \ldots, \gamma_r$ to $G \setminus H$ is a set of t pairwise orthogonal orthomorphisms of $G \setminus H$. □

A similar result holds for direct products.

Lemma 12.24. *If $G = H \times K$ and $t = \min\{\omega(H), \omega(K)\}$, then*

$$\omega(G \setminus ((H \times \{1\}) \cup (\{1\} \times K))) \geq t.$$

Proof. If θ is an orthomorphism of H and ϕ is an orthomorphism of K, then, by Theorem 8.1, the mapping $\theta \times \phi$, defined by $(\theta \times \phi)(h, k) = (\theta(h), \phi(k))$, is an orthomorphism of $H \times K$. As $\omega(H) \geq t$, there exist a pairwise orthogonal set of normalized orthomorphisms $\theta_1, \ldots, \theta_t$ of H, and, as $\omega(K) \geq t$, there exist a pairwise orthogonal set of normalized orthomorphisms $\phi_1, \ldots, \phi_t$ of K. By Theorem 8.1, $\theta_1 \times \phi_1, \ldots, \theta_t \times \phi_t$ is a pairwise orthogonal set of normalized orthomorphisms of $H \times K$, which fix $\{1\}$, $H \times \{1\}$, $\{1\} \times K$, and $G \setminus ((H \times \{1\}) \cup (\{1\} \times K))$, and induces a pairwise orthogonal set of t orthomorphisms of $G \setminus ((H \times \{1\}) \cup (\{1\} \times K))$. Hence the result. □

We will make much use of the following more general result.

Lemma 12.25. *If H, K are subgroups of an abelian group G, $G = HK$, and*

$$t = \min\{\omega(H \cap K), \omega(H/(H \cap K)), \omega(K/(H \cap K))\},$$

then $\omega(H \setminus (H \cap K)) \geq t$, $\omega(K \setminus (H \cap K)) \geq t$, and $\omega(G \setminus (H \cup K)) \geq t$.

Proof. We will use the constructions of Lemmas 12.23 and 12.24 to construct a pairwise orthogonal set of t orthomorphisms of G that induces a pairwise orthogonal set of t orthomorphisms of $H \setminus (H \cap K)$, a pairwise orthogonal set of t orthomorphisms of $K \setminus (H \cap K)$, and a pairwise orthogonal set of t orthomorphisms of $G \setminus (H \cup K)$.

Let $\alpha\colon G \to G/(H \cap K)$ be the canonical homomorphism, let $\theta_1, \ldots, \theta_t$ be a pairwise orthogonal set of orthomorphisms of $\alpha(H)$, and let $\phi_1, \ldots, \phi_t$ be a pairwise orthogonal set of orthomorphisms of $\alpha(K)$. By Lemma 12.23, $\theta_1 \times \phi_1, \ldots, \theta_t \times \phi_t$ is a pairwise orthogonal set of orthomorphisms of $\alpha(G)$ that fixes $\alpha(H)$, $\alpha(K)$, and $\alpha(H \cap K)$. Let $|G/(H \cap K)| = r$, let $1 = x_1, \ldots, x_r$ be a system of distinct coset representatives for $H \cap K$ in G, and let $\gamma_1, \ldots, \gamma_t$ be a set of pairwise orthogonal orthomorphisms of $H \cap K$. For $i = 1, \ldots, t$, define $(\theta_i \times \phi_i)'\colon \{x_1, \ldots, x_r\} \to$

$\{x_1, \ldots, x_r\}$ by $(\theta_i \times \phi_i)'(x_u) = x_v$ if $(\theta_i \times \phi_i)(x_u(H \cap K)) = x_v(H \cap K)$. By Lemma 12.23, the mappings $\nu_i : x_u g \mapsto (\theta_i \times \phi_i)'(x_u)\gamma_i(g)$, $u = 1, \ldots, r$, $g \in H \cap K$, form a set of t pairwise orthogonal orthomorphisms of G. By construction, each ν_i, $i = 1, \ldots, t$, fixes both H and K and, hence, $G \setminus (H \cap K)$. It follows that $\nu_1, \ldots, \nu_t$ induces a set of t pairwise orthogonal orthomorphisms of $G \setminus (H \cap K)$, and so $\omega(G \setminus (H \cap K)) \geq t$.

As $H \cap K$ is a normal subgroup of both H and K, $\omega(H \cap K) \geq t$, $\omega(H/(H \cap K)) \geq t$, and $\omega(K/(H \cap K)) \geq t$, by Lemma 12.23, $\omega(H \setminus (H \cap K)) \geq t$, and $\omega(K \setminus (H \cap K)) \geq t$. □

Let F be an algebraic closure of GF(q), and let F^* denote the multiplicative group of F. We will use E_m to denote the unique subfield of F of order q^m, and E_m^* to denote the multiplicative group of E_m. For $m > 1$, $m = p_1^{r_1} p_2^{r_2} \ldots p_k^{r_k}$, for some $p_1, \ldots, p_k$ distinct primes. Set $m_j = m/p_j$, $j = 1, \ldots, k$, $m_{i,j} = m/(p_i p_j)$, $i, j = 1, \ldots, k$, $i \neq j$, and, more generally, for $\pi \subseteq \{1, \ldots, k\}$, $m_\pi = m/\prod_{i \in \pi} p_i$. As E_l is a maximal subfield of E_m if and only if m/l is a prime, $E_{m_1}, \ldots, E_{m_k}$ are the maximal subfields of E_m, and, as E_m^* is cyclic, $E_{m_1}^* \cdots E_{m_k}^*$ and $E_m^*/E_{m_1}^* \cdots E_{m_k}^*$ are cyclic groups. Further

$$
\begin{aligned}
|E_{m_1}^* \cdots E_{m_k}^*| &= \operatorname{lcm}\{|E_{m_i}^*| \mid i = 1, \ldots, k\} \\
&= \operatorname{lcm}\{q^{m_i} - 1 \mid i = 1, \ldots, k\} \\
&= \operatorname{lcm}\{q^i - 1 \mid i \text{ divides } m\}.
\end{aligned}
$$

It follows that $E_m^*/(E_{m_1}^* \cdots E_{m_k}^*) \cong \mathbb{Z}_{\mathrm{quot}_m(q)}$.

Lemma 12.26. *If $m > 1$, and $m = p_1^{r_1} p_2^{r_2} \ldots p_k^{r_k}$, for some $p_1, \ldots, p_k$ distinct primes, then*

$$
\omega(E_{m_1}^* \cdots E_{m_k}^*) \geq \min\{w_i(q) \mid i \text{ divides } m\} \text{ and}
$$
$$
\omega(E_m^* \setminus (E_{m_1}^* \cdots E_{m_k}^*)) \geq \min\{w_i(q) \mid i \text{ divides } m\}.
$$

Proof. By Lemma 12.22,

$$
\prod_{i|m, i \neq m} \mathrm{quot}_i(q) = \frac{q^m - 1}{\mathrm{quot}_m(q)} = |E_{m_1}^* \cdots E_{m_k}^*|.
$$

Let $d_1, \ldots, d_l$ be the divisors of m other than m. Then, as $E_{m_1}^* \cdots E_{m_k}^*$ is a cyclic group, there exist subgroups $\{1\} = G_0 < G_1 < \cdots < G_l = E_{m_1}^* \cdots E_{m_k}^*$ satisfying G_i/G_{i-1} cyclic of order $quot_{d_i}(q)$. It follows, by repeated application of Lemma 12.23, that $\omega(E_{m_1}^* \cdots E_{m_k}^*) \geq \min\{w_i(q) \mid i \text{ divides } m, i \neq m\} \geq \min\{w_i(q) \mid i \text{ divides } m\}$. As

$$
\omega(E_m^*/(E_{m_1}^* \cdots E_{m_k}^*)) = w_m(q) \geq \min\{w_i(q) \mid i \text{ divides } m\},
$$

by Lemma 12.23, $\omega(E_m^* \setminus (E_{m_1}^* \cdots E_{m_k}^*)) \geq \min\{w_i(q) \mid i \text{ divides } m\}$. □

Lemma 12.27. *Let q be a power of 2 and let m be a positive integer. Then*

$$\omega(E_m^* \setminus \cup\{E_k^* \mid k \text{ divides } m, k \neq m\}) \geq \min\{w_k(q) \mid k|m\}.$$

Proof. Suppose this to be false and let m be the smallest positive integer for which this is false. Set $t = \min\{w_k(q) \mid k|m\}$. For $m = 1$, $E_m^* \setminus \cup\{E_k^* \mid k \text{ divides } m, k \neq m\} = \mathrm{GF}(q)^*$ and $\omega(\mathrm{GF}(q)^*) = \omega_1(q) \geq t$. Hence $m > 1$ and $m = p_1^{r_1} p_2^{r_2} \ldots p_k^{r_k}$, $p_1, \ldots, p_k$ distinct primes. If $k = 1$, then $E_m^* \setminus \cup\{E_k^* \mid k \text{ divides } m, k \neq m\} = E_m^* \setminus E_{m_1}^*$. Now $E_m^*/E_{m_1}^* \cong \mathbb{Z}_{\mathrm{quot}_m(q)}$, and so $\omega(E_m^*/E_{m_1}^*) = w_m(q) \geq t$ and, by the minimality of m, $\omega(E_{m_1}^*) \geq t$. Hence, by Lemma 12.23, $\omega(E_m^* \setminus E_{m_1}^*) \geq t$. It follows that $k > 1$.

Now, if l divides m and $l \neq m$, then $E_l^* \subseteq E_{m_i}^*$ for some $i = 1, \ldots, k$. It follows that

$$E_m^* \setminus \cup\{E_l^* \mid l \text{ divides } m, l \neq m\} = E_m^* \setminus (E_{m_1}^* \cup \cdots \cup E_{m_k}^*).$$

As $E_{m_1}^* \ldots E_{m_k}^*$ contains $E_{m_i}^*$ for $i = 1, \ldots, k$, the elements of $E_m^* \setminus (E_{m_1}^* \cup \cdots \cup E_{m_k}^*)$ can be partitioned into those that are contained in $E_{m_1}^* \ldots E_{m_k}^*$ and those that are not. Hence

$$E_m^* \setminus (E_{m_1}^* \cup \cdots \cup E_{m_k}^*) = (E_m^* \setminus (E_{m_1}^* \cdots E_{m_k}^*)) \cup ((E_{m_1}^* \cdots E_{m_k}^*) \setminus (E_{m_1}^* \cup \cdots \cup E_{m_k}^*)).$$

By Lemma 12.26, $\omega(E_m^* \setminus (E_{m_1}^* \cdots E_{m_k}^*)) \geq t$. We will show that

$$\omega((E_{m_1}^* \cdots E_{m_k}^*) \setminus (E_{m_1}^* \cup \cdots \cup E_{m_k}^*)) \geq t, \tag{12.1}$$

by induction on k.

Inequality (12.1) is true for $k = 1$ as $\omega(\{1\}) = \infty$. Let us assume that inequality (12.1) is true for $k = 1, \ldots, s$ and that $k = s + 1$. Now

$$\begin{aligned}
&(E_{m_1}^* \cdots E_{m_k}^*) \setminus (E_{m_1}^* \cup \cdots \cup E_{m_k}^*) \\
&= (((E_{m_1}^* \cdots E_{m_{k-1}}^*) E_{m_k}^*) \setminus ((E_{m_1}^* \cdots E_{m_{k-1}}^*) \cup E_{m_k}^*)) \\
&\cup (((E_{m_1}^* \cdots E_{m_{k-1}}^*) \cup E_{m_k}^*) \setminus ((E_{m_1}^* \cup \cdots \cup E_{m_{k-1}}^*) \cup E_{m_k}^*)) \\
&= (((E_{m_1}^* \cdots E_{m_{k-1}}^*) E_{m_k}^*) \setminus ((E_{m_1}^* \cdots E_{m_{k-1}}^*) \cup E_{m_k}^*)) \\
&\cup ((E_{m_1}^* \cdots E_{m_{k-1}}^*) \setminus (E_{m_1}^* \cup \cdots \cup E_{m_{k-1}}^*)),
\end{aligned}$$

and by the inductive hypothesis $\omega((E_{m_1}^* \cdots E_{m_{k-1}}^*) \setminus (E_{m_1}^* \cup \cdots \cup E_{m_{k-1}}^*)) \geq t$.

As $E_{m_{i,k}}^* \subseteq E_{m_k}^*$ for $i = 1, \ldots, k - 1$

$$E_{m_{1,k}}^* \cdots E_{m_{k-1,k}}^* \subseteq E_{m_1}^* \cdots E_{m_{k-1}}^* \cap E_{m_k}^*.$$

Using Lemmas 12.20 and 12.21, we see that

$$\begin{aligned}
|E^*_{m_1} \cdots E^*_{m_{k-1}} \cap E^*_k| &= \gcd(\mathrm{lcm}(q^{m_1} - 1, \ldots, q^{m_{k-1}} - 1), q^{m_k} - 1) \\
&= \frac{(q^{m_k} - 1)\mathrm{lcm}(q^{m_1} - 1, \ldots, q^{m_{k-1}} - 1)}{\mathrm{lcm}(q^{m_1} - 1, \ldots, q^{m_k} - 1)} \\
&= \frac{(q^{m_k} - 1) \prod_{\pi \in \{1,\ldots,k-1\}, |\pi| \text{ odd}} \gcd\{q^{m_i} - 1 \mid i \in \pi\}}{\prod_{\pi \in \{1,\ldots,k-1\}, |\pi| \text{ even}, \pi \neq \emptyset} \gcd\{q^{m_i} - 1 \mid i \in \pi\}} \\
&\quad \times \frac{\prod_{\pi \in \{1,\ldots,k\}, |\pi| \text{ even}, \pi \neq \emptyset} \gcd\{q^{m_i} - 1 \mid i \in \pi\}}{\prod_{\pi \in \{1,\ldots,k\}, |\pi| \text{ odd}} \gcd\{q^{m_i} - 1 \mid i \in \pi\}} \\
&= \frac{(q^{m_k} - 1) \prod_{\pi \in \{1,\ldots,k-1\}, |\pi| \text{ odd}} (q^{m_\pi} - 1)}{\prod_{\pi \in \{1,\ldots,k-1\}, |\pi| \text{ even}, \pi \neq \emptyset} (q^{m_\pi} - 1)} \\
&\quad \times \frac{\prod_{\pi \in \{1,\ldots,k\}, |\pi| \text{ even}, \pi \neq \emptyset} (q^{m_\pi} - 1)}{\prod_{\pi \in \{1,\ldots,k\}, |\pi| \text{ odd}} (q^{m_\pi} - 1)} \\
&= \frac{\prod_{\pi \in \{1,\ldots,k-1\}, |\pi| \text{ odd}} (q^{m_{\pi \cup \{k\}}} - 1)}{\prod_{\pi \in \{1,\ldots,k-1\}, |\pi| \text{ even}, \pi \neq \emptyset} (q^{m_{\pi \cup \{k\}}} - 1)} \\
&= \frac{\prod_{\pi \in \{1,\ldots,k-1\}, |\pi| \text{ odd}} \gcd\{q^{m_{i,k}} - 1 \mid i \in \pi\}}{\prod_{\pi \in \{1,\ldots,k-1\}, |\pi| \text{ even}, \pi \neq \emptyset} \gcd\{q^{m_{i,k}} - 1 \mid i \in \pi\}} \\
&= \mathrm{lcm}\{q^{m_{1,k}} - 1, \ldots, q^{m_{k-1,k}} - 1\} = |E^*_{m_{1,k}} \cdots E^*_{m_{k-1,k}}|.
\end{aligned}$$

Hence $E^*_{m_1} \cdots E^*_{m_{k-1}} \cap E^*_{m_k} = E^*_{m_{1,k}} \cdots E^*_{m_{k-1,k}}$.

By Lemma 12.26, $\omega(E^*_{m_1} \cdots E^*_{m_{k-1}} \cap E^*_k) = \omega(E^*_{m_{1,k}} \cdots E^*_{m_{k-1,k}}) \geq t$. By a similar argument, $\omega(E^*_{m_1} \cdots E^*_{m_{k-1}}/(E^*_{m_1} \cdots E^*_{m_{k-1}} \cap E^*_k)) \geq t$. Also

$$\omega(E^*_k/(E^*_{m_1} \cdots E^*_{m_{k-1}} \cap E^*_k)) = \omega(E^*_k/(E^*_{m_{1,k}} \cdots E^*_{m_{k-1,k}})) = w_k(q) \geq t.$$

Hence, by Lemma 12.25, $\omega((E^*_{m_1} \cdots E^*_{m_k}) \setminus (E^*_{m_1} \cup \cdots \cup E^*_{m_k})) \geq t$, and inequality (12.1) follows by induction. As $E_m \setminus \cup\{E_k \mid k \text{ divides } m, k \neq m\}$ is the disjoint union of $E^*_m \setminus (E^*_{m_1} \cdots E^*_{m_k})$ and $(E^*_{m_1} \cdots E^*_{m_k}) \setminus (E^*_{m_1} \cup \cdots \cup E^*_{m_k})$, $\omega(E^*_m \setminus (E^*_{m_1} \cdots E^*_{m_k})) \geq t$, and $\omega((E^*_{m_1} \cdots E^*_{m_k}) \setminus (E^*_{m_1} \cup \cdots \cup E^*_{m_k})) \geq t$, $\omega(E^*_m \setminus \cup\{E^*_k \mid k \text{ divides } m, k \neq m\}) \geq t$, a contradiction from which the result follows. □

We are now in a position to establish lower bounds for $\omega(S)$ when $G = \mathrm{GL}(n, q)$, q a power of 2, $q \neq 2$.

Theorem 12.28. *For $G = \mathrm{GL}(n, q)$, q a power of 2, $q \neq 2$, $n > 1$,*

$$\omega_U(S) \geq \min\{w_1(q), \ldots, w_n(q)\}.$$

Proof. Let F be an algebraic closure of GF(q) and let S be the set of odd-order elements of GL(n, q). If $A \in S$ and $B = XAX^{-1}$, $X \in$ GL(n, F), then $P_{n,q}(B) = XP_{n,q}(A)X^{-1}$. If θ is an orthomorphism of $P_{n,q}(A)$, then the mapping θ': $C \mapsto X\theta(X^{-1}CX)X^{-1}$ is an orthomorphism of $P_{n,q}(B)$, and the mapping $\theta \mapsto \theta'$ induces a bijection between the set of orthomorphisms of $P_{n,q}(A)$ and the set of orthomorphisms of $P_{n,q}(B)$: further, $\theta' \perp \phi'$ if and only if $\theta \perp \phi$. Hence $\omega(P_{n,q}(A)) = \omega(P_{n,q}(B))$ and, as A is diagonalizable by Lemma 12.18, it is thus sufficient to determine $\omega(P_{n,q}(A))$ for diagonal matrices $A \in$ GL(n, F) only.

Let $A \in S$ have characteristic polynomial $\prod_{i=1}^{r} \mathrm{f}_i^{k_i}(\lambda)$, where $\mathrm{f}_1, \ldots, \mathrm{f}_r$ are distinct irreducible polynomials over GF(q), f_i of degree m_i. Let D_i be a diagonal matrix of order m_i with nonzero entries the zeroes of f_i, and let $A_i = \mathrm{diag}(D_i, \ldots, D_i)$ consist of k_i copies of D_i. Then $XAX^{-1} = \mathrm{diag}(A_1, \ldots, A_r)$, for some $X \in GL(n, F)$, and so we are free to take A to be $\mathrm{diag}(A_1, \ldots, A_r)$. Clearly, if $\mathrm{diag}(B_1, \ldots, B_r) \equiv_p A$, then $B_i \equiv_p A_i$ for $i = 1, \ldots, r$. Now, if $B_i \equiv_p A_i$ for $i = 1, \ldots, r$, then there exist polynomials $\mathrm{p}_{i,1}$ and $\mathrm{p}_{i,2}$ over GF(q), for $i = 1, \ldots, r$, for which $\mathrm{p}_{i,1}(A_i) = B_i$ and $\mathrm{p}_{i,2}(B_i) = A_i$. By the Chinese remainder theorem, there exist polynomials p_1 and p_2 over GF(q) satisfying $\mathrm{p}_1 \equiv \mathrm{p}_{i,1} \pmod{\mathrm{f}_i}$ and $\mathrm{p}_2 \equiv \mathrm{p}_{i,2} \pmod{\mathrm{f}_i}$ for $i = 1, \ldots, r$. Then $\mathrm{p}_1(A) = \mathrm{diag}(B_1, \ldots, B_r)$ and $\mathrm{p}_2(\mathrm{diag}(B_1, \ldots, B_r)) = A$. Hence

$$P_{n,q}(A) = \{\mathrm{diag}(B_1, \ldots, B_r) \mid B_i \in P_{k_i m_i,q}(A_i), i = 1, \ldots, r\}.$$

If $\theta_{i,1}, \ldots, \theta_{i,t}$ is a pairwise orthogonal set of orthomorphisms of $P_{k_i m_i,q}(A_i)$ for $i = 1, \ldots, r$, then $\theta'_1, \ldots, \theta'_t$, defined by

$$\theta'_j(\mathrm{diag}(B_1, \ldots, B_r)) = \mathrm{diag}(\theta_{1,j}(B_1), \ldots, \theta_{r,j}(B_r))$$

for $j = 1, \ldots, t$, is a pairwise orthogonal set of orthomorphisms of $P_{n,q}(A)$. Also, if $\theta_1, \ldots, \theta_t$ is a pairwise orthogonal set of orthomorphisms of $P_{m_i,q}(D_i)$, then $\theta'_1, \ldots, \theta'_t$, defined by

$$\theta'_j(A_i) = \theta'_j(\mathrm{diag}(D_i, \ldots, D_i)) = \mathrm{diag}(\theta_j(D_i), \ldots, \theta_j(D_i))$$

for $j = 1, \ldots, t$, is a pairwise orthogonal set of orthomorphisms of $P_{m_i,q}(A_i)$. Hence

$$\begin{aligned}\omega(P_{n,q}(A)) &\geq \min\{\omega(P_{k_i m_i,q}(A_i)) \mid i = 1, \ldots, r\}\\ &\geq \min\{\omega(P_{m_i,q}(D_i)) \mid i = 1, \ldots, r\}\end{aligned}$$

and it suffices to restrict ourselves to odd-order elements of GL(m, q), $m \leq n$, whose minimal polynomials over GF(q) are irreducible.

Let $t = \min\{w_1(q), \ldots, w_n(q)\}$. To complete the proof, we need to show that $\omega(P_{m,q}(A)) \geq t$ for any odd-order $A \in$ GL(m, q), $m \leq n$, that has an irreducible minimal polynomial of degree m. Let $A \in$ GL(m, q) be of odd order and have an irreducible minimal polynomial over GF(q) of degree m. Set $E = \{\mathrm{p}(A) \mid \mathrm{p}$ a polynomial over GF$(q)\}$. Under addition and multiplication, E is a field of degree m over GF(q). For k a divisor of m, E_k will denote the unique subfield of E of degree k over GF(q), and for $B \in E$, $d(B)$, the *degree* of B, will be defined by $d(B) = \min\{k \mid B \in E_k\}$. If $B, C \in E$, $B, C \neq 0$, then $B \equiv_p C$ if and only if $d(B) = d(C)$ as, if $d(B) = k$, then $E_k = \{\mathrm{p}(B) \mid \mathrm{p}$ a polynomial over GF$(q)\}$. In particular, the elements of GL(m, q), that are polynomially equivalent to A, are precisely the elements of E of degree m, or, equivalently, if $m > 1$ the elements of E that are not contained in any proper subfield of E, or, if $m = 1$, the elements of E^*, the nonzero elements of E. By Lemma 12.27, $\omega(P_{m,q}(A)) \geq t$.

As S is partitioned into polynomial equivalence classes, the result follows from Lemma 12.19. □

12.3.4 Lower Bounds for $\omega(G)$

Using the results of Sections 12.2 and 12.3, we can establish lower bounds for $\omega(\mathrm{GL}(n, q))$, q even, $q \neq 2$.

Theorem 12.29. *If q is a power of 2, $q \neq 2$, then*

$$\omega(\mathrm{GL}(n, q)) \geq \min\{w_1(q), \ldots, w_n(q)\}.$$

Proof. By Theorem 12.14, $\omega(\mathrm{GL}(n, q)) \geq \min\{\omega_S(U), \omega_U(S)\}$. By Theorem 12.17, $\omega_S(U) \geq q-2$, and by Theorem 12.28, $\omega_U(S) \geq \min\{w_1(q), \ldots, w_n(q)\}$. It follows that

$$\begin{aligned} \omega(\mathrm{GL}(n, q)) &\geq \min\{\omega_S(U), \omega_U(S)\} \\ &\geq \min\{q-2, w_1(q), \ldots, w_n(q)\} \\ &\geq \min\{w_1(q), \ldots, w_n(q)\}, \end{aligned}$$

as $w_1(q) < q - 2$. □

If G is a nontrivial group of odd order, then $\omega(G) \geq 1$: this is a consequence of Corollary 3.6. As each quot$_n(q)$ is odd and greater than one, the following is an immediate corollary of Theorem 12.29.

Corollary 12.30. *If q is a power of 2, $q \neq 2$, then $\omega(\mathrm{GL}(n, q)) \geq 1$.*

While Corollary 12.30 is a consequence of the Hall-Paige conjecture, its proof precedes the proof of the Hall-Paige conjecture, having been proved by Dalla Volta and Gavioli [83] in 1993: their proof was described in Section 5.3.

The functions quot$_n(q)$ are polynomials in q: the polynomials quot$_n(q)$ for small n are given in Table 12.2, and the values of quot$_n(q)$ for small n and q are given in Table 12.3.

Table 12.2 The polynomials $\text{quot}_n(q)$ for small n

n	$\text{quot}_n(q)$	n	$\text{quot}_n(q)$
1	$q - 1$	6	$q^2 - q + 1$
2	$q + 1$	7	$q^6 + q^5 + q^4 + q^3 + q^2 + q + 1$
3	$q^2 + q + 1$	8	$q^4 + 1$
4	$q^2 + 1$	9	$q^6 + q^3 + 1$
5	$q^4 + q^3 + q^2 + q + 1$	10	$q^4 - q^3 + q^2 - q + 1$

Table 12.3 Values $\text{quot}_n(q)$ for small n and q

$q \backslash n$	1	2	3	4	5
4	3	5	21	17	341
8	7	9	73	65	4,681
16	15	17	273	257	69,905
32	31	33	1,057	1,025	1,082,401
64	63	65	4,161	4,097	17,043,521
$q \backslash n$	6	7	8	9	10
4	13	5,461	257	4,161	205
8	57	299,593	4,097	262,657	3,641
16	241	17,895,697	65,537	16,781,313	61,681
32	993	1,108,378,657	1,048,577	1,073,774,593	1,016,801
64	4,033	69,810,262,081	16,777,217	68,719,738,881	16,519,105

To apply Theorem 12.29, we need to know more about known lower bounds for $w_n(q) = \omega(\mathbb{Z}_{\text{quot}_n(q)})$. General bounds for $\omega(\mathbb{Z}_v)$, v odd, $v > 1$, were given in Chapters 8 and 11, and lower bounds for small v will be given in Chapter 13. These results are summarized in Lemmas 12.31, 12.32, and 12.33. Lemma 12.31 gives known values and bounds for small v.

Lemma 12.31. *$\omega(\mathbb{Z}_9) = 1$, $\omega(\mathbb{Z}_{15}) = 3$, $\omega(\mathbb{Z}_{21}) \geq 4$, $\omega(\mathbb{Z}_{33}) \geq 4$, $\omega(\mathbb{Z}_{35}) \geq 4$, $\omega(\mathbb{Z}_{39}) \geq 4$, $\omega(\mathbb{Z}_{51}) \geq 4$, and $\omega(\mathbb{Z}_{55}) \geq 5$.*

Proof. For the lower bounds for $\omega(\mathbb{Z}_9)$, $\omega(\mathbb{Z}_{15})$, and $\omega(\mathbb{Z}_{21})$ see Table 13.1. For the lower bounds for $\omega(\mathbb{Z}_{33})$, $\omega(\mathbb{Z}_{35})$, $\omega(\mathbb{Z}_{39})$, $\omega(\mathbb{Z}_{51})$, and $\omega(\mathbb{Z}_{55})$ see Table 13.14. □

Some general lower bounds are given in Lemma 12.32.

Lemma 12.32.

1. *If v is a prime, then $\omega(\mathbb{Z}_v) = v - 2$.*
2. *If $v = st$, then $\omega(\mathbb{Z}_v) \geq \min\{\omega(\mathbb{Z}_s), \omega(\mathbb{Z}_t)\}$.*
3. *If p is the smallest prime divisor of v, then $\omega(\mathbb{Z}_v) \geq p - 2$.*
4. *If v is odd, $v \neq 3m$ or $9m$, m not divisible by 3, then $\omega(\mathbb{Z}_v) \geq 3$.*
5. *If $v = 3m$, $m > 1$, m odd and not divisible by 3, then $\omega(\mathbb{Z}_v) \geq 3$.*

Proof. For (1) and (3) see Corollary 8.15; (2) is a consequence of Corollary 1.39; for (4) see Theorem 8.14; and for (5) see Corollary 11.50. □

Additional lower bounds were obtained by Jungnickel ([201] and [202]) in 1980 and 1981, using difference sets and difference families.

Lemma 12.33. *If q is a prime power, then the following hold.*

1. $\omega(\mathbb{Z}_{q^2+q+1}) \geq N(q+1) - 1$.
2. $\omega(\mathbb{Z}_{q^2-1}) \geq \omega(\mathbb{Z}_{q-1})$.
3. $\omega(\mathbb{Z}_{q^n+\cdots+q+1}) \geq \begin{cases} N(q+1) - 1 & \text{if } n \text{ is even,} \\ \omega(\mathbb{Z}_{q-1}) & \text{if } n \text{ is odd.} \end{cases}$
4. $\omega(\mathbb{Z}_{q^n-1}) \geq \omega(\mathbb{Z}_{q-1})$.

Proof. For (1) see Corollary 8.36.1; (2) is a consequence of Theorem 8.47; for (3) see Corollary 8.58 if n is even and Corollary 8.63 if n is odd; and for (4) see Corollary 8.61. □

The lower bounds for $w_n(q)$ for small n and q, given in Table 12.4, are obtained from Lemmas 12.31, 12.32, and 12.33; and from the table of lower bounds for $N(v)$ in Section III.3 of [79]. The lower bound for $w_9(64)$ requires explanation: $\text{quot}_9(64) = 68{,}719{,}738{,}881 = 1 + 2^{18} + 2^{36}$. Thus $\omega(\mathbb{Z}_{68,719,738,881}) \geq N(2^{18} + 1) - 1$ by Lemma 12.33.1. The table of lower bounds for $N(v)$ in Section III.3 of [79] covers all $v < 10{,}000$ and $2^{18} + 1 = 262{,}144$, but n_{30}, the largest v for which the existence of a set of 30 MOLS of order v is unknown, is 52,502, and hence $N(2^{18} + 1) \geq 30$.

Some consequences of Theorem 12.29 follow.

Corollary 12.34. *If q is a power of 2, $q > 4$, and $\text{quot}_i(q)$ is not of the form $9m$, m not divisible by 3, $\text{quot}_i(q) > 3$, for $i = 1, \ldots, n$, then $\omega(\text{GL}(n, q)) \geq 3$.*

Proof. This follows from Lemmas 12.32.4 and 12.32.5. □

Table 12.4 Lower bounds for $w_n(q)$ for small n and q

$q \diagdown n$	1	2	3	4	5	6	7	8	9	10
4	1^b	3^b	4^a	15^b	9^d	11^b	41^d	255^b	6^g	3^d
8	5^b	1^a	71^b	3^d	29^d	6^g	7^i	15^d	$262{,}655^b$	9^d
16	3^a	15^b	15^g	255^b	15^i	239^b	27^d	$65{,}535^b$	29^g	$61{,}679^b$
32	29^b	4^a	5^g	3^d	599^d	30^g	69^d	15^d	71^d	249^d
64	5^h	3^d	6^g	15^d	9^d	35^d	$7^{i,c}$	95^d	29^g	3^d

a: Lemma 12.31. *b*: Lemma 12.32.1. *c*: Lemma 12.32.2. *d*: Lemma 12.32.3.
e: Lemma 12.32.4. *f*: Lemma 12.32.5. *g*: Lemma 12.33.1. *h*: Lemma 12.33.2.
i: Lemma 12.33.3. *j*: Lemma 12.33.4.

Table 12.4 and Theorem 12.29 yield the following.

Corollary 12.35.

1. *If* $q = 16$ *and* $2 \leq n \leq 10$, *then* $\omega(\mathrm{GL}(n, q)) \geq 3$.
2. *If* $q = 32$ *and* $n = 2$ *or* 3, *then* $\omega(\mathrm{GL}(n, q)) \geq 4$.
3. *If* $q = 32$ *and* $4 \leq n \leq 10$, *then* $\omega(\mathrm{GL}(n, q)) \geq 3$.
4. *If* $q = 64$ *and* $2 \leq n \leq 10$, *then* $\omega(\mathrm{GL}(n, q)) \geq 3$.

Chapter 13
Groups of Small Order

In this chapter we will describe what is known about the structure of Orth(G) for G a group of small order. Much of this knowledge is the result of computer searches. In Table 13.1 we summarize known values and bounds for $|\text{Orth}(G)|$ and $\omega(G)$ when $|G| \leq 23$. We have omitted from this table groups with nontrivial, cyclic Sylow 2-subgroups as these groups do not admit orthomorphisms by Theorem 3.34. Order 23 marks a divide. We can find values of $|\text{Orth}(G)|$ in the literature for every group of order 23 or less, but we can find no such value for any group of order 24, other than the cyclic group of order 25 and those that have cyclic Sylow 2-subgroups and hence do not admit orthomorphisms. Table 13.1 also illustrates the limits to our knowledge of $\omega(G)$: we know $\omega(G)$ for all abelian groups G of order 19 or less but can only establish a lower bound for $\omega(\mathbb{Z}_2 \times \mathbb{Z}_2 \times \mathbb{Z}_5)$; and we can only establish lower bounds for $\omega(G)$ for nonabelian groups G of order 16. For groups of order greater than 23, work has concentrated on trying to improve lower bounds for $\omega(G)$ for groups G that are direct products of elementary abelian groups: a number of our lower bounds for $N(n)$ have been obtained in this way.

13.1 The Smallest Groups

The trivial group has one orthomorphism, the identity mapping, which is orthogonal to itself. By convention $\omega(1) = \infty$.

The orthomorphism graphs of the groups, $\mathbb{Z}_3$, $\mathbb{Z}_2 \times \mathbb{Z}_2$, and $\mathbb{Z}_5$, have a very simple structure. The group $\mathbb{Z}_3$ has only one normalized orthomorphism, and hence $|\text{Orth}(\mathbb{Z}_3)| = 1$ and $\omega(\mathbb{Z}_3) = 1$; the group $\mathbb{Z}_2 \times \mathbb{Z}_2$ has two normalized orthomorphisms, which are orthogonal to each other, and hence $|\text{Orth}(\mathbb{Z}_2 \times \mathbb{Z}_2)| = 2$ and $\omega(\mathbb{Z}_2 \times \mathbb{Z}_2) = 2$; and the group $\mathbb{Z}_5$ has three normalized orthomorphisms, which are pairwise orthogonal, and hence $|\text{Orth}(\mathbb{Z}_5)| = 3$ and $\omega(\mathbb{Z}_5) = 3$. These orthomorphism graphs are all complete graphs. Each of the groups, $\mathbb{Z}_3$, $\mathbb{Z}_2 \times \mathbb{Z}_2$,

A. B. Evans, *Orthogonal Latin Squares Based on Groups*, Developments in Mathematics 57, https://doi.org/10.1007/978-3-319-94430-2_13

Table 13.1 Data on orthomorphism graphs of small groups

Group G	Order G	$\|\mathrm{Orth}(G)\|$	$\omega(G)$	Group G	Order G	$\|\mathrm{Orth}(G)\|$	$\omega(G)$
1	1	1	∞	$\mathbb{Z}_2 \times \mathbb{Z}_2 \times \mathbb{Z}_4$	16	14,886,912	6
$\mathrm{GF}(3)^+$	3	1	1	$\mathrm{GF}(16)^+$	16	15,296,512	14
$\mathrm{GF}(4)^+$	4	2	2	$D_8 \times \mathbb{Z}_2$	16	7,849,984	≥ 2
$\mathrm{GF}(5)^+$	5	3	3	$Q_8 \times \mathbb{Z}_2$	16	7,571,456	≥ 2
$\mathrm{GF}(7)^+$	7	19	5	$\Gamma_2 b$	16	7,710,720	≥ 2
$\mathbb{Z}_2 \times \mathbb{Z}_4$	8	48	2	$\Gamma_2 c_1$	16	7,743,488	≥ 2
D_8	8	48	1	$\Gamma_2 c_2$	16	7,636,992	≥ 2
Q_8	8	48	1	$\Gamma_2 d$	16	7,608,320	≥ 2
$\mathrm{GF}(8)^+$	8	48	6	D_{16}	16	3,930,112	≥ 3
$\mathbb{Z}_9$	9	225	1	$\Gamma_3 a_2$	16	3,913,728	≥ 1
$\mathrm{GF}(9)^+$	9	249	7	Q_{16}	16	3,897,344	≥ 1
$\mathrm{GF}(11)^+$	11	3441	9	$\mathrm{GF}(17)^+$	17	94,471,089	15
$\mathbb{Z}_2 \times \mathbb{Z}_6$	12	16,512	4	$GF(19)^+$	19	4,613,520,889	17
D_{12}	12	6,336	2	$\mathbb{Z}_2 \times \mathbb{Z}_2 \times \mathbb{Z}_5$	20	34,864,619,520	≥ 3
A_4	12	3,840	1	D_{20}	20	7,043,328,000	≥ 2
$\mathrm{GF}(13)^+$	13	79,259	11	$\mathbb{Z}_{21}$	21	275,148,653,115	≥ 4
$\mathbb{Z}_{15}$	15	2,424,195	3	$\mathbb{Z}_7 \rtimes \mathbb{Z}_3$	21	39,372,127,200	≥ 1
$\mathbb{Z}_2 \times \mathbb{Z}_8$	16	14,735,360	3	$\mathrm{GF}(23)^+$	23	19,686,730,313,955	21
$\mathbb{Z}_4 \times \mathbb{Z}_4$	16	14,813,184	6				

and $\mathbb{Z}_5$, is an elementary abelian group and, hence, is isomorphic to the additive group of a finite field: the orthomorphisms of $\mathrm{GF}(q)^+$ were studied in Chapters 9 and 10. By Theorem 9.17, any orthomorphism of $\mathrm{GF}(q)^+$ can be represented by an orthomorphism polynomial of degree at most $q-3$, and all orthomorphism polynomials of degree 5 or less were described in Theorem 9.19. We see from this list that the only normalized orthomorphism polynomials of $\mathrm{GF}(q)^+$, $q = 3, 4$, or 5, are ax, $a \neq 0, 1$: the mapping $x \mapsto ax$ was denoted $[a]$. This same conclusion was reached, using (σ, ε)-systems, in Section 10.3. Recall that we used $\mathscr{C}_1(q)$ to denote the set, $\{[a] \mid a \neq 0, 1\}$, of linear orthomorphisms of $\mathrm{GF}(q)^+$. Thus, for $q = 3, 4$, or 5, $\mathrm{Orth}(\mathrm{GF}(q)^+) = \mathscr{C}_1(q)$, and so $|\mathrm{Orth}(\mathrm{GF}(q)^+)| = |\mathscr{C}_1(q)| = q - 2$. Further, as $[a] \perp [b]$ if $a \neq b$, $\omega(\mathrm{GF}(q)^+) = |\mathscr{C}_1(q)| = q - 2$.

13.1.1 The Group $\mathbb{Z}_7$

The orthomorphism graph of $\mathbb{Z}_7$ has a little more structure. In 1779 Euler [110] described the 19 transversals through a fixed cell of the cyclic Latin square of order 7, i.e., the Cayley table of $\mathbb{Z}_7$. By Theorem 1.14, these transversals correspond to 19 normalized complete mappings of $\mathbb{Z}_7$. Hence, the group $\mathbb{Z}_7$ has 19 normalized orthomorphisms: these orthomorphisms are listed in [184] and [185]. Five of these are pairwise orthogonal, and each of the 14 other normalized orthomorphisms is orthogonal to no orthomorphism of $\mathbb{Z}_7$. Hence $|\mathrm{Orth}(\mathbb{Z}_7)| = 19$ and $\omega(\mathbb{Z}_7) = 5$. The group $\mathbb{Z}_7$ is isomorphic to the additive group of GF(7), and by Theorem 9.17, any orthomorphism of $\mathrm{GF}(7)^+$ can be represented by an orthomorphism polynomial of degree at most 4. By Theorem 9.19 the nonlinear normalized orthomorphism polynomials of $\mathrm{GF}(7)^+$ are $\pm(x+c)^4 + 4x - c^4$, where c is arbitrary. Thus each nonlinear normalized orthomorphism polynomial of GF(7) is a translate of $x^4 + 4x$ or $-x^4 + 4x$. The mapping $x \mapsto x^4 + 4x$ maps a square a to $5a$ and a nonsquare b to $3b$: recall that we called such an orthomorphism a quadratic orthomorphism in Section 9.3. This orthomorphism was given the symbol $[5, 3]$, and we used $\mathscr{C}_2(q)$ to denote the set of quadratic orthomorphisms of $\mathrm{GF}(q)^+$. Similarly, the orthomorphism polynomial $-x^4+4x$ represents the quadratic orthomorphism $[3, 5]$. By Example 9.31, the nonlinear quadratic orthomorphisms of $\mathrm{Orth}(\mathrm{GF}(7)^+)$ are $\mathscr{C}_2(7) \setminus \mathscr{C}_1(7) = \{[3, 5], [5, 3]\}$. By Corollary 9.35, neither of these orthomorphisms is orthogonal to any linear orthomorphism, and as $(3-5)(5-3) = 3$, a nonsquare, by Corollary 9.25(3), they are not orthogonal to each other. By Theorem 10.28, the translates of the nonlinear, quadratic orthomorphisms are distinct, thus accounting for the 14 nonlinear orthomorphisms of $\mathrm{GF}(7)^+$. Also by Theorem 10.28, there are no orthogonalities between orthomorphisms from distinct translates of $\mathscr{C}_2(7) \setminus \mathscr{C}_1(7)$. Thus the data for groups of order at most 7 can be completely explained theoretically.

Two alternative descriptions of the normalized nonlinear orthomorphisms of $\mathrm{GF}(7)^+$, derived from (σ, ε)-systems, are given in Section 10.3. In Example 10.33, the normalized nonlinear orthomorphisms of $\mathrm{GF}(7)^+$ are described as

$$\theta(x) = \begin{cases} a(ab+c) & \text{if } x = c, \\ a(b+c) & \text{if } x = ab+c, \\ ac & \text{if } x = b+c, \\ ax & \text{otherwise,} \end{cases}$$

where $a^2 - a + 1 = 0$, i.e., $a = 3$ or 5, $b \neq 0$, and $c \neq 0, -b, -ab$. In Theorem 10.34, the normalized nonlinear orthomorphisms of $\mathrm{GF}(7)^+$ are described as

$$\theta(x) = \begin{cases} ab(g^{i+1} - c) & \text{if } x = b(g^i - c), \\ ax & \text{otherwise,} \end{cases}$$

where $g = 2$ or 4, a primitive cube root of unity, $a = (1-g^2)/(g-g^2)$, $b \neq 0$, and $c \neq 1, g, g^2$. We leave it to the reader to verify that the three descriptions that we have given of the normalized nonlinear orthomorphisms of $\mathrm{GF}(7)^+$, the polynomial description from Theorem 9.19, the description from Example 10.33, and the description from Theorem 10.34, are equivalent.

13.2 Groups of Order 8

There are 5 groups of order 8. The cyclic group $\mathbb{Z}_8$ has no orthomorphisms by Theorem 3.34. There are two noncyclic abelian groups, $\mathbb{Z}_2 \times \mathbb{Z}_4$ and $GF(8)^+$, and there are two nonabelian groups, the dihedral group D_8 and the quaternion group Q_8. Each of these noncyclic groups has 48 normalized orthomorphisms.

13.2.1 *The Group* $\mathbb{Z}_2 \times \mathbb{Z}_4$

The structure of the orthomorphism graph of $\mathbb{Z}_2 \times \mathbb{Z}_4$ was first studied in 1961 by Johnson, Dulmage, and Mendelsohn [196]. They partitioned the set of normalized orthomorphisms of $\mathbb{Z}_2 \times \mathbb{Z}_4$ into 24 orthogonal pairs. In 1964 Chang, Hsiang, and Tai [69], through extensive hand computations, concluded that $\omega(\mathbb{Z}_2 \times \mathbb{Z}_4) = 2$, and in 1986 Jungnickel and Grams [207], through a computer search, found the only maximal cliques in $\text{Orth}(\mathbb{Z}_2 \times \mathbb{Z}_4)$ to be 2-cliques, thus confirming that $\omega(\mathbb{Z}_2 \times \mathbb{Z}_4) = 2$.

The complete structure of $\text{Orth}(\mathbb{Z}_2 \times \mathbb{Z}_4)$ was determined by Evans and Perkel, using the computer algebra system Cayley (a forerunner of magma): the results were presented in 1992 by Evans [119]. In describing these results, we will use ij to denote the element $(i, j) \in \mathbb{Z}_2 \times \mathbb{Z}_4$, $i \in \mathbb{Z}_2$, and $j \in \mathbb{Z}_4$. We found each orthomorphism of $\mathbb{Z}_2 \times \mathbb{Z}_4$ to be orthogonal to exactly two other orthomorphisms of $\mathbb{Z}_2 \times \mathbb{Z}_4$. In fact $\text{Orth}(\mathbb{Z}_2 \times \mathbb{Z}_4)$ is the disjoint union of twelve 4-cycles. We present some of these cycles in Table 13.2: those not listed are translates of these.

Table 13.2 Some 4-cycles of $\text{Orth}(\mathbb{Z}_2 \times \mathbb{Z}_4)$

x	00	10	01	11	02	12	03	13	x	00	10	01	11	02	12	03	13
$\alpha_1(x)$	00	13	11	02	03	10	12	01	$\delta_1(x)$	00	01	10	03	12	13	02	11
$\alpha_2(x)$	00	03	12	13	01	02	11	10	$\delta_2(x)$	00	11	03	02	10	01	13	12
$\alpha_3(x)$	00	01	13	12	11	10	02	03	$\delta_3(x)$	00	03	02	13	12	11	10	01
$\alpha_4(x)$	00	11	10	03	13	02	01	12	$\delta_4(x)$	00	13	11	12	10	03	01	02
$\beta_1(x)$	00	12	13	01	03	11	10	02	$\zeta_1(x)$	00	03	13	02	12	11	01	10
$\beta_2(x)$	00	02	12	10	11	13	01	03	$\zeta_2(x)$	00	13	02	01	10	03	12	11
$\beta_3(x)$	00	12	11	03	13	01	02	10	$\zeta_3(x)$	00	01	03	10	12	13	11	02
$\beta_4(x)$	00	02	10	12	01	03	13	11	$\zeta_4(x)$	00	11	12	13	10	01	02	03
$\gamma_1(x)$	00	11	10	13	12	03	02	01	$\eta_1(x)$	00	11	13	10	12	03	01	02
$\gamma_2(x)$	00	01	03	12	10	11	13	02	$\eta_2(x)$	00	01	02	13	10	11	12	03
$\gamma_3(x)$	00	13	02	03	12	01	10	11	$\eta_3(x)$	00	13	03	02	12	01	11	10
$\gamma_4(x)$	00	03	11	02	10	13	01	12	$\eta_4(x)$	00	03	12	01	10	13	02	11

The orthomorphisms α_1, α_2, α_3, and α_4 form a 4-cycle, which we will refer to as the α-*cycle*, $\alpha_i \perp \alpha_{i+1}$, the subscripts being added modulo 4. The β-*cycle*, γ-*cycle*, δ-*cycle*, ζ-*cycle*, and η-*cycle* are defined similarly. $\text{Orth}(\mathbb{Z}_2 \times \mathbb{Z}_4)$ contains six more 4-

cycles, three translates of the α-cycle and three translates of the β-cycle. The action of translation on the normalized orthomorphisms of $\mathbb{Z}_2 \times \mathbb{Z}_4$ is easily determined. The γ-, δ-, ζ-, and η-cycles are each fixed by all translations, but not pointwise: the orbits under translations are $\{\gamma_1, \gamma_3\}$, $\{\gamma_2, \gamma_4\}$, $\{\delta_1, \delta_3\}$, $\{\delta_2, \delta_4\}$, $\{\zeta_1, \zeta_3\}$, $\{\zeta_2, \zeta_4\}$, $\{\eta_1, \eta_3\}$, and $\{\eta_2, \eta_4\}$. Any translation of an α- or β-cycle, if fixed by a translation, is fixed pointwise by the translation: any orthomorphism contained in the translate of an α- or β-cycle is fixed by a translation of order 2.

13.2.2 The Group GF(8)$^+$

The structure of the orthomorphism graph of GF(8)$^+$ was also first studied in 1961 by Johnson, Dulmage, and Mendelsohn [196]. They determined all the 6-cliques of Orth(GF(8)$^+$). Every orthomorphism of GF(8)$^+$ is also an automorphism, and the automorphism group of GF(8)$^+$ is the simple group, GL(3, 2), of order 168. This group has eight Sylow 7-subgroups, and the nonidentity elements of each of these subgroups form a 6-clique of Orth(GF(8)$^+$). In 1964 Chang, Hsiang, and Tai [69] listed representatives for each orbit of Orth(GF(8)$^+$) under the action of congruences. In 1986 Jungnickel and Grams [207] found that the only maximal cliques in Orth(GF(8)$^+$) are 2- and 6-cliques. In 1990 Stothers [302], through a computer search, found that GF(8)$^+$ has 384 complete mappings, i.e., 48 normalized complete mappings. Further details of the structure of Orth(GF(8)$^+$) were provided by Lazebnik and Thomason [224] in 2004: they found that all 48 vertices in the orthomorphism graph of GF(8)$^+$ are of degree 12; that the graph has 288 edges and eight 6-cliques, which are clearly maximal; and that the graph has maximal 2-cliques but no maximal 3-, 4-, or 5-cliques.

In trying to explain this data, we will make use of the theory of permutation polynomials and solutions to (σ, ε)-systems: these topics were covered in Chapters 9 and 10. The orthomorphism polynomials of GF(8) were listed in Theorem 9.19: they fall into two classes:

1. ax is an orthomorphism polynomial if and only if $a \neq 0, 1$. This is the class of linear orthomorphisms, and
2. $ax^4+bx^2+cx, a \neq 0$, is an orthomorphism polynomial if and only if ax^3+bx+c and $ax^3 + bx + (c + 1)$ are both irreducible over GF(8).

The nonlinear orthomorphism polynomials of GF(8) were described in Theorem 9.19 as $a^{-1}(x^4+bx^2+(a+c)x)+d, a \neq 0, d$ arbitrary, and $x^4+bx^2+cx = 0$ and $x^4 + bx^2 + (a + c)x = 0$ each have only $x = 0$ as a solution in GF(8): this is equivalent to the description that we have given above. Orthogonalities between these orthomorphisms were described by Evans [113] in 1989.

Theorem 13.1 (Evans, 1989). *For the orthomorphism graph of* $\mathrm{GF}(8)^+$, *the following hold:*

1. *If* $ax, bx \in \mathrm{Orth}(\mathrm{GF}(8)^+)$, *then* $ax \perp bx$ *if and only if* $a \neq b$.
2. *If* $ax, bx^4+cx^2+dx \in \mathrm{Orth}(\mathrm{GF}(8)^+)$, $b \neq 0$, *then* $ax \perp bx^4+cx^2+dx$ *if and only if* $bx^3+cx+(d+a)$ *is irreducible over* $\mathrm{GF}(8)$.
3. *If* ax^4+bx^2+cx *and* $ax^4+ex^2+fx \in \mathrm{Orth}(\mathrm{GF}(8)^+)$, $a \neq 0$, *then* $ax^4+bx^2+cx \perp ax^4+ex^2+fx$ *if and only if* $b = e$ *and* $c \neq f$ *or* $c = f$ *and* $b \neq e$.
4. *If* ax^4+bx^2+cx *and* $dx^4+ex^2+fx \in \mathrm{Orth}(\mathrm{GF}(8)^+)$, $a, d \neq 0$, $a \neq d$, *then* $ax^4+bx^2+cx \perp dx^4+ex^2+fx$ *if and only if* $(a+d)x^3+(b+e)x^2+(c+f)$ *is irreducible over* $\mathrm{GF}(8)$.

Proof. The normalized permutation polynomials of GF(8) of degree 4 or less, given in Theorem 9.12, are x; x^2; x^3; and x^4+ax^2+cx, where $x^4+ax^2+cx=0$ has only $x=0$ as a solution. Note that $x^4+ax^2+cx=0$ has only $x=0$ as a solution if and only if x^3+ax+c is irreducible over GF(8). Now two orthomorphism polynomials are orthogonal if and only if their difference can be written in the form $\alpha f(x+\beta)+\gamma$ for some normalized permutation polynomial f and some $\alpha, \beta, \gamma \in \mathrm{GF}(8)$, $\alpha \neq 0$. The result is then routine. □

The fact that $ax^4+bx^2+cx \perp ax^4+ex^2+fx$ if $c=f$ and $b \neq e$ was missed in [113].

An alternative description of the normalized orthomorphisms of $\mathrm{GF}(8)^+$ was given in Section 10.3 by solving (σ, ε)-systems. From these solutions we obtain 6 linear orthomorphisms and 42 nonlinear orthomorphisms. The nonlinear orthomorphims are $\phi_{a,b}$, $a \neq 0, 1$, $b \neq 0$ defined by

$$\phi_{a,b}(x) = \begin{cases} ax & \text{if } x \in \{0, b, ab, ab+b\}, \\ a(x+ab+b) & \text{otherwise.} \end{cases}$$

Some orthogonalities in $\mathrm{Orth}(\mathrm{GF}(8)^+)$ were given in Examples 10.48 and 10.49. But what matters most is that $\phi_{a,b} \not\perp \phi_{a,b'}$ for all b, b' and Corollary 10.52 that shows that $\phi_{a,b} \perp [d]$ if and only if $d = a^2$.

From solutions to (σ, ε)-systems, we showed the normalized orthomorphisms of $\mathrm{GF}(8)^+$ to consist of 6 linear orthomorphisms and 42 nonlinear orthomorphisms, $\phi_{a,b}$, $a \neq 0, 1$ and $b \neq 0$. When we look at the list of orthomorphism polynomials of GF(8), it is clear that each of the normalized orthomorphisms of $\mathrm{GF}(8)^+$ is also an automorphism. Hence $\mathrm{Orth}(\mathrm{GF}(8)^+) = \mathscr{A}(\mathrm{GF}(8)^+)$, the set of fixed-point-free automorphisms studied in Section 8.4. Now $\mathrm{Aut}(\mathrm{GF}(8)^+) \cong \mathrm{GL}(3,2)$, the simple group of order 168. Let g be a primitive element of GF(8), and let $\tau_g \in \mathrm{Aut}(\mathrm{GF}(8)^+)$ denote the mapping $x \mapsto gx$. The group $\langle \tau_g \rangle$ is a Sylow 7-subgroup of $\mathrm{Aut}(\mathrm{GF}(8)^+)$: the nonidentity elements of this subgroup form a 6-clique of $\mathrm{Orth}(\mathrm{GF}(8)^+)$. For any $\alpha \in \mathrm{Aut}(\mathrm{GF}(8)^+)$, $\langle H_\alpha \tau_g \rangle = \langle \alpha \tau_g \alpha^{-1} \rangle$ is a Sylow 7-subgroup of $\mathrm{Aut}(\mathrm{GF}(8)^+)$, and, as H_α is an automorphism of $\mathrm{Orth}(\mathrm{GF}(8)^+)$, the nonidentity elements of $\langle H_\alpha \tau_g \rangle$ form a 6-clique of

Orth(GF(8)$^+$). Now Aut(GF(8)$^+$) has 8 Sylow 7-subgroups, each of which is equal to $\langle H_\alpha \tau_g \rangle$ for some $\alpha \in$ Aut(GF(8)$^+$). In each of these Sylow 7-subgroups of Aut(GF(8)$^+$), the nonidentity elements form a 6-clique of Orth(GF(8)$^+$). Let us label these 6-cliques $\mathscr{S}_1, \ldots, \mathscr{S}_8$. As these 6-cliques are disjoint, they account for all 48 orthomorphisms of GF(8)$^+$. Now GF(8)$^+$ can be regarded as the additive group of several different but isomorphic fields, and each of $\mathscr{S}_1, \ldots, \mathscr{S}_8$ is a clique of linear orthomorphisms with respect to one of these fields. By Corollary 10.52, each nonlinear orthomorphism of GF(8)$^+$ is orthogonal to precisely one linear orthomorphism of GF(8)$^+$. It follows that each element of $\mathscr{S}_i$ is orthogonal to exactly one element of $\mathscr{S}_j$, for $i \neq j$. This implies that each vertex of the orthomorphism graph of GF(8)$^+$ has degree 12.

13.2.3 The Groups D_8 and Q_8

In 1964 Chang and Tai [70] found both D_8 and Q_8 to have 48 normalized orthomorphisms. They also reported that $\omega(D_8) = 1$ and $\omega(Q_8) = 1$: this was confirmed by Jungnickel and Grams [207] in 1986. The lack of orthogonalities in Orth(D_8) and Orth(Q_8) has not yet been explained theoretically. Using Cayley, Evans and Perkel (reported in [119]) verified that both D_8 and Q_8 have 48 normalized orthomorphisms. In 1993 Bedford [34] described the orthomorphisms of D_8 and Q_8. Using

$$D_8 = \langle a, b \mid a^4 = b^2 = 1, ab = ba^{-1} \rangle$$

and

$$Q_8 = \langle a, b \mid a^4 = 1, a^2 = b^2, ab = ba^{-1} \rangle,$$

he gave distinct representatives of orbits of the group of homologies, $\{H_\alpha \mid \alpha \in \mathrm{Aut}(G)\}$. For D_8 each orbit is of length 8, and a set of distinct representatives of these orbits is given in Table 13.3, and for Q_8 each orbit is of length 24, and a set of distinct representatives of these orbits is given in Table 13.4.

Table 13.3 Representative orthomorphisms of D_8

x	1	a	a^2	a^3	b	ab	a^2b	a^3b
$\theta_1(x)$	1	b	a	ab	a^3	a^3b	a^2	a^2b
$\theta_2(x)$	1	b	a	a^3b	a^2b	a^3	ab	a^2
$\theta_3(x)$	1	a^2	b	a^2b	a^3	a	a^3b	ab
$\theta_4(x)$	1	a^3b	a^2b	a	ab	b	a^3	a^2
$\theta_5(x)$	1	a^2	b	a^2b	ab	a^3b	a	a^3
$\theta_6(x)$	1	a^3b	a^2b	a	a^3	a^2	ab	b

Table 13.4 Representative orthomorphisms of Q_8

x	1	a	a^2	a^3	b	ab	a^2b	a^3b
$\theta_1(x)$	1	a^2	b	a^2b	a	a^3	a^3b	ab
$\theta_2(x)$	1	a^2	b	a^2b	ab	a^3b	a^3	a

13.2.4 Counting Orthomorphisms

It is an interesting fact that each of the noncyclic groups of order 8 has 48 normalized orthomorphisms. A partial explanation for this fact was given by Bedford [33] in 1991 when he established a bijection between the orthomorphisms of $\mathbb{Z}_2 \times \mathbb{Z}_4$ and those of $\mathrm{GF}(8)^+$ and a bijection between the orthomorphisms of D_8 and those of Q_8. A complete explanation was subsequently given by Bedford and Whitaker [38] in 1999. Both papers relied on a "uniform" presentation of the four groups.

$$D_8 = \langle a, b, c \mid a^2 = b, b^2 = c^2 = 1, a \cdot b = b \cdot a, b \cdot c = c \cdot b, c \cdot a = a \cdot b \cdot c\rangle$$

$$Q_8 = \langle a, b, c \mid a^2 = b = c^2, b^2 = 1, a \odot b = b \odot a, b \odot c = c \odot b, c \odot a = a \odot b \odot c\rangle$$

$$\mathrm{GF}(8)^+ = \langle a, b, c \mid a^2 = b^2 = c^2 = 1, a \times b = b \times a, a \times c = c \times a, b \times c = c \times b\rangle$$

$$\mathbb{Z}_2 \times \mathbb{Z}_4 = \langle a, b, c \mid a^2 = b^2 = 1, c^2 = b, a \otimes b = b \otimes a, a \otimes c = c \otimes a, b \otimes c = c \otimes b\rangle$$

It is clear that the elements of each group are precisely $\{a^e b^f c^g \mid e, f, g = 0, 1\}$. We have used different symbols for the group operations, $\cdot$ for D_8, $\odot$ for Q_8, $\times$ for $\mathrm{GF}(8)^+$, and $\otimes$ for $\mathbb{Z}_2 \times \mathbb{Z}_4$, to distinguish between the four groups. This is especially important when describing Bedford's bijections. Each of our four groups can be described as having the same element set, $G = \{1, a, b, c, ab, ac, bc, abc\}$, with the operation $\cdot$, $\odot$, $\times$, or $\otimes$ applied. Thus the identity mapping on G defines a bijection between the element set of any two of D_8, Q_8, $\mathrm{GF}(8)^+$, and $\mathbb{Z}_2 \times \mathbb{Z}_4$. This enables us to write equations in which the elements on opposite sides of the equal sign are from different groups. As an example:

$$c \odot a^2 = c \otimes c \otimes c \quad \text{as} \quad c \odot a^2 = bc \quad \text{and} \quad c \otimes c \otimes c = bc.$$

Let $H = \{1, a, b, ab\} \subseteq G$. In all four groups, H is a subgroup of index 2, and $G \setminus H = Hc = \{c, ac, bc, abc\}$. In D_8 and Q_8, $H \cong \mathbb{Z}_4$; and in $\mathrm{GF}(8)^+$ and $\mathbb{Z}_2 \times \mathbb{Z}_4$, $H \cong \mathbb{Z}_2 \times \mathbb{Z}_2$.

Lemma 13.2. *If $x, y \in G$, then*

$$x \cdot y = \begin{cases} x \odot y \odot b & \text{if } x, y \in Hc, \\ x \odot y & \text{otherwise;} \end{cases} \qquad x \odot y = \begin{cases} x \cdot y \cdot b & \text{if } x, y \in Hc, \\ x \cdot y & \text{otherwise;} \end{cases}$$

$$x \times y = \begin{cases} x \otimes y \otimes b & \text{if } x, y \in Hc, \\ x \otimes y & \text{otherwise;} \end{cases} \quad \text{and} \quad x \otimes y = \begin{cases} x \times y \times b & \text{if } x, y \in Hc, \\ x \times y & \text{otherwise.} \end{cases}$$

Proof. Routine. □

Bedford's correspondences were described in terms of complete mappings and shall be presented as such. First we need an elementary lemma describing the actions of any complete mapping on the cosets of H.

Lemma 13.3. *If* θ *is a complete mapping of* D_8, Q_8, $\mathrm{GF}(8)^+$, *or* $\mathbb{Z}_2 \times \mathbb{Z}_4$, *then*

$$|\{x \mid x \in H, \theta(x) \in H\}| = |\{x \mid x \in H, \theta(x) \in Hc\}|$$
$$= |\{x \mid x \in Hc, \theta(x) \in H\}| = |\{x \mid x \in Hc, \theta(x) \in Hc\}| = 2.$$

Proof. If we set $C_{ij} = |\{x \mid x \in Hi, \theta(x) \in Hj\}|$, $i, j = 1, c$, then, from the properties of a complete mapping, we compute

$$C_{i1}+C_{ic} = 4 \text{ for } i = 1, c;\ C_{ij}+C_{cj} = 4 \text{ for } j = 1, c;\ C_{11}+C_{cc} = 4;\ \text{and } C_{1c}+C_{c1} = 4.$$

Solving this system of linear equations yields $C_{ij} = 2$ for all i, j. □

Theorem 13.4 (Bedford, 1991). *There exists a natural bijection between the set of normalized complete mappings of* D_8 *and the set of normalized complete mappings of* Q_8, *and there exists a natural bijection between the set of normalized complete mappings of* $\mathrm{GF}(8)^+$ *and the set of normalized complete mappings of* $\mathbb{Z}_2 \times \mathbb{Z}_4$.

Proof. Let θ be a normalized complete mapping of D_8. Then, by Lemma 13.3,

$$\theta = \left(\begin{array}{cccc|cccc} 1 & x_1 & x_2 & x_3 & y_1 & y_2 & y_3 & y_4 \\ 1 & u_1 & v_1 & v_2 & u_2 & u_3 & v_3 & v_4 \end{array} \right),$$

where $H = \{1, x_1, x_2, x_3\} = \{1, u_1, u_2, u_3\}$ and $Hc = \{y_1, y_2, y_3, y_4\} = \{v_1, v_2, v_3, v_4\}$. Set $\eta(x) = x \cdot \theta(x)$. From θ we will construct a complete mapping θ' of Q_8 and set $\eta'(x) = x \odot \theta'(x)$. There are two cases to consider, $\eta(y_3) = b \cdot \eta(y_4)$ and $\eta(y_3) \neq b \cdot \eta(y_4)$. In the first of these cases, we will show that θ is a complete mapping of Q_8 as well as D_8; in the second we will need to modify θ to obtain a complete mapping of Q_8.

Case 1. $\eta(y_3) = b \cdot \eta(y_4)$

We will show that $\theta' = \theta$ is a complete mapping of Q_8. It suffices to show that η' is a bijection. It is easy to verify that if $h_1, h_2 \in H$, then $h_1 \cdot h_2 = h_1 \odot h_2$ and $h_1 \cdot h_2 \cdot c = h_1 \odot h_2 \odot c$. Hence $\eta'(x) = \eta(x)$ if $x \in H \cup \{y_1, y_2\}$. Now $\eta'(y_3) = y_3 \odot v_3 = b \cdot y_3 \cdot v_3 = b \cdot \eta(y_3) = \eta(y_4)$ and $\eta'(y_4) = y_4 \odot v_4 = b \cdot y_4 \cdot v_4 = b \cdot \eta(y_4) = \eta(y_3)$. Hence η' is a bijection from which it follows that θ' is a complete mapping of Q_8.

Case 2. $\eta(y_3) \neq b \cdot \eta(y_4)$

Let θ' be defined by

$$\theta'(x) = \begin{cases} \theta(x) & \text{if } x \in H \cup \{y_1, y_2\}; \\ \theta(y_4) & \text{if } x = y_3; \\ \theta(y_3) & \text{if } x = y_4. \end{cases}$$

We will show that θ' is a complete mapping of Q_8. Again, as θ' is a bijection, it suffices to show that η' is a bijection. As in Case 1, $\eta'(x) = \eta(x)$ if $x \in H \cup \{y_1, y_2\}$; thus we need to show that $\{\eta'(y_3), \eta'(y_4)\} = \{\eta(y_3), \eta(y_4)\}$. Now

$$y_3 \odot v_3 = b \cdot y_3 \cdot v_3 = b \cdot \eta(y_3) \neq \eta(y_4),$$

and

$$y_4 \odot v_4 = b \cdot y_4 \cdot v_4 = b \cdot \eta(y_4) \neq \eta(y_3).$$

Hence $\{\eta(y_3), \eta(y_4)\} = H \setminus \{y_3 \odot v_3, y_4 \odot v_4\}$. But $\{y_3 \odot v_4, y_4 \odot v_3\} = H \setminus \{y_3 \odot v_3, y_4 \odot v_4\}$, and so $\{y_3 \odot v_4, y_4 \odot v_3\} = \{\eta(y_3), \eta(y_4)\}$.

Our construction of normalized complete mappings of Q_8 from normalized complete mappings of D_8 implicitly defines a function which is invertible: to construct the inverse, swap the operations $\cdot$ and $\odot$. Therefore D_8 and Q_8 admit the same number of normalized complete mappings.

The proof that $GF(8)^+$ and $\mathbb{Z}_4 \times \mathbb{Z}_2$ admit the same number of normalized complete mappings is similar with the operations $\times$ and $\otimes$ replacing the operations $\cdot$ and $\odot$. □

In 1999, Bedford and Whitaker [38] gave a proof that each noncyclic group of order 8 admits 48 normalized complete mappings. If $K = \{1, b\}$ and G is one of $\mathbb{Z}_2 \times \mathbb{Z}_4$, $GF(8)^+$, D_8, or Q_8, then K is a subgroup of G with the following properties.

Lemma 13.5. *If G is one of $\mathbb{Z}_2 \times \mathbb{Z}_4$, $GF(8)^+$, D_8, or Q_8 and $K = \{1, b\}$, then K is a normal subgroup of G contained in the center of G, and $G/K \cong \mathbb{Z}_2 \times \mathbb{Z}_2$.*

Proof. Routine. □

From each normalized complete mapping of G/K, we will construct 24 normalized complete mappings of G. As G/K admits 2 normalized complete mappings, this yields at most 48 normalized complete mappings of G. We will then show that the 48 complete mappings constructed are distinct and that they consist of every normalized complete mapping of G. We will describe Bedford and Whitaker's construction of complete mappings of G from complete mappings of G/K.

Let θ be a normalized complete mapping of G/K, and pick $zK \in (G/K) \setminus K$. Set $T_\theta = \{(aK, \theta(aK)) \mid aK \in G/K\}$ and $T_{\theta,z} = \{(aK, \theta(aK)(zK)) \mid aK \in G/K\}$. Now $K = \theta(xK)(zK)$ for a unique choice of $xK \in G/K$, and

$G/K = \{K, xK, yK, zK\}$ for a unique choice of $yK \in G/K$. Now $(K, zK) \in T_{\theta,z}$ and $(xK, K) \in T_{\theta,z}$. As θ is normalized, $(K, K) \in T_\theta$, and, as $\theta(xK)zK = K$, $\theta(xK) = zK$, and so $(xK, zK) \in T_\theta$. As θ is a bijection, the remaining two elements of T_θ are (zK, yK) and (yK, xK), and the remaining two elements of $T_{\theta,z}$ are (zK, xK) and (yK, yK). The sets

$$S_1 = \{(K, K), (xK, K), (K, zK), (xK, zK)\}$$

and

$$S_2 = \{(yK, xK), (yK, yK), (zK, xK), (zK, yK)\}$$

partition $T_{\theta,z} \cup T_\theta$, and $\{aKbK \mid (aK, bK) \in S_i\} = G/K$ for $i = 1, 2$.

Pick $x_1 \in xK$, $y_1 \in yK$ and $z_1 \in zK$: then $xK = \{x_1, x_1'\}$, $yK = \{y_1, y_1'\}$, and $zK = \{z_1, z_1'\}$. As $y_1KyK = K = \{1, b\}$, there exists a unique $y_2 \in yK$ satisfying $y_1y_2 = b$: then $yK = \{y_2, y_2'\}$. Further, there exist unique $x_2 \in xK$ and $z_2 \in zK$ for which $1, x_1b, bz_1, x_1'z_1', b, y_1'x_2$, and z_2y_2' are distinct: then $xK = \{x_2, x_2'\}$ and $zK = \{z_2, z_2'\}$. We next set

$$S_1^* = \{(1, 1), (x_1, b), (b, z_1), (x_1', z_1')\},$$

and

$$S_2^* = \{(y_1, y_2), (y_1', x_2), (z_2, y_2'), (z_2', x_2')\}.$$

Under the canonical homomorphism $G \to G/K$, S_1^* is mapped onto S_1, and S_2^* is mapped onto S_2. We define a mapping $\theta^*: G \to G$ by $\{(g, \theta^*(g)) \mid g \in G\} = S_1^* \cup S_2^*$. We shall call the mapping $\theta^*: G \to G$ a *Bedford-Whitaker mapping* of G. The Bedford-Whitaker mappings of G are in fact normalized complete mappings of G if G is one of $GF(8)^+$, $\mathbb{Z}_2 \times \mathbb{Z}_4$, D_8, or Q_8.

Lemma 13.6. *If G is one of $GF(8)^+$, $\mathbb{Z}_2 \times \mathbb{Z}_4$, D_8, or Q_8, then any Bedford-Whitaker mapping of G is a normalized complete mapping of G.*

Proof. Let θ^* be a Bedford-Whitaker mapping of G. This mapping is well-defined as

$$\{a \mid (a, b) \in S_1^* \cup S_2^*\} = \{1, b, x_1, x_1', y_1, y_1', z_2, z_2'\}$$

are all distinct and is a bijection as

$$\{b \mid (a, b) \in S_1^* \cup S_2^*\} = \{1, b, z_1, z_1', y_2, y_2', z_2, z_2'\}$$

are all distinct. Now $1, y_1y_2 \in K$ and as $y_1y_2 = b$, these are distinct.

To show that θ^* is a complete mapping, we need to show that

$$\{ab \mid (a, b) \in S_1^* \cup S_2^*\} = \{1, x_1b, bz_1, x_1'z_1', y_1y_2, y_1'x_2, z_2y_2', z_2'x_2'\}$$

are all distinct. Now $1, y_1y_2 \in K$, and as $y_1y_2 = b$ these are distinct; $x_1b, z_2y_2' \in xK$, which are distinct by construction; and $bz_1, y_1'x_2 \in zK$, which are distinct by construction. Now $z_2y_2' \in xK$ and, by construction, $z_2y_2' \neq bx_1$, and so $z_2y_2' = x_1$. Also, $y_1'x_2 \in zK$ and, by construction, $y_1'x_2 \neq bz_1$, and so $y_1'x_2 = z_1$. As $x_1'z_1', z_2'x_2' \in yK, b \in Z(G), x_i' = bx_i, z_i' = bz_i$, and

$$z_2'x_2' = z_2x_2 = z_2\left(y_2'\right)^{-1}\left(y_1'\right)^{-1}y_1'x_2 = x_1\left(y_1'y_2'\right)^{-1}z_1 =$$

$$x_1(y_1y_2)^{-1}z_1 = bx_1z_1 = bx_1'z_1',$$

these are distinct. It follows that θ^* is a complete mapping of G.

Lastly θ^* is normalized as, by construction, $\theta^*(1) = 1$. □

In constructing Bedford-Whitaker mappings, we chose a normalized complete mapping of G/K, for which there are two choices: a coset $zK \neq K$, for which there are three choices, and $x_1 \in xK$, $y_1 \in yK$, and $z_1 \in zK$, for which there are two choices each. By Lemma 13.6, if G is one of $\mathrm{GF}(8)^+$, $\mathbb{Z}_2 \times \mathbb{Z}_4$, D_8, or Q_8, then these are all normalized complete mappings of G. In fact, they are all distinct and account for all of the normalized complete mappings of G: this was proved by Bedford and Whitaker [38] in 1999.

Lemma 13.7. *If θ is a complete mapping of G, L is a normal subgroup of G, L contains K, L is of index 2 in G, and $G/L = \{L, dL\}$, then*

$$|\{g \mid g \in L, \theta(g) \in L\}| = |\{g \mid g \in L, \theta(g) \in dL\}|$$
$$= |\{g \mid g \in dL, \theta(g) \in L\}| = |\{g \mid g \in dL, \theta(g) \in dL\}| = 2.$$

Proof. Similar to the proof of Lemma 13.3. □

Theorem 13.8 (Bedford, Whitaker, 1999). *If G is one of* $\mathrm{GF}(8)^+$, $\mathbb{Z}_2 \times \mathbb{Z}_4$, D_8, *or* Q_8, *then* $|\mathrm{Orth}(G)| = 48$.

Proof. Let θ and ϕ be normalized complete mappings of G/K, and let θ^* and ϕ^* be their corresponding Bedford-Whitaker mappings. Let $x^\theta K$, $y^\theta K$, $z^\theta K$, $x_1^\theta \in x^\theta K$, $y_1^\theta \in y^\theta K$, and $z_1^\theta \in z^\theta K$ be the choices made in constructing the Bedford-Whitaker mapping θ^* from θ, and let $x^\phi K$, $y^\phi K$, $z^\phi K$, $x_1^\phi \in x^\phi K$, $y_1^\phi \in y^\phi K$, and $z_1^\phi \in z^\phi K$ be the choices made in constructing the Bedford-Whitaker mapping ϕ^* from ϕ. Then

$$\{(g, \theta^*(g)) \mid g \in G\}$$
$$= \{(1,1), (x_1^\theta, b), (b, z_1^\theta), (x_1^{\theta'}, z_1^{\theta'}), (y_1^\theta, y_2^\theta), (y_1^{\theta'}, x_2^\theta), (z_2^\theta, y_2^{\theta'}), (z_2^{\theta'}, x_2^{\theta'})\},$$

and

$$\{(g, \phi^*(g)) \mid g \in G\}$$
$$= \{(1,1), (x_1^\phi, b), (b, z_1^\phi), (x_1^{\phi'}, z_1^{\phi'}), (y_1^\phi, y_2^\phi), (y_1^{\phi'}, x_2^\phi), (z_2^\phi, y_2^{\phi'}), (z_2^{\phi'}, x_2^{\phi'})\}.$$

Suppose that $\theta^* = \phi^*$. As $\theta^*(x_1^\theta) = b = \phi^*(x_1^\phi)$, it follows that $x_1^\theta = x_1^\phi = x_1$, that $x^\theta K = x^\phi K = xK$, and that $x_1^{\theta'} = x_1^{\phi'} = x_1'$. As $z_1^\theta = \theta^*(b) = \phi^*(b) = z_1^\phi$, it follows that $z_1^\theta = z_1^\phi = z_1$, that $z^\theta K = z^\phi K = zK$, and that $z_1^{\theta'} = z_1^{\phi'} = z_1'$. As yK is uniquely determined by xK and zK, $y^\theta K = y^\phi K = yK$. As $K = \theta(xK)(zK) = \phi(xK)(z)$, it follows that $\theta(xK) = \phi(xK)$, and as $xK \neq K$ and either $\theta \perp \phi$ or $\theta = \phi$, it must be that $\theta = \phi$. As $y^\theta K = y^\phi K$, either $y_1^\theta = y_1^\phi$ or $y_1^\theta = y_1^{\phi'}$. But, if $y_1^\theta = y_1^{\phi'}$, then $y_2^\theta = \theta^*(y_1^\theta) = \phi^*(y_1^{\phi'}) \in xK$, an impossibility. Hence $y_1^\theta = y_1^\phi$. We have thus shown that, if $\theta^* = \phi^*$, then $\theta = \phi$, $z^\theta K = z^\phi K$, $x_1^\theta = x_1^\phi$, $y_1^\theta = y_1^\phi$, and $z_1^\theta = z_1^\phi$. Hence the 48 Bedford-Whitaker mappings of G are distinct. As, by Lemma 13.6, each of these Bedford-Whitaker mappings are normalized complete mappings, G admits at least 48 normalized complete mappings.

Next, let $\mathrm{f}: G \to G/K$ be the canonical homomorphism, and let θ be a normalized complete mapping of G. We will show that θ is a Bedford-Whitaker mapping of G. Set $X = \{(\mathrm{f}(g), \mathrm{f}(\theta(g))) \mid g \in G\}$. If $(\mathrm{f}(g), \mathrm{f}(\theta(g))) = (\mathrm{f}(g'), \mathrm{f}(\theta(g')))$, for some $g, g' \in G$, then $g = h_1 k_g$ and $g' = h_1 k_{g'}$ for some $k_g, k_{g'} \in K$, and $\theta(g) = h_2 k_g'$ and $\theta(g') = h_2 k_{g'}'$ for some $k_g', k_{g'}' \in K$. It is an exercise to show that if $k_g \neq k_{g'}$ and $k_g' \neq k_{g'}'$, then $k_g k_g' = k_{g'} k_{g'}'$. Hence

$$g\theta(g) = h_1 h_2 k_g k_g' = h_1 h_2 k_{g'} k_{g'}' = g'\theta(g'),$$

and thus $g = g'$. It follows that X contains 8 distinct elements. Now each coset hK of K can be expressed in exactly two ways as $hK = \mathrm{f}(g)$, $g \in G$; exactly two ways as $hK = \mathrm{f}(\theta(g))$, $g \in G$; and exactly two ways as $hK = \mathrm{f}(g\theta(g))$, $g \in G$. In particular, $K = \mathrm{f}(1) = \mathrm{f}(g)$ for some $g \in G$, $g \neq 1$, and, as the elements of X are distinct, $\mathrm{f}(\theta(1)) = K$ and $\mathrm{f}(\theta(g)) = zK$ for some $z \notin K$. Thus $(K, K), (K, zK) \in X$. If $L = \langle K, zK \rangle$, then $L \cap X = \{(K, K), (K, zK)\}$ by Lemma 13.7. Therefore $(zK, K) \notin X$. Now $(xK, K) \in X$ for some $x \notin K \cup zK$. Hence $G/K = \{K, xK, yK, zK\}$ for some $y \in G$. As $(K, K), (xK, K) \in X$, if we set $L = \langle K, xK \rangle$, then $L \cap X = \{(K, K), (xK, K)\}$, and hence $(xK, xK) \notin X$ and $(K, xK) \notin X$. Hence $(yK, xK), (zK, xK) \in X$. Now, either $(xK, yK) \in X$ or $(xK, zK) \in X$, but not both. However, $(xK, yK) \notin X$ as then $zK = K(zK) = (xK)(yK) = (yK)(xK)$ can be expressed three times in the form $\mathrm{f}(g\theta(g))$, an impossibility. Therefore $(xK, zK) \in X$. As $(xK, yK), (K, yK) \notin X$, it must be that $(yK, yK), (zK, yK) \in X$. We have now shown that

$$X = \{(K, K), (K, zK), (xK, K), (yK, xK), (zK, xK), (xK, zK), (yK, yK), (zK, yK)\}.$$

Let the partion $T_1 \cup T_2$ of X be given by

$$T_1 = \{(K, K), (xK, zK), (yK, xK), (zK, yK)\},$$

and

$$T_2 = \{(K, zK), (xK, K), (yK, yK), (zK, xK)\}.$$

If we define $\theta'\colon G/K \to G/K$ by $\{(A, \theta'(A)) \mid A \in G/K\} = T_1$, then θ' is a normalized complete mapping of G/K, $T_{\theta'} = T_1$, and $T_{\theta',z} = T_2$. It is now straightforward to show that θ is a Bedford-Whitaker mapping constructed from θ'. The result follows. □

13.3 Groups of Order 9

There are two groups of order 9, $\mathbb{Z}_9$ and $\mathbb{Z}_3 \times \mathbb{Z}_3 \cong \mathrm{GF}(9)^+$. In 1960 Singer [295] found $\mathbb{Z}_9$ to have 225 normalized orthomorphisms. This result has since been confirmed multiple times by Johnson, Dulmage, and Mendelsohn [196] in 1961; by Chang, Hsiang, and Tai [69] in 1964; by Shieh, Hsiang, and Hsu [293] in 2000; by Lazebnik and Thomason [224] and by Hsiang, Hsu, and Shieh [183] in 2004; by McKay, McLeod, and Wanless [239] in 2006; and by Wanless [325] in 2011. It was first reported that $\omega(\mathbb{Z}_9) = 1$ by Chang, Hsiang, and Tai [69] in 1964: they used extensive hand computations to obtain this result. That $\omega(\mathbb{Z}_9) = 1$ was confirmed in 1986 by Jungnickel and Grams [207] using computers. A complete list of orthomorphisms of $\mathbb{Z}_9$ can be found in the 1980 monograph by Hsu [184] on cyclic neofields, where they are described as presentation functions of cyclic neofields of order 10.

13.3.1 The Group GF(9)$^+$

The orthomorphism graph of GF(9)$^+$ was first studied by Johnson, Dulmage, and Mendelsohn [196] in 1961. They found 21 complete sets of normalized orthomorphisms of GF(9)$^+$ and described 12 of these. They also found 27 automorphisms of GF(9)$^+$ to also be orthomorphisms. Using extensive hand computations, in 1964 Chang, Hsiang, and Tai [69] found GF(9)$^+$ to have 249 normalized orthomorphisms. This number has since been confirmed using computer searches by Stothers [302] in 1990, Hsiang, Hsu, and Shieh [183] and Lazebnik and Thomason [224] in 2004, and by Wanless [325] in 2011. In 1986 Jungnickel and Grams [207] found Orth(GF(9)$^+$) to have maximal cliques of sizes 2, 4, and 7 only. More information on the structure of Orth(GF(9)$^+$) was given by Lazebnik and Thomason [224] in 2004. They found the graph to have 1,248 edges, minimum degree 6, and twenty one 7-cliques, agreeing with the number of 7-cliques found by Johnson, Dulmage, and Mendelsohn. They also found the second largest size of a maximal clique to be 4, in agreement with Jungnickel and Grams' results.

In Theorem 9.19, all orthomorphism polynomials of degree 5 or less were described, from which we can read off the normalized orthomorphism polynomials of GF(9) of degree 5 or less.

Theorem 13.9. *The normalized orthomorphism polynomials of* GF(9) *of degree* 5 *or less are as follows:*

1. ax, $a \neq 0, 1$.
2. $-ax^3 + x$, *a a nonsquare.*
3. ax^3, *a a nonsquare.*
4. $ax^3 + bx$, *ab and $a(b-1)$ nonsquares.*
5. $\pm i(x+b)^5 - b^5$, *b arbitrary.*
6. $\pm i(x+b)^5 + x - b^5$, *b arbitrary.*
7. $\pm i(x+b)^5 - x - b^5$, *b arbitrary.*

Proof. These are described in Theorem 9.19 except for (4), which is described as $(b-a)^{-1}x^3 + \left(1 - b(b-a)^{-1}\right)x$, a and b distinct nonsquares. If we set $\alpha = (b-a)^{-1}$ and $\beta = \left(1 - b(b-a)^{-1}\right)$, then $a = -\alpha^{-1}\beta$ and $b = \alpha^{-1}(1-\beta)$. Now $-\alpha^{-1}\beta$ is a nonsquare if and only if $-\alpha\beta$ is a nonsquare if and only if, as -1 is a square, $\alpha\beta$ is a nonsquare. Similarly $\alpha^{-1}(1-\beta)$ is a nonsquare if and only if $\alpha(\beta-1)$ is a nonsquare. □

Theorem 13.9 accounts for 81 of the 249 normalized orthomorphisms of $\mathrm{GF}(9)^+$. The remaining 168 normalized orthomorphisms of $\mathrm{GF}(9)^+$ are represented by orthomorphism polynomials of degree 6. The orthomorphisms in Theorem 13.9(1) are the linear orthomorphisms, collectively $\mathscr{C}_1(9)$. The orthomorphisms in Theorem 13.9(5), (6), and (7) are the translates of the nonlinear quadratic orthomorphisms, i.e., translates of $\mathscr{C}_2(9) \setminus \mathscr{C}_1(9)$. In Section 9.3 it is shown that the nonlinear quadratic orthomorphisms of $\mathrm{GF}(9)^+$ form a 6-clique of $\mathrm{Orth}(\mathrm{GF}(9)^+)$, and each is orthogonal to the linear orthomorphism [2]. By Theorem 10.26 the orthomorphisms in Theorem 13.9(2), (3), and (4) are nonquadratic, cyclotomic orthomorphisms of index 4, i.e., $\mathscr{C}_4(9) \setminus \mathscr{C}_2(9)$. Note that the orthomorphisms in $\mathscr{C}_4(9) \setminus \mathscr{C}_2(9)$ are also automorphisms of $\mathrm{GF}(9)^+$: in fact $\mathscr{A}(\mathrm{GF}(9)^+) = \mathscr{C}_1(9) \cup (\mathscr{C}_4(9) \setminus \mathscr{C}_2(9))$. Simple counting then yields $|\mathscr{A}(\mathrm{GF}(9)^+)| = 27$. The orthogonalities among the orthomorphisms of $\mathrm{GF}(9)^+$, represented by orthomorphism polynomials of degree 5 or less, were summarized in 1989 by Evans [113].

Theorem 13.10. *The following hold for the orthomorphism graph of* $\mathrm{GF}(9)^+$:

1. *If $ax, bx \in \mathscr{C}_1(9)$ then $ax \perp bx$ if and only if $a \neq b$.*
2. *If $ax^5 + bx \in \mathscr{C}_2(9) \setminus \mathscr{C}_1(9)$ and $cx \in \mathscr{C}_1(9)$, then $ax^5 + bx \perp cx$ if and only if $c = -1$.*
3. *If $ax^5 + bx$ and $a'x^5 + b'x$ are distinct elements of $\mathscr{C}_2(9) \setminus \mathscr{C}_1(9)$, then $ax^5 + bx \perp a'x^5 + b'x$.*
4. *If $ax^3 + bx \in \mathscr{C}_4(9) \setminus \mathscr{C}_2(9)$ and $cx \in \mathscr{C}_1(9)$, then $ax^3 + bx \perp cx$ if and only if $b = c$ or at least one of $b(b-c)$, $(b-1)(b-c)$ is a nonzero square.*
5. *If $ax^3 + bx$ and $cx^3 + dx$ are two distinct elements of $\mathscr{C}_4(9) \setminus \mathscr{C}_2(9)$, then $ax^3 + bx \perp cx^3 + dx$ if and only if $(a-c)(b-d)$ is zero or a nonsquare.*
6. *No element of $\mathscr{C}_2(9) \setminus \mathscr{C}_1(9)$ is orthogonal to any element of $\mathscr{C}_4(9) \setminus \mathscr{C}_2(9)$.*

Proof. The proofs of (1), (2), and (3) are given in Section 9.3. The normalized permutation polynomials of GF(9) of degree 4 or less, given in Theorem 9.12, are x; x^3; $x^3 - ax$, a a nonsquare; x^5; and $x^5 \pm ix$. If $ax^3 + bx \in \mathscr{C}_4(9) \setminus \mathscr{C}_2(9)$ and $cx \in \mathscr{C}_1(9)$, then $ax^3 + bx \perp cx$ if and only if $ax^3 + (b - c)x$ is a permutation polynomial of GF(9) if and only if $b = c$ or $b \neq c$ and $(c - b)a^{-1}$ is a nonsquare. Now each of ab, $a(b - 1)$, is 0 or a nonsquare. It follows that if $b \neq c$, then $(c - b)a^{-1}$ is a nonsquare if and only if at least one of $(c - b)a^{-1}ab = (c - b)b$ or $(c - b)a^{-1}a(b - 1) = (c - b)(b - 1)$ is a square if and only if, as -1 is a square, at least one of $(b - c)b$ or $(b - c)(b - 1)$ is a square. If $ax^3 + bx$ and $cx^3 + dx$ are two distinct elements of $\mathscr{C}_4(9) \setminus \mathscr{C}_2(9)$, then $ax^3 + bx \perp cx^3 + dx$ if and only if $(a - c)x^3 + (b - d)x$ is a permutation polynomial of GF(9) if and only if $a = c$ or $b = d$ or $(d - b)/(a - c)$ is a nonsquare. Now $(d - b)/(a - c)$ is a nonsquare if and only if $(d - b)(a - c)$ is a nonsquare if and only if, as -1 is a square, $(b - d)(a - c)$ is a nonsquare.

If $ax^5 + bx \in \mathscr{C}_2(9) \setminus \mathscr{C}_1(9)$ and $cx^3 + dx \in \mathscr{C}_4(9) \setminus \mathscr{C}_2(9)$, then $ax^5 + bx \perp cx^3 + dx$ if and only if $ax^5 - cx^3 + (b - d)x$ is a permutation polynomial of GF(9). As $a, c \neq 0$, this is not possible, and so no element of $\mathscr{C}_2(9) \setminus \mathscr{C}_1(9)$ can be orthogonal to any element of $\mathscr{C}_4(9) \setminus \mathscr{C}_2(9)$. □

There are some nonorthogonalities that are not mentioned in Theorem 13.10. For instance, Theorem 10.28(2) tells us that elements from different translates of $\mathscr{C}_2(9) \setminus \mathscr{C}_1(9)$ cannot be orthogonal. Also, as all the elements of $\mathscr{C}_4(9) \setminus \mathscr{C}_2(9)$ are automorphisms of $\text{GF}(9)^+$ and hence are fixed by translations, no translate of an element of $\mathscr{C}_2(9) \setminus \mathscr{C}_1(9)$ can be orthogonal to any element of $\mathscr{C}_4(9) \setminus \mathscr{C}_2(9)$.

13.4 The Group $\mathbb{Z}_{11}$

Orthomorphisms of the group $\mathbb{Z}_{11}$ were first studied by Singer [295] in 1960. He found 3,441 normalized orthomorphisms. This has been confirmed many times through computer searches by Johnson, Dulmage, and Mendelsohn [196] in 1961; by Evans and McFarland [138] in 1984; by Stothers [302] in 1990; by Shieh, Hsiang, and Hsu [293] in 2000; by Shieh [292] in 2001; by Lazebnik and Thomason [224], by Hsiang, Hsu, and Shieh [183], and by Maenhaut and Wanless [236] in 2004; and by McKay, McLeod, and Wanless [239] in 2006.

In 1960 Singer [295] gave a list of representatives of orbits of the congruence group generated by the homologies, H_α, $\alpha \in \mathrm{Aut}(\mathbb{Z}_{11})$; the translations T_g, $g \in \mathbb{Z}_{11}$; the reflection R; and the inversion I: these congruences are described in Section 8.1. Singer's list of representatives for these orbits is given in Table 13.5. These orthomorphisms lie in eleven orbits, one orbit of length 3, one of length 6, two of length 66, four of length 330, and three of length 660.

Table 13.5 Representative orthomorphisms of $\mathbb{Z}_{11}$

x	0	1	2	3	4	5	6	7	8	9	10	|Orbit|
$\theta_1(x)$	0	2	4	6	8	10	1	3	5	7	9	3
$\theta_2(x)$	0	3	6	9	1	4	7	10	2	5	8	6
$\theta_3(x)$	0	2	4	8	3	9	1	10	5	7	6	66
$\theta_4(x)$	0	2	5	8	3	1	10	9	6	4	7	66
$\theta_5(x)$	0	2	4	6	9	1	10	5	3	8	7	330
$\theta_6(x)$	0	2	4	7	9	1	5	10	3	6	8	330
$\theta_7(x)$	0	2	5	8	10	9	3	6	4	7	1	330
$\theta_8(x)$	0	2	5	9	6	10	4	3	1	8	7	330
$\theta_9(x)$	0	2	4	7	1	10	9	6	3	5	8	660
$\theta_{10}(x)$	0	2	4	7	1	10	9	5	3	8	6	660
$\theta_{11}(x)$	0	2	4	7	1	10	9	3	6	8	5	660

The study of the structure of Orth($\mathbb{Z}_{11}$) was initiated by Johnson, Dulmage, and Mendelsohn [196] in 1961. They found, via a computer search, that Orth($\mathbb{Z}_{11}$) contained only one 9-clique. Computer searches by Cates and Killgrove [66] in 1981, by Evans and Mcfarland [138] in 1984, and by Lazebnik and Thomason [224] in 2004 confirmed that Orth($\mathbb{Z}_{11}$) contained only one 9-clique: Cates and Killgrove cite Parker as having obtained the same result.

In 1964 Chang, Hsiang, and Tai [69], using extensive hand computations and Singer's list in Table 13.5, claimed that the orthomorphism graph of $\mathbb{Z}_{11}$ had no edges outside of the unique 9-clique: the results of Section 9.3 make it clear that this is not correct. In 1984 Evans and Mcfarland [138], using a computer, determined the structure of Orth($\mathbb{Z}_{11}$). They showed that of the 3,441 vertices of this orthomorphism graph, 2,640 have degree 0, 660 have degree 3, 135 have degree 8, and 6 have degree 162. The vertices of degree 162 represent the linear orthomorphisms [3], [4], [5], [7], [8], and [9]. The neighborhood of a vertex of degree 3 consists of an independent set of vertices, two of degree 3, and one of degree 162. The neighborhood of a vertex of degree 8, if it represents a nonlinear orthomorphism, is the cube graph, six of its vertices having degree 8 and two of its vertices having degree 162. In Evans and Mcfarland [138], it is stated that the neighborhood of a vertex of degree 8 is the cube graph: of course, the neighborhood of a vertex of degree 8, representing a linear orthomorphism, must be a complete graph. The only other computer study of the structure of Orth($\mathbb{Z}_{11}$) was done by Lazebnik and Thomason [224] in 2004. They found this graph to have 2,016 edges; 2,640 isolated vertices; six vertices of degree 162, representing the linear orthomorphisms [3], [4], [5], [7], [8], and [9]; and all other vertices to be of degree at most 8. They also found the 8-core, the largest subgraph in which each vertex has degree at least 8, to consist of 141 vertices, and the second largest maximal clique to be a 3-clique. Lazebnik and Thomason's results are consistent with Evans and Mcfarland's results.

Regarding $\mathbb{Z}_{11}$ as the additive group of the field GF(11), all of the structure of Orth($\mathbb{Z}_{11}$) is contained in the set of normalized linear orthomorphisms, $\mathscr{C}_1(11)$; the set of quadratic orthomorphisms, $\mathscr{C}_2(11)$, and their translates; and the set of cyclotomic orthomorphisms of index 5, $\mathscr{C}_5(11)$, and their translates. Orthomorphisms not in these sets and their translates are represented by isolated vertices. There are no orthogonalities between orthomorphisms in a translate of $\mathscr{C}_2(11) \setminus \mathscr{C}_1(11)$ and orthomorphisms in a translate of $\mathscr{C}_5(11) \setminus \mathscr{C}_1(11)$, no orthogonalities between orthomorphisms in different translates of $\mathscr{C}_2(11) \setminus \mathscr{C}_1(11)$, and no orthogonalities between orthomorphisms in different translates of $\mathscr{C}_5(11) \setminus \mathscr{C}_1(11)$. The structure of $\mathrm{Orth}(\mathbb{Z}_{11})$, minus isolated vertices, is depicted in Figure 13.1.

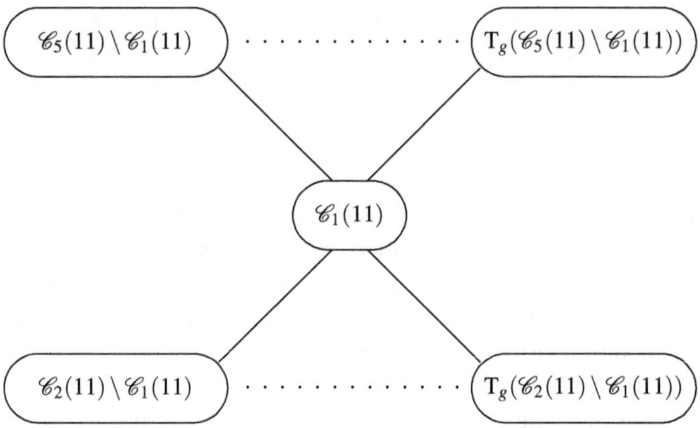

Fig. 13.1 The structure of $\mathrm{Orth}(\mathbb{Z}_{11})$

The elements of $\mathscr{C}_1(11)$ are fixed by all translates and homologies: the actions of the congruences I and R on $\mathscr{C}_1(11)$ are described in Figure 13.2. We see that Singer's orbit of length 3 consists of the linear orthomorphisms [2], [6], and [10], represented by vertices of degree 8; and Singer's orbit of length 6 consists of the

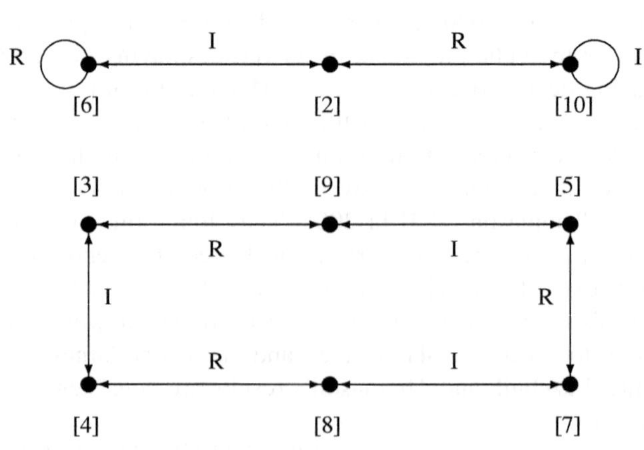

Fig. 13.2 The actions of I and R on $\mathscr{C}_1(11)$

linear orthomorphisms [3], [4], [5], [7], [8], and [9], represented by vertices of degree 162.

By Corollary 9.29, $GF(11)^+ \cong \mathbb{Z}_{11}$ has 12 nonlinear quadratic orthomorphisms, each of which, by Corollary 9.35, is orthogonal to precisely two normalized linear orthomorphisms. By Theorem 10.28, orthomorphisms in different translates of the set of nonlinear quadratic orthomorphism cannot be orthogonal. The orthogonalities between orthomorphisms within $\mathscr{C}_2(11) \setminus \mathscr{C}_1(11)$ and between orthomorphisms in $\mathscr{C}_2(11) \setminus \mathscr{C}_1(11)$ and orthomorphisms in $\mathscr{C}_1(11)$ are easily computed from Theorem 9.24 and are given in Table 13.6.

Table 13.6 Neighbors of nonlinear quadratic orthomorphisms

θ	Neighbors of θ in $\mathscr{C}_2(11) \setminus \mathscr{C}_1(11)$						In $\mathscr{C}_1(11)$	
[2, 6]	[3, 9]	[9, 3]	[4, 5]	[5, 4]	[10, 2]	[6, 10]	[7]	[8]
[6, 2]	[3, 9]	[9, 3]	[4, 5]	[5, 4]	[2, 10]	[10, 6]	[7]	[8]
[2, 10]	[4, 5]	[5, 4]	[7, 8]	[8, 7]	[6, 2]	[10, 6]	[3]	[9]
[10, 2]	[4, 5]	[5, 4]	[7, 8]	[8, 7]	[2, 6]	[6, 10]	[3]	[9]
[3, 9]	[2, 6]	[6, 2]	[6, 10]	[10, 6]	[5, 4]	[8, 7]	[5]	[7]
[9, 3]	[2, 6]	[6, 2]	[6, 10]	[10, 6]	[4, 5]	[7, 8]	[5]	[7]
[4, 5]	[2, 6]	[6, 2]	[2, 10]	[10, 2]	[9, 3]	[7, 8]	[8]	[9]
[5, 4]	[2, 6]	[6, 2]	[2, 10]	[10, 2]	[3, 9]	[8, 7]	[8]	[9]
[6, 10]	[3, 9]	[9, 3]	[7, 8]	[8, 7]	[2, 6]	[10, 2]	[4]	[5]
[10, 6]	[3, 9]	[9, 3]	[7, 8]	[8, 7]	[6, 2]	[2, 10]	[4]	[5]
[7, 8]	[2, 10]	[10, 2]	[6, 10]	[10, 6]	[9, 3]	[4, 5]	[3]	[4]
[8, 7]	[2, 10]	[10, 2]	[6, 10]	[10, 6]	[3, 9]	[5, 4]	[3]	[4]

The actions of the congruences on $\mathscr{C}_2(11) \setminus \mathscr{C}_1(11)$ is described in Theorem 9.26, from which we determine that the orbits under the homologies, H_α, I, and R, are {[2, 6], [6, 2], [2, 10], [10, 2], [6, 10], [10, 6]} and {[3, 9], [9, 3], [4, 5], [5, 4], [7, 8], [8, 7]}: under the actions of the translations, these yield the two Singer orbits of length 66. The 132 vertices, representing nonlinear quadratic orthomorphisms and their translates, are of degree 8: these and the linear orthomorphisms [2], [6], and [10] are the 135 orthomorphisms represented by vertices of degree 8. In Figure 13.3, the neighborhood of the vertex representing the quadratic orthomorphism [2, 6] is shown: this is the cube graph.

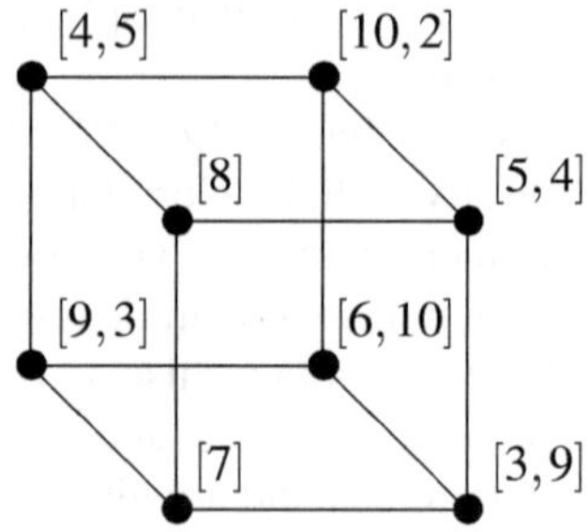

Fig. 13.3 The neighbors of [2, 6]

The complete structure of $\mathscr{C}_2(11)$ is shown in Figure 13.4. This is a collapsing of the orthomorphism graph $\mathscr{C}_2(11)$: the vertex label $\{a, b\}$ represents the pair $[a, b]$ and $[b, a]$, and the edges represent subgraphs:

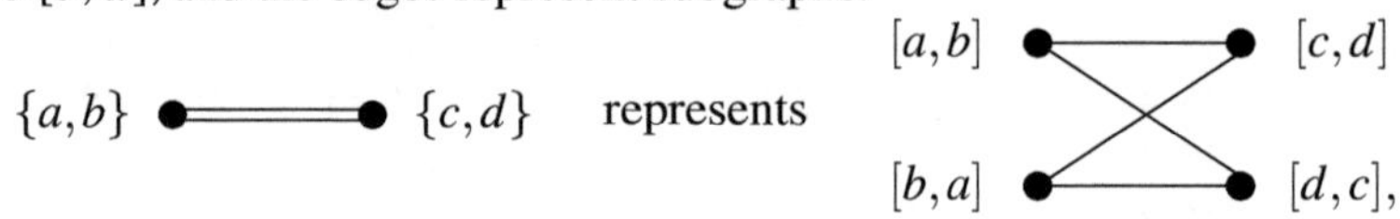

and

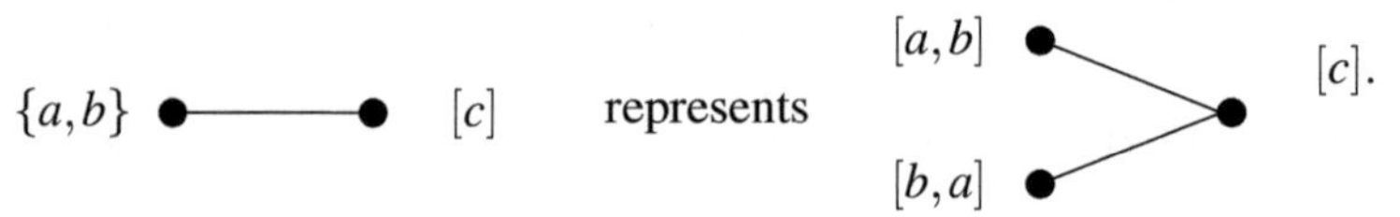

It is clear from Figure 13.4 that the full automorphism group of the graph $\mathscr{C}_2(11)$ is transitive on vertices representing nonlinear quadratic orthomorphisms. Hence, as the neighborhood of the vertex representing [2, 6] is the cube graph, the neighborhood of any vertex representing a nonlinear quadratic orthomorphism is the cube graph.

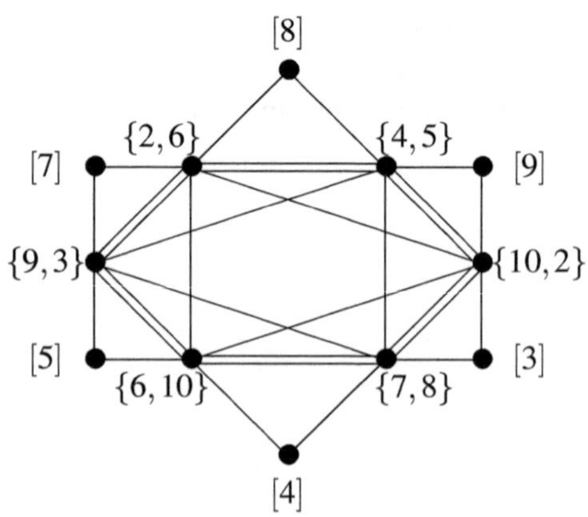

Fig. 13.4 The structure of $\mathscr{C}_2(11)$

$\mathscr{C}_5(11)$ illustrates the limitations of the formula given in Theorem 10.24 for the number of cyclotomic orthomorphisms of a given index. In the case of $\mathscr{C}_5(11)$, this formula is a sum consisting of 14,400 terms, each of which is either 0 or 1!

For $e = 5$ and $q = 11$, the cyclotomic classes are:

1. $C_0 = \{1, 10\}$,
2. $C_1 = \{2, 9\}$,
3. $C_2 = \{4, 7\}$,
4. $C_3 = \{3, 8\}$, and
5. $C_4 = \{5, 6\}$.

The following is a list of four elements of $\mathscr{C}_5(11) \setminus \mathscr{C}_1(11)$.

Theorem 13.11. *The cyclotomic mappings*

1. $\sigma_1 = [10, 4, 4, 6, 7]$,
2. $\sigma_2 = [7, 3, 3, 8, 5]$,
3. $\sigma_3 = [10, 9, 9, 2, 4]$, *and*
4. $\sigma_4 = [2, 3, 3, 10, 8]$,

are orthomorphisms in $\mathscr{C}_5(11) \setminus \mathscr{C}_1(11)$.

Proof. As $10C_0 = C_0$, $4C_1 = C_3$, $4C_2 = C_4$, $6C_3 = C_2$, and $7C_4 = C_1$, σ_1 is a bijection by Theorem 10.20. Now $\sigma_1 - [1] = [9, 3, 3, 5, 6]$ and $9C_0 = C_1$, $3C_1 = C_4, 3C_2 = C_0, 5C_3 = C_2$, and $6C_4 = C_3$. It follows, by Theorem 10.20, that σ_1 is an orthomorphism of $\mathbb{Z}_{11}$. The proof that σ_2, σ_3, and σ_4 are orthomorphisms is similar. □

Now any element of $\mathscr{C}_5(11) \setminus \mathscr{C}_1(11)$ is fixed by the congruences H_α, $\alpha(x) = ax$, $a \in C_0$, only. Hence, every element of $\mathscr{C}_5(11) \setminus \mathscr{C}_1(11)$ is in a Singer orbit of length 330. The orthomorphisms σ_1 and σ_2 are in different Singer orbits of length 330, and hence there are at least 60 orthomorphisms in $\mathscr{C}_5(11) \setminus \mathscr{C}_1(11)$. In Singer's list of representatives of the orbits of the group of congruences of Orth($\mathbb{Z}_{11}$), the orthomorphisms θ_5 and θ_8 and their translates are not in $\mathscr{C}_5(11)$. It follows that $\mathscr{C}_5(11) \setminus \mathscr{C}_1(11)$ contains exactly 60 orthomorphisms each of which is an image of either σ_1 or σ_2 under some congruence of Orth($\mathbb{Z}_{11}$). All orthogonalities in $\mathscr{C}_5(11)$ can be derived from the following list of orthogonalities by applying congruences.

Theorem 13.12. *The following hold in* $\mathscr{C}_5(11)$:

1. $\sigma_1 \perp \sigma_2$,
2. $\sigma_1 \perp \sigma_4$,
3. $\sigma_1 \perp [5]$,
4. $\sigma_2 \perp \sigma_3$, *and*
5. $\sigma_2 \perp [4]$.

Proof. Now $\sigma_1 - \sigma_2 = [3, 1, 1, 9, 2]$ and $3C_0 = C_3$, $1C_1 = C_1$, $1C_2 = C_2$, $9C_3 = C_4$, and $2C_4 = C_0$. It follows, by Theorem 10.20, that $\sigma_1 \perp \sigma_2$. The proof that $\sigma_1 \perp \sigma_4$, $\sigma_1 \perp [5]$, $\sigma_2 \perp \sigma_3$, and $\sigma_2 \perp [4]$ is similar. □

Thus, σ_1 and σ_2 are represented by vertices of degree 3 in Orth($\mathbb{Z}_{11}$) and, as they are in distinct orbits of length 330, we have accounted for all of the vertices of Orth($\mathbb{Z}_{11}$) of degree 3.

13.5 Groups of Order 12

There are five groups of order 12: two abelian groups, $\mathbb{Z}_{12}$ and $\mathbb{Z}_2 \times \mathbb{Z}_6$, and three nonabelian groups, A_4, D_{12}, and

$$T = \langle a, b | a^6 = e, b^2 = a^3, ba = a^{-1}b \rangle.$$

The Sylow 2-subgroup of $\mathbb{Z}_{12}$ is cyclic, and the Sylow 2-subgroup of T, which is $\langle b \rangle$, is also cyclic. Hence, by Theorem 3.34, neither $\mathbb{Z}_{12}$ nor T admit orthomorphisms. In what follows we will discuss what is known about cliques in orthomorphism graphs of each of the three groups of order 12 that do not have cyclic Sylow 2-subgroups. We will see, as the result of computer searches, that $\omega(G)$ cannot exceed 4 for any of these groups. The fact that $\omega(G) < 10$ for any group of order 12, i.e., no group of order 12 can admit complete sets of orthomorphisms, was established in 1973 by Baumert and Hall [31]: this was via computer searches. A theoretical proof that no group of order 12 can admit complete sets of orthomorphisms follows from a 1983 result of Brown [52]. Brown's result was proved in the context of projective planes, but we will state it in the context of affine planes: see Section 1.4 for the definitions and basic results.

Theorem 13.13 (Brown, 1983). *If an affine plane $\mathscr{A}$ of order* 12 *has a translation of order* 3*, then the collineation group of $\mathscr{A}$ is of order* 3^n *for some* n.

Proof. See [52]. □

As a corollary:

Corollary 13.14. *No group of order* 12 *admits a complete set of orthomorphisms.*

Proof. Let G be a group of order 12 that admits a complete set of orthomorphisms. By Theorem 1.52, there exists an affine plane of order 12 that has a group of translation isomorphic to G. It follows that this plane has a translation of order 3 and hence, by Theorem 13.13 that $|G| = 3^n$ for some n, a contradiction from which the result follows. □

13.5.1 *The Groups A_4 and D_{12}*

The alternating group A_4 can be defined by the presentation

$$A_4 = \langle a, b, c \mid a^2 = b^2 = c^3 = 1, ab = ba, ca = abc, cb = ac \rangle.$$

In 1964 Chang and Tai [70] found, via extensive hand computations, that A_4 admits 3,776 normalized orthomorphisms and that $\omega(A_4) = 1$. Later, in 1965 Hall and Knuth [170], using a computer, also found that $\omega(A_4) = 1$. However, Hall and Knuth found A_4 to admit 3,840 normalized complete mappings, a result that was confirmed, via computer searches, by Shieh, Hsiang, and Hsu [293] in 2000 and by Wanless [325] in 2011. In 1973 Baumert and Hall [31] searched for a complete set of mutually orthogonal Latin squares based on groups of order 12. Implicitly this meant looking for a set of ten pairwise orthogonal orthomorphisms of a group of order 12: equivalently ten pairwise orthogonal complete mappings of a group of order 12. For A_4, they found 416 normalized complete mappings of A_4 satisfying $\theta(a) = b$: an example is given in Figure 13.5. They could not extend any of the pairs of orthogonal Latin squares that they had found based on A_4 to a set of four pairwise orthogonal Latin squares based on A_4, thus proving that $\omega(A_4) < 10$.

Fig. 13.5 A complete mapping of A_4

x	1	a	b	ab	c	ac	bc	abc	c^2	ac^2	bc^2	abc^2
$\theta(x)$	1	b	ab	a	c^2	bc^2	abc^2	ac^2	c	abc	ac	bc

The dihedral group D_{12} has the presentation

$$D_{12} = \langle a, b \mid a^6 = b^2 = 1, ba = a^{-1}b \rangle.$$

In 1964 Chang and Tai [70] found, using extensive hand computations, that D_{12} admits 6,336 normalized orthomorphisms: this has been confirmed, via computer searches, by Shieh, Hsiang, and Hsu [293] in 2000 and by Wanless [325] in 2011. Chang and Tai [70] also reported that $\omega(D_{12}) = 2$: again the result of extensive hand computations. In 1965 Hall and Knuth [170] found a set of three mutually orthogonal Latin squares of order 12, obtained from the Cayley table of D_{12} by permuting rows. This corresponds to a right orthogonal pair of right orthomorphisms. Recall from Section 1.3 that a right orthomorphism of a group is a bijection, θ, for which the mapping $x \mapsto \theta(x)x^{-1}$ is a bijection and two right orthomorphisms θ and ϕ are right orthogonal if the mapping $x \mapsto \theta(x)\phi(x)^{-1}$ is a bijection. If θ is a right orthomorphism, then the mapping θ', defined by $\theta'(x) = x\theta(x)^{-1}$, is an orthomorphism. This conversion yields the orthogonal pair of orthomorphisms described in Figure 13.6. In Baumert and Hall's [31] search for a complete set of mutually orthogonal Latin squares based on the group D_{12}, they found 576 normalized complete mappings of D_{12} satisfying $\theta(a^3) = a^4$. They could not extend any of the pairs of orthogonal Latin squares that they had found based on D_{12} to a set of four pairwise orthogonal Latin squares based on D_{12}, thus proving that $\omega(D_{12}) < 10$.

x	1	a	a^2	a^3	a^4	a^5	b	ab	a^2b	a^3b	a^4b	a^5b
$\theta_1(x)$	1	a^4	a^4b	a^2b	a^5b	a^3	ab	b	a^2	a^5	a	a^3b
$\theta_2(x)$	1	ab	a^3b	a	a^2b	a^2	a^5b	a^4	b	a^4b	a^5	a^3

Fig. 13.6 An orthogonal pair of orthomorphisms of D_{12}

13.5.2 The Group $\mathbb{Z}_6 \times \mathbb{Z}_2$

The study of orthomorphisms of $\mathbb{Z}_6 \times \mathbb{Z}_2 = \{nk \mid n \in \mathbb{Z}_6, k \in \mathbb{Z}_2\}$ began in 1959 with Dulmage, Johnson, and Mendelsohn [106]. They estimated that one could obtain about 30,000 Latin squares, orthogonal to the Cayley table of $\mathbb{Z}_6 \times \mathbb{Z}_2$, by permuting rows. This would imply that $\mathbb{Z}_6 \times \mathbb{Z}_2$ admits approximately 30,000 orthomorphisms, i.e., approximately 2,500 normalized orthomorphisms. This is a significant underestimate as the authors of this study found in 1961, via an exhaustive computer search, that the number of normalized orthomorphisms of $\mathbb{Z}_6 \times \mathbb{Z}_2$ is actually 16,512: see Johnson, Dulmage, and Mendelsohn [196]. This number has been confirmed in subsequent computer searches by Shieh, Hsiang, and Hsu [293] in 2000 and by Lazebnik and Thomason [224] in 2004.

In 1959 Dulmage, Johnson, and Mendelsohn [106] found a set of three mutually orthogonal Latin squares of order 12 based on the group $\mathbb{Z}_6 \times \mathbb{Z}_2$. This yields the orthogonal pair of orthomorphisms of $\mathbb{Z}_6 \times \mathbb{Z}_2$ described in Figure 13.7. Note that one of the orthomorphisms, γ, that they found is not normalized.

Fig. 13.7 Orthogonal orthomorphisms of $\mathbb{Z}_6 \times \mathbb{Z}_2$

x	00	10	20	30	40	50	01	11	21	31	41	51
$\beta(x)$	00	30	50	31	01	21	40	20	10	11	51	41
$\gamma(x)$	20	10	50	21	11	51	41	01	31	40	00	30

Subsequently five mutually orthogonal Latin squares based on $\mathbb{Z}_6 \times \mathbb{Z}_2$ were found by Bose, Chakravarti, and Knuth [43] in 1960 and by Johnson, Dulmage, and Mendelsohn [196] in 1961. Johnson, Dulmage, and Mendelsohn found their set of mutually orthogonal Latin squares by permuting the rows of the Cayley table of $\mathbb{Z}_6 \times \mathbb{Z}_2$: seven of the orthomorphisms of $\mathbb{Z}_6 \times \mathbb{Z}_2$ that they found in the process are described in Figure 13.8. In this figure we display an orthomorphism θ_1 and six orthomorphisms, $\theta_2, \ldots, \theta_7$, that are orthogonal to θ_1. The orthomorphisms θ_1, θ_5, θ_6, and θ_7 form a 4-clique of $\mathbb{Z}_6 \times \mathbb{Z}_2$, thus establishing the existence of five mutually orthogonal Latin squares of order 12. The other orthomorphisms, θ_2, θ_3, and θ_4, are not orthogonal to any of the orthomorphisms $\theta_2, \ldots, \theta_7$.

The construction by Bose, Chakravarti, and Knuth [43] in 1960 of a 4-clique in Orth($\mathbb{Z}_6 \times \mathbb{Z}_2$) employed a somewhat different approach. By Theorem 1.21, an r-clique of Orth($\mathbb{Z}_6 \times \mathbb{Z}_2$) corresponds to a $(12, r+2; 1)$-difference matrix over $\mathbb{Z}_6 \times \mathbb{Z}_2$, and by Theorem 1.36, the canonical homomorphism $\mathbb{Z}_6 \times \mathbb{Z}_2 \rightarrow \mathbb{Z}_2$ maps this difference matrix to a $(12, r+2; 6)$-difference matrix over $\mathbb{Z}_2$. Thus, given a difference matrix over $\mathbb{Z}_2$, we might try to reconstruct a difference matrix over $\mathbb{Z}_6 \times \mathbb{Z}_2$. Bose, Chakravarti, and Knuth began with the generalized Hadamard matrix over $\mathbb{Z}_2$, A, given in Figure 13.9, and attempted to construct a difference matrix over $\mathbb{Z}_6 \times \mathbb{Z}_2$ row by row.

Fig. 13.8 Seven orthomorphisms of $\mathbb{Z}_6 \times \mathbb{Z}_2$

x	$\theta_1(x)$	$\theta_2(x)$	$\theta_3(x)$	$\theta_4(x)$	$\theta_5(x)$	$\theta_6(x)$	$\theta_7(x)$
00	00	00	00	00	00	00	00
10	01	41	51	51	30	21	40
20	21	11	41	41	01	10	51
30	20	50	21	50	10	51	41
40	11	21	20	10	31	50	20
50	10	30	40	21	51	31	11
01	31	20	10	11	20	30	21
11	51	10	31	20	21	41	01
21	40	51	50	01	50	20	31
31	41	40	30	30	40	11	10
41	50	31	11	31	11	01	30
51	30	01	01	40	41	40	50

Fig. 13.9 A generalized Hadamard matrix over $\mathbb{Z}_2$

$$A = \begin{bmatrix} 0&0&0&0&0&0&0&0&0&0&0&0 \\ 0&0&0&0&0&0&1&1&1&1&1&1 \\ 0&0&0&1&1&1&1&1&1&0&0&0 \\ 0&1&0&1&0&1&0&1&0&1&0&1 \\ 0&0&1&0&1&1&0&1&0&1&1&0 \\ 0&1&0&1&1&0&1&0&0&1&1&0 \\ 0&1&1&1&0&0&0&1&1&0&1&0 \\ 0&0&1&1&1&0&0&0&1&1&0&1 \\ 0&1&0&0&1&1&0&0&1&0&1&1 \\ 0&1&1&0&1&0&1&1&0&0&0&1 \\ 0&1&1&0&0&1&1&0&1&1&0&0 \\ 0&0&1&1&0&1&1&0&0&0&1&1 \end{bmatrix}$$

For Bose, Chakravarti, and Knuth's construction, each entry in the first row of A is converted from 0 to 00, and the entries in the second row of A are converted from $0, 0, \ldots, 0$ to $00, 10, \ldots, 50$ and from $1, 1, \ldots, 1$ to $01, 11, \ldots, 51$. Hence each entry in the first row of A becomes the identity of $\mathbb{Z}_2 \times \mathbb{Z}_6$, and the entries of the second row of A become the elements of $\mathbb{Z}_2 \times \mathbb{Z}_6$ in a natural order. We then create orthogonal sets of orthomorphisms by converting entries in rows, a 0 in a row to $k0$ for some $k \in \mathbb{Z}_6$ and a 1 in a row to $k1$ for some $k \in \mathbb{Z}_6$, so that the modified rows form a difference matrix over $\mathbb{Z}_2 \times \mathbb{Z}_6$. By so modifying rows, Bose, Chakravarti, and Knuth constructed a set of four pairwise orthogonal orthomorphisms of $\mathbb{Z}_2 \times \mathbb{Z}_6$, $\psi_1, \ldots, \psi_4$, ψ_1 by modifying row 3, ψ_2 by modifying row 4, ψ_3 by modifying row 9, and ψ_4 by modifying row 11: this 4-clique of $\mathrm{Orth}(\mathbb{Z}_2 \times \mathbb{Z}_6)$ is shown in Figure 13.10. The orthomorphisms $\phi_1, \ldots, \phi_4$, shown in Figure 13.10, form another 4-clique of $\mathrm{Orth}(\mathbb{Z}_2 \times \mathbb{Z}_6)$ constructed by Bose, Chakravarti, and Knuth in a similar manner.

In 1977 Mills [243] gave a construction of $(pq, p; 1)$-difference matrices over $\mathrm{Orth}(\mathrm{GF}(q)^+ \times \mathrm{GF}(p)^+)$, p an odd prime divisor of $q - 1$, candidates for rows to extend this difference matrix, and a simple rule for determining when

Fig. 13.10 Two 4-cliques of $\text{Orth}(\mathbb{Z}_6 \times \mathbb{Z}_2)$

x	$\phi_1(x)$	$\phi_2(x)$	$\phi_3(x)$	$\phi_4(x)$	$\psi_1(x)$	$\psi_2(x)$	$\psi_3(x)$	$\psi_4(x)$
00	00	00	00	00	00	00	00	00
10	20	41	50	01	20	11	21	31
20	10	40	31	50	10	40	30	51
30	41	21	20	11	11	51	20	10
40	31	20	41	51	01	20	41	50
50	51	01	11	40	21	31	11	41
01	30	31	21	20	31	30	50	21
11	50	10	40	21	51	41	40	20
21	40	11	51	41	41	10	51	11
31	11	50	10	30	40	21	10	01
41	01	51	30	10	30	50	01	40
51	21	30	01	31	50	01	31	30

a candidate could or could not be used to extend this difference matrix. Mills' construction is described in Theorem 13.15. Using his construction, Mills obtained a (12, 6; 1)-difference matrix over $\text{GF}(4)^+ \times \text{GF}(3)^+$, thus yielding another 4-clique of $\text{Orth}(\mathbb{Z}_6 \times \mathbb{Z}_2)$: Mills' difference matrix is given in Figure 13.11.

$$\begin{bmatrix} (0,0) & (0,2) & (0,0) & (1,2) & (1,2) & (1,1) & (w,1) & (w,1) & (w,2) & (w^2,0) & (w^2,1) & (w^2,0) \\ (0,0) & (0,0) & (0,2) & (w,1) & (w,2) & (w,2) & (w^2,2) & (w^2,1) & (w^2,1) & (1,0) & (1,0) & (1,1) \\ (0,2) & (0,0) & (0,0) & (w^2,2) & (w^2,1) & (w^2,2) & (1,1) & (1,2) & (1,1) & (w,1) & (w,0) & (w,0) \\ (1,0) & (w,0) & (w^2,0) & (1,0) & (w,0) & (w^2,0) & (1,0) & (w,0) & (w^2,0) & (1,0) & (w,0) & (w^2,0) \\ (1,0) & (w,1) & (w^2,2) & (w^2,2) & (1,0) & (w,1) & (w,1) & (w^2,2) & (1,0) & (w,2) & (w^2,0) & (1,1) \\ (w^2,0) & (1,2) & (w,1) & (w,0) & (w^2,2) & (1,1) & (1,0) & (w,2) & (w^2,1) & (w,1) & (w^2,0) & (1,2) \end{bmatrix}$$

Fig. 13.11 A (12, 6; 1)-difference matrix over $\text{GF}(4)^+ \times \text{GF}(3)^+$

In Baumert and Hall's [31] search for a complete set of mutually orthogonal Latin squares based on the group $\mathbb{Z}_2 \times \mathbb{Z}_6$, they found 1,728 normalized complete mappings of $\mathbb{Z}_2 \times \mathbb{Z}_6$ satisfying $\theta(01) = 31$. They were able to extend many of the pairs of orthogonal Latin squares that they had found based on $\mathbb{Z}_2 \times \mathbb{Z}_6$ to sets of five pairwise orthogonal Latin squares based on $\mathbb{Z}_2 \times \mathbb{Z}_6$, thus yielding yet another confirmation that $\omega(\mathbb{Z}_2 \times \mathbb{Z}_6) \geq 4$.

In 1992 Roberts [281] found via computer 24 sets of 5 MOLS of order 12. These MOLS were obtained by permuting the columns of a base square. As Roberts' base square is isotopic to the Cayley table of $\mathbb{Z}_2 \times \mathbb{Z}_6$, this showed once again that $\omega(\mathbb{Z}_2 \times \mathbb{Z}_6) \geq 4$.

The results so far show that $\omega(\mathbb{Z}_6 \times \mathbb{Z}_2) \geq 4$. In fact $\omega(\mathbb{Z}_6 \times \mathbb{Z}_2) = 4$. In 1961 Johnson, Dulmage, and Mendelsohn [196] cited Parker and Van Duren as having proved this result, and in 1973, via a computer search, Baumert and Hall [31] also showed that $\omega(\mathbb{Z}_6 \times \mathbb{Z}_2) = 4$.

13.6 Larger Groups

For groups of order 13 or more, very little is known about the structure of their orthomorphism graphs. Several researchers have used computer searches to determine the number of orthomorphisms larger groups have: this number has been determined for all groups of order 23 or less and for the cyclic group of order 25. The data for groups of order 23 or less are presented in Table 13.1. In 1990 Stothers [302] found the number of complete mappings of $GF(q)^+$ for $7 \leq q \leq 17$. In 1999 Hsu [185] listed the number of normalized orthomorphisms for cyclic groups of order 15 or less. The number of orthomorphisms of cyclic groups of order 23 or less was computed by Shieh [292] in 2001. It was reported in [239] that Shieh had found $\mathbb{Z}_{25}$ to have 1,664,382,756,757,625 normalized orthomorphisms. In 2000 Shieh, Hsiang, and Hsu [293] computed the number of orthomorphisms for all abelian groups of order 18 or less. In 2006 McKay, McLeod, and Wanless [239] computed the number of orthomorphisms for all groups of order at most 20 and for the noncyclic group of order 21. In 2004 Lazebnik and Thomason [224] computed the number of normalized orthomorphisms for abelian groups of order 16 or less. In 2004 Hsiang, Hsu, and Shieh [183] listed the number of normalized orthomorphisms for all cyclic groups of order at most 23, and in 2011 Wanless [325] listed the number of normalized orthomorphisms for all groups of order 23 or less. The hardness of the problem of counting orthomorphisms was studied by Hsiang, Hsu, and Shieh [183] in 2004.

13.6.1 The Group $\mathbb{Z}_{13}$

In 1981 Cates and Killgrove [66] determined, via a computer search, that the only 11-clique of $\mathrm{Orth}(\mathbb{Z}_{13})$ is $\mathscr{C}_1(13)$: this was confirmed by Mendelsohn and Wolk [240] in 1985 and by Lazebnik and Thomason [224] in 2004. In 1990 Stothers [302], through a computer search, found that $GF(13)^+$ has 1,030,367 complete mappings, i.e., 79,259 normalized complete mappings: this was confirmed by Shieh, Hsiang, and Hsu [293] in 2000; by Shieh [292] in 2001; by McKay, McLeod, and Wanless [239] in 2006; and by Lazebnik and Thomason [224] in 2004. More details on the structure of $\mathrm{Orth}(\mathbb{Z}_{13})$ were given by Lazebnik and Thomason [224] in 2004. They found the graph to contain 395,242 edges; one 11-clique, again confirming Cates and Killgrove's result; a maximal clique of size 5, but no maximal cliques of size $6, \ldots, 10$; and 271 vertices in its core, the largest induced subgraph in which each vertex has degree at least 10.

Very little theoretical work on $\mathrm{Orth}(\mathbb{Z}_{13})$ has been done. We know that $\mathbb{Z}_{13}$ has 20 nonlinear quadratic orthomorphisms, listed in Example 9.32, and 54 nonlinear cyclotomic orthomorphisms of index 3 by Corollary 10.25. Neighbors in $\mathscr{C}_2(13)$ of ten of the nonlinear quadratic orthomorphisms are given in Table 13.7: neighbors of the other ten are obtained by noting that $[a, b] \perp [c, d]$ implies $[b, a] \perp [d, c]$.

Table 13.7 Neighbors of nonlinear quadratic orthomorphisms of $\mathbb{Z}_{13}$

θ	Neighbors of θ in $\mathscr{C}_2(13) \setminus \mathscr{C}_1(13)$											In $\mathscr{C}_1(13)$		
[9, 12]	[2, 5]	[5, 2]	[5, 11]	[11, 5]	[12, 3]	[7, 6]	[6, 8]	[4, 10]				[4]	[7]	[8]
[2, 5]	[3, 9]	[9, 3]	[9, 12]	[12, 9]	[11, 2]	[4, 10]	[6, 8]	[8, 7]				[6]	[7]	[10]
[11, 2]	[6, 7]	[7, 6]	[6, 8]	[8, 6]	[2, 5]	[12, 3]	[3, 9]	[4, 10]				[4]	[9]	[12]
[12, 3]	[6, 8]	[8, 6]	[7, 8]	[8, 7]	[11, 2]	[4, 10]	[5, 11]	[9, 12]				[2]	[5]	[10]
[7, 6]	[2, 11]	[11, 2]	[5, 11]	[11, 5]	[3, 9]	[4, 10]	[8, 7]	[9, 12]				[3]	[10]	[12]
[8, 7]	[3, 9]	[9, 3]	[3, 12]	[12, 3]	[2, 5]	[4, 10]	[5, 11]	[7, 6]				[2]	[4]	[11]
[3, 9]	[2, 5]	[5, 2]	[4, 10]	[10, 4]	[7, 8]	[8, 7]	[11, 2]	[5, 11]	[7, 6]	[6, 8]		[6]	[11]	[12]
[5, 11]	[4, 10]	[10, 4]	[6, 7]	[7, 6]	[9, 12]	[12, 9]	[3, 9]	[12, 3]	[6, 8]	[8, 7]		[2]	[3]	[8]
[6, 8]	[2, 11]	[11, 2]	[3, 12]	[12, 3]	[4, 10]	[10, 4]	[2, 5]	[3, 9]	[5, 11]	[9, 12]		[5]	[7]	[9]
[4, 10]	[3, 9]	[9, 3]	[5, 11]	[11, 5]	[6, 8]	[8, 6]	[2, 5]	[11, 2]	[12, 3]	[7, 6]		[2]	[7]	[12]
	[8, 7]	[9, 12]												

It is clear from Table 13.7 that each vertex in the orthomorphism graph $\mathscr{C}_2(13)$ is of degree at least 10 and hence $\mathscr{C}_2(13)$ is contained in the core of Orth($\mathbb{Z}_{13}$). Counting the 11 linear orthomorphisms of $\mathbb{Z}_{13}$ and the 13 translates of the 20 nonlinear quadratic orthomorphisms of $\mathbb{Z}_{13}$ yields 271 orthomorphisms in the core of Orth($\mathbb{Z}_{13}$). Thus the core consists of the translates of $\mathscr{C}_2(13)$, and, hence, by Theorem 9.51, the core contains exactly one 11-clique, and, by Theorem 9.38, the core contains maximal 5-cliques. As there are no orthogonalities between nonlinear orthomorphisms in distinct translates of $\mathscr{C}_2(13)$ by Theorem 10.28, the structure of the core of Orth($\mathbb{Z}_{13}$) is completely determined by the structure of $\mathscr{C}_2(13)$: this structure is shown in Figures 13.12, 13.13, and 13.14. Here $\mathscr{C}_2(13)$ is depicted as the union of three subgraphs, and for simplicity, edges between linear orthomorphisms have been suppressed. In each subgraph the vertex label $\{a, b\}$ represents the pair $[a, b]$ and $[b, a]$, and the edges represent subgraphs:

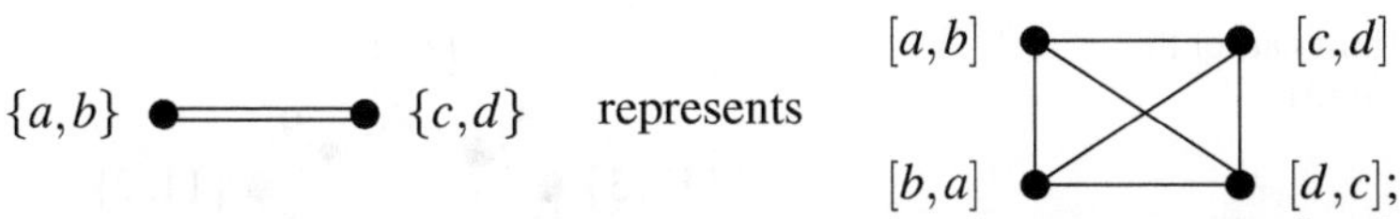

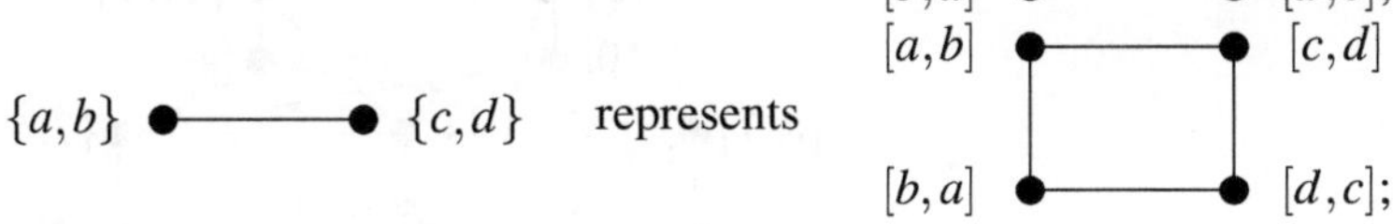

and

Fig. 13.12 First part of the structure of $\mathscr{C}_2(13)$

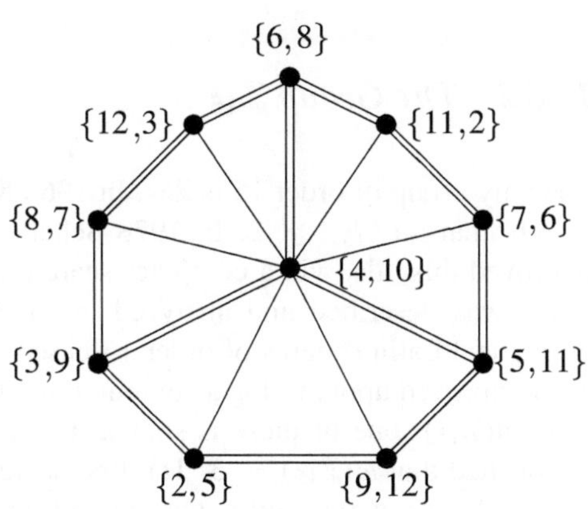

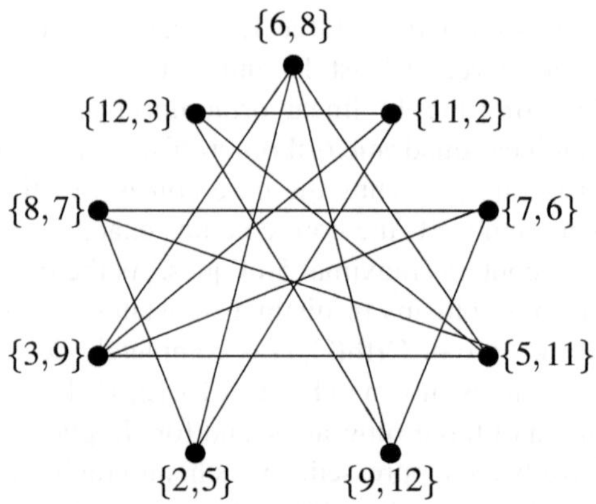

Fig. 13.13 Second part of the structure of $\mathscr{C}_2(13)$

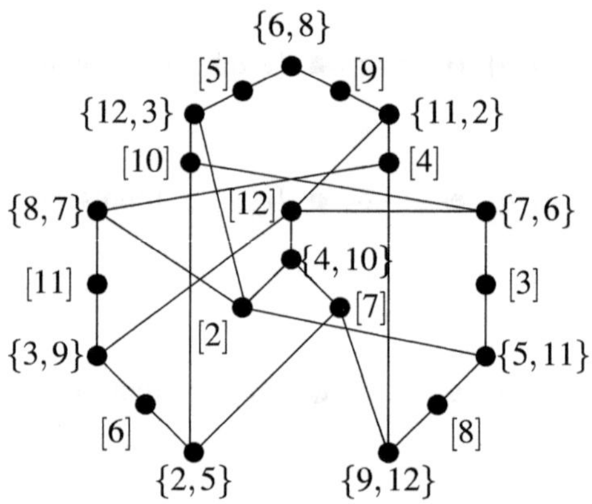

Fig. 13.14 Third part of the structure of $\mathscr{C}_2(13)$

13.6.2 The Group $\mathbb{Z}_{15}$

The only group of order 15 is $\mathbb{Z}_{15}$. In 1966 Keedwell [212] proved, via a computer search, that $\omega(\mathbb{Z}_{15}) \geq 2$. In 1978 Schellenberg, van Rees, and Vanstone [289] improved this, also via a computer search, to $\omega(\mathbb{Z}_{15}) \geq 3$: the 3-clique that they found was described and analyzed in Chapter 11. Thus there exist 4 mutually orthogonal Latin squares of order 15 based on the group $\mathbb{Z}_{15}$, a result that has not been improved upon. Using a computer in 1992, Roberts [281] found 360 3-cliques of Orth($\mathbb{Z}_{15}$): one of these is shown in Table 13.8. Two computer searches have established that $\omega(\mathbb{Z}_{15}) = 3$. The first search was by Roth and Wilson [283]: the results of their search have not been published. The second computer search to show that $\omega(\mathbb{Z}_{15}) = 3$ was conducted by Bedford and Whitaker [39] in 2000.

Table 13.8 3-clique of Orth($\mathbb{Z}_{15}$) found by Roberts

x	0	5	10	9	14	4	3	8	13	12	2	7	6	11	1
$\theta_1(x)$	0	10	5	1	11	6	12	7	2	13	8	3	9	4	14
$\theta_2(x)$	0	9	8	10	7	11	14	13	12	6	5	4	1	2	3
$\theta_3(x)$	0	14	1	4	2	3	11	12	10	8	9	5	7	13	6

13.6.3 Groups of Order 16

There are 13 noncyclic groups of order 16: four of these, GF(16)$^+$, $\mathbb{Z}_2 \times \mathbb{Z}_2 \times \mathbb{Z}_4$, $\mathbb{Z}_4 \times \mathbb{Z}_4$, and $\mathbb{Z}_2 \times \mathbb{Z}_8$, are abelian. For the orthomorphism graph of GF(16)$^+$, in 1990 Stothers [302], via a computer search, found that GF(16)$^+$ has 244,744,192 complete mappings, i.e., 15,296,512 normalized complete mappings; and Lazebnik and Thomason [224] determined some of its structure in 2004. Lazebnik and Thomason found this graph to have 15,296,512 vertices and 2,199,658,496 edges. The minimum vertex degree is 109; the maximum vertex degree is 60,160, and 112 vertices are of maximum degree; the second largest degree is 9,984; and the average vertex degree is 143.8. They also found that the graph contained 2,439,392 K_{14}s. For the other noncyclic abelian groups, Lazebnik and Thomason found $\omega(\mathbb{Z}_2 \times \mathbb{Z}_8) = 3$, $\omega(\mathbb{Z}_4 \times \mathbb{Z}_4) = 6$, and $\omega(\mathbb{Z}_2 \times \mathbb{Z}_2 \times \mathbb{Z}_4) = 6$. They found that Orth($\mathbb{Z}_2 \times \mathbb{Z}_8$) contained 256,000 3-cliques, Orth($\mathbb{Z}_4 \times \mathbb{Z}_4$) contained 21,504 6-cliques, and Orth($\mathbb{Z}_2 \times \mathbb{Z}_2 \times \mathbb{Z}_4$) contained 1,062,656 6-cliques.

There are 9 nonabelian groups of order 16, $D_8 \times \mathbb{Z}_2$, $Q_8 \times \mathbb{Z}_2$, $\Gamma_2 b$, $\Gamma_2 c_1$, $\Gamma_2 c_2$, $\Gamma_2 d$, D_{16}, $\Gamma_3 a_2$, and Q_{16}. For $D_{16} = \langle a, b \mid a^8 = b^2 = 1, bab = a^{-1} \rangle$, Quinn [273] found a 3-clique of Orth(D_{16}) in 1999: her 3-clique is shown in Figure 13.15. For any other nonabelian group G of order 16, $\omega(G) \geq 2$ if G has a normal subgroup H for which H and G/H are both isomorphic to $\mathbb{Z}_2 \times \mathbb{Z}_2$ and $\omega(G) \geq 1$ otherwise. Information about subgroups and quotient groups for each of these groups can be found in the 1980 book by Thomas and Woods [309]. In this book $D_8 \times \mathbb{Z}_2$ is denoted 16/6, $Q_8 \times \mathbb{Z}_2$ is denoted 16/7, $\Gamma_2 b$ is denoted 16/8, $\Gamma_2 c_1$ is denoted 16/9, $\Gamma_2 c_2$ is denoted 16/10, $\Gamma_2 d$ is denoted 16/11, $\Gamma_3 a_2$ is denoted 16/13, and Q_{16} is denoted 16/14. From this data we can determine that $\omega(G) \geq 2$ if G is one of $D_8 \times \mathbb{Z}_2$, $Q_8 \times \mathbb{Z}_2$, $\Gamma_2 b$, $\Gamma_2 c_1$, $\Gamma_2 c_2$, or $\Gamma_2 d$ and $\omega(G) \geq 1$ is $G = \Gamma_3 a_2$ or Q_{16}.

x	1	a	a^2	a^3	a^4	a^5	a^6	a^7	b	a	a^2b	a^3b	a^4b	a^5b	a^6b	a^7b
$\theta_1(x)$	1	b	a^6b	a^6	a^4b	a^4	a^2	a^2b	a^3	a^7b	a	a^5b	a^3b	a^7	ab	a^5b
$\theta_2(x)$	1	a^3	a^4b	a^4	a^3b	a^2b	a	a^7b	a^7	a^6	a^5b	a^5	a^6b	ab	a^2	b
$\theta_3(x)$	1	a^3b	a^3	a^7b	a^5b	b	a^4	a^2	ab	a^4b	a^6b	a^6	a^2b	a^5	a^7	a

Fig. 13.15 A 3-clique of Orth(D_{16})

13.6.4 *The Groups* $GF(4)^+ \times GF(5)^+$, $\mathbb{Z}_{21}$, *and* $GF(3)^+ \times GF(8)^+$

Work on groups of order greater than 16 has mostly focused on improving lower bounds for $\omega(G)$. Cyclic groups were dealt with in Section 11.5, dihedral groups in Section 12.1, and linear groups of characteristic 2 in Sections 12.2 and 12.3. For the remainder of this section, we will only study lower bounds for $\omega(G)$ when $G \cong GF(q_1)^+ \times \cdots \times GF(q_n)^+$, $n \geq 2$, $q_1 < \cdots < q_n$ powers of distinct primes, $q_1 \neq 2$: lower bounds for such groups of order 21 or less are included in Table 13.1 and for such groups of orders 24 through 84 in Figure 13.14. As $\omega(GF(q_i)^+) = q_i - 2$, by Corollary 8.5, $\omega(G) \geq q_1 - 2$. MacNeish's conjecture, if it were true, would have implied that $\omega(G) = q_1 - 2$. We have already demonstrated counterexamples to MacNeish's conjecture, $\omega(\mathbb{Z}_2 \times \mathbb{Z}_6) = 4$ and $\omega(\mathbb{Z}_{15}) = 3$, and in each of these cases, the MacNeish's conjecture would have predicted a value of 2. These results established that $N(12) \geq 5$ and $N(15) \geq 4$, values that have not been improved upon.

For order 20, in 1989 Todorov [312] proved, using transversal designs, that $N(20) \geq 4$, exceeding the value predicted by the MacNeish conjecture by 1. In 1992 Roberts [281] gave a new proof of this result using orthomorphisms of $GF(4)^+ \times GF(5)^+$: one of the 3-cliques that he found is depicted in Table 13.9, where the elements of $GF(4)^+ \times GF(5)^+$ are described as $\{ijk \mid i \in \mathbb{Z}_5, j \in \mathbb{Z}_2, k \in \mathbb{Z}_2\}$.

Table 13.9 A 3-clique of $\mathrm{Orth}(GF(4)^+ \times GF(5)^+)$ found by Roberts

x	000	001	010	011	100	101	110	111	200	201
$\theta_1(x)$	000	010	011	001	400	410	411	401	300	310
$\theta_2(x)$	000	011	001	010	300	311	301	310	110	101
$\theta_3(x)$	000	100	210	411	301	110	401	201	311	410
x	210	211	300	301	310	311	400	401	410	411
$\theta_1(x)$	311	301	201	211	210	200	101	111	110	100
$\theta_2(x)$	111	100	410	401	411	400	211	200	210	201
$\theta_3(x)$	101	010	310	111	211	001	200	400	300	011

In 1959 Parker [264] constructed 4 MOLS of order 21. This provided the first counterexample to the MacNeish conjecture: according to this conjecture, $N(21) = 2$. In his 1995 survey of orthomorphisms and near orthomorphisms, Bedford [35] showed that Parker's construction was based on the group $\mathbb{Z}_{21}$, thus proving that $\omega(\mathbb{Z}_{21}) \geq 3$. A corrected version of the 3-clique of $\mathrm{Orth}(\mathbb{Z}_{21})$, derived from Parker's construction, is described in Table 13.10: in Bedford's depiction, the images of $\theta_3(1)$, $\theta_3(6)$, $\theta_3(8)$, and $\theta_3(18)$ were incorrect.

Other researchers have since found 3-cliques of $\mathrm{Orth}(\mathbb{Z}_{21})$. In 1978 Schellenberg, van Rees, and Vanstone [289] found such a clique: their clique is described in Example 11.5. In 1980 Jungnickel [201] constructed a $(21, 5; 1)$-difference matrix over $\mathbb{Z}_{21}$, implying that $\omega(\mathbb{Z}_{21}) \geq 3$: Jungnickel's example is described

Table 13.10 Parker's 3-clique of Orth($\mathbb{Z}_{21}$)

x	0	1	2	3	4	5	6	7	8	9	10	11	12	13	14	15	16	17	18	19	20
$\theta_1(x)$	0	6	12	4	11	7	18	20	1	3	13	9	16	19	10	2	15	5	8	14	17
$\theta_2(x)$	0	18	16	11	9	20	8	17	6	4	19	3	15	14	13	12	2	7	1	10	5
$\theta_3(x)$	0	8	15	9	3	17	1	5	18	11	14	4	2	10	19	16	12	20	6	13	7

in Figure 8.3. Several 3-cliques of Orth($\mathbb{Z}_{21}$) were discovered by Roberts [281] in 1992. In 1991 Nazarok [251] improved on these results by finding a 4-clique of Orth($\mathbb{Z}_{21}$) thus proving that $N(21) \geq 5$, the best lower bound to date. Nazarok's 4-clique of Orth($\mathbb{Z}_{21}$) is shown in Table 13.11.

Table 13.11 Nazarok's 4-clique of Orth($\mathbb{Z}_{21}$)

x	0	1	2	3	4	5	6	7	8	9	10	11	12	13	14	15	16	17	18	19	20
$\theta_1(x)$	0	19	15	12	10	1	13	2	9	11	3	14	20	7	18	6	5	16	8	17	4
$\theta_2(x)$	0	17	6	20	3	19	7	15	11	14	12	18	4	2	8	13	1	5	9	16	10
$\theta_3(x)$	0	16	13	4	12	17	2	6	14	18	20	8	10	15	9	7	19	1	11	5	3
$\theta_4(x)$	0	5	7	10	20	16	15	13	18	8	4	9	3	6	11	2	17	19	14	1	12

In 1987 Roth and Peters [282] found several 3-cliques of $\mathbb{Z}_2 \times \mathbb{Z}_2 \times \mathbb{Z}_6$ using a computer. The orthomorphisms they found are listed in the Table 13.12: in this table the elements of $\mathbb{Z}_2 \times \mathbb{Z}_2 \times \mathbb{Z}_6$ are described as $\{ijk \mid i \in \mathbb{Z}_6, j \in \mathbb{Z}_2, k \in \mathbb{Z}_2\}$. Of these orthomorphisms the following sets form 3-cliques: $\{\sigma_1, \sigma_2, \sigma_3\}$, $\{\sigma_1, \sigma_2, \sigma_4\}$, $\{\sigma_1, \sigma_2, \sigma_5\}$, $\{\sigma_1, \sigma_2, \sigma_6\}$, $\{\sigma_1, \sigma_2, \sigma_7\}$, $\{\sigma_1, \sigma_8, \sigma_9\}$, $\{\sigma_1, \sigma_8, \sigma_{10}\}$. This implies the existence of 4 mutually orthogonal Latin squares of order 24.

In 1992 Roberts [281] improved on this result by proving the existence of 5 mutually orthogonal Latin squares of order 24, obtained by constructing a 4-clique of $\mathbb{Z}_2 \times \mathbb{Z}_2 \times \mathbb{Z}_6$. Each of Roberts' orthomorphisms is a cyclic permutation of order 23 of the nonzero elements of $\mathbb{Z}_2 \times \mathbb{Z}_2 \times \mathbb{Z}_6$. Roberts' 4-clique is shown in Table 13.13: we have used the description of the elements of $\mathbb{Z}_2 \times \mathbb{Z}_2 \times \mathbb{Z}_6$ as $\{ijk \mid i \in \mathbb{Z}_6, j \in \mathbb{Z}_2, k \in \mathbb{Z}_2\}$. This result was further improved by Abel, Colbourn, and Wojta [7] in 2004 by constructing a (24, 8; 1)-difference matrix over $GF(8)^+ \times GF(3)^+$, yielding $N(24) \geq 7$, the current best lower bound: their construction is described in [6].

13.6.5 Further Results

We have already seen some constructions of cliques of Orth(G), $|G| > 24$, G a direct product of elementary abelian groups: many cyclic groups are of this form. In Example 11.3, we described Schellenberg, van Rees, and Vanstone's [289] construction of a 2-clique of Orth($\mathbb{Z}_{33}$) and in Example 11.4, their construction of a 2-clique

Table 13.12 Orthomorphisms of $\mathbb{Z}_2 \times \mathbb{Z}_2 \times \mathbb{Z}_6$ found by Roth and Peters

x	$\sigma_1(x)$	$\sigma_2(x)$	$\sigma_3(x)$	$\sigma_4(x)$	$\sigma_5(x)$	$\sigma_6(x)$	$\sigma_7(x)$	$\sigma_8(x)$	$\sigma_9(x)$	$\sigma_{10}(x)$
000	000	000	000	000	000	000	000	000	000	000
100	200	300	210	511	001	510	310	300	311	011
200	400	510	211	010	100	211	111	501	401	101
300	501	201	411	510	101	400	100	211	411	400
400	311	210	501	411	501	301	211	200	301	510
500	410	301	110	001	111	501	501	311	510	501
010	011	001	401	410	400	410	411	001	400	411
110	211	411	010	011	300	310	210	400	200	401
210	511	101	500	110	510	401	201	110	410	200
310	110	400	011	501	301	111	300	401	010	500
410	100	511	301	400	201	011	311	510	201	010
510	301	100	400	211	500	500	400	101	511	001
001	010	011	101	101	110	110	110	011	100	100
101	210	401	310	300	010	010	511	410	500	110
201	510	111	200	401	200	101	500	100	110	511
301	111	410	311	210	011	411	001	411	310	211
401	101	501	001	111	511	311	010	500	501	301
501	300	110	100	500	210	200	101	111	211	310
011	001	010	300	311	310	300	301	010	300	311
111	201	310	510	200	311	210	011	310	011	300
211	401	500	511	301	410	511	410	511	101	410
311	500	211	111	201	411	100	401	201	111	111
411	310	200	201	100	211	001	510	210	001	201
511	411	311	410	310	401	201	200	301	210	210

of Orth$(\mathbb{Z}_{39})$. Both of these groups are isomorphic to direct products of elementary abelian groups, $\mathbb{Z}_{33} \cong \mathbb{Z}_3 \times \mathbb{Z}_{11}$ and $\mathbb{Z}_{39} \cong \mathbb{Z}_3 \times \mathbb{Z}_{13}$. Jungnickel's [201, 202] constructions in 1980 and 1981 of difference matrices over cyclic groups, using difference sets and families, were described in Section 8.3. Some explicit lower bounds for $\omega(\mathbb{Z}_n)$, derived from Jungnickel's constructions are given in Tables 8.1 and 8.3. One example from Table 8.1: for $\mathbb{Z}_{57} \cong \mathbb{Z}_3 \times \mathbb{Z}_{19}$, $\omega(\mathbb{Z}_{57}) \geq 6$. Also, in Section 11.5, we proved that $\omega(\mathbb{Z}_{3m}) \geq 3$ if $m > 3$ is relatively prime to 6.

In Table 13.14 we list the current best lower bounds for $\omega(G)$, $24 \leq |G| \leq 84$, G a direct product of elementary abelian groups. Several of the bounds in this table also yield the current lower bounds for $N(n)$. Constructions of difference matrices that establish the current lower bounds for $N(n)$ can be found in [6]. In 2004 Abel, Colbourn, and Wojta [7] 2004 constructed a $(24, 8; 1)$-difference matrix over GF$(8)^+ \times$ GF$(3)^+$, establishing that $N(24) \geq 7$; in 2008 Abel [2] constructed a $(28, 6; 1)$-difference matrix over GF$(7)^+ \times$ GF$(4)^+$, establishing that $N(28) \geq 5$; in 2005 Abel and Ge [9] constructed a $(33, 6; 1)$-difference matrix over GF$(11)^+ \times$ GF$(3)^+$, establishing that $N(33) \geq 5$; in 2004 Abel, Colbourn, and Wojta [7] constructed a $(36, 9; 1)$-difference matrix over GF$(9)^+ \times$ GF$(4)^+$, establishing that $N(36) \geq 8$; in 2005 Abel and Ge [9] constructed a

Table 13.13 Roberts' 4-clique of $\mathrm{Orth}(\mathbb{Z}_2 \times \mathbb{Z}_2 \times \mathbb{Z}_6)$

x	$\sigma_1(x)$	$\sigma_2(x)$	$\sigma_3(x)$	$\sigma_4(x)$
000	000	000	000	000
100	011	210	511	401
200	400	211	410	011
300	411	401	501	210
400	200	001	401	201
500	211	400	300	411
010	300	301	311	001
110	311	511	200	400
210	100	510	101	010
310	111	100	210	211
410	500	300	110	200
510	511	101	011	410
001	010	311	310	301
101	001	501	201	100
201	410	500	100	310
301	401	110	211	511
401	210	310	111	500
501	201	111	010	110
011	310	010	001	300
111	301	200	510	101
211	110	201	411	311
311	101	411	500	510
411	510	011	400	501
511	501	410	301	111

(39, 6; 1)-difference matrix over $\mathrm{GF}(13)^+ \times \mathrm{GF}(3)^+$, establishing that $N(39) \geq 5$; in 1994 Abel and Cheng [5] constructed a (40, 8; 1)-difference matrix over $\mathrm{GF}(8)^+ \times \mathrm{GF}(5)^+$, establishing that $N(40) \geq 7$; in 2008 Abel [2] constructed a (44, 6; 1)-difference matrix over $\mathrm{GF}(11)^+ \times \mathrm{GF}(4)^+$, establishing that $N(44) \geq 5$; in 2002 Abel and Bennett [3] constructed a (45, 7; 1)-difference matrix over $\mathrm{GF}(9)^+ \times \mathrm{GF}(5)^+$, establishing that $N(45) \geq 6$; in 2007 Abel and Cavenagh [4] constructed a (48, 9; 1)-difference matrix over $\mathrm{GF}(16)^+ \times \mathrm{GF}(3)^+$, establishing that $N(48) \geq 8$; in 2005 Abel and Ge [9] constructed a (51, 6; 1)-difference matrix over $\mathrm{GF}(17)^+ \times \mathrm{GF}(3)^+$, establishing that $N(51) \geq 5$; in 2008 Abel [2] constructed a (52, 6; 1)-difference matrix over $\mathrm{GF}(13)^+ \times \mathrm{GF}(4)^+$, establishing that $N(52) \geq 5$; in 2000 Wojta [334] constructed a (55, 7; 1)-difference matrix over $\mathrm{GF}(11)^+ \times \mathrm{GF}(5)^+$, establishing that $N(55) \geq 6$; and in 2004 Abel, Colbourn, and Wojta [7] constructed a (75, 8; 1)-difference matrix over $\mathrm{GF}(25)^+ \times \mathrm{GF}(3)^+$, establishing that $N(75) \geq 7$. In 1996 Wojta [333] constructed a (35, 6; 1)-difference matrix over $\mathrm{GF}(7)^+ \times \mathrm{GF}(5)^+$, establishing that $N(35) \geq 5$: the construction described in [6] is an alternate construction, also by Wojta. The constructions in [6] of a (56, 8; 1)-difference matrix over $\mathrm{GF}(8)^+ \times \mathrm{GF}(7)^+$ and a (80, 9; 1)-difference matrix over $\mathrm{GF}(16)^+ \times \mathrm{GF}(5)^+$ from Abel's [1] 1995 PhD thesis are given instead: the original constructions are Mills' [243] 1977 construction of a

Table 13.14 Bounds for $\omega(G)$ for some groups of order at least 24

Group G	Order G	$\omega(G) \geq$	Group G	Order G	$\omega(G) \geq$
$GF(8)^+ \times GF(3)^+$	24	6	$GF(8)^+ \times GF(7)^+$	56	6
$GF(7)^+ \times GF(4)^+$	28	4	$GF(3)^+ \times GF(19)^+$	57	6
$GF(11)^+ \times GF(3)^+$	33	4	$GF(4)^+ \times GF(3)^+ \times GF(5)^+$	60	3
$GF(7)^+ \times GF(5)^+$	35	4	$GF(9)^+ \times GF(7)^+$	63	5
$GF(9)^+ \times GF(4)^+$	36	7	$GF(5)^+ \times GF(13)^+$	65	3
$GF(13)^+ \times GF(3)^+$	39	4	$GF(4)^+ \times GF(17)^+$	68	2
$GF(8)^+ \times GF(5)^+$	40	6	$GF(3)^+ \times GF(23)^+$	69	3
$GF(11)^+ \times GF(4)^+$	44	4	$GF(8)^+ \times GF(9)^+$	72	7
$GF(9)^+ \times GF(5)^+$	45	5	$GF(3)^+ \times GF(25)^+$	75	6
$GF(16)^+ \times GF(3)^+$	48	7	$GF(4)^+ \times GF(19)^+$	76	2
$GF(17)^+ \times GF(3)^+$	51	4	$GF(7)^+ \times GF(11)^+$	77	5
$GF(13)^+ \times GF(4)^+$	52	4	$GF(16)^+ \times GF(5)^+$	80	8
$GF(11)^+ \times GF(5)^+$	55	5	$GF(4)^+ \times GF(3)^+ \times GF(7)^+$	84	4

(56, 8; 1)-difference matrix over $GF(8)^+ \times GF(7)^+$, establishing that $N(56) \geq 7$, and Abel and Cheng's [5] 1994 construction of a (80, 9; 1)-difference matrix over $GF(16)^+ \times GF(5)^+$, establishing that $N(80) \geq 8$.

Several other entries in Table 13.14 yield bounds for $N(n)$ that match the current best lower bounds. Jungnickel's [201] 1980 construction of a (57, 8; 1)-difference matrix over $GF(19)^+ \times GF(3)^+$, yielding $N(57) \geq 7$, was described in Section 8.3. For the group $G = GF(5)^+ \times GF(4)^+ \times GF(3)^+$, of order 60, as G has a normal subgroup $H \cong GF(4)^+ \times GF(3)^+$, $\omega(H) = 4$, and $\omega(G/H) = 3$, it follows from Corollary 1.39 that $\omega(G) \geq 3$ and hence that $N(60) \geq 4$. For three of the groups $GF(9)^+ \times GF(7)^+$ of order 63, $GF(9)^+ \times GF(8)^+$ of order 72, and $GF(11)^+ \times GF(7)^+$ of order 77, the lower bounds are from Corollary 1.39, and the corresponding lower bounds for $N(n)$, $N(63) \geq 6$, $N(72) \geq 7$, and $N(77) \geq 6$ do not exceed the bound in Conjecture 1.11, the MacNeish conjecture.

Other entries in Table 13.14 yield lower bounds for $N(n)$ that are less than the current best lower bounds. The bound, $\omega(GF(23)^+ \times GF(3)^+) \geq 3$, is a special case of Corollary 11.50: the current lower bound for $N(69)$ is 6. For the group $G = GF(7)^+ \times GF(4)^+ \times GF(3)^+$, of order 84, as G has a normal subgroup $H \cong GF(4)^+ \times GF(3)^+$, $\omega(H) = 4$, and $\omega(G/H) = 5$, it follows from Corollary 1.39 that $\omega(G) \geq 4$: the current lower bound for $N(84)$ is 6. For three of the groups $GF(11)^+ \times GF(5)^+$ of order 65, $GF(17)^+ \times GF(4)^+$ of order 68, and $GF(19)^+ \times GF(4)^+$ of order 76, the lower bounds are from Corollary 1.39: the corresponding lower bounds for $N(n)$, $N(65) \geq 7$, $N(68) \geq 5$, and $N(76) \geq 6$ exceed the bound in Conjecture 1.11, the MacNeish conjecture.

In 1977 Mills [243] gave a construction of a $(pq, p; 1)$-difference matrix over $GF(p)^+ \times GF(q)^+$, p an odd prime divisor of $q - 1$, which, in many instances, could be extended to a larger difference matrix by adding rows. The rows of the matrix, which we will denote $D_{p,q}$, are indexed by the elements of $GF(p)$ and the columns by the elements of $GF(p) \times GF(q)$. We let w be a primitive pth root of unity in

GF(q), and, for each $g \in \mathrm{GF}(q)$, we choose $\mathrm{m}_g(x) = a_g x^2 + b_g x + c_g$, a polynomial over GF($p$), $a_g \neq 0$. Then, for $h \in \mathrm{GF}(p)$ and each $(k, g) \in \mathrm{GF}(p) \times \mathrm{GF}(q)$, we set the $h,(k, g)$th entry of $D_{p,q}$ equal to $(\mathrm{m}_g(k-h), gw^h)$. We will use R_h to denote the hth row of $D_{p,q}$. $D_{p,q}$ is a $(pq, p; 1)$-difference matrix over $\mathrm{GF}(q)^+ \times \mathrm{GF}(p)^+$.

Theorem 13.15 (Mills, 1977). *The matrix $D_{p,q}$ is a $(pq, p; 1)$-difference matrix over* $\mathrm{GF}(q)^+ \times \mathrm{GF}(p)^+$.

Proof. Let $h, h' \in \mathrm{GF}(p)$, $h \neq h'$, and $(k, g), (k', g') \in \mathrm{GF}(p) \times \mathrm{GF}(q)$. Then

$$(\mathrm{m}_g(k-h), gw^h) - (\mathrm{m}_g(k-h'), gw^{h'}) = (\mathrm{m}_{g'}(k'-h), g'w^h) - (\mathrm{m}_{g'}(k'-h'), g'w^{h'}),$$

if and only if

$$\mathrm{m}_g(k-h) - \mathrm{m}_g(k-h') = \mathrm{m}_{g'}(k'-h) - \mathrm{m}_{g'}(k'-h') \text{ and } gw^h - gw^{h'} = g'w^h - g'w^{h'},$$

if and only if

$$\mathrm{m}_g(k-h) - \mathrm{m}_g(k-h') = \mathrm{m}_{g'}(k'-h) - m_{g'}(k'-h') \text{ and } g = g'.$$

Now

$$\mathrm{m}_g(k-h) - \mathrm{m}_g(k-h') = \mathrm{m}_g(k'-h) - \mathrm{m}_g(k'-h')$$

if and only if

$$(h'-h)(a_g(2k-h-h') + b_g) = (h'-h)(a_g(2k'-h-h') + b_g)$$

if and only if

$$2k - h - h' = 2k' - h - h'$$

if and only if $k = k'$. Hence $D_{p,q}$ is a difference matrix. □

For $d \in \mathrm{GF}(p)$ and $g \in \mathrm{GF}(q)$, we choose $e_{g,d}, f_{g,d} \in \mathrm{GF}(p)$. We will use D_d to denote the $1 \times pq$ row vector whose entries are indexed by the elements of $\mathrm{GF}(p) \times \mathrm{GF}(q)$ and whose (k, g)th entry is $((f_{g,d} + kd), w^k e_{g,d})$. Row vectors of the form D_p are the candidates for extending $D_{p,q}$ to a larger difference matrix. There is a simple rule for determining when D_d can extend $D_{p,q}$.

Theorem 13.16 (Mills, 1977). *The matrix*

$$\begin{bmatrix} D_{p,q} \\ D_d \end{bmatrix}$$

is a $(pq, p+1; 1)$-difference matrix over $\mathrm{GF}(p)^+ \times \mathrm{GF}(q)^+$ *if and only if the matrix*

$$\begin{bmatrix} R_0 \\ D_d \end{bmatrix}$$

is a $(pq, 2; 1)$*-difference matrix over* $\mathrm{GF}(p)^+ \times \mathrm{GF}(q)^+$.

Proof. For $(k, g), (k', g') \in \mathrm{GF}(p) \times \mathrm{GF}(q)$,

$$((f_{g,d} + kd), w^k e_{g,d}) - (\mathrm{m}_g(k), g) = ((f_{g',d} + k'd), w^{k'} e_{g',d}) - (\mathrm{m}_{g'}(k'), g')$$

if and only if

$$f_{g,d} + kd - \mathrm{m}_g(k) = f_{g',d} + k'd - \mathrm{m}_{g'}(k') \text{ and } w^k e_{g,d} - g = w^{k'} e_{g',d} - g'.$$

Now, for $h \in \mathrm{GF}(p)$,

$$((f_{g,d} + kd), w^k e_{g,d}) - (\mathrm{m}_g(k-h), gw^h)$$
$$= ((f_{g',d} + k'd), w^{k'} e_{g',d}) - (\mathrm{m}_{g'}(k'-h), g'w^h)$$

if and only if

$$f_{g,d}+kd-\mathrm{m}_g(k-h) = f_{g',d}+k'd-\mathrm{m}_{g'}(k'-h) \text{ and } w^k e_{g,d}-gw^h = w^{k'} e_{g',d}-g'w^h$$

if and only if

$$f_{g,d} + (k-h)d - \mathrm{m}_g(k-h) = f_{g',d} + (k'-h)d - \mathrm{m}_{g'}(k'-h) \text{ and}$$
$$w^{k-h} e_{g,d} - g = w^{k'-h} e_{g',d} - g'.$$

It follows that

$$\begin{bmatrix} R_h \\ D_d \end{bmatrix}$$

is a $(pq, 2; 1)$-difference matrix over $\mathrm{GF}(p)^+ \times \mathrm{GF}(q)^+$ if and only if

$$\begin{bmatrix} R_0 \\ D_d \end{bmatrix}$$

is a $(pq, 2; 1)$-difference matrix over $\mathrm{GF}(p)^+ \times \mathrm{GF}(q)^+$. The result follows. □

By Theorem 13.15, $D_{3,4}$ is a $(12, 3; 1)$-difference matrix over $\mathrm{GF}(4)^+ \times \mathrm{GF}(3)^+$. Using Theorem 13.16, Mills was able to extend this difference matrix by adding 3 extra rows, yielding a $(12, 6; 1)$-difference matrix over $\mathrm{GF}(4)^+ \times \mathrm{GF}(3)^+$: the difference matrix was depicted in Figure 13.11. This construction yielded one more proof that $\omega(\mathrm{GF}(4)^+ \times \mathrm{GF}(3)^+) \geq 4$.

Mills also extended $D_{3,13}$ by adding 2 extra rows, yielding a (39, 5; 1)-difference matrix over $\mathrm{GF}(13)^+ \times \mathrm{GF}(3)^+$; $D_{3,16}$ by adding 2 extra rows, yielding a (48, 5; 1)-difference matrix over $\mathrm{GF}(16)^+ \times \mathrm{GF}(3)^+$; $D_{5,11}$ by adding 1 extra row, yielding a (55, 6; 1)-difference matrix over $\mathrm{GF}(11)^+ \times \mathrm{GF}(5)^+$; and $D_{7,8}$ by adding 1 extra row, yielding a (56, 8; 1)-difference matrix over $\mathrm{GF}(8)^+ \times \mathrm{GF}(7)^+$.

Part IV
Additional Topics

Chapter 14
Projective Planes from Complete Sets of Orthomorphisms

We saw in Chapter 1 that, from a complete set of orthomorphisms of a group G of order n, we can construct a complete set of mutually orthogonal Latin squares of order n, and a projective plane of order n. In this chapter we will describe the relationship between the complete sets of orthomorphisms and the corresponding projective planes.

In Section 14.1 we will introduce Cartesian projective planes and show that these are the projective planes constructed from complete sets of orthomorphisms. In Section 14.2 we will describe the known classes of Cartesian projective planes. In Section 14.3 we will discuss nonexistence results for Cartesian projective planes. In particular we will prove the nonexistence of certain generalized Hadamard matrices, from which the nonexistence of complete sets of orthomorphisms of certain groups can be derived. In Section 14.4 we will study Cartesian projective planes of prime order.

14.1 Introduction

Projective and affine planes were introduced in Section 1.4. We showed that a projective plane of order $n \geq 2$ has n^2+n+1 points, n^2+n+1 lines, each point is on $n+1$ lines, each line passes through $n+1$ points, two distinct points are incident with exactly one line, and two distinct lines meet in exactly one point. Removing any line and the points on it from a projective plane yields an affine plane. Starting with an affine plane, if, for each parallel class we add a new point incident with every line in the parallel class, a *point at infinity*, and a new line through the points at infinity, the *line at infinity*, we obtain a projective plane. In light of this relationship, we will, for the most part, restrict ourselves in what follows to projective planes.

We will use capital letters for points and small letters for lines except that we will reverse this notation when dealing with the points and lines of the dual of a

A. B. Evans, *Orthogonal Latin Squares Based on Groups*, Developments in Mathematics 57, https://doi.org/10.1007/978-3-319-94430-2_14

projective plane. If A, B are distinct points of a projective plane, then AB denotes the unique line that passes through A and B, and if l and m are distinct lines of a projective plane, then $l \cap m$ denotes the unique point that is on both lines.

14.1.1 Collineations

A *collineation* of a projective or affine plane is a bijection from the point set to the point set that maps lines to lines. Under composition, the collineations form a group, the *collineation group* of the plane. Any subgroup of the collineation group of a plane is a *collineation group* of the plane. One class of collineations of projective planes will be particularly important to us, perspectivities. Let $\mathscr{P}$ be a projective plane, l a line of $\mathscr{P}$, and P a point of $\mathscr{P}$. A (P, l)-*perspectivity* of $\mathscr{P}$ with *axis* l and *center* P is a collineation of $\mathscr{P}$ that fixes every point on l and every line through P. A (P, l)-perspectivity is a (P, l)-*elation* if P is on l and a (P, l)-*homology* if P is not on l.

We have already seen elations in Section 1.4 in the guise of translations. Recall that a translation of an affine plane is a collineation that fixes all parallel classes and all lines of one parallel class. Let ϕ be a (P, l)-elation of a projective plane $\mathscr{P}$. The restriction, ϕ', of ϕ to the affine plane $\mathscr{A} = \pi \setminus \{l\}$ is a collineation that, as ϕ fixes all points on l, ϕ' fixes all parallel classes of $\mathscr{A}$, and, as ϕ fixes all lines through P, ϕ' fixes all lines in the parallel class of lines through P in $\mathscr{P}$. Thus ϕ' is a translation of $\mathscr{A}$. Similarly, any translation of an affine plane induces an elation of the corresponding projective plane.

Lemma 14.1. *Let ϕ be a (P, l)-perspectivity of a projective plane $\mathscr{P}$, and let $Q \neq P$ be a point not on l. The perspectivity ϕ is uniquely determined by $\phi(Q)$.*

Proof. Let R be a point of $\mathscr{P}$. We may assume that R is not on l and that $R \neq P, Q$. If R is not on PQ, let m be the unique line through $\phi(Q)$ that meets RQ in l. As ϕ fixes all points on l and so maps RQ to m, $\phi(R)$ must be on m and, as ϕ fixes all lines through P, ϕ fixes PR and, hence, $\phi(R)$ must be on PR. It follows that $\phi(R)$ is uniquely determined as the point of intersection of m and PR. If R is on PQ, let T be a point of π that is not on PQ or l. By the above argument, $\phi(T)$ is uniquely determined and, hence, repeating the above argument with T in place of Q, $\phi(R)$ is also uniquely determined. □

Lemma 14.2. *Let $\mathscr{P}$ be a projective plane of order n, P a point of $\mathscr{P}$, and l a line of $\mathscr{P}$. If G is the group of (P, l)-perspectivities of $\mathscr{P}$, then*

$$|G| \leq \begin{cases} n & \textit{if } P \textit{ is on } l, \\ n-1 & \textit{if } P \textit{ is not on } l. \end{cases}$$

Proof. Let Q be a point of $\mathscr{P}$ not on l, $Q \neq P$. If ϕ is a (P, l)-perspectivity of $\mathscr{P}$, then, as ϕ fixes all lines through P, $\phi(Q)$ must be on PQ. Further, $\phi(Q) \neq P$ and

$\phi(Q)$ cannot be on l. By Lemma 14.1, if R is any point on PQ not on l, $R \neq P$, then there can be at most one (P, l)-perspectivity mapping Q to R. As there are $n + 1$ points on PQ, there can be at most n images under (P, l)-perspectivites for Q if P is on l and at most $n - 1$ images under (P, l)-perspectivites for Q if P is not on l, hence the result. □

If $\mathscr{P}$ is a projective plane, P is a point of $\mathscr{P}$, and l is a line of $\mathscr{P}$, then the (P, l)-perspectivities form a group G under composition: $\mathscr{P}$ is (P, l)-*transitive* if G acts transitively on the points of any line m through P, $m \neq l$, except for P and the intersection of m and l. If $\mathscr{P}$ is (P, l)-transitive with G being the group of (P, l)-perspectivities, then any collineation group of $\mathscr{P}$ containing G is also said to be (P, l)-transitive. If P is on l, as (P, l)-elations correspond to translations of the affine plane $\pi \setminus \{l\}$, $\mathscr{P}$ can be constructed from a complete set of orthomorphisms.

Theorem 14.3. *There exists a (P, l)-transitive projective plane, P on l, with G the group of (P, l)-elations, if and only if G admits a complete set of orthomorphisms.*

Proof. This follows from Theorem 1.52. □

In 1954 Lenz [226] classified projective planes by the (P, l)-transitivities, P on l, that they admitted. In 1957 Barlotti [28] extended Lenz's classification to include (P, l)-transitivities with P not on l. The resulting classification of projective planes is known as the Lenz-Barlotti classification. For projective planes corresponding to complete sets of orthomorphisms, by Theorem 14.3, these planes should be (P, l)-transitive for some P on l. Some of the planes in the original Lenz-Barlotti classification have been shown not to exist. The complete classification can be found in Dembowski's 1968 book [88].

Theorem 14.4. *Let $\mathscr{P}$ be a finite projective plane of order at least four with point set $\mathbb{P}$ and line set $\mathbb{L}$, that is (P, l)-transitive for at least one $(P, l) \in \mathbb{P} \times \mathbb{L}$, P on l, and let $T(\mathscr{P}) \subseteq \mathbb{P} \times \mathbb{L}$ be defined by*

$$T(\mathscr{P}) = \{(P, l) \mid P \in \mathbb{P}, l \in \mathbb{L}, \pi \textit{ is } (P, l)\textit{-transitive}\}.$$

Then $T(\mathscr{P})$ must be one of the following.

II.1. $T(\mathscr{P}) = \{(P, l)\}$ *for some $P \in \mathbb{P}$ and some $l \in \mathbb{L}$, P on l.*
II.2. $T(\mathscr{P}) = \{(P, l), (Q, m)\}$, *for some $P, Q \in \mathbb{P}$ and some $l, m \in \mathbb{L}$, $P \neq Q$, $l \neq m$, Q on l, and P on both l and m.*
III.1. $T(\mathscr{P}) = \{(P, PQ) \mid P \in \mathbb{P}, P \textit{ on } l\}$ *for some $Q \in \mathbb{P}$ and some $l \in \mathbb{L}$, Q not on l.*
IVa.1. $T(\mathscr{P}) = \{(P, l) \mid P \in \mathbb{P}, P \textit{ on } l\}$ *for some $l \in \mathbb{L}$.*
IVb.1. $T(\mathscr{P}) = \{(P, l) \mid l \in \mathbb{L}, l \textit{ passes through } P\}$ *for some $P \in \mathbb{P}$.*
IVa.2. $T(\mathscr{P}) = \{(R, l) \mid R \in \mathbb{P}, R \textit{ on } l\} \cup \{(P, m) \mid m \in \mathbb{L}, m \textit{ passes through } Q\} \cup \{(Q, w) \mid w \in \mathbb{L}, w \textit{ passes through } P\}$ *for some $P, Q \in \mathbb{P}$, $P \neq Q$, and some $l \in \mathbb{L}$, l passes through both P and Q.*

IVb.2. $T(\mathscr{P}) = \{(P, w) \mid w \in \mathbb{L}, w$ *passes through* $P\} \cup \{(Q, l) \mid Q \in \mathbb{P}, Q$ *on* $m\} \cup \{(R, m) \mid R \in \mathbb{P}, R$ *on* $l\}$, *for some* $l, m \in \mathbb{L}$, $l \neq m$, *and for some* $P \in \mathbb{P}$, P *the intersection of* l *and* m.

V.1. $T(\mathscr{P}) = \{(P, l) \mid P \in \mathbb{P}, P$ *on* $l\} \cup \{(Q, m) \mid m \in \mathbb{L}, m$ *passes through* $Q\}$ *for some* $Q \in \mathbb{P}$ *and some* $l \in \mathbb{L}$, Q *on* l.

VII.2, $T(\mathscr{P}) = \mathbb{P} \times \mathbb{L}$.

Proof. See Table 1 of Chapter 3 in [88]. □

14.1.2 Coordinates

In 1947 Hall [166] coordinatized projective planes. Let $\mathscr{P}$ be a projective plane of order n and let S be a set of n symbols containing the symbols 0 and 1, but not ∞. To coordinatize $\mathscr{P}$, using the symbols $S \cup \{\infty\}$, first pick a line, which we will designate l_∞, and three distinct points X, Y, and I on l_∞, and one point O not on l_∞. Points on l_∞ will be assigned coordinates (m), $m \in S \cup \{\infty\}$, and points not on l_∞ will be assigned coordinates (a, b), $a, b \in S$.

To assign coordinates to points not on l_∞, we first assign distinct coordinates (a, a), $a \in S$, $a \neq 0$, to the points on OI other than I, with O assigned the coordinates $(0, 0)$. To a point P, not on l_∞, we assign the coordinates (a, b) if YP passes through (a, a) and XP passes through (b, b).

To assign coordinates to points on l_∞, we assign Y the coordinate (∞) and to any other point P on l_∞ the coordinate (m) if OP passes through the point $(1, m)$. Note that this assigns (0) to X and (1) to I.

This coordinatization is depicted in Figure 14.1. Removing l_∞, and the points on l_∞, yields a coordinatization of the corresponding affine plane.

We can use this coordinate system to define binary operations $+$ and $\times$ on S. We set $am = b$ if the line through $(0, 0)$ and (m) passes through (a, b) and $a + b = c$ if the line through (1) and $(0, b)$ passes through (a, c). These operations are depicted

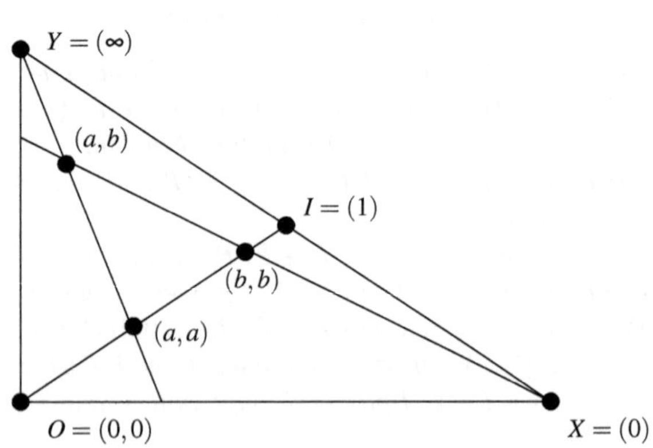

Fig. 14.1 The coordinatization of a projective plane

in Figure 14.2. In defining multiplication on S some authors write ma where we have written am: see [190] for instance.

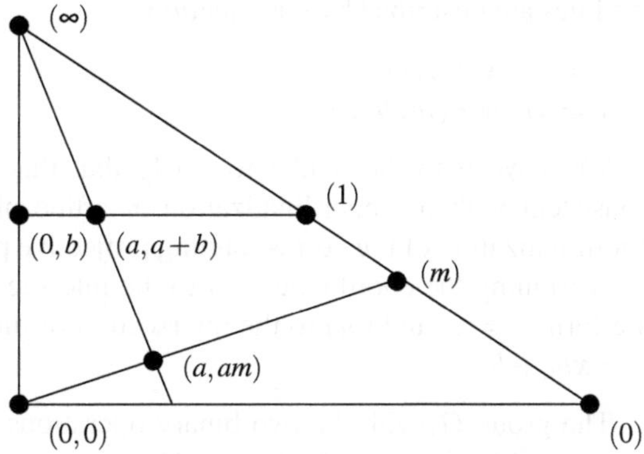

Fig. 14.2 Multiplication and addition in S

Theorem 14.5.

1. *$a \times 0 = 0 \times a = 0$ for all $a \in S$,*
2. *S is a loop under addition with identity* 0,
3. *$S \setminus \{0\}$ is a loop under multiplication with identity* 1.

Proof. As the point $(a, a \times 0)$ is on the line joining $(0, 0)$ and (0), $a \times 0 = 0$ and, as the point $(0, 0 \times a)$ is on the line joining $(0, 0)$ to (a), $0 \times a = 0$.

As the point $(a, a+0)$ is on the line joining $(0, 0)$ and (1), $a+0 = a$ and, as the point $(0, 0+a)$ is on the line joining $(0, 0)$ to (∞), $0+a = a$. Now if $a+b = c$, then the point (a, c) is the intersection of three lines, one through (0), one through (1), and one through (∞). As any two of these lines determine the third, any two of a, b, c determine the third in the equation $a+b = c$. If follows that, under addition, S is a loop with identity 0.

As the point $(a, a \times 1)$ is on the line joining $(0, 0)$ and (1), $a \times 1 = a$ and, as the point $(1, 1 \times a)$ is on the line joining $(0, 0)$ to (a), $1 \times a = a$. Now if $a \times b = c$, $a, b, c \neq 0$, then the point (a, c) is the intersection of three lines, one through (0), one through (1), and one through (∞). As any two of these lines determine the third, any two of a, b, c determine the third in the equation $a \times b = c$. If follows that, under addition, S is a loop with identity 0. □

Example 14.6. Modifying the construction of nets from orthomorphisms given in Theorem 1.48, let G be a group, written additively whether abelian or not, with identity 0 and let $1 \neq 0$ be a specified element of G. Let $\{\theta_m \mid m \in G \setminus \{0, 1\}\}$ be a complete set of normalized orthomorphisms of G, where $\theta_m(1) = m$. We next define multiplication on G by setting

- $x \times 0 = 0$ for all $x \in G$,
- $x \times 1 = x$ for all $x \in G$, and
- $x \times m = \theta_m(x)$ for all $x \in G$, $m \neq 0, 1$.

Using this multiplication, the construction of Theorem 1.48 can be written as follows.

The points of the affine plane $\mathscr{A}$ are ordered pairs (x, y) of elements of G and the lines are described by the equations:

- $x = c, c \in G$, and
- $y = xm + b, m, b \in G$.

We leave it to the reader to verify that this description of an affine plane is consistent with our coordinatization of affine planes. This can be extended to a coordinatization of the corresponding projective plane by adding the line at infinity, l_∞, assigning the coordinate (∞) to the intersection of the lines with equations of the form $x = c$, and (m) to the intersection of the lines with equations of the form $y = xm + b$.

The group G, with the two binary operations of addition and multiplication, in Example 14.6 is an example of a Cartesian group. Let $\mathscr{P}$ be a projective plane coordinatized by $S \cup \{\infty\}$. We call S a *Cartesian group* if $(S, +)$ is a group and the lines of $\pi \setminus \{l_\infty\}$ are described by the equations:

- $x = c, c \in S$, and
- $y = xm + b, m, b \in S$.

A projective or affine plane coordinatized by a Cartesian group is called a *Cartesian plane*.

Theorem 14.7. *If S is a Cartesian group, then the set $\{x \mapsto xm \mid m \in S, m \neq 0, 1\}$ is a complete set of orthomorphisms of $(S, +)$.*

Proof. It is sufficient to show that, if $m \neq n$, then the equation $-xn + xm = b$ has a unique solution. This is true as the lines $y = xn$ and $y = xm + b$ meet in exactly one point. □

It follows from Theorem 14.7 that Example 14.6 characterizes Cartesian planes.

14.1.3 Desargues' Configuration

A *triangle* in a projective plane is a 3-tuple of noncollinear points (P_1, P_2, P_3), which we call the *vertices* of the triangle, together with the lines P_1P_2, P_1P_3, and P_2P_3, which we call the *sides* of the triangle. Two triangles (P_1, P_2, P_3) and (Q_1, Q_2, Q_3) are *perspective from* a point P if the lines P_1Q_1, P_2Q_2, and P_3Q_3 all pass through P, and are *perspective from* a line l if corresponding sides meet on l, i.e., the intersections $P_1P_2 \cap Q_1Q_2$, $P_1P_3 \cap Q_1Q_3$, and $P_2P_3 \cap Q_2Q_3$ are all on l. If two triangles are perspective from both a point P and a line l, then the configuration consisting of P, l, the triangles, and the three lines from P through the vertices of the triangles is called the *Desargues configuration*. Figure 14.3 shows the Desargues configuration. The special case of the Desargues configuration with P on l is called the *little Desargues configuration*.

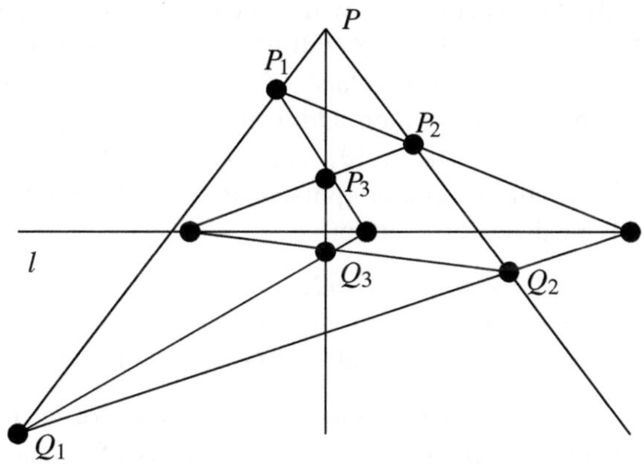

Fig. 14.3 The Desargues configuration

We say that a projective plane is (P, l)*-Desargesian* if for any pair of triangles (P_1, P_2, P_3) and (Q_1, Q_2, Q_3) perspective from P, if $P_1P_2 \cap Q_1Q_2$ and $P_1P_3 \cap Q_1Q_3$ are on l, then $P_2P_3 \cap Q_2Q_3$ is also on l. The connection between (P, l)-transitive projective planes and (P, l)-Desarguesian projective planes was established by Baer [24] in 1942.

Theorem 14.8 (Baer, 1942). *A projective plane is* (P, l)*-transitive if and only if it is* (P, l)*-Desarguesian.*

Proof. See Theorem 4.29 in [190] or Theorem 5.11 in [210] for instance. □

A projective plane is *Desarguesian* if it is (P, l)-Desarguesian for all point-line pairs (P, l).

Theorem 14.9. *A finite projective plane is Desarguesian if and only if it can be coordinatized by a field.*

Proof. Routine computations show that a projective plane coordinatized by a field is Desarguesian.

By Theorem 6.9 in [190], if a projective plane is Desarguesian, then it is coordinatized by a division ring, and any finite division ring is a field. □

14.2 Known Cartesian Planes

In this section we describe the known classes of finite Cartesian projective planes, translation planes, dual translation planes, derived dual translation planes, and Coulter-Matthews planes. We will make no attempt to be exhaustive: a comprehensive listing with descriptions of these planes can be found in the 2007 *Handbook of finite translation planes* by Johnson, Jha, and Bilioti [198].

In this section $\mathscr{P}$ will denote a finite Cartesian plane coordinatized as in Example 14.6. Thus $\mathscr{P}$ will be $((\infty), l_\infty)$-transitive. We will use $\mathscr{A}$ to denote the corresponding affine plane $\mathscr{P} \setminus l_\infty$, and G will denote the group of $((\infty), l_\infty)$-elations of $\mathscr{P}$, written additively with identity 0, and $1 \neq 0$ will be a specified element of G. The set $\{\theta_m \mid m \in G \setminus \{0, 1\}\}$ will be the corresponding complete set of normalized orthomorphisms of G, where $\theta_m(1) = m$, and multiplication on G will be defined by setting

- $x \times 0 = 0$ for all $x \in G$,
- $x \times 1 = x$ for all $x \in G$, and
- $x \times m = \theta_m(x)$ for all $x \in G, m \neq 0, 1$.

From the construction of Theorem 1.48 we see that the points of the affine plane $\mathscr{A}$ are ordered pairs (x, y) of elements of G and the lines are described by the equations:

- $x = c, c \in G$, and
- $y = xm + b, m, b \in G$.

14.2.1 Translation Planes

Let us consider the possibility that $\mathscr{P}$ is (P, l_∞)-transitive for some P on l_∞, $P \neq (\infty)$. Without loss of generality we may assume that $P = (0)$. In 1965 Repphun [278] gave orthomorphism characterizations of several classes of projective planes.

Theorem 14.10 (Repphun, 1965). *$\mathscr{P}$ is $((0), l_\infty)$-transitive if and only if $\theta_m \in$ Aut(G) for all $m \in G$, $m \neq 0, 1$.*

Proof. For $b \in G$ let σ_b denote the $((\infty), l_\infty)$-elation that maps $(0, 0)$ to $(0, b)$. By Lemma 14.1, σ_b is uniquely determined. As the mapping $(x, y) \mapsto (x, y+b)$ maps the line $x = c$ into itself and the line $y = xm + c$ into the line $y = xm + (c + b)$, it must be that $\sigma_b(x, y) = (x, y + b)$ for all $x, y \in G$.

Let us suppose that $\mathscr{P}$ is $((0), l_\infty)$-transitive and, for $a \in G$, let τ_a denote the $((0), l_\infty)$-elation that maps $(0, 0)$ to $(a, 0)$. Following the construction in the proof of Lemma 14.1, we can show that $\tau_a(x, y) = (x + a, y)$.

The points $(a, \theta_m(a))$, $(b, \theta_m(b))$, and $(a + b, \theta_m(a + b))$ are on the line $y = \theta_m(x)$. The mapping $\sigma_{\theta_m(b)}\tau_b$ fixes all points on l_∞ and maps $(0, 0)$ to $(b, \theta_m(b))$, and so fixes the line $y = \theta_m(x)$. Thus

$$\sigma_{\theta_m(b)}\tau_b(a, \theta_m(a)) = (a + b, \theta_m(a) + \theta_m(b)) = (a + b, \theta_m(a + b)).$$

Hence $\theta_m(a + b) = \theta_m(a) + \theta_m(b)$ for all $a, b \in G$, and so $\theta_m \in$ Aut(G).

Next, let us suppose that $\theta_m \in$ Aut(G) for all $m \in G$, $m \neq 0, 1$. For $a \in G$ define τ_a by $\tau_a(x, y) = (x+a, y)$. Clearly τ_a permutes the points of $\mathscr{A}$. Further, τ_a

fixes all lines of the form $y = c$, maps the line $x = c$ to the line $x = c+a$, and maps the line $y = x+b$ to the line $y = x+(-a+b)$; and τ_a maps the line $y = \theta_m(x)+b$ to the line $y = \theta_m(x) + (-\theta_m(a) + b)$, and so τ_a fixes all points on l_∞. Thus, for all $a \in G$, τ_a is a $((0), l_\infty)$-elation and so $\mathscr{P}$ is $((0), l_\infty)$-transitive. □

As a corollary:

Corollary 14.11. *If $\mathscr{P}$ is $((0), l_\infty)$-transitive, then G is elementary abelian.*

Proof. This follows from Corollary 8.93 and Theorem 14.10. □

We can show, using the Lenz-Barlotti classification of projective planes given in Theorem 14.4, that any projective plane that is (P, l)-transitive for more than one point P on l must be (P, l)-transitive for all points P on l. However this result is easily derived as a corollary to Theorem 14.10.

Corollary 14.12. *If $\mathscr{P}$ is (P, l_∞)-transitive, for some P on l_∞, $P \neq (\infty)$, then $\mathscr{P}$ is (Q, l_∞)-transitive for all Q on l_∞.*

Proof. Without loss of generality we may assume that $P = (0)$. Let σ_a and τ_b be as in the proof of Theorem 14.10. Note that by Corollary 14.11, G is abelian.

Now $\tau_a\sigma_a(x, x+b) = (x+a, x+a+b)$ and so the line $y = x+b$ is fixed by $\tau_a\sigma_a$ for all $a, b \in G$. Hence, as $\tau_a\sigma_a$ fixes all points on l_∞, $\mathscr{P}$ is $((1), l_\infty)$-transitive.

As $\theta_m \in \mathrm{Aut}(G)$ for all $m \neq 0, 1$ by Theorem 14.10,

$$\tau_a\sigma_{\theta_m(a)}(x, \theta_m(x) + b) = (x + a, \theta_m(x) + \theta_m(a) + b) = (x + a, \theta_m(x + a) + b)$$

and so the line $y = \theta_m(x) + b$ is fixed by $\tau_a\sigma_{\theta_m(a)}$ for all $a, b \in G$. Hence, as $\tau_a\sigma_{\theta_m(a)}$ fixes all points on l_∞, $\mathscr{P}$ is $((m), l_\infty)$-transitive, hence the result. □

An affine plane $\mathscr{A}$ is a *translation plane* if for all points A, B of $\mathscr{A}$ there exists a translation mapping A to B. Equivalently, the corresponding projective plane $\mathscr{P}$ is a *translation plane* if $\mathscr{P}$ is (P, l_∞)-transitive for all P on l_∞. By Corollary 14.12, π is a translation plane if and only if $\mathscr{P}$ is (P, l_∞)-transitive for at least two points P on l_∞, and by Corollary 14.11, if $\mathscr{P}$ is a translation plane, then the additive group of its Cartesian group is elementary abelian. We see that translation planes are of Lenz-Barlotti type IVa.1, IVa.2, V.1, or VII.2. By Theorem 14.10, $\mathscr{A}$ is a translation plane if and only if $\theta_m \in \mathrm{Aut}(G)$ for all $m \in G \setminus \{0, 1\}$. Thus translation planes are translation nets: these were discussed in Section 8.4. While translation planes can always be constructed from orthomorphisms that are also automorphisms, constructions of translation planes in the literature employ quasifields, spreads, and net replacement.

A *(right) quasifield* is a set Q, $0, 1 \in Q$, $0 \neq 1$, with two binary operations, $+$ and $\times$, satisfying the following:

- Under addition Q is a group with identity 0.
- Under multiplication the nonzero elements of Q form a loop with identity 1.
- The right distributive law holds, i.e., $(a + b)c = ac + bc$ for all $a, b, c \in Q$.
- $a0 = 0a = 0$ for all $a \in Q$.
- For all $a, b, c \in Q$, $a \neq b$, the equation $xa = xb + c$ has a unique solution.

A *left quasifield* is defined similarly with the left distributive law in place of the right distributive law. A *(right) semifield* is a quasifield that satisfies the left distributive law, i.e., a semifield is both a right and a left quasifield. A *(right) nearfield* is a quasifield in which multiplication is associative. Left semifields and left nearfields are defined similarly. There is a very strong connection between finite translation planes and finite quasifields.

Theorem 14.13. *Any finite translation plane is coordinatized by a finite quasifield and any finite quasifield coordinatizes a finite translation plane.*

Proof. Let $\mathscr{A}$ be a finite translation plane coordinatized by a set Q. By Theorem 14.5, Q is a group under addition with identity 0, the nonzero elements of Q form a loop with identity 1, and $a0 = 0a = 0$ for all $a \in Q$. By Theorem 14.10, $\theta_m \in \text{Aut}(G)$ for $m \neq 0, 1$. Hence, if $c \neq 0$, then

$$(a+b)c = \theta_c(a+b) = \theta_c(a) + \theta_c(b) = ac + bc.$$

Clearly $(a+b)c = ac + bc$ if $c = 0$ or $c = 1$.

It remains to show that the equation $xa = xb + c$ has a unique solution if $a \neq b$. This is clearly true if $\{a, b\} = \{0, 1\}$. If $b = 0$ and $a \neq 0, 1$, then $xa = c$ has a unique solution as θ_a is a bijection. If $b = 1$ and $a \neq 0, 1$, then $xa = x + c$ has a unique solution as θ_a is an orthomorphism. If $a, b \neq 0, 1$, then $xa = xb + c$ has a unique solution as $\theta_a \perp \theta_b$. It follows that $\mathscr{A}$ is coordinatized by a quasifield.

Let Q be a finite quasifield, let $\mathscr{A}$ be the affine plane coordinatized by Q, and, and for $m \neq 0, 1$, define $\theta_m : G \to G$ by $\theta_m(x) = xm$.

Let $m \in G \setminus \{0, 1\}$. As the equation $xa = xb + c$ has a unique solution for all $a, b, c \in Q$, $a \neq b$, $\theta_m(x) = c$ has a unique solution for all $c \in Q$, and so θ_m is a bijection. Similarly $\theta_m(x) = x + c$ has a unique solution for all $c \in Q$, and so θ_m is an orthomorphism. Further if $m, n \in G \setminus \{0, 1\}$, $m \neq n$, then $\theta_m(x) = \theta_n(x) + c$ has a unique solution for all $c \in Q$, and so $\theta_m \perp \theta_n$. Hence $\{\theta_m \mid m \in G \setminus \{0, 1\}\}$ is a complete set of normalized orthomorphisms of G.

Further, for $m \neq 0, 1$,

$$\theta_m(a+b) = (a+b)m = am + bm = \theta_m(a) + \theta_m(b),$$

and hence $\theta_m \in \text{Aut}(G)$. It follows from Theorem 14.10 that $\mathscr{A}$ is a translation plane. □

Translation planes that are coordinatized by semifields are called *semifield planes* and translation planes that are coordinatized by nearfields are called *nearfield planes*.

Example 14.14. Let us construct a quasifield by starting with the field of order nine, $\text{GF}(9) = \{a + bi \mid a, b \in \text{GF}(3), i^2 + 1 = 0\}$, and modifying multiplication. Let Q be the set of elements of GF(9) with addition as in GF(9) and multiplication defined by

$$(a+bi)(c+di) = \begin{cases} (ad) + (bc)i & \text{if } d = 0; \\ (-bd) + (ad)i & \text{if } d \neq 0 \text{ and } c = 0; \\ (ac+bd) + (ad-bc)i & \text{if } c, d \neq 0. \end{cases}$$

With addition and multiplication so defined, it is straightforward to show that Q is a quasifield, that multiplication in Q is nonassociative, and that the left distributive law holds. Hence Q is a nearfield. The affine (projective) plane constructed from Q is a nearfield plane, called the *Hall plane* of order nine. This is the smallest nondesarguesian translation plane and is the only nondesarguesian translation plane of order nine.

As corollaries to Theorem 14.13 we obtain Repphun's [278] 1965 characterizations of nearfield and semifield planes by the complete sets of normalized orthomorphisms from which they can be constructed. Let us extend the set $\{\theta_m \mid m \in G, m \neq 0, 1\}$ to include θ_0 defined by $\theta_0(x) = 0$, and θ_1 defined by $\theta_1(x) = x$.

Corollary 14.15 (Repphun, 1965). *$\mathscr{P}$ is a nearfield plane if and only if $\{\theta_m \mid m \in G, m \neq 0\}$ is a subgroup of* $\mathrm{Aut}(G)$.

Corollary 14.16 (Repphun, 1965). *$\mathscr{P}$ is a semifield plane if and only if $\{\theta_m \mid m \in G\}$ forms a group under the operation $\theta_a + \theta_b \colon x \mapsto \theta_a(x) + \theta_b(x)$, and $\{\theta_m \mid m \in G, m \neq 0\}$ is a subset of* $\mathrm{Aut}(G)$.

Given a vector space V of dimension $2n$ over a field $\mathrm{GF}(q)$ a *spread* is a set S of n-dimensional subspaces of V such that any nonzero vector of V is contained in precisely one element of S. It is easy to see that we can construct an affine plane from this spread by taking the points of the plane to be the vectors of V and the lines to be the cosets of the elements of S. This affine plane is a translation plane.

Theorem 14.17. *The affine plane constructed from a spread is a translation plane.*

Proof. Let $\mathscr{A}$ be the affine plane constructed from a spread S of a vector space V and let $u, v \in V$. The mapping $w \mapsto w + v - u$ is a translation that maps u to v. It follows that $\mathscr{A}$ is a translation plane. □

In fact, any translation plane can be constructed from a spread. A proof of this fact can be found in Chapter 3, Result 23 of [88]: note that the term used there is congruence partition instead of spread.

Another construction of translation planes uses replaceable nets: see Section 1.4 for information on nets. Given an affine plane $\mathscr{A}$, we can form a net $\mathscr{N}$ whose points are the points of $\mathscr{A}$ and whose lines are all of the lines of some of the parallel classes of $\mathscr{A}$. $\mathscr{N}$ is a *replaceable net* if there exists another net $\mathscr{N}'$ on the same point set with the property that two distinct points are joined by a line in $\mathscr{N}$ if and only if they are joined by a line in $\mathscr{N}'$. Clearly, if we replace $\mathscr{N}$ in $\mathscr{A}$ by $\mathscr{N}'$, we obtain an affine plane $\mathscr{A}'$. Many translation planes have been constructed in this fashion: see [258], for example. There is an extensive literature on translation planes: see [198] for instance.

14.2.2 Dual Translation Planes

Recall that in Section 1.4 we defined the dual of a projective plane $\mathscr{P}$ to be the projective plane $\mathscr{P}'$ whose points are the lines of $\mathscr{P}$ and whose lines are the points of $\mathscr{P}$, the point l of $\mathscr{P}'$ is on the line P of $\mathscr{P}'$ if and only if the point P of $\mathscr{P}$ is on the line l of $\mathscr{P}$.

Lemma 14.18. *For a point P of $\mathscr{P}$ and a line l of $\mathscr{P}$, $\mathscr{P}$ is (P, l)-transitive if and only if the dual $\mathscr{P}'$ is (l, P)-transitive.*

Proof. Any (P, l)-perspectivity of $\mathscr{P}$ fixes all points on l and all lines through P. This corresponds to a (l, P)- perspectivity of $\mathscr{P}'$. By Lemma 14.1, a (P, l)-perspectivity of $\mathscr{P}$ is uniquely determined by the image of a point Q, $Q \neq P$, Q not on l. Similarly a (l, P)-perspectivity of $\mathscr{P}'$ is uniquely determined by the image of a point m, $m \neq l$, m not on P. Thus the number of (P, l)-perspectivities of $\mathscr{P}$ is equal to the number of (l, P)-perspectivities of $\mathscr{P}'$. Hence, $\mathscr{P}$ is (P, l)-transitive if and only if the dual $\mathscr{P}'$ is (l, P)-transitive. □

Theorem 14.19. *The dual $\mathscr{P}'$ of a translation plane $\mathscr{P}$ is (l_∞, P)-transitive for all lines P through the point l_∞.*

Proof. A (P, l)-elation of $\mathscr{P}$ fixes all points Q on the line l and all lines m through the point P. The corresponding collineation of $\mathscr{P}'$ is an (l, P)-elation fixing all lines Q through the point l and all points m on the line P. □

It follows from Theorem 14.19 that a dual translation plane is of Lenz-Barlotti type IVb.1, IVb.2, V.1, or VII.2. As an immediate corollary of Theorem 14.19, we obtain the following:

Corollary 14.20. *The dual $\mathscr{P}'$ of a translation plane $\mathscr{P}$ is a Cartesian plane.*

Let Q be a quasifield with operations $+$ and $\times$. The *dual* of Q is the left quasifield with operations $+'$ and $\times'$ such that $a +' b = a + b$ and $a \times' b = b \times a$. Let π be the projective plane coordinatized by Q, and let $\mathscr{P}'$ be the incidence structure with point set

$$\{(x, y)' \mid x, y \in Q'\} \cup \{(m)' \mid m \in Q'\} \cup \{(\infty)'\},$$

and lines l'_∞; $x = c, c \in Q'$; and $y = x \times' m +' b, m, b \in Q'$. The line l'_∞ is incident with the points $(m)'$, $m \in Q'$, and $(\infty)'$; the line $x = c$ is incident with the points $(c, y)'$, $c \in Q'$, and $(\infty)'$; and the line $y = x \times' m +' b$, $m, b \in Q'$, is incident with the points $(x, x \times' m +' b)'$, $x \in Q'$, and $(m)'$. The incidence structure $\mathscr{P}'$ is essentially the dual of $\mathscr{P}$.

Theorem 14.21. *$\mathscr{P}'$ is a projective plane isomorphic to the dual of $\mathscr{P}$.*

Proof. Let us establish a correspondence between the lines of $\mathscr{P}$ and the points of $\mathscr{P}'$.

Lines in $\mathscr{P}$	Points in $\mathscr{P}'$
l_∞	$(\infty)'$
$x = c$	$(c)'$
$y = x \times m + b$	$(m, -b)'$

Let us also establish a correspondence between the points of $\mathscr{P}$ and the lines of $\mathscr{P}'$.

Points in $\mathscr{P}$	Lines in $\mathscr{P}'$
(∞)	l'_∞
(m)	$x = m$
(m, b)	$y = x \times' m +' (-b)$

It is routine to show that these correspondences establish an isomorphism between the dual of $\mathscr{P}$ and $\mathscr{P}'$ and hence that $\mathscr{P}'$ is a projective plane. □

An immediate corollary:

Corollary 14.22. *A projective plane is a dual translation plane if and only if it can be coordinatized by a left quasifield.*

As nearfields are both left and right quasifields, we obtain the following:

Corollary 14.23. *A projective plane is a nearfield plane if and only if it is both a dual translation plane and a translation plane.*

Theorem 14.21 yields a characterization of complete sets of normalized orthomorphisms from which dual translation planes can be constructed. As before, let us define θ_0 and θ_1 by $\theta_0(x) = 0$ for all $x \in GF(q)$, and $\theta_1(x) = x$ for all $x \in GF(q)$.

Corollary 14.24 (Repphun, 1965). *$\mathscr{P}$ is a dual translation plane if and only if*

$$\theta_{a+b}(x) = \theta_a(x) + \theta_b(x) \textit{ for all } a, b, x \in GF(q)^+.$$

Proof. If $\mathscr{P}$ is a dual translation plane, then, by Corollary 14.22, $\mathscr{P}$ is coordinatized by a left quasifield and so

$$\theta_{a+b}(x) = x(a + b) = xa + xb = \theta_a(x) + \theta_b(x).$$

If $\theta_{a+b}(x) = \theta_a(x) + \theta_b(x)$ for all $a, b, x \in GF(q)^+$, $a, b, a + b \neq 0, 1$, then π is coordinatized by a left quasifield and so is a dual translation plane by Corollary 14.22. □

An alternative version of Corollary 14.24:

Corollary 14.25. *$\mathscr{P}$ is a dual translation plane if and only if $\{\theta_m \mid m \in GF(q)\}$ is a group with the operation addition defined by $\theta_{a+b}(x) = \theta_a(x) + \theta_b(x)$.*

If $\mathscr{P}$ is a translation plane constructed from the orthomorphisms θ_a, $a \in GF(q)$, $a \neq 0, 1$, then corresponding complete set of orthomorphisms, ϕ_a, $a \in GF(q)$, $a \neq 0, 1$, from which the dual $\mathscr{P}'$ can be constructed are easily determined.

Corollary 14.26. *For $a, b \in GF(q)$, $a, b \neq 0, 1$, $\phi_a(b) = \theta_b(a)$.*

Proof. Suppose that $\mathscr{P}$ is coordinatized by the quasifield Q with operation $+$ and $\times$ and let Q' be the dual of Q with operation $+'$ and $\times'$. Then $\phi_b(a) = a \times' b = b \times a = \theta_a(b)$. □

14.2.3 Derived Dual Translation Planes

In the early 1960s Ostrom [256, 257] developed a powerful method called *derivation* for constructing new projective planes of square order from known planes of square order. In derivation, lines in a replaceable net are replaced by Baer subplanes.

A *subplane* of a projective plane $\mathscr{P}$ with point set $\mathbb{P}$ and line set $\mathbb{L}$ is a projective plane $\mathscr{P}'$ with point set $\mathbb{P}' \subseteq \mathbb{P}$, and line set $\mathbb{L}' \subseteq \mathbb{L}$, where $P \in \mathbb{P}'$ is on $l \in \mathbb{L}'$ in $\mathscr{P}'$ if and only if P is on l in $\mathscr{P}$. If $\mathscr{P}$ is a projective plane of order n^2, then a *Baer subplane* of $\mathscr{P}$ is a subplane of order n.

To describe derivation, let $\mathscr{P}$ be a projective plane of order n^2 and let $\mathscr{A}$ be the affine plane obtained from $\mathscr{P}$ by removing a line l_∞ and all points on l_∞. Let $\mathscr{N}$ be a net whose points are the points of $\mathscr{A}$ and whose lines are the lines of $n+1$ parallel classes of $\mathscr{A}$. We say that a Baer subplane $\mathscr{B}$ is contained in $\mathscr{N}$ if $\mathscr{B}$ contains the line l_∞ and all other lines of $\mathscr{B}$ are lines of $\mathscr{N}$.

Theorem 14.27 (Derivation). *If P and Q are points of $\mathscr{N}$ and PQ is a line of $\mathscr{N}$, then there is at most one Baer subplane that contains the line PQ and is contained in $\mathscr{N}$.*

If, for all points P and Q of $\mathscr{N}$ for which PQ is a line of $\mathscr{N}$, there is a Baer subplane that contains the line PQ, then the incidence structure obtained from $\mathscr{A}$ by removing the lines of $\mathscr{N}$ and replacing them with the Baer subplanes contained in $\mathscr{N}$ is an affine plane.

Proof. See [88, 198, 210] for instance. □

If, for all points P and Q of $\mathscr{N}$ for which PQ is a line of $\mathscr{N}$, there is a Baer subplane that contains the line PQ, then we say that $\mathscr{A}$ is *derivable*, and the affine plane obtained from $\mathscr{A}$ by removing the lines of $\mathscr{N}$ and replacing them with the Baer subplanes contained in $\mathscr{N}$ is called a *derived plane*.

We will discuss two classes of affine/projective planes constructed by derivation from dual translation planes; the derived dual Lüneberg planes; and the derived dual Walker planes. Each of these planes is of Lenz-Barlotti type II.1.

Let $q = (2^n)^2$, $n = 2k + 1 \geq 3$, and let $Q = \{(a, b) \mid a, b \in \mathrm{GF}(q)\}$. Let us define addition in Q by

$$(a,b)+(c,d)=(a+c,b+d),$$

and multiplication in Q by

$$(a,b)\cdot(c,d)=\left(a\left(c^{2k+1}-dd^{2k+1}\right)+bd, ac+bd\right).$$

Each such Q is a right quasifield and so, by Theorem 14.13, coordinatizes a translation plane. These translation planes were discovered by Lüneberg [233] in 1965. In 1966 Ostrom [259] showed the duals of the Lüneberg planes to be derivable and proved that the derived dual Lüneberg planes are of Lenz-Barlotti type II.1.

Theorem 14.28. *If $q=(2^n)^2$, $n\geq 3$ odd, then there exist a projective plane of order q that is of Lenz-Barlotti type II.1.*

Proof. See [210] or [259]. □

Let $q\equiv 5 \pmod 6$ be a prime power and let $Q=\{(a,b)\mid a,b\in \mathrm{GF}(q)\}$. Let us define addition in Q by

$$(a,b)+(c,d)=(a+c,b+d),$$

and multiplication in Q by

$$(a,b)\cdot(c,d)=\left(a\left(d-c^2\right)+bc, \left(\frac{q+1}{3}\right)ac^3+bd\right).$$

Each such Q is a right quasifield and so, by Theorem 14.13, coordinatizes a translation plane. These translation planes were discovered by Walker [319] in 1976. In 1974 Johnson and Piper [199] showed the duals of the Walker planes to be derivable and proved that the derived dual Walker planes are of Lenz-Barlotti type II.1.

Theorem 14.29. *If q is a prime power, $q\equiv 5 \pmod 6$, then there exist a projective plane of order q^2 that is of Lenz-Barlotti type II.1.*

Proof. See [199] or [210]. □

14.2.4 The Coulter-Matthews Planes

A mapping f: $\mathrm{GF}(q)\to \mathrm{GF}(q)$ is a *planar function* over $\mathrm{GF}(q)$ if the mapping $x\mapsto \mathrm{f}(x+a)-\mathrm{f}(x)$ is a bijection for all $a\neq 0$. Planar functions were introduced by Dembowski and Ostrom [89] in 1968. Section 9.5 of [247] covers planar functions in more generality. Here we are only concerned with planar functions that give rise to Cartesian planes that are not translation planes or dual translation planes.

If f is a planar function over $\mathrm{GF}(q)$, set $\theta_a(x)=\mathrm{f}(x+a)-\mathrm{f}(x)$. Then, for $a\neq b$, $a,b\neq 0$,

$$\theta_a(x) - \theta_b(x) = \mathrm{f}(x+a) - \mathrm{f}(x+b) = \mathrm{f}((x+b)+(a-b)) - \mathrm{f}(x+b) = \theta_{a-b}(x+b).$$

Hence $\theta_a \perp \theta_b$ if $a \neq b$. By Theorem 1.30, it follows that, if f is a planar function over $GF(q)$, then the set $\{\theta_a\theta_1^{-1} \mid a \in \mathrm{GF}(q), a \neq 0, 1\}$ is a complete set of orthomorphisms of $\mathrm{GF}(q)^+$. Hence, from planar functions we can construct Cartesian projective planes.

A simple example.

Example 14.30. Let q be odd and let $\mathrm{f}(x) = x^2$. If $a \neq 0$, then

$$\theta_a(x) = \mathrm{f}(x+a) - \mathrm{f}(x) = (x+a)^2 - x^2 = 2ax + a^2.$$

This is clearly a bijection and so f is a planar function. If, further, $a \neq 1$, then $\theta_a\theta_1^{-1}(x) = ax + (a^2 - a)$. Normalizing yields the set $\{x \mapsto ax \mid a \in \mathrm{GF}(q), a \neq 0, 1\}$ of linear orthomorphims of $\mathrm{GF}(q)^+$. The corresponding projective plane is the Desarguesian projective plane of order q.

In 2004 Pierce and Kallaher [268] determined the possible Lenz-Barlotti types of planes constructed from planar functions.

Theorem 14.31 (Pierce, Kallaher, 2004). *If $\mathscr{P}$ is a finite projective plane constructed from a planar function, then $\mathscr{P}$ is of Lenz-Barlotti type II.1, V.1, or VII.2.*

Planes of Lenz-Barlotti type V.1 are translation planes and planes of Lenz-Barlotti type VII.2 are Desarguesian planes. The only planes of Lenz-Barlotti type II.1 that have been constructed from planar functions are the Coulter-Matthews planes constructed by Coulter and Matthews [82] in 1997.

Theorem 14.32 (Coulter and Matthews, 1997). *If* $\gcd(m, 2n) = 1$, *then* $x^{(3^m+1)/2}$ *is a planar function over* $\mathrm{GF}(3^n)$. *The corresponding projective planes are of Lenz-Barlotti type II.1.*

Proof. See [82]. □

14.3 Some Nonexistence Results

In Theorem 14.7 we saw that the additive group of a Cartesian group must admit a complete set of orthomorphisms, and in Example 14.6 we saw that, from a complete set of orthomorphisms of a group G, we can construct a Cartesian group with additive group G. There is only one class of groups that are known to admit complete sets of orthomorphisms, the class of elementary abelian groups. In this section we will summarize what is known about groups that do not admit complete sets of orthomorphisms.

In 1949 Bruck and Ryser [58] established a nonexistence result for finite projective planes.

Theorem 14.33 (Bruck-Ryser, 1949). *If $n \equiv 1$ or $2 \pmod 4$ and n is not the sum of two squares, then no projective plane of order n exists.*

An immediate corollary.

Corollary 14.34. *Let G be a group of order n. If $n \equiv 1$ or $2 \pmod 4$, and n is not the sum of two squares, then G cannot admit a complete set of orthomorphisms.*

As an immediate consequence of the Hall-Paige Theorem (Theorem 3.34), we obtain the following.

Theorem 14.35. *A group with a nontrivial, cyclic Sylow 2-subgroup cannot admit a complete set of orthomorphisms.*

In 1972 Studnicka [303] proved that Cartesian projective planes of order $2p^n$, p an odd prime did not exist, a special case of Theorem 14.35. Studnicka's proof is essentially the same as the proof of the Hall-Paige Theorem.

Data for groups of small order was given in Chapter 13, where we saw that no abelian group of order less than 20, that is not elementary abelian, and no nonabelian group of order less than 16, can admit a complete set of orthomorphisms.

14.3.1 Generalized Hadamard Matrices

In Section 1.3, we showed that, for a group G of order n, the existence of a generalized Hadamard matrix of index 1, $\mathrm{GH}(n, 1)$, over G corresponded to the existence of a complete set of orthomorphisms of G. As a consequence of Theorem 1.19, if H is a normal subgroup of G of order m, then there exists a $\mathrm{GH}(n, m)$ over G/H. Thus nonexistence results for generalized Hadamard matrices give rise to nonexistence results for complete sets of orthomorphisms.

In 1984 De Launey [86] established the nonexistence of several classes of generalized Hadamard matrices: his results are given in Theorems 14.36, 14.38, and 14.40.

Let $\mathbb{Q}[z]$ denote the ring of polynomials with rational coefficients and let $R[z]$ denote the quotient ring

$$R[z] = \mathbb{Q}[z] / \left(\sum_{i=0}^{m-1} z^i \right).$$

Thus, each element of $R[z]$ can be represented by a polynomial of the form $a_0 + a_1 z + \cdots + a_{m-2} z^{m-2}$. In $R[z]$, $z^{m-1} = -1 - z - \cdots - z^{m-2}$, and, under multiplication, $G = \{1, z, z^2, \ldots, z^{m-1}\}$ is a group isomorphic to $\mathbb{Z}_m$. Let $H = (h_{ij})$ be a generalized Hadamard matrix of order n over G and set $H^* = (h_{ji}^{-1})$. By the properties of generalized Hadamard matrices,

$$HH^* = nI,$$

where I is the $n \times n$ identity matrix. Thus H is nonsingular.

$$\det(H) = \mathrm{f}(z) = \sum_{i=0}^{m-2} a_i z^i,$$

a polynomial in z with integer coefficients, and

$$\det(H^*) = \mathrm{f}(z^{-1}) = \sum_{i=0}^{m-2} a_i z^{-i}.$$

Hence

$$\mathrm{f}(z)\mathrm{f}(z^{-1}) = \left(\sum_{i=0}^{m-2} a_i z^i\right)\left(\sum_{i=0}^{m-2} a_i z^{-i}\right) = \det(H)\det(H^*) = n^n. \qquad (14.1)$$

De Launey used Equation 14.1 to establish the nonexistence of certain generalized Hadamard matrices over $\mathbb{Z}_m$, $m = 3, 5, 7$.

Theorem 14.36 (De Launey, 1984). *Let $G = \mathbb{Z}_3$ and*

$$n = 3^t p_1^{k_1} \cdots p_m^{k_m},$$

where $p_1, \ldots, p_m$ are distinct odd primes. If $p_i \equiv 5 \pmod 6$ and k_i is odd, for some i, then no GH$(n, n/3)$ *over G exists.*

Proof. If $m = 3$, then $\mathrm{f}(z) = a_0 + a_i z$. Setting $v_0 = a_0$, $v_1 = a_1 - a_0$, and $v_2 = -a_1$, we see that

$$v_0 + v_1 + v_2 = 0,$$

and, Equation 14.1 becomes

$$v_0^2 + v_1^2 + v_2^2 = 2n^n.$$

Suppose that for a given i, $p_i \equiv 5 \pmod 6$ and k_i is odd. We will show that p_i divides each of v_0, v_1, and v_2. If this is not the case, then we may assume, without loss of generality, that $v_0 \not\equiv 0 \pmod{p_i}$. Setting $x = v_1 v_0^{-1}$, we see that

$$v_2 v_0^{-1} \equiv -(1 + x) \pmod{p_i}$$

and

$$1 + x + x^2 \equiv 0 \pmod{p_i}.$$

But this equation has no solution modulo p_i as -3 is not a quadratic residue modulo p_i. It follows that p_i divides each of v_0, v_1, and v_2. Thus p_i^2 divides $2n^n = v_0^2+v_1^2+v_2^2$. Repeating this argument leads to the conclusion that k_i is even: a contradiction from which the result follows. □

The following is an immediate consequence:

Corollary 14.37. *Let G admit a homomorphism onto* $\mathbb{Z}_3$, *and*

$$|G| = 3^t p_1^{k_1} \cdots p_m^{k_m},$$

where $p_1, \ldots, p_m$ *are distinct odd primes. If* $p_i \equiv 5 \mod 6$ *and* k_i *is odd, for some* i, *then G does not admit a complete set of orthomorphisms.*

In particular the groups $\mathbb{Z}_{15}$, $\mathbb{Z}_{33}$, $\mathbb{Z}_{45}$, $\mathbb{Z}_{99}$, $\mathbb{Z}_{105}$, and any group of these orders, with a normal subgroup of index 3, cannot admit a complete set of orthomorphisms.

The proofs of Theorems 14.38 and 14.40 are similar to the proof of Theorem 14.36, though more complex. The reader should consult [86] for the proofs of these theorems.

Theorem 14.38 (De Launey, 1984). *Let* $G = \mathbb{Z}_5$ *and* $n = 3^t 7^k m$. *If m is odd, m is relatively prime to* 21, *and one of t, k is odd, then no* GH$(n, n/5)$ *over G exists.*

Proof. See Theorem 4.2 in [86]. □

The following is an immediate consequence:

Corollary 14.39. *Let G admit a homomorphism onto* $\mathbb{Z}_5$, *and* $|G| = 3^t 7^k m$. *If m is odd, m is relatively prime to* 21, *and one of t, k is odd, then G does not admit a complete set of orthomorphisms.*

Theorem 14.40 (De Launey, 1984). *Let* $G = \mathbb{Z}_7$ *and* $n = 3^t m$. *If m is relatively prime to* 3, *and t and m are both odd, then no* GH$(n, n/7)$ *over G exists.*

Proof. See Theorem 4.3 in [86]. □

The following is an immediate consequence:

Corollary 14.41. *Let G admit a homomorphism onto* $\mathbb{Z}_7$, *and* $|G| = 3^t m$. *If m is relatively prime to* 3, *and t and m are both odd, then G does not admit a complete set of orthomorphisms.*

Also in 1984 De Launey [87] established the nonexistence of GH$(n, 1)$ for an infinite class of abelian groups of order n. This nonexistence proof applied to a larger class of matrices, generalized weighing matrices, and employed the same techniques as were used to prove Theorems 14.36, 14.38, and 14.40.

Theorem 14.42 (De Launey, 1984). *Let G be an abelian group of order n and let $p \neq 2$ be a prime that divides n. If there exists an integer $m \not\equiv 0 \pmod{p}$, m dividing the square free part of n, and the order of m modulo p being even, then no* $\mathrm{GH}(n, 1)$ *over G exists.*

Proof. See Theorem 5.1 in [87]. □

As a corollary we obtain the following:

Corollary 14.43 (De Launey, 1984). *Let G be an abelian group of order n and let $p \neq 2$ be a prime that divides n. If there exists an integer $m \not\equiv 0 \pmod{p}$, m dividing the square free part of n, and the order of m modulo p being even, then G does not admit a complete set of orthomorphisms.*

14.3.2 A Special Case

In 1986 Woodcock [335] established the nonexistence of a complete set of orthomorphisms for an infinite number of cyclic groups of odd order. Woodcock's result can be derived from Theorem 14.36. However Woodcock's proof is quite different from De Launey's proof. While both proofs start with generalized Hadamard matrices, in De Launey's proof, the groups are multiplicative and in Woodcock's proof additive.

Theorem 14.44. *If $n \equiv 15 \pmod{18}$ then $\mathbb{Z}_n$ does not admit a complete set of orthomorphisms.*

Proof. Let $\theta_1, \ldots, \theta_{n-2}$ be a complete set of orthomorphisms of $\mathbb{Z}_n$ which we may assume to be normalized, and let $\pi\colon \mathbb{Z}_n \to \mathbb{Z}_3$ be an epimorphism.

Let $H = (h_{ij}) = (\pi(\theta_i(j)))$: H is essentially a generalized Hadamard matrix over $\mathbb{Z}_3$ with one column and two rows removed. The inner products of rows of H are ± 1.

$$\sum_{j=1}^{n-1} h_{ij}^2 = \frac{n(n-1)(2n-1)}{6} = 1,$$

and, for $k \neq i$, as

$$\sum_{j=1}^{n-1}(h_{ij} - h_{kj})^2 = \sum_{j=1}^{n-1} h_{ij}^2 - 2\sum_{j=1}^{n-1} h_{ij}h_{kj} + \sum_{j=1}^{n-1} h_{kj}^2,$$

$$\sum_{j=1}^{n-1} h_{ij}h_{kj} = -1.$$

It follows that $HH^t = -J - I$, where J is the $(n-2) \times (n-2)$ matrix of all ones and I is the $(n-2) \times (n-2)$ identity matrix. We leave it to the reader to verify that $\det(HH^t) = 1$. It follows, as each column of HH^t is a linear combination of the columns of H and H has $n-1$ rows, that $\text{rank}(H) = n-2$. Now $H = (K \mid c)$, where K is an $(n-2) \times (n-2)$ matrix. As

$$\sum_{j=1}^{n-1} h_i j = \frac{n(n-1)}{2} = 0.$$

for all $i = 1, \ldots, n-2$, c is a linear combination of the columns of K. Hence K is nonsingular. Set $M = K^{-1}H$. Simple computation shows that $M = (I \mid e)$, where I is the $(n-2) \times (n-2)$ identity matrix and e is a column vector, each of whose entries is -1. Computing MM^t in two different ways yields

$$MM^t = -HH^t = K^{-1}H(K^{-1})^t.$$

But then $\det(MM^t) = -1 = \det(K)^2$, a contradiction as -1 is not a square in $\mathbb{Z}_3$, hence the result. □

We leave it to the reader to verify that Theorem 14.44 is really a special case of Theorem 14.36. The following are the smallest cases of Theorem 14.44.

Corollary 14.45. *$\mathbb{Z}_{15}$, $\mathbb{Z}_{33}$, and $\mathbb{Z}_{51}$, do not admit complete sets of orthomorphisms.*

Note that $\mathbb{Z}_{15}$, $\mathbb{Z}_{33}$, and $\mathbb{Z}_{51}$ are the only groups of these orders.

14.4 Cartesian Projective Planes of Prime Order

It has long been conjectured that the only projective planes of prime order are the Desarguesian projective planes. By Theorem 14.9, these are precisely the projective planes coordinatized by fields of prime order.

Conjecture 14.46. Any projective plane of prime order is Desarguesian.

Conjecture 14.46 has been proven true for all primes less than or equal to 7. In 1907 Macinnes [234] proved this conjecture true for orders 2, 3, and 5. In 1941 Bose and Nair [44] gave a proof that this conjecture was true for order 7 based on a list of Latin squares of order 7 determined by Norton [254] in 1939. However, Norton's list was incomplete. In 1951 Sade [285, 286] completed the list of Latin squares of order 7. In 1953 Hall [168, 169] gave a geometric proof of this conjecture for order 7. Conjecture 14.46 remains open for all primes greater than or equal to 11. A weaker conjecture is the following.

Conjecture 14.47. Any Cartesian projective plane of prime order is Desarguesian.

In this section we will discuss attempts to prove or disprove Conjecture 14.47.

By Theorem 14.7 the study of Cartesian projective planes of prime order p reduces to the study of $(p-2)$-cliques of $\mathrm{Orth}(\mathbb{Z}_p)$. The only $(p-2)$-clique of $\mathrm{Orth}(\mathbb{Z}_p)$ we know is $\mathscr{C}_1(p)$, the complete set of orthomorphisms $\{x \mapsto ax \mid a \in \mathrm{GF}(p), a \neq 0, 1\}$. A crucial result for us is a 1965 result of Repphun [278] that characterizes complete sets of normalized orthomorphisms from which Desarguesian planes can be constructed.

Theorem 14.48 (Repphun, 1965). *The affine plane $\mathscr{A}$ is Desarguesian if and only if $G = \mathrm{GF}(q)^+$ and $\theta_m(x) = mx$, the multiplication being multiplication in* GF(q).

Proof. See [278]. □

Theorem 14.48 was rediscovered in 1984 by Evans and Mcfarland [138] for the special case of planes of prime order. From Theorem 14.48 it is immediate that the existence of any other $(p-2)$-clique of $\mathrm{Orth}(\mathbb{Z}_p)$ would imply the existence of a nondesarguesian projective plane of order p.

Corollary 14.49 (Evans and McFarland, 1984). *Let p be a prime. If* $\mathrm{Orth}(\mathbb{Z}_p)$ *contains more than one $(p-2)$-clique, then there exists a nondesarguesian projective plane of order p.*

Data on the orthomorphism graphs of groups of small order was given in Chapter 13: in particular data on $(p-2)$-cliques of $\mathrm{Orth}(\mathbb{Z}_p)$ for $p \leq 13$, p a prime. To recap, $\mathrm{Orth}(\mathbb{Z}_{11})$ was first shown to contain only one 9-clique by Johnson, Dulmage and Mendelsohn [196] in 1961. This has since been confirmed by Cates and Killgrove [66] in 1981, by Evans and Mcfarland [138] in 1984, and by Lazebnik and Thomason [224] in 2004. All of these proofs required the use of computers. It follows, by Corollary 14.49, that the only Cartesian projective plane of order 11 is the Desarguesian projective plane. The orthomorphism graph $\mathrm{Orth}(\mathbb{Z}_{13})$ was shown to contain only one 11-clique by Cates and Killgrove [66] in 1981, and by Lazebnik and Thomason [224] in 2004. These proofs also required the use of computers. It follows, by Corollary 14.49, that the only Cartesian projective plane of order 13 is the Desarguesian projective plane.

We studied complete sets of orthomorphisms in $\mathscr{C}_2(q)$, the orthomorphism graph consisting of quadratic orthomorphisms of $\mathrm{GF}(q)^+$ in Section 9.4. In particular in Theorem 9.51, we proved that the only $(p-2)$-clique in $\mathscr{C}_2(p)$, p a prime, is $\mathscr{C}_1(p)$. This result has the following geometric consequence.

Theorem 14.50. *Let p be a prime and let $\mathscr{P}$ be a projective plane of order p that is (P, l)-transitive for some point P and some line l, P on l. If $\mathscr{P}$ has a homology with axis l and order $(p-1)/2$, then $\mathscr{P}$ is Desarguesian.*

Proof. Let η be a homology of $\mathscr{P}$ of order $(p-1)/2$ with axis l and center Q. Let us coordinatize $\mathscr{P}$ so that l is l_∞, P is (∞), and Q is $(0,0)$. As $\mathscr{P}$ is $((\infty), l_\infty)$-transitive, by Theorem 14.3, $\mathscr{P}$ can be constructed from a complete set of orthomorphisms, $\theta_1, \ldots, \theta_{p-2}$, of $\mathbb{Z}_p$. As in Example 14.6, the points of $\mathscr{P}$, not on l_∞, can be coordinatized by ordered pairs of elements of $\mathbb{Z}_p$ and the lines of the affine plane $\mathscr{A} = \mathscr{P} \setminus \{l_\infty\}$ can be represented by the equations

- $x = c$, $c \in \mathbb{Z}_p$ a constant,
- $y = b$, $b \in \mathbb{Z}_p$ a constant,
- $y = x + b$, $b \in \mathbb{Z}_p$ a constant,
- $y = \theta_i(x) + b$, $b \in \mathbb{Z}_p$ a constant, $i = 1, \ldots, p-2$.

With multiplication modulo p, $\mathbb{Z}_p$ is a field. As η fixes all lines through $(0, 0)$, η maps $(1, 1)$ to (g, g) for some $g \in \mathbb{Z}_p$, $g \neq 0$. We will show that η maps the point (x, y) to the point (gx, gy) and that $|g| = (p-1)/2$.

As η fixes the point (0), it maps the line $y = 1$ to the line $y = g$, and as η fixes the line $x = 0$, the point $(0, 1)$, the intersection of the lines $x = 0$ and $y = 1$, is moved to the point $(0, g)$, the intersection of the lines $x = 0$ and $y = g$. Similarly η maps the point $(1, 0)$ to the point $(g, 0)$. To show that $\eta(x, y) = (gx, gy)$, for all $x, y \in \mathbb{Z}_p$, it is sufficient to show that $\eta(x, x) = (gx, gx)$, for all $x \in \mathbb{Z}_p$. Suppose that $\eta(a, a) = (ga, ga)$. Then $\eta(a, 0) = (ga, 0)$ and $\eta(0, a) = (0, ga)$. The point $(a, a+1)$, the intersection of the line through (∞) and $(a, 0)$ and the line through (1) and $(0, 1)$, is moved by η to the point $(ga, ga+g)$, the intersection of the line through (∞) and $(ga, 0)$ and the line through (1) and $(0, g)$. The point (a, a), the intersection of the line through (0) and $(a, a+1)$ and the line through (1) and $(0, 0)$, is moved by η to the point $(g(a+1), g(a+1))$, the intersection of the line through (0) and $(ga, ga+q)$ and the line through (1) and $(0, 0)$. By induction we see that $\eta(x, x) = (gx, gx)$ for all $x, y \in \mathbb{Z}_p$ and so $\eta(x, y) = (gx, gy)$ for all $x, y \in \mathbb{Z}_p$. As η has order $(p-1)/2$, it follows that $|g| = (p-1)/2$.

Now η fixes the lines $y = \theta_i(x)$ for $i = 1, \ldots, p-2$ and so the point $(x, \theta_i(x))$ is mapped to the point $(gx, g\theta_i(x)) = (gx, \theta_i(gx))$ for $i = 1, \ldots, p-2$. Hence $\theta_i(gx) = g\theta_i(x)$ for $i = 1, \ldots, p-2$, and so $\theta_i(g^r x) = g^r\theta_i(x)$ for $i = 1, \ldots, p-2$. Let us pick a nonsquare $w \in \mathbb{Z}_p$, set $A_i = \theta_i(1)$, and $B_i = \theta_i(w)$ for $i = 1, \ldots, p-2$. As $\{g^r \mid r = 1, \ldots, (p-1)/2\}$ is the set of nonzero squares of $\mathbb{Z}_p$, and $\{g^r w \mid r = 1, \ldots, (p-1)/2\}$ is the set of nonsquares of $\mathbb{Z}_p$, θ_i is the quadratic orthomorphism $[A_i, B_i]$: quadratic orthomorphisms were covered in Section 9.3. Thus $\theta_1, \ldots, \theta_{p-2}$ are in $\mathscr{C}_2(p)$. Hence, by Theorem 9.51, $\theta_1, \ldots, \theta_{p-2}$ are in $\mathscr{C}_1(p)$, and $\mathscr{P}$ is thus Desarguesian. □

Other studies of possible nondesarguesian projective planes of prime order have concentrated on the possible collineation groups of such planes: see [179] and [197], for instance.

Chapter 15
Related Topics

In this chapter we will introduce several topics that have not been covered elsewhere in this book but that are closely related to complete mappings and orthomorphisms of groups. In our discussion of these topics, we will outline the work that has been done, presenting many of the results without proofs. Our emphasis will be on the role played by orthomorphisms and related mappings. In Section 15.1 we will introduce classes of complete mappings and orthomorphisms as well as mappings that are closely related to complete mappings and orthomorphisms, in particular near complete mappings and near orthomorphisms. In Section 15.2 we will discuss neofields; in Section 15.3 group sequencings and its variations; and in Section 15.4 we will introduce starters.

15.1 More Mappings

There are several variants and special classes of complete mappings and orthomorphisms. Included among those covered in this chapter are near complete mappings, near orthomorphisms, quasi-complete mappings, quasi-orthomorphisms, and antisymmetric mappings.

We saw in Section 1.3 right orthomorphisms and left orthomorphisms. Throughout this book the term orthomorphism has been used to refer to left orthomorphisms, that is bijections θ for which $x \mapsto x^{-1}\theta(x)$ is a bijection as opposed to bijections θ for which $x \mapsto \theta(x)x^{-1}$ is a bijection. Similarly we can define left and right complete mappings and left and right versions of any variant of complete mappings and orthomorphisms. As it is fairly easy to convert the right version of a mapping into its left version, we will only consider left versions of any mappings covered in this section, and we will omit the adjective "left." Many authors use the term complete mappings to refer to what we have called orthomorphisms. To maintain

A. B. Evans, *Orthogonal Latin Squares Based on Groups*, Developments in Mathematics 57, https://doi.org/10.1007/978-3-319-94430-2_15

consistency, in such instances, we will refer to orthomorphisms instead. While many of these mappings can be defined for quasigroups, we will restrict ourselves to groups.

15.1.1 Near Complete Mappings and Near Orthomorphisms

There is more than one definition for near complete mappings The most general definition is given in [132]. A *near complete mapping* of a group G is a bijection $\theta\colon G \setminus \{a\} \to G \setminus \{b\}$ for which the mapping $\sigma\colon x \mapsto x\theta(x)$ is a bijection $G \setminus \{a\} \to G \setminus \{c\}$ for some $a, b, c \in G$. For convenience, we will call the ordered triple (a, b, c) the *ex elements* of θ. Any other definition of near complete mappings can be obtained from this general definition by imposing restrictions on the choice of a, b, and c. In other definitions, it is most often assumed that $b = 1$ and very often that $b = 1$ and $a \neq 1$. In some cases, a near complete mapping θ of a group G can be extended to a complete mapping of G. If θ has ex elements (a, b, c) and $ab = c$, then the mapping $\theta'\colon G \to G$ defined by

$$\theta'(g) = \begin{cases} b & \text{if } g = a, \\ \theta(g) & \text{if } g \neq a, \end{cases}$$

is a complete mapping of G. Note that this extension is only possible if $ab = c$. Let θ be a near complete mapping of a group G with ex elements (a, b, c) and define θ^u and θ_v by $\theta^u(g) = \theta(u^{-1}g)$ if $g \neq ua$ and $\theta_v(g) = \theta(g)v$ if $g \neq a$. Then θ^u is a near complete mapping of G with ex elements (ua, b, uc), and θ_v is a near complete mapping of G with ex elements (a, bv, cv). Using these transformations any near complete mapping of G can be transformed into a near complete mapping with ex elements $(h, 1, 1)$. A near complete mapping with ex elements $(h, 1, 1)$ is said to be *normalized* or in *canonical form*, in which case h is called the *exdomain* element of θ. A normalized near complete mapping with exdomain element h can be extended to a complete mapping if and only if $h = 1$. To distinguish between near complete mappings that can be extended to complete mappings and those that cannot, we will say that a near complete mapping in canonical form is *proper* if its exdomain element is not the identity.

By Theorem 1.14, there is a one-to-one correspondence between complete mappings of a group and transversals of its Cayley table. A similar result holds for near complete mappings of a group and near transversals of its Cayley table. A *near transversal* of a Latin square L of order n is a set T of $n-1$ cells of L, at most one cell in each row, at most one cell in each column, and each symbol occurring at most once in T.

Theorem 15.1. *Let $G = \{g_1, \ldots, g_n\}$ be a group and let L be its Cayley table. For $T = \{(1, j_1), \ldots, (n, j_n)\} \setminus \{(k, j_k)\}$, $\{j_1, \ldots, j_n\} = \{1, \ldots, n\}$, a near transversal of L, define θ_T by $\theta_T(g_i) = g_{j_i}$, $i \neq k$. Then θ_T is a near complete mapping of G,* *and the mapping $T \mapsto \theta_T$ defines a one-to-one correspondence between the near transversals of L and the near complete mappings of G.*

Proof. Similar to the proof of Theorem 1.14. □

Note that the ex elements of the near complete mapping, constructed in Theorem 15.1, are (g_k, g_{j_k}, c), where c is the unique element of G not in the set $\{g_i g_{j_i} \mid i = 1, \ldots, n, i \neq k\}$. Note also that the near transversal T can be extended to a transversal if and only if $g_k g_{j_k} = c$ if and only if θ_T can be extended to a complete mapping of G. To illustrate Theorem 15.1, consider the Latin square L of order 6 shown in Figure 15.1: L is the Cayley table of $\mathbb{Z}_6 = \{0, 1, \ldots, 5\}$. The entries of a near transversal of L are shown in bold, and the corresponding near complete mapping is shown in Figure 15.2: this is a proper near complete mapping with exdomain element 3. By Corollary 3.2, $\mathbb{Z}_6$ does not admit complete mappings, and, hence, by Theorem 1.14, the Latin square L of Figure 15.1 does not possess transversals. It follows that the near transversal shown in Figure 15.1 cannot be extended to a transversal and the near complete mapping in Figure 15.2 cannot be extended to a complete mapping.

Fig. 15.1 A near transversal of the Cayley table of $\mathbb{Z}_6$

$$L = \begin{pmatrix} 0\,\mathbf{1}\,2\,3\,4\,5 \\ 1\,2\,\mathbf{3}\,4\,5\,0 \\ 2\,3\,4\,\mathbf{5}\,0\,1 \\ 3\,4\,5\,0\,1\,2 \\ 4\,5\,0\,1\,\mathbf{2}\,3 \\ 5\,0\,1\,2\,3\,\mathbf{4} \end{pmatrix}$$

Fig. 15.2 A near complete mapping of $\mathbb{Z}_6$

g	0	1	2	3	4	5
$\theta(g)$	1	2	3	·	4	5
$g+\theta(g)$	1	3	5	·	2	4

The question arises as to which finite groups admit proper near complete mappings. In 1947 Paige [261] characterized admissible finite abelian groups: see Theorem 3.9. Recall that a group is admissible if it admits complete mappings. In the process, he implicitly characterized finite abelian groups that admit proper near complete mappings.

Theorem 15.2. *A finite abelian group admits proper near complete mappings if and only if its Sylow 2-subgroup is nontrivial and cyclic.*

Proof. See Corollary 3.10. □

Thus any finite abelian group must admit complete mappings or proper near complete mappings, but not both. The situation is quite different for nonabelian groups. Recall that we used D_{2n} to denote the dihedral group of order $2n$ with the presentation

$$D_{2n} = \langle a, b \mid a^n = b^2 = 1, ba = a^{-1}b \rangle$$

and Q_{4n} to denote the quaternion group of order $4n$ with the presentation

$$Q_{4n} = \langle a, b \mid a^{2n} = 1, b^2 = a^n, ba = a^{-1}b \rangle.$$

A proper near complete mapping of D_6 with exdomain element ab is shown in Figure 15.3, a proper near complete mapping of D_8 with exdomain element a^2 is shown in Figure 15.4, and a proper near complete mapping of Q_8 with exdomain element a^2 is shown in Figure 15.5. Thus, every nonabelian group of order less than 10 admits proper near complete mappings: two of these, D_8 and Q_8, also admit complete mappings. In Section 15.3 we will discuss sequenceable groups, which can be characterized as groups that admit a special kind of proper near complete mapping. Keedwell has conjectured (see Conjecture 15.55) that all nonabelian groups of order at least 10 are sequenceable. The following is a weaker version of Keedwell's conjecture.

Conjecture 15.3. Every finite nonabelian group admits proper near complete mappings.

Fig. 15.3 A proper near complete mapping of D_6

g	1	a	a^2	ab	a^2b	b
$\theta(g)$	ab	a	b	$\cdot$	a^2	a^2b
$g\theta(g)$	ab	a^2	a^2b	$\cdot$	b	a

Fig. 15.4 A proper near complete mapping of D_8

g	1	a	a^2	a^3	b	ab	a^2b	a^3b
$\theta(g)$	a^2	b	$\cdot$	a^3b	a	a^2b	ab	a^3
$g\theta(g)$	a^2	ab	$\cdot$	a^2b	a^3b	a^3	a	b

Fig. 15.5 A proper near complete mapping of Q_8

g	1	a	a^2	a^3	b	ab	a^2b	a^3b
$\theta(g)$	a^2	a^3b	$\cdot$	b	a^3	a^2b	ab	a
$g\theta(g)$	a^2	b	$\cdot$	a^3b	ab	a	a^3	a^2b

Closely related to near complete mappings are near orthomorphisms. A *near orthomorphism* of a group G is a bijection $\theta\colon G \setminus \{a\} \to G \setminus \{b\}$ for which the mapping $\delta\colon x \mapsto x^{-1}\theta(x)$ is a bijection $G \setminus \{a\} \to G \setminus \{c\}$ for some $a, b, c \in G$. As with near complete mappings, for convenience, we will call the ordered triple

(a, b, c) the *ex elements* of θ. Note that θ is a near orthomorphism with ex elements (a, b, c) if and only if δ is a near complete mapping with ex elements (a, c, b). A near orthomorphism θ of a group G can be extended to an orthomorphism of G if and only if the ex elements (a, b, c) of θ satisfy $a^{-1}b = c$. Let θ be a near orthomorphism of a group G with ex elements (a, b, c), and define θ^u and θ_v by $\theta^u(g) = \theta(gu)$ if $g \neq au^{-1}$ and $\theta_v(g) = \theta(g)v$ if $g \neq a$. Then θ^u is a near orthomorphism of G with ex elements (au^{-1}, b, uc), and θ_v is a near orthomorphism of G with ex elements (a, bv, cv). Using these transformations any near orthomorphism of G can be transformed into a near orthomorphism with ex elements $(h, 1, 1)$. Such a near orthomorphism is said to be *normalized* or in *canonical form*, and h is called the *exdomain* element of θ: this near orthomorphism is *proper* if $h \neq 1$. Note that, analogous to near complete mappings, a normalized near orthomorphism can be extended to an orthomorphism if and only if it is not proper. The definition given here differs subtly from that of Hsu [185]. Hsu defines a near orthomorphism of a group G to be a bijection $\theta\colon G \to G$ for which the set $\{x^{-1}\theta(x) \mid x \in G\} = G - \{1\}$. If $t \neq 1$ is the exdomain element of a normalized near orthomorphism θ, then setting $\theta(t) = 1$ yields a mapping that satisfies Hsu's definition of a near orthomorphism.

In 1996 Keedwell and Wang [220] defined a class of near orthomorphisms that they called *special near orthomorphisms*. A special near orthomorphism is a normalized near orthomorphism, with exdomain element h, such that $\theta(1) = h$. They gave constructions of special near orthomorphisms for all finite abelian groups with nontrivial cyclic Sylow 2-subgroups, except for those groups whose Sylow 2-subgroup is either $\mathbb{Z}_2$ or $\mathbb{Z}_4$.

Theorem 15.4 (Keedwell and Wang, 1996). *A finite abelian group, whose order is divisible by 8, that has a cyclic Sylow 2-subgroup admits a special near orthomorphism.*

Cycle notation has proved to be a useful way to represent permutations. The cycle $(a_0\ a_1\ \dots\ a_{k-1})$ is used to denote the mapping $a_i \mapsto a_{i+1}$, indices being added modulo k. It is well-known that any permutation can be written as a product of disjoint cycles. Cycle notation can be modified to represent near complete mappings and near orthomorphisms. The sequence $[a_0\ a_1\ \dots\ a_{k-1}]$ is used to denote the mapping $a_i \mapsto a_{i+1}$, $i = 0, \dots, k-2$. It is easily seen that any near complete mapping or near orthomorphism can be written as a product of disjoint cycles/sequences and that this product is unique up to the order of the cycles and sequences. As examples, the near complete mapping of $\mathbb{Z}_6$ depicted in Figure 15.2 can be written as $[0\ 1\ 2\ 3](4)(5)$ or as $[0\ 1\ 2\ 3]$; the near complete mapping of D_6 depicted in Figure 15.3 can be written as $[1\ ab](a^2\ b\ a^2b)$; the near complete mapping of D_8 depicted in Figure 15.4 can be written as $[1\ a^2](a\ b)(a^3\ a^3b)(ab\ a^2b)$; and the near complete mapping of Q_8 depicted in Figure 15.5 can be written as $[1\ a^2](a\ a^3b)(a^3\ b)(ab\ a^2b)$. The cycle/sequence structures of near complete mappings, etc., will prove to be useful in several applications: they will prove particularly useful in Section 15.3 in constructing sequencings and R-sequencings of groups.

15.1.2 Products of k-Cycles

We next turn our attention to orthomorphisms and complete mappings that can be written as products of disjoint k-cycles. For a group G of order n, we will consider two types of such mappings: normalized orthomorphisms that are products of k-cycles, where k divides $n-1$, and complete mappings that are products of k-cycles, where k divides n.

In 1981 Friedlander, Gordon, and Tannenbaum [147] studied k-regular orthomorphisms, which they called k-regular complete mappings. Let G be a group of order n and let $k > 1$ be a divisor of $n-1$. A *k-regular orthomorphism* of G is a normalized orthomorphism of G that can be written as a product of $(n-1)/k$ disjoint cycles, each of length k. Note that we need the condition $k > 1$ as 1-regular orthomorphisms cannot exist for nontrivial groups. As examples, we leave it to the reader to verify that (1 6)(2 5)(3 4) is a 2-regular orthomorphism of $\mathbb{Z}_7$; that (1 2 4)(3 6 5) is a 3-regular orthomorphism of $\mathbb{Z}_7$; and that (1 8 12 5)(2 3 11 10)(4 6 9 7) is a 4-regular orthomorphism of $\mathbb{Z}_{13}$.

There are two cases of special interest. The case $k = n-1$ will be treated in Sect, 15.3, where we will see that a group G of order n admits an R-sequencing if and only if G admits an $(n-1)$-regular orthomorphism. The case $k = 2$ will be covered in Section 15.4, where we will see that starters and 2-regular orthomorphisms are equivalent. There are other instances in which k-regular orthomorphisms arise: 6-regular orthomorphisms are related to cyclic Steiner triple systems of order $n \equiv 1 \pmod 6$ (see [194]), and k-regular orthomorphisms are related to I-matrices (see [184]). In both of these cases, the relationship is implicitly given in the context of neofields: neofields will be discussed in Section 15.2.

Which finite groups admit k-regular orthomorphisms? For abelian groups Friedlander, Gordon, and Tannenbaum conjectured the following.

Conjecture 15.5. If G is an abelian group of order n whose Sylow 2-subgroup is either trivial or noncyclic and $k > 1$ is a divisor of $n-1$, then G admits a k-regular orthomorphism.

We will deal with the special case $k = 2$ in Section 15.4 and the special case $k = n-1$ in Section 15.3. Other special cases of Conjecture 15.5 that were proved by Friedlander, Gordon, and Tannenbaum are given in Theorems 15.6, 15.7, 15.8, and 15.9.

Theorem 15.6. *If G is an abelian group of order at most* 15, *$k > 1$ is a divisor of $|G|-1$, and the Sylow* 2*-subgroup of G is either trivial or noncyclic, then G admits k-regular orthomorphisms.*

Theorem 15.7. *If $G = \mathrm{GF}(q)^+$, $q > 2$, and $k > 1$ divides $q-1$, then G admits k-regular orthomorphisms.*

The proof of Theorem 15.7 uses the fact that if a is a primitive element of $GF(q)$, then the linear orthomorphism $x \mapsto a^d x, d \neq n-1$ a divisor of $q-1$, is a $(q-1)/d$-regular orthomorphism of $\mathrm{GF}(q)^+$.

Theorem 15.8. *If G is an abelian p-group, p is an odd prime, and $k > 1$ is a divisor of $p-1$, then G admits k-regular orthomorphisms.*

Theorem 15.9. *If $G = \mathbb{Z}_n$, $n \geq 5$ odd, then G admits $(n-1)/2$-regular orthomorphisms.*

In 1994 Wang [323] introduced the concept of uniformly harmonious groups. Let G be a group of order n. The group G is said to be *uniformly harmonious* if, for each divisor k of n, G admits complete mappings that are products of disjoint cycles of length of k.

Theorem 15.10 (Wang, 1994). *Every abelian group of odd order is uniformly harmonious.*

Clearly the only mapping that is a product of 1-cycles is the identity mapping. Hence 1-regular orthomorphisms of nontrivial groups cannot exist. As the identity mapping is a complete mapping of any group of odd order (see Corollary 3.6), and of no group of even order (as $g^2 = 1$ would then have at least two solutions), uniformly harmonious groups must be of odd order.

15.1.3 Generalized Orthomorphisms and Near Orthomorphisms

In 1984 Hsu and Keedwell [186] gave a generalization of orthomorphisms and near orthomorphisms. Hsu and Keedwell's generalized mappings are not mappings, but arrangements of the elements of a group, possibly repeated, into cycles and sequences. These cycles and sequences need not be disjoint though it is implicit in [186] that a group element may appear at most once in a cycle or sequence. Let G be a group of order n, λ a positive integer, and $K = \{k_1, \ldots, k_r\}$ be positive integers that sum to $\lambda(n-1)$. A (K, λ)-*orthomorphism* is an arrangement of the nonidentity elements of G, each used λ times, into r cycles

$$(g_{1,0} \ \cdots \ g_{1,k_1-1})(g_{2,0} \ \cdots \ g_{2,k_2-1}) \cdots (g_{r,0} \ \cdots \ g_{r,k_r-1})$$

such that each nonidentity element of G appears exactly λ times in the form $g_{i,j}^{-1} g_{i,j+1}$, the subscripts being added modulo k_i for $i = 0, \ldots, r$. If $k_1 = \cdots = k_r = k$ and $r = (n-1)/k$, then we call a (K, λ)-orthomorphism a (k, λ)-orthomorphism. It is clear that a $(K, 1)$-orthomorphism corresponds to an orthomorphism with cycle lengths in K and that a $(k, 1)$-orthomorphism corresponds to a k-regular orthomorphism.

Example 15.11. (1 2 6 5)(1 3 6 4)(2 4 5 3) is a (4, 2)-orthomorphism of $\mathbb{Z}_7$: the differences are 1, 4, 6, 3, 2, 3, 5, 4, 2, 1, 5, 6.

Let G be a group of order n, λ a positive integer, and $K = \{h_1, \ldots, h_\lambda; k_1, \ldots, k_r\}$ be positive integers that sum to λn. A (K, λ)-*near orthomorphism* is an arrangement of the elements of G, each used λ times, into λ sequences and r cycles

$$[\bar{g}_{1,0} \cdots \bar{g}_{1,h_1-1}] \cdots [\bar{g}_{\lambda,0} \cdots \bar{g}_{\lambda.h_\lambda-1}](g_{1,0} \ \cdots \ g_{1,k_1-1}) \cdots (g_{r,0} \ \cdots \ g_{r,k_r-1})$$

such that the list of the quotients $\bar{g}_{i,j}^{-1}\bar{g}_{i,j+1}$, $j = 0, \ldots, h_i - 1$, $i = 0, \ldots, \lambda$ and $g_{i,j}^{-1} g_{i,j+1}$, $j = 0, \ldots, k_i - 1$, the subscripts being added modulo k_i, $i = 0, \ldots, r$, contain each nonidentity element of G exactly λ times. If $k_1 = \cdots = k_r = k = h + 1$ and $r = (n-1)/k$, then we call a (K, λ)-near orthomorphism a (k, λ)-near orthomorphism. As an example [3 1 0][3 5 0](1 4 5 2)(2 4) is a (3, 3; 4, 2)-near orthomorphism of $\mathbb{Z}_6$: the differences are 4, 5, 2, 1, 3, 1, 3, 5, 2, 4. A $(K, 1)$-near orthomorphism is clearly a near orthomorphism. Some constructions of (K, λ)-orthomorphisms and near orthomorphisms are to be found in [187] and [215].

In their 1984 paper, Hsu and Keedwell [186] concentrated on applications of (K, λ)-orthomorphisms and near orthomorphisms when $\lambda = 1$: these applications will be considered in Sections 15.2 and 15.3. In a subsequent 1985 paper [187], they applied (K, λ)-orthomorphisms and near orthomorphisms, $\lambda \geq 1$, in the construction of Mendelsohn designs. Recall from Section 1.4 that an incidence structure is a set of blocks and a set of points with an incidence relation, point P incident with block l. When blocks are uniquely determined by the points incident with them, i.e., there are no repeated blocks, then the blocks may be represented by sets of points: then point P incident with block l becomes $P \in l$. In a Mendelsohn design, there are no repeated blocks and the points in a block are cyclically ordered. If $K = \{k_1, \ldots, k_r\}$ is a set of positive integers, a (v, K, λ)- *Mendelsohn design* consists of a set $\mathscr{P}$ of v points and a set of cyclically ordered subsets of $\mathscr{P}$ called blocks, where K is the set of cardinalities of these blocks, and each ordered pair of distinct elements of $\mathscr{P}$ appear consecutively in these blocks exactly λ times. A (v, k, λ)-*Mendelsohn design* is a (v, K, λ)-Mendelsohn design in which $K = \{k\}$.

Hsu and Keedwell [187] gave a construction of Mendelsohn designs from generalized orthomorphisms. If G is a group of order n, $K = \{k_1, \ldots, k_r\}$, and

$$(g_{1,0} \ \cdots \ g_{1,k_1-1})(g_{2,0} \ \cdots \ g_{2,k_2-1}) \cdots (g_{r,0} \ \cdots \ g_{r,k_r-1})$$

is a (K, λ)-orthomorphism of G, then we can construct an $(n, \bar{K}, \lambda)$-Mendelsohn design $\mathscr{M}$, where $\bar{K}$ is the set of distinct elements of K, by taking the points to be the elements of G and the blocks to be the cyclically ordered sets $(gg_{i,0} \ \cdots \ gg_{i,k_i-1})$, $g \in G$, $i = 1, \ldots, r$. Note that the mapping $\tau_g \colon x \mapsto gx$ permutes the points of $\mathscr{M}$ and the blocks of $\mathscr{M}$. The set $\{\tau_g \mid g \in G\}$ is a group of automorphisms of $\mathscr{M}$. Hsu and Keedwell characterized those Mendelsohn designs that can be constructed in this way. More information on Mendelsohn designs can be found in [79].

15.1.4 Other Variants

In 1998 Bedford [37] introduced quasicomplete mappings. A list $a_1, a_2, \ldots, a_n$ of elements of a finite group G of order n is a *quasi-ordering* of G if the list contains exactly one occurrence of each element z of G such that $z^2 = 1$; and, for every element g of G such that $g^2 \neq 1$, there are either two occurrences of g and none of g^{-1} or one occurrence of g and one of g^{-1} or two occurrences of g^{-1} and none of g. A *quasi-complete mapping* of a group or quasigroup G is a mapping $g \mapsto \theta(g)$ of the elements of G onto a quasi-ordering of G such that the mapping $\phi\colon g \mapsto g\theta(g)$ is a permutation of the elements of G. The mapping $g \mapsto \phi(g)$ is a *quasi-orthomorphism* of G. From a quasi-complete mapping of a group G of order n, we can construct a quasi-orthogonal pair of Latin squares of order n. Two Latin squares of the same order on the same symbol set are *quasi-orthogonal* if when superimposed each unordered pair of distinct symbols occurs exactly twice.

Theorem 15.12. *The Cayley table of a finite group G has a quasi-orthogonal mate if and only if G admits quasi-complete mappings.*

If θ is a quasi-complete mapping of a group G, L the Cayley table of G, and ϕ the quasi-orthomorphism corresponding to θ, then L and L_ϕ are quasi-orthogonal. Recall that if the xyth entry of L is xy, then the xyth entry of L_ϕ is $x\phi(y)$.

Example 15.13. The following defines a quasi-complete mapping θ of $\mathbb{Z}_4$ as well as the corresponding quasi-orthomorphism ϕ.

x	0	1	2	3
$\theta(x)$	0	1	1	2
$\phi(x)$	0	2	3	1

The corresponding quasi-orthogonal pair L and L_ϕ is shown in Figure 15.6, and the square of unordered pairs obtained by superimposing L and L_ϕ is shown in Figure 15.7.

Fig. 15.6 Two quasi-orthogonal Latin squares of order 4

$$L = \begin{pmatrix} 0\,1\,2\,3 \\ 1\,2\,3\,0 \\ 2\,3\,0\,1 \\ 3\,0\,1\,2 \end{pmatrix}, \quad L_\phi = \begin{pmatrix} 0\,2\,3\,1 \\ 1\,3\,0\,2 \\ 2\,0\,1\,3 \\ 3\,1\,2\,0 \end{pmatrix}$$

Fig. 15.7 L and L_ϕ superimposed

$$\begin{pmatrix} \{0,0\} & \{1,2\} & \{2,3\} & \{1,3\} \\ \{1,1\} & \{2,3\} & \{0,3\} & \{0,2\} \\ \{2,2\} & \{0,3\} & \{0,1\} & \{1,3\} \\ \{3,3\} & \{0,1\} & \{1,2\} & \{0,2\} \end{pmatrix}$$

Bedford [37] characterized abelian groups that admit quasi-complete mappings.

Theorem 15.14 (Bedford, 1998). *An abelian group admits a quasi-complete mapping if and only if it is not of the form* $\mathbb{Z}_{4m+2} \times G$, $|G|$ *odd.*

By contrast, there are very few results for finite nonabelian groups.

Theorem 15.15 (Bedford, 1998). D_{2n}*, the dihedral group of order* $2n$*, does not admit quasi-complete mappings if* n *is odd.*

For G a finite group, a bijection $\theta: G \to G$ is an *antisymmetric mapping* of G if $\theta(g)h = \theta(h)g$ implies that $g = h$. These mappings have been used in error detection schemes: see [161] for instance. Antisymmetric mappings are closely related to orthomorphisms. If G is abelian, then θ is an antisymmetric mapping of G if and only if θ is an orthomorphism of G. Hence, we can characterize finite abelian groups that admit antisymmetric mappings.

Theorem 15.16. *A finite abelian group admits antisymmetric mappings if and only if its Sylow* 2*-subgroup is either trivial or noncyclic.*

Proof. This is an immediate consequence of Theorem 3.9. □

Thus the determination of those finite groups that admit antisymmetric mappings reduces to determining those finite nonabelian groups that admit antisymmetric mappings. A number of finite nonabelian groups have been shown to admit antisymmetric mappings: see [161, 290, 317], and [332].

Example 15.17. Recall that $D_6 = \langle a, b \mid a^3 = b^2 = 1, bab = a^2 \rangle$ denotes the dihedral group of order 6. The following defines an antisymmetric mapping of D_6.

x	1	a	a^2	b	ba	ba^2
$\theta(x)$	a^2	a	1	b	ba	ba^2

Table 15.1 shows that θ is an antisymmetric mapping of D_6. The xyth entry of Table 15.1 is $\theta(x)y$, and $\theta(y)x$ is the xyth entry of the transpose of Table 15.1. It can be readily seen that the xyth and yxth entries of this table match if and only if $x = y$. Hence, θ is an antisymmetric mapping of D_6.

Table 15.1 Proof that θ is an antisymmetric mapping of D_6

	1	a	a^2	b	ba	ba^2
1	a^2	1	a	ba	ba^2	b
a	a	a^2	1	ba^2	b	ba
a^2	1	a	a^2	b	ba	ba^2
b	b	ba	ba^2	1	a	a^2
ba	ba	ba^2	b	a^2	1	a
ba^2	ba^2	b	ba	a	a^2	1

As, by Theorem 3.34, D_6 does not admit complete mappings, this example shows that nonabelian groups that do not admit complete mappings may still admit antisymmetric mappings. In 1995 Gallian and Mullin [150] conjectured that all nonabelian groups admit antisymmetric mappings.

Conjecture 15.18. All nonabelian groups admit antisymmetric mappings.

Gallian and Mullin [150] proved this conjecture true for a number of finite groups. Some of their results are shown in Theorems 15.19, 15.20, 15.21, and 15.22.

Theorem 15.19 (Gallian and Mullin, 1995). *All groups of odd order admit antisymmetric mappings.*

Theorem 15.20 (Gallian and Mullin, 1995). *Finite simple groups, other than $\mathbb{Z}_2$, admit antisymmetric mappings.*

Theorem 15.21 (Gallian and Mullin, 1995). *p-groups, other than cyclic 2-groups, admit antisymmetric mappings.*

Theorem 15.22 (Gallian and Mullin, 1995). *All nonabelian groups of order less than 36 admit antisymmetric mappings.*

There are a number of other closely related mappings in the literature. ψ-complete mappings, defined in 1982 by Niederreiter and Robinson [252], generalize both complete mappings and orthomorphisms. While they were originally defined for the additive groups of finite fields, they can be defined for groups in general. Let G be a finite group and let $\psi\colon G \to G$ be a bijection. A bijection $\theta\colon G \to G$ is a *ψ-complete mapping* of G if the mapping $g \mapsto \psi(g)\theta(g)$ is a bijection. This generalizes complete mappings and orthomorphisms as if ψ is the identity mapping, then ψ-complete mappings are synonymous with complete mappings; and if $\psi(g) = g^{-1}$ for all $g \in G$, then ψ-complete mappings are synonymous with orthomorphisms. R-orthomorphisms were defined in 2001 by Cohen, Niederreiter, Shparlinski, and Zieve [77] for groups in general and studied in [75] and [253] in the context of elementary abelian groups. Let R be a nonempty set of positive integers. The bijection $\theta\colon G \to G$ is an *R-orthomorphism* of G if θ^r, θ composed with itself r times, is an orthomorphism of G for all $r \in R$. A closely related class of mappings, K-complete mappings, was defined by Winterhof [331] in 2014. For q a prime power and K a set of positive integers, a *K-complete mapping* is a permutation polynomial f of GF(q) for which the mapping

$$x \mapsto x + \sum_{k \in K} \mathrm{f}^k(x)$$

is a permutation polynomial of GF(q). *K-orthomorphisms* are defined similarly. Compound and compatible orthomorphisms were defined in 2010 by Stones and Wanless [300]. These are a class of orthomorphisms of cyclic groups. An orthomorphism θ of $\mathbb{Z}_n$ is *d-compound* if $\theta(i) \equiv \theta(j) \pmod{d}$ whenever $i \equiv j \pmod{d}$. An orthomorphism of $\mathbb{Z}_n$ is *compatible* if it is d-compound for all d

that divide n. Quasi-near complete mappings and quasi-near orthomorphisms were defined in 1998 by Bedford [37], where they were used in constructions of quasi-sequencings of groups. A *quasi-near complete mapping* of a group G is a mapping $g \mapsto \theta(g)$ of the elements of $G \setminus \{h\}$ onto a quasi-ordering of $G \setminus \{1\}$ such that the mapping $\phi: g \mapsto g\theta(g)$ is a one-to-one mapping from $G \setminus \{h\}$ to $G \setminus \{k\}$ say. Here h and k are fixed elements of G. The mapping $g \mapsto \phi(g)$ is a *quasi-near orthomorphism* of G.

15.2 Neofields

There are a number of algebraic structures with two binary operations, addition and multiplication: for example, rings, in particular fields and division rings, and quasifields. In each of these examples, the additive structure is that of a group. In this section we will discuss neofields, another algebraic structure with two binary operations. Neofields differ in that under addition the elements of a neofield form a loop. Recall from Section 1.1 that a loop has one binary operation, an identity, and both left and right cancellation holds. A group is an associative loop. Neofields were introduced by Paige [260] in his 1947 Ph.D. thesis.

We will see that there is a natural one-to-one correspondence between left neofields and normalized near orthomorphisms. Hence, properties of a left neofield can, in principle, be determined by properties of its corresponding normalized near orthomorphism.

15.2.1 Left Neofields and Neofields

A *left neofield* is a set N with two binary operations, addition and multiplication, satisfying the following:

1. The elements of N form a loop under addition, with identity 0.
2. The nonzero elements of N form a group under multiplication, with identity 1.
3. The left distributive law holds: $a(b+c) = ab + ac$ for all $a, b, c \in N$.
4. $0 \times a = 0$ for all $a \in N$.

A left neofield is called a *neofield* if the right distributive law is also satisfied. For a neofield or left neofield, we will use N^+ to denote the additive loop and N^* to denote the multiplicative group of nonzero elements. A left neofield N is said to be *abelian* if N^* is abelian and *cyclic* if N^* is cyclic. Clearly, if N is abelian, then the terms neofield and left neofield are synonymous. The *order* of the left neofield N, denoted $|N|$, is the number of elements of N.

Lemma 15.23. *Let N be a left neofield. Then $a \times 0 = 0$ for all $a \in N$.*

Proof. Note that $0 + a \times 0 = a \times 0 = a(0 + 0) = a \times 0 + a \times 0$ for all $a \in N$. Hence, by cancellation on the right, $0 = a \times 0$ for all $a \in N$. □

The axiom, $0 \times a = 0$ for all $a \in N$, is not to be found in the literature. It is a "nondegeneracy" condition that cannot be derived from the first three axioms of a left neofield. Without this axiom, the mapping $x \mapsto 0 \times x$ can be any endomorphism of N^+, e.g., $0 \times a = a$ for all $a \in N$. Thus we have added this axiom to avoid these "degenerate" possibilities. Note that this axiom is not needed for neofields as it can be derived from the right distributive law.

As a consequence of Lemma 15.23, multiplication in a left neofield N is associative for all elements of N, not just for elements of N^*.

An *automorphism* of a neofield N is a bijection $\alpha\colon N \to N$ satisfying:

1. $\alpha(x + y) = \alpha(x) + \alpha(y)$ and
2. $\alpha(xy) = \alpha(x)\alpha(y)$.

Thus an automorphism of N is both an automorphism of N^+ and an automorphism of N^*. We can consider $\alpha\colon N^* \to N^*$ to be a bijection $N \to N$ by setting $\alpha(0) = 0$. With this convention, it follows that $\mathrm{Aut}(N) = \mathrm{Aut}(N^+) \cap \mathrm{Aut}(N^*)$. Loops that can be the additive loop of a left neofield can be characterized by their automorphism groups.

Theorem 15.24. *A loop can be the additive loop of a left neofield if and only if it admits an automorphism group that acts sharply transitively on its nonidentity elements.*

Proof. Let N be a left neofield and, for each $g \in N^*$, define $\tau_g\colon N \to N$ by $\tau_g(a) = ga$. As $\tau_g(x) + \tau_g(y) = gx + gy = g(x + y) = \tau_g(x + y)$, $\tau_g \in \mathrm{Aut}(N^+)$ for all $g \in N^*$. As $\tau_g \circ \tau_h(x) = \tau_g(hx) = ghx = \tau_{gh}(x)$, the set $\{\tau_g \mid g \in N^*\}$ is a group of automorphisms of N^*. That this group acts sharply transitively on N^* can be seen by noting that if $x, y \in N^*$, then $\tau_g(x) = y$ if and only if $g = yx^{-1}$.

Conversely, let L be a loop written additively with identity 0. Let us assume that G is an automorphism group of L that acts sharply transitively on the nonzero elements of L. We will use G to define multiplication on L. Pick a nonzero element of L and denote it 1, and for each nonzero element $a \in L$, let τ_a denote the unique element of G satisfying $\tau_a(1) = a$. Define multiplication on L by:

$$ab = \begin{cases} 0 & \text{if } a = 0 \text{ or } b = 0, \\ \tau_a(b) & \text{if } a, b \neq 0. \end{cases}$$

Clearly, under multiplication, the nonzero elements of L form a group isomorphic to G. If we set $\tau_a(b) = 0$ if a or b is 0, then $ab = \tau_a(b)$ for all $a, b \in L$. Also, with this convention, $\tau_a(b + c) = \tau_a(b) + \tau_a(c)$ for all $a, b, c \in L$. Then

$$a(b + c) = \tau_a(b + c) = \tau_a(b) + \tau_a(c) = ab + ac.$$

Hence, with this multiplication, L is a left neofield. □

In 1951 Bruck [54] proved that any loop with an automorphism group that is transitive on its nonzero elements is either a simple loop or an elementary abelian group. See [54], for instance, for the definition of a simple loop. Thus, as a corollary to Theorem 15.24, we obtain the following.

Corollary 15.25. *The additive loop of a left neofield, whose additive loop is not an elementary abelian group, is a simple loop.*

Here is an example of a neofield of order 6.

Example 15.26. Let $G = \{1, g, g^2, g^3, g^4\}$ be the cyclic group of order 5, and let the addition and multiplication tables of N be given as follows:

$+$	0	1	g	g^2	g^3	g^4
0	0	1	g	g^2	g^3	g^4
1	1	0	g^4	g^3	g^2	g
g	g	g^2	0	1	g^4	g^3
g^2	g^2	g^4	g^3	0	g	1
g^3	g^3	g	1	g^4	0	g^2
g^4	g^4	g^3	g^2	g	1	0

$\times$	0	1	g	g^2	g^3	g^4
0	0	0	0	0	0	0
1	0	1	g	g^2	g^3	g^4
g	0	g	g^2	g^3	g^4	1
g^2	0	g^2	g^3	g^4	1	g
g^3	0	g^3	g^4	1	g	g^2
g^4	0	g^4	1	g	g^2	g^3

It is routine, but tedious, to verify that N is a neofield.

Let us take a closer look at the neofield in Example 15.26, specifically at the second line, that gives the values of $1 + x$, of the addition table for N.

x	0	1	g	g^2	g^3	g^4
$1+x$	1	0	g^4	g^3	g^2	g
$x^{-1}(1+x)$	$\cdot$	$\cdot$	g^3	g	g^4	g^2

We see that the mapping $x \mapsto 1+x$, restricted to $G \setminus \{1\}$, is a near orthomorphism of G that easily extends to an orthomorphism of G. We next examine the relationship between left neofields and near orthomorphisms.

15.2.2 Left Neofields and Near Orthomorphisms

The *presentation function* of a left neofield N is the mapping $\theta \colon N \to N$ defined by $\theta(x) = 1 + x$. A left neofield N is completely determined by its multiplicative group N^* and its presentation function θ as, if $a, b \neq 0$, then

$$a + b = a(1 + a^{-1}b) = a\theta(a^{-1}b).$$

As we shall see, the presentation function of a left neofield is essentially a near orthomorphism. In the first paper on neofields in 1949, Paige [262] implicitly used near orthomorphisms in the construction of neofields, citing Bruck for the implicit characterization of those near orthomorphisms that can be used to construct neofields. In 1984 Hsu and Keedwell [186] established a correspondence between left neofields with multiplicative group G and normalized near orthomorphisms of G: when we impose the condition, $0 \times a = 0$ for all $a \in N$, this correspondence becomes a one-to-one correspondence. But this correspondence was actually first discovered by Johnson [195] in her 1981 Ph.D. thesis: this was pointed out in [216].

Theorem 15.27. *Let G be a group, written multiplicatively with identity 1. Let θ be a normalized near orthomorphism of G, with exdomain element t, and extend θ to a bijection $G \cup \{0\} \to G \cup \{0\}$ by setting $\theta(t) = 0$ and $\theta(0) = 1$. Then θ is the presentation function of a left neofield. Conversely, Let θ be the presentation function of a left neofield with multiplicative group G. Let t be the unique solution to the equation $1 + x = 0$. Then, θ restricted to $G - \{t\}$ is a normalized near orthomorphism of G with exdomain element t.*

Proof. Let θ be a near orthomorphism of a group G, with exdomain element t, and extend θ to a bijection $G \cup \{0\} \to G \cup \{0\}$ by setting $\theta(t) = 0$ and $\theta(0) = 1$. Let $N = G \cup \{0\}$ and define addition and multiplication in N as follows. Multiplication is as in G except that $0a = a0 = 0$ for all $a \in N$. To define addition,

$$x + y = \begin{cases} y & \text{if } x = 0, \\ x\theta(x^{-1}y) & \text{if } x \neq 0. \end{cases}$$

We see that $0 + y = y$ for all $y \in N$ and if $y \neq 0$ then $y + 0 = y\theta(0) = y$. Suppose that $a + b = c$. Given a and b, the value of c is uniquely determined. Given a and c, then $b = c$ if and only if $a = 0$, and b is uniquely determined by the equation $\theta(a^{-1}b) = a^{-1}c$ if $a \neq 0$. Given b and c, then $a = 0$ if and only if $b = c$. Otherwise a is uniquely determined by the equation $(a^{-1}b)^{-1}\theta(a^{-1}b) = b^{-1}c$. Thus N is a loop under addition. Next consider $a(b + c)$. This must equal $ab + ac$ if any of a, b, or c is 0. If $a, b, c \neq 0$, then

$$a(b + c) = ab(1 + b^{-1}c) = (ab)\theta(b^{-1}c) = (ab)\theta((ab)^{-1}ac) = ab + ac,$$

and so the left distributive law holds. Thus N is a left neofield. Conversely, if θ is the presentation function of a left neofield N with multiplicative group G, then $\theta(t) = 0$ for a unique element t of N. It is easily seen that θ is a bijection from $G - \{t\}$ onto $G - \{1\}$, and the mapping $x \to x^{-1}\theta(x) = x^{-1} + 1$ is a bijection from $G - \{t\}$ onto $G - \{1\}$. Thus the restriction of θ to $G - \{t\}$ is a near orthomorphism of G. □

Corollary 15.28. *There is a one–one correspondence between the set of normalized orthomorphisms of a group G and the set of left neofields with multiplicative group G in which $1 + 1 = 0$.*

Corollary 15.29. *There is a one–one correspondence between the set of proper near orthomorphisms of a group G and the set of left neofields with multiplicative group G in which $1 + 1 \neq 0$.*

In Section 8.1 we defined the homology H_α for orthomorphisms of a group G by $\mathrm{H}_\alpha[\theta] = \alpha\theta\alpha^{-1}$ for $\alpha \in \mathrm{Aut}(G)$. Similarly we define the *homology* H_α for near orthomorphisms of a group G by $\mathrm{H}_\alpha[\theta] = \alpha\theta\alpha^{-1}$ for $\alpha \in \mathrm{Aut}(G)$.

Theorem 15.30. *Let θ be a near orthomorphism of G with exdomain element t, let N be the left neofield constructed from θ, and let α be an automorphism of G.*

1. *$\mathrm{H}_\alpha[\theta]$ is a near orthomorphism with exdomain element $\alpha(t)$.*
2. *α extends to an automorphism of N, by setting $\alpha(0) = 0$, if and only if $\mathrm{H}_\alpha[\theta] = \theta$.*
3. *If $\alpha(x) = c^{-1}xc$, then $\mathrm{H}_\alpha[\theta] = \theta$ if and only if $(a + b)c = ac + bc$ for all $a, b \in N$.*
4. *N is a neofield if and only if $\mathrm{H}_\beta[\theta] = \theta$ for all $\beta \in \mathrm{Inn}(G)$.*

Proof.

1. This follows as $x^{-1}\alpha\theta\alpha^{-1}(x) = \alpha(\alpha^{-1}(x)^{-1}\theta\alpha^{-1}(x))$.
2. If $a \neq 0$ then $\alpha(a + b) = \alpha(a) + \alpha(b)$ if and only if

$$\alpha(a\theta(a^{-1}b)) = \alpha(a)\theta(\alpha(a^{-1}b)).$$

By setting $x = \alpha(a^{-1}b)$, this is seen to be true if and only if $\alpha\theta\alpha^{-1}(x) = \theta(x)$. Hence the result.

3. If any of a, b, or c is zero, then $(a+b)c = ac+bc$. If $a, b, c \neq 0$ then $(a+b)c = a\theta(a^{-1}b)c$ and $ac + bc = ac\theta(c^{-1}a^{-1}bc)$ and $a\theta(a^{-1}b)c = ac\theta(c^{-1}a^{-1}bc)$ if and only if $\mathrm{H}_\alpha[\theta] = \theta$.
4. This is an immediate consequence of 2. □

Corollary 15.31. *If θ is a near orthomorphism of G corresponding to a neofield, then the exdomain element of θ is in the center of G.*

Proof. Let θ be a near orthomorphism of G corresponding to a neofield. By Theorem 15.30, $\mathrm{H}_\alpha(\theta) = \theta$ for all $\alpha \in \mathrm{Inn}(G)$. But then, by Theorem 15.30, if t is the exdomain element of θ, then $\alpha(t) = t$ for all $\alpha \in \mathrm{Inn}(G)$. It follows that t is in the center of G. □

15.2.3 Classes of Left Neofields and Neofields

There are several properties of neofields and left neofields that have been studied. A (left) neofield N is *commutative* if addition is commutative. A (left) neofield N has the *right inverse property* if for all $a \in N$ there exists $b \in N$ such that $(x+a)+b = x$ for all $x \in N$, the *left inverse property* if for all $a \in N$ there exists $b \in N$ such that $b + (a + x) = x$ for all $x \in N$, the *inverse property* if it has both the left and right inverse properties, and the *exchange inverse property* if for all $a \in N$ there exists $b \in N$ such that $b + (x + a) = x$ for all $x \in N$.

In 1980 the above properties were used by Hsu [184] to classify cyclic neofields: recall that a cyclic neofield is a neofield whose multiplicative group is cyclic. In 1949 Paige [262] showed that for a finite abelian neofield, i.e., a neofield whose multiplicative group is abelian, the element t satisfying $1 + t = 0$ is uniquely determined.

Lemma 15.32 (Paige, 1949). *If N is a finite abelian neofield with multiplicative group G and $1 + t = 0$, then one of the following holds.*

1. *If $|G|$ is odd or G has more than one involution, then $t = 1$.*
2. *If G has exactly one involution u, then $t = u$.*

Proof. See Theorem II.2 in [262]. □

The implications for finite cyclic neofields are immediate.

Corollary 15.33. *If N is a cyclic neofield of order n and g is a generator of N^* and $1 + t = 0$, then*

$$t = \begin{cases} 1 & \text{if } n \text{ is even,} \\ g^{(n-1)/2} & \text{if } n \text{ is odd.} \end{cases}$$

Corollary 15.34. *If N is a finite cyclic neofield and $1 + t = 0$, then $t + 1 = 0$. Further, if $a, b \in N$, then $a + b = 0$ if and only if $b + a = 0$.*

For a cyclic neofield N, we will use $-a$ to denote the unique element of N that satisfies $a + (-a) = (-a) + a = 0$ and $a - b$ to denote $a + (-b)$: note that $-a = (-1)a$. A cyclic neofield N satisfies the *exchange minus property* if $-(a + b) = -a - b$ for all $a, b \in N$. In Hsu's classification of cyclic neofields, there are six classes of neofields. A cyclic neofield is a *XMP-neofield* if it satisfies the exchange minus property, a *CIP-neofield* if it is commutative and satisfies the inverse property, an *XIP-neofield* if it satisfies the exchange inverse property, an *LXP-neofield* if it satisfies the left inverse property, a *CMP-neofield* if it is commutative, and an *RXP-neofield* if it satisfies the right inverse property.

Any cyclic neofield is an XMP-neofield as

$$-(a + b) = (-1)(a + b) = (-1)a + (-1)b = -a - b.$$

Ordered by inclusion, these classes form a lattice as shown in Figure 15.8. Hsu [184] gave a number of constructions for each class of cyclic neofield.

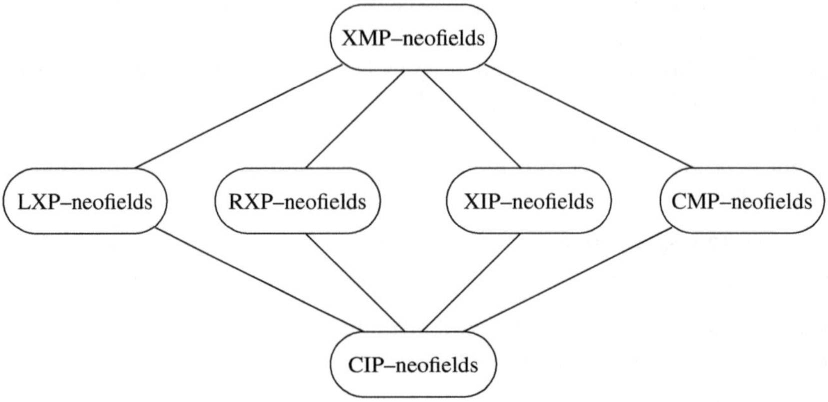

Fig. 15.8 The six classes of cyclic neofields

We can extend the congruences R and I, introduced in Section 8.1, of orthomorphisms to the domain of near orthomorphisms as follows.

Lemma 15.35. *Let θ be a near orthomorphism of a group G with exdomain element t.*

1. $\mathrm{R}[\theta]: x \to x\theta(x^{-1})$ *is a near orthomorphism of a group G with exdomain element t.*
2. $\mathrm{I}[\theta]: x \to t^{-1}\theta^{-1}(tx)$ *is a near orthomorphisms of G with exdomain element t^{-1}.*

As before, R is called a *reflection* and I an *inversion*.

Lemma 15.36. *Let G be a finite group, $\alpha, \beta \in \mathrm{Aut}(G)$, and let θ be a near orthomorphism of G with exdomain element t.*

1. $\mathrm{H}_\alpha \mathrm{H}_\beta[\theta] = \mathrm{H}_{\alpha\beta}[\theta]$.
2. $\mathrm{R}^2[\theta] = \theta$.
3. $\mathrm{H}_\alpha \mathrm{R}[\theta] = \mathrm{R}\mathrm{H}_\alpha[\theta]$.
4. $\mathrm{I}^2[\theta] = \theta$.
5. $\mathrm{H}_\alpha I[\theta] = \mathrm{I}\mathrm{H}_\alpha[\theta]$.
6. *If $t \in Z(G)$ and $t^2 = 1$, then* $(\mathrm{IR})^3[\theta] = \theta$.

Proof. Similar to the proof of Theorem 8.8. □

As, by Lemma 15.36, H_α commutes with both R and I for all $\alpha \in \mathrm{Aut}(G)$, we obtain the following as a consequence of Theorem 15.30.

Corollary 15.37. *If θ is a near orthomorphism that defines a neofield, then* $\mathrm{R}[\theta]$ *and* $\mathrm{I}[\theta]$ *are near orthomorphisms that define neofields.*

Correspondences between properties of left neofields and mappings that fix the corresponding near orthomorphisms are established in the next theorem.

Theorem 15.38. *Let θ be a near orthomorphism of G with exdomain element t, and let N be the left neofield constructed from θ.*

1. N is commutative if and only if $R[\theta] = \theta$.
2. If $t \in Z(G)$, *then N has the right inverse property if and only if* $IRI[\theta] = \theta$.
3. N has the left inverse property if and only if $I[\theta] = \theta$.
4. If $t \in Z(G)$, *then N has the inverse property if and only if* $I[\theta] = \theta$ *and* $IRI[\theta] = \theta$.
5. N has the exchange inverse property if and only if $RI[\theta] = \theta$.

Proof. Let t be the exdomain element of θ. Thus $1 + t = 0$. We will first establish that a necessary condition for N to be commutative or have the left or right inverse property is that $t^2 = 1$. Note that $t + 1 = 0$ if and only if $t(1+t^{-1}) = 0$ if and only if $t^{-1} = t$. Thus N commutative implies that $t + 1 = 1 + t = 0$ and so $t^2 = 1$. If N has the right inverse property, then there exists $b \in N$ satisfying $(x + t) + b = x$ for all $x \in N$. But then $b = 1$ as $(1 + t) + b = 1$ and then $(0 + t) + 1 = 0$ again implies that $t^2 = 1$. Similarly, if N has the left inverse property, then there exists $b \in N$ satisfying $b + (1 + x) = x$ for all $x \in N$. But then $b = t$ as $b + (1 + t) = t$ and then $t + (1 + 0) = 0$ again implies that $t^2 = 1$. It is also important to note that $T[\theta] = \theta$, for T = I, R, and IRI, only if $t^2 = 1$. Thus, in the proofs of $1, \cdots, 4$, we shall assume that $t^2 = 1$.

1. If either a or $b = 0$, or $a = bt$, or $b = at$, then $a + b = b + a$. Otherwise $a + b = b + a$ if and only if $a\theta(a^{-1}b) = b\theta(b^{-1}a)$ if and only if $\theta(a^{-1}b) = (a^{-1}b)\theta((a^{-1}b)^{-1})$ if and only if $R[\theta](a^{-1}b) = \theta(a^{-1}b)$.
2. If N has the right inverse property, then there exists $b \in N$ satisfying $(x+a)+b = x$ for all $x \in N$: setting $x = 0$ shows that $b = at$.
   If $a = 0$, $x = 0$, or $x = at$, then $(x + a) + at = x$. Otherwise $(x + a) + at = x$ if and only if $x\theta(x^{-1}a) + at = x$, if and only if $x\theta(x^{-1}a)\theta(u\theta(x^{-1}a)^{-1}x^{-1}at) = x$, if and only if $\theta(\theta(x^{-1}a)^{-1}x^{-1}at) = \theta(x^{-1}a)^{-1}$, if and only if $\theta(y\theta^{-1}(ty^{-1})) = ty$, where $y = \theta(x^{-1}a)^{-1}t$, if and only if $ty\theta^{-1}(ty^{-1}) = t\theta^{-1}(ty)$ if and only if $RI[\theta](y) = I[\theta](y)$ if and only if $IRI[\theta] = \theta$.
3. If N has the left inverse property, then there exists $b \in N$ satisfying $b+(a+x) = x$ for all $x \in N$: setting $x = 0$ shows that $b = at$.
   If $a = 0, x = 0$, or $x = at$, then $at+(a+x) = x$. Otherwise $at+(a+x) = x$ if and only if $t\theta(t\theta(a^{-1}x)) = a^{-1}x$ if and only if $I[\theta](a^{-1}x) = \theta(a^{-1}x)$.
4. This follows from 2 and 3.
5. If N has the exchange inverse property, then there exists $b \in N$ satisfying $b + (x + a) = x$ for all $x \in N$: setting $x = 0$ shows that $b = at^{-1}$.
   If $a = 0, x = 0$, or $x = at^{-1}$, then $at^{-1}+(x+a) = x$. Otherwise $at^{-1}+(x+a) = x$ if and only if $at^{-1}\theta(ta^{-1}x\theta(x^{-1}a)) = x$ if and only if $RI[\theta](a^{-1}x) = \theta(a^{-1}x)$. □

Let us revisit Hsu's classification of cyclic neofields. By the relations in Theorem 15.38, the group $\langle I, R\rangle$ acting on a cyclic group is formally isomorphic to the dihedral group of order six. This group has precisely six subgroups, the trivial subgroup $\langle 1\rangle$, $\langle I\rangle$, $\langle R\rangle$, $\langle RI\rangle$, $\langle IRI\rangle$, and the whole group $\langle I, R\rangle$. Each class in Hsu's classification corresponds to one of these subgroups. Near orthomorphisms that define CIP-neofields are fixed by every element of $\langle \mathrm{I}, \mathrm{R}\rangle$, those that define CMP-neofields are fixed by every element of $\langle \mathrm{R}\rangle$, those that define XIP-neofields are fixed by every element of $\langle \mathrm{RI}\rangle$, those that define LXP-neofields are fixed by every element of $\langle \mathrm{I}\rangle$, those that define RXP-neofields are fixed by every element of $\langle \mathrm{IRI}\rangle$, and those that define XMP-neofields are fixed by every element of $\langle 1\rangle$.

Further correspondences between the properties of neofields and properties of the corresponding near orthomorphisms can be found in [214].

Let us introduce one more class of neofields: planar neofields. Recall that, in Chapter 14, we described coordinatizing projective planes using the symbols $S \cup \{\infty\}$, and we used the structure of projective planes to define addition and multiplication in S. In Theorem 14.5, we showed that S formed a loop with identity 0 under addition, and the nonzero elements of S formed a loop with identity 1 under multiplication. We might ask, can a neofield coordinatize a projective plane? We will call such neofields *planar neofields*. Planar neofields were first studied by Paige in his 1947 Ph.D. thesis [260] and in his 1949 paper [262]. Paige gave a characterization of planar neofields.

Theorem 15.39 (Paige, 1947). *A neofield N is a planar neofield if and only if*

1. *the equation $1 + xa = b + xb$ has a unique solution for $x \in N$ for all $a, b \in N$, $a, b \neq 0, 1$, $a \neq b$; and*
2. *the equation $1 + ax = b + bx$ has a unique solution for $x \in N$ for all $a, b \in N$, $a, b \neq 0, 1$, $a \neq b$.*

Proof. See Theorem III.1 in [262]. □

Examples of planar neofields are fields and semifields. To date, no examples of planar neofields of non-prime power order are known. Neofields were generalized to division neo-rings by Hughes in his 1955 Ph.D. thesis [188] and in his 1955 paper [189]. The definition of a *division neo-ring* is the same as the definition of a neofield except that the nonzero elements form a loop under multiplication, not necessarily a group. A division neo-ring is a *planar division neo-ring* if it can be used to coordinatize a projective plane. Hughes gave a characterization of planar division neo-rings.

Theorem 15.40 (Hughes, 1955). *A division neo-ring R is a planar division neo-ring if and only if*

1. *the equation $xa + b = xc + d$ has a unique solution for $x \in R$ for all $a, b, c, d \in N$, $a \neq c$; and*
2. *the equation $ax + b = cx + d$ has a unique solution for $x \in R$ for all $a, b, c, d \in N$, $a \neq c$.*

Proof. See [189]. □

As in the case of neofields, to date, no examples of planar division neo-rings of non-prime power order are known. For later work on planar division neo-rings, see the 1974 paper by Kantor [211] or the 2003 paper by Ghinelli and Jungnickel [154]: note that, in the latter paper, planar division neo-rings are called neofields.

15.2.4 Multiplicative Groups of Neofields

Which groups can be the multiplicative group of a neofield? This question was posed in 1949 by Paige [262], who called such groups admissible: this is not to be confused with the current usage of this term. Paige answered this question for abelian groups.

Theorem 15.41 (Paige, 1949). *Any finite abelian group can be the multiplicative group of a neofield.*

Proof. Let G be a finite abelian group. If the Sylow 2-subgroup of G is trivial or noncyclic, then G admits complete mappings by Theorem 3.9. If the Sylow 2-subgroup of G is nontrivial and cyclic, then G admits proper near complete mappings by Theorem 15.2. In either case G can be the multiplicative group of a left neofield by Theorem 15.27. But, as G is abelian, any left neofield with multiplicative group G is also a neofield. □

Thus, the question as to which finite groups can be multiplicative groups of neofields reduces to: which nonabelian groups can be the multiplicative groups of neofields? In 1983 Keedwell [215] gave a list of problems which included a closely related problem: for which orders do there exist neofields with nonabelian multiplicative groups? A simple proof, given by Paige [262] in 1949, shows that groups of odd order can be multiplicative groups of neofields.

Theorem 15.42 (Paige, 1949). *Any group of odd order can be the multiplicative group of a neofield.*

Proof. Let G be a group of odd order. The mapping $x \mapsto x^2$ is an orthomorphism that is fixed by H_α for all $\alpha \in \mathrm{Aut}(G)$. Hence, by Theorem 15.30, G can be the multiplicative group of a neofield. □

A *subneofield* of a neofield N is a subset of the elements of N that form a neofield with the same addition and multiplication as in N. In determining whether a group can be the multiplicative group of a neofield or not, one subneofield is of particular importance.

Lemma 15.43. *If N be a finite neofield with multiplicative group G, then $M = Z(G) \cup \{0\}$ is a subneofield of N.*

Proof. Clearly M is closed under multiplication. If $a, b \in M$ and $g \in G$, then

$$g(a+b) = ga + gb = ag + bg = (a+b)g.$$

Hence M is closed under addition. It is now easy to show that M is a subneofield of N. □

In 1981 Johnson [195] showed the importance of this subneofield in determining whether a finite group could be the multiplicative group of a neofield or not.

Theorem 15.44 (Johson, 1981). *Let G be a finite group. If G is the multiplicative group of a neofield, then either $1+1=0$ for all neofields with multiplicative group G or $1+1 \neq 0$ for all neofields with multiplicative group G.*

Proof. Let N be a neofield with multiplicative group G. By Lemma 15.43, $M = Z(G) \cup \{0\}$ is a subneofield of N.

If the Sylow 2-subgroup of $Z(G)$ is trivial or noncyclic, then, by Theorems 3.9 and 15.2, $Z(G)$ admits orthomorphisms but not proper near orthomorphisms. It follows that N must be constructed from an orthomorphism and so, by Theorem 15.27, $1 + 1 = 0$.

If the Sylow 2-subgroup of $Z(G)$ is nontrivial and cyclic, then, by Theorems 3.9 and 15.2, $Z(G)$ admits proper near orthomorphisms but not orthomorphisms. It follows that N must be constructed from a proper near orthomorphism and so, by Theorem 15.27, $1 + 1 \neq 0$. □

Let G be a finite group. Theorem 15.44 tells us that there are two possibilities to consider. If the Sylow 2-subgroup of $Z(G)$ is trivial or noncyclic, then any neofield with multiplicative group G must be constructed from an orthomorphism. In this case, determining whether G can be the multiplicative group of a neofield or not is tantamount to determining if G admits orthomorphisms that are fixed by H_α for all $\alpha \in \mathrm{Aut}(G)$ or not. Similarly, if the Sylow 2-subgroup of $Z(G)$ is nontrivial and cyclic, then any neofield with multiplicative group G must be constructed from a proper near orthomorphism. In this case, determining whether G can be the multiplicative group of a neofield or not is tantamount to determining if G admits proper near orthomorphisms that are fixed by H_α for all $\alpha \in \mathrm{Aut}(G)$ or not. In 1981 Johnson [195] showed that in neither case could a dihedral group be the multiplicative group of a neofield.

Theorem 15.45 (Johnson, 1981). *No dihedral group can be the multiplicative group of a neofield.*

Proof. See Theorems 4.3 and 4.6 in [195]. □

In 2014 certain linear groups were shown by Evans [131] to be multiplicative groups of neofields.

Theorem 15.46 (Evans, 2014). *If q is a power of 2, $q \neq 2$, then $\mathrm{GL}(n, q)$ can be the multiplicative group of a neofield.*

Proof. The orthomorphisms of $\mathrm{GL}(n, q)$, q is a power of 2, $q \neq 2$, constructed in Section 12.3 are fixed by H_α for all $\alpha \in \mathrm{GL}(n, q)$. Hence, by Theorem 15.27, these orthomorphisms correspond to neofields in which $1 + 1 = 0$. □

The result of Theorem 15.46 was extended to other linear groups by Evans [133] in 2016.

Theorem 15.47 (Evans, 2016). *If q is a power of 2, $q \neq 2$, then each of $\mathrm{SL}(n, q)$, $\mathrm{PG}(n, q)$, and $\mathrm{PSL}(n, q)$ can be the multiplicative group of a neofield.*

Proof. See [133]. □

15.3 Complete Latin Squares and Group Sequencings

A *sequencing* of a group G of order n is an ordering $a_0 = 1, a_1, a_2, \ldots, a_{n-1}$ of the elements of G such that the *partial products* $b_0 = a_0 = 1$, $b_1 = a_0a_1$, $b_2 = a_0a_1a_2, \ldots, b_{n-1} = a_0a_1a_2 \cdots a_{n-1}$ are distinct. We say that a group is *sequenceable* if it possesses a sequencing.

Group sequencings were introduced by Gordon [155] in 1961 in the construction of complete Latin squares. A Latin square $L = \{l_{ij}\}$ of order n is *row complete* if the $n(n-1)$ ordered pairs $(l_{ij}, l_{i,j+1})$, $i = 1, \ldots, n$ and $j = 1, \ldots, n-1$, are distinct, *column complete* if the $n(n-1)$ ordered pairs $(l_{ij}, l_{i+1,j})$, $i = 1, \ldots, n-1$ and $j = 1, \ldots, n$, are distinct, and *complete* if it is both row complete and column complete.

Theorem 15.48 (Gordon, 1961). *Let $a_0, a_1, a_2, \ldots, a_{n-1}$ be a sequencing of a group G of order n, and let $b_0, b_1, b_2, \ldots, b_{n-1}$ be the corresponding sequence of partial products. Then the $n \times n$ matrix with ijth entry $\{b_i^{-1}b_j\}$ is a complete Latin square of order n.*

Proof. See Theorem 2 in [155]. □

Example 15.49. Let 0, 1, 8, 3, 6, 5, 4, 7, 2, 9 be an ordering of the elements of $\mathbb{Z}_{10}$. As the partial sums 0, 1, 9, 2, 8, 3, 7, 4, 6, 5 are distinct, this is a sequencing of $\mathbb{Z}_{10}$. The associated complete Latin square is shown in Figure 15.9.

The sequencing of Example 15.49 can be generalized: the ordering

$$0, 1, -2, 3, -4, \ldots, 2n-3, -(2n-2), 2n-1$$

is a sequencing of $\mathbb{Z}_{2n}$ as the partial sums are

$$0, 1, -1, 2, -2, \ldots, n-1, -(n-1), n.$$

Fig. 15.9 A complete Latin square of order 10

$$\begin{pmatrix}
0&1&9&2&8&3&7&4&6&5\\
9&0&8&1&7&2&6&3&5&4\\
1&2&0&3&9&4&8&5&7&6\\
8&9&7&0&6&1&5&2&4&3\\
2&3&1&4&0&5&9&6&8&7\\
7&8&6&9&5&0&4&1&3&2\\
3&4&2&5&1&6&0&7&9&8\\
6&7&5&8&4&9&3&0&2&1\\
4&5&3&6&2&7&1&8&0&9\\
5&6&4&7&3&8&2&9&1&0
\end{pmatrix}$$

It should be noted that the complete Latin square in Figure 15.9 can be obtained from the Cayley table of $\mathbb{Z}_{10}$ by permuting rows and columns. This was observed by Keedwell [213] in 1976.

Theorem 15.50. *A complete Latin square can be obtained from the Cayley table of a finite group G, by permuting rows and columns, if and only if G is sequenceable.*

From a sequencing of a group, we can construct a near complete mapping of the group.

Theorem 15.51. *Let $a_0, a_1, a_2, \ldots, a_{n-1}$ be a sequencing of a group G of order n, and let $b_0, b_1, b_2, \ldots, b_{n-1}$ be the partial products. Define $\theta\colon G\setminus\{b_{n-1}\} \to G\setminus\{1\}$ by*

$$\theta(b_i) = a_{i+1}, i = 0, \ldots, n-2.$$

Then θ is a near complete mapping of G with exdomain element b_{n-1}.

Proof. First note that $\{b_0, \ldots, b_{n-2}\} = G \setminus \{b_{n-1}\}$.

Now

$$\{\theta(b_0), \ldots, \theta(b_{n-2})\} = \{a_1, \ldots, a_{n-1}\} = G \setminus \{1\}$$

and

$$\{b_0\theta(b_0), \ldots, b_{n-2}\theta(b_{n-2})\} = \{b_1, \ldots, b_{n-1}\} = G \setminus \{1\},$$

from which the result follows. □

As an example, the near complete mapping derived from the sequencing of $\mathbb{Z}_{10}$, described in Example 15.49, is shown in Figure 15.10. The exdomain element of this near complete mapping is 5.

Fig. 15.10 A near complete mapping from a sequencing of $\mathbb{Z}_{10}$

g	0	1	2	3	4	5	6	7	8	9
$\theta(g)$	1	8	6	4	2	.	9	7	5	3
$g+\theta(g)$	1	9	8	7	6	.	5	4	3	2

The near orthomorphism, $g \mapsto g + \theta(g)$, associated with the near complete mapping in Figure 15.10 can be written as the sequence [0 1 9 2 8 3 7 4 6 5].

In 1984 Hsu and Keedwell [186] characterized the normalized near orthomorphisms from which group sequencings can be constructed.

Theorem 15.52 (Hsu, Keedwell, 1984). *A group G of order n is sequenceable if and only if it admits a normalized near orthomorphism that consists of one sequence of length n.*

Proof. Let $a_0, a_1, a_2, \ldots, a_{n-1}$ be a sequencing of a group G of order n, and let $b_0, b_1, b_2, \ldots, b_{n-1}$ be the partial products. Then $[b_0\ b_1\ \cdots\ b_{n-1}]$ is a normalized near orthomorphism of G.

If $[b_0\ b_1\ \cdots\ b_{n-1}]$ is a normalized near orthomorphism of G, then setting

$$a_i = \begin{cases} 1 & \text{if } i = 0, \\ b_{i-1}^{-1} b_i & \text{if } i = 1, \dots, n-1, \end{cases}$$

yields a sequencing $a_0, \dots, a_{n-1}$ of G. □

Which groups are sequenceable? This question was answered for abelian groups by Gordon [155] in 1961.

Theorem 15.53 (Gordon, 1961). *An abelian group is sequenceable if and only if it has a unique element of order* 2.

Proof. See Theorem 1 in [155]. □

The situation is different for nonabelian groups. Order 10 appears to be a dividing line.

Theorem 15.54. *No nonabelian group of order less than* 10 *is sequenceable.*

Proof. See Gordon [155]. □

However, the nonabelian group of order 10, the dihedral group $D_{10} = \langle a, b \mid a^5 = b^2 = 1, ab = ba^{-1} \rangle$ is sequenceable. $1, ba, a^4, ba^2, b, ba^4, a^2, a, ba^3, a^3$ is a sequencing for this group. In 1983 Keedwell [215] conjectured that nonabelian groups of order less that 10 were the only nonsequenceable nonabelian groups.

Conjecture 15.55 (Keedwell). All nonabelian groups of order at least 10 are sequenceable.

Keedwell's conjecture has been proved true for many classes of groups.

Theorem 15.56 (Anderson, 1987). *All nonabelian groups of order n,* $10 \le n \le 32$ *are sequenceable.*

Proof. See [17] and [18]. □

Theorem 15.57 (Anderson, 1987). A_5 *and* S_5 *are sequenceable.*

Proof. See [17]. □

The proof that the dihedral groups satisfy Keedwell's conjecture is the result of work by several mathematicians, whose work is described in the dynamic survey [255] by Ollis.

Theorem 15.58. *The dihedral group of order* 2*n,* D_{2n}*,* $n \ge 5$*, is sequenceable.*

There are a number of results for binary groups: a group is *binary* if it has exactly one involution. Theorem 15.53 can be restated as a finite abelian group which is sequenceable if and only if it is a binary group. Keedwell's conjecture has been proved for binary solvable groups.

Theorem 15.59 (Anderson and Ihrig, 1993). *All binary solvable groups, except the quaternion group of order* 8*, are sequenceable.*

Proof. See [19]. □

Anderson and Ihrig actually proved the stronger result that solvable groups with a unique element of order 2 are symmetrically sequenceable. A *symmetric sequencing* of a group G of order $2n$, with a unique element u of order 2, is a sequencing $a_0 = 1, a_1, a_2, \ldots, a_{2n-1}$ of G for which $a_n = u$ and $a_{n-i} = a_{n+i}^{-1}$, $i = 1, 2, \ldots, n-1$. A group is *symmetrically sequenceable* if it possesses a symmetric sequencing. A number of other groups have been shown to be sequenceable including many binary groups and groups of odd order: see the dynamic survey by Ollis [255] for details.

15.3.1 R-Sequencings

An *R-sequencing* of a group G of order n is an ordering $a_0 = 1, a_1, a_2, \ldots, a_{n-1}$ of the elements of G such that the *partial products* $b_0 = a_0 = 1$, $b_1 = a_0a_1$, $b_2 = a_0a_1a_2$, ..., $b_{n-2} = a_0a_1a_2 \cdots a_{n-2}$ are distinct and $a_0a_1a_2 \cdots a_{n-1} = 1$. A group is *R-sequenceable* if it possesses an R-sequencing. R-sequencings were introduced by Paige [263] in 1951 as a sufficient condition for a group to admit complete mappings, equivalently orthomorphisms. They were also used by Ringel [280] in 1974 in his solution of the map-coloring problem for all compact *two*-dimensional manifolds except the sphere. Note that in an R-sequencing of a finite group G, exactly one element of G does not appear as a partial product.

Theorem 15.60. *Let $a_0, a_1, a_2, \ldots, a_{n-1}$ be an R-sequencing of a group G of order n, let $b_0, b_1, b_2, \ldots, b_{n-2}$ be the corresponding sequence of partial products, and let c be the element of G that is not in the list of partial products. Then, the mapping $\theta: G \to G$ defined by*

$$\theta(g) = \begin{cases} b_{i+1} & \text{if } g = b_i, i = 0, 1, \ldots, n-3, \\ b_0 & \text{if } g = b_{n-2}, \\ c & \text{if } g = c, \end{cases}$$

is an orthomorphism of G.

Proof. Routine. □

An immediate consequence of Theorems 3.34 and 15.60:

Corollary 15.61. *If G is a finite R-sequenceable group, then its Sylow 2 subgroup is either trivial or noncyclic.*

As an example $0, 12, 2, 10, 4, 8, 6, 5, 9, 3, 11, 1, 7$ is an R-sequencing of $\mathbb{Z}_{13}$. The partial sums are $0, 12, 1, 11, 2, 10, 3, 8, 4, 7, 5, 6$, missing 9. The associated orthomorphism is shown in Figure 15.11.

g	0	1	2	3	4	5	6	7	8	9	10	11	12
$\theta(g)$	12	11	10	8	7	6	0	5	4	9	3	2	1
$\theta(g)-g$	12	10	8	5	3	1	7	11	9	0	6	4	2

Fig. 15.11 An orthomorphism of $\mathbb{Z}_{13}$

The orthomorphism in Figure 15.11 is the cycle (0 12 1 11 2 10 3 8 4 7 5 6). In 1984 Hsu and Keedwell [186] characterized the normalized orthomorphisms from which R-sequencings can be constructed.

Theorem 15.62 (Hsu, Keedwell, 1984). *A group G of order n is R-sequenceable if and only if it admits a normalized orthomorphism that consists of one cycle of length $n-1$.*

Proof. Similar to the proof of Theorem 15.52. □

Cyclic groups of odd order were shown to be R-sequenceable groups by Friedlander, Gordon, and Miller [146] in 1978.

Theorem 15.63 (Friedlander, Gordon and Miller, 1978). *If n is odd, then $\mathbb{Z}_n$ is R-sequenceable.*

Proof.

$$0, -1, 2, -3, 4, \ldots, -(2n-1), 2n, 2n-1, -(2n-2),$$
$$2n-3, -(2n-4), \ldots, 3, -2, 1, -2n$$

is an R-sequencing of $\mathbb{Z}_{4n+1}$, and

$$0, -1, 2, -3, 4, \ldots, -(2n-1), 2n, -(2n+2), 2n+3,$$
$$-(2n+4), \ldots, -4n, 4n+1, -(4n+2), 2n+2$$

is an R-sequencing of $\mathbb{Z}_{4n+3}$. □

Friedlander, Gordon, and Miller [146] conjectured that any finite abelian group with a trivial or noncyclic Sylow 2-subgroup is R-sequenceable. This conjecture was proved by Alspach, Kreher, and Pastine [15] in 2017.

Theorem 15.64 (Alspach, Kreher, and Pastine, 2017). *A finite abelian group with a trivial or noncyclic Sylow 2-subgroup is R-sequenceable.*

Note that, a finite group with a nontrivial, cyclic Sylow 2-subgroup is not admissible and, hence, cannot be R-sequenceable. It follows that Theorem 15.64 characterizes finite abelian R-sequenceable groups.

There are many other classes of R-sequenceable groups known: see the dynamic survey by Ollis [255].

15.3.2 Harmonious Groups

A *harmonious ordering* of a group G of order n is an ordering $a_0 = 1, a_1, a_2, \ldots, a_{n-1}$ of the elements of G such that the products $a_0a_1,\ a_1a_2,\ a_2a_3, \ldots,\ a_{n-1}a_0$ are distinct. G is a *harmonious group* if it possesses a harmonious ordering. Harmonious groups were introduced by Beals, Gallian, Headley, and Jungreis [32] in 1991.

Theorem 15.65. *If $a_0 = 1, a_1, a_2, \ldots, a_{n-1}$ is a harmonious ordering of a group G, of order n, then the mapping $a_i \mapsto a_ia_{i+1}$, indices added modulo n, is an orthomorphism of G.*

Proof. Routine. □

As an example, $0, 1, 2, \ldots, n-1$ is a harmonious ordering of $\mathbb{Z}_n$ if n is odd. The associated orthomorphism is $i \mapsto 2i + 1$. Note that this orthomorphism is not normalized and that its associated complete mapping $i \mapsto i+1$ is a cycle of length n. Beals, Gallian, Headley, and Jungreis characterized complete mappings from which harmonious orderings can be constructed.

Theorem 15.66. *A group G of order n is harmonious if and only if it admits a complete mapping that consists of one cycle of length n.*

Proof. Routine. □

An immediate corollary of Theorems 3.34 and 15.65:

Corollary 15.67. *Finite groups with nontrivial cyclic* 2*-groups are not harmonious.*

Beals, Gallian, Headley, and Jungreis [32] discovered an additional class of nonharmonious groups.

Theorem 15.68 (Beals, Gallian, Headley, and Jungreis, 1991). *The additive group of the finite field* $\mathrm{GF}(2^n)$ *is not harmonious.*

Beals, Gallian, Headley, and Jungreis [32] completely characterized finite abelian harmonious groups and showed all groups of odd order to be harmonious.

Theorem 15.69 (Beals, Gallian, Headley, and Jungreis, 1991). *Groups of odd order are harmonious.*

Theorem 15.70 (Beals, Gallian, Headley, and Jungreis, 1991). *Abelian groups, except $GF(2^n)^+$, with trivial or noncyclic* 2*-groups, are harmonious.*

In addition, several dihedral and quaternion groups have been shown to be harmonious: see the dynamic survey by Ollis [255].

15.4 Starters

Let G be an abelian group of odd order n. A *starter* of G is a set of unordered pairs $S = \{\{x_i, y_i\} \mid i = 1, \ldots, (n-1)/2\}$ satisfying the following:

- $\{x_i \mid i = 1, \ldots, (n-1)/2\} \cup \{y_i \mid i = 1, \ldots, (n-1)/2\} = G \setminus \{0\}$.
- $\{x_i - y_i \mid i = 1, \ldots, (n-1)/2\} \cup \{y_i - x_i \mid i = 1, \ldots, (n-1)/2\} = G \setminus \{0\}$.

Starters are closely related to orthomorphisms and have several applications in design theory. We will describe two of these applications, the construction of Room squares and the construction of orthogonal-symmetric Latin squares. For more on starters and their applications, see the 1992 survey paper by Dinitz and Stinson [103] or the 2007 article by Dinitz [101]. While starters can be defined for nonabelian groups, we will define starters for abelian groups only and use additive notation.

15.4.1 Basic Results

There is a natural correspondence between the set of starters of a group and the set of normalized orthomorphisms of a group fixed by the congruence I, i.e., the set of normalized orthomorphisms θ for which $\theta = \theta^{-1}$.

Lemma 15.71. *Let $S = \{\{x_i, y_i\} \mid i = 1, \ldots, (n-1)/2\}$ be a starter of a group G of order n. Then the mapping $\theta\colon G \to G$ defined by*

$$\theta(g) = \begin{cases} 1 & \text{if } g = 1; \\ y_i & \text{if } g = x_i,\ i = 1, \ldots, (n-1)/2; \\ x_i & \text{if } g = y_i,\ i = 1, \ldots, (n-1)/2; \end{cases}$$

is an orthomorphism of G that is fixed by the congruence I. *Further, this establishes a one-to-one correspondence between the set of starters of G and the set of normalized orthomorphisms of G that are fixed by the congruence* I.

Proof. Routine. □

With S and θ as in Lemma 15.71, we will call θ the orthomorphism *associated* with S and S the starter *associated* with θ. Note that the orthomorphisms associated with starters are the 2-regular orthomorphisms: see Section 15.1. Clearly, no abelian group of even order can admit starters. However, every abelian group G of odd order admits starters, e.g., the *patterned starter*, $P = \{\{x, -x\} \mid x \in G\}$, associated with the orthomorphism $x \mapsto -x$.

Let $S = \{\{x_i, y_i\} \mid i = 1, \ldots, (n-1)/2\}$ be a starter of an abelian group G of order n. We say that S is a *strong starter* if $x_i + y_i \neq 0$ for $i = 1, \ldots, (n-1)/2$ and $x_i + y_i \neq x_j + y_j$ if $i \neq j$ and $i, j = 1, \ldots, (n-1)/2$.

Example 15.72. In $\mathbb{Z}_{13}$

$$S = \{\{1,2\},\{3,6\},\{4,11\},\{5,9\},\{7,12\},\{8,10\}\}$$

is a strong starter. The sums $x_i + y_i$ are

$$1, 2, 3, 5, 6, 9.$$

The associated orthomorphism is given in Figure 15.12.

Fig. 15.12 The associated orthomorphism for a strong starter of $\mathbb{Z}_{13}$

x	0	1	2	3	4	5	6	7	8	9	10	11	12
$\theta(x)$	0	2	1	6	11	9	3	12	10	5	8	4	7
$\theta(x) - x$	0	1	12	3	7	4	10	5	2	9	11	6	8

Example 15.73. In $\mathbb{Z}_{15}$

$$S = \{\{1,11\},\{2,3\},\{4,12\},\{5,14\},\{6,8\},\{7,10\},\{9,13\}\}$$

is a strong starter. The sums $x_i + y_i$ are

$$1, 2, 4, 5, 7, 12, 14.$$

The associated orthomorphism is given in Figure 15.13.

x	0	1	2	3	4	5	6	7	8	9	10	11	12	13	14
$\theta(x)$	0	11	3	2	12	14	8	10	6	13	7	1	4	9	5
$\theta(x) - x$	0	10	1	14	8	9	2	3	13	4	12	5	7	11	6

Fig. 15.13 The associated orthomorphism for a strong starter of $\mathbb{Z}_{15}$

Let $S = \{\{x_i, y_i\} \mid i = 1, \ldots, (n-1)/2\}$ and $T = \{\{w_i, z_i\} \mid i = 1, \ldots, (n-1)/2\}$ be starters of an abelian group G of order n. For each $i = 1, \ldots, (n-1)/2$, $x_i - y_i = \pm(w_j - z_j)$ for a unique $j \in \{1, \ldots, (n-1)/2\}$. Thus, by reordering the unordered pairs of T and swapping w_i with z_i for some i if needed, we may ensure that $x_i - y_i = w_i - z_i$ for $i = 1, \ldots, (n-1)/2$. With this condition, we say that S and T are *orthogonal* if $w_i \neq x_i$ for all i and $w_i - x_i \neq w_j - x_j$ if $i \neq j$.

Example 15.74. For the group $\mathbb{Z}_{15}$, the starters S, T, and P, ordered so that the differences $x_i - y_i$ match, are shown below.

S	$\{1, 11\}$	$\{2, 3\}$	$\{4, 12\}$	$\{5, 14\}$	$\{6, 8\}$	$\{7, 10\}$	$\{9, 13\}$
T	$\{4, 14\}$	$\{12, 13\}$	$\{3, 11\}$	$\{1, 10\}$	$\{7, 9\}$	$\{5, 8\}$	$\{2, 6\}$
P	$\{10, 5\}$	$\{7, 8\}$	$\{11, 4\}$	$\{3, 12\}$	$\{14, 1\}$	$\{6, 9\}$	$\{13, 2\}$
$x_i - y_i$	5	14	7	6	13	12	11

We leave it to the reader to verify that these three starters are pairwise orthogonal.

If $S = \{\{x_i, y_i\} \mid i = 1, \ldots, (n-1)/2\}$ is a starter of an abelian group G of order n, then it is easy to show that $-S = \{\{-x_i, -y_i\} \mid i = 1, \ldots, (n-1)/2\}$ is also a starter of G. Note that, if we rewrite $-S$ as $-S = \{\{-y_i, -x_i\} \mid i = 1, \ldots, (n-1)/2\}$, then the differences for S and $-S$ match. In Example 15.74 the starter S is the strong starter from Example 15.73, $T = -S$, and P is the patterned starter. From a strong starter, we can construct a set of three pairwise orthogonal starters.

Theorem 15.75. *If S is a strong starter, then S, $-S$, and P are pairwise orthogonal starters.*

Proof. Let $S = \{\{x_i, y_i\} \mid i = 1, \ldots, (n-1)/2\}$ be a strong starter of an abelian group G of order n. Then $-S = \{\{-y_i, -x_i\} \mid i = 1, \ldots, (n-1)/2\}$ and $P = \{\{(x_i - y_i)[(n-1)/2], (y_i - x_i)[(n-1)/2]\} \mid i = 1, \ldots, (n-1)/2\}$. It is routine to show that these three starters are pairwise orthogonal. □

One approach to constructing orthogonal pairs of starters is the starter-adder construction. If $S = \{\{x_i, y_i\} \mid i = 1, \ldots, (n-1)/2\}$ is a starter of an abelian group G of order n, then an *adder* for S is a set $A = \{a_i \mid i = 1, \ldots, (n-1)/2\}$, $a_i \neq 0$ for all i, $a_i \neq a_j$ if $i \neq j$, for which $S + A = \{\{x_i + a_i, y_i + a_i\} \mid i = 1, \ldots, (n-1)/2\}$ is a starter.

Theorem 15.76. *If S is a starter for an abelian group G of order n and A is an adder for S, then S and $S + A$ are orthogonal. Further, if T is a starter of G orthogonal to S, then $T = S + A$ for some adder A for S.*

Proof. Let $S = \{\{x_i, y_i\} \mid i = 1, \ldots, (n-1)/2\}$ be a starter of an abelian group G of order n, let $A = \{a_i \mid i = 1, \ldots, (n-1)/2\}$ be an adder for S, and let $T = \{\{w_i, z_i\} \mid i = 1, \ldots, (n-1)/2\} = S + A$, $w_i = x_i + a_i$ and $z_i = y_i + a_i$ for all i. Now $z_i - w_i = y_i - x_i$ for all i, and, as A is an adder, $w_i = x_i + a_i \neq x_i$ for all i and $w_i - x_i = a_i = a_j = w_j - x_j$ only if $i = j$. Hence, S and T are orthogonal.

Let $S = \{\{x_i, y_i\} \mid i = 1, \ldots, (n-1)/2\}$ and $T = \{\{w_i, z_i\} \mid i = 1, \ldots, (n-1)/2\}$ be an orthogonal pair of starters. Without loss of generality, we may assume that $z_i - w_i = y_i - x_i$ for all i. Set $a_i = w_i - x_i$ for all i. As S and T are orthogonal, $a_i \neq 0$ for all i and $a_i = a_j$ only if $i = j$. Further $T = S + A$ is a starter and so A is an adder for S. □

Let us revisit the three pairwise orthogonal starters of $\mathbb{Z}_{15}$ described in Example 15.74. The three sets $A = \{3, 10, 14, 11, 1, 13, 8\}$, $B = \{9, 5, 7, 13, 8, 14, 4\}$, and $C = \{6, 10, 8, 2, 7, 1, 11\}$ are adders; A and B are both adders for S, and C is an adder for T. In fact $T = S + A$, $P = S + B$, and $P = T + C$, from which it follows, by Theorem 15.76, that S, T, and P are pairwise orthogonal.

15.4.2 Room Squares

A *Room square of side n* on a symbol set of order $n+1$ is an $n \times n$ array in which

- every cell is empty or contains an unordered pair of symbols,
- each symbol occurs exactly once in each row and exactly once in each column, and
- every unordered pair of symbols occurs once.

Orthogonal pairs of starters can be used to construct Room squares.

Theorem 15.77. *If there exists an orthogonal pair of starters for an abelian group of order n, then there exists a Room square of side n.*

Proof. Let $S = \{\{x_i, y_i\} \mid i = 1, \ldots, (n-1)/2\}$ be a starter of an abelian group G of order n and suppose that there exists a starter orthogonal to S. Then, by Theorem 15.76, there exists an adder $A = \{a_i \mid i = 1, \ldots, (n-1)/2\}$ for S. We construct an $n \times n$ array $R = \{R_{gh}\}$, using the elements of G to index the rows and columns of R, and ordered pairs of elements from $G \cup \{\infty\}$ as possible entries of R, setting

$$R_{gh} = \begin{cases} \{x_i + g, y_i + g\} & \text{if } h = g - a_i, \\ \{\infty, g\} & \text{if } h = g, \end{cases}$$

all other cells being empty. R is a Room square of side n. □

The following is an example of this construction.

Example 15.78. The sets $S = \{\{2, 3\}, \{4, 6\}, \{1, 5\}\}$ and $T = \{\{4, 5\}, \{1, 3\}, \{2, 6\}\}$ are orthogonal starters for the group $\mathbb{Z}_7$. The corresponding adder for S is $A = \{2, 4, 1\}$. The construction of Theorem 15.77 yields the Room square of side 7 shown in Figure 15.14.

Fig. 15.14 A Room square of side 7

$$\begin{pmatrix}
\{\infty,0\} & - & - & \{4,6\} & - & \{2,3\} & \{1,5\} \\
\{2,6\} & \{\infty,1\} & - & - & \{5,0\} & - & \{3,4\} \\
\{4,5\} & \{3,0\} & \{\infty,2\} & - & - & \{6,1\} & - \\
- & \{5,6\} & \{4,1\} & \{\infty,3\} & - & - & \{0,2\} \\
\{1,3\} & - & \{6,0\} & \{5,2\} & \{\infty,4\} & - & - \\
- & \{2,4\} & - & \{0,1\} & \{6,3\} & \{\infty,5\} & - \\
- & - & \{3,5\} & - & \{1,2\} & \{0,4\} & \{\infty,6\}
\end{pmatrix}$$

A number of constructions can be found in the literature, as outlined in [103], culminating in a complete determination of those n for which a Room square of side n exists.

Theorem 15.79. *A Room square of side n exists if and only if n is odd and $n \neq 3, 5$.*

Proof. Two proofs are given in [103]. □

15.4.3 The Existence of Strong Starters

As we have seen in Theorem 15.75, from a strong starter of an abelian group of order n, we can construct a pairwise orthogonal set of three starters of order n, and, by Theorem 15.77, from an orthogonal pair of starters of order n, we can construct a Room square of side n. Thus from a strong starter of an abelian group of order n, we can construct a Room square of side n. Indeed most of the early existence results for Room squares were derived from constructions of strong starters. It is natural to ask which finite abelian groups admit strong starters. While the existence problem for Room squares has been settled (see Theorem 15.79), the existence problem for strong starters is still open. Other than groups of even order, which cannot admit starters, only four abelian groups have been shown not to admit strong starters.

Theorem 15.80. *The groups $\mathbb{Z}_3$, $\mathbb{Z}_5$, $\mathbb{Z}_9$, and $\mathbb{Z}_3 \times \mathbb{Z}_3$ do not admit strong starters.*

A number of finite abelian groups have been shown to admit strong starters. In 1990 Horton [181] conjectured that the converse to Theorem 15.80 held for finite abelian groups of odd order.

Conjecture 15.81 (Horton, 1990). All finite abelian groups of odd order admit strong starters, except for $\mathbb{Z}_3$, $\mathbb{Z}_5$, $\mathbb{Z}_9$, and $\mathbb{Z}_3 \times \mathbb{Z}_3$.

Conjecture 15.81 generalizes an older conjecture for cyclic groups.

Conjecture 15.82. For all odd $n > 9$, the group $\mathbb{Z}_n$ admits strong starters.

In what follows, we will describe some of the progress made toward settling Conjectures 15.81 and 15.82. Many small groups have been dealt with via computer searches. As an example, cited in [79], $\mathbb{Z}_n$ admits strong starters for $7 \leq n \leq 999$, n odd, $n \neq 9$. A number of small groups, whose orders are divisible by three, were shown to admit strong starters by Horton [181] in 1990.

Theorem 15.83 (Horton, 1990). *All abelian groups of orders 3^3, 3^4, 3^5, 5×3^2, 5×3^3, $5^2 \times 3$, and $5^2 \times 3^2$ admit strong starters.*

Proof. The proof is via a computer search using a hill-climbing algorithm of Dinitz and Stinson [102]. □

The elementary abelian groups that admit strong starters have been determined.

Theorem 15.84. *If q is odd, $q \neq 3, 5, 9$, then $\mathrm{GF}(q)^+$ admits strong starters.*

Proof. See [73, 231], and [248]. □

A 1986 result of Rees and Wallis [276] shows that the only groups we need to be concerned with are those whose orders are divisible by three.

Theorem 15.85 (Rees and Wallis, 1986). *If G is an abelian group of order n, $n > 5$ and $\gcd(n, 6) = 1$, then G admits strong starters.*

Theorem 15.85 was rediscovered by Horton [181] in 1990. Horton also showed that, to prove Conjecture 15.81, it is sufficient to construct strong starters for four classes of groups.

Theorem 15.86 (Horton, 1990). *To prove Conjecture 15.81, it is sufficient to construct strong starters for each of the following groups.*

1. $\mathbb{Z}_{3p}$, $p > 111$ *prime.*
2. $\mathbb{Z}_3 \times \mathbb{Z}_3 \times \mathbb{Z}_p$, $p > 111$ *prime.*
3. $\mathbb{Z}_{3^n}$, $n \geq 7$.
4. $\mathbb{Z}_{5\times 3^n}$, $n \geq 4$.

15.4.4 Orthogonal Symmetric Latin Squares

Let $L = \{l_{ij}\}$ be a Latin square of order n on the symbol set $\{s_1, \ldots, s_n\}$. L is *idempotent* if $l_{ii} = s_i$ for $i = 1, \ldots, n$ and is *symmetric* if $l_{ij} = l_{ji}$ for $i, j = 1, \ldots, n$. If L and L' are two idempotent, symmetric Latin squares of order n, on the same symbol set, we say that L and L' are *orthogonal-symmetric* if for all $a, b \in \{1, \ldots, n\}$, there is at most one ordered pair (i, j) for which $l_{ij} = s_a$ and $l'_{ij} = s_b$. Note that, in 1973 Gross, Mullin, and Wallis [158] showed that the existence of a pair of orthogonal-symmetric idempotent, symmetric Latin squares of order n was equivalent to the existence of a Room square side n. Idempotent, symmetric Latin squares and orthogonal-symmetric idempotent, symmetric Latin squares can be constructed from starters and orthogonal starters.

Theorem 15.87. *Let $G = \{g_1, \ldots, g_n\}$ be an abelian group of order n, and let $S = \{\{x_i, y_i\} \mid i = 1, \ldots, (n-1)/2\}$ be a starter for G. The $n \times n$ array $L = \{l_{ij}\}$ defined by*

$$l_{ij} = \begin{cases} g_i & \text{if } j = i, \\ g_k & \text{if } j \neq i \text{ and } \{g_i, g_j\} = \{x_r + g_k, y_r + g_k\} \\ & \text{for some } r \in \{1, \ldots, (n-1)/2\}, \end{cases}$$

is an idempotent, symmetric Latin square. Further, if L and L' are idempotent, symmetric Latin squares constructed from starters S and S' of G, then L and L' are orthogonal-symmetric if and only if S and S' are orthogonal.

Proof. Routine. □

As an example.

Example 15.88. For the group $\mathbb{Z}_7$, let S be the starter in Example 15.74. It is easily seen that S is a strong starter and that the starter T in Example 15.74 is $-S$. Thus, by Theorem 15.75, S, $T = -S$, and P, the patterned starter are three pairwise orthogonal starters of $\mathbb{Z}_7$. These starters, ordered so that the differences $x_i - y_i$ match, are shown below.

S	$\{2,3\}$	$\{4,6\}$	$\{1,5\}$
$-S$	$\{4,5\}$	$\{1,3\}$	$\{2,6\}$
P	$\{3,4\}$	$\{6,1\}$	$\{5,2\}$
$x_i - y_i$	1	2	4

Let us label the elements of $\mathbb{Z}_7$ in their natural order, $\{0, 1, \ldots, 6\}$. The corresponding pairwise orthogonal-symmetric Latin squares, L_S, L_{-S}, and L_P, of order 7, are shown in Figure 15.15.

$$L_S = \begin{pmatrix} 0&5&3&2&6&1&4\\ 5&1&6&4&3&0&2\\ 3&6&2&0&5&4&1\\ 2&4&0&3&1&6&5\\ 6&3&5&1&4&2&0\\ 1&0&4&6&2&5&3\\ 4&2&1&5&0&3&6 \end{pmatrix}, L_{-S} = \begin{pmatrix} 0&3&6&1&5&4&2\\ 3&1&4&0&2&6&5\\ 6&4&2&5&1&3&0\\ 1&0&5&3&6&2&4\\ 5&2&1&6&4&0&3\\ 4&6&3&2&0&5&1\\ 2&5&0&4&3&1&6 \end{pmatrix}, L_P = \begin{pmatrix} 0&4&1&5&2&6&3\\ 1&1&5&2&6&3&0\\ 4&5&2&6&3&0&4\\ 5&2&6&3&0&4&1\\ 3&6&3&0&4&1&5\\ 6&3&0&4&1&5&2\\ 3&0&4&1&5&2&6 \end{pmatrix}.$$

Fig. 15.15 Three pairwise orthogonal-symmetric idempotent, symmetric Latin squares of order 7

Let $s(G)$ denote the maximum possible number of pairwise orthogonal starters for an abelian group G, let $s(n)$ denote the maximum possible number of pairwise orthogonal starters for any abelian group of order n, and let $\nu(n)$ denote the maximum possible number of pairwise symmetric-orthogonal idempotent, symmetric Latin squares of order n.

Theorem 15.89. *For $n \geq 3$ odd, $1 \leq s(n) \leq \nu(n) \leq n - 2$.*

Proof. $1 \leq s(n)$ as every abelian group of odd order admits starters, e.g., the patterned starter. By Theorem 15.87, $s(n) \leq \nu(n)$, and the bound $\nu(n) \leq n - 2$, was established by Horton [180] in 1981. □

Some lower bounds for $s(\mathrm{GF}(q)^+)$ can be found in [99] and [158]; and a table of lower bounds for $\nu(n)$, n odd, $n \leq 103$ can be found in [100].

Chapter 16
Problems

In the 1992 book [119], we gave a list of 74 problems. Some of these have since been solved. Notably Problems 73 and 74 deal with the existence of complete mappings. These problems were solved when the Hall-Paige conjecture was proved: see Chapter 7. In this chapter we will reorganize, update, and amend the list of problems in [119] and will include several additional problems. The problems will range from very specific to general problems that indicate directions for further research.

In [119] we listed the following general problems. Given a group G and an orthomorphism graph $\mathscr{H}$ of G:

1. Determine exact values of, or bounds for, $\omega(\mathscr{H})$.
2. What can we say about the structure of $\mathscr{H}$?
3. Can $\mathscr{H}$ contain a complete set of orthomorphisms?
4. Given a clique of $\mathscr{H}$, can we extend it to a larger clique of $\mathscr{H}$ or $\mathrm{Orth}(G)$?
5. Is $\mathscr{H}$ nonempty?
6. Does there exist an r-clique in $\mathscr{H}$ that is maximal in $\mathscr{H}$ or $\mathrm{Orth}(G)$?

Most of these problems are concerned with cliques of orthomorphism graphs, i.e., pairwise orthogonal sets of orthomorphisms. The problems that do not deal with cliques of orthomorphism graphs concern the structure of orthomorphism graphs and the existence of types of orthomorphisms. Problems on the constructions of orthomorphisms and classes of orthomorphisms will be given in Section 16.1: each class of orthomorphisms induces an orthomorphism graph. In Section 16.2 we will give problems dealing with cliques of orthomorphism graphs and in Section 16.3 problems dealing with the structure of orthomorphism graphs. Some related and additional problems will be given in Section 16.4.

A. B. Evans, *Orthogonal Latin Squares Based on Groups*, Developments in Mathematics 57, https://doi.org/10.1007/978-3-319-94430-2_16

16.1 Constructions and Classes of Orthomorphisms

In this section we will be interested in finding new, interesting orthomorphism graphs. We should expect new orthomorphism graphs to have simple rules for membership and orthogonality: as an example, the membership and orthogonality rules for $\mathscr{P}(G)$, described in Section 8.2, are simple as are those for $\mathscr{C}_2(q)$, described in Section 9.3. Most of the problems in this section are concerned with orthomorphisms of elementary abelian groups.

16.1.1 (σ, ε)-Systems

In Section 10.3 we constructed orthomorphisms of $\mathrm{GF}(q)^+$ by solving systems of equations of the form $ax_{\sigma(i)} - x_i = (a-1)x_{\varepsilon(i)}$, σ and ε permutations of $\{1, \cdots, q-1\}$, for $a, x_1, \cdots, x_{q-1}$: these systems of equations were called (σ, ε)-systems. Several problems dealing with (σ, ε)-systems were presented in [119]. None of these problems have been solved.

Problem 16.1 (Problem 1 from [119]). Can Theorem 10.35 be strengthened? This theorem concerns the (σ, ε)-system for $\mathrm{GF}(q)$ with $\sigma = (12\cdots n)$ and $\varepsilon = \sigma^r$, $2 \le r < n \le q-1$. Is it true that this system yields orthomorphisms if and only if n divides $q-1$? If n divides $q-1$ are the orthomorphisms yielded by this system precisely those described in Theorem 10.34? Can the rank of the solution space of this (σ, ε)-system ever be greater than 2?

Problem 16.2 (Problem 2 from [119]). Let R be a finite commutative ring with unity, and let $\sigma = (12\cdots n)$ and $\varepsilon = \sigma^r$, $2 \le r < n$. Under what conditions will this (σ, ε)-system yield orthomorphisms of R^+? Describe the orthomorphisms yielded by this system. In Section 10.3, when studying this type of (σ, ε)-system, we restricted ourselves to additive groups of finite fields.

Problem 16.3 (Problem 3 from [119]). Find new interesting classes of (σ, ε)-systems, determine the orthomorphisms yielded by these systems, and study the orthomorphism graphs these determine over the additive groups of finite fields or rings.

We might, for instance, define new classes by splicing together (σ, ε)-systems. As an example, the system defined by $\sigma = (123)$ and $\varepsilon = (132)$ could be spliced with itself to define a new system with $\sigma' = (123)(456)$ and $\varepsilon' = (132)(465)$. The solutions to the new system can be obtained from solutions to the original system.

In Chapters 9 and 10, we used three approaches in studying orthomorphism graphs of $\mathrm{GF}(q)^+$; (σ, ε)-systems, cyclotomy, and permutation polynomials. In Theorem 10.26 we characterized permutation polynomials that represent cyclotomic orthomorphisms. However we have no similar theorems relating (σ, ε)-systems to cyclotomic orthomorphisms and permutation polynomials. Problems 16.4, 16.5, and 16.6 ask for ways to address this deficiency.

Problem 16.4 (Problem 4 from [119]). Can we determine a (σ, ε)-system that yields $H_\alpha[\theta]$ from a (σ, ε)-system that yields the orthomorphism θ? More generally, if F is a congruence of Orth($GF(q)^+$), can we determine a (σ, ε)-system that yields $F[\theta]$ from a (σ, ε)-system that yields the orthomorphism θ?

Problem 16.5 (Problem 5 from [119]). Suppose that $e|q-1$. Characterize those (σ, ε)-systems of $GF(q)$ that yield cyclotomic orthomorphisms of $GF(q)^+$ of index e.

Problem 16.6 (Problem 6 from [119]). Given a (σ, ε)-system of $GF(q)$, characterize those permutation polynomials of $GF(q)$ that represent orthomorphisms of $GF(q)^+$ yielded by this system.

Problem 16.7 (Problem 7 from [119]). Let R be a commutative ring with unity, and let $a \in R$, with a and $a-1$ both units of R. If θ is an orthomorphism of $R^+ = \{r_1, \cdots, r_n\}$, then $\theta(r_i) = ar_{\sigma(i)}$ and $\theta(r_i) - r_i = (a-1)r_{\varepsilon(i)}$, for some σ and ε permutations of $\{1, \cdots, n\}$. This defines a (σ, ε)-system. Use this approach to construct new (σ, ε)-systems, and hence new orthomorphism graphs, from known cliques of orthomorphism graphs of abelian groups.

More generally.

Problem 16.8 (Problem 8 from [119]). Let $G = \{g_1, \cdots, g_n\}$ be a group. Any orthomorphism of a group G can be written in the form $\theta(g_i) = \alpha(g_{\sigma(i)})$, for some $\alpha \in \mathrm{Aut}(G)$ and some permutation σ of $\{1, \cdots, n\}$, and $g_i^{-1}\theta(g_i) = g_i^{-1}\alpha(g_{\sigma(i)}) = \beta(g_{\varepsilon(i)})$, for some $\beta \in \mathrm{Aut}(G)$ and some permutation ε of $\{1, \cdots, n\}$. Try to solve systems of equations of the form $x_i^{-1}\alpha(x_{\sigma(i)}) = \beta(x_{\varepsilon(i)})$, $\alpha, \beta \in \mathrm{Aut}(G)$, σ, ε permutations of $\{1, \cdots, n\}$. If G is abelian and α is a fixed-point-free automorphism of G, then we might set $\beta(g) = g^{-1}\alpha(g)$ as this will also be an automorphism of G.

16.1.2 Cyclotomic Orthomorphisms and Their Generalizations

In Chapters 9 and 10, we introduced quadratic and cyclotomic orthomorphisms. In contrast to (σ, ε)-systems, which yield large classes of orthomorphisms with relative ease, but offer little information on orthogonalities, the use of cyclotomy yields simultaneously constructions of orthomorphisms and simple rules for orthogonality between these orthomorphisms. In addition cyclotomic orthomorphisms can be viewed in several different ways, as permutations of cyclotomy classes, as fixed point of congruences, and as permutation polynomials, allowing several generalizations to arbitrary groups.

Problem 16.9 (Problem 9 from [119]). Let $\{X_1, \cdots, X_m\}$ be a partition of the set of nonidentity elements of a group G. Study classes of orthomorphisms θ of G for which $\theta(X_i) = X_{\sigma(i)}$, for some permutation σ of $\{1, \cdots, m\}$, and the corresponding orthomorphism graph.

Problem 16.10 (Problem 10 from [119]). or K, a group of congruences of $\mathrm{Orth}(G)$, study the orthomorphism graph induced by $\{\theta \in \mathrm{Orth}(G) \mid \mathrm{T}[\theta] = \theta \text{ for all } \mathrm{T} \in K\}$.

Many orthomorphisms and their corresponding orthomorphism graphs can be defined by the congruences that fix them. Cyclotomic orthomorphisms are characterized by the homologies that fix them. Elements of $\mathscr{A}(G)$ are fixed by each element of $\{\mathrm{T}_g \mid g \in G\}$, and this characterizes elements of $\mathscr{A}(G)$.

Problem 16.11 (Problem 11 from [119]). In studying orthomorphisms of $\mathbb{Z}_p \times \cdots \times \mathbb{Z}_p$, we have used the fact that this group can be thought of as the additive group of a finite field and have used field multiplication to define linear, quadratic, and cyclotomic orthomorphisms. There are many ways to define multiplication on $\mathbb{Z}_p \times \cdots \times \mathbb{Z}_p$ so that the field axioms are satisfied. Extend the class of linear (respectively quadratic or cyclotomic) orthomorphisms to include orthomorphisms that are linear (respectively quadratic or cyclotomic) with respect to some field multiplication, and study the corresponding orthomorphism graphs. As an example, every orthomorphism of $\mathbb{Z}_2 \times \mathbb{Z}_2 \times \mathbb{Z}_2$ is linear with respect to some field multiplication.

16.1.3 Strong Complete Mappings

Strong complete mappings, mappings that are both complete mappings and orthomorphisms, were discussed in Section 8.5. There were two problems, Problems 70 and 71, on the existence of strong complete mappings in [119].

- Is it true that if the Sylow 3-subgroup of a finite group G is nontrivial and cyclic, then G cannot admit strong complete mappings? This is the case if $|G|$ is odd as there must then exist a homomorphism from G onto its Sylow 3-subgroup.
- If G is a finite abelian group and its Sylow 2 and Sylow 3-subgroups are each either trivial or noncyclic, does G admit strong complete mappings?

Both of these problems have now been solved. In a preprint Shieh, Hsiang, and Hsu [294] exhibited a strong complete mapping for D_{12}, the dihedral group of order 12; and in 2012 Evans [129] proved that a finite abelian group admits strong complete mappings if and only if its Sylow 2 and Sylow 3-subgroups are each either trivial or noncyclic.

Problem 16.12. Find more examples of finite groups with nontrivial, cyclic Sylow 3-subgroups that admit strong complete mappings.

Problem 16.13. Does there exist a group G, $|G| \neq 8$, with a noncyclic Sylow 2-subgroup and a nontrivial, cyclic Sylow 3-subgroup that is not a homomorphic image of G and that does not admit strong complete mappings?

Problem 16.14. In Section 4.1 we used HP-systems, and in Section 7.1 W-systems, to extend complete mappings of a subgroup to complete mappings of a group. Find similar ways to extend strong complete mappings of a subgroup to strong complete mappings of a group.

Problem 16.15. Characterize nonabelian 2-groups that admit strong complete mappings.

Problem 16.16. Characterize nonabelian 3-groups that admit strong complete mappings.

More generally.

Problem 16.17. Characterize finite solvable groups that admit strong complete mappings.

16.1.4 More Constructions

Here are some miscellaneous problems on constructions of orthomorphisms.

Problem 16.18 (Problem 12 from [119]). Let R be a ring, and let

$$P(R) = \{\mathrm{f} \mid \mathrm{f} \in R(x), r \to \mathrm{f}(r) \text{ is a permutation of } R\},$$

the set of permutation polynomials of R.

Study the orthomorphism graph $\mathrm{Orth}(R^+) \cap P(R)$. In general not all permutations of the elements of R can be represented by permutation polynomials of R.

Problem 16.19 (Problem 13 from [119]). Find more classes of permutation polynomials and orthomorphism polynomials.

Problem 16.20 (Problem 14 from [119]). If q is even, $q \neq 2$, then $\mathrm{Orth}(\mathrm{GF}(q)^*) \neq \emptyset$, and any orthomorphism of $\mathrm{GF}(q)^*$ can be represented by a permutation polynomial of $\mathrm{GF}(q)$ with constant term 0. Use permutation polynomials to study $\mathrm{Orth}(\mathrm{GF}(q)^*)$. The first case of interest is $\mathrm{Orth}(\mathrm{GF}(16)^*) \cong \mathrm{Orth}(\mathbb{Z}_{15})$.

Problem 16.21 (Problem 15 from [119]). Find new interesting orthomorphism graphs by studying subgraphs of the orthomorphism graphs $\mathscr{D}_D(G)$ and $\mathscr{J}_{D,H}(G)$, described in Section 8.3. As an example, if K is a group of order $q+1$, then each orthomorphism graph of K can be used to define a subgraph of the orthomorphism graph $\mathscr{D}_D(\mathbb{Z}_n)$, $n = q^2 + q + 1$.

Problem 16.22 (Problem 16 from [119]). Theorem 13.15 is stated in difference matrix form. Study the new orthomorphism graph obtained when this theorem is rewritten in orthomorphism form.

Problem 16.23 (Problem 18 from [119]). If H is a normal subgroup of G, find new ways to extend orthomorphisms of H or G/H to orthomorphisms of G and hence extend orthomorphism graphs of H or G/H to orthomorphism graphs of G. More generally if H is a (not necessarily normal) subgroup of G, find new ways to extend orthomorphisms of H to orthomorphisms of G.

16.2 Cliques of Orthomorphism Graphs

Our problems on cliques of orthomorphism graphs fall into four general categories: determining $\omega(G)$ or bounds on $\omega(G)$, determining $\omega(\mathscr{H})$ or bounds on $\omega(\mathscr{H})$ for an orthomorphism graph $\mathscr{H}$, finding cliques that are maximal in $\mathscr{H}$ or $\mathrm{Orth}(G)$, and finding complete set of orthomorphisms, i.e., cliques of size $|G| - 2$.

16.2.1 *Bounds on $\omega(G)$*

By Theorem 1.21, from an r-clique of the orthomorphism graph of a group of order n, we can construct a set of $r + 1$ mutually orthogonal Latin squares of order n. A number of lower bounds for $N(n)$ have been obtained in this way. The following problems ask for $\omega(G)$, or bounds on $\omega(G)$, for several groups.

Problem 16.24 (Problem 38 from [119]). Determine $\omega(D_{16})$. What can we say about $\omega(D_q)$, $q = 2^n$, or $\omega(D_{4n})$?

Progress towards solving Problem 16.24 was described in Section 12.1.

Problem 16.25 (Problem 41 from [119]). Determine $\omega(\mathbb{Z}_{27})$.

Problem 16.26 (Problem 42 from [119]). If $q = p^n$, p a prime, then $\omega(\mathbb{Z}_q) \geq p - 2$: see Section 8.2. When do we have equality? When is $\omega(\mathbb{Z}_q) > p - 2$?

Problem 46 in [119] was to determine clique numbers for the orthomorphism graphs of groups of orders 16, 17, 19, 20, 21, 23, and 24. But 17, 19, and 23 are primes, and for prime orders, the clique numbers were known: by the construction of Theorem 1.6, $\omega(\mathbb{Z}_p) = p - 2$ if p is a prime. Progress towards solving this problem was described in Chapter 13. In particular $\omega(G)$ has been determined for all abelian groups of order 16. We have divided the remaining cases of this problem into Problems 16.27–16.30.

Problem 16.27. Determine clique numbers for the orthomorphism graphs of the nine nonabelian groups of order 16.

Problem 16.28. Determine clique numbers for the orthomorphism graphs of the two groups of order 20 that have noncyclic Sylow 2-subgroups, $\mathbb{Z}_2 \times \mathbb{Z}_{10}$ and D_{20}.

Problem 16.29. Determine clique numbers for the orthomorphism graphs of the two groups of order 21.

Problem 16.30. Determine clique numbers for the orthomorphism graphs of the 13 groups of order 24 that have noncyclic Sylow 2-subgroups.

Problem 16.31 (Problem 47 from [119]). If $p < q$ are odd primes, then $\omega(\mathbb{Z}_{pq}) \geq p - 2$. This is the best available lower bound for many values of p and q. Try to improve this bound.

16.2.2 Cliques of Orthomorphism Graphs

While we are more interested in determining $\omega(G)$ for a group G, it is often easier to determine $\omega(\mathscr{H})$ for an orthomorphism graph $\mathscr{H}$ of G. The following problems concern cliques of these orthomorphism graphs.

Problem 16.32 (Problem 49 from [119]). Recall that $\mathscr{S}(G)$ is the orthomorphism graph induced by the strong complete mappings of G studied in Section 8.5. If $|G|$ is odd, when is $\omega(\mathscr{S}(G)) = \omega(\mathrm{Orth}(G)) - 1$? When is $\omega(\mathscr{S}(G)) < \omega(\mathrm{Orth}(G)) - 1$? If $|G|$ is even, when is $\omega(\mathscr{S}(G)) = \omega(\mathrm{Orth}(G))$? When is $\omega(\mathscr{S}(G)) < \omega(\mathrm{Orth}(G))$?

In Section 8.4 we studied the orthomorphism graphs $\mathscr{A}(G)$, $\mathscr{A}^{-}(G)$, and their union. These orthomorphism graphs consist of orthomorphisms constructed from automorphisms. Problems 16.33, 16.34, 16.35, and 16.36 concern these orthomorphism graphs.

Problem 16.33 (Problem 50 from [119]). In Theorem 8.86 we proved that

$$\omega(A_{\mathscr{H}} \cap \mathrm{Orth}(G)) \leq \omega(A_H \cap \mathrm{Orth}(G)).$$

When does equality hold?

Problem 16.34 (Problem 51 from [119]). In Corollary 8.76 we proved that

$$\omega(\mathscr{A}(G)) \leq \min\{\omega(\mathscr{A}(H)), \omega(\mathscr{A}(G/H))\}$$

if H char G. When does equality hold?

Problem 16.35 (Problem 52 from [119]). In Theorem 8.77 we proved that

$$\omega(\mathscr{A}(G \times H)) = \min\{\omega(\mathscr{A}(H)), \omega(\mathscr{A}(G))\}$$

if $|G|$ and $|H|$ are relatively prime. What happens if $|G|$ and $|H|$ are not relatively prime?

Problem 16.36 (Problem 53 from [119]). Determine $\omega(\mathscr{A}(G))$ for (a class of) nonabelian p-groups. Some upper bounds can be found in Section 8.4.

16.2.3 Maximal Cliques

In Theorem 1.32 we showed that a maximal set of $r+1$ mutually orthogonal Latin squares of order n could be constructed from a maximal r-clique of Orth(G), G a group of order n. The following problems deal with maximal cliques of orthomorphism graphs, in particular maximal cliques of Orth(G).

Problem 16.37 (Problem 56 from [119]). In Section 8.2 we studied the orthomorphism graph $\mathscr{P}(G)$ induced by orthomorphisms of G of the form $\phi_r : x \mapsto x^r$. Let p be the smallest prime divisor of $|G|$, and let S be a Sylow p-subgroup of G. By Theorem 8.14, $\omega(\mathscr{P}(G)) = p - 2$, and the only maximal cliques of $\mathscr{P}(G)$ are $(p-2)$-cliques. When are maximal cliques of $\mathscr{P}(G)$ also maximal in Orth(G)?

If S is cyclic, then there exists a homomorphism from G onto S, and Corollary 8.24 then implies that maximal cliques of $\mathscr{P}(G)$ are maximal in Orth(G). Thus this question is only of interest when S is noncyclic. On the other hand, if $G = S = \mathrm{GF}(q)^+, q > p$, then no maximal clique of $\mathscr{P}(G)$ is maximal in Orth(G). It is also of interest to know, when these cliques are not maximal, how far they can be extended.

Problem 16.38 (Problem 58 from [119]). For which $(n, r; \lambda)$-difference matrices D and which positive integers m is mD a maximal difference matrix. Necessary conditions are that D be a maximal difference matrix and n does not divide m. These conditions are not sufficient in general. Find sufficient conditions.

Such difference matrices with $\lambda = 1$ were used in Section 8.2 to construct maximal sets of mutually orthogonal Latin squares.

Problem 16.39 (Problem 59 from [119]). Find new pairs (n, r) for which there exists a maximal r-clique of a group of order n. By Theorem 1.32 such a pair implies the existence of a maximal set of $r+1$ mutually orthogonal Latin squares of order n.

Problem 16.40 (Problem 60 from [119]). In Theorem 9.38 we established the maximality of certain cliques of $\mathscr{C}_2(p)$ in Orth$(\mathrm{GF}(p)^+)$, p an odd prime. For which odd prime powers does this result hold? Some computer-generated results were described in Section 9.4.

Problem 16.41 (Problem 61 from [119]). Find maximal cliques of $\mathscr{C}_e(q)$, $e \neq 2$, that are maximal in Orth$(\mathrm{GF}(q)^+)$. Some computer-generated results were described in Section 10.2.

16.2.4 Complete Sets of Orthomorphisms

The construction of projective planes of order n from complete sets of orthomorphisms of groups of order n is described in Chapter 14. We give here a few problems on complete sets of orthomorphisms.

Problem 16.42 (Problem 62 from [119]). If p is a prime, is there only one $(p-2)$-clique in $\mathrm{Orth}(\mathbb{Z}_p)$?

The smallest unsolved case is $p = 17$.

Problem 16.43 (Problem 65 from [119]). Is there only one 15-clique in $\mathrm{Orth}(\mathbb{Z}_{17})$?

Problem 16.44 (Problem 66 from [119]). In Theorem 9.51, we proved that the only $(p-2)$-clique of $\mathscr{C}_2(p)$ is $\mathscr{C}_1(p)$ if p is a prime. Is $\mathscr{C}_1(p)$ the only $(p-2)$-clique of $\mathscr{C}_e(p)$, e a proper divisor of $p-1$?

Problem 16.45 (Problem 67 from [119]). In Theorem 9.53 the $(q-2)$-cliques of $\mathscr{C}_2(q)$ were characterized for q an odd prime power. Characterize the $(q-2)$-cliques of $\mathscr{C}_e(q)$, e a proper divisor of $q-1$.

Problem 16.46. Find new classes of Cartesian projective planes.

Problem 16.47. Which groups can be Cartesian groups?

16.2.5 Theoretical Proofs of Computer Generated Results

In Chapter 13 we gave data for small groups. Many of these results were obtained via computer searches. In [119] we asked for theoretical proofs of many of these results. To date, none of these problems has been solved. Here are those problems related to cliques of orthomorphism graphs.

Problem 16.48 (Problem 37 from [119]). Prove theoretically that $\omega(D_8) = 1$. Recall that we have used D_n to denote the dihedral group of order n.

Problem 16.49 (Problem 39 from [119]). Prove theoretically that $\omega(Q_8) = 1$.

Problem 16.50 (Problem 40 from [119]). Prove theoretically that $\omega(\mathbb{Z}_9) = 1$.

Problem 16.51 (Problem 43 from [119]). Prove theoretically that $\omega(A_4) = 1$.

Problem 16.52 (Problem 44 from [119]). Prove theoretically that $\omega(\mathbb{Z}_2 \times \mathbb{Z}_6) = 4$.

Problem 16.53 (Problem 45 from [119]). Prove theoretically that $\omega(\mathbb{Z}_{15}) = 3$.

Problem 16.54 (Problem 57 from [119]). In Section 13.4 we showed that [5] and [10, 4, 4, 6, 7], an element of $\mathscr{C}_5(11)$, together form a maximal 2-clique of

$\mathrm{Orth}(\mathbb{Z}_{11})$. Prove this theoretically and so obtain a theoretical proof that there exists a maximal set of *three* mutually orthogonal Latin squares of order 11. If possible generalize your argument.

In Problem 16.42 we asked whether there is only one $(p-2)$-clique in $\mathrm{Orth}(\mathbb{Z}_p)$, p a prime. We showed in Chapter 13 that the answer is yes for $p \leq 13$, but this has only been shown theoretically for $p \leq 7$.

Problem 16.55 (Problem 63 from [119]). Prove theoretically that there is only one 9-clique in $\mathrm{Orth}(\mathbb{Z}_{11})$.

Problem 16.56 (Problem 64 from [119]). Prove theoretically that there is only one 11-clique in $\mathrm{Orth}(\mathbb{Z}_{13})$.

16.3 The Structure of Orthomorphism Graphs

Our problems on the structure of orthomorphism graphs are determining the structure of $\mathrm{Orth}(G)$, determining the structure of $\mathscr{H}$ for an orthomorphism graph $\mathscr{H}$, and determining possible orthogonalities within and between orthomorphism graphs.

16.3.1 General Structure Problems

Here are some problems on the structure of orthomorphism graphs that do not consist of cyclotomic orthomorphisms and their translates.

Problem 16.57 (Problem 17 from [119]). In Section 8.2 we introduced the orthomorphism graph $\mathscr{P}(G)$ consisting of those orthomorphisms of the form $\phi_r(x) = x^r$, and in Section 8.5, we defined strong complete mappings to be orthomorphisms that are orthogonal to ϕ_{-1}. Study the orthomorphism graphs consisting of those orthomorphisms of G that are orthogonal to each element of S, $S \subseteq \mathscr{P}(G)$.

Problem 16.58 (Problem 20 from [119]). Explain the structure of $\mathrm{Orth}(GF(9)^+)$. The description given in Theorem 13.10 is incomplete.

Problem 16.59 (Problem 23 from [119]). Determine the structure of $\mathrm{Orth}(\mathbb{Z}_2 \times \mathbb{Z}_6)$.

Problem 16.60 (Problem 24 from [119]). Determine the structure of $\mathrm{Orth}(\mathbb{Z}_{13})$.

In Problem 25 from [119], we asked for a theoretical determination of the structure of $\mathrm{Orth}(\mathbb{Z}_{15})$. We first need to know the structure of this orthomorphism graph.

Problem 16.61. Determine the structure of $\mathrm{Orth}(\mathbb{Z}_{15})$.

It would also be of interest to know the structures of the orthomorphism graphs of groups of orders 16, 17, 19, 20, 21, 23, and 24.

Problem 16.62 (Problem 34 from [119]). Determine the structure of $\mathscr{A}(G)$, or $\mathscr{A}^-(G)$, or $\mathscr{A}(G) \cup \mathscr{A}^-(G)$ for some class of groups other than the elementary abelian groups. These orthomorphism graphs were studied in Section 8.4.

16.3.2 Cyclotomic Orthomorphisms

In Chapter 9 we discussed the orthomorphism graphs $\mathscr{C}_1(q)$ and $\mathscr{C}_2(q)$, and in Chapter 10 the more general class of orthomorphism graphs $\mathscr{C}_e(q)$, $e|q-1$. Here are some problems on the structure of these graphs and their translates.

Problem 16.63 (Problem 26 from [119]). For p an odd prime, what types of orthomorphisms are allowed as neighbors of a nonlinear quadratic orthomorphism of $GF(p)^+$? For $p \leq 11$ we showed in Chapter 13 that only linear and quadratic orthomorphisms are allowed.

Problem 16.64 (Problem 27 from [119]). For p an odd prime, what types of orthomorphisms are allowed as neighbors of a nonlinear cyclotomic orthomorphism of index e, $2 < e < p-1$? For $p \leq 11$ we showed in Chapter 13 that only linear orthomorphisms and cyclotomic orthomorphisms of the same index are allowed.

In a recent paper, Fear and Wanless [140] gave partial solutions to Problems 16.63 and 16.64. In this same paper, they gave a complete solution to Problem 28 from [119]. This problem asked, "In Theorem 10.23 we showed that if $a, b|q-1$ and $a|b$, then $\mathscr{C}_a(q) \subseteq \mathscr{C}_b(q)$. When do we have equality? When do we have inequality?" Fear and Wanless proved that $\mathscr{C}_a(q) \neq \mathscr{C}_b(q)$ whenever $a, b|q-1$ and $a|b$, $a \neq b$, except for the following cases: $\mathscr{C}_1(3) = \mathscr{C}_2(3)$, $\mathscr{C}_1(4) = \mathscr{C}_3(4)$, $\mathscr{C}_1(5) = \mathscr{C}_2(5) = \mathscr{C}_4(5)$, and $\mathscr{C}_1(7) = \mathscr{C}_3(7)$.

Problem 16.65 (Problem 29 from [119]). In Theorem 10.24 we gave a formula for $|\mathscr{C}_e(q)|$ which is only practical for small values of e. Find a better formula.

Problem 16.66 (Problem 30 from [119]). In Theorem 10.27 we showed that if q is prime, then $T_g[\mathscr{C}_a(q)] \cap T_h[\mathscr{C}_b(q)] = \mathscr{C}_1(q)$. If q is not prime, describe $T_g[\mathscr{C}_a(q)] \cap T_h[\mathscr{C}_b(q)]$, where $a, b|q-1$, $a, b \neq q-1$, and $g, h \in \mathrm{GF}(q)$, $g \neq h$.

Problem 16.67 (Problem 31 from [119]). If $[A, B]$ is a nonlinear quadratic orthomorphism of $\mathrm{GF}(q)^+$, what can we say about the structure of the orthomorphism graph induced by its neighbors in $\mathscr{C}_2(q)$?

Problem 16.68 (Problem 32 from [119]). If θ is a nonlinear cyclotomic orthomorphism of $\mathrm{GF}(q)^+$ of index e, $2 < e < q-1$, what can we say about the structure of the orthomorphism graph induced by its neighbors in $\mathscr{C}_e(q)$?

Problem 16.69 (Problem 33 from [119]). Theorem 10.30 gives rules for membership and orthogonality in the set of cyclotomic orthomorphism polynomials discovered by Niederreiter and Robinson. What can we say about the structure of the orthomorphism graph induced by this set of cyclotomic orthomorphisms?

16.3.3 Theoretical Proofs of Computer-Generated Results

In Chapter 13 we gave data for small groups, mostly obtained via computer searches. Here are some problems asking for theoretical explanations for computer-generated results on the structure of orthomorphism graphs.

Problem 16.70 (Problem 19 from [119]). In Section 13.2 we described Orth($\mathbb{Z}_2 \times \mathbb{Z}_4$) to consist of 12 disjoint 4-cycles. This was determined through a computer search. Give a theoretical proof of this. Does this give us any insight into the structure of Orth($\mathbb{Z}_2 \times \mathbb{Z}_8$) or more generally Orth($\mathbb{Z}_2 \times \mathbb{Z}_q$), $q = 2^n$?

Problem 16.71 (Problem 21 from [119]). Explain, without the use of computer generated data, the structure of Orth($\mathbb{Z}_{11}$).

Problem 16.71 can be broken down into several smaller problems.

Problem 16.72. Why is no element of $\mathscr{C}_2(11) \setminus \mathscr{C}_1(11)$ orthogonal to any orthomorphism outside of $\mathscr{C}_2(11)$? Can we extend this explanation to obtain information about possible orthogonalities between elements of $\mathscr{C}_2(p) \setminus \mathscr{C}_1(p)$ and orthomorphisms outside of $\mathscr{C}_2(p)$, p a prime?

Problem 16.73. Similarly, why is no element of $\mathscr{C}_5(11) \setminus \mathscr{C}_1(11)$ orthogonal to any orthomorphism outside of $\mathscr{C}_5(11)$? Can we extend this explanation to obtain information about possible orthogonalities between elements of $\mathscr{C}_e(p) \setminus \mathscr{C}_1(p)$ and orthomorphisms outside of $\mathscr{C}_e(p)$, p a prime?

Problem 16.74. Prove theoretically that all edges of Orth($\mathbb{Z}_{11}$) are within $\mathscr{C}_1(11)$, $\mathscr{C}_2(11)$, $\mathscr{C}_5(11)$, and their translates.

Problem 16.75. Why is each element of $\mathscr{C}_5(11) \setminus \mathscr{C}_1(11)$ orthogonal to exactly one element of $\mathscr{C}_1(11)$? Can elements of $\mathscr{C}_e(p) \setminus \mathscr{C}_1(p)$, p a prime, $e \neq p-1$, be characterized by the number of elements of $\mathscr{C}_1(p)$ they are orthogonal to?

Problem 16.76 (Problem 22 from [119]). Give a theoretical proof that $|\text{Orth}(A_4)| = 3776$ and $|\text{Orth}(D_6)| = 6336$.

16.4 Related and Additional Problems

In this section we will give a number of problems that are closely related to orthomorphisms, near orthomorphisms, complete mappings, and near complete mappings of groups, plus some additional problems. The reader will find other relevant lists of problems in the literature. In their 1974 book, Dénes and Keedwell [91] gave a list of problems dealing with Latin squares. They updated this list in the second edition [95], published in 2015. In 1984 Jungnickel [203] gave a list of problems concerning Latin squares and difference matrices. He translated this paper into English and updated the list in1990: see [206].

16.4.1 General Problems

Here are some problems concerning Latin squares, difference matrices, and generalized Hadamard matrices.

Problem 16.77 (Problem 48 from [119]). In Section 1.3 we defined difference matrices and generalized Hadamard matrices, and in Section 13.5, we described the 1960 construction by Bose, Chakravarti, and Knuth of four pairwise orthogonal orthomorphisms of $\mathbb{Z}_2 \times \mathbb{Z}_6$. They began with a generalized Hadamard matrix $M =$ GH(6, 2) over $\mathbb{Z}_2$, i.e., a Hadamard matrix of order 12. They then used the canonical homomorphism $\pi\colon \mathbb{Z}_2 \times \mathbb{Z}_6 \to \mathbb{Z}_2$ to replace entries of M, row by row, entry x by an element of $\pi^{-1}(x)$ until they obtained a $(12, 6; 1)$-difference matrix over $\mathbb{Z}_2 \times \mathbb{Z}_6$, yielding $five$ mutually orthogonal Latin squares of order 12.

Try to adapt this construction to obtain new mutually orthogonal sets of Latin squares. Specifically, given a group H of order n, an epimorphism $G \to H$, $|G|/|H| = \lambda$, and a $(n, r; \lambda)$-difference matrix over H or a GH(n, λ) over H, try to construct an $(n\lambda, t; 1)$-difference matrix over G for large t, yielding $t - 1$ mutually orthogonal Latin squares of order $n\lambda$.

Problem 16.78 (Problem 54 from [119]). In Section 1.2 we defined $N(n)$ to be the maximum number of squares possible in a set of mutually orthogonal Latin squares of order n. Let us define $R(n)$ to be the maximum number of squares possible in a set of mutually orthogonal Latin squares of order n constructed from pairwise orthogonal orthomorphisms of a group of order n. Thus $R(n) = 1 + \max\{\omega(G) \mid G \text{ a group}, |G| = n\}$.

For what values of n is $R(n) < N(n)$? For what values of n is $R(n) = N(n)$? We know, by Theorem 3.22, that, if $n \equiv 2 \pmod 4$ and $n > 6$, then $R(n) = 1$. From the disproof of Euler's conjecture (see Section 1.2), we know that $N(n) \geq 2$ for all $n > 6$. It follows that $R(n) < N(n)$ for all $n \equiv 2 \pmod 4$, $n > 6$. We also know, by the construction of Theorem 1.6, that $R(n) = N(n)$ if n is a power of a prime. This problem has not been answered for any other value of $n > 6$.

Recall that if $G = \{g_1, \ldots, g_n\}$ is a group, the Cayley table of G is the Latin square $L = \{g_i g_j\}$, and for a bijection $\theta\colon G \to G$, in Section 1.3, we defined L_θ to be the Latin square with ijth entry $g_i\theta(g_j)$. In Problem 55 from [119], we asked, for L the Cayley table for a group G and $\theta_1, \cdots, \theta_s$ a clique of Orth(G), if $L, L_{\theta_1}, \cdots, L_{\theta_s}, M_{s+1}, \cdots, M_r$ is a mutually orthogonal set of Latin squares can

$\theta_1, \cdots, \theta_s$ be extended to an r-clique of Orth(G)? Examples 1.33 and 1.34 show that the answer to this question is no in general: this prompts us to ask when the answer would be yes.

Problem 16.79. Let L be a Cayley table for a group G, and let $\theta_1, \cdots, \theta_s$ be a clique of Orth(G). If $L, L_{\theta_1}, \cdots, L_{\theta_s}, M_{s+1}, \cdots, M_r$ is a mutually orthogonal set of Latin squares, under what conditions can $\theta_1, \cdots, \theta_s$ be extended to an r-clique of Orth(G)? In Theorem 1.32 we showed the answer to be yes if $r = s + 1$. The next case of interest is $r = s + 2$.

Problem 16.80 (Problem 68 from [119]). In Theorems 14.36, 14.38, and 14.40, we established the nonexistence of generalized Hadamard matrices, GH(n, λ) over $\mathbb{Z}_p$, $p = 3$, 5, or 7, for many values of n and λ. Try to improve these results. We might for instance try to extend these results to the case $p = 11$.

Problem 16.81. For what values of n and λ and for what groups G of order n do there exist generalized Hadamard matrices GH(n, λ) over G?

Problem 16.82 (Problem 69 from [119]). In Theorem 14.42, G is assumed to be abelian. Does Theorem 14.42 still hold if this assumption is dropped?

16.4.2 Related Mappings

In Section 15.1 we introduced special classes of orthomorphisms and complete mappings of groups as well as variants of these mappings. We give here some problems dealing with these mappings.

Problem 16.83. Find more examples of finite nonabelian groups that admit proper near complete mappings.

Problem 16.84. Find more examples of abelian groups of order n, whose Sylow 2-subgroups are either trivial or noncyclic, that admit k-regular orthomorphisms for $k > 1$ a divisor of $n - 1$.

Problem 16.85. Find more finite nonabelian groups that admit antisymmetric mappings.

16.4.3 Neofields

In Section 15.2 we discussed neofields and left neofields: these are closely related to orthomorphisms and near orthomorphisms of groups. We present here some problems dealing with neofields and left neofields.

Problem 16.86 (Problem 72 from [119]). Suppose that we are given a property P of (left) neofields. Which groups can be the multiplicative group of a (left) neofield, with property P, in which $1+1=0$? Which groups can be the multiplicative group of a (left) neofield, with property P, in which $1+1\neq 0$?

Problems 16.87, 16.88, 16.89, 16.90, and 16.91 are special cases of Problem 16.86. By Theorem 15.38, solving these problems is equivalent to solving existence problems for special classes of orthomorphisms and near orthomorphisms.

Problem 16.87. Which groups can be the multiplicative group of a commutative (left) neofield in which $1+1=0$? Which groups can be the multiplicative group of a commutative (left) neofield in which $1+1\neq 0$?

Problem 16.88. Which groups can be the multiplicative group of a (left) neofield with the right inverse property in which $1+1=0$? Which groups can be the multiplicative group of a (left) neofield with the right inverse property in which $1+1\neq 0$?

Problem 16.89. Which groups can be the multiplicative group of a (left) neofield with the left inverse property in which $1+1=0$? Which groups can be the multiplicative group of a (left) neofield with the left inverse property in which $1+1\neq 0$?

Problem 16.90. Which groups can be the multiplicative group of a (left) neofield with the inverse property in which $1+1=0$? Which groups can be the multiplicative group of a (left) neofield with the inverse property in which $1+1\neq 0$?

Problem 16.91. Which groups can be the multiplicative group of a (left) neofield with the exchange inverse property in which $1+1=0$? Which groups can be the multiplicative group of a (left) neofield with the exchange inverse property in which $1+1\neq 0$?

Problem 16.92. Which nonabelian groups can be the multiplicative groups of neofields.

In 1983 Keedwell [215] gave a list of problems which included a problem very closely related to Problem 16.92.

Problem 16.93. For which orders do there exist neofields with nonabelian multiplicative groups.

16.4.4 Group Labeling Problems

In Section 15.3 we covered sequencings, R-sequencings, and harmonious labelings of groups: these correspond to special classes of orthomorphisms, near orthomorphisms, and complete mappings of groups. We list here some problems on group labelings.

Problem 16.94. Which groups are sequenceable? This is equivalent to asking which finite groups admit near orthomorphisms that consist of one sequence of length n, where n is the order of the group.

Problem 16.95. Which finite groups are R-sequenceable? This is equivalent to asking which finite groups admit normalized orthomorphisms that consist of one cycle of length $n - 1$, where n is the order of the group.

Problem 16.96. Which finite groups are harmonious? This is equivalent to asking which finite groups admit complete mappings that consist of one cycle of length n, where n is the order of the group.

Problem 16.97. Prove that all nonabelian groups of order at least 10 are sequenceable or find a nonabelian group of order at least 10 that is not sequenceable.

Solving Problem 16.97 would settle Conjecture 15.55.

16.4.5 Starters

In Section 15.4 we covered starters: these are associated with a class of orthomorphisms. We present here some problems on starters.

Problem 16.98. For which n does there exist an abelian group of order n that admits strong starters?

More particularly, we might ask.

Problem 16.99. Which finite abelian groups admit strong starters?

Problem 16.100. Prove the existence of strong starters for $\mathbb{Z}_{3p}$, $p > 111$ prime.

Problem 16.101. Prove the existence of strong starters for $\mathbb{Z}_3 \times \mathbb{Z}_3 \times \mathbb{Z}_p$, $p > 111$ prime.

Problem 16.102. Prove the existence of strong starters for $\mathbb{Z}_{3^n}$, $n \geq 7$.

Problem 16.103. Prove the existence of strong starters for $\mathbb{Z}_{5\times 3^n}$, $n \geq 4$.

Solving Problems 16.100, 16.101, 16.102, and 16.103 would settle Conjecture 15.81.

Problem 16.104. Prove the existence of strong starters for some abelian group of order $3p$ for some prime $p > 111$.

Problem 16.105. Prove the existence of strong starters for some abelian group of order $9p$ for some prime $p > 111$.

Solving Problems 16.104 and 16.105 would settle Conjecture 15.82. A list of problems on starters was given by Dinitz and Stinson [103] in 1992.

16.4.6 Graph Representations

In Section 8.2 questions about the structure of $\mathscr{P}(G)$ led naturally to the study of graph representations. Two problems on graph representations were presented in [119].

Problem 16.106 (Problem 35 from [119]). Given a graph G, determine $\text{rep}(G)$, the representation number of G, which is defined to be the least n for which G can be represented modulo n.

Problem 16.107 (Problem 36 from [119]). Characterize those graphs that can be represented modulo pqr, for some p, q, and r distinct primes.

A list of problems on graph representations was given by Evans, Narayan, and Urick [139] in 2009.

16.4.7 Additional Problems

We conclude this chapter with some further problems that were suggested for inclusion, most by Ian Wanless and the others by anonymous reviewers.

Problem 16.108. In Theorem 8.92 we proved that $\omega(\mathscr{A}(G) \cup \mathscr{A}^{-}(G)) = |G| - 2$ if and only if G is elementary abelian. Under what conditions can we strengthen this result to $\omega(\mathscr{A}(G) \cup \mathscr{A}^{-}(G)) \geq |G| - r$ if and only if G is elementary abelian, for some $r \geq 3$?

In Section 15.2 a number of classes of neofields were characterized in terms of properties of their associated near orthomorphisms. We gave no such characterization for planar neofields.

Problem 16.109. Find a nice characterization of planar neofields in terms of properties of their associated near orthomorphisms.

In Example 1.33 we described a set of *three* mutually orthogonal Latin squares of order 8, one of which was the Cayley table of D_8, and in Section 13.2 we saw that $\omega(D_8) = 1$. In Example 1.34 we described a set of *four* mutually orthogonal Latin squares of order 9, one of which was the Cayley table of $\mathbb{Z}_9$, and in Section 13.3, we saw that $\omega(\mathbb{Z}_9) = 1$. This suggests the following problem:

Problem 16.110. Which groups G have the property that their Cayley tables are contained in a set of t mutually orthogonal Latin squares for some $t > \omega(G) + 1$? For a given group, how large can t be?

In 2004 Grüttmüller [159] studied Latin squares and partial Latin squares with cyclically generated diagonals. He posed two problems: his first problem can be reworded in terms of partial orthomorphisms. A *partial orthomorphism* of $\mathbb{Z}_n$ of *deficiency* r is a one-one mapping θ from a subset of order $n - r$ of $\mathbb{Z}_n$ to $\mathbb{Z}_n$ for which the mapping $x \mapsto \theta(x) - x$ is also one-one.

Problem 16.111. Can a partial orthomorphism, of deficiency $n - r$, of $\mathbb{Z}_n$, n odd, be completed to an orthomorphism if $r \leq (n + 1)/3$? For any r, can a partial orthomorphism of $\mathbb{Z}_n$ of deficiency $n - r$, n odd, be completed to an orthomorphism if n is sufficiently large relative to r?

In a 2006 paper, McKay, McLeod, and Wanless [239] observed that setting z_n equal to the number of normalized orthomorphisms of $\mathbb{Z}_n$, modulo 8

$$(z_1, z_3, z_5, z_7, z_9, z_{11}, z_{13}, z_{15}, z_{17}, z_{19}, z_{21}, z_{23}, z_{25}) \\ \equiv (1, 1, 3, 3, 1, 1, 3, 3, 1, 1, 3, 3, 1).$$

This suggests the following problem.

Problem 16.112. Is the sequence $\{z_{2k+1} \mod 8\}$ periodic?

In their 2017 paper, Fear and Wanless [140] defined a normalized orthomorphism θ of $\mathrm{GF}(q)^+$ to be *irregular* if $\mathrm{T}_g(\theta)$ is not a cyclotomic orthomorphism for any $g \in \mathrm{GF}(q)$. They conjectured that irregular orthomorphisms exist for all sufficiently large q.

Problem 16.113. Do there exist irregular orthomorphisms of $\mathrm{GF}(q)^+$ for all sufficiently large q? If so, what is the largest q for which no irregular orthomorphism of $\mathrm{GF}(q)^+$ exists?

Also in [140], Fear and Wanless studied the question, for d a divisor of $q - 1$, does there exist a cyclotomic orthomorphism of index d, and no smaller index, that is orthogonal to some linear orthomorphism? They answered this question except for $d > 50$ and $q = 2d + 1 \equiv 3 \pmod 4$ and for $d > 50$ and $q = 3d + 1$.

Problem 16.114. For $d > 50$ and $q = 2d + 1 \equiv 3 \pmod 4$, does there exist a cyclotomic orthomorphism of index d, and no smaller index, that is orthogonal to some linear orthomorphism?

Problem 16.115. For $d > 50$ and $q = 3d + 1$, does there exist a cyclotomic orthomorphism of index d, and no smaller index, that is orthogonal to some linear orthomorphism?

In 2005 Wanless [324] gave a construction of Latin squares using cyclotomic orthomorphisms. Via a computer search, he found a number of these squares to be atomic Latin squares of composite order: see [80] or [324] for the definition of atomic Latin squares.

Problem 16.116. Are there finitely many or infinitely many cyclotomic orthomorphisms that produce atomic Latin squares of composite order by the construction in [324]?

References

1. Abel, R.J.R.: On the existence of balanced incomplete block designs and transversal designs. PhD thesis, University of New South Wales (1995)
2. Abel, R.J.R.: Some $V(12, t)$ vectors and other designs from difference and quasi-difference matrices. Australas. J. Combin. **40**, 69–85 (2008)
3. Abel, R.J.R., Bennett, F.E.: The existence of 2-SOLSSOMs. In: Wallis, W.D. (ed.) Designs 2002, pp. 1–21. Kluwer, Boston (2003)
4. Abel, R.J.R., Cavenagh, N.J.: Concerning eight mutually orthogonal Latin squares. J. Combin. Des. **15**, 255–261 (2007)
5. Abel, R.J.R., Cheng, Y.W.: Some new MOLS of order $2^n p$ for p a prime power. Australas. J. Combin. **10**, 175–186 (1994)
6. Abel, R.J.R., Colbourn, C.J., Dinitz, J.H.: Mutually orthogonal Latin squares (MOLS). In: Colbourn, C.J., Dinitz, J.H. (eds) Handbook of Combinatorial Designs, 2nd. edn., pp. 160–193. Chapman & Hall/CRC, Florida (2007)
7. Abel, R.J.R., Colbourn, C.J., Wojtas, M.: Concerning seven and eight mutually orthogonal Latin squares. J. Combin. Des. **12**, 123–131 (2004)
8. Abel, R.J.R., Finizio, N.J., Ge, G., Greig, M.: New $\mathbb{Z}$-cyclic triplewhist frames and triplewhist tournament designs. Discrete Appl. Math. **154**, 1649–1673 (2006)
9. Abel, R.J.R., Ge, G.: Some difference matrix constructions and an almost completion for the existence of triplewhist tournaments TWh(v). European J. Combin. **26**, 1094–1104 (2005)
10. Aczél, J.: Kvázicsoportok–hálózatok–nomogramok. Mat. Lapok **15**, 114–162 (1964)
11. Aczél, J.: Quasigroups, nets and nomograms. Adv. Math. **1**, 383–450 (1965)
12. Aczél, J.: Conditions for a loop to be a group and for a groupoid to be a semigroup. Amer. Math. Monthly **76**, 547–549 (1969)
13. Afsarinejad, K.: On mutually orthogonal Knut Vik designs. Statist. Probab. Lett. **5**, 323–324 (1987)
14. Albert, A.A.: Quasigroups I. Trans. Amer. Math. Soc. **54**, 507–519 (1943)
15. Alspach, B., Kreher, D.L., Pastine, A.: The Friedlander-Gordon-Miller conjecture is true. Australas. J. Combin. **67(1)**, 11–24 (2017)
16. Anderson, B.A.: Sequencings and houses. Congr. Numer. **43**, 23–43 (1984)
17. Anderson, B.A.: S_5, A_5 and all nonabelian groups of order 32 are sequenceable. Congr. Numer. **58**, 53–68 (1987)
18. Anderson, B.A.: A fast method for sequencing low order non-abelian groups. Ann. Discrete Math. **34**, 27–42 (1987)
19. Anderson, B.A., Ihrig, E.C.: Every finite solvable group with a unique element of order two, except the quaternion group, has a symmetric sequencing. J. Combin. Des. **1**, 3–14 (1993)

A. B. Evans, *Orthogonal Latin Squares Based on Groups*, Developments in Mathematics 57, https://doi.org/10.1007/978-3-319-94430-2

20. Anderson, I., Colbourn, C.J., Dinitz, J.H., Griggs, T.S.: Design theory: antiquity to 1950. In: Colbourn, C.J., Dinitz, J.H. (eds) Handbook of Combinatorial Designs, 2nd. edn., pp. 11–22, Chapman & Hall/CRC, Florida (2007)
21. Aschbacher, M.: Lectures delivered at NSA (1990)
22. Aschbacher, M.: Sporadic groups. Cambridge Tracts in Mathematics **104**, Cambridge University Press, Cambridge (2000)
23. Aschbacher, M.: Finite group theory. Cambridge Studies in Advanced Mathematics **10**, 2nd edn., Cambridge University Press, Cambridge (2000)
24. Baer, R.: Homogeneity of projective planes. Amer. J. Math. **64**, 137–152 (1942)
25. Bailey, R.A., Jungnickel, D.: Translation nets and fixed-point-free group automorphisms. J. Combin. Theory Ser. A **55**, 1–13 (1990)
26. Bambah, R.P.: On complete primitive residue sets. Bull. Calcutta Math. Soc. **38**, 113–116 (1946)
27. Baofeng W., Dongdai L.: On constructing complete permutation polynomials over finite fields of even characteristic. Discrete Appl. Math. **184**, 213–222 (2015)
28. Barlotti, A.: Le possibili configurazioni del sistema delle coppie punto-retta (A, a) per cui un piano grafico risulta (A, a)-transitivo. Boll. Unione Mat. Ital. **12**, 212–226 (1957)
29. Bateman, P.T.: A remark on infinite groups. Amer. Math. Monthly **57**, 623–624 (1950)
30. Baumert, L.D., Gordon, D.M.: On the existence of cyclic difference sets with small parameters. In: High primes and misdemeanours, pp. 61–68. Amer. Math. Soc., Providence RI (2004)
31. Baumert, L., Hall Jr., M.: Nonexistence of certain planes of order 10 and 12. J. Combin. Theory Ser. A **14**, 273–280 (1973)
32. Beals, R., Gallian, J.A., Headley, P., Jungreis, D.: Harmonious groups. J. Combin. Theory Ser. A **56**, 223–238 (1991)
33. Bedford, D.: Transversals in the Cayley tables of the non-cyclic groups of order 8. European J. Combin. **12**, 455–458 (1991)
34. Bedford, D.: Construction of orthogonal Latin squares using left neofields. Discrete Math. **115**, 17–38 (1993)
35. Bedford, D.: Orthomorphisms and near orthomorphisms of groups and orthogonal Latin squares: a survey. Bull. Inst. Combin. Appl. **15**, 13–33 (1995)
36. Bedford, D.: Addendum to: "Orthomorphisms and near orthomorphisms of groups and orthogonal Latin squares: a survey". Bull. Inst. Combin. Appl. **18**, 86 (1996)
37. Bedford, D.: Quasi-orthogonal Latin squares and related designs. J. Combin. Math. Combin. Comput. **26**, 213–224 (1998)
38. Bedford, D., Whitaker, R.M.: Enumeration of transversals in the Cayley tables of the non-cyclic groups of order 8. Discrete Math. **197/198**, 77–81 (1999)
39. Bedford, D., Whitaker, R.M.: New and old values for maximal MOLS(n). Ars Combin. **54**, 255–258 (2000)
40. Bell, J.: Cyclotomic orthomorphisms of finite fields. Discrete Appl. Math. **161**, 294–300 (2013)
41. Beth, T., Jungnickel, D., Lenz, H.: Design Theory, 2nd. edn. Cambridge University Press, Cambridge (1999)
42. Bondesen, A.: Er det en gruppetavle? Nordisk Mat. Tidskr. **17**, 132–136 (1969)
43. Bose, R.C., Chakravarti, I.M., Knuth, D.E.: On methods of constructing sets of mutually orthogonal Latin squares using a computer I. Technometrics **2**, 507–516 (1960)
44. Bose, R.C., Nair, K.R.: On complete sets of Latin squares. Sankhya **5**, 361–382 (1941)
45. Bose, R.C., Shrikhande, S.S.: On the falsity of Euler's conjecture about the non-existence of two orthogonal Latin squares of order $4t+2$. Proc. Natl. Acad. Sci. USA **45**, 734–737 (1959)
46. Bose, R.C., Shrikhande, S.S., Parker, E.T.: Further results on the construction of mutually orthogonal Latin squares and the falsity of Euler's conjecture. Canad. J. Math. **12**, 189–203 (1960)
47. Bosma, W., Cannon, J., Playoust, C.: The Magma algebra system. I. The user language. J. Symbolic Comput. **24**, 235–265 (1997) http://magma.maths.usyd.edu.au/magma/

48. Bowler, A.: Orthomorphisms of dihedral groups. Discrete Math. **167/168**, 141–144 (1997)
49. Brandt, H.: Verallgemeinierung des Gruppenbegriffs. Math. Ann. **96**, 360–366 (1927)
50. Bray, J.N.: personal communication.
51. Breuer, T., Müller, J.: Character tables of endomorphism rings of multiplicity-free permutation modules in GAP. http://www.math.rwth-aachen.de/~Juergen.Mueller/mferctbl/mferctbl.html
52. Brown, J.M.N.: Elations of order 3 in projective planes of order 12. In: Finite Geometries (Pullman, 1981), pp. 61–65. Lect. Notes in Pure And Appl. Math. **82**. Dekker, New York (1983)
53. Bruck, R.H.: Finite nets I, numerical invariants. Canad. J. Math. **3**, 94–107 (1951)
54. Bruck, R.H.: Loops with transitive automorphism groups. Pacific J. Math. **1**, 481–483 (1951)
55. Bruck, R.H.: Difference sets in a finite group. Trans. Amer. Math. Soc. **78**, 464–481 (1955)
56. Bruck, R.H.: A survey of binary systems. Springer-Verlag, Berlin-Göttingen-Heidelberg (1958)
57. Bruck, R.H.: Finite nets II, uniqueness and imbedding. Pacific J. Math. **13**, 421–457 (1963)
58. Bruck R.H., Ryser, H.J.: The nonexistence of certain finite projective planes. Canad. J. Math. **1**, 88–93 (1949)
59. Burn, R.P.: Cayley tables and associativity. Math. Gaz. **62**, 278–281 (1978)
60. Burton, D.M.: Elementary number theory, 6th edn. McGraw Hill, New York (2007)
61. Carlitz, L.: A note on abelian groups. Proc. Amer. Math. Soc. **4**, 937–938 (1953)
62. Carlitz, L.: A theorem on permutations in a finite field. Proc. Amer. Math. Soc. **11**, 456–459 (1960)
63. Carlitz, L.: Some theorems on permutation polynomials. Bull. Amer. Math. Soc. **68**, 120–122 (1962)
64. Carlitz, L.: A note on permutation functions over a finite field. Duke Math. J. **29**, 325–332 (1962)
65. Carter, R.W.: Simple Groups of Lie Type. Wiley Classics Lib., John Wiley & Sons Ltd., New York (1972)
66. Cates, M.L., Killgrove, R.B.: One-directional translation planes of order 13. Congr. Numer. **32**, 173–180 (1981)
67. Cayley, A.: On the theory of groups as depending on the symbolical equation $\theta^n = 1$. Philos. Mag. **7**, 40–47 (1854)
68. Cayley, A.: On the theory of groups. Proc. Lond. Math. Soc. **9**, 126–133 (1877/78)
69. Chang, L.Q., Hsiang, K., Tai, S.: Congruent mappings and congruence classes of orthomorphisms of groups. Acta Math. Sinica **14**, 747–756 (1964), Chinese. Translated as: Chinese Math. Acta **6**, 141–152 (1965)
70. Chang, L.Q., Tai, S.S.: On the orthogonal relations among orthomorphisms of non-commutative groups of small orders. Acta Math. Sinica **14**, 471–480 (1964), Chinese: translated as: Chinese Math. Acta **5**, 506–515 (1964)
71. Chein, O., Pflugfelder, H.O., Smith, J.D.H. (eds): Quasigroups and Loops: Theory and Applications. Sigma Series in Pure Math. **8**, Heldermann Verlag, Berlin (1990)
72. Chevalley, C.: Sur certains groupes simples. Tohoku Math. J. (2) **7**, 14–66 (1955)
73. Chong, B.C., Chan, K.M.: On the existence of normalized Room squares. Nanta Math. **7**, 8–17 (1974)
74. Chou, W.-S.: Set-complete mappings on finite fields. In: Mullen G.L., Shiue, P.J. (eds.) Finite Fields, Coding Theory, and Advances in Communications and Computing, pp. 33–41, Dekker, New York (1993)
75. Chou, W.-S., Niederreiter, H.: Monomials and binomials over finite fields as R-orthomorphisms. Publ. Math. Debrecen **61**, 511–521 (2002)
76. Clifford, A.H., Preston, G.B.: The Algebraic Theory of Semigroups, Volume **1**. American Math. Soc., Providence (1961)
77. Cohen, S.D., Niederreiter, H., Shparlinski, I.E., Zieve, M.: Incomplete character sums and a special class of permutations. In: 21st Journées Arithmétiques (Rome, 2001), pp. 53–63. J. Théor. Nombres Bordeaux **13**, (2001)

78. Colbourn, C.J., Dinitz, J.H.: Mutually orthogonal Latin squares: a brief survey of constructions. J. Statist. Plann. Inference **95**, 9–48 (2001)
79. Colbourn, C.J., Dinitz, J.H. (eds): Handbook of Combinatorial Designs, 2nd. edn. Chapman & Hall/CRC, Florida (2007)
80. Colbourn, C.J., Dinitz, J.H., Wanless, I.M.: Latin squares. In: Colbourn, C.J., Dinitz, J.H. (eds) Handbook of Combinatorial Designs, 2nd. edn., pp. 135–160. Chapman & Hall/CRC, Florida (2007)
81. Conway, J.H., Curtis, R.T., Norton, S.P., Parker, R.A., Wilson, R.A.: Atlas of finite groups. Clarendon Press, Oxford (1985)
82. Coulter, R.S., Matthews, R.W.: Planar functions and planes of Lenz-Barlotti class II. Des. Codes Cryptogr. **10**, 167–184 (1997)
83. Dalla Volta, F., Gavioli, N.: Complete mappings in some linear and projective groups. Arch. Math. (Basel) **61**, 111–118 (1993)
84. Dalla Volta, F., Gavioli, N.: On the admissibility of some linear and projective groups in odd characteristic. Geom. Dedicata **66**, 245–254 (1997)
85. Dalla Volta F., Gavioli, N: Minimal counterexamples to a conjecture of Hall and Paige. Arch. Math. (Basel) **77**, 209–214 (2001)
86. De Launey, W.: On the nonexistence of generalised Hadamard matrices. J. Statist. Plann. Inference **10**, 385–396 (1984)
87. De Launey, W.: On the non-existence of certain generalised weighing matrices. Ars Combin. **17A** , 117–132 (1984)
88. Dembowski, H.P.: Finite geometries. Springer-Verlag, Berlin - Heidelberg - New York (1968)
89. Dembowski, P., Ostrom, T.G.: Planes of order n with collineation groups of order n^2. Math. Z. **103**, 239–258 (1968)
90. Dénes, J., Hermann, P.: On the product of all the elements of a finite group. Discrete Math. **15**, 105–109 (1982)
91. Dénes, J., Keedwell, A.D.: Latin Squares and Their Applications. English Universities Press, London (1974)
92. Dénes, J., Keedwell, A.D.: A new conjecture concerning admissibility of groups. European J. Combin. **10**, 171–174 (1989)
93. Dénes, J., Keedwell, A.D.: Latin Squares: New Developments in the Theory and Applications. Annals of Discrete Mathematics **46**, North Holland, Amsterdam (1991)
94. Dénes, J., Keedwell, A.D.: On two conjectures related to admissible groups and quasigroups. Period. Polytech. Transportation Engrg. **19**, 33–35 (1991)
95. Dénes, J., Keedwell, A.D.: Latin Squares and Their Applications, 2nd. edn. North Holland, Amsterdam (2015)
96. Di Vincenzo, O.M.: On the existence of complete mappings of finite groups. Rend. Mat. Appl. (7) **9**, 189–198 (1989)
97. Dickson, L.E.: The analytic representation of substitutions on a power of a prime number of letters with a discussion of the linear group. Ann. of Math. **11**, 65–120, 161–183 (1897)
98. Dickson, L.E.: Linear groups with an exposition of the Galois field theory. Teubner, Leibzig (1901)
99. Dinitz, J.H.: New lower bounds for the number of pairwise orthogonal symmetric Latin squares. Congr. Numer. **23**, 393–398 (1979)
100. Dinitz, J.H.: Room squares. In: Colbourn, C.J., Dinitz, J.H. (eds) Handbook of Combinatorial Designs, 2nd. edn., pp. 584–590. Chapman & Hall/CRC, Florida (2007)
101. Dinitz, J.H.: Starters. In: Colbourn, C.J., Dinitz, J.H. (eds) Handbook of Combinatorial Designs, 2nd. edn., pp. 622–628. Chapman & Hall/CRC, Florida (2007)
102. Dinitz, J.H., Stinson, D.R.: A hill-climbing algorithm for the construction of one-factorizations and Room squares. SIAM J. Algebraic Discrete Math. **8**, 430–438 (1987)
103. Dinitz, J.H., Stinson, D.R.: Room squares and related designs, in: Dinitz, J.H., Stinson, D.R. (eds.) Contemporary design theory: a collection of surveys, pp. 137–204. John Wiley & Sons, New York (1992)

104. Dinitz, J.H., Williams, H.C.: Number theory and finite fields. In: Colbourn, C.J., Dinitz, J.H. (eds) Handbook of Combinatorial Designs, 2nd. edn., pp. 791–818. Chapman & Hall/CRC, Florida (2007)
105. Drake, D.A.: Maximal sets of Latin squares and partial transversals. J. Statist. Plann. Inference **1**, 143–149 (1977)
106. Dulmage, A.L., Johnson, D., Mendelsohn, N.S.: Orthogonal Latin squares: preliminary report. Canad. Math. Bull. **2**, 211–216 (1959)
107. Egan, J., Wanless, I.M.: Enumeration of MOLS of small order. Math. Comp. **85**, 799–824 (2016).
108. Erdős, P., Evans, A.B.: Representations of graphs and orthogonal Latin square graphs. J. Graph Theory **13**, 593–595 (1989)
109. Erdős, P., Evans, A.B.: Sets of prime numbers satisfying a divisibility condition. J. Number Theory **61**, 39–43 (1996)
110. Euler, L.: Recherche sur une nouvelle espèce de quarrès magiques. Leonardi Euleri Opera Omnia series 1 **7**, 291–392 (1923)
111. Evans, A.B.: Orthomorphisms of $\mathbb{Z}_p$. Discrete Math. **64**, 147–156 (1987)
112. Evans, A,B,: Generating orthomorphisms of GF$(q)^+$. Discrete Math. **63**, 21–26 (1987)
113. Evans, A.B.: Orthomorphisms of GF$(q)^+$. Ars Combin. **27**, 121–132 (1989)
114. Evans, A.B.: On planes of prime order with translations and homologies. J. Geom. **34**, 36–41 (1989)
115. Evans, A.B.: Orthomorphism graphs of groups. J. Geom. **35**, 66–74 (1989)
116. Evans, A.B.: On strong complete mappings. Congr. Numer. **70**, 241–248 (1990)
117. Evans, A.B.: Maximal sets of mutually orthogonal Latin squares I. European J. Combin. **12**, 477–482 (1991)
118. Evans, A.B.: Maximal sets of mutually orthogonal Latin squares II. European J. Combin. **13**, 345–350 (1992)
119. Evans, A.B.: Orthomorphism graphs of groups. Lecture Notes in Mathematics **1535**, Springer-Verlag, Berlin (1992)
120. Evans, A.B.: The existence of complete mappings of finite groups. Congr. Numer. **90**, 65–75 (1992)
121. Evans, A.B.: Mutually orthogonal Latin squares based on linear groups. In: Coding Theory, Design Theory, Group Theory (Burlington, VT, 1990), pp. 171–175. Wiley-Intersci. publ., New York (1993)
122. Evans, A.B.: Latin squares based on the cyclic group of order $3p$. In: Proc. Eighth Internat. Conf. on Graph Theory, Combinatorics, Algorithms and Applications (Kalamazoo, Michigan, 1996), pp. 379–386. New Issues Press, Michigan (1999)
123. Evans, A.B.: The existence of complete mappings of SL$(2, q)$, $q \equiv 1 \pmod 4$. Finite Fields Appl. **7**, 373–381 (2001)
124. Evans, A.B.: On orthogonal orthomorphisms of cyclic and non-abelian groups. Discrete Math. **243**, 229–233 (2002)
125. Evans, A.B.: The existence of complete mappings of $SL(2, q)$, $q \equiv 3 \pmod 4$. Finite Fields Appl. **11**, 151–155 (2005)
126. Evans, A.B.: Latin squares without orthogonal mates. Des. Codes Cryptogr. **40**, 121–130 (2006)
127. Evans, A.B.: On orthogonal orthomorphisms of cyclic and non-abelian groups. II. J. Combin. Des. **15**, 195–209 (2007)
128. Evans, A.B.: The admissibility of sporadic simple groups. J. Algebra **321**, 105–116 (2009)
129. Evans, A.B.: The existence of strong complete mappings. Electron. J. Combin. **19**, # P34 (2012)
130. Evans, A.B.: The existence of strong complete mappings of finite groups: a survey. Discrete Math. **313**, 1191–1196 (2013)
131. Evans, A.B.: Mutually orthogonal Latin squares based on general linear groups. Des. Codes Cryptogr. **71**, 479–492 (2014)

132. Evans, A.B.: A class of Latin squares derived from finite abelian groups. Comment. Math. Univ. Carolin. **55,3**, 401–409 (2014)
133. Evans, A.B.: Linear groups that are the multiplicative groups of neofields. Bul. Acad. Ştiinţe Repub. Mold. Mat. **1000**, 1–6 (2016)
134. Evans, A.B.: The orthomorphism graph $\mathscr{L}_3(q)$. Electron. J. Combin. **25(1)**, #P1.43 (2018)
135. Evans, A.B.: Orthomorphism graphs of GF$(q)^+ \times$ GF$(3)^+$. Preprint
136. Evans, A.B., Fricke, G.H., Maneri, C.C., McKee, T.A., Perkel, M.: Representations of graphs modulo n. J. Graph Theory **18**, 801–815 (1994)
137. Evans, A.B., Isaak, G., Narayan, D.A.: Representations of graphs modulo n. Discrete Math. **223**, 109–123 (2000)
138. Evans, A.B., Mcfarland, R.L.: Planes of prime order with translations. Congr. Numer. **44**, 41–46 (1984)
139. Evans, A.B., Narayan, D.A.. Urick, J.: Representations of graphs modulo n: Some problems. Bull. Inst. Combin. Appl. **56**, 85–97 (2009)
140. Fear, D., Wanless, I.M.: Existence results for cyclotomic orthomorphisms. J. Algebraic Combin. **46**, 1–14 (2017).
141. Feit, W., Thompson, J.G.: Solvability of groups of odd order. Pacific J. Math. **13**, 775–1029 (1963)
142. Ferrero, G., Ferrero Cotti, C.C.: Come verificare la proprietà associativa. Boll. Unione Mat. Ital. (4) **11**, 322–329 (1975)
143. Fiala, N.: Short identities implying a quasigroup is a loop or group. Quasigroups Related Systems **15**, 263–271 (2007)
144. Fleisher, E.: Eulerian squares. PhD thesis, New York University (1934)
145. Ford, D., Johnson, K.W.: Determinants of Latin squares of order 8. Exp. Math. **5**, 317–325 (1996)
146. Friedlander, R.J., Gordon, B., Miller, M.D.: On a group sequencing problem of Ringel. Congr. Numer. **21**, 307–321 (1978)
147. Friedlander, R.J., Gordon, B., Tannenbaum, P.: Partitions of groups and complete mappings. Pacific J. Math. **92**, 283–293 (1981)
148. Frisinger, H.H.: The solution of a famous two-centuries-old problem. The Leonhard Euler Latin square conjecture. Hist. Math. **8**, 56–60 (1981)
149. Frolov, M.: Recherches sur les permutations carrées. J. Math. Spéc. (3) **4**, 8–11 (1890)
150. Gallian, J.A., Mullin, M.D.: Groups with antisymmetric mappings. Arch. Math. **65**, 273–280 (1995)
151. Ganley, M.J., Spence, E.: Relative difference sets and quasiregular collineation groups. J. Combin. Theory Ser. A **19**, 134–153 (1975)
152. Gao, S., Wei, W.: On non-abelian group difference sets. Discrete Math. **112**, 93–102 (1993)
153. The GAP group: GAP: Groups, algorithms, and programming. Version 4.4.10, http://www.gap-system.org (2007)
154. Ghinelli, D., Jungnickel, D.: On finite projective planes in Lenz-Barlotti class at least I.3. Adv. Geom. **suppl.**, S28–S48 (2003)
155. Gordon, B.: Sequences in Groups with Distinct Partial Products. Pacific J. Math. **11**, 1309–1313 (1961)
156. Gordon, D.M.: The prime power conjecture is true for $n < 2,000,000$. Electron. J. Combin. **1**, R6 (1994)
157. Gorenstein, D.: Finite groups. Harper & Row, New York (1968)
158. Gross, K.B., Mullin, R.C., Wallis, W.D.: The number of pairwise orthogonal symmetric Latin squares. Util. Math. **4**, 239–251 (1973)
159. Grüttmüller, M.: Completing partial Latin squares with prescribed diagonals. Discrete Appl. Math. **138**, 89–97 (2004)
160. Guangkui, X., Cao, X.: Complete permutation polynomials over finite fields of odd characteristic. Finite Fields Appl. **31**, 228–240 (2015)
161. Gumm, H.P.: Encoding of numbers to detect typing errors. Internat. J. Appl. Engin. **2**, 61–65 (1986)

162. Hachenberger, D.: Constructions of large translation nets with nonabelian translation groups. Des. Codes Cryptogr. **1**, 219–236 (1991)
163. Hachenberger, D.: On the existence of translation nets. J. Algebra **152**, 207–229 (1992)
164. Hachenberger, D., Jungnickel, D.: Bruck nets with a transitive direction. Geom. Dedicata **36**, 287–313 (1990)
165. Hachenberger, D., Jungnickel, D.: Translation nets: a survey. Discrete Math. **106/107**, 231–242 (1992)
166. Hall, M.: Projective planes. Trans. Amer. Math. Soc. **54** , 229–277 (1943)
167. Hall, M.: A combinatorial problem on abelian groups. Proc. Amer. Math. Soc. **3**, 584–587 (1952)
168. Hall, M.: Uniqueness of the projective plane with 57 points. Proc. Amer. Math. Soc. **4**, 912–916 (1953)
169. Hall, M.: Correction to "Uniqueness of the projective plane with 57 points". Proc. Amer. Math. Soc. **5**, 994–997 (1954)
170. Hall, M., Knuth, D.E.: Combinatorial analysis and computing. Amer. Math. Monthly **72**, 21–28 (1965)
171. Hall, M., Paige, L.J.: Complete mappings of finite groups. Pacific J. Math. **5**, 541–549 (1955)
172. Hall, P.: A characteristic property of soluble groups. London Math. Soc. **12**, 198–200 (1937)
173. Hammel, A.: Verifying the associative property for finite groups. Math. Teacher **61**, 136–139 (1968)
174. Hedayat, A.: A complete solution to the existence and non-existence of Knut Vic designs and orthogonal Knut Vic designs. J. Combin. Theory Ser. A **22**, 331–337 (1977)
175. Hedayat, A., Federer, W.T.: An application of group theory to the existence and nonexistence of orthogonal Latin squares. Biometrika **56**, 547–551 (1969)
176. Hedayat, A., Federer, W.T.: On the non-existence of Knut Vic designs for all even orders. Ann. Statist. **3**, 445–447 (1975)
177. Hering, C.: On subgroups with trivial normalizer intersection. J. Algebra **20**, 622–629 (1972)
178. Hermite, C.: Sur les fonctions de sept lettres. C.R. Acad. Sci. Paris **57**, 750–757 (1863)
179. Ho, C.Y., Goncalves, A.: On collineation groups of a projective plane of prime order. Geom. Dedicata **20**, 357–366 (1986)
180. Horton, J.D.: Room designs and one-factorizations. Aequationes Math. **22**, 56–63 (1981)
181. Horton, J.D.: Orthogonal starters in finite abelian groups. Discrete Math. **79**, 265–278 (1990)
182. Hou, X-d.: Permutation polynomials over finite fields – A survey of recent advances. Finite Fields Appl. **32**, 82–119 (2015)
183. Hsiang, J., Hsu, D.F., Shieh, Y.-P.: On the hardness of computing problems of complete mappings. Discrete Math. **277**, 87–100 (2004)
184. Hsu, D.F.: Cyclic neofields and combinatorial designs. Lecture Notes in Mathematics **824**, Springer-Verlag, Berlin-Heidelberg-New York (1980)
185. Hsu, D.F.: Orthomorphisms and near orthomorphisms. In: Alavi,Y., Chartrand, G., Oellermann, O.R., Schwenk, A.J. (eds.) Graph Theory, Combinatorics, and Applications, pp. 667–679. Wiley, New York (1991)
186. Hsu, D.F., Keedwell, A.D.: Generalized complete mappings, neofields, sequenceable groups and block designs. I. Pacific J. Math. **111**, 317–332 (1984)
187. Hsu, D.F., Keedwell, A.D.: Generalized complete mappings, neofields, sequenceable groups and block designs. II. Pacific J. Math. **117(2)**, 291–312 (1985)
188. Hughes, D.R.: Planar division neo-rings, PhD dissertation, University of Wisconsin-Madison (1955)
189. Hughes, D.R.: Planar division neo-rings. Trans. Amer. Math. Soc. **80**, 502–527 (1955)
190. Hughes, D.R., Piper, F.C.: Projective planes. Springer, Berlin-Heidelberg-New York (1973)
191. Huppert, B.: Endliche Gruppen I. Springer, Berlin-Heidelberg-New York (1967)
192. Huppert, B., Blackburn, N.: Finite groups III. Springer-Verlag, Berlin-Heidelberg-New York (1982)
193. Hurwitz, A.: ____. Nouv. Ann. Serie **31**, 389 (1882)

194. Johnsen, E.C., Storer, T.: Combinatorial Structures in Loops, IV. Steiner Triple Systems in Neofields, Math. Z. **138**, 1–14 (1974)
195. Johnson, C.P.: Constructions of neofields and right neofields, PhD thesis, Emory University (1981)
196. Johnson, D.M., Dulmage, A.L., Mendelsohn, N.S.: Orthomorphisms of groups and orthogonal Latin squares. I. Canad. J. Math. **13**, 356–372 (1961)
197. Johnson, N.L.: Projective planes of prime order p that admit collineation groups of order p^2. J. Geom. **30**, 49–68 (1987)
198. Johnson, N.L., Jha, V., Bilioti, M.: Handbook of finite translation planes. Chapman and Hall, Boca Raton-London-New York (2007)
199. Johnson, N.L., Piper, F.C.: On planes of Lenz-Barlotti class II-1. Bull. Lond. Math. Soc. **6**, 152–154 (1974)
200. Jungnickel, D.: On regular sets of Latin squares. In: Problemes combinatoires et theorie des graphes, pp. 255–256. Colloq. Internat. CNRS **260**, Paris (1978)
201. Jungnickel, D.: On difference matrices and regular Latin squares. Abh. Math. Semin. Univ. Hambg. **50**, 219–231 (1980)
202. Jungnickel, D.: Einige neue Differenzenmatrizen. Mitt. Math. Semin. Giessen **149**, 47–57 (1981)
203. Jungnickel, D.: Lateinsche Quadrate, ihre Geometrien und ihre Gruppen. Jahresber. Dtsch Math.-Ver. **86**, 69–108 (1984)
204. Jungnickel, D.: Existence results for translation nets II. J. Algebra **122**, 288–298 (1989)
205. Jungnickel, D.: Partial spreads over $\mathbb{Z}_q$. Linear Algebra Appl. **114**, 95–102 (1989)
206. Jungnickel, D.: Latin squares, their geometries and their groups. A survey. In: Coding Theory and design Theory, part II, pp. 166–225. IMA Vol. Math. Appl. **21**. Springer-Verlag, New York (1990)
207. Jungnickel, D., Grams, G.: Maximal difference matrices of order $\leq$ 10. Discrete Math. **58**, 199–203 (1986)
208. Jungnickel, D., Pott, A.: Computational non-existence results for abelian affine difference sets. Congr. Numer. **68**, 91–98 (1989)
209. Jungnickel, D., Pott, A., Smith K.W.: Difference sets. In: Colbourn, C.J., Dinitz, J.H. (eds) Handbook of Combinatorial Designs, 2nd. edn., pp. 419–435. Chapman & Hall/CRC, Florida (2007)
210. Kallaher, M.J.: Affine Planes with Transitive Collineation Groups. North-Holland, New York-Amsterdam (1982)
211. Kantor, W.M.: Projective planes of type I-4. Geom. Dedicata **3**, 335–346 (1974)
212. Keedwell, A.D.: On orthogonal Latin squares and a class of neofields. Rend. Mat. Appl. **(5)25**, 519–561 (1966)
Erratum: Math. Reviews 37(1969), p.1469.
213. Keedwell, A.D.: Latin squares P-quasigroups and graph decompositions. Recueil des Travaux de l'Institute Mathématique, Belgrade, N.S. **1(9)**, 41–48 (1976)
214. Keedwell, A.D.: The existence of pathological left neofields. Ars Combin. **16B**, 161–170 (1983)
215. Keedwell, A.D.: Sequenceable groups, generalized complete mappings, neofields and block designs. In: Combinatorial Mathematics **X** (Adelaide, 1982), pp. 49–71. Lecture Notes in Math. **1036**. Springer, Berlin (1983)
216. Keedwell, A.D.: Construction, properties and applications of finite neofields. Comment. Math. Univ. Carolin. **41**, 283–297 (2000)
217. Keedwell, A. D.: Tests for loop nuclei and a new criterion for a Latin square to be group-based. European J. Combin. **26**, 111–116 (2005)
218. Keedwell, A. D.: Realizations of loops and groups defined by short identities. Comment. Math. Univ. Carolin. **50**(3), 373–383 (2009)
219. Keedwell, A. D.: Corrigendum to "Realizations of loops and groups defined by short identities". Comment. Math. Univ. Carolin. **50**, 639–640 (2009)

220. Keedwell, A.D., Wang, C.D.: On special near orthomorphisms. J. Combin. Math. Combin. Comput. **21**, 187–192 (1996)
221. Klotz, W., Sander, T.: Some properties of unitary Cayley graphs. Electron. J. Combin. **14**, #R45 (2007)
222. Klyve, D., Stemkosi, L.: Graeco-Latin squares and a mistaken conjecture of Euler. College Math. J. **37**, 2–15 (2006)
223. Laywine, C.F., Mullen, G.L.: Discrete Mathematics using Latin squares. Wiley, New York (1998)
224. Lazebnik, F., Thomason, A.: Orthomorphisms and the construction of projective planes. Math. Comp. **73**, 1547–1557 (2004)
225. Lemmermeyer, F.: Reciprocity laws. From Euler to Eisenstein. Springer, Berlin (2000)
226. Lenz, H.: Kleiner Desarguessche Satz und Dualität in projektiven Ebenen. Jahresber. Dtsch Math.-Ver. **57**, 20–31 (1954)
227. Li, J., Chandler, D.B., Xiang, Q.: Permutation polynomials of degree 6 or 7 over finite fields of characteristic 2. Finite Fields Appl. **16**, 406–419 (2010)
228. Li, Z.H.: Nonexistence of a special kind of orthomorphic permutation polynomials over finite fields with characteristic 2. J. Shaanxi Normal Univ. Nat. Sci. Ed. **36**, 5–10 (2008)
229. Lidl, R., Niederreiter, H.: Finite fields, 2nd. edn. Encyclopedia of Mathematics and its Applications **20**, Cambridge University Press, Cambridge (1997)
230. Lindner, C.C., Mendelsohn, E., Mendelsohn, N.S., Wolk, B.: Orthogonal Latin square graphs. J. Graph Theory **3**, 325–338 (1979)
231. Lins, S., Schellenberg, P.J.: The existence of strong starters in $\mathbb{Z}_{16k^2+1}$: a simpler proof. Ars Combin. **11**, 123–129 (1985)
232. Lovász, L., Schrijver, A.: Remarks on a theorem of Rédei. Studia Sci. Math. Hungar. **16**, 449–454 (1981)
233. Lüneberg, H.: Über projektive Ebenen, in denen jede Fahne von einer nicht-trivalen Elation invariant gelassen wird. Abh. Math. Semin. Univ. Hambg. **29** (1965), 37–76 (1965)
234. Macinnes, C.R.: Finite planes with less than eight points on a line. Amer. Math. Monthly **14**, 171–174 (1907)
235. MacNeish, H.F.: Euler squares. Ann. of Math. **23**, 221–227 (1922)
236. Maenhaut, B.M., Wanless, I.M.: Atomic Latin squares of order eleven. J. Combin. Des. **12**, 12–34 (2004)
237. Mann, H.B.: The construction of orthogonal Latin squares. Ann. Math. Statist. **13**, 418–423 (1942)
238. Mann, H.B.: On orthogonal Latin squares. Bull. Amer. Math. Soc. **50**, 249–257 (1944)
239. McKay, B.D., McLeod, J.C., Wanless, I.M.: The number of transversals in a Latin square. Des. Codes Crytogr. **40**, 269–284 (2006)
240. Mendelsohn, N.S., Wolk, B.: A search for a nondesarguesian plane of prime order. In: Lecture Notes in Pure and Applied Mathematics **103**, pp. 199–208. Marcel Dekker, New York (1985)
241. Michler, G.O.: Theory of finite simple groups. Cambridge University Press, Cambridge (2006).
242. Miller, G.A.: A new proof of the generalized Wilson's theorem. Ann. of Math. (2) **4**, 188–190 (1903)
243. Mills, W.H.: Some mutually orthogonal Latin squares. Congr. Numer. **19**, 473–487 (1977)
244. Moghaddamfar, A.R., Zokayi, A.R.: On the admissibility of finite groups. Southeast Asian Bull. Math. **33**, no. 3, 485–489 (2009).
245. Mullen, G.L.: Permutation polynomials over finite fields. In: Finite Fields, Coding Theory and Advances in Communications and Computing (Las Vegas, NV, 1991), pp. 131–151. Lecture Notes in Pure Appl. Math. **141**, Dekker, New York (1993)
246. Mullen, G.L., Niederreiter, H.: Dickson polynomials over finite fields and complete mappings. Canad. Math. Bull. **30**, 19–27 (1987)
247. Mullen, G.L., Panario, D.: Handbook of finite fields. CRC Press, Boca Raton-London-New York (2013)

248. Mullin, R.C., Nemeth, E.: An existence theorem for Room squares. Canad. Math. Bull. **12**, 493–497 (1969)
249. Muratović-Ribić, A., Pasalic, E.: A note on complete polynomials over finite fields and their applications in cryptography. Finite Fields Appl. **25**, 306–315 (2014)
250. Narayan, D.A.: An upper bound for representation number of graphs with fixed order. Integers **A12**, 1–4 (2003)
251. Nazarok, A.V.: Five pairwise orthogonal Latin squares of order 21. Issled. Oper. i ASU, 54–56 (1991)
252. Niederreiter, H., Robinson, K.H.: Complete mappings of finite fields. J. Aust. Math. Soc. Ser. A **33**, 197–212 (1982)
253. Niederreiter, H., Winterhof, A.: Cyclotomic R-orthomorphisms of finite fields. Discrete Math. **295**, 161–171 (2005)
254. Norton, H.W.: The 7×7 squares. Ann. Eugenics **9**, 269–307 (1939)
255. Ollis, M.A.: Sequenceable groups and related topics, Dynamic survey. Electron. J. Combin. **20**(2), #DS10v2 (2013)
256. Ostrom, T.G.: Translation planes and configurations in Desarguesian planes. Arch. Math. **11**, 457–464 (1960)
257. Ostrom, T.G.: Semi-translation planes. Trans. Amer. Math. Soc. **111**, 1–18 (1964)
258. Ostrom, T.G.: Replaceable nets, net collineations, and net extensions. Canad. J. Math. **18**, 666–672 (1966)
259. Ostrom, T.G.: The dual Lüneburg planes. Math. Z. **92**, 201–209 (1966)
260. Paige, L.J.: Neofields. PhD thesis, University of Wisconsin (1947)
261. Paige, L.J.: A note on finite abelian groups. Bull. Amer. Math. Soc. **53**, 590–593 (1947)
262. Paige, L.J.: Neofields. Duke Math. J. **16**, 39–60 (1949)
263. Paige, L.J.: Complete mappings of finite groups. Pacific J. Math. **1**, 111–116 (1951)
264. Parker, E.T.: Construction of some sets of mutually orthogonal Latin squares. Proc. Amer. Math. Soc. **10**, 946–949 (1959)
265. Parker, F.D.: When is a loop a group? Amer. Math. Monthly **72**, 765–766 (1965)
266. Passman, D.S.: Permutation groups. W.A. Benjamin, New York (1968)
267. Pflugfelder, H.O.: Quasigroups and loops: introduction. Sigma Series in Pure Mathematics, **7**, Heldermann Verlag, Berlin (1990)
268. Pierce, D., Kallaher, M.J.: A note on planar functions and their planes. Bull. Inst. Combin. Appl. **42**, 53–75 (2004)
269. Pott, A.: Maximal difference matrices of order q. J. Combin. Des. **1**, 171–176 (1993)
270. Pott, A: Finite geometry and character theory. Lecture Notes in Mathematics **1601**, Springer-Verlag, Berlin (1995)
271. Praeger, C.E., Soicher, L.H.: Low rank representations and graphs for sporadic groups. Cambridge University Press, Cambridge (1997).
272. Pula, K.: Products of all elements in a loop and a framework for non-associative analogues of the Hall-Paige conjecture. Electron. J. Combin. **16**, no. 1, Research Paper 60, 15 pp. (2009).
273. Quinn, K.A.S.: Difference matrices and orthomorphisms over non-abelian groups. Ars Combin. **52**, 289–295 (1999)
274. Ramanathan, K.G.: On the product of the elements of a finite abelian group. J. Indian Math. Soc. (N. S.) **11**, 44–48 (1947)
275. Rédei, L.: Lacunary polynomials over finite fields. North-Holland, Amsterdam-London (1973)
276. Rees, R., Wallis, W.D.: A class of resolvable pairwise balanced designs. Congr. Numer. **55**, 211–220 (1986)
277. Reidemeister, K.: Topoogische Frage der Differentialgeometrie, V, Gewebe und Gruppen. Math. Z. **29**, 427–435 (1929)
278. Repphun, K.: Geometrische Eigenschaften vollständiger Orthomorphismsysteme von Gruppen. Math. Z. **89**, 206–212 (1965)
279. Rhemtulla, A.R.: On a problem of L. Fuchs. Studia Sci. Math. Hungar. **4**, 195–200 (1969)

280. Ringel, G.: Cyclic arrangements of the elements of a group. Notices Amer. Math. Soc. **21**, A95–96 (1974)
281. Roberts, C.E. Jr.: Sets of mutually orthogonal Latin squares with "like subsquares". J. Combin. Theory Ser. A **61**, 50–63 (1992)
282. Roth, R., Peters, M.: Four pairwise orthogonal Latin squares of order 24. J. Combin. Theory Ser. A **44**, 152–155 (1987)
283. Roth, R., Wilson, R.M.: Personal communication.
284. Rotman, J.J.: An introduction to the theory of groups. Springer, New York (1995)
285. Sade, A.: Omission dans les listes de Norton pour les carrés 7 × 7. J. Reine Angew. Math. **189**, 190–191 (1951)
286. Sade, A.: An omission in Norton's list of 7 × 7 squares. Ann. Math. Statist. **22**, 306–307 (1951)
287. Sade, A.: Isotopies d'un groupoide avec son conjoint. Rend. Circ. Mat. Palermo (2) **12**, 357–381 (1963)
288. Saeli, D.: Complete mappings and difference ratio in double loops. Riv. Math. Univ. Parma (4) **15**, 111–117 (1989)
289. Schellenberg, P.J., Van Rees, G.H.J., Vanstone, S.A.: Four pairwise orthogonal Latin squares of order 15. Ars Combin. **6**, 141–150 (1978)
290. Schulz, R.-H.: Some check digit systems over non-abelian groups. Mitt. Math. Ges. Hamburg **12**, 819–827 (1991)
291. Shallue, C.J., Wanless, I.M.: Permutation polynomials and orthomorphism polynomials of degree six. Finite Fields Appl. **20**, 84–92 (2013)
292. Shieh, Y.P.: Partition Strategies for # P-complete problem with applications to enumerative combinatorics. PhD thesis, National Taiwan University, (2001)
293. Shieh, Y.P., Hsiang, J., Hsu, D.F.: On the enumeration of abelian K-complete mappings. Congr. Numer. **144**, 67–88 (2000)
294. Shieh, Y.-P., Hsiang, J., Hsu, D.F.: On the existence problems of complete mappings. Preprint.
295. Singer, J.: A class of groups associated with Latin squares. Amer. Math. Monthly **67**, 235–240 (1960)
296. Siu, M.-K.: Which Latin squares are Cayley tables? Amer. Math. Monthly **98**, 625–627 (1991)
297. Speiser, A.: Die Theorie der Gruppen von endlicher Ordnung, 2nd edn. Springer, Berlin (1927)
298. Sprague, A.P.: Translation nets. Mitt. Math. Semin. Giessen **157**, 46–68 (1982)
299. Steinberg, R.: Variations on a theme of Chevalley. Pacific J. Math. **9**, 875–891 (1959).
300. Stones, D., Wanless, I.M.: Compound orthomorphisms of the cyclic group. Finite Fields Appl. **16**, 277–289 (2010)
301. Storer, T.: Cyclotomy and difference sets. Markham, Chicago (1967)
302. Stothers, W.W.: On permutation polynomials whose difference is linear. Glasg. Math. J. **32**, 165–171 (1990)
303. Studnicka, I.: Non-existence of Cartesian groups of order $2p^m$. Comment. Math. Univ. Carolin. **13**, 721–725 (1972)
304. Sun, Q., Zhang, Q.F.: A simple proof of a conjecture about complete mappings over finite fields. Sichuan Daxue Xuebao **35**, 840–842 (1998)
305. Suschkewitsch, A.: On a generalization of the associative law. Trans. Amer. Math. Soc. **31**, 204–214 (1929)
306. Suzuki, M.: Group theory I. Springer-Verlag, Berlin-New York (1982)
307. Tarry, G.: Le problème des 36 officiers. Assoc. Franc. Paris **29**, 170–203 (1900)
308. Tecklenburg, H.: Latin squares and configuration axioms in affine planes. Ars Combin. **29**, 33–47 (1990)
309. Thomas, A.D., Woods, G.V.: Group Tables. Devon Print Group, Exeter, Devon (1980)
310. Thompson, J.: Finite groups with fixed-point-free automorphisms of prime order. Proc. Natl. Acad. Sci. USA **45**, 578–581 (1959).
311. Thomsen, G.: Topologische Fragen der Differentialgeometrie XII, Schnittepunktsätze in ebenen Gewebe. Abh. Math. Seminar Univ. Hambg. **7**, 99–106 (1939)

312. Todorov, D.: Four mutually orthogonal Latin squares of order 20. Ars. Combin. **27**, 63–65 (1989)
313. Ullrich, P.: An Eulerian square before Euler and an experimental design before R.A. Fisher: on the early history of Latin squares. Chance **12**, 22–27 (1999)
314. Ullrich, P.: Officers, playing cards, and sheep. On the history of Eulerian squares and the design of experiments. Metrika **56**, 189–204 (2002)
315. van Rees, G.H.J.: Subsquares and transversals in Latin squares. Ars Combin. 29B, 193–204 (1990)
316. Vaughan-Lee, M., Wanless, I.M.: Latin squares and the Hall-Paige conjecture. Bull. Lond. Math. Soc. **35**, 191–195 (2003).
317. Vereoff, J.: Error detecting decimal codes. Math. Centre Tracts **29**, Amsterdam (1969)
318. Vijayaraghavan, T., Chowla, S.: On complete residue sets. Quart. J. Math. (Oxford Ser.) **19**, 193–199 (1948)
319. Walker, M.: A class of translation planes. Geom. Dediicata **5**, 135–146 (1976)
320. Wan, D.: On a problem of Niederreiter and Robinson about finite fields. J. Aust. Math. Soc. Ser. A **41**, 336–338 (1986)
321. Wan, D., Mullen, G.L., Shiue, P.J.-S.: The number of permutation polynomials of the form $f(x) + cx$ over a finite field. Proc. Edinb. Math. Soc. **(2)38**, 133–149 (1995)
322. Wan, D., Mullen, G.L., Shiue, P.J.-S.: Erratum "The number of permutation polynomials of the form $f(x) + cx$ over a finite field". Proc. Edinb. Math. Soc. **38**, i (1995).
323. Wang, C.-D.: On harmoniousness and complete mappings decomposable into disjoint cycles of the same length. In: Combinatorics, Graph Theory, Algorithms and Applications (Beijing, 1993), pp. 347–353. World Sci. Publ., River Edge, NJ (1994)
324. Wanless, I.M.: Atomic Latin squares based on cyclotomic orthomorphisms. Electron. J. Combin. **12**, R22 (2005)
325. Wanless, I.M.: Transversals in Latin squares: a survey. In: Surveys in combinatorics 2011, pp. 403–437. London Math. Soc. Lecture Note Ser. **392**, Cambridge Univ. Press, Cambridge (2011)
326. Wanless, I.M., Webb, B.S.: The existence of Latin squares without orthogonal mates. Des. Codes Cryptogr. **40**, 131–135 (2006)
327. Wanless, I.M., Webb, B.S.: Small partial Latin squares that cannot be embedded in a Cayley table. Australas. J. Combin. **67(2)**, 352–363 (2017).
328. Watson, D.: Condition for a loop to be a group. Amer. Math. Monthly **74**, 843–844 (1967)
329. Wilcox, S.: Reduction of the Hall-Paige conjecture to sporadic simple groups. J. Algebra **321**, 1407–1428 (2009)
330. Wilson, R., Walsh, P., Tripp, J., Suleiman, I., Parker, R., Norton, S., Nickerson, S., Linton, S., Bray, J., Abbott, R.: Atlas of finite group representations - Version 3. http://brauer.maths.qmul.ac.uk/Atlas/v3/
331. Winterhof, A.: Generalizations of complete mappings of finite fields and some applications. J. Symbolic Comput. **64**, 42–52 (2014)
332. Winters, S.: Error-detecting schemes using dihedral groups. UMAP J. **11**, 299–308 (1990)
333. Wojtas, M.: Five mutually orthogonal Latin squares of order 35. J. Combin. Des. **4**, 153–154 (1996)
334. Wojtas, M.: Three new constructions of mutually orthogonal Latin squares. J. Combin. Des. **8**, 218–220 (2000)
335. Woodcock, C.F.: On Orthogonal Latin Squares. J. Combin. Theory Ser. A **43**, 146–148 (1986)
336. Wu, G., Li, N., Helleseth, T., Zhang, Y.: Some classes of monomial complete permutation polynomials over finite fields of characteristic two. Finite Fields Appl. **28**, 148–165 (2014)
337. Xiang, Q.: Recent progress in algebraic design theory. Finite Fields Appl. **11**, 622–653 (2005)
338. Zassenhaus, H.J.: The theory of groups. 2nd edn., Chelsea, New York (1958).
339. Zassenhaus, H.: What makes a loop a group? Amer. Math. Monthly **75**, 139–142 (1968)
340. Zha, Z., Hu, L., Cao, X.: Constructing permutations and complete permutations over finite fields via subfield-valued polynomials. Finite Fields Appl. **31**, 162–177 (2015)

Index

A. B. Evans, *Orthogonal Latin Squares Based on Groups*, Developments in Mathematics 57, https://doi.org/10.1007/978-3-319-94430-2

MIX
Papier aus verantwortungsvollen Quellen
Paper from responsible sources
FSC® C105338

If you have any concerns about our products,
you can contact us on
ProductSafety@springernature.com

In case Publisher is established outside the EU,
the EU authorized representative is:
Springer Nature Customer Service Center GmbH
Europaplatz 3, 69115 Heidelberg, Germany

Printed by Libri Plureos GmbH
in Hamburg, Germany